THIRD EDITION

Matter & Interactions

VOLUME I
Modern Mechanics

VOLUME II
Electric and Magnetic Interactions

THIRD EDITION

Matter & Interactions

VOLUME I
Modern Mechanics

VOLUME II
Electric and Magnetic Interactions

RUTH W. CHABAY

BRUCE A. SHERWOOD

North Carolina State University

WILEY

JOHN WILEY & SONS, INC.

EXECUTIVE EDITOR   *Stuart Johnson*

ASSISTANT EDITOR   *Aly Rentrop*

ASSOCIATE MARKETING DIRECTOR   *Christine Kushner*

SENIOR PRODUCTION EDITOR   *Elizabeth Swain*

ART DIRECTOR   *Jeof Vita*

TEXT DESIGN   *Laura Ierardi*

SENIOR MEDIA EDITOR   *Thomas Kulesa*

SENIOR ILLUSTRATION EDITORS   *Sigmund Malinowski and Anna Melhorn*

COVER IMAGE   *Ruth Chabay*

Cover Description

> Volume 1: Ball-and-spring model of a solid
> Volume 2: Magnetic field of a moving charge

This book was set in 10/12 Times Ten Roman in LaTex by Aptara®, Inc. and printed and bound by Courier Kendallville. The cover was printed by Courier Kendallville.

This book is printed on acid-free paper.

Library of Congress Cataloging-in-Publication Data

Chabay, Ruth W.
    Matter & interactions / Ruth W. Chabay, Bruce A. Sherwood.—3rd ed.
      p. cm.
    Includes index.
    Volume 1 ISBN 978-0-470-50345-4 (pbk.)
    Volume 2 ISBN 978-0-470-50346-1 (pbk.)
    Complete ISBN 978-0-470-50347-8 (cloth)
    1. Physics—Textbooks. 2. Mechanics—Textbooks. I. Sherwood, Bruce A. II. Title. III.
Title: Matter and interactions.

    QC23.2.C43 2011
    530—dc22

                                                    2009034010

Printed in the United States of America

10 9 8 7 6 5 4 3 2

Preface

This textbook emphasizes a 20th-century perspective on introductory physics with the goal of involving students in the contemporary physics enterprise:

- Apply fundamental principles to a wide range of systems (from nuclei to stars; unify mechanics and thermal physics, and electrostatics and circuits)
- Integrate contemporary physics throughout the curriculum (atomic models of matter, quantized energy, relativistic dynamics)
- Engage students in physical modeling (idealization, approximation, assumptions, estimation)
- Integrate computational physics (now a partner of theory and experiment)

Modern Mechanics (volume 1, Chapters 1–13) focuses on the atomic structure of matter and interactions between material objects and emphasizes the wide applicability of a small number of fundamental principles: the Momentum Principle, the Energy Principle, and the Angular Momentum Principle. We can use these principles to explain and predict the behavior of systems as different as molecules and galaxies.

Electric & Magnetic Interactions (volume 2, Chapters 14–25) emphasizes the concepts of electric and magnetic fields and extends the study of the atomic structure of matter to include the role of electrons. The principles of electricity and magnetism are the foundation for much of today's technology, from cell phones to medical imaging.

Prerequisites

This book is intended for introductory calculus-based college physics courses taken by science and engineering students. You will need a basic knowledge of derivatives and integrals, which can be obtained by studying calculus concurrently.

Modeling

Matter & Interactions places a major emphasis on constructing and using physical models. A central aspect of science is the modeling of complex real-world phenomena. A physical model is based on what we believe to be fundamental principles; its intent is to predict or explain the most important aspects of an actual situation. Modeling necessarily involves making approximations and simplifying assumptions in order that the model can be analyzed in detail.

Computational Modeling

Computational modeling is now as important as theory and experiment in contemporary science and engineering. We introduce you to serious computer modeling right away to help you build a strong foundation in the use of this important tool.

Computational modeling allows us to analyze complex systems that would otherwise require very sophisticated mathematics or that could not be analyzed at all without a computer. Numerical calculations based on the Momentum Principle give us the opportunity to watch the dynamical evolution of the behavior of a system. Simple models frequently need to be refined and extended. This can be done straightforwardly with a computer model but is often impossible with a purely analytical (non-numerical) model.

Questions

As you read the text, you will frequently come to a question that invites you to stop and think, to make a prediction, to carry out a step in a derivation or analysis, or to apply a principle. These questions are answered in the following paragraphs, but it is important that you make a serious effort to answer the questions on your own before reading further. Be honest in comparing your answers to those in the text. Paying attention to surprising or counterintuitive results can be a useful learning strategy.

Exercises

Small exercises that require you to apply new concepts are found at the end of many sections of the text. These may involve qualitative reasoning or simple calculations. You should work these exercises when you come to them, to consolidate your understanding of the material you have just read. Answers to these exercises are found at the end of each chapter.

Conventions Used in Diagrams

Force

Component of force

Velocity

Momentum

Electric field

Component of electric field

Magnetic field

Component of Magnetic field

Position

Angular momentum

Torque

Distance

A path

The conventions most commonly used to represent vectors and scalars in diagrams in this text are shown in the margin. In equations and text, a vector will be written with an arrow above it: $\vec{p}$.

TO THE INSTRUCTOR

The approach to introductory physics in this textbook differs significantly from that in most textbooks. Key emphases of the approach are these:

- Starting from fundamental principles rather than secondary formulas
- Atomic-level description and analysis
- Modeling the real world through idealizations and approximations
- Computational modeling of physical systems
- Unification of mechanics and thermal physics
- Unification of electrostatics and circuits
- The use of 3D vectors throughout

Instructor Web Site

Much useful information may be found on the Matter & Interactions public web site:

http://www.matterandinteractions.org

This public web site includes information of a general character, including articles about M&I and useful educational software.

Also at this public web site are links to the Wiley web site for instructors, where you will find useful resources that include the following:

- Lecture notes
- Videos of lectures, and videos that are useful to show during lectures

- Additional lecture demo software
- Figures from the textbook in jpeg format
- Sample course calendar
- Test questions
- Clicker questions
- Lab activities including experiments, computational modeling, and white-board problems for group problem solving

Articles Dealing with Matter & Interactions

Five key articles dealing with this curriculum are the following:

- Chabay, R. & Sherwood, B. (1999). Bringing atoms into first-year physics. *American Journal of Physics* 67, 1045–1050.
- Chabay, R. W. & Sherwood, B. (2004). Modern mechanics. *American Journal of Physics* 72, 439–445.
- Chabay, R. W. & Sherwood, B. (2006). Restructuring Introductory E&M. *American Journal of Physics* 74, 329–336.
- Chabay, R. & Sherwood, B. (2008) Computational physics in the introductory calculus-based course. *American Journal of Physics* 76(4&5), 307–313.
- Chabay, R. & Sherwood, B. (2007) Matter & Interactions. A chapter in the on-line book *Reviews in PER Volume 1: Research-Based Reform of University Physics*, E. F. Redish & P. J. Cooney, editors, http://www.compadre.org/per/per reviews/volume1.cfm.

These articles are available on the Matter & Interactions web site. The last article includes implementation aspects, evaluation information, and so on.

Computational Homework Problems

Some important homework problems require the student to write a computer program. We strongly recommend VPython as the best tool for student use. An adequate subset of the underlying Python programming language can be taught in an hour or two, even to students who have never written a program before, and instructional materials are available to instructors.

Real-time 3D animations are generated as a side effect of student computations, and these animations provide powerfully motivating and instructive visualizations of fields and motions. VPython supports true vector computations, which encourages students to begin thinking about vectors as much more than mere components. VPython can be obtained at no cost for Windows, Macintosh, and Linux at http://vpython.org.

Homework Problems

In addition to the exercises and problems included at the end of a chapter, there are many exercises distributed throughout the chapter. We have prepared versions of the textbook exercises and problems for the WebAssign computer homework system (http://www.webassign.net).

Desktop Experiment Kit

On the public web site indicated above is information on a desktop experiment kit for E&M that is distributed by PASCO that makes it possible for students to make key observations of electrostatic, circuit, and magnetic phenomena in the classroom, tightly integrated with the theory (http://www.pasco.com, search for EM-8675). Several chapters contain optional experiments that can be done with this kit. This does not preclude having other, more complex laboratory experiences associated with the curriculum.

For example, one such lab that we use deals with Faraday's law and requires signal generators, large coils, and oscilloscopes. You may have lab experiments already in place that will go well with this textbook.

What's New in the 3rd Edition

The 3rd edition of this text includes the following new features:

- Full color, which helps clarify many diagrams and allows examples to stand out more clearly.
- A larger set of exercises and problems at the end of each chapter.
- More numerical examples and problems to complement symbolic ones.
- Improvements in the treatments of computational modeling, angular momentum, the effect of electric fields on matter, and electromagnetic radiation.
- A section on geometric optics in Chapter 24.

Details of Changes from the 2nd Edition

The iterative approach to predicting motion, which is the foundation for computational modeling, has been moved from Chapter 3 to Chapter 2 in recognition of its central importance in modern mechanics. This makes it possible to practice with the iterative approach in the context of simple force laws before applying it to more complicated $1/r^2$ vector force laws in Chapter 3.

Chapter 3 has been renamed "Fundamental Interactions" and includes examples of all four fundamental interactions: gravitational, electromagnetic, weak, and strong.

What had been a very long Chapter 4 has been split into Chapter 4 on the ball-and-spring model of solids and spring-mass oscillations and Chapter 5 on statics and dynamics free-body problems, including dynamics problems involving curving motion. Included in the new Chapter 5 is extended and more consistent instruction on statics and dynamics free-body problems. The Momentum Principle in the form $d\vec{p}/dt = \vec{F}_{net}$ is still primary, but more consistent connections are made to acceleration.

In the first chapter on energy (Chapter 6) potential energy is introduced in a more concrete way in the context of a particular force law (gravitation).

The second energy chapter (Chapter 7) deals a bit more extensively with the full range of types of energy inputs, not just work and thermal energy transfer but also radiative transfer, mass transfer, sound, and electrical inputs. It also introduces explicitly the concept of a steady state, where there are several energy transfers but no change in the energy of the system.

In Chapter 8 on quantized energy there is explicit mention of the existence of selection rules, which in this introductory treatment are not actually taken into account.

Chapter 9 on multiparticle systems now introduces moment of inertia and $K_{rot} = \frac{1}{2}I\omega^2$. There is an example of using integral calculus to find the moment of inertia for a rod.

Chapter 11 on angular momentum builds on the concept of moment of inertia and rotational kinetic energy which are now introduced in Chapter 9. There are many more numerical examples to make the concepts more concrete. There is a basic treatment of rotational kinematics.

Chapter 12 on quantum statistical mechanics now uses k_B instead of k for the Boltzmann constant.

Chapter 15 on the effects of electric field on matter has been reorganized and strengthened, and $v = uE$ is introduced to help students understand with a deeper sense of mechanism why $E = 0$ in a conductor in equilibrium. The case study on sparks has been moved to Chapter 21 on magnetic force and has been rewritten to use the concept of potential and to capitalize on the discussion of Jack and Jill and Einstein in discussing the extended muon lifetime.

Chapter 17 on electric potential includes an explicit procedure for computing potential difference from a known electric field.

Chapter 22 now contains the discussion of p-n junctions, formerly in a separate chapter, as a case study in the use of Gauss's law.

Chapter 23 on Faraday's law is now careful to say that a curly electric field and a time-varying magnetic field accompany each other, rather than asserting that a time-varying magnetic field "causes" a curly electric field.

Chapter 24 on electromagnetic radiation has been reorganized for greater clarity and includes sections on geometrical optics. There is now an optional section that derives the equation for the radiative field of an accelerated charge, based on a derivation by Purcell.

What Can Be Omitted?

In a large course for engineering and science students with three 50-minute lectures and one 110-minute small-group studio lab per week, or in a studio format with five 50-minute sessions per week, it is possible to complete most but not all of the mechanics and E&M material in two 15-week semesters. In an honors course, or a course for physics majors, it is possible to do almost everything. You may be able to go further or deeper if your course has a weekly recitation session in addition to lecture and lab.

What can be omitted if there is not enough time to do everything? In mechanics, the one thing we feel should not be omitted is the introduction to entropy in terms of quantum statistical mechanics (Chapter 12). This is a climax of the theme of integrating mechanics and thermal physics. One way to decide what mechanics topics can be omitted is to be guided by what foundation is required for Chapter 12. See detailed suggestions below.

In E&M, one should not omit electromagnetic radiation and its effects on matter (Chapter 24). This is the climax of the whole E&M enterprise. One way to decide what E&M topics can be omitted is to be guided by what foundation is required for Chapter 24. See detailed suggestions below.

Any starred section (*) can safely be omitted. Material in these sections is not referenced in later work.

Chapter 1 (Interactions and Motion): Unless your students have a particularly strong background in vectors, all of this chapter is important for later work (except for starred sections).

Chapter 2 (The Momentum Principle): None of the unstarred sections of this foundational chapter should be omitted.

Chapter 3 (The Fundamental Interactions): The section on determinism may be omitted.

Chapter 4 (Contact Interactions): Buoyancy and pressure may be omitted (one can return to these topics during Chapter 13 on gases).

Chapter 5 (Rate of Change of Momentum): All of this chapter is important for later work.

Chapter 6 (The Energy Principle): This is the foundation for the energy concept, and nothing should be omitted (other than starred sections, of course).

Chapter 7 (Internal Energy): If you are pressed for time, you might choose to omit the second half of the chapter on energy dissipation, beginning with Section 7.10.

Chapter 8 (Energy Quantization): None of the unstarred sections of this short chapter should be omitted.

Chapter 9 (Multiparticle Systems): The formalism of finding the center of mass may be skipped, because the important applications have obvious locations of the center of mass.

Chapter 10 (Collisions): A good candidate for omission is the analysis of collisions in the center-of-mass frame. Since there is a basic introduction to collisions in Chapter 3 (before energy is introduced), one could omit all of Chapter 10. On the other hand, the combined use of the Momentum Principle and the Energy Principle can illuminate both fundamental principles.

Chapter 11 (Angular Momentum): The main content of this chapter should not be omitted, as it introduces the third fundamental principle of mechanics, the Angular Momentum Principle. One might choose to omit most applications involving nonzero torque.

Chapter 12 (Entropy: Limits on the Possible): It is strongly recommend to do at least the first half of this chapter, on the Einstein solid. To make this work well, exposition and student computing should be interwoven. It might seem odd to ask the students simply to reproduce the results shown in the textbook, but in our experience and that of our students it is enormously educational to be required to make the ideas concrete in a computer program. You find yourself asking "What exactly is q_1?" and answering very concretely by actually calculating with the initially unfamiliar quantities. It is the difference between writing an essay about entropy and calculating the actual value of the entropy. Despite its scientific importance, one might omit the second half of the chapter, on the Boltzmann distribution. However, if you cover Chapter 13 you will need a few key results from this analysis.

Chapter 13 (Gases and Engines): This entire chapter may be omitted. If you include this chapter, you should not omit the initial sections, including the relationship of pressure to the results from Chapter 12. The important formula for particle flow ($nA\bar{v}$) and the concept of mean free path are both used in E&M, though they are rederived there. You might choose to omit the sections on macroscopic energy transfers (isothermal and adiabatic processes). The sections on heat engines represent additional applications of the second law of thermodynamics and may be omitted. Another possibility is to omit just the sections on nonzeropower engines.

Chapter 14 (Electric Field): The field concept is the backbone of the treatment of E&M, so none of this brief chapter should be omitted. Electric dipoles, which are introduced here, are central to what follows.

Chapter 15 (Electric Fields and Matter): Nothing in this foundational chapter should be omitted.

Chapter 16 (Electric Field of Distributed Charges): It is important that students acquire a good working knowledge of the patterns of electric field around some standard charged objects (rod, ring, disk, capacitor, sphere). If however they themselves are to acquire significant expertise in setting up physical integrals, they need extensive practice, and you might decide that the amount of time necessary for acquiring this expertise is not an appropriate use of the available course time.

Chapter 17 (Electric Potential): The section on dielectric constant can be omitted if necessary, as can the section on the potential of distributed charges.

Chapter 18 (Magnetic Field): In the sections on the atomic structure of magnets, you might choose to discuss only the first part, in which one finds that the magnetic moment of a bar magnet is consistent with an atomic model. Omitting the remaining sections on spin and domains will not cause significant difficulties later.

Chapter 19 (Electric Field and Circuits): It is best to do everything because it solidifies the field concept. However, you may choose not to do all of the final applications.

Chapter 20 (Circuit Elements): The discussion of field in capacitor circuits helps significantly in solidifying the students' understanding of electric field, and offers an opportunity to think about transient phenomena on an ordinary time scale. The derivation of resistance, and the connection between macro and micro circuit analysis, is useful to most students of science and engineering. The sections on series and parallel resistors and on internal resistance, meters, quantitative analysis of RC circuits, and multi-loop circuits can be omitted. Physics and engineering students who need to analyze complex multiloop circuits will later take specialized courses on the topic; in the introductory physics course the emphasis should be on giving all students a good grounding in the fundamental mechanisms underlying circuit behavior.

Chapter 21 (Magnetic Force): We recommend discussing Jack and Jill and Einstein, but it is safe to omit the sections on relativistic field transformations. However, students often express high interest in the relationship between electric fields and magnetic fields, and here is an opportunity to satisfy some aspects of their curiosity. Motors and generators may be omitted or downplayed. The case study on sparks in air can be omitted, because nothing later depends critically on this topic, though it provides an introductory-level example of a phenomenon where an intuitively appealing model fails utterly, while a different model predicts several key features of the phenomenon. Another possibility is to discuss sparks near the end of the course, because it can be a useful review of many aspects of E&M.

Chapter 22 (Patterns of Field in Space): Gauss's law and Ampere's law are important foundations for the following chapters on Faraday's law and electromagnetic radiation. It would in principle be possible to do Gauss's law just after Chapter 16, and Ampere's law just after Chapter 18, but it is not a good idea. Students need extensive experience with patterns of field in space before they can make sense of these topologically complex relations. Moreover, studying Gauss's law just before Faraday's law makes the concept of flux fresh and unproblematic in Chapter 23.

Chapter 23 (Faraday's Law): Though it can safely be omitted, we recommend retaining the section on superconductors, because students are curious about this topic. The section on inductance can also be omitted.

Chapter 24 (Electromagnetic Radiation): This is the climax toward which much of the textbook has been heading. We strongly recommend omitting nothing on radiation. However, you may wish to omit the treatment of geometrical optics.

Chapter 25 (Waves and Particles): It may not be possible to get to this chapter at all, and its inclusion is optional. It contrasts the classical wave model of light developed in Chapter 24 with the particle model that was used in mechanics, and it attempts to show the relationship between these two very different models. It also develops key concepts in physical optics. Students who are interested in physics often find these sections interesting.

Acknowledgments

We owe much to the unusual working environment provided by the Department of Physics and the former Center for Innovation in Learning at Carnegie Mellon, which made it possible during the 1990s to carry out the research and development leading to the first edition of this textbook in 2002. We are grateful for the open-minded attitude of our colleagues in the Carnegie Mellon physics department toward curriculum innovations.

We are grateful to the support of our colleagues Robert Beichner and John Risley in the Physics Education Research and Development group at North Carolina State University, and to other colleagues in the NCSU physics department.

We thank Fred Reif for emphasizing the role of the three fundamental principles of mechanics, and for his view on the reciprocity of electric and gravitational forces. We thank Robert Bauman, Gregg Franklin, and Curtis Meyer for helping us think deeply about energy.

Much of Chapter 12 on quantum statistical mechanics is based on an article by Thomas A. Moore and Daniel V. Schroeder, "A different approach to introducing statistical mechanics," American Journal of Physics vol. 65, pp. 26–36 (January 1997). We have benefited from many stimulating conversations with Thomas Moore, author of another introductory textbook that takes a contemporary view of physics, "Six Ideas that Shaped Physics." Michael Weissman and Robert Swendsen provided particularly helpful critiques on some aspects of our implementation of Chapter 12.

We thank Hermann Haertel for opening our eyes to the fundamental mechanisms of electric circuits. Robert Morse, Priscilla Laws, and Mel Steinberg stimulated our thinking about desktop experiments. Bat-Sheva Eylon offered important guidance at an early stage. Ray Sorensen provided deep analytical critiques that influenced our thinking in several important areas. Randall Feenstra taught us about semiconductor junctions. Thomas Moore showed us a useful way to present the differential form of Maxwell's equations. Fred Reif helped us devise an assessment of student learning of basic E&M concepts. Uri Ganiel suggested the high-voltage circuit used to demonstrate the reality of surface charge. The unusual light bulb circuits at the end of Chapter 23 are based on an article by P. C. Peters, "The role of induced emf's in simple circuits," American Journal of Physics 52, 1984, 208-211. Thomas Ferguson gave us unusually detailed and useful feedback on the E&M chapters. Discussions with John Jewett about energy transfers were helpful.

We thank David Andersen, David Scherer, and Jonathan Brandmeyer for the development of tools that enabled us and our students to write associated software.

We thank our colleagues David Brown, Krishna Chowdary, Laura Clarke, Norman Derby, Thomas Foster, Chris Gould, Mark Haugan, Joe Heafner, Robert Hilborn, Eric Hill, Andrew Hirsch, Leonardo Hsu, Matthew Kohlmyer, Barry Luokkala, Sara Majetich, Jonathan Mitschele, Arjendu Pattanayak, Jeff Polak, Prabha Ramakrishnan, Vidhya Ramachandran, Michael Schatz, Robert Swendsen, Aaron Titus, Michael Weissman, and Hugh Young.

We thank a group of reviewers assembled by the publisher, who gave us useful critiques on the second edition of this volume: Kelvin Chu, Michael Dubson, Tom Furtak, David Goldberg, Javed Iqbal, Shawn Jackson, Craig Ogilvie, Michael Politano, Norris Preyer, Rex Ramsier, Tycho Sleator, Robert Swendsen, Larry Weinstein, and Michael Weissman.

We have benefited greatly from the support and advice of Stuart Johnson of John Wiley & Sons. Elizabeth Swain of John Wiley & Sons was exceptionally skilled in managing the 3rd edition project. Helen Walden did a superb job of careful copyediting; any remaining errors are ours.

This project was supported, in part, by the National Science Foundation (grants MDR-8953367, USE-9156105, DUE-9554843, DUE-9972420, DUE-0320608, DUE-0237132, and DUE-0618504). We are grateful to the National Science Foundation and its reviewers for the long-term support of this challenging twenty-year project. Opinions expressed are those of the authors, and not necessarily those of the Foundation.

Ruth Chabay
Bruce Sherwood
Department of Physics
North Carolina State University
July 2009

Image Credits

Randall Feenstra (Figure 1.3 and Figure 4.5); Stacey Benson (Figure 7.39 and Figure 7.40); Judith Harrison (Figure 7.45); Barry Luokkala (Figure 25.18); Jeremy Henriksen (Figure 25.20); NASA/JPL/Space Science Institute (Jupiter in Figure 1.4); NASA, ESA, K. Kuntz (JHU), F. Bresolin (University of Hawaii), J. Trauger (Jet Propulsion Lab), J. Mould (NOAO), Y.-H. Chu (University of Illinois, Urbana), STScI, Canada-France-Hawaii Telescope/ J.-C. Cuillandre/Coelum, G. Jacoby, B. Bohannan, M. Hanna/NOAO/ AURA/NSF (M101 Pinwheel Galaxy in Figure 1.4); NASA, ESA, the Hubble Heritage Team (STScI/AURA), D. Carter (Liverpool John Moores University) and the Coma HST ACS Treasury Team (Coma Cluster of galaxies in Figure 1.4); NASA (Figures 3.69 and 3.70).

Contents

VOLUME I Modern Mechanics

CHAPTER 1 *Interactions and Motion*   1

1.1 KINDS OF MATTER   1
1.2 DETECTING INTERACTIONS   3
1.3 NEWTON'S FIRST LAW OF MOTION   6
1.4 OTHER INDICATORS OF INTERACTION   7
1.5 DESCRIBING THE 3D WORLD: VECTORS   9
1.6 SI UNITS   20
1.7 VELOCITY   20
1.8 MOMENTUM   28
1.9 CHANGE OF MOMENTUM   32
1.10 *THE PRINCIPLE OF RELATIVITY   35
1.11 *UPDATING POSITION AT HIGH SPEED   38
SUMMARY   39
EXERCISES AND PROBLEMS   40
ANSWERS TO EXERCISES   46

CHAPTER 2 *The Momentum Principle*   48

2.1 SYSTEM AND SURROUNDINGS   48
2.2 THE MOMENTUM PRINCIPLE   48
2.3 APPLYING THE MOMENTUM PRINCIPLE   53
2.4 MOMENTUM CHANGE WITH CHANGING FORCE   60
2.5 ITERATIVE PREDICTION OF MOTION   62
2.6 SPECIAL CASE: CONSTANT FORCE   72
2.7 ESTIMATING INTERACTION TIMES   78
2.8 PHYSICAL MODELS   82
2.9 *DERIVATION: SPECIAL-CASE AVERAGE VELOCITY   83
2.10 *INERTIAL FRAMES   84
2.11 *MEASUREMENTS AND UNITS   85
SUMMARY   87
EXERCISES AND PROBLEMS   87
ANSWERS TO EXERCISES   94

CHAPTER 3 *The Fundamental Interactions*   95

3.1 THE FUNDAMENTAL INTERACTIONS   95
3.2 THE GRAVITATIONAL FORCE   96
3.3 APPROXIMATE GRAVITATIONAL FORCE NEAR THE EARTH'S SURFACE   101
3.4 RECIPROCITY   103
3.5 PREDICTING MOTION OF GRAVITATIONALLY INTERACTING OBJECTS   104
3.6 THE ELECTRIC FORCE   109
3.7 THE STRONG INTERACTION   111

3.8 NEWTON AND EINSTEIN 112
3.9 PREDICTING THE FUTURE OF COMPLEX SYSTEMS 113
3.10 DETERMINISM 115
3.11 CONSERVATION OF MOMENTUM 118
3.12 THE MULTIPARTICLE MOMENTUM PRINCIPLE 123
3.13 COLLISIONS: NEGLIGIBLE EXTERNAL FORCES 126
3.14 POINTS AND SPHERES 128
3.15 MEASURING THE GRAVITATIONAL CONSTANT G 128
SUMMARY 128
EXERCISES AND PROBLEMS 129
ANSWERS TO EXERCISES 137

CHAPTER 4 *Contact Interactions* 138

4.1 TARZAN AND THE VINE 138
4.2 A MODEL OF A SOLID: BALLS CONNECTED BY SPRINGS 138
4.3 TENSION FORCES 140
4.4 LENGTH OF AN INTERATOMIC BOND 141
4.5 THE STIFFNESS OF AN INTERATOMIC BOND 143
4.6 STRESS, STRAIN, AND YOUNG'S MODULUS 146
4.7 COMPRESSION (NORMAL) FORCES 148
4.8 FRICTION 149
4.9 SPEED OF SOUND IN A SOLID AND INTERATOMIC BOND STIFFNESS 152
4.10 DERIVATIVE FORM OF THE MOMENTUM PRINCIPLE 154
4.11 ANALYTICAL SOLUTION: SPRING–MASS SYSTEM 156
4.12 ANALYTICAL EXPRESSION FOR SPEED OF SOUND 161
4.13 CONTACT FORCES DUE TO GASES 163
4.14 *A VERTICAL SPRING–MASS SYSTEM 167
4.15 *GENERAL SOLUTION FOR THE MASS–SPRING SYSTEM 168
SUMMARY 170
EXERCISES AND PROBLEMS 171
ANSWERS TO EXERCISES 177

CHAPTER 5 *Rate of Change of Momentum* 178

5.1 IDENTIFYING FORCES ON A SYSTEM 178
5.2 MOMENTUM NOT CHANGING (STATICS) 181
5.3 FINDING THE RATE OF CHANGE OF MOMENTUM 186
5.4 CURVING MOTION 189
5.5 RATE OF CHANGE OF DIRECTION 193
5.6 WHY DOES THE VINE BREAK? 200
5.7 PROBLEM SOLVING 204
SUMMARY 208
EXERCISES AND PROBLEMS 208
ANSWERS TO EXERCISES 218

CHAPTER 6 *The Energy Principle* 220

6.1 THE ENERGY PRINCIPLE 220
6.2 THE SIMPLEST SYSTEM: A SINGLE PARTICLE 221
6.3 WORK: MECHANICAL ENERGY TRANSFER 226
6.4 UPDATE FORM OF THE ENERGY PRINCIPLE 233
6.5 CHANGE OF REST ENERGY 236
6.6 PROOF OF THE ENERGY PRINCIPLE FOR A PARTICLE 238
6.7 WORK DONE BY A NONCONSTANT FORCE 240
6.8 POTENTIAL ENERGY IN MULTIPARTICLE SYSTEMS 241
6.9 GRAVITATIONAL POTENTIAL ENERGY 246

6.10 GENERAL PROPERTIES OF POTENTIAL ENERGY   251
6.11 PLOTTING ENERGY vs. SEPARATION   253
6.12 APPLYING GRAVITATIONAL POTENTIAL ENERGY   255
6.13 GRAVITATIONAL POTENTIAL ENERGY NEAR THE EARTH'S SURFACE   259
6.14 ELECTRIC POTENTIAL ENERGY   260
6.15 THE MASS OF A MULTIPARTICLE SYSTEM   262
6.16 REFLECTION: WHY ENERGY?   266
6.17 IDENTIFYING INITIAL AND FINAL STATES   267
6.18 *A PUZZLE   271
6.19 *GRADIENT OF POTENTIAL ENERGY   272
6.20 *INTEGRALS AND ANTIDERIVATIVES   273
6.21 *APPROXIMATION FOR KINETIC ENERGY   273
6.22 *FINDING THE FORMULA FOR PARTICLE ENERGY   274
S U M M A R Y   276
E X E R C I S E S   A N D   P R O B L E M S   277
A N S W E R S   T O   E X E R C I S E S   286

CHAPTER 7 *Internal Energy*   288

7.1 POTENTIAL ENERGY OF MACROSCOPIC SPRINGS   288
7.2 POTENTIAL ENERGY OF A PAIR OF NEUTRAL ATOMS   293
7.3 PATH INDEPENDENCE OF POTENTIAL ENERGY   295
7.4 INTERNAL ENERGY AND THERMAL ENERGY   296
7.5 ENERGY TRANSFER DUE TO A TEMPERATURE DIFFERENCE   301
7.6 REFLECTION: FORMS OF ENERGY   304
7.7 POWER: ENERGY PER UNIT TIME   304
7.8 OPEN AND CLOSED SYSTEMS   305
7.9 THE CHOICE OF SYSTEM AFFECTS ENERGY ACCOUNTING   306
7.10 ENERGY DISSIPATION   310
7.11 POTENTIAL ENERGY AND "CONSERVATIVE" FORCES   315
7.12 *RESONANCE   316
S U M M A R Y   318
E X E R C I S E S   A N D   P R O B L E M S   319
A N S W E R S   T O   E X E R C I S E S   325

CHAPTER 8 *Energy Quantization*   326

8.1 PHOTONS   326
8.2 ELECTRONIC ENERGY LEVELS   327
8.3 THE EFFECT OF TEMPERATURE   334
8.4 VIBRATIONAL ENERGY LEVELS   335
8.5 ROTATIONAL ENERGY LEVELS   338
8.6 OTHER ENERGY LEVELS   339
8.7 COMPARISON OF ENERGY LEVEL SPACINGS   340
8.8 *CASE STUDY: HOW A LASER WORKS   340
8.9 *WAVELENGTH OF LIGHT   342
S U M M A R Y   342
E X E R C I S E S   A N D   P R O B L E M S   343
A N S W E R S   T O   E X E R C I S E S   346

CHAPTER 9 *Multiparticle Systems*   347

9.1 THE MOTION OF THE CENTER OF MASS   347
9.2 SEPARATION OF MULTIPARTICLE SYSTEM ENERGY   350
9.3 ROTATIONAL KINETIC ENERGY   354
9.4 THE "POINT PARTICLE SYSTEM"   360
9.5 ANALYZING POINT PARTICLE AND REAL SYSTEMS   365

9.6 *MODELING FRICTION IN DETAIL 367

9.7 *A PHYSICAL MODEL FOR DRY FRICTION 369

9.8 *DERIVATION: KINETIC ENERGY OF A MULTIPARTICLE SYSTEM 372

9.9 *DERIVATION: THE POINT PARTICLE ENERGY EQUATION 373

SUMMARY 375

EXERCISES AND PROBLEMS 375

ANSWERS TO EXERCISES 381

CHAPTER 10   *Collisions*   382

10.1 INTERNAL INTERACTIONS IN COLLISIONS 382

10.2 ELASTIC AND INELASTIC COLLISIONS 383

10.3 A HEAD-ON COLLISION OF EQUAL MASSES 385

10.4 HEAD-ON COLLISIONS BETWEEN UNEQUAL MASSES 388

10.5 FRAME OF REFERENCE 389

10.6 SCATTERING: COLLISIONS IN 2-D AND 3-D 391

10.7 DISCOVERING THE NUCLEUS INSIDE ATOMS 394

10.8 *DISTRIBUTION OF SCATTERING ANGLES 396

10.9 RELATIVISTIC MOMENTUM AND ENERGY 398

10.10 INELASTIC COLLISIONS AND QUANTIZED ENERGY 401

10.11 COLLISIONS IN OTHER REFERENCE FRAMES 402

SUMMARY 407

EXERCISES AND PROBLEMS 407

ANSWERS TO EXERCISES 413

CHAPTER 11   *Angular Momentum*   414

11.1 TRANSLATIONAL ANGULAR MOMENTUM 414

11.2 ROTATIONAL ANGULAR MOMENTUM 421

11.3 TRANSLATIONAL PLUS ROTATIONAL ANGULAR MOMENTUM 425

11.4 THE ANGULAR MOMENTUM PRINCIPLE 426

11.5 MULTIPARTICLE SYSTEMS 430

11.6 THREE FUNDAMENTAL PRINCIPLES OF MECHANICS 435

11.7 SYSTEMS WITH ZERO TORQUE 436

11.8 SYSTEMS WITH NONZERO TORQUES 439

11.9 PREDICTING POSITIONS WHEN THERE IS ROTATION 441

11.10 ANGULAR MOMENTUM QUANTIZATION 443

11.11 *GYROSCOPES 448

11.12 *MORE COMPLEX ROTATIONAL SITUATIONS 454

11.13 *RATE OF CHANGE OF A ROTATING VECTOR 454

SUMMARY 455

EXERCISES AND PROBLEMS 456

ANSWERS TO EXERCISES 471

CHAPTER 12   *Entropy: Limits on the Possible*   472

12.1 STATISTICAL ISSUES 472

12.2 A STATISTICAL MODEL OF SOLIDS 474

12.3 THERMAL EQUILIBRIUM OF BLOCKS IN CONTACT 480

12.4 THE SECOND LAW OF THERMODYNAMICS 484

12.5 WHAT IS TEMPERATURE? 485

12.6 SPECIFIC HEAT CAPACITY OF A SOLID 489

12.7 THE BOLTZMANN DISTRIBUTION 493

12.8 THE BOLTZMANN DISTRIBUTION IN A GAS 496

SUMMARY 506

EXERCISES AND PROBLEMS 506

ANSWERS TO EXERCISES 511

CHAPTER 13   *Gases and Engines*   512

13.1 GASES, SOLIDS, AND LIQUIDS   512
13.2 GAS LEAKS THROUGH A HOLE   513
13.3 MEAN FREE PATH   516
13.4 PRESSURE AND TEMPERATURE   518
13.5 ENERGY TRANSFERS   524
13.6 FUNDAMENTAL LIMITATIONS ON EFFICIENCY   532
13.7 A MAXIMALLY EFFICIENT PROCESS   534
13.8 *WHY DON'T WE ATTAIN THE THEORETICAL EFFICIENCY?   543
13.9 *APPLICATION: A RANDOM WALK   546
13.10 *DERIVATION: MAXIMUM-POWER EFFICIENCY   548
SUMMARY   550
EXERCISES AND PROBLEMS   550
ANSWERS TO EXERCISES   552

VOLUME II   Electric and Magnetic Interactions

CHAPTER 14   *Electric Field*   553

14.1 NEW CONCEPTS   553
14.2 ELECTRIC CHARGE AND FORCE   554
14.3 THE CONCEPT OF "ELECTRIC FIELD"   555
14.4 THE ELECTRIC FIELD OF A POINT CHARGE   559
14.5 SUPERPOSITION OF ELECTRIC FIELDS   562
14.6 THE ELECTRIC FIELD OF A DIPOLE   564
14.7 CHOICE OF SYSTEM   573
14.8 IS ELECTRIC FIELD REAL?   573
SUMMARY   575
EXERCISES AND PROBLEMS   576
ANSWERS TO EXERCISES   583

CHAPTER 15   *Electric Fields and Matter*   584

15.1 CHARGED PARTICLES IN MATTER   584
15.2 HOW INSULATORS BECOME CHARGED   586
15.3 POLARIZATION   590
15.4 POLARIZATION OF INSULATORS   595
15.5 POLARIZATION OF CONDUCTORS   597
15.6 A MODEL OF A METAL   599
15.7 CHARGING AND DISCHARGING   606
15.8 WHEN THE FIELD CONCEPT IS LESS USEFUL   608
SUMMARY   609
BASIC EXPERIMENTS   610
ADDITIONAL EXPERIMENTS   614
EXERCISES AND PROBLEMS   616
ANSWERS TO EXERCISES   627

CHAPTER 16   *Electric Field of Distributed Charges*   628

16.1 OVERVIEW   628
16.2 A UNIFORMLY CHARGED THIN ROD   629
16.3 PROCEDURE FOR CALCULATING ELECTRIC FIELD   636
16.4 A UNIFORMLY CHARGED THIN RING   638
16.5 A UNIFORMLY CHARGED DISK   640
16.6 TWO UNIFORMLY CHARGED DISKS: A CAPACITOR   643

16.7 A SPHERICAL SHELL OF CHARGE   645
16.8 A SOLID SPHERE CHARGED THROUGHOUT ITS VOLUME   648
16.9 INFINITESIMALS AND INTEGRALS IN SCIENCE   650
16.10 *UNIFORM THIN ROD AT AN ARBITRARY LOCATION   650
16.11 *INTEGRATING THE SPHERICAL SHELL   651
S U M M A R Y   652
E X E R C I S E S   A N D   P R O B L E M S   654
A N S W E R S   T O   E X E R C I S E S   663

CHAPTER 17   *Electric Potential*   664

17.1 A REVIEW OF POTENTIAL ENERGY   664
17.2 SYSTEMS OF CHARGED OBJECTS   667
17.3 POTENTIAL DIFFERENCE IN A UNIFORM FIELD   671
17.4 SIGN OF POTENTIAL DIFFERENCE   674
17.5 POTENTIAL DIFFERENCE IN A NONUNIFORM FIELD   675
17.6 PATH INDEPENDENCE   682
17.7 THE POTENTIAL AT ONE LOCATION   686
17.8 POTENTIAL DIFFERENCE IN AN INSULATOR   690
17.9 ENERGY DENSITY AND ELECTRIC FIELD   692
17.10 *POTENTIAL OF DISTRIBUTED CHARGES   694
17.11 *INTEGRATING THE SPHERICAL SHELL   695
S U M M A R Y   697
E X E R C I S E S   A N D   P R O B L E M S   697
A N S W E R S   T O   E X E R C I S E S   709

CHAPTER 18   *Magnetic Field*   710

18.1 ELECTRON CURRENT   710
18.2 DETECTING MAGNETIC FIELDS   711
18.3 BIOT-SAVART LAW: SINGLE MOVING CHARGE   713
18.4 RELATIVISTIC EFFECTS   715
18.5 ELECTRON CURRENT & CONVENTIONAL CURRENT   716
18.6 THE BIOT-SAVART LAW FOR CURRENTS   719
18.7 THE MAGNETIC FIELD OF CURRENT DISTRIBUTIONS   720
18.8 A CIRCULAR LOOP OF WIRE   723
18.9 MAGNETIC DIPOLE MOMENT   725
18.10 THE MAGNETIC FIELD OF A BAR MAGNET   726
18.11 THE ATOMIC STRUCTURE OF MAGNETS   729
18.12 *ESTIMATE OF ORBITAL ANGULAR MOMENTUM OF AN ELECTRON IN AN ATOM   734
18.13 *MAGNETIC FIELD OF A SOLENOID   735
S U M M A R Y   737
E X P E R I M E N T S   738
E X E R C I S E S   A N D   P R O B L E M S   742
A N S W E R S   T O   E X E R C I S E S   750

CHAPTER 19   *Electric Field and Circuits*   751

19.1 OVERVIEW   751
19.2 CURRENT IN DIFFERENT PARTS OF A CIRCUIT   752
19.3 ELECTRIC FIELD AND CURRENT   754
19.4 WHAT CHARGES MAKE THE ELECTRIC FIELD IN THE WIRES?   757
19.5 CONNECTING A CIRCUIT: THE INITIAL TRANSIENT   762
19.6 FEEDBACK   764
19.7 SURFACE CHARGE AND RESISTORS   765
19.8 ENERGY IN A CIRCUIT   767

19.9 APPLICATIONS OF THE THEORY 772
19.10 DETECTING SURFACE CHARGE 777
SUMMARY 779
EXPERIMENTS 779
EXERCISES AND PROBLEMS 782
ANSWERS TO EXERCISES 791

CHAPTER 20   *Circuit Elements*   793

20.1 CAPACITORS 793
20.2 RESISTORS 800
20.3 WORK AND POWER IN A CIRCUIT 807
20.4 BATTERIES 808
20.5 AMMETERS, VOLTMETERS, AND OHMMETERS 810
20.6 QUANTITATIVE ANALYSIS OF AN RC CIRCUIT 813
20.7 REFLECTION: THE MACRO-MICRO CONNECTION 816
20.8 *WHAT ARE AC AND DC? 816
20.9 *ELECTRONS IN METALS 818
20.10 *A COMPLICATED RESISTIVE CIRCUIT 819
SUMMARY 821
EXPERIMENTS 822
EXERCISES AND PROBLEMS 823
ANSWERS TO EXERCISES 834

CHAPTER 21   *Magnetic Force*   835

21.1 MAGNETIC FORCE ON A MOVING CHARGE 836
21.2 MAGNETIC FORCE ON A CURRENT-CARRYING WIRE 840
21.3 COMBINING ELECTRIC AND MAGNETIC FORCES 843
21.4 THE HALL EFFECT 845
21.5 MOTIONAL EMF 850
21.6 MAGNETIC FORCE IN A MOVING REFERENCE FRAME 853
21.7 MAGNETIC TORQUE 857
21.8 POTENTIAL ENERGY FOR A MAGNETIC DIPOLE 858
21.9 MOTORS AND GENERATORS 863
21.10 *CASE STUDY: SPARKS IN AIR 866
21.11 *RELATIVISTIC FIELD TRANSFORMATIONS 875
SUMMARY 879
EXPERIMENTS 880
EXERCISES AND PROBLEMS 881
ANSWERS TO EXERCISES 896

CHAPTER 22   *Patterns of Field in Space*   898

22.1 PATTERNS OF ELECTRIC FIELD: GAUSS'S LAW 898
22.2 DEFINITION OF "ELECTRIC FLUX" 900
22.3 GAUSS'S LAW 902
22.4 REASONING FROM GAUSS'S LAW 908
22.5 GAUSS'S LAW FOR MAGNETISM 913
22.6 PATTERNS OF MAGNETIC FIELD: AMPERE'S LAW 914
22.7 MAXWELL'S EQUATIONS 920
22.8 *THE DIFFERENTIAL FORM OF GAUSS'S LAW 921
22.9 *THE DIFFERENTIAL FORM OF AMPERE'S LAW 926
22.10 *SEMICONDUCTOR DEVICES 928
SUMMARY 942
EXERCISES AND PROBLEMS 943
ANSWERS TO EXERCISES 947

CHAPTER 23 *Faraday's Law* 948

23.1 CURLY ELECTRIC FIELDS 948
23.2 FARADAY'S LAW 951
23.3 FARADAY'S LAW AND MOTIONAL EMF 959
23.4 MAXWELL'S EQUATIONS 962
23.5 SUPERCONDUCTORS 962
23.6 INDUCTANCE 965
23.7 *SOME PECULIAR CIRCUITS 973
23.8 *THE DIFFERENTIAL FORM OF FARADAY'S LAW 975
23.9 *LENZ'S RULE 975
SUMMARY 976
EXERCISES AND PROBLEMS 977
ANSWERS TO EXERCISES 985

CHAPTER 24 *Electromagnetic Radiation* 986

24.1 MAXWELL'S EQUATIONS 986
24.2 FIELDS TRAVELING THROUGH SPACE 989
24.3 ACCELERATED CHARGES PRODUCE RADIATION 995
24.4 SINUSOIDAL ELECTROMAGNETIC RADIATION 999
24.5 ENERGY AND MOMENTUM IN RADIATION 1002
24.6 EFFECTS OF RADIATION ON MATTER 1007
24.7 LIGHT PROPAGATION THROUGH A MEDIUM 1011
24.8 REFRACTION: BENDING OF LIGHT 1013
24.9 LENSES 1016
24.10 IMAGE FORMATION 1020
24.11 *THE FIELD OF AN ACCELERATED CHARGE 1030
24.12 *DIFFERENTIAL FORM OF MAXWELL'S EQUATIONS 1032
SUMMARY 1033
EXERCISES AND PROBLEMS 1033
ANSWERS TO EXERCISES 1039

CHAPTER 25 *Waves and Particles* 1041

25.1 WAVE PHENOMENA 1042
25.2 MULTI-SOURCE INTERFERENCE: DIFFRACTION 1050
25.3 THE WAVE MODEL vs. THE PARTICLE MODEL OF LIGHT 1054
25.4 FURTHER APPLICATIONS OF THE WAVE MODEL 1061
25.5 ANGULAR RESOLUTION 1066
25.6 STANDING WAVES 1070
25.7 *DERIVATION: TWO SLITS ARE LIKE TWO SOURCES 1076
SUMMARY 1077
EXERCISES AND PROBLEMS 1078
ANSWERS TO EXERCISES 1080

Index I-1

Interactions and Motion

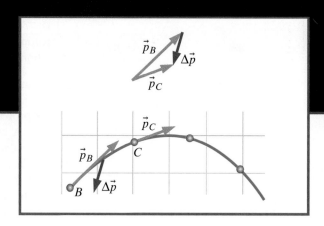

This textbook deals with the nature of matter and its interactions. The variety of phenomena that we will be able to explain and understand is very wide, including the orbit of stars around a black hole, nuclear fusion, and the speed of sound in a solid.

> The main goal of this textbook is to have you engage in a process central to science: the attempt to explain in detail a broad range of phenomena using a small set of powerful fundamental principles.

> The specific focus is on learning how to model the nature of matter and its interactions in terms of a small set of physical laws that govern all mechanical interactions, and in terms of the atomic structure of matter.

KEY IDEAS

Chapter 1 introduces the notion of interactions between material objects and the changes they produce.

- **Fundamental physics principles apply to all kinds of matter, from galaxies to subatomic particles.**
- **Change is an indication of an interaction.**
- **Position and motion in 3D space can be described precisely by vectors.**
- **The momentum of an object depends on both mass and velocity.**

■ Reading a science textbook efficiently and productively requires "active" reading. To encourage active reading, you will find "Stop and Think" questions, indicated by QUESTION . Trying to answer a question by using what you already know, as well as material you have just read, helps you learn more than if you just read passively. You might cover the rest of the page while you're thinking.

1.1 KINDS OF MATTER

We will deal with material objects of many sizes, from subatomic particles to galaxies. All of these objects have certain things in common.

Atoms and Nuclei

Ordinary matter is made up of tiny atoms. An atom isn't the smallest type of matter, for it is composed of even smaller objects (electrons, protons, and neutrons), but many of the ordinary everyday properties of ordinary matter can be understood in terms of atomic properties and interactions. As you probably know from studying chemistry, atoms have a very small, very dense core, called the nucleus, around which is found a cloud of electrons. The nucleus contains protons and neutrons, collectively called nucleons. Electrons are kept close to the nucleus by electric attraction to the protons (the neutrons hardly interact with the electrons).

Hydrogen
1 electron

1×10^{-10} m

Carbon
6 electrons

Iron
26 electrons

Uranium
92 electrons

Figure 1.1 Atoms of hydrogen, carbon, iron, and uranium. The gray blur represents the electron cloud surrounding the nucleus. The black dot shows the location of the nucleus. On this scale, however, the nucleus would be much too small to see.

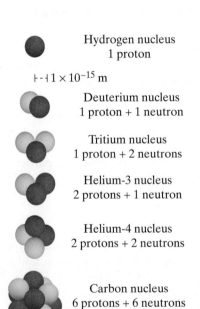

Hydrogen nucleus
1 proton

1×10^{-15} m

Deuterium nucleus
1 proton + 1 neutron

Tritium nucleus
1 proton + 2 neutrons

Helium-3 nucleus
2 protons + 1 neutron

Helium-4 nucleus
2 protons + 2 neutrons

Carbon nucleus
6 protons + 6 neutrons

Figure 1.2 Nuclei of hydrogen, helium, and carbon. Note the *very* much smaller scale than in Figure 1.1!

QUESTION   Recall your previous studies of chemistry. How many protons and electrons are there in a hydrogen atom? In a helium or carbon atom?

If you don't remember the properties of these atoms, see the periodic table on the inside front cover of this textbook. Hydrogen is the simplest atom, with just one proton and one electron. A helium atom has two protons and two electrons. A carbon atom has six protons and six electrons. Near the other end of the chemical periodic table, a uranium atom has 92 protons and 92 electrons. Figure 1.1 shows the relative sizes of the electron clouds in atoms of several elements but cannot show the nucleus to the same scale; the tiny dot marking the nucleus in the figure is much larger than the actual nucleus.

The radius of the electron cloud for a typical atom is about 1×10^{-10} meter. The reason for this size can be understood using the principles of quantum mechanics, a major development in physics in the early 20th century. The radius of a proton is about 1×10^{-15} meter, very much smaller than the radius of the electron cloud.

Nuclei contain neutrons as well as protons (Figure 1.2). The most common form or "isotope" of hydrogen has no neutrons in the nucleus. However, there exist isotopes of hydrogen with one or two neutrons in the nucleus (in addition to the proton). Hydrogen atoms containing one or two neutrons are called deuterium or tritium. The most common isotope of helium has two neutrons (and two protons) in its nucleus, but a rare isotope has only one neutron; this is called helium-3.

The most common isotope of carbon has six neutrons together with the six protons in the nucleus (carbon-12), whereas carbon-14 with eight neutrons is an isotope that plays an important role in dating archaeological objects.

Near the other end of the periodic table, uranium-235, which can undergo a fission chain reaction, has 92 protons and 143 neutrons, whereas uranium-238, which does not undergo a fission chain reaction, has 92 protons and 146 neutrons.

Molecules and Solids

When atoms come in contact with each other, they may stick to each other ("bond" to each other). Several atoms bonded together can form a molecule—a substance whose physical and chemical properties differ from those of the constituent atoms. For example, water molecules (H_2O) have properties quite different from the properties of hydrogen atoms or oxygen atoms.

An ordinary-sized rigid object made of bound-together atoms and big enough to see and handle is called a solid, such as a bar of aluminum. A new kind of microscope, the scanning tunneling microscope (STM), is able to map the locations of atoms on the surface of a solid, which has provided new techniques for investigating matter at the atomic level. Two such images appear in Figure 1.3. You can see that atoms in a crystalline solid are arranged in a regular three-dimensional array. The arrangement of atoms on the surface depends on the direction along which the crystal is cut. The irregularities in the bottom image reflect "defects," such as missing atoms, in the crystal structure.

Liquids and Gases

When a solid is heated to a higher temperature, the atoms in the solid vibrate more vigorously about their normal positions. If the temperature is raised high enough, this thermal agitation may destroy the rigid structure of the solid. The atoms may become able to slide over each other, in which case the substance is a liquid.

At even higher temperatures the thermal motion of the atoms or molecules may be so large as to break the interatomic or intermolecular bonds completely, and the liquid turns into a gas. In a gas the atoms or molecules are quite free to move around, only occasionally colliding with each other or the walls of their container.

We will learn how to analyze many aspects of the behavior of solids and gases. We won't have much to say about liquids, because their properties are much harder to analyze. Solids are simpler to analyze than liquids because the atoms stay in one place (though with thermal vibration about their usual positions). Gases are simpler to analyze than liquids because between collisions the gas molecules are approximately unaffected by the other molecules. Liquids are the awkward intermediate state, where the atoms move around rather freely but are always in contact with other atoms. This makes the analysis of liquids very complex.

Figure 1.3 Two different surfaces of a crystal of pure silicon. The images were made with a scanning tunneling microscope. Images courtesy of Randall Feenstra, Carnegie Mellon University.

Planets, Stars, Solar Systems, and Galaxies

In our brief survey of the kinds of matter that we will study, we make a giant leap in scale from atoms all the way up to planets and stars, such as our Earth and Sun. We will see that many of the same principles that apply to atoms apply to planets and stars. By making this leap we bypass an important physical science, geology, whose domain of interest includes the formation of mountains and continents. We will study objects that are much bigger than mountains, and we will study objects that are much smaller than mountains, but we don't have time to apply the principles of physics to every important kind of matter.

Our Sun and its accompanying planets constitute our Solar System. It is located in the Milky Way galaxy, a giant rotating disk-shaped system of stars. On a clear dark night you can see a band of light (the Milky Way) coming from the huge number of stars lying in this disk, which you are looking at from a position in the disk, about two-thirds of the way out from the center of the disk. Our galaxy is a member of a cluster of galaxies that move around each other much as the planets of our Solar System move around the Sun (Figure 1.4). The Universe contains many such clusters of galaxies.

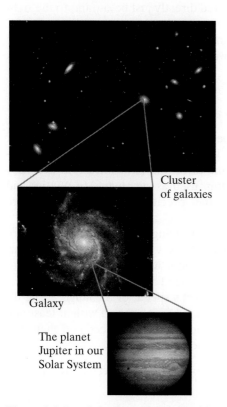

Cluster of galaxies

Galaxy

The planet Jupiter in our Solar System

Figure 1.4 Our Solar System exists inside a galaxy, which itself is a member of a cluster of galaxies.

1.2 | DETECTING INTERACTIONS

Objects made of different kinds of matter interact with each other in various ways: gravitationally, electrically, magnetically, and through nuclear interactions. How can we detect that an interaction has occurred? In this section we consider various kinds of observations that indicate the presence of interactions.

QUESTION Before you read further, take a moment to think about your own ideas of interactions. How can you tell that two objects are interacting with each other?

Change of Direction of Motion

Suppose that you observe a proton moving through a region of outer space, far from almost all other objects. The proton moves along a path like the one shown in Figure 1.5. The arrow indicates the initial direction of the proton's motion, and the "×'s" in the diagram indicate the position of the proton at equal time intervals.

QUESTION Do you see evidence in Figure 1.5 that the proton is interacting with another object?

Figure 1.5 A proton moves through space, far from almost all other objects. The initial direction of the proton's motion is upward, as indicated by the arrow. The ×'s represent the position of the proton at equal time intervals.

Figure 1.6 An electron moves through space, far from almost all other objects. The initial direction of the electron's motion is upward and to the left, as indicated by the arrow. The ×'s represent the position of the electron at equal time intervals.

Figure 1.7 Two successive positions of a particle (indicated by a dot), with arrows indicating the velocity of the particle at each location. The shorter arrow indicates that the speed of the particle at location 2 is less than its speed at location 1.

Evidently a change in direction is a vivid indicator of interactions. If you observe a change in direction of the motion of a proton, you will find another object somewhere that has interacted with this proton.

> QUESTION   Suppose that the only other object nearby was another proton. What was the approximate initial location of this second proton?

Since two protons repel each other electrically, the second proton must have been located to the right of the bend in the first proton's path.

Change of Speed

Suppose that you observe an electron traveling in a straight line through outer space far from almost all other objects (Figure 1.6). The path of the electron is shown as though a camera had taken multiple exposures at equal time intervals.

> QUESTION   Where is the electron's speed largest? Where is the electron's speed smallest?

The speed is largest at the upper left, where the ×'s are farther apart, which means that the electron has moved farthest during the time interval between exposures. The speed is smallest at the bottom right, where the ×'s are closer together, which means that the electron has moved the least distance during the time interval between exposures.

> QUESTION   Suppose that the only other object nearby was another electron. What was the approximate initial location of this other electron?

The other electron must have been located directly just below and to the right of the starting location, since electrons repel each other electrically.

Evidently a change in speed is an indicator of interactions. If you observe a change in speed of an electron, you will find another object somewhere that has interacted with the electron.

Change of Velocity: Change of Speed or Direction

In physics, the word "velocity" has a special technical meaning that is different from its meaning in everyday speech. In physics, the quantity called "velocity" denotes a combination of speed and direction. (In contrast, in everyday speech, "speed" and "velocity" are often used as synonyms. In physics and other sciences, however, words have rather precise meanings and there are few synonyms.)

For example, consider an airplane that is flying with a speed of 1000 kilometers/hour in a direction that is due east. We say the velocity is 1000 km/hr, east, where we specify both speed and direction. An airplane flying west with a speed of 1000 km/hr would have the same speed but a different velocity.

We have seen that a change in an object's speed, or a change in the direction of its motion, indicates that the object has interacted with at least one other object. The two indicators of interaction, change of speed and change of direction, can be combined into one compact statement:

A change of velocity (speed or direction or both) indicates the existence of an interaction.

Pushing forward or back, parallel to the direction of the motion, will make an object speed up or slow down but cannot make it turn. To change the direction of the motion you have to push sideways, perpendicular to the motion.

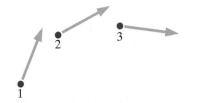

Figure 1.8 Three successive positions of a particle (indicated by a dot), with arrows indicating the velocity of the particle at each location. The arrows are the same length, indicating the same speed, but they point in different directions, indicating a change in direction and therefore a change in velocity.

Figure 1.9 "Uniform motion"—no change in speed or direction.

Diagrams Showing Changes in Velocity

In physics diagrams, the velocity of an object is represented by an arrow: a line with an arrowhead. The tail of the arrow is placed at the location of the object, and the arrow points in the direction of the motion of the object. The length of the arrow is proportional to the speed of the object. Figure 1.7 shows two successive positions of a particle at two different times, with velocity arrows indicating a change in speed of the particle (it's slowing down). Figure 1.8 shows three successive positions of a different particle at three different times, with velocity arrows indicating a change in direction but no change in speed.

We will see a little later in the chapter that velocity is only one example of a physical quantity that has a "magnitude" (an amount or a size) and a direction.

Other examples of such quantities are position relative to an origin in 3D space, force, and magnetic field. Quantities having magnitude and direction can be usefully described as "vectors." Vectors are mathematical quantities that have their own special rules of algebra, similar (but not identical) to the rules of ordinary algebra. Arrows are commonly used in diagrams to denote vector quantities. We will use vectors extensively.

Uniform Motion

Suppose that you observe a rock moving along in outer space far from all other objects. We don't know what made it start moving in the first place; presumably a long time ago an interaction gave it some velocity and it has been coasting through the vacuum of space ever since.

It is an observational fact that such an isolated object moves at constant, unchanging speed, in a straight line. Its velocity does not change (neither its direction nor its speed changes). We call such motion with unchanging velocity "uniform motion" (Figure 1.9).

An Object at Rest

A special case of uniform motion is the case in which an object's speed is zero and remains zero—the object remains at rest. In this case the object's speed is constant (zero) and the direction of motion, while undefined, is not changing.

Uniform Motion Implies No Net Interaction

When we observe an object in uniform motion, we conclude that since its velocity is not changing, either it is not interacting significantly with any other object, or else it is undergoing multiple interactions that cancel each other out. In either case, we can say that there is no "net" (total) interaction.

■ **Exercises**
At the end of a section you will usually find short exercises. It is important to work through exercises as you come to them, to make sure that you can apply what you have just read. Simply reading about concepts in physics is not enough—you must be able to use the concepts in answering questions and solving problems. Make a serious attempt to do an exercise before checking the answer at the end of the chapter. This will help you assess your own understanding.

1.X.1 Which of the following do you see moving with constant velocity?
(a) A ship sailing northeast at a speed of 5 meters per second
(b) The Moon orbiting the Earth
(c) A tennis ball traveling across the court after having been hit by a tennis racket
(d) A can of soda sitting on a table
(e) A person riding on a Ferris wheel that is turning at a constant rate
1.X.2 In which of the following situations is there observational evidence for significant interaction between two objects? How can you tell?
(a) A ball bounces off a wall with no change in speed.
(b) A baseball that was hit by a batter flies toward the outfield.
(c) A communications satellite orbits the Earth.

(d) A space probe travels at constant speed toward a distant star.

(e) A charged particle leaves a curving track in a particle detector.

1.3   NEWTON'S FIRST LAW OF MOTION

The basic relationship between change of velocity and interaction is summarized qualitatively by Newton's "first law of motion":

NEWTON'S FIRST LAW OF MOTION

**An object moves in a straight line and at constant speed
except to the extent that it interacts with other objects.**

The words "to the extent" imply that the stronger the interaction, the more change there will be in direction and/or speed. The weaker the interaction, the less change. If there is no net interaction at all, the direction doesn't change and the speed doesn't change (uniform motion). This case can also be called "uniform velocity" or "constant velocity," since velocity refers to both speed and direction. It is important to remember that if an object is not moving at all, its velocity is not changing, so it too may be considered to be in uniform motion.

Newton's first law of motion is only qualitative, because it doesn't give us a way to calculate quantitatively how much change in speed or direction will be produced by a certain amount of interaction, a subject we will take up in the next chapter. Nevertheless, Newton's first law of motion is important in providing a conceptual framework for thinking about the relationship between interaction and motion.

The English physicist Isaac Newton was the first person to state this law clearly. Newton's first law of motion represented a major break with ancient tradition, which assumed that constant pushing was required to keep something moving. This law says something radically different: no interactions at all are needed to keep something moving!

Does Newton's First Law Apply in Everyday Life?

Superficially, Newton's first law of motion may seem at first not to apply to many everyday situations. To push a chair across the floor at constant speed, you have to keep pushing all the time.

QUESTION   Doesn't Newton's first law of motion say that the chair should keep moving at constant speed without anyone pushing it? In fact, shouldn't the speed or direction of motion of the chair change due to the interaction with your hands? Does this everyday situation violate Newton's first law of motion? Try to answer these questions before reading farther.

The complicating factor here is that your hands aren't the only objects that are interacting with the chair. The floor also interacts with the chair, in a way that we call friction. If you push just hard enough to compensate exactly for the floor friction, the sum of all the interactions is zero, and the chair moves at constant speed as predicted by Newton's first law. (If you push harder than the floor does, the chair's speed does increase.)

Motion without Friction

It is difficult to observe motion without friction in everyday life, because objects almost always interact with many other objects, including air, flat surfaces, and so on. This explains why it took people such a long time to understand clearly the relationship between interaction and change (Newton was born in 1642).

You may be able to think of situations in which you have seen an object keep moving at constant (or nearly constant) velocity, without being pushed or pulled. One example of a nearly friction-free situation is a hockey puck sliding on ice. The puck slides a long way at nearly constant speed in a straight line (constant velocity) because there is little friction with the ice. An even better example is the uniform motion of an object in outer space, far from all other objects.

1.X.3 Apply Newton's first law to each of the following situations. In which situations can you conclude that the object is undergoing a net interaction with one or more other objects?

  (a) A book slides across the table and comes to a stop.

  (b) A proton in a particle accelerator moves faster and faster.

  (c) A car travels at constant speed around a circular race track.

  (d) A spacecraft travels at a constant speed toward a distant star.

  (e) A hydrogen atom remains at rest in outer space.

1.X.4 A spaceship far from all other objects uses its rockets to attain a speed of 1×10^4 m/s. The crew then shuts off the power. According to Newton's first law, which of the following statements about the motion of the spaceship after the power is shut off are correct? (Choose all statements that are correct.)

  (a) The spaceship will move in a straight line.

  (b) The spaceship will travel on a curving path.

  (c) The spaceship will enter a circular orbit.

  (d) The speed of the spaceship will not change.

  (e) The spaceship will gradually slow down.

  (f) The spaceship will stop suddenly.

1.4   OTHER INDICATORS OF INTERACTION

Change of Identity

Change of velocity (change of speed and/or direction) is not the only indicator of interactions. Another is change of identity, such as the formation of water (H_2O) from the burning of hydrogen in oxygen. A water molecule behaves very differently from the hydrogen and oxygen atoms of which it is made.

Change of Shape or Configuration

Another indicator of interaction is change of shape or configuration (arrangement of the parts). For example, slowly bend a pen or pencil, then hold it in the bent position. The speed hasn't changed, nor is there a change in the direction of motion (it's not moving!). The pencil has not changed identity. Apparently a change of shape can be evidence for interactions, in this case with your hand.

Other changes in configuration include "phase changes" such as the freezing or boiling of a liquid, brought about by interactions with the surroundings. In different phases (solid, liquid, gas), atoms or molecules are arranged differently. Changes in configuration at the atomic level are another indication of interactions.

Change of Temperature

Another indication of interaction is change of temperature. Place a pot of cold water on a hot stove. As time goes by, a thermometer will indicate a change in the water due to interaction with the hot stove.

Other Indications of Interactions

Is a change of position an indicator of an interaction? That depends. If the change of position occurs simply because a particle is moving at constant speed and direction, then a mere change of position is not an indicator of an interaction, since uniform motion is an indicator of no net interaction.

> QUESTION However, if you observe an object at rest in one location, and later you observe it again at rest but in a different location, did an interaction take place?

Yes. You can infer that there must have been an interaction to give the object some velocity to move the object toward the new position, and another interaction to slow the object to a stop in its new position.

In later chapters we will consider interactions involving change of identity, change of shape, and change of temperature, but for now we'll concentrate on interactions that cause a change of velocity (speed and/or direction).

Indirect Evidence for an Additional Interaction

Sometimes there is indirect evidence for an additional interaction. When something doesn't change although you would normally expect a change due to a known interaction, this indicates that another interaction is present. Consider a balloon that hovers motionless in the air despite the downward gravitational pull of the Earth. Evidently there is some other kind of interaction that opposes the gravitational interaction. In this case, interactions with air molecules have the net effect of pushing up on the balloon ("buoyancy"). The lack of change implies that the effect of the air molecules exactly compensates for the gravitational interaction with the Earth.

When you push a chair across the floor and it moves with constant velocity despite your pushing on it (which ought to change its speed), that means that something else must also be interacting with it (the floor).

The stability of the nucleus of an atom is another example of indirect evidence for an additional interaction. The nucleus contains positively charged protons that repel each other electrically, yet the nucleus remains intact. We conclude that there must be some other kind of interaction present, a non-electric attractive interaction that overcomes the electric repulsion. This is evidence for an interaction called the "strong interaction" that acts between protons and neutrons in the nucleus.

Summary: Changes as Indicators of Interactions

Here then are the most common indicators of interactions:

- Change of velocity (change of direction and/or change of speed)
- Change of identity
- Change of shape

- Change of temperature
- Lack of change when change is expected (indirect evidence)

The important point is this: **Interactions cause change.**

In the absence of interactions, there is no change, which is usually uninteresting. The exception is the surprise when nothing changes despite our expectations that something should change. This is indirect evidence for some interaction that we hadn't recognized was present, that more than one interaction is present and the interactions cancel each others' effects.

For the next few chapters we'll concentrate on change of velocity as evidence for an interaction (or lack of change of velocity, which can give indirect evidence for additional interactions).

1.X.5 You slide a coin across the floor, and observe that it travels in a straight line, slowing down and eventually stopping. A sensitive thermometer shows that the coin's temperature increased. What can we conclude? (Choose all statements that are correct.)

(a) Because the coin traveled in a straight line, we conclude that it did not interact with anything.

(b) Because the coin did not change shape, we conclude that it did not interact with anything.

(c) Because the coin slowed down, we conclude that Newton's first law does not apply to objects in everyday life, such as coins.

(d) Because the coin's speed changed, we conclude that it interacted with one or more other objects.

(e) Because the coin got hot, we conclude that it interacted with one or more other objects.

1.5   DESCRIBING THE 3D WORLD: VECTORS

Physical phenomena take place in the 3D world around us. In order to be able to make quantitative predictions and give detailed, quantitative explanations, we need tools for describing precisely the positions and velocities of objects in 3D, and the changes in position and velocity due to interactions. These tools are mathematical entities called 3D "vectors."

3D Coordinates

Figure 1.10 Right-handed 3D coordinate system.

We will use a 3D coordinate system to specify positions in space and other vector quantities. Usually we will orient the axes of the coordinate system as shown in Figure 1.10: $+x$ axis to the right, $+y$ axis upward, and $+z$ axis coming out of the page, toward you. This is a "right-handed" coordinate system: if you hold the thumb, first, and second fingers of your right hand perpendicular to each other, and align your thumb with the x axis and your first finger with the y axis, your second finger points along the z axis. (In some math and physics textbook discussions of 3D coordinate systems, the x axis points out, the y axis points to the right, and the z axis points up, but we will also use a 2D coordinate system with y up, so it makes sense always to have the y axis point up.)

Basic Properties of Vectors: Magnitude and Direction

A vector is a quantity that has a magnitude and a direction. For example, the velocity of a baseball is a vector quantity. The magnitude of the baseball's velocity is the speed of the baseball—for example, 20 meters/second. The direction of the baseball's velocity is the direction of its motion at a particular instant—for example, "up" or "to the right" or "west" or "in the $+y$ direction."

A symbol denoting a vector is written with an arrow over it:

$$\vec{v} \text{ is a vector}$$

Position

A position in space can also be considered to be a vector, called a *position vector*, pointing from an origin to that location. Figure 1.11 shows a position vector that might represent your final position if you started at the origin and walked 4 meters along the x axis, then 2 meters parallel to the z axis, then climbed a ladder so you were 3 meters above the ground. Your new position relative to the origin is a vector that can be written like this:

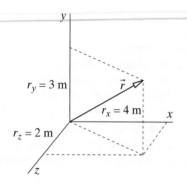

Figure 1.11 A position vector $\vec{r} = \langle 4, 3, 2 \rangle$ m and its x, y, and z components.

$$\vec{r} = \langle 4, 3, 2 \rangle \text{ m}$$
$$x \text{ component } r_x = 4 \text{ m}$$
$$y \text{ component } r_y = 3 \text{ m}$$
$$z \text{ component } r_z = 2 \text{ m}$$

In three dimensions a vector is a triple of numbers $\langle x, y, z \rangle$. Quantities like the position of an object and the velocity of an object can be represented as vectors:

$$\vec{r} = \langle x, y, z \rangle \text{ (a position vector)}$$
$$\vec{r}_1 = \langle 3.2, -9.2, 66.3 \rangle \text{ m (a position vector)}$$
$$\vec{v} = \langle v_x, v_y, v_z \rangle \text{ (a velocity vector)}$$
$$\vec{v}_1 = \langle -22.3, 0.4, -19.5 \rangle \text{ m/s (a velocity vector)}$$

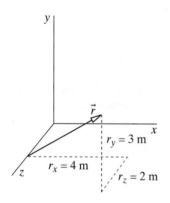

Figure 1.12 The arrow represents the vector $\vec{r} = \langle 4, 3, 2 \rangle$ m, drawn with its tail at location $\langle 0, 0, 2 \rangle$.

Components of a Vector

Each of the numbers in the triple is referred to as a *component* of the vector. The x component of the vector $\vec{v}$ is the number v_x. The z component of the vector $\vec{v}_1 = \langle -22.3, 0.4, -19.5 \rangle$ m/s is -19.5 m/s. A component such as v_x is not a vector, since it is only one number.

It is important to note that the x component of a vector specifies the difference between the x coordinate of the tail of the vector and the x coordinate of the tip of the vector. It does not give any information about the location of the tail of the vector (compare Figure 1.11 and Figure 1.12).

Drawing Vectors

In Figure 1.11 we represented your position vector relative to the origin graphically by an arrow whose tail is at the origin and whose arrowhead is at your position. The length of the arrow represents the distance from the origin, and the direction of the arrow represents the direction of the vector, which is the direction of a direct path from the initial position to the final position (the "displacement"; by walking and climbing you "displaced" yourself from the origin to your final position).

Since it is difficult to draw a 3D diagram on paper, when working on paper you will usually be asked to draw vectors that all lie in a single plane. Figure 1.13 shows an arrow in the xy plane representing the vector $\langle -3, -1, 0 \rangle$.

Figure 1.13 The position vector $\langle -3, -1, 0 \rangle$, drawn at the origin, in the xy plane. The components of the vector specify the displacement from the tail to the tip. The z axis, which is not shown, comes out of the page, toward you.

Vectors and Scalars

A quantity that is represented by a single number is called a *scalar*. A scalar quantity does not have a direction. Examples include the mass of an object, such as 5 kg, or the temperature, such as $-20°$C. Vectors and scalars are very different entities; a vector can never be equal to a scalar, and a scalar cannot be added to a vector. Scalars can be positive or negative:

$$m = 50\,\text{kg}$$
$$T = -20°\text{C}$$

Although a component of a vector such as v_x is not a vector, it's not a scalar either, despite being only one number. An important property of a true scalar is that its value doesn't change if we orient the xyz coordinate axes differently. Rotating the axes doesn't change an object's mass, or the temperature, but it does change what we mean by the x component of the velocity since the x axis now points in a different direction.

1.X.6 How many numbers are needed to specify a 3D position vector?

1.X.7 How many numbers are needed to specify a scalar?

Magnitude of a Vector

In Figure 1.14 we again show the vector from Figure 1.11, showing your displacement from the origin. Using a 3D extension of the Pythagorean theorem for right triangles (Figure 1.15), the net distance you have moved from the starting point is

$$\sqrt{(4\,\text{m})^2 + (3\,\text{m})^2 + (2\,\text{m})^2} = \sqrt{29}\,\text{m} = 5.39\,\text{m}$$

We say that the *magnitude* $|\vec{r}|$ of the position vector $\vec{r}$ is

$$|\vec{r}| = 5.39\,\text{m}$$

The magnitude of a vector is written either with absolute-value bars around the vector as $|\vec{r}|$, or simply by writing the symbol for the vector without the little arrow above it, r.

The magnitude of a vector can be calculated by taking the square root of the sum of the squares of its components (see Figure 1.15).

Figure 1.14 A vector representing a displacement from the origin.

Figure 1.15 The magnitude of a vector is the square root of the sum of the squares of its components (3D version of the Pythagorean theorem).

MAGNITUDE OF A VECTOR

If the vector $\vec{r} = \langle r_x, r_y, r_z \rangle$ then $|\vec{r}| = \sqrt{r_x^2 + r_y^2 + r_z^2}$ (a scalar).

The magnitude of a vector is always a positive number. The magnitude of a vector is a single number, not a triple of numbers, and it is a scalar, not a vector.

The magnitude of a vector is a true scalar, because its value doesn't change if you rotate the coordinate axes. Rotating the axes changes the individual components, but the length of the arrow representing the vector doesn't change.

Can a *Vector* Be Positive or Negative?

QUESTION Consider the vector $\vec{v} = \langle 8 \times 10^6, 0, -2 \times 10^7 \rangle$ m/s. Is this vector positive? Negative? Zero?

None of these descriptions is appropriate. The x component of this vector is positive, the y component is zero, and the z component is negative. Vectors aren't positive, or negative, or zero. Their components can be positive or negative or zero, but these words just don't mean anything when used with the vector as a whole.

On the other hand, the *magnitude* of a vector such as $|\vec{v}|$ is always positive.

> **1.X.8** If $\vec{r} = \langle -3, -4, 1 \rangle$ m, find $|\vec{r}|$.
>
> **1.X.9** Can the magnitude of a vector be a negative number?

Mathematical Operations Involving Vectors

Although the algebra of vectors is similar to the scalar algebra with which you are very familiar, it is not identical. There are some algebraic operations that cannot be performed on vectors.

Algebraic operations that *are* legal for vectors include the following operations, which we will discuss in this chapter:

- Adding one vector to another vector: $\vec{a} + \vec{w}$
- Subtracting one vector from another vector: $\vec{b} - \vec{d}$
- Finding the magnitude of a vector: $|\vec{r}|$
- Finding a unit vector (a vector of magnitude 1): $\hat{r}$
- Multiplying (or dividing) a vector by a scalar: $3\vec{v}$ or $\vec{w}/2$
- Finding the rate of change of a vector: $\Delta \vec{r}/\Delta t$ or $d\vec{r}/dt$.

In later chapters we will also see that there are two more ways of combining two vectors:

> The vector dot product, whose result is a scalar
>
> The vector cross product, whose result is a vector

Operations That Are *Not* Legal for Vectors

Although vector algebra is similar to the ordinary scalar algebra you have used up to now, there are certain operations that are not legal (and not meaningful) for vectors:

> A vector cannot be set equal to a scalar.
>
> A vector cannot be added to or subtracted from a scalar.
>
> A vector cannot occur in the denominator of an expression. (Although you can't divide by a vector, note that you can legally divide by the *magnitude* of a vector, which is a scalar.)

> **1.X.10** The vector $\vec{g} = \langle 2, -7, 3 \rangle$ and the scalar $h = -2$. What is $h + \vec{g}$?
> - **(a)** $\langle 0, -9, 1 \rangle$
> - **(b)** $\langle 4, -5, 5 \rangle$
> - **(c)** $\langle 4, 9, 5 \rangle$
> - **(d)** This is a meaningless expression.
>
> **1.X.11** Is $4/\langle 6, -7, 4 \rangle$ a meaningful expression? If so, what is its value?

Multiplying a Vector by a Scalar

A vector can be multiplied (or divided) by a scalar. If a vector is multiplied by a scalar, each of the components of the vector is multiplied by the scalar:

If $\vec{r} = \langle x, y, z \rangle$, then $a\vec{r} = \langle ax, ay, az \rangle$

If $\vec{v} = \langle v_x, v_y, v_z \rangle$, then $\dfrac{\vec{v}}{b} = \left\langle \dfrac{v_x}{b}, \dfrac{v_y}{b}, \dfrac{v_z}{b} \right\rangle$

$$\left\langle \frac{1}{2} \right\rangle \langle 6, -20, 9 \rangle = \langle 3, -10, 4.5 \rangle$$

Multiplication by a scalar "scales" a vector, keeping its direction the same but making its magnitude larger or smaller (Figure 1.16). Multiplying by a negative scalar reverses the direction of a vector.

1.X.12 The vector $\vec{a} = \langle 0.03, -1.4, 26.0 \rangle$ and the scalar $f = -3.0$. What is $f\vec{a}$?
1.X.13 If $\vec{r} \langle 2, -3, 5 \rangle$ m, what is $\vec{r}/2$?

Figure 1.16 Multiplying a vector by a scalar changes the magnitude of the vector. Multiplying by a negative scalar reverses the direction of the vector.

Magnitude of a Scalar

You may wonder how to find the magnitude of a quantity like $-3\vec{r}$, which involves the product of a scalar and a vector. This expression can be factored:

$$|-3\vec{r}| = |-3| \cdot |\vec{r}|$$

The magnitude of a scalar is its absolute value, so:

$$|-3\vec{r}| = |-3| \cdot |\vec{r}| = 3\sqrt{r_x^2 + r_y^2 + r_z^2}$$

1.X.14 If $\vec{v} = \langle 2, -3, 5 \rangle$ m/s, what is the magnitude of $3\vec{v}$?
1.X.15 How does the magnitude of the vector $-\vec{a}$ compare to the magnitude of the vector $\vec{a}$?

Direction of a Vector: Unit Vectors

One way to describe the direction of a vector is by specifying a *unit vector*. A unit vector is a vector of magnitude 1, pointing in some direction. A unit vector is written with a "hat" (caret) over it instead of an arrow. The unit vector $\hat{a}$ is called "a-hat."

QUESTION   Is the vector $\langle 1, 1, 1 \rangle$ a unit vector?

The magnitude of $\langle 1, 1, 1 \rangle$ is $\sqrt{1^2 + 1^2 + 1^2} = 1.73$, so this is not a unit vector. The vector $\langle 1/\sqrt{3}, 1/\sqrt{3}, 1/\sqrt{3} \rangle$ is a unit vector, since its magnitude is 1:

$$\sqrt{\left(\frac{1}{\sqrt{3}}\right)^2 + \left(\frac{1}{\sqrt{3}}\right)^2 + \left(\frac{1}{\sqrt{3}}\right)^2} = 1$$

Note that every component of a unit vector must be less than or equal to 1.

In our 3D Cartesian coordinate system, there are three special unit vectors, oriented along the three axes. They are called i-hat, j-hat, and k-hat, and they point along the x, y, and z axes, respectively (Figure 1.17):

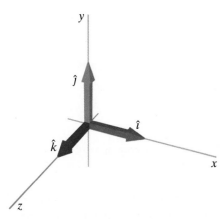

Figure 1.17 The unit vectors $\hat{i}$, $\hat{j}$, $\hat{k}$.

$$\hat{i} = \langle 1, 0, 0 \rangle$$
$$\hat{j} = \langle 0, 1, 0 \rangle$$
$$\hat{k} = \langle 0, 0, 1 \rangle$$

One way to express a vector is in terms of these special unit vectors:

$$\langle 0.02, -1.7, 30.0 \rangle = 0.02\hat{\imath} + (-1.7)\hat{\jmath} + 30.0\hat{k}$$

We will usually use the $\langle x, y, z \rangle$ form rather than the $\hat{\imath}\hat{\jmath}\hat{k}$ form in this book, because the familiar $\langle x, y, z \rangle$ notation, used in many calculus textbooks, emphasizes that a vector is a single entity.

Not all unit vectors point along an axis, as shown in Figure 1.18. For example, the vectors

$$\hat{g} = \langle 0.5774, 0.5774, 0.5774 \rangle \quad \text{and} \quad \hat{F} = \langle 0.424, 0.566, 0.707 \rangle$$

are both unit vectors, since the magnitude of each is equal to 1. Note that every component of a unit vector is less than or equal to 1.

$\vec{v} = \langle 1.5, 1.5, 0 \rangle$ m/s

$\hat{v} = \langle \frac{\sqrt{2}}{2}, \frac{\sqrt{2}}{2}, 0 \rangle$

Figure 1.18 The unit vector $\hat{v}$ has the same direction as the vector $\vec{v}$, but its magnitude is 1, and it has no physical units.

Calculating Unit Vectors

Any vector may be factored into the product of a unit vector in the direction of the vector, multiplied by a scalar equal to the magnitude of the vector.

$$\vec{w} = |\vec{w}| \cdot \hat{w}$$

For example, a vector of magnitude 5, aligned with the y axis, could be written as:

$$\langle 0, 5, 0 \rangle = 5\langle 0, 1, 0 \rangle$$

Therefore, to find a unit vector in the direction of a particular vector, we just divide the vector by its magnitude:

CALCULATING A UNIT VECTOR

$$\hat{r} = \frac{\vec{r}}{|\vec{r}|} = \frac{\langle x, y, z \rangle}{\sqrt{(x^2 + y^2 + z^2)}}$$

$$\hat{r} = \left\langle \frac{x}{\sqrt{(x^2 + y^2 + z^2)}}, \frac{y}{\sqrt{(x^2 + y^2 + z^2)}}, \frac{z}{\sqrt{(x^2 + y^2 + z^2)}} \right\rangle$$

EXAMPLE   Unit Vector

If $\vec{v} = \langle -22.3, 0.4, -19.5 \rangle$ m/s, then

$$\hat{v} = \frac{\vec{v}}{|\vec{v}|} = \frac{\langle -22.3, 0.4, -19.5 \rangle \text{ m/s}}{\sqrt{(-22.3)^2 + (0.4)^2 + (-19.5)^2} \text{ m/s}} = \langle -0.753, 0.0135, -0.658 \rangle$$

Remember that to divide a vector by a scalar, you divide each component of the vector by the scalar. The result is a new vector. Note also that a unit vector has no physical units (such as meters per second), because the units in the numerator and denominator cancel.

1.X.16 What is the unit vector in the direction of $\langle 0, 6, 0 \rangle$?

1.X.17 What is the unit vector $\hat{a}$ in the direction of $\vec{a}$, where $\vec{a} = \langle 400, 200, -100 \rangle$ m/s^2?

Equality of Vectors

EQUALITY OF VECTORS

A vector is equal to another vector if and only if all the components of the vectors are equal.

$$\vec{w} = \vec{r} \quad \text{means that}$$

$$w_x = r_x \quad \text{and} \quad w_y = r_y \quad \text{and} \quad w_z = r_z$$

The magnitudes and directions of two equal vectors are the same:

$$|\vec{w}| = |\vec{r}| \quad \text{and} \quad \hat{w} = \hat{r}$$

EXAMPLE   **Equal Vectors**

$$\vec{r} = \langle 4, 3, 2 \rangle$$

$$|\vec{r}| = \sqrt{(4^2 + 3^2 + 2^2)} = 5.39$$

$$\hat{r} = \frac{\langle 4, 3, 2 \rangle}{5.39} = \langle 0.742, 0.557, 0.371 \rangle$$

If $\vec{w} = \vec{r}$, then

$$\vec{w} = \langle 4, 3, 2 \rangle$$

$$|\vec{w}| = 5.39$$

$$\hat{w} = \langle 0.742, 0.557, 0.371 \rangle$$

1.X.18 If $\vec{a} = \langle -3, 7, 0.5 \rangle$, and $\vec{b} = \vec{a}$, what is the y component of $\vec{b}$?

1.X.19 Consider the vectors $\vec{r}_1$ and $\vec{r}_2$ represented by arrows in Figure 1.19. Are these two vectors equal?

Vector Addition

Figure 1.19 Are these two vectors equal? (Exercise 1.X.19.)

ADDING VECTORS

The sum of two vectors is another vector, obtained by adding the components of the vectors.

$$\vec{A} = \langle A_x, A_y, A_z \rangle$$

$$\vec{B} = \langle B_x, B_y, B_z \rangle$$

$$\vec{A} + \vec{B} = \langle (A_x + B_x), (A_y + B_y), (A_z + B_z) \rangle$$

EXAMPLE   **Adding Vectors**

$$\langle 1, 2, 3 \rangle + \langle -4, 5, 6 \rangle = \langle -3, 7, 9 \rangle$$

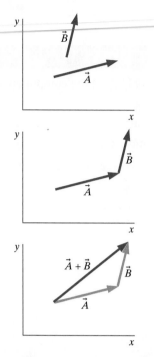

Figure 1.20 The procedure for adding two vectors graphically: draw vectors tip to tail. To add $\vec{A} + \vec{B}$ graphically, move $\vec{B}$ so the tail of $\vec{B}$ is at the tip of $\vec{A}$, then draw a new arrow starting at the tail of $\vec{A}$ and ending at the tip of $\vec{B}$.

Warning: Don't Add Magnitudes!

The magnitude of a vector is *not* in general equal to the sum of the magnitudes of the two original vectors! For example, the magnitude of the vector $\langle 3, 0, 0 \rangle$ is 3, and the magnitude of the vector $\langle -2, 0, 0 \rangle$ is 2, but the magnitude of the vector $(\langle 3, 0, 0 \rangle + \langle -2, 0, 0 \rangle)$ is 1, not 5!

Adding Vectors Graphically: Tip to Tail

The sum of two vectors has a geometric interpretation. In Figure 1.20 you first walk along displacement vector $\vec{A}$, followed by walking along displacement vector $\vec{B}$. What is your net displacement vector $\vec{C} = \vec{A} + \vec{B}$? The x component C_x of your net displacement is the sum of A_x and B_x. Similarly, the y component C_y of your net displacement is the sum of A_y and B_y.

GRAPHICAL ADDITION OF VECTORS

To add two vectors $\vec{A}$ and $\vec{B}$ graphically (Figure 1.20):

- Draw the first vector $\vec{A}$.
- Move the second vector $\vec{B}$ (without rotating it) so its tail is located at the *tip* of the first vector.
- Draw a new vector from the tail of vector $\vec{A}$ to the tip of vector $\vec{B}$.

1.X.20 If $\vec{F}_1 = \langle 300, 0, -200 \rangle$ and $\vec{F}_2 = \langle 150, -300, 0 \rangle$, what is the magnitude of $\vec{F}_1$? What is the magnitude of $\vec{F}_2$? What is the magnitude of $\vec{F}_1 + \vec{F}_2$? What is the magnitude of $\vec{F}_1$ plus the magnitude of $\vec{F}_2$? Is $|\vec{F}_1 + \vec{F}_2| = |\vec{F}_1| + |\vec{F}_2|$?

1.X.21 $\vec{A} = \langle 3 \times 10^3, -4 \times 10^3, -5 \times 10^3 \rangle$ and $\vec{B} = \langle -3 \times 10^3, 4 \times 10^3, 5 \times 10^3 \rangle$. Calculate the following:

    (a) $\vec{A} + \vec{B}$ **(b)** $|\vec{A} + \vec{B}|$ **(c)** $|\vec{A}|$ **(d)** $|\vec{B}|$ **(e)** $|\vec{A}| + |\vec{B}|$

Vector Subtraction

The difference of two vectors will be very important in this and subsequent chapters. To subtract one vector from another, we subtract the components of the second vector from the components of the first:

$$\vec{A} - \vec{B} = \langle (A_x - B_x), (A_y - B_y), (A_z - B_z) \rangle$$
$$\langle 1, 2, 3 \rangle - \langle -4, 5, 6 \rangle = \langle 5, -3, -3 \rangle$$

1.X.22 If $\vec{F}_1 = \langle 300, 0, -200 \rangle$ and $\vec{F}_2 = \langle 150, -300, 0 \rangle$, what is the sum $\vec{F}_1 + \vec{F}_2$? What is the difference $\vec{F}_1 - \vec{F}_2$? What is $\vec{F}_2 - \vec{F}_1$?

Subtracting Vectors Graphically: Tail to Tail

To subtract one vector $\vec{B}$ from another vector $\vec{A}$ graphically:

- Draw the first vector $\vec{A}$.
- Move the second vector $\vec{B}$ (without rotating it) so its tail is located at the *tail* of the first vector.
- Draw a new vector from the tip of vector $\vec{B}$ to the tip of vector $\vec{A}$.

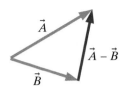

Figure 1.21 The procedure for subtracting vectors graphically: draw vectors tail to tail; draw a new vector from the tip of the second vector to the tip of the first vector.

Figure 1.22 Exercise 1.X.23.

Note that you can check this algebraically and graphically. As shown in Figure 1.21, since the tail of $\vec{A} - \vec{B}$ is located at the tip of $\vec{B}$, then the vector $\vec{A}$ should be the sum of $\vec{B}$ and $\vec{A} - \vec{B}$, as indeed it is:

$$\vec{B} + (\vec{A} - \vec{B}) = \vec{A}$$

1.X.23 Which of the following statements about the three vectors in Figure 1.22 are correct?

(a) $\vec{s} = \vec{t} - \vec{r}$ (b) $\vec{r} = \vec{t} - \vec{s}$ (c) $\vec{r} + \vec{t} = \vec{s}$ (d) $\vec{s} + \vec{t} = \vec{r}$
(e) $\vec{r} + \vec{s} = \vec{t}$

Commutativity and Associativity

Vector addition is commutative:

$$\vec{A} + \vec{B} = \vec{B} + \vec{A}$$

Vector subtraction is *not* commutative:

$$\vec{A} - \vec{B} \neq \vec{B} - \vec{A}$$

The associative property holds for vector addition and subtraction:

$$(\vec{A} + \vec{B}) - \vec{C} = \vec{A} + (\vec{B} - \vec{C})$$

The Zero Vector

It is convenient to have a compact notation for a vector whose components are all zero. We will use the symbol $\vec{0}$ to denote a zero vector, in order to distinguish it from a scalar quantity that has the value 0.

$$\vec{0} = \langle 0, 0, 0 \rangle$$

For example, the sum of two vectors $\vec{B} + (-\vec{B}) = \vec{0}$.

Change in a Quantity: The Greek Letter Δ

Frequently we will want to calculate the change in a quantity. For example, we may want to know the change in a moving object's position or the change in its velocity during some time interval. The Greek letter Δ (capital delta suggesting "d for difference") is used to denote the change in a quantity (either a scalar or a vector).

We use the subscript i to denote an *initial* value of a quantity, and the subscript f to denote the *final* value of a quantity. If a vector $\vec{r}_i$ denotes the initial position of an object relative to the origin (its position at the beginning of a time interval), and $\vec{r}_f$ denotes the final position of the object, then

$$\Delta \vec{r} = \vec{r}_f - \vec{r}_i$$

$\Delta \vec{r}$ means "change of $\vec{r}$" or $\vec{r}_f - \vec{r}_i$ (displacement)
Δt means "change of t" or $t_f - t_i$ (time interval)

The symbol Δ (delta) always means "final minus initial," not "initial minus final." For example, when a child's height changes from 1.1 m to 1.2 m, the change is $\Delta y = +0.1$ m, a positive number. If your bank account dropped from

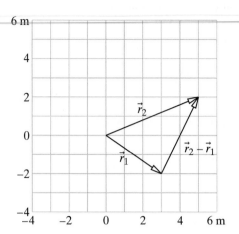

Figure 1.23 Relative position vector.

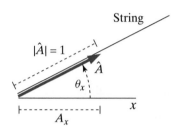

Figure 1.24 A unit vector whose direction is at a known angle from the +x axis.

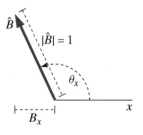

Figure 1.25 A unit vector in the second quadrant from the +x axis.

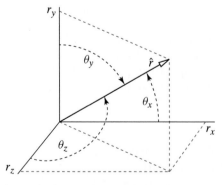

Figure 1.26 A 3D unit vector and its angles to the x, y, and z axes.

$150 to $130, what was the change in your balance? Δ (bank account) = −20 dollars.

Relative Position Vectors

Vector subtraction is used to calculate relative position vectors, vectors that represent the position of an object relative to another object. In Figure 1.23, object 1 is at location $\vec{r}_1$ and object 2 is at location $\vec{r}_2$. We want the components of a vector that points from object 1 to object 2. This is the vector obtained by subtraction: $\vec{r}_{2\text{ relative to }1} = \vec{r}_2 - \vec{r}_1$. Note that the form is always "final" minus "initial" in these calculations.

> **1.X.24** At 10:00 AM you are at location $\langle -3, 2, 5 \rangle$ m. By 10:02 AM you have walked to location $\langle 6, 4, 25 \rangle$ m.
>   **(a)** What is $\Delta\vec{r}$, the change in your position?
>   **(b)** What is Δt, the time interval during which your position changed?
> **1.X.25** A snail is initially at location $\vec{r}_1 = \langle 3, 0, -7 \rangle$ m. At a later time the snail has crawled to location $\vec{r}_2 = \langle 2, 0, -8 \rangle$ m. What is $\Delta\vec{r}$, the change in the snail's position?

Unit Vectors and Angles

Suppose that a taut string is at an angle θ_x to the +x axis, and we need a unit vector in the direction of the string. Figure 1.24 shows a unit vector $\hat{A}$ pointing along the string. What is the x component of this unit vector? Consider the triangle whose base is A_x and whose hypotenuse is $|\hat{A}| = 1$. From the definition of the cosine of an angle we have this:

$$\cos\theta = \frac{\text{adjacent}}{\text{hypotenuse}} = \frac{A_x}{1}, \text{ so } A_x = \cos\theta_x$$

In Figure 1.24 the angle θ_x is shown in the first quadrant (θ_x less than 90°), but this works for larger angles as well. For example, in Figure 1.25 the angle from the +x axis to a unit vector $\hat{B}$ is in the second quadrant (θ_x greater than 90°) and $\cos\theta_x$ is negative, which corresponds to B_x being negative.

What is true for x is also true for y and z. Figure 1.26 shows a 3D unit vector $\hat{r}$ and indicates the angles between the unit vector and the x, y, and z axes. Evidently we can write

$$\hat{r} = \langle \cos\theta_x, \cos\theta_y, \cos\theta_z \rangle$$

These three cosines of the angles between a vector (or unit vector) and the coordinate axes are called the "direction cosines" of the vector. The cosine function is never greater than 1, just as no component of a unit vector can be greater than 1.

A common special case is that of a unit vector lying in the xy plane, with zero z component (Figure 1.27). In this case $\theta_x + \theta_y = 90°$, so that $\cos\theta_y = \cos(90° - \theta_x) = \sin\theta_x$, therefore you can express the cosine of θ_y as the sine of θ_x, which is often convenient. However, in the general 3D case shown in Figure 1.26 there is no such simple relationship among the direction angles or among their cosines.

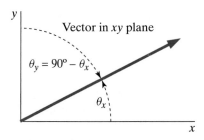

Figure 1.27 If a vector lies in the xy plane, $\cos\theta_y = \sin\theta_x$.

FINDING A UNIT VECTOR FROM ANGLES

To find a unit vector if angles are given:
- Redraw the vector of interest with its tail at the origin, and determine the angles between this vector and the axes.
- Imagine the vector $\langle 1, 0, 0 \rangle$, which lies on the $+x$ axis. θ_x is the angle through which you would rotate the vector $\langle 1, 0, 0 \rangle$ until its direction matched that of your vector. θ_x is positive, and $\theta_x \le 180°$.
- θ_y is the angle through which you would rotate the vector $\langle 0, 1, 0 \rangle$ until its direction matched that of your vector. θ_y is positive, and $\theta_y \le 180°$.
- θ_z is the angle through which you would rotate the vector $\langle 0, 0, 1 \rangle$ until its direction matched that of your vector. θ_z is positive, and $\theta_z \le 180°$.

EXAMPLE

From Angle to Unit Vector

A rope lying in the xy plane, pointing up and to the right, supports a climber at an angle of 20° to the vertical (Figure 1.28). What is the unit vector pointing up along the rope?

Solution

Figure 1.28 A climber supported by a rope.

Follow the procedure given above for finding a unit vector from angles. In Figure 1.29 we redraw the vector with its tail at the origin, and we determine the angles between the vector and the axes. If we rotate the unit vector $\langle 1, 0, 0 \rangle$ from along the $+x$ axis to the vector of interest, we see that we have to rotate through an angle $\theta_x = 70°$. To rotate the unit vector $\langle 0, 1, 0 \rangle$ from along the $+y$ axis to the vector of interest, we have to rotate through an angle of $\theta_y = 20°$. The angle from the $+z$ axis to our vector is $\theta_z = 90°$. Therefore the unit vector that points along the rope is this:

$$\langle \cos 70°, \cos 20°, \cos 90° \rangle = \langle 0.342, 0.940, 0 \rangle$$

FURTHER DISCUSSION   You may have noticed that the y component of the unit vector can also be calculated as $\sin 70° = 0.940$, and it is often useful to recognize that a vector component can be obtained using sine instead of cosine. There is, however, some advantage always to calculate in terms of direction cosines. This is a method that always works, including in 3D, and that avoids having to decide whether to use a sine or a cosine. Just use the cosine of the angle from the relevant positive axis to the vector.

EXAMPLE

From Unit Vector to Angles

A vector $\vec{r}$ points from the origin to the location $\langle -600, 0, 300 \rangle$ m. What is the angle that this vector makes to the x axis? To the y axis? To the z axis?

Solution

$$\hat{r} = \frac{\langle -600, 0, 300 \rangle}{\sqrt{(-600)^2 + (0)^2 + (300)^2} \text{ m}} = \langle -0.894, 0, 0.447 \rangle$$

But we also know that $\hat{r} = \langle \cos\theta_x, \cos\theta_y, \cos\theta_z \rangle$, so $\cos\theta_x = -0.894$, and the arccosine gives $\theta_x = 153.4°$. Similarly,

$$\cos\theta_y = 0, \theta_y = 90° \text{ (which checks; no } y \text{ component)}$$
$$\cos\theta_z = 0.447, \theta_z = 63.4°$$

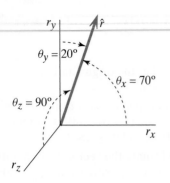

Figure 1.29 Redraw the vector with its tail at the origin. Identify the angles between the positive axes and the vector. In this example the vector lies in the xy plane.

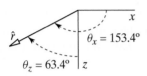

Figure 1.30 Look down on the xz plane. The difference in the two angles is $90°$, as it should be.

FURTHER DISCUSSION   Looking down on the xz plane in Figure 1.30, you can see that the difference between $\theta_x = 153.4°$ and $\theta_z = 63.4°$ is $90°$, as it should be.

> **1.X.26** A unit vector lies in the xy plane, at an angle of $160°$ from the $+x$ axis, with a positive y component. What is the unit vector? (It helps to draw a diagram.)
>
> **1.X.27** A string runs up and to the left in the xy plane, making an angle of $40°$ to the vertical. Determine the unit vector that points along the string.

1.6   SI UNITS

In this book we use the SI (Système Internationale) unit system, as is customary in technical work. The SI unit of mass is the kilogram (kg), the unit of distance is the meter (m), and the unit of time is the second (s). In later chapters we will encounter other SI units, such as the newton (N), which is a unit of force.

It is essential to use SI units in physics equations; this may require that you convert from some other unit system to SI units. If mass is known in grams, you need to divide by 1000 and use the mass in kilograms. If a distance is given in centimeters, you need to divide by 100 to convert the distance to meters. If the time is measured in minutes, you need to multiply by 60 to use a time in seconds. A convenient way to do such conversions is to multiply by factors that are equal to 1, such as $(1\,\text{min})/(60\,\text{s})$ or $(100\,\text{cm})/(1\,\text{m})$. As an example, consider converting 60 miles per hour to SI units, meters per second. Start with the 60 mi/hr and multiply by factors of 1:

$$\left(60\frac{\text{mi}}{\text{hr}}\right)\left(\frac{1\,\text{hr}}{60\,\text{min}}\right)\left(\frac{1\,\text{min}}{60\,\text{s}}\right)\left(\frac{5280\,\text{ft}}{1\,\text{mi}}\right)\left(\frac{12\,\text{in}}{1\,\text{ft}}\right)\left(\frac{2.54\,\text{cm}}{1\,\text{in}}\right)\left(\frac{1\,\text{m}}{100\,\text{cm}}\right) = 26.8\frac{\text{m}}{\text{s}}$$

Observe how most of the units cancel, leaving final units of m/s.

> **1.X.28** A snail moved 80 cm (80 centimeters) in 5 minutes. What was its average speed in SI units? Write out the factors as was done above.

1.7   VELOCITY

We use vectors not only to describe the position of an object but also to describe velocity (speed and direction). If we know an object's present speed in meters per second and the object's direction of motion, we can predict where it will be a short time into the future. As we have seen, *change* of velocity is an indication of interaction. We need to be able to work with velocities of objects in 3D, so we need to learn how to use 3D vectors to represent velocities. After learning how to describe velocity in 3D, we will also learn how to describe *change* of velocity, which is related to interactions.

Average Speed

The concept of speed is a familiar one. Speed is a single number, so it is a scalar quantity (speed is the magnitude of velocity). A world-class sprinter can run 100 meters in 10 seconds. We say the sprinter's average speed is

(100 m)/(10 s) = 10 m/s. In SI units speed is measured in meters per second, abbreviated "m/s."

A car that travels 100 miles in 2 hours has an average speed of (100 mi)/ (2 hr) = 50 miles per hour (about 22 m/s). In symbols,

$$v_{\text{avg}} = \frac{d}{t}$$

where v_{avg} is the "average speed," d is the distance the car has traveled, and t is the elapsed time.

There are other useful versions of the basic relationship among average speed, distance, and time. For example,

$$d = v_{\text{avg}}t$$

expresses the fact that if you run 5 m/s for 7 seconds you go 35 meters. Or you can use

$$t = \frac{d}{v_{\text{avg}}}$$

to calculate that to go 3000 miles in an airplane that flies at 600 miles per hour will take 5 hours.

Units

While it is easy to make a mistake in one of the formulas relating speed, time interval, and change in position, it is also easy to catch such a mistake by looking at the units. If you had written $t = v_{\text{avg}}/d$, you would discover that the right-hand side has units of (m/s)/m, or 1/s, not s. Always check units!

Instantaneous Speed Compared to Average Speed

If a car went 70 miles per hour for the first hour and 30 miles an hour for the second hour, it would still go 100 miles in 2 hours, with an average speed of 50 miles per hour. Note that during this 2-hour interval, the car was almost never actually traveling at its average speed of 50 miles per hour.

To find the "instantaneous" speed—the speed of the car at a particular instant—we should observe the short distance the car goes in a very short time, such as a hundredth of a second: If the car moves 0.3 meters in 0.01 s, its instantaneous speed is 30 meters per second.

Vector Velocity

Earlier we calculated vector differences between two different objects. The vector difference $\vec{r}_2 - \vec{r}_1$ represented a *relative* position vector—the position of object 2 relative to object 1 at a particular time. Now we will be concerned with the change of position of *one* object during a time interval, and $\vec{r}_f - \vec{r}_i$ will represent the "displacement" of this single object during the time interval, where $\vec{r}_i$ is the initial 3D position and $\vec{r}_f$ is the final 3D position (note that as with relative position vectors, we always calculate "final minus initial"). Dividing the (vector) displacement by the (scalar) time interval $t_f - t_i$ (final time minus initial time) gives the average (vector) velocity of the object:

DEFINITION: AVERAGE VELOCITY

$$\vec{v}_{avg} = \frac{\vec{r}_f - \vec{r}_i}{t_f - t_i}$$

Another way of writing this expression, using the "Δ" symbol (Greek capital delta, defined in Section 1.5) to represent a change in a quantity, is

$$\vec{v}_{avg} = \frac{\Delta \vec{r}}{\Delta t}$$

Remember that this is a compact notation for

$$\vec{v}_{avg} = \frac{\Delta \vec{r}}{\Delta t} = \left\langle \frac{\Delta x}{\Delta t}, \frac{\Delta y}{\Delta t}, \frac{\Delta z}{\Delta t} \right\rangle$$

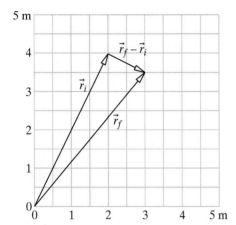

Figure 1.31 The displacement vector points from initial position to final position.

■ Note that the displacement $\vec{r}_f - \vec{r}_i$ refers to the positions of one object (the bee) at two different times, *not* the position of one object relative to a second object at one particular time. However, the vector subtraction is the same kind of operation for either kind of situation.

Determining Average Velocity from Change in Position

Consider a bee in flight (Figure 1.31). At time $t_i = 15\,\text{s}$ after 9:00 AM, the bee's position vector was $\vec{r}_i = \langle 2, 4, 0 \rangle$ m. At time $t_f = 15.1\,\text{s}$ after 9:00 AM, the bee's position vector was $\vec{r}_f = \langle 3, 3.5, 0 \rangle$ m. On the diagram, we draw and label the vectors $\vec{r}_i$ and $\vec{r}_f$.

Next, on the diagram, we draw and label the vector $\vec{r}_f - \vec{r}_i$, with the tail of the vector at the bee's initial position. One useful way to think about this graphically is to ask yourself what vector needs to be added to the initial vector $\vec{r}_1$ to make the final vector $\vec{r}_f$, since $\vec{r}_f$ can be written in the form $\vec{r}_f = \vec{r}_i + (\vec{r}_f - \vec{r}_i)$.

The vector we just drew, the change in position $\vec{r}_f - \vec{r}_i$, is called the "displacement" of the bee during this time interval. This displacement vector points from the initial position to the final position, and we always calculate displacement as "final minus initial."

We calculate the bee's displacement vector numerically by taking the difference of the two vectors, final minus initial:

$$\vec{r}_f - \vec{r}_i = \langle 3, 3.5, 0 \rangle \text{ m} - \langle 2, 4, 0 \rangle \text{ m} = \langle 1, -0.5, 0 \rangle \text{ m}$$

This numerical result should be consistent with our graphical construction. Look at the components of $\vec{r}_f - \vec{r}_i$ in Figure 1.31. Do you see that this vector has an x component of $+1$ and a y component of $-0.5\,\text{m}$? Note that the (vector) displacement $\vec{r}_f - \vec{r}_i$ is in the direction of the bee's motion.

The average velocity of the bee, a vector quantity, is the (vector) displacement $\vec{r}_f - \vec{r}_i$ divided by the (scalar) time interval $t_f - t_i$. Calculate the bee's average velocity:

$$\vec{v}_{avg} = \frac{\vec{r}_f - \vec{r}_i}{t_f - t_i} = \frac{\langle 1, -0.5, 0 \rangle \text{ m}}{(15.1 - 15)\,\text{s}} = \frac{\langle 1, -0.5, 0 \rangle \text{ m}}{0.1\,\text{s}} = \langle 10, -5, 0 \rangle \text{ m/s}$$

Since we divided $\vec{r}_f - \vec{r}_i$ by a scalar $(t_f - t_i)$, the average velocity $\vec{v}_{avg}$ points in the direction of the bee's motion, if the bee flew in a straight line.

What is the speed of the bee?

$$\text{speed of bee} = |\vec{v}_{avg}| = \sqrt{10^2 + (-5)^2 + 0^2} \text{ m/s} = 11.18 \text{ m/s}$$

What is the direction of the bee's motion, expressed as a unit vector?

$$\text{direction of bee: } \hat{v}_{avg} = \frac{\vec{v}_{avg}}{|\vec{v}_{avg}|} = \frac{\langle 10, -5, 0 \rangle \text{ m/s}}{11.18 \text{ m/s}} = \langle 0.894, -0.447, 0 \rangle$$

Note that the "m/s" units cancel; the result is dimensionless. We can check that this really is a unit vector:

$$\sqrt{0.894^2 + (-0.447)^2 + 0^2} = 0.9995$$

This is not quite 1.0 due to rounding the velocity coordinates and speed to three significant figures.

Put the pieces back together and see what we get. The original vector factors into the product of the magnitude times the unit vector:

$$|\vec{v}|\hat{v} = (11.18\,\text{m/s})\langle 0.894, -0.447, 0\rangle = \langle 10, -5, 0\rangle\,\text{m/s}$$

This is the same as the original vector $\vec{v}$.

1.X.29 At a time 0.2 seconds after it has been hit by a tennis racket, a tennis ball is located at $\langle 5, 7, 2\rangle$ m, relative to an origin in one corner of a tennis court. At a time 0.7 seconds after being hit, the ball is located at $\langle 9, 2, 8\rangle$ m.

  (a) What is the average velocity of the tennis ball?

  (b) What is the average speed of the tennis ball?

  (c) What is the unit vector in the direction of the ball's velocity?

1.X.30 A spacecraft is observed to be at a location $\langle 200, 300, -400\rangle$ m relative to an origin located on a nearby asteroid, and 5 seconds later is observed at location $\langle 325, 25, -550\rangle$ m.

  (a) What is the average velocity of the spacecraft?

  (b) What is the average speed of the spacecraft?

  (c) What is the unit vector in the direction of the spacecraft's velocity?

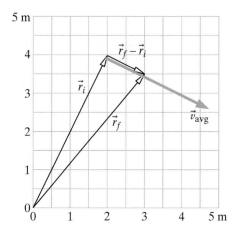

Figure 1.32 Average velocity vector: displacement divided by time interval.

Scaling a Vector to Fit on a Graph

We can plot the average velocity vector on the same graph that we use for showing the vector positions of the bee (Figure 1.32). However, note that velocity has units of meters per second whereas positions have units of meters, so we're mixing apples and oranges.

Moreover, the magnitude of the vector, 11.18 m/s, doesn't fit on a graph that is only 5 units wide (in meters). It is standard practice in such situations to scale down the arrow representing the vector to fit on the graph, preserving the correct direction. In Figure 1.32 we've scaled down the velocity vector by about a factor of 3 to make the arrow fit on the graph. Of course if there is more than one velocity vector we use the same scale factor for all the velocity vectors. The same kind of scaling is used with other physical quantities that are vectors, such as force and momentum, which we will encounter later.

Predicting a New Position

We can rewrite the velocity relationship in the form

$$(\vec{r}_f - \vec{r}_i) = \vec{v}_{\text{avg}}(t_f - t_i)$$

That is, the (vector) displacement of an object is its average (vector) velocity times the time interval. This is just the vector version of the simple notion that if you run at a speed of 7 m/s for 5 s you move a distance of $(7\,\text{m/s})(5\,\text{s}) = 35\,\text{m}$, or that a car going 50 miles per hour for 2 hours goes $(50\,\text{mi/hr})(2\,\text{hr}) = 100$ miles.

Is $(\vec{r}_f - \vec{r}_i) = \vec{v}_{avg}(t_f - t_i)$ a valid vector relation? Yes, multiplying a vector $\vec{v}_{avg}$ times a scalar $(t_f - t_i)$ yields a vector. We make a further rearrangement to obtain a relation for updating the position when we know the velocity:

THE POSITION UPDATE FORMULA

$$\vec{r}_f = \vec{r}_i + \vec{v}_{avg}(t_f - t_i)$$

or

$$\vec{r}_f = \vec{r}_i + \vec{v}_{avg}\Delta t$$

This equation says that if we know the starting position, the average velocity, and the time interval, we can predict the final position. This equation will be important throughout our work.

Using the Position Update Formula

The position update formula $\vec{r}_f = \vec{r}_i + \vec{v}_{avg}\,\Delta t$ is a vector equation, so we can write out its full component form:

$$\langle x_f, y_f, z_f \rangle = \langle x_i, y_i, z_i \rangle + \langle v_{avg,x}, v_{avg,y}, v_{avg,z} \rangle \,\Delta t$$

Because the x component on the left of the equation must equal the x component on the right (and similarly for the y and z components), this compact vector equation represents three separate component equations:

$$x_f = x_i + v_{avg,x}\,\Delta t$$
$$y_f = y_i + v_{avg,y}\,\Delta t$$
$$z_f = z_i + v_{avg,z}\,\Delta t$$

EXAMPLE

Updating the Position of a Ball

At time $t_i = 12.18\,\text{s}$ after 1:30 PM a ball's position vector is $\vec{r}_i = \langle 20, 8, -12 \rangle$ m. The ball's velocity at that moment is $\vec{v} = \langle 9, -4, 6 \rangle$ m/s. At time $t_f = 12.21\,\text{s}$ after 1:30 PM, where is the ball, assuming that its velocity hardly changes during this short time interval?

Solution

$$\vec{r}_f = \vec{r}_i + \vec{v}(t_f - t_i) = \langle 20, 8, -12 \rangle \text{ m} + (\langle 9, -4, 6 \rangle \text{ m/s})(12.21 - 12.18)\,\text{s}$$
$$\vec{r}_f = \langle 20, 8, -12 \rangle \text{ m} + \langle 0.27, -0.12, 0.18 \rangle \text{ m}$$
$$\vec{r}_f = \langle 20.27, 7.88, -11.82 \rangle \text{ m}$$

■ Note that if the velocity changes significantly during the time interval, in either magnitude or direction, our prediction for the new position may not be very accurate. In this case the velocity at the initial time could differ significantly from the average velocity during the time interval.

1.X.31 A proton traveling with a velocity of $\langle 3 \times 10^5, 2 \times 10^5, -4 \times 10^5 \rangle$ m/s passes the origin at a time 9.0 seconds after a proton detector is turned on. Assuming that the velocity of the proton does not change, what will its position be at time 9.7 seconds?

1.X.32 How long does it take a baseball with velocity $\langle 30, 20, 25 \rangle$ m/s to travel from location $\vec{r}_1 = \langle 3, 7, -9 \rangle$ m to location $\vec{r}_2 = \langle 18, 17, 3.5 \rangle$ m?

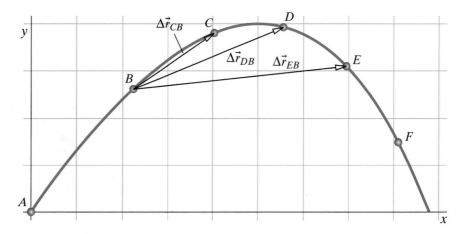

Figure 1.33 The trajectory of a ball through air. The axes represent the *x* and *y* distance from the ball's initial location; each square on the grid corresponds to 10 meters. Three different displacements, corresponding to three different time intervals, are indicated by arrows on the diagram.

loc.	t (s)	position (m)
A	0.0	$\langle 0, 0, 0 \rangle$
B	1.0	$\langle 22.3, 26.1, 0 \rangle$
C	2.0	$\langle 40.1, 38.1, 0 \rangle$
D	3.0	$\langle 55.5, 39.2, 0 \rangle$
E	4.0	$\langle 69.1, 31.0, 0 \rangle$
F	5.0	$\langle 80.8, 14.8, 0 \rangle$

Figure 1.34 Table showing elapsed time and position of the ball at each location marked by a dot in Figure 1.33.

Instantaneous Velocity

The curved colored line in Figure 1.33 shows the path of a ball through the air. The colored dots mark the ball's position at time intervals of one second. While the ball is in the air, its velocity is constantly changing, due to interactions with the Earth (gravity) and with the air (air resistance).

Suppose we ask: What is the velocity of the ball at the precise instant that it reaches location *B*? This quantity would be called the "instantaneous velocity" of the ball. We can start by approximating the instantaneous velocity of the ball by finding its average velocity over some larger time interval.

The table in Figure 1.34 shows the time and the position of the ball for each location marked by a colored dot in Figure 1.33. We can use these data to calculate the average velocity of the ball over three different intervals, by finding the ball's displacement during each interval, and dividing by the appropriate Δt for that interval:

$$\vec{v}_{EB} = \frac{\Delta \vec{r}_{EB}}{\Delta t} = \frac{\vec{r}_E - \vec{r}_B}{t_E - t_B} = \frac{(\langle 69.1, 31.0, 0 \rangle - \langle 22.3, 26.1, 0 \rangle)\,\text{m}}{(4.0 - 1.0)\,\text{s}}$$

$$= \langle 15.6, 1.6, 0 \rangle \frac{\text{m}}{\text{s}}$$

$$\vec{v}_{DB} = \frac{\Delta \vec{r}_{DB}}{\Delta t} = \frac{\vec{r}_D - \vec{r}_B}{t_D - t_B} = \frac{(\langle 55.5, 39.2, 0 \rangle - \langle 22.3, 26.1, 0 \rangle)\,\text{m}}{(3.0 - 1.0)\,\text{s}}$$

$$= \langle 16.6, 6.55, 0 \rangle \frac{\text{m}}{\text{s}}$$

$$\vec{v}_{CB} = \frac{\Delta \vec{r}_{CB}}{\Delta t} = \frac{\vec{r}_C - \vec{r}_B}{t_C - t_B} = \frac{(\langle 40.1, 38.1, 0 \rangle - \langle 22.3, 26.1, 0 \rangle)\,\text{m}}{(2.0 - 1.0)\,\text{s}}$$

$$= \langle 17.8, 12.0, 0 \rangle \frac{\text{m}}{\text{s}}$$

Not surprisingly, the average velocities over these different time intervals are not the same, because both the direction of the ball's motion and the speed of the ball were changing continuously during its flight. The three average velocity vectors that we calculated are shown in Figure 1.35.

QUESTION Which of the three average velocity vectors depicted in Figure 1.35 best approximates the instantaneous velocity of the ball at location *B*?

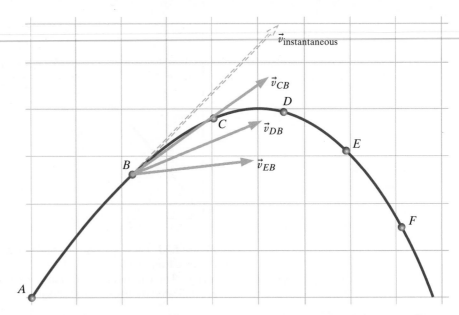

Figure 1.35 The three different average velocity vectors calculated above are shown by three arrows, each with its tail at location *B*. Note that since the units of velocity are m/s, these arrows use a different scale from the distance scale used for the path of the ball. The three arrows representing average velocities are drawn with their tails at the location of interest. The dashed arrow represents the actual instantaneous velocity of the ball at location *B*.

Simply by looking at the diagram, we can tell that $\vec{v}_{CB}$ is closest to the actual instantaneous velocity of the ball at location *B*, because its direction is closest to the direction in which the ball is actually traveling. Because the direction of the instantaneous velocity is the direction in which the ball is moving at a particular instant, the instantaneous velocity is tangent to the ball's path. Of the three average velocity vectors we calculated, $\vec{v}_{CB}$ best approximates a tangent to the path of the ball. Evidently $\vec{v}_{CB}$, the velocity calculated with the shortest time interval, $t_C - t_B$, is the best approximation to the instantaneous velocity at location *B*. If we used even smaller values of Δt in our calculation of average velocity, such as 0.1 second, or 0.01 second, or 0.001 second, we would presumably have better and better estimates of the actual instantaneous velocity of the object at the instant when it passes location *B*.

Two important ideas have emerged from this discussion:

- The direction of the instantaneous velocity of an object is tangent to the path of the object's motion.
- Smaller time intervals yield more accurate estimates of instantaneous velocity.

1.X.33 How does average velocity differ from instantaneous velocity?

1.X.34 A comet travels in an elliptical path around a star, in the direction shown in Figure 1.36. Which arrow best indicates the direction of the comet's velocity vector at each of the numbered locations in the orbit?

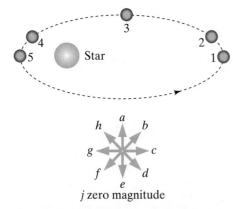

Figure 1.36 A comet goes around a star.

Connection to Calculus

You may already have learned about derivatives in calculus. The instantaneous velocity is a derivative, the limit of $\Delta \vec{r}/\Delta t$ as the time interval Δt used in the calculation gets closer and closer to zero:

$$\vec{v} = \lim_{\Delta t \to 0} \frac{\Delta \vec{r}}{\Delta t}, \text{ which is written as } \vec{v} = \frac{d\vec{r}}{dt}$$

In Figure 1.35, the process of taking the limit is illustrated graphically. As smaller values of Δt are used in the calculation, the average velocity vectors approach the limiting value: the actual instantaneous velocity.

A useful way to see the meaning of the derivative of a vector is to consider the components:

$$\vec{v} = \frac{d\vec{r}}{dt} = \frac{d}{dt}\langle x, y, z \rangle = \left\langle \frac{dx}{dt}, \frac{dy}{dt}, \frac{dz}{dt} \right\rangle = \langle v_x, v_y, v_z \rangle$$

■ The rate of change of a vector (the derivative) is itself a vector.

The derivative of the position vector $\vec{r}$ gives components that are the components of the velocity, as we should expect.

Informally, you can think of $d\vec{r}$ as a very small ("infinitesimal") displacement, and dt as a very small ("infinitesimal") time interval. It is as though we had continued the process illustrated in Figure 1.35 to smaller and smaller time intervals, down to an extremely tiny time interval dt with a correspondingly tiny displacement $d\vec{r}$. The ratio of these tiny quantities is the instantaneous velocity.

The ratio of these two tiny quantities need not be small. For example, suppose that an object moves in the x direction a tiny distance of 1×10^{-15} m, the radius of a proton, in a very short time interval of 1×10^{-23} s:

$$\vec{v} = \frac{\langle 1 \times 10^{-15}, 0, 0 \rangle \text{ m}}{1 \times 10^{-23} \text{ s}} = \langle 1 \times 10^{8}, 0, 0 \rangle \text{ m/s}$$

which is one-third the speed of light (3×10^{8} m/s)!

Acceleration

Velocity is the time rate of change of position: $\vec{v} = d\vec{r}/dt$. Similarly, we define "acceleration" as the time rate of change of velocity: $\vec{a} = d\vec{v}/dt$. Acceleration, which is itself a vector quantity, has units of meters per second per second, written as m/s/s or m/s^2.

DEFINITION: ACCELERATION

Instantaneous acceleration is the time rate of change of velocity:

$$\vec{a} = \frac{d\vec{v}}{dt}$$

Average acceleration can be calculated from a change in velocity:

$$\vec{a}_{\text{avg}} = \frac{\Delta \vec{v}}{\Delta t}$$

The units of acceleration are m/s^2.

If a car traveling in a straight line speeds up from 20 m/s to 26 m/s in 3 seconds, we say that the magnitude of the acceleration is $(26 - 20)/3 = 2$ m/s/s. If you drop a rock, its speed increases 9.8 m/s every second, so its acceleration is 9.8 m/s/s, as long as air resistance is negligible.

1.X.35 Powerful sports cars can go from zero to 25 m/s (about 60 mph) in 5 seconds. What is the magnitude of the acceleration? How does this compare with the acceleration of a falling rock?

1.X.36 Suppose the position of an object at time t is $\langle 3 + 5t, 4t^2, 2t - 6t^3 \rangle$. What is the instantaneous velocity at time t? What is the acceleration at time t? What is the instantaneous velocity at time $t = 0$? What is the acceleration at time $t = 0$?

Change of Magnitude and/or Change in Direction

There are two parts to the acceleration, the time rate of change of the velocity $\vec{v} = |\vec{v}|\hat{v}$:

$$\text{rate of change of the magnitude of the velocity (speed)} \; \frac{d|\vec{v}|}{dt}$$

$$\text{rate of change of the direction of the velocity} \; \frac{d\hat{v}}{dt}$$

As we'll see in later chapters, these two parts of the acceleration are associated with pushing or pulling parallel to the motion (changing the speed) or perpendicular to the motion (changing the direction).

1.8 MOMENTUM

Newton's first law of motion,

An object moves in a straight line and at constant speed except to the extent that it interacts with other objects

gives us a conceptual connection between interactions and their effects on the motion of objects. However, this law does not allow us to make quantitative (numerical) predictions or explanations—we could not have used this law to predict the exact trajectory of the ball shown in Figure 1.33, and we could not use this law alone to figure out how to send a rocket to the Moon.

In order to make quantitative predictions or explanations of physical phenomena, we need a quantitative measure of interactions and a quantitative measure of effects of those interactions.

Newton's first law of motion does contain the important idea that if there is no interaction, a moving object will continue to move in a straight line, with no change of direction or speed, and an object that is not moving will remain at rest. A quantitative version of this law would provide a means of predicting the motion of an object, or of deducing how it must have moved in the past, if we could list all of its interactions with other objects.

Changes in Velocity

QUESTION What factors make it difficult or easy to change an object's velocity?

You have probably noticed that if two objects have the same velocity but one is much more massive than the other, it is more difficult to change the heavy object's speed or direction. It is easier to stop a baseball traveling at a hundred miles per hour than to stop a car traveling at a hundred miles per hour! It is easier to change the direction of a canoe than to change the direction of a large, massive ship such as the *Titanic* (which couldn't change course quickly enough after the iceberg was spotted).

Momentum Involves Both Mass and Velocity

To take into account both an object's mass and its velocity, we define a vector quantity called "momentum" that involves the product of mass (a scalar) and velocity (a vector). Instead of saying "the stronger the interaction, the bigger the change in the velocity," we now say "the stronger the interaction, the bigger the change in the momentum."

Momentum, a vector quantity, is usually represented by the symbol $\vec{p}$. We might expect that the mathematical expression for momentum would be simply $\vec{p} = m\vec{v}$, and indeed this is almost, but not quite, correct.

Experiments on particles moving at very high speeds, close to the speed of light $c = 3 \times 10^8$ m/s, show that changes in $m\vec{v}$ are not really proportional to the strength of the interactions. As you keep applying a force to a particle near the speed of light, the speed of the particle barely increases, and it is not possible to increase a particle's speed beyond the speed of light.

Through experiments it has been found that changes in the following quantity are proportional to the amount of interaction:

DEFINITION OF MOMENTUM

Momentum is defined as the product of mass times velocity, multiplied by a proportionality factor gamma:

$$\vec{p} = \gamma m\vec{v}$$

The proportionality factor γ (lower-case Greek gamma) is defined as

$$\gamma = \frac{1}{\sqrt{1 - \left(\frac{|\vec{v}|}{c}\right)^2}}$$

In these equations $\vec{p}$ represents momentum, $\vec{v}$ is the velocity of the object, m is the mass of the object, $|\vec{v}|$ is the magnitude of the object's velocity (the speed), and c is the speed of light (3×10^8 m/s). Momentum has units of kg·m/s. To calculate momentum in these units, you must specify mass in kg and velocity in meters per second.

This is the "relativistic" definition of momentum. Albert Einstein in 1905 in his Special Theory of Relativity predicted that this would be the appropriate definition for momentum at high speeds, a prediction that has been abundantly verified in a wide range of experiments.

EXAMPLE

Momentum of a Fast-Moving Proton

Suppose that a proton (mass 1.7×10^{-27} kg) is traveling with a velocity of $\langle 2 \times 10^7, 1 \times 10^7, -3 \times 10^7 \rangle$ m/s. (a) What is the momentum of the proton? (b) What is the magnitude of the momentum of the proton?

Solution

(a) $|\vec{v}| = \sqrt{(2 \times 10^7)^2 + (1 \times 10^7)^2 + (-3 \times 10^7)^2} = 3.7 \times 10^7 \dfrac{\text{m}}{\text{s}}$

$\dfrac{|\vec{v}|}{c} = \dfrac{3.7 \times 10^7 \frac{\text{m}}{\text{s}}}{3 \times 10^8 \frac{\text{m}}{\text{s}}} = 0.12$

$\gamma = \dfrac{1}{\sqrt{1 - (0.12)^2}} = 1.007$

$\vec{p} = (1.007)(1.7 \times 10^{-27} \text{ kg})\langle 2 \times 10^7, 1 \times 10^7, -3 \times 10^7 \rangle \dfrac{\text{m}}{\text{s}}$

$\vec{p} = \langle 3.4 \times 10^{-20}, 1.7 \times 10^{-20}, -5.1 \times 10^{-20} \rangle \dfrac{\text{kg·m}}{\text{s}}$

$|\vec{p}| = \sqrt{(3.4 \times 10^{-20})^2 + (1.7 \times 10^{-20})^2 + (-5.1 \times 10^{-20})^2} \dfrac{\text{kg·m}}{\text{s}}$

(b) $= 6.4 \times 10^{-20} \dfrac{\text{kg·m}}{\text{s}}$

Approximate Expression for Momentum

In the example above, we found that $\gamma = 1.007$. Since in that calculation we used only two significant figures, we could have used the approximation that $\gamma \approx 1.0$ without affecting our answer. Let's examine the expression for γ to see whether we can come up with a guideline for when it is reasonable to use the approximate expression

$$\vec{p} \approx 1 \cdot m\vec{v}$$

Looking at the expression for γ

$$\gamma = \frac{1}{\sqrt{1 - \left(\dfrac{|\vec{v}|}{c}\right)^2}}$$

we see that it depends only on the ratio of the speed of the object to the speed of light (the object's mass doesn't appear in this expression).

If $|\vec{v}|/c$ is a very small number, then $1 - (|\vec{v}|/c)^2 \approx 1 - 0 \approx 1$, so $\gamma \approx 1$.

$\|\vec{v}\|$ **m/s**	$\|\vec{v}\|/c$	γ
0	0	1.0000
3	1×10^{-8}	1.0000
300	1×10^{-6}	1.0000
3×10^6	1×10^{-8}	1.0001
3×10^7	0.1	1.0050
1.5×10^8	0.5	1.1547
2.997×10^8	0.999	22.3663
2.9997×10^8	0.9999	70.7124
3×10^8	1	Infinite! Impossible!

Figure 1.37 Values of γ calculated for some speeds. γ is shown to four decimal places, which is more accuracy than we will usually need.

APPROXIMATION: MOMENTUM AT LOW SPEEDS

$$\vec{p} \approx m\vec{v} \text{ when } |\vec{v}| \ll c$$
$$\text{When } \vec{p} \approx m\vec{v}, \vec{v} \approx \vec{p}/m$$

Some values of $(|\vec{v}|/c)$ and γ are displayed in Figure 1.37. From this table you can see that even at the very high speed where $|\vec{v}|/c = 0.1$, which means that $|\vec{v}| = 3 \times 10^7$ m/s, the relativistic factor γ is only slightly different from 1. For large-scale objects such as a space rocket, whose speed is typically only about 1×10^4 m/s, we can ignore the factor γ, and momentum is to a good approximation $\vec{p} \approx m\vec{v}$. It is only for high-speed cosmic rays or particles produced in high-speed particle accelerators that we need to use the full relativistic definition for momentum, $\vec{p} = \gamma m\vec{v}$.

From this table you can also get a sense of why it is not possible to exceed the speed of light. As you make a particle go faster and faster, approaching the speed of light, additional increases in the speed become increasingly difficult, because a tiny increase in speed means a huge increase in momentum, requiring huge amounts of interaction. In fact, for the speed to equal the speed of light, the momentum would have to increase to be infinite! There is a cosmic speed limit in the Universe, 3×10^8 m/s.

We will repeatedly emphasize the role of momentum throughout this textbook because of its fundamental importance not only in classical (prequantum) mechanics but also in relativity and quantum mechanics. The use of momentum clarifies the physics analysis of certain complex processes such as collisions, including collisions at speeds approaching the speed of light.

1.X.37 A good sprinter can run 100 meters in 10 seconds. What is the magnitude of the momentum of a sprinter whose mass is 65 kg and who is running at a speed of 10 m/s?

1.X.38 What is the momentum of an electron traveling at a velocity of $\langle 0, 0, -2 \times 10^8 \rangle$ m/s? (Masses of particles are given on the inside back cover of this textbook.) What is the magnitude of the momentum of the electron?

Direction of Momentum

Like velocity, momentum is a vector quantity, so it has a magnitude and a direction. The direction is the same as the direction of the velocity.

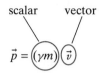

Figure 1.38 The expression for momentum is the product of a scalar times a vector. The scalar factor is positive, so the direction of an object's momentum is the same as the direction of its velocity.

> QUESTION A leaf is blown by a gust of wind, and at a particular instant is traveling straight upward, in the $+y$ direction. What is the direction of the leaf's momentum?

The mathematical expression for momentum can be looked at as the product of a scalar part times a vector part. Since the mass m must be a positive number, and the factor gamma (γ) must be a positive number, this scalar factor cannot change the direction of the vector (Figure 1.38). Therefore the direction of the leaf's momentum is the same as the direction of its velocity: straight up (the $+y$ direction).

j zero magnitude

Figure 1.39 A comet's momentum is in the direction of the arrow labeled *b*.

1.X.39 A comet's momentum is in the direction of the arrow labeled *b* in Figure 1.39. What arrow best describes the direction of the comet's velocity?

Using Momentum to Update Position

If you know the momentum of an object, you can calculate the change in position of the object over a given time interval. This is straightforward if the object is traveling at a speed low enough that the approximate expression for momentum can be used, since in this case $\vec{v} \approx \vec{p}/m$. For speeds near the speed of light, we must use the exact expression

$$\vec{v} = \frac{\vec{p}/m}{\sqrt{1 + \left(\frac{|\vec{p}|}{mc}\right)^2}}$$

This expression was derived simply by rearranging the equation $\vec{p} = \gamma m \vec{v}$. The detailed derivation is given at the end of the chapter.

EXAMPLE **Displacement of a Fast Proton**

A proton with constant momentum $\langle 0, 0, 2.72 \times 10^{-19}\rangle$ kg $\cdot$ m/s leaves the origin 10.0 seconds after an accelerator experiment is started. What is the location of the proton 2 ns later? (ns = nanosecond = 1×10^{-9} s)

Solution

$$\vec{v} = \frac{\vec{p}/m}{\sqrt{1 + \left(\frac{|\vec{p}|}{mc}\right)^2}} = \frac{\langle 0, 0, 2.72 \times 10^{-19}\rangle \text{ kg} \cdot \text{m/s}}{(1.7 \times 10^{-27} \text{ kg})\sqrt{1 + \left(\frac{2.72 \times 10^{-19} \text{ kg} \cdot \text{m/s}}{(1.7 \times 10^{-27} \text{ kg})(3 \times 10^8 \text{ m/s})}\right)^2}}$$

$$= \langle 0, 0, 1.4 \times 10^8\rangle \text{ m/s}$$

$$\vec{r}_f = \vec{r}_i + \vec{v}_{\text{avg}}\,\Delta t$$

$$= \langle 0, 0, 0\rangle \text{ m} + \langle 0, 0, 1.4 \times 10^8 \text{ m/s}\rangle (2 \times 10^{-9} \text{ s})$$

$$= \langle 0, 0, 0.28\rangle \text{ m}$$

The proton traveled 28 cm in 2 ns.

EXAMPLE	Displacement of an Ice Skater

An ice skater whose mass is 50 kg moves with constant momentum $\langle 400, 0, 300 \rangle$ kg·m/s. At a particular instant in her skating program she passes location $\langle 0, 0, 3 \rangle$ m. What was her location at a time 3 seconds earlier?

Solution

$$\vec{v} \approx \frac{\vec{p}}{m} = \frac{\langle 400, 0, 300 \rangle \text{ kg} \cdot \text{m/s}}{50 \text{ kg}} = \langle 8, 0, 6 \rangle \text{ m/s}$$

$$(\vec{r}_f - \vec{r}_i) = \vec{v} \, \Delta t$$

$$\vec{r}_i = \vec{r}_f - \vec{v} \, \Delta t$$

$$= \langle 0, 0, 3 \rangle \text{ m} - (\langle 8, 0, 6 \rangle \text{ m/s})(3 \text{ s})$$

$$= \langle -24, 0, -15 \rangle \text{ m}$$

1.X.40 A 4.5 kg bowling ball rolls down an alley with nearly constant momentum of $\langle 0.9, 0, 22.5 \rangle$ kg·m/s, starting from the origin. What will be the location of the ball after 2 seconds?

1.X.41 At time $t_1 = 12$ s, a car with mass 1300 kg is located at $\langle 94, 0, 30 \rangle$ m and has momentum $\langle 4500, 0, -3000 \rangle$ kg·m/s. The car's momentum is not changing. At time $t_2 = 17$ s, what is the position of the car?

1.9   CHANGE OF MOMENTUM

In the next chapter we will introduce "the Momentum Principle," which quantitatively relates change in momentum $\Delta \vec{p}$ to the strength and duration of an interaction. In order to be able to use the Momentum Principle we need to know how to calculate changes in momentum.

Momentum is a vector quantity and proportional to the velocity, so just as was the case with velocity, there are two aspects of momentum that can change: magnitude and direction. A mathematical description of change of momentum must include either a change in the magnitude of the momentum, or a change in the direction of the momentum, or both. As we will see in later chapters, pushing parallel to the motion changes the magnitude of the momentum, but to change the direction of the momentum requires pushing perpendicular (sideways) to the motion.

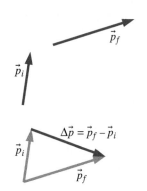

Figure 1.40 Calculation of $\Delta \vec{p}$.

Change of the Vector Momentum

The change in the momentum $\Delta \vec{p}$ during a time interval is a vector: $\Delta \vec{p} = \vec{p}_f - \vec{p}_i$. This vector expression captures both changes in magnitude and changes in direction. Figure 1.40 is a graphical illustration of a change from an initial momentum $\vec{p}_i$ to a final momentum $\vec{p}_f$. Place the initial and final momentum vectors tail to tail, then draw a vector from initial to final. This is the vector representing $\Delta \vec{p} = \vec{p}_f - \vec{p}_i$. This is the same procedure you used to calculate relative position vectors by subtraction, or displacement vectors by subtraction. The rule for subtracting vectors is always the same: Place the vectors tail to tail, then draw from the tip of the initial vector to the tip of the final vector. This resultant vector is "final minus initial."

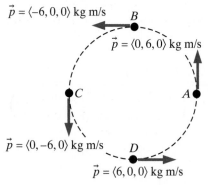

Figure 1.41 The momentum of an object traveling in a circle changes, even if the magnitude of momentum does not change.

Change in Magnitude of Momentum

If an object's speed changes (that is, the magnitude of its velocity changes), the magnitude of the object's momentum also changes. Pushing in the direction of the momentum (and velocity) makes the magnitude of the momentum

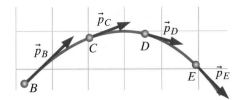

Figure 1.42 A portion of the trajectory of a ball moving through air, subject to gravity and air resistance. The arrows represent the momentum of the ball at the locations indicated by letters.

increase (the speed increases); pushing backwards makes the magnitude decrease (the object slows down).

Change of Direction of Momentum

There are various ways to specify a change in the direction of motion. For example, if you used compass directions, you could say that an airplane changed its direction from 30° east of north to 45° east of north: a 15° clockwise change. One can imagine various other schemes, involving other kinds of coordinate systems. The standard way to deal with this is to use vectors, as in Figure 1.41, which shows the changing momentum of an object traveling at constant speed along a circular path.

EXAMPLE **Change in Momentum of a Ball**

Figure 1.42 shows a portion of the trajectory of a ball in air, subject to gravity and air resistance. At location B, the ball's momentum is $\vec{p}_B = \langle 3.03, 2.83, 0 \rangle$ kg·m/s. At location C, the ball's momentum is $\vec{p}_C = \langle 2.55, 0.97, 0 \rangle$ kg·m/s. (a) Find the change in the ball's momentum between these locations, and show it on the diagram. (b) Find the change in the magnitude of the ball's momentum.

Solution **(a)**

$$\Delta\vec{p} = \vec{p}_C - \vec{p}_B = \langle 2.55, 0.97, 0 \rangle \text{ kg·m/s} - \langle 3.03, 2.83, 0 \rangle \text{ kg·m/s}$$
$$= \langle -0.48, -1.86, 0 \rangle \text{ kg·m/s}$$

Both the x and y components of the ball's momentum decreased, so $\Delta\vec{p}$ has negative x and y components. This is consistent with the graphical subtraction shown in Figure 1.43.

(b)

$$|\vec{p}_B| = \sqrt{(3.03)^2 + (2.83)^2 + (0)^2} = 4.15 \text{ kg·m/s}$$
$$|\vec{p}_C| = \sqrt{(2.55)^2 + (0.97)^2 + (0)^2} = 2.73 \text{ kg·m/s}$$

Figure 1.43 Graphical calculation of $\Delta\vec{p}$. The result is also shown superimposed on the ball's path, midway between the initial and final locations.

FURTHER DISCUSSION The magnitude of the ball's momentum decreased, which means that there must have been a component of gravitational and air resistance forces acting on the ball in a direction opposite to the ball's motion, which slowed it down. The direction of the ball's momentum also changed, as is seen in Figure 1.42, which means that there must have been components of the forces acting perpendicular to the ball's motion.

EXAMPLE **Change in Momentum and in Magnitude of Momentum**

Suppose you are driving a 1000 kilogram car at 20 m/s in the $+x$ direction. After making a 180 degree turn, you drive the car at 20 m/s in the $-x$ direction. (20 m/s is about 45 miles per hour or 72 km per hour.) (a) What is the change of magnitude of the momentum of the car $\Delta|\vec{p}|$? (b) What is the magnitude of the change of momentum of the car $|\Delta\vec{p}|$?

Solution **(a)**

$$\Delta|\vec{p}| = |\vec{p}_2| - |\vec{p}_1| \approx |m\vec{v}_2| - |m\vec{v}_1| = |m||\vec{v}_2| - |m||\vec{v}_1|$$
$$\Delta|\vec{p}| = (1000 \text{ kg})(20 \text{ m/s}) - (1000 \text{ kg})(20 \text{ m/s})$$
$$\Delta|\vec{p}| = 0 \text{ kg·m/s}$$

(b) $\Delta \vec{p} = \vec{p}_2 - \vec{p}_1 = 1000\,\text{kg}\langle 20, 0, 0 \rangle\,\text{m/s} - 1000\,\text{kg}\langle -20, 0, 0 \rangle\,\text{m/s}$

$$= \langle 4 \times 10^4, 0, 0 \rangle\,\text{kg} \cdot \text{m/s}$$

$$|\Delta \vec{p}| = 4 \times 10^4\,\text{kg} \cdot \text{m/s}$$

So $|\Delta \vec{p}| \neq \Delta |\vec{p}|$.

FURTHER DISCUSSION This calculation shows something important. Change of magnitude, $\Delta |\vec{p}|$, isn't the same as magnitude of change, $|\Delta \vec{p}|$. Here's an even more dramatic example:

QUESTION   A ball with momentum $\langle 5, 0, 0 \rangle$ kg·m/s rebounds from a wall with nearly the same speed, so its final momentum is approximately $\langle -5, 0, 0 \rangle$ kg·m/s. What is $\Delta |\vec{p}|$? What is $|\Delta \vec{p}|$?

The change of the magnitude of the momentum, $\Delta |\vec{p}|$, is zero ($5 - 5 = 0$), but the magnitude of the change of the momentum, $|\Delta \vec{p}|$, is $|\langle -5, 0, 0 \rangle - \langle 5, 0, 0 \rangle| = |\langle -10, 0, 0 \rangle| = 10$ kg·m/s.

1.X.42 A tennis ball of mass 57 g travels with velocity $\langle 50, 0, 0 \rangle$ m/s toward a wall. After bouncing off the wall, the tennis ball is observed to be traveling with velocity $\langle -48, 0, 0 \rangle$ m/s.

  (a) Draw a diagram showing the initial and final momentum of the tennis ball.

  (b) What is the change in the momentum of the tennis ball?

  (c) What is the change in the magnitude of the tennis ball's momentum?

1.X.43 The planet Mars has a mass of 6.4×10^{23} kg, and travels in a nearly circular orbit around the Sun, as shown in Figure 1.44. When it is at location A, the velocity of Mars is $\langle 0, 0, -2.5 \times 10^4 \rangle$ m/s. When it reaches location B, the planet's velocity is $\langle -2.5 \times 10^4, 0, 0 \rangle$ m/s. We're looking down on the orbit from above the north poles of the Sun and Mars, with $+x$ to the right and $+z$ down the page.

  (a) What is $\Delta \vec{p}$, the change in the momentum of Mars between locations A and B?

  (b) On a copy of the diagram in Figure 1.44, draw two arrows representing the momentum of Mars at locations C and D, paying attention to both the length and direction of each arrow.

  (c) What is the direction of the change in the momentum of Mars between locations C and D? Draw the vector $\Delta \vec{p}$ on your diagram.

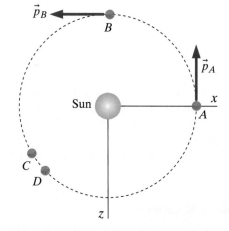

Figure 1.44 The nearly circular orbit of Mars around the Sun, viewed from above the orbital plane ($+x$ to the right, $+z$ down the page). Not to scale: the sizes of the Sun and Mars are exaggerated.

Average Rate of Change of Momentum

The rate of change of the vector position is such an important quantity that it has a special name: "velocity." In Section 1.7 we discussed how to find both the average rate of change of position (average velocity) and the instantaneous rate of change of position (instantaneous velocity) of an object.

The average rate of change of momentum and the instantaneous rate of change of momentum are also extremely important quantities. In some situations, we will only be able to find an average rate of change of momentum:

AVERAGE RATE OF CHANGE OF MOMENTUM

$$\frac{\Delta \vec{p}}{\Delta t} = \frac{\vec{p}_f - \vec{p}_i}{t_f - t_i}$$

This quantity is a vector, and points in the direction of $\Delta \vec{p}$.

EXAMPLE

Average Rate of Change of Momentum

If the momentum of a ball changes from $\langle 1, 2, 0 \rangle$ kg $\cdot$ m/s to $\langle 0.5, 0, 0.5 \rangle$ kg $\cdot$ m/s in half a second, the average rate of change of momentum of the ball is

$$\frac{(\langle 0.5, 0, 0.5 \rangle - \langle 1, 2, 0 \rangle)\, \text{kg} \cdot \text{m/s}}{0.5\,\text{s}} = \langle -1, -4, 1 \rangle \frac{\text{kg} \cdot \text{m}}{\text{s}^2}$$

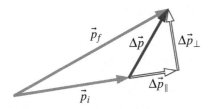

Figure 1.45 The parallel and perpendicular components of the momentum change, $\Delta \vec{p}$.

Parallel and Perpendicular Change

In Figure 1.45 we see the parallel and perpendicular components of a change of momentum, $\Delta \vec{p} = \vec{p}_f - \vec{p}_i$. To make a nonzero perpendicular component $\Delta \vec{p}_\perp$, involving a change in direction, requires pushing perpendicular (sideways) to the initial momentum $\vec{p}_i$.

1.X.44 A hockey puck was sliding on ice with momentum $\langle 0, 0, 10 \rangle$ kg $\cdot$ m/s when it was kicked, and the new momentum was $\langle 7, 0, 12 \rangle$ kg $\cdot$ m/s. (a) What was $\Delta \vec{p}_\parallel$? (b) What was $\Delta \vec{p}_\perp$? It helps to draw a diagram.

1.10   *THE PRINCIPLE OF RELATIVITY

Sections marked with a "*" are optional. They provide additional information and context, but later sections of the textbook don't depend critically on them. This optional section deals with some deep issues about the "reference frame" from which you observe motion. Newton's first law of motion only applies in an "inertial reference frame," which we will discuss here in the context of the principle of relativity.

A great variety of experimental observations has led to the establishment of the following principle:

THE PRINCIPLE OF RELATIVITY

Physical laws work in the same way for observers in uniform motion as for observers at rest.

This principle is called "the principle of relativity." (Einstein's extensions of this principle are known as "special relativity" and "general relativity.") Phenomena observed in a room in uniform motion (for example, on a train moving with constant speed on a smooth straight track) obey the same physical laws in the same way as experiments done in a room that is not moving. According to this principle, Newton's first law of motion should be true both for an observer moving at constant velocity and for an observer at rest.

For example, suppose that you're riding in a car moving with constant velocity, and you're looking at a map lying on the dashboard. As far as you're concerned, the map isn't moving, and no interactions are required to hold it still on the dashboard. Someone standing at the side of the road sees the car go by, sees the map moving at a high speed in a straight line, and can see that no interactions are required to hold the map still on the dashboard. Both you and the bystander agree that Newton's first law of motion is obeyed: the bystander sees the map moving with constant velocity in the absence of interactions, and you see the map not moving at all (a zero constant velocity) in the absence of interactions.

On the other hand, if the car suddenly speeds up, it moves out from under the map, which ends up in your lap. To you it looks like "the map sped up in the backwards direction" without any interactions to cause this to happen, which looks like a violation of Newton's first law of motion. The problem is that you're strapped to the car, which is an accelerated reference frame, and Newton's first law of motion applies only to nonaccelerated reference frames, called "inertial" reference frames. Similarly, if the car suddenly turns to the right, moving out from under the map, the map tends to keep going in its original direction, and to you it looks like "the map moved to the left" without any interactions. So a change of speed or a change of direction of the car (your reference frame) leads you to see the map behave in a strange way.

The bystander, who is in an inertial (nonaccelerating) reference frame, doesn't see any violation of Newton's first law of motion. The bystander's reference frame is an inertial frame, and the map behaves in an understandable way, tending to keep moving with unchanged speed and direction when the car changes speed or direction.

The Cosmic Microwave Background

The principle of relativity, and Newton's first law of motion, apply only to observers who have a constant speed and direction (or zero speed) relative to the "cosmic microwave background," which provides the only backdrop and frame of reference with an absolute, universal character. It used to be that the basic reference frame was loosely called "the fixed stars," but stars and galaxies have their own individual motions within the Universe and do not constitute an adequate reference frame with respect to which to measure motion.

The cosmic microwave background is low-intensity electromagnetic radiation with wavelengths in the microwave region, which pervades the Universe, radiating in all directions. Measurements show that our galaxy is moving through this microwave radiation with a large, essentially constant velocity, toward a cluster of a large number of other galaxies. The way we detect our motion relative to the microwave background is through the "Doppler shift" of the frequencies of the microwave radiation, toward higher frequencies in front of us and lower frequencies behind. This is essentially the same phenomenon as that responsible for a fire engine siren sounding at a higher frequency when it is approaching us and a lower frequency when it is moving away from us.

The discovery of the cosmic microwave background provided major support for the "Big Bang" theory of the formation of the Universe. According to the Big Bang theory, the early Universe must have been an extremely hot mixture of charged particles and high-energy, short-wavelength electromagnetic radiation (visible light, x-rays, gamma rays, etc.). Electromagnetic radiation interacts strongly with charged particles, so light could not travel very far without interacting, making the Universe essentially opaque. Also, the Universe was so hot that electrically neutral atoms could not form without the electrons immediately being stripped away again by collisions with other fast-moving particles.

As the Universe expanded, the temperature dropped. Eventually the temperature was low enough for neutral atoms to form. The interaction of electromagnetic radiation with neutral atoms is much weaker than with individual charged particles, so the radiation was now essentially free, dissociated from the matter, and the Universe became transparent. As the Universe continued to expand (the actual space between clumps of matter got bigger), the wavelengths of the electromagnetic radiation got longer, until today this fossil radiation has wavelengths in the relatively low-energy, long-wavelength microwave portion of the electromagnetic spectrum.

Inertial Frames of Reference

It is an observational fact that in reference frames that are in uniform motion with respect to the cosmic microwave background, far from other objects (so that interactions are negligible), an object maintains uniform motion. Such frames are called "inertial frames" and are reference frames in which Newton's first law of motion is valid. All of these reference frames are equally valid; the cosmic microwave background simply provides a concrete example of such a reference frame.

QUESTION Is the surface of the Earth an inertial frame?

No! The Earth is rotating on its axis, so the velocity of an object sitting on the surface of the Earth is constantly changing direction, as is a coordinate frame tied to the Earth (Figure 1.46). Moreover, the Earth is orbiting the Sun, and the Solar System itself is orbiting the center of our Milky Way galaxy, and our galaxy is moving toward other galaxies. So the motion of an object sitting on the Earth is actually quite complicated and definitely not uniform with respect to the cosmic microwave background.

However, for many purposes the surface of the Earth can be considered to be (approximately) an inertial frame. For example, it takes 6 hours for the rotation of the Earth on its axis to make a 90° change in the direction of the velocity of a "fixed" point. If a process of interest takes only a few minutes, during these few minutes a "fixed" point moves in nearly a straight line at constant speed due to the Earth's rotation, and velocity changes in the process of interest are typically much larger than the very small velocity change of the approximate inertial frame of the Earth's surface.

Similarly, although the Earth is in orbit around the Sun, it takes 365 days to go around once, so for a period of a few days or even weeks the Earth's orbital motion is nearly in a straight line at constant speed. Hence for many purposes the Earth represents an approximately inertial frame despite its motion around the Sun.

The Special Theory of Relativity

Einstein's Special Theory of Relativity (published in 1905) built on the basic principle of relativity but added the conjecture that the speed of a beam of light must be the same as measured by observers in different frames of reference in uniform motion with respect to each other. In Figure 1.47, observers on each spaceship measure the speed of the light c emitted by the ship at the top to be the same ($c = 3 \times 10^8$) m/s, despite the fact that they are moving at different velocities.

This additional condition seems peculiar and has far-reaching consequences. After all, the map on the dashboard of your car has different speeds relative to different observers, depending on the motion of the observer. Yet a wide range of experiments has confirmed Einstein's conjecture: all observers measure the same speed for the same beam of light, $c = 3 \times 10^8$ m/s. (The color of the light is different for the different observers, but the speed is the same.)

On the other hand, if someone on the ship at the top throws a ball or a proton or some other piece of matter, the speed of the object will be different for observers on the three ships; it is only light whose speed is independent of the observer.

Einstein's theory has interesting consequences. For example, it predicts that time will run at different rates in different frames of reference. These

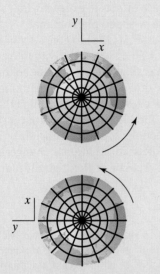

Figure 1.46 Axes tied to the Earth rotate through 90° in a quarter of a day (6 hours).

$c = 3 \times 10^8$ m/s

$\vec{v}_1$

$\vec{v}_2 = 0$

$c = 3 \times 10^8$ m/s

$\vec{v}_3$

$c = 3 \times 10^8$ m/s

Figure 1.47 Light emitted by the top spaceship is measured to have the same speed by observers in all three ships.

predictions have been confirmed by many experiments. These unusual effects are large only at very high speeds (a sizable fraction of the speed of light), which is why we don't normally observe these effects in everyday life, and why we can use nonrelativistic calculations for low-speed phenomena.

However, for the Global Positioning System (GPS) to give adequate accuracy it is necessary to take Einstein's Special Theory of Relativity into account. The atomic clocks on the satellites run slower than our clocks, due to the speed of the satellites, and the difference in clock rate depends on γ. Although γ for the GPS satellites orbiting the Earth is nearly 1, it differs just enough that making the approximation $\gamma \approx 1$ would make the GPS hopelessly inaccurate and useless. It is even necessary to apply corrections based on Einstein's General Theory of Relativity, which correctly predicts an additional change in clock rate due to the gravity of the Earth being weaker at the high altitude of the GPS satellites.

1.X.45 A spaceship at rest with respect to the cosmic microwave background emits a beam of red light. A different spaceship, moving at a speed of 2.5×10^8 m/s toward the first ship, detects the light. Which of the following statements are true for observers on the second ship? (More than one statement may be correct.)

(a) They observe that the light travels at 3×10^8 m/s.

(b) They see light that is not red.

(c) They observe that the light travels at 5.5×10^8 m/s.

(d) They observe that the light travels at 2.5×10^8 m/s.

1.11 *UPDATING POSITION AT HIGH SPEED

If $v \ll c$, $\vec{p} \approx m\vec{v}$ and $\vec{v} \approx \vec{p}/m$. But at high speed it is more complicated to determine the velocity from the (relativistic) momentum. Here is a way to solve for $\vec{v}$ in terms of $\vec{p}$:

$$|\vec{p}| = \frac{1}{\sqrt{1 - (|\vec{v}|/c)^2}} m|\vec{v}|$$

Divide by m and square: $\dfrac{|\vec{p}|^2}{m^2} = \dfrac{|\vec{v}|^2}{1 - (|\vec{v}|/c)^2}$

Multiply by $(1 - (|\vec{v}|/c)^2)$: $\dfrac{|\vec{p}|^2}{m^2} - \left(\dfrac{|\vec{p}|^2}{m^2 c^2}\right)|\vec{v}|^2 = |\vec{v}|^2$

Collect terms: $\left(1 + \dfrac{|\vec{p}|^2}{m^2 c^2}\right)|\vec{v}|^2 = \dfrac{|\vec{p}|^2}{m^2}$

$$|\vec{v}| = \frac{|\vec{p}|/m}{\sqrt{1 + \left(\dfrac{|\vec{p}|}{mc}\right)^2}}$$

The expression above gives the magnitude of $\vec{v}$, in terms of the magnitude of $\vec{p}$. To get an expression for the vector $\vec{v}$, recall that any vector can be factored into its magnitude times a unit vector in the direction of the vector, so

$$\vec{p} = |\vec{p}|\hat{p} \quad \text{and} \quad \vec{v} = |\vec{v}|\hat{v}$$

But since $\vec{p}$ and $\vec{v}$ are in the same direction, $\hat{v} = \hat{p}$, so

$$\vec{v} = |\vec{v}|\hat{p} = \frac{|\vec{p}|/m}{\sqrt{1 + \left(\dfrac{|\vec{p}|}{mc}\right)^2}}\hat{p} = \frac{(|\vec{p}|\hat{p})/m}{\sqrt{1 + \left(\dfrac{|\vec{p}|}{mc}\right)^2}}$$

$$\vec{v} = \frac{\vec{p}/m}{\sqrt{1 + \left(\dfrac{|\vec{p}|}{mc}\right)^2}}$$

THE RELATIVISTIC POSITION UPDATE FORMULA

$$\vec{r}_f = \vec{r}_i + \frac{1}{\sqrt{1 + \left(\dfrac{|\vec{p}|}{mc}\right)^2}}\left(\frac{\vec{p}}{m}\right)\Delta t \text{ (for small } \Delta t\text{)}$$

Note that at low speeds $|\vec{p}| \approx m|\vec{v}|$, and the denominator is

$$\sqrt{1 + \left(\frac{|\vec{v}|}{c}\right)^2} \approx 1$$

so the formula becomes the familiar $\vec{r}_f = \vec{r}_i + (\vec{p}/m)\Delta t$.

SUMMARY

Interactions (Section 1.2 and Section 1.4)
Interactions are indicated by
- Change of velocity (change of direction and/or change of speed)
- Change of identity
- Change of shape of multiparticle system
- Change of temperature of multiparticle system
- Lack of change when change is expected

Newton's first law of motion (Section 1.3)
An object moves in a straight line and at constant speed, except to the extent that it interacts with other objects.

Vectors (Section 1.5)
A 3D *vector* is a quantity with magnitude and a direction, which can be expressed as a triple $\langle x, y, z\rangle$. A vector is indicated by an arrow: $\vec{r}$.
A *scalar* is a single number.
Legal mathematical operations involving vectors include:
- Adding one vector to another vector
- Subtracting one vector from another vector
- Multiplying or dividing a vector by a scalar

- Finding the magnitude of a vector
- Taking the derivative of a vector

Operations that are *not* legal with vectors include:
- A vector cannot be added to a scalar.
- A vector cannot be set equal to a scalar.
- A vector cannot appear in the denominator (you can't divide by a vector).

A unit vector $\hat{r} = \vec{r}/|\vec{r}|$ has magnitude 1.
A vector can be factored using a unit vector: $\vec{F} = |\vec{F}|\hat{F}$.
Direction cosines: $\hat{r} = \langle\cos\theta_x, \cos\theta_y, \cos\theta_z\rangle$

The symbol Δ
The symbol Δ (delta) means "change of": $\Delta t = t_f - t_i$, $\Delta\vec{r} = \vec{r}_f - \vec{r}_i$.
Δ always means "final minus initial."

Velocity and change of position (Section 1.7)
Definition of average velocity

$$\vec{v}_{\text{avg}} = \frac{\Delta\vec{r}}{\Delta t} = \frac{\vec{r}_f - \vec{r}_i}{t_f - t_i}$$

Velocity is a vector. $\vec{r}$ is the position of an object (a vector). t is the time. Average velocity is equal to the change in position divided by the time elapsed. SI units of velocity are meters per second (m/s).

The position update formula

$$\vec{r}_f = \vec{r}_i + \vec{v}_{avg}\,\Delta t$$

The final position (vector) is the vector sum of the initial position plus the product of the average velocity and the elapsed time.

Definition of instantaneous velocity

$$\vec{v} = \lim_{\Delta t \to 0}\frac{\Delta \vec{r}}{\Delta t} = \frac{d\vec{r}}{dt}$$

The instantaneous velocity is the limiting value of the average velocity as the time elapsed becomes very small.

Velocity in terms of momentum

$$\vec{v} = \frac{\vec{p}/m}{\sqrt{1 + \left(\dfrac{|\vec{p}|}{mc}\right)^2}}\ \text{or}\ \vec{v} \approx \vec{p}/m \ \text{at low speeds}$$

Momentum (Section 1.8)
Definition of momentum

$$\vec{p} = \gamma m\vec{v}$$

where $\gamma = \dfrac{1}{\sqrt{1 - (|\vec{v}|/c)^2}}$ (lower-case Greek gamma)

Momentum (a vector) is the product of the relativistic factor "gamma" (a scalar), mass, and velocity.

Combined into one equation: $\vec{p} = \dfrac{1}{\sqrt{1 - (|\vec{v}|/c)^2}}\, m\vec{v}$.

Approximation for momentum at low speeds

$$\vec{p} \approx m\vec{v} \ \text{at speeds such that}\ |\vec{v}| \ll c$$

Useful numbers:
   Radius of a typical atom: about 1×10^{-10} meter
   Radius of a proton or neutron: about 1×10^{-15} meter
   Speed of light: 3×10^8 m/s

These and other useful data and conversion factors are given on the inside back cover of the textbook.

EXERCISES AND PROBLEMS

Exercises ("X") are small-scale questions that deal with basic definitions and quantities and require few steps in reasoning. Their purpose is to give you practice in dealing with the basic concepts. Problems ("P") are larger applications that require some thought and may involve several steps in reasoning. The problems give you practice in applying what you are learning.

Section 1.2

1.X.46 Give two examples (other than those discussed in the text) of interactions that may be detected by observing:

  (a) Change in velocity
  (b) Change in temperature
  (c) Change in shape
  (d) Change in identity
  (e) Lack of change when change is expected

1.X.47 Which of the following observations give conclusive evidence of an interaction? (Choose all that apply.)

  (a) Change of velocity, either change of direction or change of speed
  (b) Change of shape or configuration without change of velocity
  (c) Change of position without change of velocity
  (d) Change of identity without change of velocity
  (e) Change of temperature without change of velocity. Explain your choice.

1.X.48 Moving objects left the traces labeled *A–F* in Figure 1.48 and Figure 1.49. The dots were deposited at equal time intervals (for example, one dot each second). In each case the object starts from the square. Which trajectories show evidence that the

moving object was interacting with another object somewhere? If there is evidence of an interaction, what is the evidence?

Figure 1.48 Exercise 1.X.48. Also see Figure 1.49.

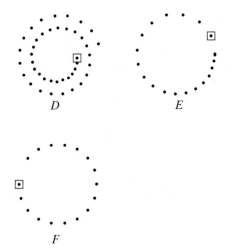

D   *E*

F

Figure 1.49 Exercise 1.X.48. Also see Figure 1.48.

Section 1.3

1.X.49 Which of the following observers might observe something that appears to violate Newton's first law of motion? Explain why.

(a) A person standing still on a street corner

(b) A person riding on a roller coaster

(c) A passenger on a starship traveling at $0.75c$ toward the nearby star Alpha Centauri

(d) An airplane pilot doing aerobatic loops

(e) A hockey player coasting across the ice

1.X.50 Why do we use a spaceship in outer space, far from other objects, to illustrate Newton's first law? Why not a car or a train? (More than one of the following statements may be correct.)

(a) A car or train touches other objects, and interacts with them.

(b) A car or train can't travel fast enough.

(c) The spaceship has negligible interactions with other objects.

(d) A car or train interacts gravitationally with the Earth.

(e) A spaceship can never experience a gravitational force.

1.X.51 A spaceship far from all other objects uses its thrusters to attain a speed of 1×10^4 m/s. The crew then shuts off the power. According to Newton's first law, what will happen to the motion of the spaceship from then on?

1.X.52 Some science museums have an exhibit called a Bernoulli blower, in which a volleyball hangs suspended in a column of air blown upward by a strong fan. If you saw a ball suspended in the air but didn't know the blower was there, why would Newton's first law suggest that something must be holding the ball up?

1.X.53 Place a ball on a book and walk with the book in uniform motion. Note that you don't really have to do anything to the ball to keep the ball moving with constant velocity (relative to the ground) or to keep the ball at rest (relative to you). Then stop suddenly, or abruptly change your direction or speed. What does Newton's first law of motion predict for the motion of the ball (assuming that the interaction between the ball and the book is small)? Does the ball behave as predicted? It may help to take the point of view of a friend who is standing still, watching you.

Section 1.5

Vector equality; components

1.X.54 Consider a vector $\vec{u} = \langle u_x, u_y, u_z \rangle$, and another vector $\vec{p} = \langle p_x, p_y, p_z \rangle$. If $\vec{u} = \vec{p}$, then which of the following statements must be true? Some, all, or none of the following may be true:

(a) $u_x = p_x$

(b) $u_y = p_y$

(c) $u_z = p_z$

(d) The direction of $\vec{u}$ is the same as the direction of $\vec{p}$.

Vectors and scalars

1.X.55 Does the symbol $\vec{a}$ represent a vector or a scalar?

1.X.56 Which of the following are vectors?

(a) 5 m/s

(b) $\langle -11, 5.4, -33 \rangle$ m

(c) $\vec{r}$

(d) v_z

1.X.57 Does the symbol $|\vec{v}|$ represent a vector or a scalar?

1.X.58 Which of the following are vectors? **(a)** 3.5 **(b)** 0 **(c)** $\langle 0.7, 0.7, 0.7 \rangle$ **(d)** $\langle 0, 2.3, -1 \rangle$ **(e)** -3×10^6 **(f)** $3 \cdot \langle 14, 0, -22 \rangle$

1.X.59 Which of the following are vectors? **(a)** $\vec{r}/2$ **(b)** $|\vec{r}|/2$ **(c)** $\langle r_x, r_y, r_z \rangle$ **(d)** $5 \cdot \vec{r}$

Magnitude of a vector

1.X.60 What is the magnitude of the vector $\vec{v}$, where $\vec{v} = \langle 8 \times 10^6, 0, -2 \times 10^7 \rangle$ m/s?

1.X.61 Figure 1.50 shows several arrows representing vectors in the xy plane.

(a) Which vectors have magnitudes equal to the magnitude of $\vec{a}$?

(b) Which vectors are equal to $\vec{a}$?

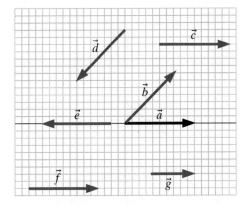

Figure 1.50 Exercise 1.X.61.

1.X.62 In Figure 1.51 three vectors are represented by arrows in the xy plane. Each square in the grid represents one meter. For each vector, write out the components of the vector, and calculate the magnitude of the vector.

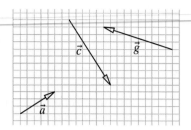

Figure 1.51 Exercise 1.X.62.

Illegal operations with vectors

1.X.63 Is $3 + \langle 2, -3, 5 \rangle$ a meaningful expression? If so, what is its value?

Multiplying a vector by a scalar

1.X.64 On a piece of graph paper, draw arrows representing the following vectors. Make sure the tip and tail of each arrow you draw are clearly distinguishable.

(a) Placing the tail of the vector at $\langle 5, 2, 0 \rangle$, draw an arrow representing the vector $\vec{p} = \langle -7, 3, 0 \rangle$. Label it $\vec{p}$.

(b) Placing the tail of the vector at $\langle -5, 8, 0 \rangle$, draw an arrow representing the vector $-\vec{p}$. Label it $-\vec{p}$.

1.X.65 A proton is located at $\vec{r}_p = \langle 2, 6, -3 \rangle$ m. An electron is located at $\vec{r}_e = \langle 4, 12, -6 \rangle$ m. Which of the following statements are true?
(a) $2\vec{r}_p = \vec{r}_e$ (b) $2\hat{r}_p = \hat{r}_e$ (c) $|2\vec{r}_p| = |\vec{r}_e|$

1.X.66 The following questions refer to the vectors depicted by arrows in Figure 1.52.

(a) What are the components of the vector $\vec{a}$? (Note that since the vector lies in the xy plane, its z component is zero.)

(b) What are the components of the vector $\vec{b}$?

(c) Is this statement true or false? $\vec{a} = \vec{b}$

(d) What are the components of the vector $\vec{c}$?

(e) Is this statement true or false? $\vec{c} = -\vec{a}$

(f) What are the components of the vector $\vec{d}$?

(g) Is this statement true or false? $\vec{d} = -\vec{c}$

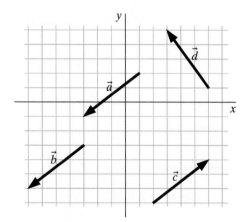

Figure 1.52 Exercise 1.X.66.

1.X.67 (a) In Figure 1.53, what are the components of the vector $\vec{d}$?
(b) If $\vec{e} = -\vec{d}$, what are the components of $\vec{e}$?

(c) If the tail of vector $\vec{d}$ were moved to location $\langle -5, -2, 4 \rangle$ m, where would the tip of the vector be located?

(d) If the tail of vector $-\vec{d}$ were placed at location $\langle -1, -1, -1 \rangle$ m, where would the tip of the vector be located?

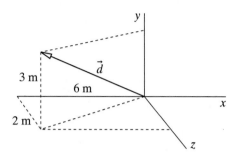

Figure 1.53 Exercise 1.X.67.

1.X.68 What is the result of multiplying the vector $\vec{a}$ by the scalar f, where $\vec{a} = \langle 0.02, -1.7, 30.0 \rangle$ and $f = 2.0$?

1.X.69 (a) On a piece of graph paper, draw the vector $\vec{f} = \langle -2, 4, 0 \rangle$, putting the tail of the vector at $\langle -3, 0, 0 \rangle$. Label the vector $\vec{f}$.

(b) Calculate the vector $2\vec{f}$, and draw this vector on the graph, putting its tail at $\langle -3, -3, 0 \rangle$, so you can compare it to the original vector. Label the vector $2\vec{f}$.

(c) How does the magnitude of $2\vec{f}$ compare to the magnitude of $\vec{f}$?

(d) How does the direction of $2\vec{f}$ compare to the direction of $\vec{f}$?

(e) Calculate the vector $\vec{f}/2$, and draw this vector on the graph, putting its tail at $\langle -3, -6, 0 \rangle$, so you can compare it to the other vectors. Label the vector $\vec{f}/2$.

(f) How does the magnitude of $\vec{f}/2$ compare to the magnitude of $\vec{f}$?

(g) How does the direction of $\vec{f}/2$ compare to the direction of $\vec{f}$?

(h) Does multiplying a vector by a scalar change the magnitude of the vector?

(i) The vector $a(\vec{f})$ has a magnitude three times as great as that of $\vec{f}$, and its direction is opposite to the direction of $\vec{f}$. What is the value of the scalar factor a?

Unit vectors

1.X.70 What is the unit vector in the direction of $\langle -300, 0, 0 \rangle$?

1.X.71 What is the unit vector in the direction of $\langle 2, 2, 2 \rangle$? What is the unit vector in the direction of $\langle 3, 3, 3 \rangle$?

1.X.72 Write the vector $\vec{a} = \langle 400, 200, -100 \rangle$ m/s² as the product $|\vec{a}| \cdot \hat{a}$.

1.X.73 (a) On a piece of graph paper, draw the vector $\vec{g} = \langle 4, 7, 0 \rangle$ m. Put the tail of the vector at the origin.
(b) Calculate the magnitude of $\vec{g}$.
(c) Calculate $\hat{g}$, the unit vector pointing in the direction of $\vec{g}$.
(d) On the graph, draw $\hat{g}$. Put the tail of the vector at $\langle 1, 0, 0 \rangle$ m so you can compare $\hat{g}$ and $\vec{g}$.
(e) Calculate the product of the magnitude $|\vec{g}|$ times the unit vector $\hat{g}$, $(|\vec{g}|)(\hat{g})$.

1.X.74 A proton is located at $\langle 3 \times 10^{-10}, -3 \times 10^{-10}, 8 \times 10^{-10} \rangle$ m.
(a) What is $\vec{r}$, the vector from the origin to the location of the proton?

(b) What is $|\vec{r}|$?

(c) What is $\hat{r}$, the unit vector in the direction of $\vec{r}$?

1.X.75 Write each of these vectors as the product of the magnitude of the vector and the appropriate unit vector:

(a) $\langle 0, 0, 9.5 \rangle$ (b) $\langle 0, -679, 0 \rangle$

(c) $\langle 3.5 \times 10^{-3}, 0, -3.5 \times 10^{-3} \rangle$ (d) $\langle 4 \times 10^{6}, -6 \times 10^{6}, 3 \times 10^{6} \rangle$

1.X.76 A cube is 3 cm on a side, with one corner at the origin. What is the unit vector pointing from the origin to the diagonally opposite corner at location $\langle 3, 3, 3 \rangle$ cm? What is the angle from this diagonal to one of the adjacent edges of the cube?

1.X.77 Which of the following are unit vectors? (Numerical values are given to only 3 significant figures.)

(a) $\langle 0, 0, -1 \rangle$

(b) $\langle 0.5, 0.5, 0 \rangle$

(c) $\langle 0.333, 0.333, 0.333 \rangle$

(d) $\langle 0.9, 0, 0.1 \rangle$

(e) $\langle 0, 3, 0 \rangle$

(f) $\langle 1, -1, 1 \rangle$

(g) $\langle 0.577, 0.577, 0.577 \rangle$

(h) $\langle 0.949, 0, -0.316 \rangle$

1.X.78 Two vectors, $\vec{f}$ and $\vec{g}$, are equal: $\vec{f} = \vec{g}$. Which of the following statements are true?

(a) $\hat{f} = \hat{g}$

(b) $g_x = f_x$

(c) $f_z = g_z$

(d) The directions of $\vec{f}$ and $\vec{g}$ may be different.

(e) The magnitudes of $\vec{f}$ and $\vec{g}$ may be different.

Vector addition and subtraction

1.X.79 Imagine that you have a baseball and a tennis ball at different locations. The center of the baseball is at $\langle 3, 5, 0 \rangle$ m, and the center of the tennis ball is at $\langle -3, -1, 0 \rangle$ m. On a piece of graph paper, do the following:

(a) Draw dots at the locations of the center of the baseball and the center of the tennis ball.

(b) Draw the position vector of the baseball, which is an arrow whose tail is at the origin and whose tip is at the location of the baseball. Label this position vector $\vec{B}$. Clearly show the tip and tail of each arrow.

(c) Complete this equation: $\vec{B} = <$ ___ , ___ , ___ $>$ m.

(d) Draw the position vector of the tennis ball. Label it $\vec{T}$.

(e) Complete this equation: $\vec{T} = <$ ___ , ___ , ___ $>$ m.

(f) Draw the relative position vector for the tennis ball relative to the baseball. The tail of this vector is at the center of the baseball, and the tip of the vector is at the center of the tennis ball. Label this relative position vector $\vec{r}$.

(g) Complete the following equation by reading the coordinates of $\vec{r}$ from your graph: $\vec{r} = <$ ___ , ___ , ___ $>$ m.

(h) Calculate this difference: $\vec{T} - \vec{B} = <$ ___ , ___ , ___ $>$ m.

(i) Is the following statement true? $\vec{r} = \vec{T} - \vec{B}$ _____.

(j) Write two other equations relating the vectors $\vec{B}$, $\vec{T}$, and $\vec{r}$.

(k) Calculate the magnitudes of the vectors $\vec{B}$, $\vec{T}$, and $\vec{r}$.

(l) Calculate the difference of the magnitudes $|\vec{T}| - |\vec{B}|$.

(m) Does $|\vec{T}| - |\vec{B}| = |\vec{T} - \vec{B}|$?

Change of a vector; relative position

1.X.80 In Figure 1.54, $\vec{r}_1 = \langle 3, -2, 0 \rangle$ m and $\vec{r}_2 = \langle 5, 2, 0 \rangle$ m. Calculate the position of object 2 relative to object 1, as a relative position vector. Before checking the answer, see whether your answer is consistent with the appearance of the vector $\vec{r}_{2 \text{ relative to } 1} = \vec{r}_2 - \vec{r}_1$ shown in the diagram. What is the position of object 1 relative to object 2, as a vector?

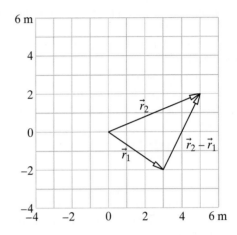

Figure 1.54 Exercise 1.X.80.

1.X.81 (a) What is the vector whose tail is at $\langle 9.5, 7, 0 \rangle$ m and whose head is at $\langle 4, -13, 0 \rangle$ m?

(b) What is the magnitude of this vector?

1.X.82 A man is standing on the roof of a building with his head at the position $\langle 12, 30, 13 \rangle$ m. He sees the top of a tree, which is at the position $\langle -25, 35, 43 \rangle$ m.

(a) What is the relative position vector that points from the man's head to the top of the tree?

(b) What is the distance from the man's head to the top of the tree?

1.X.83 A star is located at $\langle 6 \times 10^{10}, 8 \times 10^{10}, 6 \times 10^{10} \rangle$ m. A planet is located at $\langle -4 \times 10^{10}, -9 \times 10^{10}, 6 \times 10^{10} \rangle$ m.

(a) What is the vector $\vec{r}_{sp}$ pointing from the star to the planet?

(b) What is the vector $\vec{r}_{ps}$ pointing from the planet to the star?

1.X.84 A planet is located at $\langle -1 \times 10^{10}, 8 \times 10^{10}, -3 \times 10^{10} \rangle$. A star is located at $\langle 6 \times 10^{10}, -5 \times 10^{10}, 1 \times 10^{10} \rangle$.

(a) What is $\vec{r}_{sp}$, the vector from the star to the planet?

(b) What is the magnitude of $\vec{r}_{sp}$?

(c) What is $\hat{r}_{sp}$, the unit vector (vector with magnitude 1) in the direction of $\vec{r}_{sp}$?

1.X.85 A proton is located at $\langle x_p, y_p, z_p \rangle$. An electron is located at $\langle x_e, y_e, z_e \rangle$. What is the vector pointing from the electron to the proton? What is the vector pointing from the proton to the electron?

Section 1.7

1.X.86 In the expression $\Delta \vec{r} / \Delta t$, what is the meaning of $\Delta \vec{r}$? What is the meaning of Δt?

1.X.87 A "slow" neutron produced in a nuclear reactor travels from location $\langle 0.2, -0.05, 0.1 \rangle$ m to location $\langle -0.202, 0.054, 0.098 \rangle$ m in 2 microseconds ($1 \mu s = 1 \times 10^{-6}$ s).

(a) What is the average velocity of the neutron?

(b) What is the average speed of the neutron?

1.X.88 The position of a baseball relative to home plate changes from $\langle 15, 8, -3 \rangle$ m to $\langle 20, 6, -1 \rangle$ m in 0.1 second. As a vector, write the average velocity of the baseball during this time interval.

1.P.89 A spacecraft traveling at a velocity of $\langle -20, -90, 40 \rangle$ m/s is observed to be at a location $\langle 200, 300, -500 \rangle$ m relative to an origin located on a nearby asteroid. At a later time the spacecraft is at location $\langle -380, -2310, 660 \rangle$ m.

  (a) How long did it take the spacecraft to travel between these locations?

  (b) How far did the spacecraft travel?

  (c) What is the speed of the spacecraft?

  (d) What is the unit vector in the direction of the spacecraft's velocity?

1.X.90 Start with the definition of average velocity and derive the position update formula from it. Show all steps in the derivation.

1.X.91 At time $t_1 = 12$ s, a car is located at $\langle 84, 78, 24 \rangle$ m and has velocity $\langle 4, 0, -3 \rangle$ m/s. At time $t_2 = 18$ s, what is the position of the car? (The velocity is constant in magnitude and direction during this time interval.)

1.X.92 At a certain instant a ball passes location $\langle 7, 21, -17 \rangle$ m. In the next 3 seconds, the ball's average velocity is $\langle -11, 42, 11 \rangle$ m/s. At the end of this 3 second time interval, what is the height y of the ball?

1.X.93 Here are the positions at three different times for a bee in flight (a bee's top speed is about 7 m/s).

Time	6.3 s	6.8 s	7.3 s
Position	$\langle -3.5, 9.4, 0 \rangle$ m	$\langle -1.3, 6.2, 0 \rangle$ m	$\langle 0.5, 1.7, 0 \rangle$ m

  (a) Between 6.3 s and 6.8 s, what was the bee's average velocity? Be careful with signs.

  (b) Between 6.3 s and 7.3 s, what was the bee's average velocity? Be careful with signs.

  (c) Of the two average velocities you calculated, which is the best estimate of the bee's instantaneous velocity at time 6.3 s?

  (d) Using the best information available, what was the displacement of the bee during the time interval from 6.3 s to 6.33 s?

1.X.94 The position of a golf ball relative to the tee changes from $\langle 50, 20, 30 \rangle$ m to $\langle 53, 18, 31 \rangle$ m in 0.1 second. As a vector, write the velocity of the golf ball during this short time interval.

1.X.95 The crew of a stationary spacecraft observe an asteroid whose mass is 4×10^{17} kg. Taking the location of the spacecraft as the origin, the asteroid is observed to be at location $\langle -3 \times 10^3, -4 \times 10^3, 8 \times 10^3 \rangle$ m at a time 18.4 seconds after lunchtime. At a time 21.4 seconds after lunchtime, the asteroid is observed to be at location $\langle -1.4 \times 10^3, -6.2 \times 10^3, 9.7 \times 10^3 \rangle$ m. Assuming that the velocity of the asteroid does not change during this time interval, calculate the vector velocity $\vec{v}$ of the asteroid.

1.X.96 An electron passes location $\langle 0.02, 0.04, -0.06 \rangle$ m, and $2\,\mu s$ later is detected at location $\langle 0.02, 1.84, -0.86 \rangle$ m (1 microsecond is 1×10^{-6} s).

  (a) What is the average velocity of the electron?

(b) If the electron continues to travel at this average velocity, where will it be in another $5\,\mu s$?

1.P.97 Figure 1.55 shows the trajectory of a ball traveling through the air, affected by both gravity and air resistance. Here are the positions of the ball at several successive times:

location	t(s)	position (m)
A	0.0	$\langle 0, 0, 0 \rangle$
B	1.0	$\langle 22.3, 26.1, 0 \rangle$
C	2.0	$\langle 40.1, 38.1, 0 \rangle$

  (a) What is the average velocity of the ball as it travels between location A and location B?

  (b) If the ball continued to travel at the same average velocity during the next second, where would it be at the end of that second? (That is, where would it be at time $t = 2$ seconds?)

  (c) How does your prediction from part (b) compare to the actual position of the ball at $t = 2$ seconds (location C)? If the predicted and observed locations of the ball are different, explain why.

Figure 1.55 Problem 1.P.97.

1.P.98 At 6 seconds after 3:00, a butterfly is observed leaving a flower whose location is $\langle 6, -3, 10 \rangle$ m relative to an origin on top of a nearby tree. The butterfly flies until 10 seconds after 3:00, when it alights on a different flower whose location is $\langle 6.8, -4.2, 11.2 \rangle$ m relative to the same origin. What was the location of the butterfly at a time 8.5 seconds after 3:00? What assumption did you have to make in calculating this location?

Section 1.8

1.X.99 Which of the following statements about the velocity and momentum of an object are correct?

  (a) The momentum of an object is always in the same direction as its velocity.

  (b) The momentum of an object can be either in the same direction as its velocity or in the opposite direction.

  (c) The momentum of an object is perpendicular to its velocity.

  (d) The direction of an object's momentum is not related to the direction of its velocity.

  (e) The direction of an object's momentum is tangent to its path.

1.X.100 In which of these situations is it reasonable to use the approximate formula for the momentum of an object, instead of the full relativistically correct formula?

(a) A car traveling on an interstate highway

(b) A commercial jet airliner flying between New York and Seattle

(c) A neutron traveling at 2700 meters per second

(d) A proton in outer space traveling at 2×10^8 m/s

(e) An electron in a television tube traveling 3×10^6 m/s

1.X.101 Answer the following questions about the factor γ (gamma) in the full relativistic formula for momentum:

(a) Is γ a scalar or a vector quantity?

(b) What is the minimum possible value of γ?

(c) Does γ reach its minimum value when an object's speed is high or low?

(d) Is there a maximum possible value for γ?

(e) Does γ become large when an object's speed is high or low?

(f) Does the approximation $\gamma \approx 1$ apply when an object's speed is low or when it is high?

1.X.102 A baseball has a mass of 0.155 kg. A professional pitcher throws a baseball 90 miles per hour, which is 40 m/s. What is the magnitude of the momentum of the pitched baseball?

1.X.103 A proton in an accelerator attains a speed of $0.88c$. What is the magnitude of the momentum of the proton?

1.X.104 A hockey puck with a mass of 0.4 kg has a velocity of $\langle 38, 0, -27 \rangle$ m/s. What is the magnitude of its momentum, $|\vec{p}|$?

1.X.105 A baseball has a mass of about 155 g. What is the magnitude of the momentum of a baseball thrown at a speed of 100 miles per hour? (Note that you need to convert mass to kilograms and speed to meters/second. See the inside back cover of the textbook for conversion factors.)

1.X.106 What is the magnitude (in kg · m/s) of the momentum of a 1000 kg airplane traveling at a speed of 500 miles per hour? (Note that you need to convert speed to meters per second.)

1.X.107 If a particle has momentum $\vec{p} = \langle 4, -5, 2 \rangle$ kg · m/s, what is the magnitude $|\vec{p}|$ of its momentum?

1.X.108 An electron with a speed of $0.95c$ is emitted by a supernova, where c is the speed of light. What is the magnitude of the momentum of this electron?

1.X.109 A "cosmic-ray" proton hits the upper atmosphere with a speed $0.9999c$, where c is the speed of light. What is the magnitude of the momentum of this proton? Note that $|\vec{v}|/c = 0.9999$; you don't actually need to calculate the speed $|\vec{v}|$.

1.X.110 A proton in an accelerator is traveling at a speed of $0.99c$. (Masses of particles are given on the inside back cover of this textbook.)

(a) If you use the approximate nonrelativistic formula for the magnitude of momentum of the proton, what answer do you get?

(b) What is the magnitude of the correct relativistic momentum of the proton?

(c) The approximate value (the answer to part a) is significantly too low. What is the ratio of magnitudes you calculated (correct/approximate)? Such speeds are attained in particle accelerators.

1.X.111 An object with mass 1.6 kg has momentum $\langle 0, 0, 4 \rangle$ kg · m/s.

(a) What is the magnitude of the momentum?

(b) What is the unit vector corresponding to the momentum?

(c) What is the speed of the object?

1.X.112 When the speed of a particle is close to the speed of light, the factor γ, the ratio of the correct relativistic momentum $\gamma m \vec{v}$ to the approximate nonrelativistic momentum $m \vec{v}$, is quite large. Such speeds are attained in particle accelerators, and at these speeds the approximate nonrelativistic formula for momentum is a very poor approximation. Calculate γ for the case where $|\vec{v}|/c = 0.9996$.

1.X.113 An electron travels at speed $|\vec{v}| = 0.996c$, where $c = 3 \times 10^8$ m/s is the speed of light. The electron travels in the direction given by the unit vector $\hat{v} = \langle 0.655, -0.492, -0.573 \rangle$. The mass of an electron is 9×10^{-31} kg.

(a) What is the value of $\gamma = 1/\sqrt{1 - (|\vec{v}|/c)^2}$? You can simplify the calculation if you notice that $(|\vec{v}|/c)^2 = (0.996)^2$.

(b) What is the speed of the electron?

(c) What is the magnitude of the electron's momentum?

(d) What is the vector momentum of the electron? Remember that any vector can be "factored" into its magnitude times its unit vector, so that $\vec{v} = |\vec{v}|\hat{v}$.

1.X.114 If p/m is $0.85c$, what is v in terms of c?

Section 1.9

1.X.115 A tennis ball of mass m traveling with velocity $\langle v_x, 0, 0 \rangle$ hits a wall and rebounds with velocity $\langle -v_x, 0, 0 \rangle$.

(a) What was the change in momentum of the tennis ball?

(b) What was the change in the magnitude of the momentum of the tennis ball?

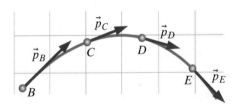

Figure 1.56 Problem 1.P.117.

1.X.116 A 50 kg child is riding on a carousel (merry-go-round) at a constant speed of 5 m/s. What is the magnitude of the change in the child's momentum $|\Delta \vec{p}|$ in going all the way around (360°)? In going halfway around (180°)? Draw a diagram showing the initial vector momentum and the final vector momentum, then subtract, then find the magnitude.

1.P.117 Figure 1.56 shows a portion of the trajectory of a ball traveling through the air. Arrows indicate its momentum at several locations.

At various locations, the ball's momentum is:

$\vec{p}_B = \langle 3.03, 2.83, 0 \rangle$ kg · m/s

$\vec{p}_C = \langle 2.55, 0.97, 0 \rangle$ kg · m/s

$\vec{p}_D = \langle 2.24, -0.57, 0 \rangle$ kg · m/s

$\vec{p}_E = \langle 1.97, -1.93, 0 \rangle$ kg · m/s

$\vec{p}_F = \langle 1.68, -3.04, 0 \rangle$ kg · m/s

(a) Calculate the change in the ball's momentum between each pair of adjacent locations.

(b) On a copy of the diagram, draw arrows representing each $\Delta \vec{p}$ you calculated in part (a).

(c) Between which two locations is the magnitude of the change in momentum greatest?

Computational Problems

These problems are intended to introduce you to using a computer to model matter, interactions, and motion. You will build on these small calculations to build models of physical systems in later chapters.

Some parts of these problems can be done with almost any tool (spreadsheet, math package, etc.). Other parts are most easily done with a programming language. We recommend the free 3D programming environment VPython, obtainable at http://vpython.org. Your instructor will introduce you to an available computational tool and assign problems, or parts of problems, that can be addressed using the chosen tool.

1.P.118 (a) Write a program that makes an object move from left to right across the screen with velocity $\langle v_x, 0, 0 \rangle$. Make v_x a variable, so you can change it later. Let the time interval for each step of the computation be a variable dt, so that the position x increases by an amount $v_x dt$ each time.
(b) Modify a copy of your program to make the object run into a wall and reverse its direction.
(c) Make a modification so that v_x is no longer a constant but changes smoothly with time. Is the change of the speed of the object clearly visible to an observer? Try to make one version in which the speed change is clearly noticeable, and another in which it is not noticeable.
(d) Corresponding to part (c), make a computer graph of x vs. t, where t is the time.
(e) Corresponding to part (c), make a computer graph of v_x vs. t, where t is the time.

1.P.119 (a) Write a program that makes an object move at an angle.
(b) Change the component of velocity of the object in the x direction but not in the y direction, or vice versa. What do you observe?
(c) Start the object moving at an angle and make it bounce off at an appropriate angle when it hits a wall.

1.P.120 Write a program that makes an object move smoothly from left to right across the screen with speed v, leaving a trail of dots on the screen at equal time intervals. If the dots are too close together, leave a dot every N steps, and adjust N to give a nice display.

1.P.121 This problem requires a 3D programming environment. Assume SI units, and give the objects reasonable and distinguishable attributes (such as sizes and colors). Run the program after each part, to make sure your display is correct.
   (a) Create a sphere representing a basketball, at location $\langle -5, 2, -3 \rangle$ m.
   (b) Create an arrow to represent the position of the basketball, relative to the origin.
   (c) Create a second sphere to represent a volleyball, at location $\langle -3, -1, 3.5 \rangle$ m.

   (d) Create an arrow to represent the position of the volleyball relative to the origin.
   (e) Using symbolic names (no numbers) for the positions and attributes of these balls, create an arrow to represent the position of the volleyball relative to the basketball.
   (f) Change the positions of the basketball and the volleyball to $\langle -4, 2, 5 \rangle$ m and $\langle 3, 1, -2 \rangle$ m, respectively. If you correctly used symbolic names in the previous part, the arrow representing the relative position vector should automatically adjust. Does it?
   (g) Add to your program instructions to print the position of each ball, and the position of the tip and the tail of the arrow (all 3D vectors).

1.P.122 Write a program to display the motion of a cart traveling on a track at constant velocity.
   (a) Create a box to represent a track 2 meters long, 10 cm wide, and 5 cm high. Orient the long axis of the track along the x axis.
   (b) Create a second box to represent a low-friction cart that is 10 cm long, 4 cm high, and 6 cm wide. Give this box a symbolic name so you can refer to it later in the program.
   (c) Position the cart so it is 1 cm above the track, and its left edge lines up with the left end of the track.
   (d) Create a vector variable to represent the velocity of the cart. Give the cart an initial velocity of $\langle 0.2, 0, 0 \rangle$ m/s.
   (e) Using a time step of 0.01 seconds, write a loop to use the position update formula to animate the motion of the cart as it travels at constant velocity from one end of the track to the other.
   (f) Change your program to start the cart at the right end of the track and move to the left.
   (g) What would happen if the cart had a nonzero y velocity component? Try it.
   (h) Have your cart leave a trail behind it as it moves.

1.P.123 Write a program to display the motion of a hockey puck sliding at constant velocity on ice.
   (a) Represent the ice surface by a box that is large in the x and z dimensions and thin in the y dimension. Represent the puck by a small cylinder, positioned just above the ice surface. (If you do not have a 3D programming environment, represent the puck by a circle rather than a cylinder.) Give the puck a symbolic name so you can refer to it later in the program.
   (b) Position the puck at one corner of the ice surface.
   (c) Create a vector variable to represent the velocity of the puck. Assign velocity components so the puck will slide diagonally across the ice.
   (d) Using a time step of 0.001 second, write a loop to use the position update formula to animate the motion of the puck as it slides from one corner of the ice surface to another.

ANSWERS TO EXERCISES

1.X.1 a, d
1.X.2 a, b, c, e
1.X.3 a, b, c
1.X.4 a, d

1.X.5 d, e
1.X.6 three
1.X.7 one
1.X.8 5.10 m

1.X.9 no

1.X.10 d

1.X.11 no

1.X.12 $\langle -0.09, 4.2, -78.0 \rangle$

1.X.13 $\langle 1, -1.5, 2.5 \rangle$

1.X.14 18.5 m/s

1.X.15 They are the same.

1.X.16 $\langle 0, 1, 0 \rangle$

1.X.17 $\langle 0.873, 0.436, -0.218 \rangle$

1.X.18 7

1.X.19 no

1.X.20 361, 335, 577, 696, no

1.X.21 $\langle 0, 0, 0 \rangle$, 0, 7.07×10^3, 7.07×10^3, 1.41×10^4

1.X.22 $\langle 450, -300, -200 \rangle$, $\langle 150, 300, -200 \rangle$, $\langle -150, -300, 200 \rangle$

1.X.23 a, b, e

1.X.24 (a) $\langle 9, 2, 20 \rangle$ m, (b) 120 s

1.X.25 $\langle -1, 0, -1 \rangle$ m

1.X.26 $\langle -0.940, 0.342, 0 \rangle$

1.X.27 $\langle -0.643, 0.766, 0 \rangle$

1.X.28 2.67×10^{-3} m/s

1.X.29 (a) $\langle 8, -10, 12 \rangle$ m/s, (b) 17.55 m/s, (c) $\langle 0.456, -0.570, 0.684 \rangle$

1.X.30 (a) $\langle 25, -55, -30 \rangle$ m/s, (b) 67.45 m/s, (c) $\langle 0.371, -0.815, -0.445 \rangle$

1.X.31 $\langle 2.1 \times 10^5, 1.4 \times 10^5, -2.8 | 5 \times 10 \rangle$ m

1.X.32 0.5 s

1.X.33 Average velocity is the displacement (change in position) divided by the total time; instantaneous velocity is the limit of the average velocity as the time interval gets very small.

1.X.34 1a, 2h, 3g, 4f, 5e

1.X.35 5 m/s/s, about half as big

1.X.36 $\langle 5, 8t, 2 - 18t^2 \rangle$, $\langle 0, 8, -16t \rangle$, $\langle 5, 0, 2 \rangle$, $\langle 0, 0, -16 \rangle$

1.X.37 650 kg·m/s

1.X.38 2.415×10^{-22} kg·m/s

1.X.39 b

1.X.40 $\langle 0.4, 0, 10 \rangle$ m

1.X.41 $\langle 111.31, 0, 41.54 \rangle$ m

1.X.42 (a) The diagram should show the ball bouncing back to the left with almost its initial speed, (b) $\langle -5.586, 0, 0 \rangle$ kg·m/s, (c) -0.114 kg·m/s

1.X.43 (a) $\langle -1.6 \times 10^{28}, 0, 1.6 \times 10^{28} \rangle$ kg·m/s; (b) and (c) if you make a careful diagram, with the two arrows tail to tail, you'll see that the arrow representing $\Delta \vec{p}$ from C to D points up and to the right on the page (toward the Sun).

1.X.44 (a) $\langle 7, 0, 0 \rangle$, (b) $\langle 0, 0, 2 \rangle$

1.X.45 a, b

The Momentum Principle

KEY IDEAS

- **One or more objects can be considered to be a "system."**
  - Everything not in the system is part of the "surroundings."
- **The momentum of a system can be changed only by interactions with the surroundings.**
  - Force is a quantitative measure of interaction.
  - Force is a vector (it has magnitude and direction).
- **The change in momentum of a system is equal to the net impulse applied, which involves:**
  - The total force exerted on it by all objects in the surroundings, and
  - The time interval over which the interaction occurs.
- **The motion of an object can be predicted by iteratively updating its momentum and position.**

■ A fundamental principle relates the change in a quantity to the interaction causing the change, and it is powerful because it applies in absolutely every situation.

The Momentum Principle is the first of three fundamental principles of mechanics that together make it possible to predict and explain a very broad range of real-world phenomena (the other two are the Energy Principle and the Angular Momentum Principle). The Momentum Principle makes a quantitative connection between amount of interaction and change of momentum.

2.1   SYSTEM AND SURROUNDINGS

One or more objects can be considered to be a "system." Everything that is not included in the system is part of the "surroundings." The Momentum Principle relates the change in momentum of a system to the amount of interaction with its surroundings. No matter what system we choose, the Momentum Principle will correctly predict the behavior of the system.

In this chapter we will always consider a single object to be the system, and everything else in the Universe to be the surroundings. However, in later chapters we will see that it is possible, and often useful, to choose a system comprising several objects. A fundamental principle such as the Momentum Principle applies to any system, no matter how complex.

2.2   THE MOMENTUM PRINCIPLE

The Momentum Principle is a fundamental principle that is also known as Newton's second law. It restates and extends Newton's first law of motion in a quantitative, causal form that can be used to predict the behavior of objects. The validity of the Momentum Principle has been verified through a very wide variety of observations and experiments, involving large and small objects, moving slowly or at speeds near the speed of light. It is a summary of the way interactions affect motion in the real world.

■ The Momentum Principle is a fundamental principle because:
It applies to every possible system, no matter how large or small (from clusters of galaxies to subatomic particles), and no matter how fast it is moving.
It is true for every kind of interaction (electric, gravitational, etc.)
It relates an effect (change in momentum) to a cause (an interaction).

THE MOMENTUM PRINCIPLE

$$\Delta \vec{p} = \vec{F}_{\text{net}} \, \Delta t$$

The change of momentum of a system (the effect) is equal to the net force acting on the system times the duration of the interaction (the cause).
The time interval Δt must be small enough that the net force is nearly constant during this time interval.

As usual, the capital Greek letter delta (Δ) means "change of" (something), or "final minus initial." We will see that the time interval Δt must be small enough that the net force on the system does not change significantly during this time interval.

To begin to understand this equation, we will consider each quantity involved.

$$\Delta \vec{p} = \vec{F}_{\text{net}} \, \Delta t$$

Change of Momentum $\Delta \vec{p}$

As we saw in the previous chapter, the change in momentum of a system can involve

- a change in the magnitude of momentum
- a change in the direction of momentum
- change in both magnitude and direction

$$\Delta \vec{p} = \vec{p}_f - \vec{p}_i$$

2.X.1 A ball's momentum changes from $\langle 1.5, -1.2, 0.7 \rangle$ kg·m/s to $\langle 1.6, -0.9, 1.1 \rangle$ kg·m/s. What is $\Delta \vec{p}$?

$$\Delta \vec{p} = \vec{F}_{\text{net}} \, \Delta t$$

Force $\vec{F}$

Scientists and engineers employ the concept of "force" to quantify interactions between two objects. Force is a vector quantity because a force has a magnitude and is exerted in a particular direction. Examples of forces include the following:

- The repulsive electric force a proton exerts on another proton
- The attractive gravitational force the Earth exerts on you
- The force that a compressed spring exerts on your hand
- The force on a spacecraft of expanding gases in a rocket engine
- The force of the surrounding air on the propeller of an airplane or airboat

Measuring a Force

Measuring the magnitude of the velocity of an object (in other words, measuring its speed) is a familiar task, but how do we measure the magnitude of a force?

Figure 2.1 Stretching of a spring is a measure of force.

Figure 2.2 Compression of a spring is also a measure of force.

$$\Delta \vec{p} = \vec{F}_{net}\, \Delta t$$

Figure 2.3 The magnitude of the gravitational force by the Earth on a small apple is about 1 newton.

■ Since the Momentum Principle relates force to change in momentum over a time interval, we see that a newton can be expressed in terms of kilograms, meters, and seconds.

$$1\ \text{N} = 1\ \text{kg} \cdot \text{m/s}^2$$

A simple way to measure force is to use the stretch or compression of a spring. In Figure 2.1 we hang a block from a spring, and note that the spring is stretched a distance s. Then we hang two such blocks from the spring, and we see that the spring is stretched twice as much. By experimentation, we find that any spring made of the same material and produced to the same specifications behaves in the same way.

Similarly, we can observe how much the spring compresses when the same blocks are supported by it (Figure 2.2). We find that one block compresses the spring by the same distance $|s|$, and two blocks compress it by $2|s|$. (Compression can be considered negative stretch, because the length of the spring decreases.)

Unit of Force

We can use a spring to make a scale for measuring forces, calibrating it in terms of how much force is required to produce a given stretch. The SI unit of force is the "newton," abbreviated as "N." One newton is a rather small force. A newton is approximately the downward gravitational force of the Earth on a small apple, or about a quarter of a pound (Figure 2.3). If you hold a small apple at rest in your hand, you apply an upward force of about one newton, compensating for the downward pull of the Earth.

The Meaning of *net*

In physics the word *net* has a precise meaning: it is the sum of all contributions to a quantity, both pluses and minuses.

Because a force has a magnitude and a direction, it is a vector, so the *net* force on a system is the vector sum of all the forces exerted on the system by all the objects in the surroundings. It is the net force $\vec{F}_{net}$ on a system, acting for some time Δt, that causes a change of momentum.

DEFINITION OF NET FORCE

$$\vec{F}_{net} \equiv \vec{F}_1 + \vec{F}_2 + \cdots$$

The "net" force acting on a system at an instant is the vector sum of *all* the forces exerted on the system by all the objects in the surroundings at that instant.

The net force acting on a system at an instant is the vector sum of all of the forces exerted on the system by all the objects in the surroundings, which are called "external" forces. There may be forces internal to the system, exerted by one object in the system on another object in the system, but such internal forces cannot change the momentum of the system. We will see in detail why this is so in Chapter 3, but the basic reason is that internal forces cancel each other's effects. A force that object 1 in a system exerts on object 2 in the system does change the momentum of object 2, but object 2 exerts a force in the opposite direction on object 1 that changes the momentum of object 1 in the opposite way, so that the changes in momenta of the two objects add up to zero.

2.X.2 Two external forces, $\langle 40, -70, 0 \rangle$ N and $\langle 20, 10, 0 \rangle$ N, act on a system. What is the net force acting on the system?

QUESTION A ball falling toward the Earth consists of many atoms. Each atom in the ball exerts forces on neighboring atoms in the ball, the Earth exerts forces on each atom in the ball, and the air exerts forces on atoms at the surface of the ball. Taking the ball as the system, and the Earth and air as the surroundings, which of these forces are external forces and which are internal forces?

The Earth and the air are part of the surroundings, so the force exerted by the Earth and the force exerted by the air are both external forces. They are forces exerted on the chosen system (the ball) by objects in the surroundings (the Earth and the air). Forces that atoms in the ball exert on neighboring atoms in the ball are internal forces. The internal forces don't contribute to the net force $\vec{F}_{net}$ that changes the ball's momentum as predicted by the Momentum Principle, $\Delta \vec{p} = \vec{F}_{net} \Delta t$.

We find experimentally that the magnitude of the net force acting on an object affects the magnitude of the change in its momentum. Many introductory physics laboratories have air tracks like the one illustrated in Figure 2.4. The long triangular base has many small holes in it, and air under pressure is blown out through these holes. The air forms a cushion under the glider, allowing it to coast smoothly with very little friction.

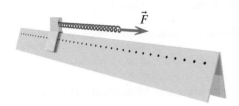

Figure 2.4 A block mounted on a nearly frictionless air track, pulled by a spring.

Suppose that we place a block on a glider on a long air track, and attach a calibrated spring to it (Figure 2.4). We pull on the spring so that it is stretched a distance s, so that it exerts a force F on the block, and we pull for a short time. Choose the block and glider as the system, so that the net force on the system is just the spring, since friction with the track is very small due to the air cushion.

We observe that the momentum of the system of block and glider increases from zero to an amount mv (since $v \ll c$ so $\gamma \approx 1$). If instead we pull on the spring so that it stretches a distance $2s$, the spring exerts a force $2F$ on the system, and we observe that the system's momentum increases to $2mv$ in the same amount of time. Apparently the magnitude of the change in momentum is proportional to the magnitude of the net force applied to the object.

2.X.3 You exert a force of magnitude 3 N on an object for one second, and observe that the magnitude of the object's momentum decreases by 3 kg·m/s. If instead you had exerted a force of 1.5 N on the object for one second, what would the magnitude of the object's momentum change have been?

$$\Delta \vec{p} = \vec{F}_{net} \Delta t$$

Duration of Interaction Δt

Another experiment will show us that not only the magnitude of the force, but also the length of time during which it acts on an object, affects the change of momentum of the object. Suppose that we again place a block on an air track and attach a spring to it. We pull on the spring for a time Δt with a force F, and we observe that the momentum of the system of block plus glider increases from zero to an amount mv. However, if we repeat the experiment but pull for a time twice as long with the same force F, the system's momentum increases to $2mv$; the change in its momentum is twice as great. We observe that the magnitude of the change in momentum is directly proportional to the length of time $\Delta t = t_2 - t_1$ during which the force acts on the object.

$$\Delta \vec{p} = \vec{F}_{net} \Delta t$$

Impulse $\vec{F}_{net} \Delta t$

The amount of interaction affecting an object includes both the strength of the interaction (expressed as the net force $\vec{F}_{net}$) and the duration Δt of the interaction. Either a bigger force or applying the force for a longer time will cause more change of momentum.

The product of a force and a time interval is called "impulse."

DEFINITION OF IMPULSE

$$\text{Impulse} \equiv \vec{F}\Delta t$$

Impulse has units of $N \cdot s$ (newton-seconds)

The time interval Δt must be small enough that the net force is nearly constant during this time interval.

With this definition of impulse we can state the Momentum Principle in words like this:

The change of momentum of a system
is equal to the net impulse applied to it.

QUESTION   A constant net force $\langle 3, -5, 4 \rangle$ N acts on an object for 10 s. What is the net impulse applied to the object? What was the change in momentum of the object?

$$\text{impulse} = \vec{F}_{net}\,\Delta t = \langle 3, -5, 4 \rangle \text{ N} \cdot (10\,\text{s}) = \langle 30, -50, 40 \rangle \text{ N} \cdot \text{s}$$

The change in the object's momentum is equal to the net impulse, so

$$\Delta \vec{p} = \langle 30, -50, 40 \rangle \text{ kg} \cdot \text{m/s}$$

Evidently $1\,\text{N} \cdot \text{s} = 1\,\text{kg} \cdot \text{m/s}$, so $1\,\text{N} = 1\,\text{kg} \cdot \text{m/s}^2$.

EXAMPLE   **Impulse Applied to a Baseball**

Just before it touches the bat, the momentum of a baseball is $\langle 0, 0, 3 \rangle$ $\text{kg} \cdot \text{m/s}$. After the batter has hit the ball, and the ball is no longer in contact with the bat, the momentum of the baseball is $\langle -2, 2, -1 \rangle$ $\text{kg} \cdot \text{m/s}$. What was the impulse applied to the ball during its contact with the bat?

Solution

$$\Delta \vec{p} = \vec{F}_{net}\,\Delta t$$
$$\langle -2, 2, -1 \rangle \text{ kg} \cdot \text{m/s} - \langle 0, 0, 3 \rangle \text{ kg} \cdot \text{m/s} = \vec{F}_{net}\,\Delta t$$
$$\langle -2, 2, -4 \rangle \text{ kg} \cdot \text{m/s} = \vec{F}_{net}\,\Delta t$$

FURTHER DISCUSSION   Note that without more information we cannot find the values of $\vec{F}_{net}$ or Δt separately.

EXAMPLE   **Direction of Impulse**

A hockey puck is sliding across the ice with momentum $\vec{p}_1$ (shown in Figure 2.5) when a hockey player hits it sharply with a hockey stick. Afterward, the puck has momentum $\vec{p}_2$. Both $\vec{p}_1$ and $\vec{p}_2$ lie in the xz plane.

(a) What was the direction of the impulse applied to the puck by the stick?

(b) What was the direction of the net force exerted on the puck during its short contact with the hockey stick?

Figure 2.5 Initial and final momenta of a hockey puck. Both momenta lie in the xz plane; this is a top view, looking down on the ice from above.

Figure 2.6 To find $\Delta\vec{p}$, the vectors $\vec{p}_1$ and $\vec{p}_2$ are placed tail to tail. The net force on the puck, shown by the red arrow, was in the direction of the observed change in momentum $\Delta\vec{p}$.

Solution

(a) We can find the direction of $\Delta\vec{p}$ graphically, as shown in Figure 2.6. From the vector equation

$$\Delta\vec{p} = \vec{F}_{\text{net}}\,\Delta t$$

we see that $\Delta\vec{p}$ and $\vec{F}_{\text{net}}\,\Delta t$ must have the same directions.

(b) Since Δt is a positive scalar (2 seconds), the direction of the net force is the same as the direction of the impulse, because multiplying the vector $\vec{F}_{\text{net}}$ by the positive scalar Δt can't change the direction of the vector.

2.X.4 A constant net force of $\langle -0.5, -0.2, 0.8 \rangle$ N acts on an object for 2 minutes. **(a)** What is the impulse applied to the object, in SI units? **(b)** What is the change in the momentum of the object?

2.X.5 A hockey puck initially has momentum $\langle 0, 2, 0 \rangle$ kg·m/s. It slides along the ice, gradually slowing down, until it comes to a stop. **(a)** What was the impulse applied by the ice to the hockey puck? **(b)** It took 3 seconds for the puck to come to a stop. During this time interval, what was the net force on the puck by the ice and the air (assuming that this force was constant)?

| 2.3 | **APPLYING THE MOMENTUM PRINCIPLE** |

Update Form of the Momentum Principle

Since $\Delta\vec{p} = \vec{p}_f - \vec{p}_i = \vec{F}_{\text{net}}\,\Delta t$ ("final minus initial"), we can rearrange the Momentum Principle to the update form:

THE MOMENTUM PRINCIPLE (UPDATE FORM)

$$\vec{p}_f = \vec{p}_i + \vec{F}_{\text{net}}\,\Delta t$$

The time interval Δt must be small enough that the net force is nearly constant during this time interval.

■ The "update form" of the Momentum Principle can be used to predict the new momentum of a system to which a known force has been applied for a known time interval.

This update version of the Momentum Principle emphasizes the fact that if you know the initial momentum, and you know the net force acting during a "short enough" time interval, you can predict the final momentum.

Separation of Components

The Momentum Principle written in terms of vectors can be interpreted as three ordinary scalar equations, for components of the motion along the x, y, and z axes:

$$p_{fx} = p_{ix} + F_{\text{net},x}\,\Delta t$$
$$p_{fy} = p_{iy} + F_{\text{net},y}\,\Delta t$$
$$p_{fz} = p_{iz} + F_{\text{net},z}\,\Delta t$$

An important aspect of this is the fact that the x component of an object's momentum cannot be affected by forces in the y or z directions.

This fact can be very useful in solving problems. In some situations, for example, if we know that the y and z components of an object's momentum are not changing, we may choose to work only with the x component of the momentum update equation.

Steps in Applying the Momentum Principle

The equation for the Momentum Principle provides an outline of how to solve a problem. To predict a final momentum $\vec{p}_f$ (left side of the equation), we need to identify the system whose momentum is changing, and we need to know the values of all the quantities on the right side of the equation: initial momentum $\vec{p}_i$, the net force $\vec{F}_{net}$, and the time interval Δt. Therefore, to apply the Momentum Principle to a real-world situation, several steps are required:

- Choose a system, consisting of a portion of the Universe. The rest of the Universe is called the surroundings.
- Make a list of the objects in the surroundings that exert significant forces on the chosen system.
  - Then make a labeled diagram showing the external forces exerted by the objects in the surroundings. This is called a "free-body diagram" in which the effects of the surroundings on the system are indicated as arrows representing the external forces. It's often useful to indicate initial momenta or velocities as well.

■ Divide and conquer: Focus on one term at a time when evaluating terms of the Momentum Principle.

- Choose initial and final times (often these are obvious; in examples we will comment on this when necessary).
- Apply the Momentum Principle to the chosen system:

$$\vec{p}_f = \vec{p}_i + \vec{F}_{net}\, \Delta t$$

Substitute known values into the terms of the Momentum Principle.

- Check units; check the reasonableness of your answer (direction of momentum, change in magnitude of momentum), based on the physical situation.

EXAMPLE

Ball and String: 2D, Single Force

Inside a spaceship in outer space there is a small steel ball of mass 0.25 kg. At a particular instant, the ball has momentum $\langle -8, 3, 0 \rangle$ kg·m/s. At that instant the ball is being pulled by a string, which exerts a net force $\langle 10, 25, 0 \rangle$ N on the ball. What is the ball's approximate momentum 0.5 seconds later?

Solution

System: the steel ball
Surroundings: the string
Free-body diagram: Figure 2.7

Momentum Principle:

$$\vec{p}_f = \vec{p}_i + \vec{F}_{net}\, \Delta t$$
$$\vec{p}_f = (\langle -8, 3, 0 \rangle \text{ kg·m/s}) + (\langle 10, 25, 0 \rangle \text{ N})(0.5 \text{ s})$$
$$\vec{p}_f = (\langle -8, 3, 0 \rangle \text{ kg·m/s}) + (\langle 5, 12.5, 0 \rangle \text{ N·s})$$
$$\vec{p}_f = \langle -3, 15.5, 0 \rangle \text{ kg·m/s}$$

Figure 2.7 A free-body diagram for the steel ball.

Check: units correct (momentum: kg·m/s and position: m); direction of final momentum makes sense (Figure 2.8).

2.X.6 A hockey puck is sliding along the ice with nearly constant momentum $\langle 10, 0, 5 \rangle$ kg·m/s when it is suddenly struck by a hockey stick with a force $\langle 0, 0, 2000 \rangle$ N that lasts for only 4 milliseconds (4×10^{-3}) s. What is the new momentum of the puck?

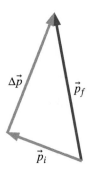

Figure 2.8 Initial momentum, change in momentum, and final momentum for a ball pulled by a string.

Figure 2.9 A fan cart on a track. The fan directs air to the left, with the result that the surrounding air pushes the car to the right.

Figure 2.10 Forces on a moving fan cart. The circle denotes the system.

Another Kind of Force

An easy way to apply a nearly constant force is to mount a battery-powered fan on a cart (Figure 2.9). If the fan directs air backwards, the interaction with the air pushes the cart forward with a nearly constant force. Airboats used in the very shallow Florida Everglades are built in a similar way, with large fans on top of the boats propelling them through the swamp.

Air Resistance and Friction

When an object gets going very fast, air resistance exerted on the front of the object, sometimes called "drag," becomes important and at high speeds can be as big as the propelling force. At that point the momentum doesn't increase any more, and the object travels at constant speed. When the speed of an object is low, air resistance is small, so we can make the approximation that the net force on the fan cart is associated solely with the push of air on the back of the fan cart, in reaction to the fan directing air backwards.

Friction between the object and the surface on which it moves can also be important. Friction is usually independent of speed, so this is a constant force. The friction between the cart and the track is relatively low, so it is reasonable to say that it is approximately zero in this case.

Other Interacting Objects

If we choose the fan cart as the system, the air is not the only external object that interacts with the system. The Earth exerts a downward gravitational force on the cart, and the track exerts an upward force, as shown in Figure 2.10. We must consider these forces as well, because they contribute to the net external force on the fan cart.

If we see the cart rolling horizontally along the track, it is clear that the y component of its momentum remains zero and does not change:

$$\Delta p_y = 0$$

QUESTION Using the Momentum Principle, what can we conclude about the relative magnitudes of the force on the cart by the Earth and the force on the cart by the track?

According to the Momentum Principle, the y component of the net force on the cart is zero. Evidently these two forces must be exactly equal in magnitude and opposite in direction.

You may wonder how the track "knows" exactly how much upward force to exert on the cart. Basically, the cart compresses the track slightly, and the compressed track pushes up, like a spring. Understanding this in detail requires a microscopic view, discussed in Chapter 4.

EXAMPLE

A Fan Cart: 1D, Several Forces, Constant Net Force

Suppose that you have a fan cart whose mass is 0.5 kg, and with the fan turned on, the force acting on the cart due to the air and friction with the track is $\langle 0.2, 0, 0 \rangle$ N and constant. You give the cart a shove, and you release the cart at position $\langle 0.5, 0, 0 \rangle$ m with initial velocity $\langle 0.3, 0, 0 \rangle$ m/s. What is the momentum of the cart 0.6 seconds later? What is its velocity at that time?

Solution

System: the cart (including the fan)
Surroundings: the Earth, the track, the air

■ We choose the initial time to be just after you release the cart, so your hand no longer exerts a force on the cart.

Figure 2.11 Free-body diagram for a fan cart. The circle denotes the system.

Free-body diagram: Figure 2.11
Initial time: see comment in margin
Momentum Principle:

$$\vec{p}_f = \vec{p}_i + \vec{F}_{net}\,\Delta t = \vec{p}_i + (\vec{F}_{track} + \vec{F}_{Earth} + \vec{F}_{air})(\Delta t)$$

$$\vec{p}_f = \vec{p}_i + \langle 0.2, (|\vec{F}_{track}| - |\vec{F}_{Earth}|), 0\rangle \ \text{N}(0.3\ \text{s})$$

$$\vec{p}_f = \langle (0.5\ \text{kg})(0.3\ \text{m/s}), 0, 0\rangle + \langle 0.2, 0, 0\rangle \ \text{N}(0.6\ \text{s})$$

$$\vec{p}_f = \langle 0.27, 0, 0\rangle \ \text{kg}\cdot\text{m/s}$$

$$\vec{v}_f \approx \frac{\vec{p}_f}{m} = \frac{\langle 0.27, 0, 0\rangle \ \text{kg}\cdot\text{m/s}}{0.5\ \text{kg}} = \langle 0.54, 0, 0\rangle \ \text{m/s}$$

Check: The speed increased; this is reasonable since force was in same direction as momentum.
Check: Only the x component of momentum changed; this is reasonable since net force was in the $+x$ direction.

FURTHER DISCUSSION In the preceding example, p_y and p_z were not changing. From this observation we could deduce that $F_{net,y} = 0$ and $F_{net,z} = 0$ (see the comment on separation of components at the beginning of this section). Knowing that $F_{net,y} = 0$ allowed us to deduce that the track must push up on the cart with a force equal to the downward gravitational force exerted by the Earth. We used a large time interval.

2.X.7 You were driving a car with velocity $\langle 25, 0, 15\rangle$ m/s. You quickly turned and braked, and your velocity became $\langle 10, 0, 18\rangle$ m/s. The mass of the car was 1000 kg.

  (a) What was the (vector) change in momentum $\Delta\vec{p}$ during this maneuver? Pay attention to signs.
  (b) What was the (vector) impulse applied to the car by the ground?
  (c) If the maneuver took 3 seconds, what was the average (vector) force exerted on the car? (The net force is due to the ground and the Earth; the y components of these forces cancel.)

2.X.8 At a certain instant, a particle is moving in the $+x$ direction with momentum $+10$ kg $\cdot$ m/s. During the next 0.1 s, a constant force $\langle -6, 3, 0\rangle$ N acts on the particle. What is the momentum of the particle at the end of this 0.1 s interval?

Relativistic Motion

The Momentum Principle can also be applied to an object moving at speeds near the speed of light:

EXAMPLE

Fast Proton: 1D, Constant Net Force, Relativistic

A proton in a particle accelerator is moving with velocity $\langle 0.960c, 0, 0\rangle$, so the speed is $0.960 \times 3 \times 10^8$ m/s $= 2.88 \times 10^8$ m/s. A constant electric force is applied to the proton to speed it up, $\vec{F}_{net} = \langle 5 \times 10^{-12}, 0, 0\rangle$ N. What is the proton's speed as a fraction of the speed of light after 20 nanoseconds (1 ns $= 1 \times 10^{-9}$ s)?

Solution

System: the proton
Surroundings: electric charges in the accelerator
Free-body diagram: Figure 2.12

Proton

Figure 2.12 The free-body diagram for the proton.

Momentum Principle:

$$\vec{p}_f = \vec{p}_i + \vec{F}_{net}\,\Delta t$$

$$\langle p_{fx}, 0, 0 \rangle = \langle \gamma_i m v_{ix}, 0, 0 \rangle + \langle F_{net,x}, 0, 0 \rangle\,\Delta t$$

Find the x component of the initial momentum:

$$\gamma = \frac{1}{\sqrt{1 - \left(\dfrac{0.960c}{c}\right)^2}} = 3.57$$

$$p_{ix} = \gamma_i m v_{ix} = 3.57 \cdot (1.7 \times 10^{-27}\text{ kg})(0.96 \cdot 3 \times 10^8\text{ m/s})$$

$$= 1.75 \times 10^{-18}\text{ kg} \cdot \text{m/s}$$

Find the x component of the impulse:

$$F_{net,x}\,\Delta t = 5 \times 10^{-12}\text{ N} \cdot (20 \times 10^{-9}\text{ s}) = 1 \times 10^{-19}\text{ N} \cdot \text{s}$$

Apply the Momentum Principle:

$$p_{fx} = (1.75 \times 10^{-18}\text{ kg} \cdot \text{m/s}) + (1 \times 10^{-19}\text{ N} \cdot \text{s})$$

$$= 1.85 \times 10^{-18}\text{ kg} \cdot \text{m/s}$$

Find the final speed:

■ See Chapter 1 for the derivation of this equation for speed in terms of momentum for a fast-moving particle.

$$\frac{v_{fx}}{c} = \frac{\dfrac{p_{fx}}{mc}}{\sqrt{1 + \left(\dfrac{p_{fx}}{mc}\right)^2}}$$

$$\frac{p_{fx}}{mc} = \frac{1.85 \times 10^{-18}\text{ kg} \cdot \text{m/s}}{(1.7 \times 10^{-27}\text{ kg})(3 \times 10^8\text{ m/s})} = 3.62 \text{ (no units)}$$

$$\frac{v_{fx}}{c} = \frac{3.62}{\sqrt{1 + 3.62^2}} = 0.964$$

$$v_{fx} = 0.964c$$

FURTHER DISCUSSION Although the magnitude of the momentum increased from 1.75×10^{-18} kg·m/s to 1.85×10^{-18} kg·m/s, the speed didn't increase very much, because the proton's initial speed, $0.960c$, was already close to the cosmic speed limit, c. Because the speed hardly changed, the distance the proton moved during the 20 ns was approximately equal to

$$(0.964 \times 3 \times 10^8\text{ m/s})(20 \times 10^{-9}\text{ s}) = 5.8\text{ m}$$

Updating Position If Momentum Is Changing

In the fan cart example discussed earlier the momentum, and therefore the velocity, of the fan cart is changing with time. If we want to predict where the fan cart will be at the end of the 0.6 second time interval, we need to use an appropriate value for the average velocity of the cart.

We can't calculate the average velocity directly from the definition

$$\vec{v}_{avg} \equiv \frac{\Delta \vec{r}}{\Delta t} = \frac{\vec{r}_f - \vec{r}_i}{\Delta t}$$

because the final location $\vec{r}_f$ of the cart is unknown. Since in this case we do not have any other information that would allow us to determine the average velocity, we need to find an approximate value for the average velocity in this situation. There are two possible approaches:

- If the change in velocity is very small over the time interval of interest, we can use the final velocity (or the initial velocity) as an approximation to the average velocity.
- We can try to estimate an average value for the velocity.

In the preceding fan cart example, the final velocity of the cart is $\langle 0.54, 0, 0 \rangle$ m/s, which is nearly twice the initial velocity of $\langle 0.3, 0, 0 \rangle$ m/s. Neither the final nor the initial velocity is a good approximation to the average velocity.

QUESTION What might be a better approximation for $\vec{v}_{avg}$ in this case?

Let's consider using the arithmetic average:

$$v_{avg,x} \approx \frac{(v_{ix} + v_{fx})}{2} \text{ is an approximation for } v_{avg,x} = \frac{\Delta x}{\Delta t}$$

The arithmetic average lies between the two extremes. For example, the arithmetic average of 6 and 8 is $(6 + 8)/2 = 14/2 = 7$, halfway between 6 and 8.

The arithmetic average is often a good approximation, but it is not necessarily equal to the true average $v_{avg,x} = \Delta x/\Delta t$. The arithmetic average does not give the true average unless v_x is changing at a constant rate, which is the case only if the net force is constant, as it happens to be for a fan cart (see Figure 2.13). The proof that $v_{avg,x} = (v_{ix} + v_{fx})/2$ when v_x changes at a constant rate is given in optional Section 2.9 at the end of this chapter. Although the result seems reasonable, the detailed proof is more complicated than one might expect.

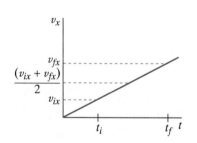

Figure 2.13 v_x is changing at a constant rate (linearly with time), so the arithmetic average is equal to $v_{avg,x}$.

QUESTION When is the arithmetic average a poor approximation for $\vec{v}_{avg}$?

The arithmetic average is a poor approximation for $\vec{v}_{avg}$ if the rate of change of velocity (the acceleration $\vec{a} = d\vec{v}/dt$) is not constant. At low speeds ($v \ll c$), the acceleration is directly proportional to the rate of change of momentum, so we can conclude that $d\vec{v}/dt$ will be constant if the net force on the system is constant.

EXAMPLE **The Arithmetic Average Can Be a Poor Approximation**

If you drive 50 mi/hr in the $+x$ direction for four hours, and then 20 mi/hr in the $+x$ direction for an hour, how far have you traveled?

Solution Using the arithmetic average velocity, you would calculate that you have traveled 175 miles:

$$v_{avg,x} \approx \frac{50 + 20}{2} \text{ mph} = 35 \text{ mph}$$

$$x_f - x_i \approx (35 \text{ mph})(5 \text{ hr}) = 175 \text{ mi}$$

However, you have actually traveled significantly farther:

$$x_f - x_i = (50 \text{ mph})(4 \text{ hr}) + (20 \text{ mph})(1 \text{ hr}) = 220 \text{ mi}$$

The x component of your actual average velocity was:

$$v_{\text{avg},x} = \frac{220 \text{ mi}}{5 \text{ hr}} = 44 \text{ mph}$$

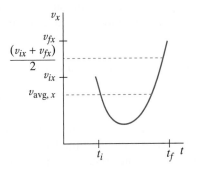

Figure 2.14 In this example, v_x is not changing linearly with time. In this case the arithmetic average is significantly different from the true average value of $v_{\text{avg},x}$.

FURTHER DISCUSSION See Figure 2.14 for another such example. In situations where the rate of change of velocity is not constant, we have to choose short enough time intervals that the velocity is nearly constant during the brief Δt.

APPROXIMATE AVERAGE VELOCITY

$$v_{\text{avg},x} \approx \frac{(v_{ix} + v_{fx})}{2} \text{ is an approximation for } v_{\text{avg},x} = \frac{\Delta x}{\Delta t}$$

This is exactly true only if v_x changes linearly with time (acceleration is constant). It may be a poor approximation if the net force is not constant.

EXAMPLE

Final Position of a Fan Cart

In the previous fan cart example, we used the Momentum Principle to find the final velocity of a fan cart that was subjected to a constant force over a time interval of 0.6 s. Now that we have a way to calculate the x component of the average velocity in this special situation (constant x component of force), we can predict the position of the fan cart at the end of the time interval:

Solution

$$v_{\text{avg},x} = \frac{(v_{ix} + v_{fx})}{2} = \frac{(0.3 + 0.54)}{2} \text{ m/s}$$

$$\vec{v}_{\text{avg}} = \langle 0.42, 0, 0 \rangle \text{ m/s}$$

$$\vec{r}_f = \vec{r}_i + \vec{v}_{\text{avg}} \Delta t = \langle 0.5, 0, 0 \rangle \text{ m} + \langle 0.42, 0, 0 \rangle \frac{\text{m}}{\text{s}} (0.6 \text{ s})$$

$$\vec{r}_f = \langle 0.752, 0, 0 \rangle \text{ m}$$

FURTHER DISCUSSION The net force on the cart was constant, so this calculation of average velocity gives the correct value.

2.X.9 On a straight road (taken to be in the $+x$ direction) you drive for an hour at 50 km per hour, then quickly speed up to 100 km per hour and drive for an additional two hours. How far do you go (Δx)? What is your average x component of velocity ($v_{\text{avg},x}$)? Why isn't $v_{\text{avg},x}$ equal to the arithmetic average of your initial and final values of v_x, $(50 + 100)/2 = 75$ km per hour?

2.X.10 You're driving on a straight road (in the $+x$ direction) at a constant speed of 30 m/s. In 10 seconds, you speed up to 40 m/s to pass a truck. Assuming that your car speeds up at a constant rate (constant force by the road on the tires), what is your average x component of velocity $v_{\text{avg},x}$ during this maneuver? How far do you go during this maneuver?

A common mistake is to try to calculate $v_{avg,x}$ as equal to $(v_{f,x} - v_{i,x})/2$, confusing average velocity with half of the *change* of velocity. But this mistake would lead to $v_{avg,x}$ only 5 m/s, which makes no sense if you're speeding up from 30 m/s to 40 m/s. Do check that the average velocity makes sense.

2.4 **MOMENTUM CHANGE WITH CHANGING FORCE**

In real-world situations, forces due to interactions with the surroundings may not be constant—it is common for both the direction and magnitude of forces to change.

QUESTION How should we take into account changing forces when calculating changes in momentum?

If the net force on a system is not constant over a time interval, we can divide the time interval into several smaller intervals, such that the net force is approximately constant over each smaller interval. We can find the total change in the system's momentum by adding up the momentum changes in each interval:

$$\Delta \vec{p} = \Delta \vec{p}_1 + \Delta \vec{p}_2 + \Delta \vec{p}_3 + \cdots$$

EXAMPLE **A Leaf in the Wind**

Suppose that on a windy day a dry leaf is drifting through the air with momentum $\vec{p}_1$ when it is hit by several gusts of wind. The net force on the leaf changes from moment to moment, causing different amounts of momentum change, as shown in the following table.

Time interval	Net force
0.0 − 2.0 s	$\langle 0.2, 0.2, 0 \rangle$ N
2.0 − 4.0 s	$\langle 0.0, -0.3, 0 \rangle$ N
4.0 − 6.0 s	$\langle -0.3, 0.4, 0 \rangle$ N

What is the total change in momentum of the leaf over this 3-second interval?

Solution

$$\Delta \vec{p}_1 = \langle 0.4, 0.4, 0 \rangle \text{ kg} \cdot \text{m/s}$$

$$\Delta \vec{p}_2 = \langle 0, -0.6, 0 \rangle \text{ kg} \cdot \text{m/s}$$

$$\Delta \vec{p}_3 = \langle -0.6, 0.8, 0 \rangle \text{ kg} \cdot \text{m/s}$$

$$\Delta \vec{p}_{total} = \Delta \vec{p}_1 + \Delta \vec{p}_2 + \Delta \vec{p}_3 = \langle -0.2, 0.6, 0 \rangle \text{ kg} \cdot \text{m/s}$$

FURTHER DISCUSSION Another way to look at this kind of sum of momentum changes is this:

$$\Delta \vec{p}_1 = \vec{p}_1 - \vec{p}_0$$

$$\Delta \vec{p}_2 = \vec{p}_2 - \vec{p}_1$$

$$\Delta \vec{p}_3 = \vec{p}_3 - \vec{p}_2$$

Add up the left side of these three equations and you get the total change in the momentum, $\Delta \vec{p}_{total}$. Adding up the right side of these three equations, note

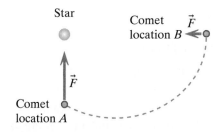

Figure 2.15 The direction and magnitude of the gravitational force on a comet by a star change as the comet's position changes. The size of the comet is exaggerated in this diagram.

- Δt must be small enough that the force can be considered to be approximately constant over the time interval.

---

that the sum includes terms $\vec{p}_1 - \vec{p}_1 = 0$ and $\vec{p}_2 - \vec{p}_2 = 0$. The only terms that remain are $\vec{p}_3 - \vec{p}_0 = \vec{p}_{final} - \vec{p}_{initial}$. Equating the sum of the left side of the three equations to the sum of the right side, we find that the total change in the momentum is as expected—just the final value minus the initial value:

$$\Delta \vec{p}_{total} = \vec{p}_{final} - \vec{p}_{initial}$$

In the air track experiment described earlier, we were able to keep the force constant during the time interval Δt, by keeping the stretch of the spring constant. In most real-world cases the force applied to an object is not constant, but changes as the object moves. For example, the magnitude of the force exerted on an object by a spring attached to the object changes when the stretch of the spring changes. As a comet moves, both the direction and the magnitude of the force exerted on the comet by a star change (Figure 2.15).

If the magnitude or the direction of a force changes during a time interval Δt, what value of the force should we use in calculating the impulse? There are two possibilities:

- Use an approximate average value of the force.
- Divide the time interval into several time intervals small enough that the force remains approximately constant over each smaller time interval.

Each of these possibilities requires that we make an approximation. This is fine, but it is important to be aware that we are making approximations!

We met the same issue with the position update relation, $\vec{r}_f = \vec{r}_i + \vec{v}_{avg} \Delta t$, where we need to use a short enough time interval that the velocity isn't changing very much, or else we need to know the average velocity during the time interval.

Momentum Change on an Air Track

We can use the Momentum Principle to predict the change in momentum of an object subject to a changing net force, in an experiment using the air track apparatus described earlier. We will choose the x axis to point in the direction of the motion. We stretch a calibrated spring, which exerts a force of 20 N on the system consisting of the glider plus block, when the spring is stretched 4 cm (Figure 2.16). Suppose that the glider and block start from rest, and the spring pulls for 1 second in the $+x$ direction with a force of magnitude 20 N (you will have to move forward in order to keep the spring stretched).

Figure 2.16 Apply a constant force to a glider and block on a low-friction air track.

QUESTION The glider and block start from rest, so $\vec{p}_i = \langle 0, 0, 0 \rangle$ kg·m/s. What would the Momentum Principle predict the new momentum of the glider and block to be after 1 second?

Since the friction force on the glider is negligibly small, and the y components of the forces exerted by the Earth and track cancel, the net force on the system is just the force exerted by the spring. The Momentum Principle (update form) applied to the glider and block is:

$$\vec{p}_f = \langle 0, 0, 0 \rangle \text{ kg·m/s} + \langle 20, 0, 0 \rangle \text{ N}(1 \text{ s})$$

Since the net force is in the x direction, we know that the y and z components of the momentum of the glider plus block will not change, and we can work with just the x component of the momentum.

$$p_{fx} = p_{ix} + F_{net,x} \Delta t = 0 + (20 \text{ N})(1 \text{ s}) = 20 \text{ kg·m/s}$$

If you do the experiment, this is what you will observe.

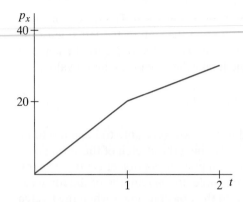

Figure 2.17 *x* component of momentum of the glider plus block (kg · m/s) vs. time (s), when pulled for 1 second with a constant force of 20 N, then for 1 second with a force of 10 N.

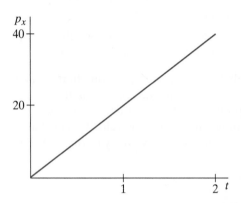

Figure 2.18 *x* component of momentum of the glider plus block (kg · m/s) vs. time (s), when pulled for 2 seconds with a force of 20 N.

Suppose the spring keeps exerting a force for another second, but now with the spring stretched half as much, 2 cm, so you know that the spring is now exerting a force of (10 N), half the original force.

QUESTION   What would the Momentum Principle predict the new *x* component of the momentum of the glider plus block to be now?

The Momentum Principle would predict the following, where we take the final momentum from the first pull and consider that to be the initial momentum for the second pull:

$$p_{fx} = p_{ix} + F_{net,x} \, \Delta t = (20 \text{ kg} \cdot \text{m/s}) + (10 \text{ N})(1 \text{ s}) = 30 \text{ kg} \cdot \text{m/s}$$

If you do the experiment, this is what you will observe (Figure 2.17). Note that the effects of the interactions in the two 1-second intervals add; we add the two momentum changes.

Instead of varying the net force, we could try varying the duration of the interaction. Start over with the glider and block initially at rest. Pull for 2 seconds with the spring stretched 4 cm, so the force of the spring on the system is 20 N.

QUESTION   The glider and block start from rest ($p_{ix} = 0$). What would the Momentum Principle predict the new *x* component of the momentum of the system to be?

$$p_{fx} = p_{ix} + F_{net,x} \, \Delta t = 0 + (20 \text{ N})(2 \text{ s}) = 40 \text{ kg} \cdot \text{m/s}$$

Here (Figure 2.18) the final *x* component of momentum was 40 kg · m/s after applying a force of 20 N for 2 s, whereas in the previous experiments we got 30 kg · m/s after the spring applied a force of 20 N for 1 s plus a force of 10 N for 1 s. In our calculations we can use big values of Δt as long as the force isn't changing much. When the force changed from 20 N for a second to 10 N for a second we had to treat the two time intervals separately.

Many different experiments have shown the validity of the Momentum Principle. If we use two springs in different directions to move the glider and block, we find that it is indeed the vector sum of the two spring forces, the "net" force, that accounts for the change in momentum.

2.X.11 The *x* component of the momentum of an object is observed to increase with time:

At $t = 0$ s, = 30 kg · m/s

At $t = 1$ s, = 40 kg · m/s

At $t = 2$ s, = 50 kg · m/s

At $t = 3$ s, = 60 kg · m/s

What can you conclude about the *x* component of the net force acting on the object during this time?

(a) $F_{net,x}$ is zero.

(b) $F_{net,x}$ is constant.

(c) $F_{net,x}$ is decreasing with time.

(d) $F_{net,x}$ is increasing with time.

(e) Not enough information is given.

| 2.5 | **ITERATIVE PREDICTION OF MOTION** |

To "iterate" means "to repeat" or "to perform again"; the word "iteration" implies repeating the same process many times. The basic idea we will use in predicting the motion of systems subject to changing forces is to divide up the time interval of interest into many very small time intervals, and iteratively apply the Momentum Principle over successive intervals. Each time interval must be short enough that we may safely make the approximation that the force on the system does not change significantly during the time interval. We will update momentum and position iteratively like this:

ITERATIVE PREDICTION OF MOTION

- Calculate the (vector) forces acting on the system.
- Update the momentum of the system: $\vec{p}_f = \vec{p}_i + \vec{F}_{net}\,\Delta t$.
- Update the position: $\vec{r}_f = \vec{r}_i + \vec{v}_{avg}\,\Delta t$.
- Repeat.

This procedure need not be confined to situations in which forces are changing. It can be used even in a constant-force situation to predict the entire path along which an object will travel. For example, we could apply this procedure to calculate the trajectory of a ball moving without air resistance. A graphical representation of three successive steps in the process might look like Figure 2.19:

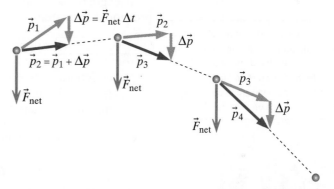

Figure 2.19 Three steps in the iterative prediction of the path of a ball in the absence of air resistance. In each step the new momentum of the ball is calculated, then this new momentum is used to update the position of the ball. In the calculation represented here, the final velocity was used as an approximation to the average velocity in each step.

At each location we calculate the net impulse $\vec{F}_{net}\,\Delta t$, the change in momentum $\Delta\vec{p}$, and the new momentum $\vec{p} + \Delta\vec{p}$. We use the new momentum to predict the new position, then iterate. (Here we have used $\vec{v}_f \approx \vec{p}_f/m$ to approximate $\vec{v}_{avg}$.) Even with only three steps the trajectory looks roughly like what is observed in the real world, and you can see the increasing displacement of the ball during each time step when headed downward, corresponding to increasing speed. If we used smaller time steps, the straight line segments representing displacement would be shorter, and the trajectory would look more and more like the actual curving path of the ball.

The Spring Force

A "force law" describes mathematically how a force depends on the situation. For a spring, it is determined experimentally that the force exerted by a spring on an object attached to the spring is given by the following equation:

Figure 2.20 The magnitude of the force exerted by a spring is proportional to the absolute value of the stretch of the spring. For an elongated spring, stretch is positive.

■ Because of the "reciprocity" of the electric forces between the protons and electrons in the spring and those in your hand, as we'll study in Chapter 3, the force you exert on the spring is equal in magnitude and opposite in direction to the force the spring exerts on your hand.

Figure 2.21 The magnitude of the force exerted by a spring is proportional to the absolute value of the stretch of the spring. For a compressed spring, stretch is negative.

THE SPRING FORCE

$$|\vec{F}_{spring}| = k_s|s|$$

$|s|$ is the absolute value of the stretch: $s = L - L_0$
L_0 is the length of the relaxed spring.
L is the length of the spring when stretched or compressed.
k_s is the "spring stiffness" (also called "spring constant").
The force acts in a direction to restore the spring to its relaxed length.

The constant k_s is a positive number, and is a property of the particular spring: the stiffer the spring, the larger the spring stiffness, and the larger the force needed to stretch the spring. Note that s is positive if the spring is stretched ($L > L_0$, Figure 2.20) and negative if the spring is compressed ($L < L_0$, Figure 2.21). It is possible to write this equation as a vector equation by using unit vectors. This equation is sometimes called "Hooke's law."

QUESTION   Suppose a certain spring has been calibrated so that we know that its spring stiffness k_s is 500 N/m. You pull on the spring and observe that it is 0.01 m (1 cm) longer than it was when relaxed. What is the magnitude of the force exerted by the spring on your hand?

The force law gives $|\vec{F}_{spring}| = (500\,\text{N/m})(|+0.01\,\text{m}|) = 5\,\text{N}$. Note that the total length of the spring doesn't matter; it's just the amount of stretch or compression that matters.

QUESTION   Suppose that instead of pulling on the spring, you push on it, so the spring becomes shorter than its relaxed length. If the relaxed length of the spring is 10 cm, and you compress the spring to a length of 9 cm, what is the magnitude of the force exerted by the spring on your hand?

The stretch of the spring in SI units is

$$s = L - L_0 = (0.09\,\text{m} - 0.10\,\text{m}) = -0.01\,\text{m}$$

The force law gives

$$|\vec{F}_{spring}| = (500\,\text{N/m})(|-0.01\,\text{m}|) = 5\,\text{N}$$

The magnitude of the force is the same as in the previous case. Of course the direction of the force exerted by the spring on your hand is now different, but we would need to write a full vector equation to incorporate this information.

2.X.12 You push on a spring whose stiffness is 11 N/m, compressing it until it is 2.5 cm shorter than its relaxed length. What is the magnitude of the force the spring now exerts on your hand?

2.X.13 A spring is 0.17 m long when it is relaxed. When a force of magnitude 250 N is applied, the spring becomes 0.24 m long. **(a)** What is the stiffness of this spring? **(b)** Next, this spring is compressed so that its length is 0.15 m. What magnitude of force is required to do this?

The Spring Force as a Vector

To write an equation for the spring force as a vector, combining magnitude and direction into one expression, we need to consider the general case, in which the spring may be stretched both vertically and horizontally. Since the force exerted by the spring will be directed along the axis of the spring, we

Figure 2.22 The vector $\vec{L}$ points from the location where the spring is attached to the ceiling to the location of the mass. The arrow is offset for clarity in the diagram. L_0 is the length of the relaxed spring.

need a unit vector in that direction. We will get this unit vector from the vector $\vec{L}$, shown in Figure 2.22, which is a relative position vector: it specifies the position of the movable end of the spring relative to the fixed end of the spring. $\vec{L}$ extends from the point at which the spring is attached to a support to the mass at the other end of the spring. We can factor the relative position vector $\vec{L}$ into the product of a magnitude (the length of the spring) and a unit vector:

$$\vec{L} = |\vec{L}|\hat{L}$$

The stretch of the spring can now be written as

$$s = |\vec{L}| - L_0$$

- When the spring is longer than its relaxed length, the value of s will be positive, but the direction of the force will be given by $-\hat{L}$.
- When the spring is shorter than its relaxed length, the value of s will be negative, but the direction of the force will be given by $+\hat{L}$.

The general expression for the vector spring force is therefore:

THE VECTOR SPRING FORCE

$$\vec{F} = -k_s s \hat{L}$$

where $s = |\vec{L}| - L_0$.
L_0 is the length of the relaxed spring.
The vector $\vec{L}$ extends from the point of attachment of the spring
    to the mass at the other end.
The stretch s may be positive or negative.

EXAMPLE   **Force Due to a Stretched Spring**

Suppose that the stiffness of the rightmost spring shown in Figure 2.22 is 9 N/m, and its relaxed length is 21 cm. At the instant shown, the location of the green mass is $\langle .07, -0.33, 0 \rangle$ m relative to an origin at the point of attachment of the spring. What is the force exerted by the spring on the green mass at this instant?

Solution

$$\vec{L} = \langle 0.07, -0.33, 0.0 \rangle \, \text{m} - \langle 0, 0, 0 \rangle \, \text{m} = \langle 0.07, -0.33, 0.0 \rangle \, \text{m}$$

$$|\vec{L}| = \sqrt{(.07\,\text{m})^2 + (-.33\,\text{m})^2} = 0.337\,\text{m}$$

$$\hat{L} = \frac{\langle 0.07, -0.33, 0.0 \rangle \, \text{m}}{0.337\,\text{m}} = \langle 0.208, -.979, 0 \rangle$$

$$s = 0.337\,\text{m} - 0.21\,\text{m} = 0.127\,\text{m}$$

$$\vec{F} = -(9\,\text{N/m})(0.127\,\text{m})\langle 0.208, -.979, 0 \rangle = \langle -.238, 1.12, 0 \rangle \, \text{N/m}$$

Check: The x component of the force is negative, and the y component is positive, as they should be.

2.X.14 A spring of stiffness 95 N/m, and with relaxed length 15 cm, stands vertically on a table, as shown in Figure 2.21. With your hand you push straight down on the spring until your hand is only 11 cm above the table.

  (a) What is the vector $\vec{L}$?

  (b) What is the magnitude of $\vec{L}$?

(c) What is the unit vector $\hat{L}$?

(d) What is the stretch s?

(e) What is the force $\vec{F}$ exerted on your hand by the spring?

Gravitational Force

In Chapter 3 we will discuss the general form of the gravitational force law that describes the gravitational attraction that all objects in the Universe have for each other. Near the Earth's surface it is approximately true that a 1 kg object is attracted to the Earth by a gravitational force of magnitude 9.8 N, and a 2 kg object is attracted by a force of magnitude 2×9.8 N. We define $g = +9.8$ N/kg to be the strength of this interaction in newtons per kilogram, so we can say that a mass m near the surface of the Earth is pulled toward the Earth by a force whose magnitude is mg.

APPROXIMATE GRAVITATIONAL FORCE

$$|\vec{F}_{\text{grav}}| \approx mg$$

where $g = +9.8$ N/kg for an object of mass m near the surface of the Earth.

2.X.15 In our usual coordinate system, $+x$ to the right, $+y$ up (away from the center of the Earth), $+z$ out of the page toward you, what is the *vector* gravitational force on a 5 kg object sitting on the ground? (From the Momentum Principle you can conclude that a force of the same magnitude is exerted upward by the ground on the object if the momentum isn't changing. In Chapter 4 we'll see more about the origin of this force.)

2.X.16 The magnitude of the gravitational force on a small apple is about 1 N. What is the mass of a small apple?

Motion of a Block–Spring System

If we attach a block to the top of a spring, push down on the block, and then release it, the block will oscillate up and down. (This repetitive motion is described as "periodic.") As the spring stretches and compresses, the force exerted on the block by the spring changes in magnitude and direction. There is also a constant gravitational force on the block. Because the net force on the block is continually changing, we can't use a one-step calculation to predict its motion. We need to apply the Momentum Principle iteratively to predict the location and velocity of the block at any instant.

In outline, here is the calculation we will do. We need two initial values:

- The initial position of the block
- The initial momentum of the block

We will do the following calculation iteratively (repeatedly):

- Find the net force on the block.
- Update the momentum of the block (by applying the Momentum Principle).
- Update the position of the block.

Block on Spring: 1D, Nonconstant Net Force

A spring has a relaxed length of 20 cm (0.2 m) and its spring stiffness is 8 N/m (Figure 2.23). You glue a 60 gram block (0.06 kg) to the top of the spring, and push the block down, compressing the spring so its total length is 10 cm (Figure 2.24). You make sure the block is at rest, then you quickly move your hand away. The block begins to move upward, because the upward force on the block by the spring is greater than the downward force on the block by the Earth. Make a graph of y vs. time for the block during a 0.3-second interval after you release the block.

Figure 2.23 A relaxed vertical spring.

Solution

System: Block
Surroundings: Spring, Earth
Free-body diagram: Figure 2.24
For convenience we place the origin at the base of the spring.

$$\vec{L} = \langle 0, 0.1, 0 \rangle - \langle 0, 0, 0 \rangle = \langle 0, 0.1, 0 \rangle$$

$$|\vec{L}| = 0.1$$

$$\hat{L} = \langle 0, 1, 0 \rangle$$

$$\vec{F}_{\text{spring}} = -k_s(|\vec{L}| - L_0)\langle 0, 1, 0 \rangle = \langle 0, -k_s(|\vec{L}| - L_0), 0 \rangle$$

$$\vec{F}_{\text{Earth}} = \langle 0, -mg, 0 \rangle$$

The initial momentum of the block is zero, since it is at rest when you release it:

$$\vec{p}_i = \langle 0, 0, 0 \rangle$$

Because both the force by the spring and the gravitational force by the Earth act in the $\pm y$ direction, and the initial x and z components of the block's momentum are zero, we need to consider only the y components of forces:

$$F_{\text{spring},y} = -k_s(|\vec{L}| - L_0)$$

$$F_{\text{Earth},y} = -mg$$

$$F_{\text{net},y} = F_{\text{spring},y} + F_{\text{Earth},y}$$

To get an approximate answer, let's divide the 0.3-second time interval into three intervals each 0.1 second long. (It would be better to use even shorter intervals, but this would be unduly tedious if done by hand.)

Figure 2.24 You compress the spring, make sure the block is at rest, then release the block. The red arrows show the forces on the block, due to the surroundings, at the instant just after you release the block.

The first time step (Figure 2.25):

$$|\vec{L}| = 0.1 \, \text{m}$$

$$s = 0.1 \, \text{m} - 0.2 \, \text{m} = -0.1 \, \text{m}$$

$$F_{\text{spring},y} = -8 \, \text{N/m}(-0.1 \, \text{m}) = +0.8 \, \text{N}$$

$$F_{\text{Earth},y} = -0.06 \, \text{kg} \cdot 9.8 \, \text{N/kg} = -0.588 \, \text{N}$$

$$F_{\text{net},y} = +0.212 \, \text{N}$$

$$p_{fy} = 0 + (0.212 \, \text{N})(0.1 \, \text{s}) \quad \text{(Momentum Principle)}$$

$$p_{fy} = 0.0212 \, \text{kg} \cdot \text{m/s}$$

$$v_{\text{avg},y} \approx v_{fy}$$

$$v_{fy} = \frac{p_{fy}}{m} = \frac{(0.0212 \, \text{kg} \cdot \text{m/s})}{0.06 \, \text{kg}} = +0.353 \, \text{m/s}$$

$$y_f = 0.1 \, \text{m} + (0.353 \, \text{m/s})(0.1 \, \text{s}) \quad \text{(position update)}$$

$$y_f = 0.135 \, \text{m}$$

Figure 2.25 During time step 1, the net force on the block is in the $+y$ direction, so the y component of the block's momentum is increasing.

Figure 2.26 During time step 2, the net force on the block is now in the $-y$ direction. The block is still moving upward, but its upward momentum is decreasing.

Figure 2.27 During time step 3, the net force on the block is in the $-y$ direction, and the block is now moving downward.

The second time step (Figure 2.26):

$$|\vec{L}| = 0.135\,\text{m}$$
$$s = 0.135\,\text{m} - 0.2\,\text{m} = -0.0647\,\text{m}$$
$$F_{spring,y} = +0.520\,\text{N}$$
$$F_{net,y} = -0.0707\,\text{N}$$
$$p_{fy} = (0.0212\ \text{kg} \cdot \text{m/s}) + (-0.0707\,\text{N})(0.1\,\text{s})$$
$$p_{fy} = 0.0141\ \text{kg} \cdot \text{m/s}$$
$$v_{fy} = 0.236\,\text{m/s}$$
$$y_f = 0.159\,\text{m}$$

The third time step (Figure 2.27):

$$|\vec{L}| = 0.159\,\text{m}$$
$$s = -0.411\,\text{m}$$
$$F_{net,y} = (-0.259\,\text{N})$$
$$p_{fy} = -0.118\ \text{kg} \cdot \text{m/s}$$
$$v_{fy} = -0.196\,\text{m/s}$$
$$y_f = 0.139\,\text{m}$$

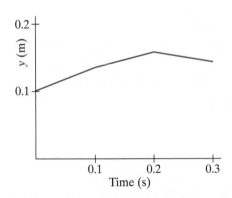

Figure 2.28 A graph of the y component of the block's position vs. time for the three-step iterative calculation.

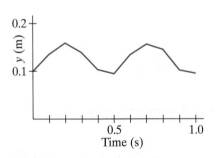

Figure 2.29 A graph of the y component of the block's position vs. time for an iterative calculation carried out for 10 steps of 0.1 second.

FURTHER DISCUSSION   The graph of y vs. time is shown in Figure 2.28.

QUESTION   Is this result reasonable?

Yes, because we expect the block to oscillate up and down on the spring.

QUESTION   What approximations were made in this calculation?

We made the approximation that the net force did not change significantly over each small time step. During each time step, we also used the final velocity as an approximation for the average velocity. (This is simpler than computing the arithmetic average, and not necessarily worse in a situation where the force is changing. In this case, it turns out that using the arithmetic average would actually have given a less accurate answer.)

Figure 2.29 shows the graph of y vs. t produced when the iterative calculation above is carried out for 10 time steps of 0.1 second.

QUESTION   What features of the graph reflect inaccuracies in the calculations?

Although the graph does show the oscillatory motion of the block, its jagged lines reflect the fact that 0.1 second is too large a step size to produce an accurate result. There are two issues:

- The graph is jagged because we're connecting temporally distant computed points by long straight lines.
- The computed points themselves are inaccurate because we used a rather long time step of 0.1 s, which is clearly large compared to the time scale of the changes occurring in the motion.

2.X.17 In step 3 above, only the results of the calculations are given. Carry out the calculations for this step yourself, to be sure you understand in detail the procedure used. Compare your results to the values given above.

2.X.18 In step 2 above, the net force on the block was downward, but the block moved upward. Explain why this was possible, and what the effect of the net force was during this time step.

Periodic Motion

Since the motion of the block–spring system repeats over and over, this motion is called *periodic* motion. The time interval between maxima in the plot of y vs. t is called the *period* of the motion, and is usually represented by the symbol T. From the graph in Figure 2.29 we can see that for this particular mass–spring system the period T appears to be around half a second. To get a better estimate of the period, we would need a better graph.

QUESTION   How could we improve the accuracy of our calculations?

Improving the Calculation: Smaller Δt

In general, decreasing the time step size will produce more accurate results, because the assumption that $\vec{F}_{net}$ is constant during the time Δt will be more valid. Using a step size of 0.01 second produces a smoothly oscillating graph of y vs. t, as shown in Figure 2.30. Now the straight lines connecting the computed points are short, making the graph look more smooth, and the computed points themselves are more accurate thanks to using a shorter time step, so that the force and velocity are more nearly constant during each time interval.

QUESTION   Suppose that we had chosen to use a Δt of one second. What effect would this have had on our prediction of the block's motion?

The periodicity of the system's motion is due to the fact that the net force on the block is changing with time. Since one second is actually longer than the period of the motion, assuming that the net force is constant over this interval will lead to an extremely inaccurate result. The resulting graph of y vs. t, shown in Figure 2.31, shows that the predicted y position of the mass after 1 second would be more than 3 meters above the floor!

Guidelines for Choosing a Time Step

In any calculation based on the Momentum Principle, we need to choose a value for the time step Δt short enough that the net force on the object of interest is approximately constant during this time interval. How do we do this? There is not a simple formula for choosing Δt, but there are some useful guidelines.

Distance-Dependent Forces

In many cases the force on the system depends on the distance of the object from some other object. For example, in the preceding mass–spring example, the distance is the stretch: the displacement of the free end of the spring from its equilibrium position. In such a case we need to pick a Δt small enough that this distance won't change very much during this interval. How do we do this?

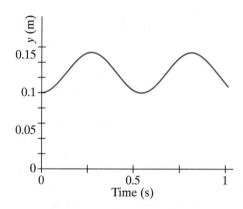

Figure 2.30 A graph of the y component of the block's position vs. time for an iterative calculation using a step size of 0.01 second.

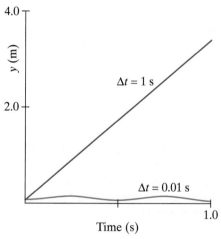

Figure 2.31 If we use a time step of 1 second to predict the motion of the mass–spring system, we predict that the mass will be more than 3 meters above the floor! Note that the vertical scale of this graph is very different from the scale of the preceding graphs.

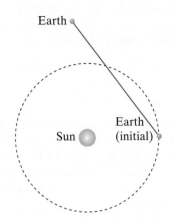

Earth

Sun ●   Earth
(initial)

Figure 2.32 A time step of 3 months is too large in this situation.

(a) If the initial velocity of the system is not zero, use $|\vec{v}_i \Delta t|$ to estimate the displacement of the system during one time step. Make sure this displacement is small compared to the relevant distance.

For example, we'll see in Chapter 3 that if we try to predict the motion of the Earth around the Sun using a time step of 3 months, we get the displacement shown in Figure 2.32, which is very large compared to the distances in the situation. The problem of course is that in 3 months the direction of the Sun's gravitational force on the Earth changes a lot, and the velocity changes a lot, so approximating force and velocity as constant during the 3-month interval is very inaccurate.

(b) In a case in which the initial velocity is zero, as in the preceding block and spring example, we may need to do a quick guess and check estimate, by completing one iteration.

In the block and spring example, the initial stretch of the spring was −10 cm. We guessed that a 0.1-second time step would be reasonable, and after calculating v_{yf} and y_f, we found that the block had moved from 0.1 m above the floor to 0.135 m above the floor: a displacement of 3.5 cm and a 35% change in the length of the spring. For a rough estimate this was acceptable, and did give us a general idea about the periodic motion of the block and spring, as shown in Figure 2.29. To make more accurate predictions, it was necessary to use a significantly shorter time step, as shown in Figure 2.30. Because the period of the system's motion (the time between successive maximum displacements) was about half a second, we could have missed the periodicity of the motion altogether if we had chosen a time step of 0.5 seconds or larger!

Speed-Dependent Forces

Some forces, like air resistance, depend on an object's speed instead of its position. In such a case, we need to make sure that the object's speed does not change very much during the time interval Δt.

Periodic Motion

For a system in periodic motion, such as a mass and spring, or an orbiting body, if we know the approximate period, we need to choose Δt to be significantly smaller than the period (the time for the motion to repeat). For a mass–spring system with a period of 2 seconds, Δt should be much less than 2 seconds. For an Earth–Sun orbit, we want Δt to be much smaller than a year (and we saw in Figure 2.32 that 3 months was a poor choice for Δt).

Telling a Computer What to Do

While this iterative scheme is very general, doing it by hand is tedious. It is not difficult to program a computer to do these calculations repetitively. Computers are now fast enough that it is possible to get high accuracy simply by taking very short time steps, so that during each step the net force and velocity aren't changing much.

In a few very special cases it is possible to derive an analytical (algebraic) solution for the motion of an object, but in most situations this is not possible (see the next section, "Why Not Just Use Calculus"). Therefore, this iterative approach for predicting motion can be very valuable. Doing iterative calculations with very small time steps is practical only if one uses a computer. This is how a computer program can be structured in order to apply the Momentum Principle repetitively to predict the future:

STRUCTURING ITERATIVE CALCULATIONS ON A COMPUTER

- Define the values of constants such as g or k_s to use in the program.
- Specify the masses, initial positions, and initial momenta of the interacting objects.
- Specify an appropriate value for Δt, small enough that the objects don't move very far during one update.
- Create a "loop" structure for repetitive calculations:
  - Calculate the (vector) forces acting on each object.
  - Update the momentum of each object: $\vec{p}_f = \vec{p}_i + \vec{F}_{net}\,\Delta t$.
  - Update the positions: $\vec{r}_f = \vec{r}_i + \vec{v}_{avg}\,\Delta t$, where $\vec{v}_{avg} \approx \vec{p}_f/m$.
  - Repeat.

Figure 2.33 shows a display created by such a computer program. The details of the actual program statements depend on which programming language or system you use, and your instructor will provide you with the details of the particular tool you will use.

Figure 2.33 Display created by a computer program to calculate the motion of a mass–spring system in 3D.

Time Step Size in a Computer Calculation

When doing a calculation by hand, there is a tradeoff between accuracy and time required to do many iterations (the smaller the time step, the more calculations must be done). Since computers are quite fast, it is reasonable to use much smaller time steps in a computer calculation than one would use by hand. However, even a fast computer can take a very long time to do calculations that use an unnecessarily tiny time step—it would not be reasonable to use a time step of 1×10^{-20} s in a prediction of the Earth's motion around the Sun.

A standard method of checking the accuracy of a computer calculation is to decrease the time step and repeat the calculation. If the results do not change significantly, the original time step was adequately small.

CHECKING THAT Δt IS SMALL ENOUGH

In an iterative computer calculation, always make sure that decreasing the step size does not significantly change the predictions. If it does, keep decreasing the step size until further decreases make no significant change.

2.X.19 A block is hung from a spring, and the spring is stretched 5 cm before the block is released from rest. You need to choose a time step to use in predicting the block's motion. Which of the following would be a reasonable distance for the block to move in the first time step, doing an iterative calculation *by hand*?

(a) 10 cm (b) 5 cm (c) 0.5 cm (d) 0.005 cm

2.X.20 Jupiter goes around the Sun in 4333 Earth days. Which of the following would be a reasonable value to try for Δt in a *computer* calculation of the orbit?

(a) 1 day (b) 4333 days (c) 0.01 second (d) 800 days

Why Not Just Use Calculus?

You might wonder why we don't simply use calculus to predict the motion of physical systems. In fact, we actually are using calculus. Step by step, we add up a large number of small increments of the momentum of an object, and a large number of small displacements of the object, to calculate a large change in its momentum and position over a long time, and this corresponds to an approximate evaluation of an integral:

$$\Delta \vec{p}_1 + \Delta \vec{p}_2 + \cdots \approx \int_i^f d\vec{p} = \int_i^f \vec{F}_{\text{net}} \, dt$$

$$\Delta \vec{r}_1 + \Delta \vec{r}_2 + \cdots \approx \int_i^f d\vec{r} = \int_i^f \vec{v}_{\text{avg}} \, dt$$

The integral sign is a distorted "S" meaning "sum" of an infinite number of infinitesimal quantities, from the initial time i to the final time f.

A more interesting answer is that the motion of most physical systems actually *cannot* be predicted using calculus in any way other than by this step-by-step, calculus-based approach. In a few special cases calculus does give a general result without going through this procedure. For example, an object subjected to a constant force has a constant rate of change of momentum and velocity, and calculus can be used to obtain a prediction for the position as a function of time, as we will see later in this chapter in the example of a ball thrown through a vacuum. An analytical solution can be derived for the motion of a mass attached to a very low mass spring, in the absence of air resistance, as we will see in Chapter 4. The elliptical orbits of two stars around each other can be predicted mathematically without an iterative approach, although the math is quite challenging.

However, in only slightly more complicated situations, an analytical solution may not be possible. For example, the general motion of three stars around each other has never been successfully analyzed in this way. The basic problem is that it is usually relatively easy to take the derivative of a known function, but it is often impossible to determine in algebraic form the integral of a known function, which is what would be involved in long-term prediction (adding up a large number of small momentum increments due to known forces).

In contrast, a step-by-step procedure of the kind we carried out for the mass–spring system can easily be extended to three or more bodies. It is also possible to include the effects of various kinds of friction and damping in an iterative calculation. This is why we study the step-by-step prediction method in detail, because it is a powerful technique of increasing importance in modern science and engineering, thanks to the availability of powerful computers to do the repetitive work for us.

2.6   SPECIAL CASE: CONSTANT FORCE

In most real situations the net force acting on a system does not remain constant over time. However, in a few special situations the net force on a system may be constant, or approximately constant, for some period of time. In such a situation the task of predicting the motion of the system is simplified.

In particular, we'll see that if we can assume that the net force on an object is constant, we not only can predict where it will be and what its velocity will be after a given time has passed, but because the object's momentum is increasing at a constant rate we can also easily answer questions such as "How long will it take the object to move a given distance?"

EXAMPLE   **1 D Motion, Constant Net Force**

Suppose a fan cart of mass 0.5 kg is coasting in the $+x$ direction at 0.3 m/s. Then the fan is turned on, and the cart experiences a constant net force of .2 N in the $+x$ direction for 0.6 seconds.

(a) What will the cart's velocity be at the end of this interval?

(b) How far will the cart have traveled?

Solution System: Cart
Surroundings: Everything else
Free-body diagram: Figure 2.34

Figure 2.34 Free-body diagram for a fan cart subject to a constant net force in the +x direction.

We can begin as before by dividing the time interval into smaller intervals of 0.1 second. We only need to consider the x component of force and momentum. We will use the symbol p_{x1} to represent the x momentum of the system after 0.1 s, and so on.

(a) Initial momentum:

$$p_{x0} = (0.5\,\text{kg} \cdot 0.3\,\text{m/s}) = 0.15\ \text{kg} \cdot \text{m/s}$$

Momentum Principle:

$$p_{x1} = 0.15\ \text{kg} \cdot \text{m/s} + 0.2\,\text{N} \cdot 0.1\,\text{s} = 0.17\ \text{kg} \cdot \text{m/s}$$
$$p_{x2} = 0.17\ \text{kg} \cdot \text{m/s} + 0.2\,\text{N} \cdot 0.1\,\text{s} = 0.19\ \text{kg} \cdot \text{m/s}$$
$$p_{x3} = 0.19\ \text{kg} \cdot \text{m/s} + 0.2\,\text{N} \cdot 0.1\,\text{s} = 0.21\ \text{kg} \cdot \text{m/s}$$
$$p_{x4} = \ldots$$

We see that the net x impulse applied to the cart in each time step is constant. Because the rate of change of momentum is constant, we can combine all the small time steps into one big time step:

$$p_{x6} = 0.15\ \text{kg} \cdot \text{m/s} + 0.2\,\text{N} \cdot 0.6\,\text{s} = 0.27\ \text{kg} \cdot \text{m/s}$$

Since at low speeds $\gamma \approx 1$ and $p_x \approx mv_x$, the final x velocity of the cart is

$$v_{x6} = \frac{0.27\ \text{kg} \cdot \text{m/s}}{0.5\,\text{kg}} = 0.54\ \text{m/s}$$

(b) Final position:
The rate of change of v_x (the x acceleration) is constant in this situation, so the arithmetic average velocity is equal to the real average velocity:

$$v_{\text{avg},x} = \frac{0.3\,\text{m/s} + 0.54\,\text{m/s}}{2} = 0.42\ \text{m/s}$$
$$x_f - x_i = 0.42\,\text{m/s} \cdot 0.6\,\text{s} = 0.25\ \text{m}$$

The cart has traveled 25 cm.

Constant Net Force: An Algebraic Solution

It is straightforward to generalize the solution to the previous example.
If the net force on a system is constant during the time Δt, and $v \ll c$:

$$p_{xf} = p_{xi} + F_{\text{net},x}\,\Delta t$$
$$v_{xf} = v_{xi} + \frac{F_{\text{net},x}}{m}\,\Delta t$$
$$v_{\text{avg},x} = \frac{\left(v_{xi} + \dfrac{F_{\text{net},x}}{m}\,\Delta t\right) + v_{xi}}{2}$$
$$v_{\text{avg},x} = v_{xi} + \frac{1}{2}\frac{F_{\text{net},x}}{m}\,\Delta t$$
$$x_f = x_i + v_{\text{avg},x}\,\Delta t$$
$$x_f = x_i + v_{xi}\,\Delta t + \frac{1}{2}\frac{F_{\text{net},x}}{m}\,\Delta t^2$$

The quantity $F_{\text{net},x}/m$ is the rate of change of v_x, so it is equal to a_x, the x component of the system's acceleration. If the net force had acted in the y or z direction the solution would be the same, except for the subscripts indicating direction. Δt can be arbitrarily large as long as the net force on the system is constant during this time interval.

Analytical Solutions

The term *analytical* describes a solution that can be written symbolically (algebraically). The final equation for x_f above is an analytical solution to the problem of motion in one dimension under the influence of a constant force.

> **2.X.21** A land yacht is a low-mass, three-wheeled wind-powered vehicle with a sail. Suppose that you are in the desert, observing a land yacht race. Consider your location to be the origin. When a particular land yacht passes you, it is traveling 10 m/s in the $-z$ direction. If the mass of the vehicle plus driver is 163 kg, and net force on the vehicle remains constant at $\langle 0, 0, -35 \rangle$ N, what is the position of the yacht one minute after it passes you?

Determining Δt

Since the equation

$$x_f = x_i + v_{xi}\Delta t + \frac{1}{2}\frac{F_{\text{net},x}}{m}\Delta t^2$$

applies to a constant force situation, we can use it to determine how long it would take an object to travel a given distance, if it were acted on during its travel time by a constant net force.

The best-known example of a system subject to an approximately constant force is a object near the Earth's surface, falling freely toward the ground. Even in this simple situation the actual force acting on the object is not exactly constant.

> **QUESTION** What factors cause the net force on a falling object to vary?

The air resistance, which is due to the collisions between the falling object and air molecules, depends on the shape of the moving object, and also on both the magnitude and direction of the velocity of the object. Air resistance is small at low speeds and big at high speeds, as you may have experienced if you put your hand out the window of a moving car. This force can be quite large. A low-density object such as a Styrofoam™ ball experiences air resistance that is comparable to the small gravitational force on the ball, so air resistance is important unless the Styrofoam ball is moving very slowly. At low speeds a baseball, which has a fairly high density, moves with negligible air resistance. But at the speed that a professional pitcher can throw a baseball (about 100 mi/hr or 44 m/s), the baseball goes only about half as far in air as it would in a vacuum, because air resistance is large at this high speed (Figure 2.35). We will discuss air resistance further in a later chapter.

It is also the case that the gravitational force itself is not exactly constant; it increases slightly as the object approaches the ground. Nevertheless, it is sometimes acceptable to treat the net force on a falling object near the Earth's surface as approximately constant over a limited time interval.

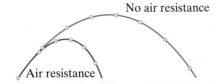

No air resistance

Air resistance

Figure 2.35 The trajectory of a baseball thrown at high speed (around 100 mph, or 44 m/s), ignoring air resistance (top curve) and including the effect of air resistance (bottom curve). The dots indicate the ball's position at equal time intervals. The ball travels about half as far in air as it would in a vacuum.

EXAMPLE

Falling Objects: 1D, Constant Net Force

You simultaneously drop two objects, a 5 kg bowling ball and a 0.2 kg billiard ball, from the top of the leaning tower of Pisa (Figure 2.36). If air resistance were negligible, how long would it take each ball to fall 56 meters to the ground?

Solution

System: Bowling ball
Surroundings: Earth
Approximations: Neglect air resistance

Figure 2.36 The leaning tower of Pisa.

Figure 2.37 Free-body diagram for a ball dropped from a tower.

Initial time: The instant the hand is no longer in contact with ball
Final time: The instant before the ball hits the ground
Free-body diagram: Figure 2.37

From the previous discussion:

$$y_f - y_i = v_{yi}\Delta t + \frac{1}{2}\frac{F_{net,y}}{m}\Delta t^2$$

$$y_f - y_i = 0 + \frac{1}{2}\frac{-mg}{m}\Delta t^2$$

$$-56\,\text{m} = \frac{1}{2}\frac{(-9.8\,\text{N/kg}\cdot 5\,\text{kg})}{5\,\text{kg}}\Delta t^2$$

Note that since the gravitational force on the ball is proportional to the mass of the ball, the mass cancels.

$$56\,\text{m} = \frac{1}{2}(9.8\,\text{N/kg})\,\Delta t^2$$

$$\Delta t = \sqrt{\frac{2\cdot 56\,\text{m}}{9.8\,\text{N/kg}}} = 3.38\,\text{s}$$

Because the result is independent of mass, the solution is the same for the billiard ball. Each ball falls to the ground in 3.38 seconds. Because the gravitational force on an object is proportional to its mass, the y acceleration does not depend on the mass of the object! This is not true for other forces, such as the spring force we discussed in previous sections.

2.X.22 Explain why the quantity $y_f - y_i$ has a negative value (-56 m) in the preceding example. Also explain why there is a negative sign associated with the quantity (-9.8 N/kg $\cdot$ 5 kg), when $g = +9.8$ N/kg is a positive constant.

2.X.23 A 20 gram apple falls from a branch that is 4 meters above the ground.

    (a) How much time elapses before the apple hits the ground?

    (b) Just before the impact, what is the velocity of the apple?

At What Point Does a Contact Force Stop Acting?

Forces due to contact between two objects, such as a bat and a baseball, or a foot and a soccer ball, are called "contact forces." In the previous example, the force by your hand on the bowling ball was a contact force. This force stopped acting the instant your hand was no longer in contact with the ball. Because we chose this instant as the initial time in our analysis, your hand did not exert a force on the ball during the period of interest.

The following example involves a ball that has been kicked into the air. We choose as the initial time the instant at which the kicker's foot is no longer in contact with the ball. From this instant on, the foot does not exert a force on the ball. Why not? In a later chapter we will see that forces involving contact between objects such as a foot and a ball occur because springlike interatomic bonds in both objects are compressed. This compression persists only as long as the objects are actually in contact. When the foot and the ball lose contact, the foot can no longer compress interatomic bonds in the ball, so the force exerted by the foot on the ball ceases at this moment.

EXAMPLE	A Ball with Negligible Air Resistance: 2D, Constant Force

A ball of mass 500 g is initially on the ground, at location $\langle 0, 0, 0 \rangle$ m, and you kick it with initial velocity $\langle 3, 7, 0 \rangle$ m/s.

(a) Where will the ball be half a second later?

(b) At what time will the ball hit the ground?

Make the approximation that air resistance is negligible.

Solution

System: Ball
Surroundings: Earth
Free body diagram: Figure 2.38
Initial time: The instant your foot is no longer touching the ball

■ Pick initial and final times so that during the time interval Δt only the Earth exerts a force on the ball.

(a) A net force in the $-y$ direction can only change the y component of the ball's momentum—the x component is not affected.

$$\vec{F}_{net} = \langle 0, -mg, 0 \rangle$$

x components:

$$F_{net,x} = 0$$
$$x_f = x_i + v_{ix} \Delta t$$
$$x_f = 0 + (3 \text{ m/s})(0.5 \text{ s}) = 1.5 \text{ m}$$

y components:

$$F_{net,y} = -mg$$
$$y_f = y_i + v_{iy} \Delta t + \frac{1}{2} \frac{(-mg)}{m} \Delta t^2$$
$$y_f = 0 + (7 \text{ m/s})(0.5 \text{ s}) - \frac{1}{2}(9.8 \text{ N/kg})(0.5 \text{ s})^2 = 2.275 \text{ m}$$

Figure 2.38 Free-body diagram for the ball moving with negligible air resistance.

Final displacement vector:

$$\vec{r}_f = \langle 1.5, 2.275, 0 \rangle \text{ m}$$

■ In this case, the mass of the ball cancels, because the gravitational force is proportional to mass.

Check: The units are correct, and the ball has moved in an appropriate direction.

(b) Final time: The instant just before the ball hits the ground. At this instant we know $y_f = 0$, so we can find Δt:

$$0 = 0 + v_{iy} \Delta t + \frac{1}{2}(-g) \Delta t^2$$

Solving this quadratic equation for the unknown time Δt, we find two possible values:

$$\Delta t = 0 \text{ and } \Delta t = \frac{2v_{iy}}{g}$$

The first value, $\Delta t = 0$, corresponds to the initial situation, when the ball is near the ground, just after the kick. The second value is the time when the ball returns to the ground, just before hitting:

$$\Delta t = \frac{2(7 \text{ m/s})}{(9.8 \text{ N/kg})} = 1.43 \text{ s}$$

QUESTION	Could we use these equations for x and y as a function of time to find the location of the ball 10 seconds after you kick it?

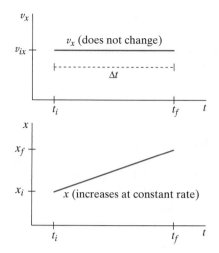

Figure 2.39 Motion graphs for the thrown ball. Top: v_x vs. t; bottom: x vs. t. Note that v_x does not change because the net force acted only in the y direction.

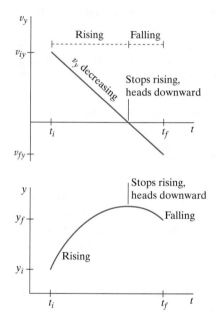

This is y vs. time t. It is not the path.

Figure 2.40 Motion graphs for the thrown ball. Top: v_y vs. t; bottom: y vs. t.

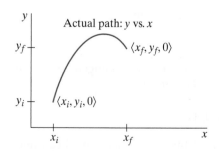

Figure 2.41 The actual trajectory of the thrown ball, with negligible air resistance (y vs. x).

No. Our result would be that the ball was far underground, since

$$y_f = (7 \text{ m/s})(10 \text{ s}) - \frac{1}{2}(9.8 \text{ N/kg})(10 \text{ s})^2 = -420 \text{ m}$$

which is not physically reasonable! (The ball would have hit the ground and stopped before 10 seconds had passed, due to other interactions not included in our model.)

Graphs of Motion

Figure 2.39 and Figure 2.40 show graphs of position and velocity components vs. time for the ball in the preceding example. In Figure 2.39 the first graph, v_x vs. t, is simply a horizontal line, because v_x doesn't change, since there is no x component of force. The graph of x is a straight line (the second graph), rising if v_x is positive. Note that the slope of the x vs. t graph is equal to v_x.

In Figure 2.40 the graph of v_y is a falling straight line (the first graph), because the y component of the force is $-mg$, which constantly makes the y component of momentum decrease. At some point the y component of momentum decreases to zero, at the top of the motion, after which the ball heads downward, with negative v_y. The graph of y vs. time t is an inverted parabola (the second graph), since the equation for y is a quadratic function in the time.

Note that the slope of the y vs. t graph (Figure 2.40) at any time is equal to v_y at that time. In particular, when the slope is zero (at the maximum y), v_y is momentarily equal to zero. Before that point the slope is positive, corresponding to $v_y > 0$, and after that point the slope is negative, corresponding to $v_y < 0$.

The actual path of the ball, the graph of y vs. x (Figure 2.41), is also an inverted parabola (the bottom graph). Since x increases linearly with t, whether we plot y vs. t or y vs. x we'll see a similar curve. The scale factor along the horizontal axis is different, of course (meters instead of seconds).

> **2.X.24** A ball is kicked from a location $\langle 9, 0, -5 \rangle$ m (on the ground) with initial velocity $\langle -10, 13, -5 \rangle$ m/s.
> **(a)** What is the velocity of the ball 0.6 seconds after being kicked?
> **(b)** What is the location of the ball 0.6 seconds after being kicked?
> **(c)** How much time does it take for the ball to reach its maximum height, where $v_y = 0$ momentarily?
> **(d)** What is the maximum height?
> **(e)** At what time does the ball hit the ground?
> **(f)** What is the location of the ball when it hits the ground?

Parallel and Perpendicular Components of the Net Force

The component of the net force that is parallel to the momentum, $F_\parallel$, changes the magnitude of the momentum and therefore changes the speed. If $F_\parallel$ is in the direction of the momentum, the object will speed up; if $F_\parallel$ is opposite to the direction of the momentum, the object will slow down.

The component of the net force that is perpendicular to the momentum, $F_\perp$, changes the direction of the momentum.

We can see these effects in the motion of a projectile. At location A in Figure 2.42, the net (gravitational) force is split into components parallel to and perpendicular to the momentum. Because the parallel component is opposite the momentum, the projectile is slowing down. The perpendicular component is turning the direction to be pointing less steeply upward.

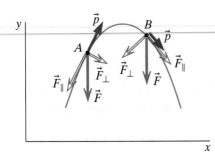

Figure 2.42 The parallel component of the force changes the magnitude of the momentum (and the speed). The perpendicular component changes the direction of the momentum.

At location B in Figure 2.42, the parallel component of the net force is in the direction of the momentum, and the projectile is speeding up. The perpendicular component is turning the direction to be pointing more steeply downward.

2.7 ESTIMATING INTERACTION TIMES

So far all of the examples we have considered have involved finding a change in momentum (and position), given a known force acting over a known time interval. The following problems require you to find either the duration of an interaction (time interval), or the net force exerted during an interaction. These problems involve several steps in reasoning.

EXAMPLE **Striking a Hockey Puck**

In a video of a hockey game, we see a 0.16 kg hockey puck sliding across the ice, as shown in Figure 2.43. A player hits the puck sharply with a hockey stick; the stick breaks, and the direction of motion of the puck changes. From examining the video frame by frame, we find that before it was hit the velocity of the puck was $\langle 20, 0, 0 \rangle$ m/s, and after the impact the velocity of the puck was $\langle 20, 0, 31.7 \rangle$ m/s. When we pile weights on the side of a similar hockey stick we find that the stick breaks under a force of about 1000 N (this is roughly 250 pounds, since a force of one newton is equivalent to a force of about a quarter of a pound, approximately the weight of a small apple).

(a) What was the (vector) force on the puck during the collision with the stick?
(b) How long was the puck in contact with the hockey stick?
(c) What assumptions were made in finding these quantities?

Solution

Plan: Find the change in momentum of the puck, and apply the Momentum Principle to find the net impulse applied to the puck. The direction of the net force is the same as the direction of the impulse, and the magnitude of the force can be estimated from other information given. Knowing both the net impulse and the net force allows us to find Δt.

System: the hockey puck
Surroundings: Earth, ice, hockey stick, air (free-body diagram: Figure 2.44)
Initial time: when stick first makes contact with puck
Final time: when stick breaks

$$\vec{p}_f - \vec{p}_i = \vec{F}_{net} \, \Delta t$$
$$\vec{p}_i = 0.16 \, \text{kg} \cdot \langle 20, 0, 0 \rangle \text{ m/s}$$
$$\vec{p}_f = 0.16 \, \text{kg} \cdot \langle 20, 0, 31.7 \rangle \text{ m/s}$$
$$\vec{p}_f - \vec{p}_i = \langle 0, 0, 5.072 \rangle \text{ m/s}$$

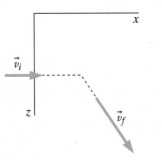

Figure 2.43 Top view of the path of a hockey puck that is hit by a hockey stick.

(a) Only the z component of the puck's momentum changed, so the net impulse must have been in the z direction. Therefore

$$\hat{F} = \langle 0, 0, 1 \rangle$$

Because the stick broke, the magnitude of the force must have been about 1000 N. Therefore

$$\vec{F}_{net} = \langle 0, 0, 1000 \rangle \text{ N}$$

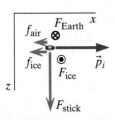

Figure 2.44 Free-body diagram for the hockey puck during the impact.

(b) From the Momentum Principle:

$$p_{fz} - p_{iz} = F_{\text{net},z}\, \Delta t$$

$$\Delta t = \frac{p_{fz} - p_{iz}}{F_{\text{net},z}} = \frac{5.072 \ \text{kg} \cdot \text{m/s}}{1000\,\text{N}} = 0.005 \ \text{s}$$

(c) We assumed that:

- The friction force due to the ice surface was negligible during the contact time.
- Air resistance was negligible during the contact time.
- The force by the hockey stick on the puck was constant during the contact time.

QUESTION How good were our assumptions?

The neglect of sliding friction and air resistance is probably pretty good, since a hockey puck slides for long distances on ice with nearly constant speed.

We know the hockey stick exerts a maximum force of $F_{\text{stick}} = 1000$ N, because we observe that the stick breaks. We approximate the force as nearly constant during contact. Actually, this force grows quickly from zero at first contact to 1000 N, then abruptly drops to zero when the stick breaks.

Even though our analysis of the stick contact time (0.005 s) isn't exact, it is adequate to get a reasonably good determination of the short interaction time, something that we wouldn't know without using the Momentum Principle and the position update formula. The very short duration of the impact explains why we hear a sharp, short crack.

Although 0.005 s is a short time, the puck moves (20 m/s)(0.005 s) = 0.1 m (about 4 inches) in the x direction during this time. Also during this time v_z increases from 0 to 31.7 m/s, with an average value of about 15.8 m/s, so the z displacement is about (15.8 m/s)(0.005 s) = 0.08 m during contact. A more accurate sketch of the path of the puck should show a bend as in Figure 2.45.

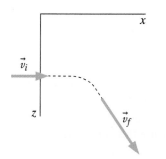

Figure 2.45 A more accurate overhead view of the path of the hockey puck, showing the bend during impact.

Estimating Times

Real-world problems often require the estimation of one or more quantities. Estimating masses or distances is usually not difficult, but most people find it more challenging to estimate time durations, especially if the times are very short. It is common to guess what seems a rather short time, such as a second, or half a second. However, we found in the previous example that the contact time between a puck and a hockey stick was significantly shorter than this—about a hundredth of a second.

A useful, systematic way to estimate a short time is to use the relationship between velocity and position (a.k.a. the position update formula).

QUESTION Suppose we had guessed that the contact time between the stick and the puck was about one second. We know that the original speed of the puck was 20 m/s. At this speed, how far would the puck have traveled in 1 s?

$$(20 \ \text{m/s})(1 \ \text{s}) = 20 \ \text{m}$$

It is clear that the puck did not move 20 meters during the impact. We must conclude that the impact took much less than 1 second.

Figure 2.46 Two students running fast collide head-on.

Suppose we guess that the puck may have slid 20 cm along the stick during the impact:

$$\Delta t \approx \frac{(0.20 \text{ m})}{(20 \text{ m/s})} = 0.01 \text{ s}$$

This estimate of the contact time differs from our result by only a factor of around 2, instead of a factor of 100, so it is a much better estimate. Using an object's speed and estimating the distance traveled during an interval allows us to come up with a much better estimate of interaction times than we would otherwise get.

EXAMPLE   **Colliding Students**

Two students who are late for tests are running to classes in opposite directions as fast as they can. They turn a corner, run into each other head-on, and crumple into a heap on the ground (Figure 2.46). Using physics principles, estimate the force that one student exerts on the other during the collision. You will need to estimate some quantities; give reasons for your choices and provide checks showing that your estimates are physically reasonable.

Solution

■ This problem is rather ill-defined and doesn't seem much like a "textbook" problem. No numbers have been given, yet you're asked to estimate the force of the collision. This kind of problem is typical of the kinds of problems engineers and scientists encounter in their professional work.

■ We could have chosen the student on the right as the system; the analysis would be very similar.

■ The student's heels are pushed to the left along the ground due to the collision, and the resisting ground pushes to the right. We'll assume that this force is much smaller than the force exerted by the other student, because the student's shoes can slip.

■ *Notes on estimates*

A 1 kg mass weighs 2.2 lb.

An Olympic sprinter can run the 100-meter dash in less than 10 seconds, so 10 m/s is an upper limit. A brisk walking speed is around 2 m/s, so 6 m/s is an intermediate value.

During the collision the student's speed decreases from 6 m/s to 0 m/s, so his or her average speed is about 3 m/s.

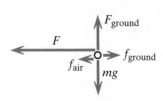

Figure 2.47 Free-body diagram for colliding students.

System: the student on the left
Surroundings: Earth, ground, other student, air
Free-body diagram: Figure 2.47
Initial time: just before impact
Final time: when speeds become zero
The symbol F represents the *magnitude* of the force we wish to find.
Simplifying assumptions:
The students have the same mass and same speed.
Friction and air resistance are negligible during this short Δt.

Momentum Principle, x component (neglecting small forces):

$$0 = p_{ix} + (-F)\Delta t$$
$$p_{ix} = F\Delta t$$

To go further we must estimate some quantities. See the comments in the margin for supporting information.

Mass of the student: about 60 kg
Initial speed of student: about 6 m/s
Average speed of student: about 3 m/s

In order to estimate Δt, we need to estimate the distance traveled during the collision—the amount by which the student's body was compressed. Perhaps the student's body is compressed by about 5 cm during the collision, which will be painful.

Approximate duration of collision:

$$\Delta x = v_{\text{avg},x} \Delta t$$
$$\Delta t = \frac{\Delta x}{v_{\text{avg},x}} = \frac{0.05 \text{ m}}{3 \text{ m/s}} = 0.017 \text{ s}$$

Solve for F:

$$F = \frac{p_{ix}}{\Delta t} \approx \frac{(60 \text{ kg})(6 \text{ m/s})}{0.017 \text{ s}} = 21000 \text{ N}$$

■ Force has units (kg)(ms)/s, which has the units of momentum (kg·m/s) divided by s, which is correct for a force (change of momentum divided by time).

Check:

■ The units check (force is in newtons, collision time is in seconds).
■ Is the result reasonable? The contact time is very short, as expected. Is the force reasonable or not? See the discussion below.

FURTHER DISCUSSION 21,000 N is a very large force. For example, the gravitational force on a 60 kg student (the "weight") is only about (60 kg)(9.8 N/kg) ≈ 600 N. The force of the impact is about 35 times the weight of the student! It's like having a stack of 35 students sitting on you. If the students hit heads instead of stomachs, the squeeze might be less than 1 cm, and the force would be over 5 times as large! This is why heads can break in such a collision.

Our result of 21,000 N shows why collisions are so dangerous. Collisions involve very large forces acting for very short times, giving impulses of ordinary magnitude.

How Good Were Our Approximations?

We made the following approximations and simplifying assumptions:

■ We estimated the running speed from known 100 m dash records.
■ We estimated the masses of the students.
■ We assumed that the horizontal component of the force of the ground on the bottom of the student's shoe was small compared to the force exerted by the other student. Now that we find that the impact force is huge, this assumption seems quite good.
■ We made the approximation that the impact force was nearly constant during the impact, so what we've really determined is an average force.
■ We assumed that the students had similar masses and similar running speeds, to simplify the analysis. If this is not the case, the analysis is significantly more complicated, but we would still find that the impact force is huge.

You might object that with all these estimates and simplifying assumptions the final result of 21,000 N for the impact force is not useful. It is certainly the case that we don't have a very accurate result. But nevertheless we gained valuable information: that the impact force is *very* large. Before doing this analysis based on the Momentum Principle, we had no idea of whether the force was small compared to the student's weight, comparable, or much bigger. Now we have a quantitative result that the force is about 35 times the weight of a student, and we can appreciate why collisions are so dangerous.

2.X.25 Study the colliding students example. The following questions review this analysis.
- **(a)** What was chosen as the system to be analyzed?
- **(b)** Which components ($F_{\text{net},x}$, $F_{\text{net},y}$, or $F_{\text{net},z}$) of the net force on the system were zero during the collision?
- **(c)** What is the correct form of the x component of the Momentum Principle for this situation?
- **(d)** Which quantities were used in estimating the duration of the collision, Δt?
- **(e)** How was the force exerted by one student on the other calculated?

(f) About how large was the estimated force exerted by one student on the other, compared to the weight mg of the student?

2.8 PHYSICAL MODELS

Our model of the colliding student situation is good enough for many purposes. However, we left out some aspects of the actual motion. For example, we mostly ignored the flexible structure of the students, how much their shoes slip on the ground during the collision, and so on. We have also quite sensibly neglected the gravitational force of Mars on the students, because it is so tiny compared to the force of one student on the other.

Making and using models is an activity central to physics, and the criteria for a "good" physical model will depend on how we intend to use the model. In fact, one of the most important problems a scientist or engineer faces is deciding what interactions must be included in a model of a real physical, chemical, or biological system, and what interactions can reasonably be ignored.

Order of Magnitude Estimates

If we neglect some effects, we say that we are constructing a simplified model of the situation. A useful model should omit extraneous detail but retain the most important features of the real-world situation. When we make many estimates, our goal is often to determine the order of magnitude of the answer: is the force on a colliding student closer to 0.0001 newton or 10,000 newtons? Knowing the order of magnitude to expect in an answer can be critically important in solving a real problem, designing an experiment, or designing a crash helmet. The designer of a crash helmet needs an approximate value for the maximum force it must withstand. Actual collisions will vary, so there is no "right" or simple answer.

Idealized Models

Some models are "idealized," by which we mean they involve simple, clean, stripped-down situations, free of messy complexities (Figure 2.48). "Ideally," a ball will roll forever on a level floor, but a real ball rolling on a real floor eventually comes to a stop. An "ideal" gas is a fictitious gas in which the molecules don't interact at all with each other, as opposed to a real gas whose molecules do interact, but only when they come close to each other. A model of a single Earth orbiting a Sun is an idealized model because it leaves out the effects of the other bodies in the Solar System.

The behavior of idealized models allows us to investigate simple patterns of motion, and learn what factors are important in determining these patterns. Once we understand these factors, we can revise and extend our models, including more interactions and complexities, to see what effects these have.

An important aspect of physical modeling is that we will engage in making appropriate approximations to simplify the messy, real-world situation enough to permit (approximate) analysis using Newton's laws. Actually, using Newton's laws is itself an example of modeling and making approximations, because we are neglecting the effects of quantum mechanics and of General Relativity (Einstein's treatment of gravitation). Newton's laws are only an approximation to the way the world works, though frequently an extremely good one.

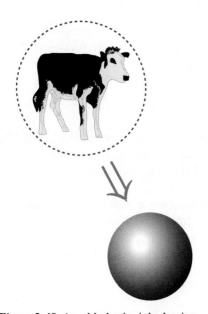

Figure 2.48 An old physics joke begins, "Consider a spherical cow ..." Sometimes this degree of idealization is actually appropriate.

*DERIVATION: SPECIAL-CASE AVERAGE VELOCITY

Here are two proofs, one geometric and one algebraic (using calculus), for the following special-case result concerning average velocity:

$$v_{\text{avg},x} = \frac{(v_{ix} + v_{fx})}{2} \ \textit{only if } v_x \text{ changes at a constant rate}$$

The results are similar for $v_{\text{avg},y}$ and $v_{\text{avg},z}$.

Geometric Proof

If $F_{\text{net},x}$ is constant, $p_{fx} = p_{ix} + F_{\text{net},x}\,\Delta t$ implies that p_x changes at a constant rate. At speeds small compared to the speed of light, $v_x \approx p_x/m$, so a graph of v_x vs. time is a straight line, as in Figure 2.49. Using this graph, we form narrow vertical slices, each of height v_x and narrow width Δt.

Within each narrow slice v_x changes very little, so the change in position during the brief time Δt is approximately $\Delta x = v_x\,\Delta t$. Therefore the change in x is approximately equal to the area of the slice of height v_x and width Δt (Figure 2.50).

If we add up the areas of all these slices, we get approximately the area under the line in Figure 2.49, and this is also equal to the total displacement $\Delta x_1 + \Delta x_2 + \Delta x_3 + \cdots = x_f - x_i$. If we go to the limit of an infinite number of slices, each with infinitesimal width, the sum of slices really *is* the area, and this area we have shown to be equal to the change in position. This kind of sum of an infinite number of infinitesimal pieces is called an "integral" in calculus.

The area under the line is a trapezoid, and from geometry we know that the area of a trapezoid is the average of the two bases times the altitude:

$$\text{area} = \frac{(\text{top} + \text{bottom})}{2}(\text{altitude})$$

Figure 2.49 Graph of v_x vs. t (constant force), divided into narrow vertical slices each of height v_x and width Δt.

Figure 2.50 One narrow slice has an area given approximately by $v_x\,\Delta t$. This is equal to Δx, the displacement of the object.

Turn Figure 2.50 on its side, as in Figure 2.51, and you see that the top and bottom have lengths v_{ix} and v_{fx}, while the altitude of the trapezoid is the total time $(t_f - t_i)$. Therefore we have the following result:

$$\text{trapezoid area } = x_f - x_i = \frac{(v_{ix} + v_{fx})}{2}(t_f - t_i)$$

Dividing by $(t_f - t_i)$, we have this:

$$\frac{x_f - x_i}{t_f - t_i} = \frac{(v_{ix} + v_{fx})}{2}$$

By definition, however, the x component of average velocity is the change in x divided by the total time, so we have proved that

$$v_{\text{avg},x} = \frac{(v_{ix} + v_{fx})}{2} \ \textit{only if } v_x \text{ changes linearly with time}$$

Figure 2.51 The area of the whole trapezoid is equal to the total displacement. The figure in Figure 2.50 has been rotated 90 degrees clockwise.

The proof depended critically on the straight-line ("linear") change in velocity, which occurs if $F_{\text{net},x}$ is constant (and $v \ll c$). Otherwise we wouldn't have a trapezoidal area. That's why the result isn't true in general; it's only true in this important but special case.

Algebraic Proof Using Calculus

An algebraic proof using calculus can also be given. We will use the x component of the derivative version of the Momentum Principle (more about this in a later chapter):

$$\Delta \vec{p} = \vec{F}_{net} \, \Delta t \text{ implies that } \frac{\Delta \vec{p}}{\Delta t} = \vec{F}_{net}$$

In the limit we have

$$\lim_{\Delta t \to 0} \frac{\Delta \vec{p}}{\Delta t} = \frac{d\vec{p}}{dt} = \vec{F}_{net} \text{ and } \frac{dp_x}{dt} = F_{net,x}$$

If $F_{net,x}$ is a constant, the time derivative of p_x is a constant, so we have

$$p_x = F_{net,x} t + p_{xi} \text{ since } p_x = p_{ix} \text{ when } t = 0$$

You can check this by taking the derivative with respect to time t, which gives the original equation $dp_x/dt = F_{net,x}$. At speeds small compared to the speed of light, $v_x \approx p_x/m$, so we can write

$$v_x = \frac{F_{net,x}}{m} t + v_{ix} \text{ since } v_x = v_{ix} \text{ when } t = 0$$

However, the x component of velocity is the rate at which x is changing:

$$v_x = \frac{dx}{dt} = \frac{F_{net,x}}{m} t + v_{ix}$$

Now the question is, can you think of a function of x that has this time derivative? Since the time derivative of t^2 is $2t$, the following formula for x has the appropriate derivative:

$$x = \frac{1}{2} \frac{F_{net,x}}{m} t^2 + v_{ix} t + x_i \text{ since } x = x_i \text{ when } t = 0$$

You can check this by taking the derivative with respect to t, which gives the equation for v_x, since $d(\frac{1}{2} t^2)/dt = t$ and $d(t)/dt = 1$.

The average velocity that we seek is the change in position divided by the total time:

$$v_{avg,x} = \frac{x - x_i}{t} = \frac{1}{2} \frac{F_{net,x}}{m} t + v_{ix} = \frac{1}{2} (v_{fx} - v_{ix}) + v_{xi}$$

where we have used the equation we previously derived for the velocity:

$$v_{fx} = v_x = \frac{F_{net,x}}{m} t + v_{ix}$$

Simplifying the expression for $v_{avg,x}$ we have the proof:

$$v_{avg,x} = \frac{(v_{ix} + v_{fx})}{2} \text{ only if } v_x \text{ changes at a constant rate}$$

| 2.10 | *INERTIAL FRAMES |

Newton's first law is valid only in an "inertial frame" of reference, one in uniform motion (or at rest) with respect to the pervasive "cosmic microwave background" (see the optional discussion at the end of Chapter 1). Since the Momentum Principle is a quantitative version of Newton's first law, we expect the Momentum Principle to be valid in an inertial reference frame, but not in a reference frame that is not in uniform motion. Let's check that this is true.

If you view some objects from a space ship that is moving uniformly with velocity $\vec{v}_s$ with respect to the cosmic microwave background, all of the velocities of those objects have the constant $\vec{v}_s$ subtracted from them, as far as you are concerned. For example, a rock moving at the same velocity as your spacecraft would have $\vec{v}_{\text{rock}} = (\vec{v} - \vec{v}_s) = \vec{0}$ in your reference frame: it would appear to be stationary as it coasted along beside your spacecraft.

With a constant space ship velocity, we have $\Delta \vec{v}_s = \vec{0}$, and the change of momentum of the moving object reduces to the following (for speeds small compared to c):

$$\Delta[m(\vec{v} - \vec{v}_s)] = \Delta(m\vec{v}) - \Delta(m\vec{v}_s) = \Delta(m\vec{v}) \text{ since}$$
$$\Delta(m\vec{v}_s) = \vec{0}$$

Therefore,

$$\Delta[m(\vec{v} - \vec{v}_s)] = \Delta(m\vec{v}) = \vec{F}_{\text{net}}\, \Delta t$$

If the velocity $\vec{v}_s$ of the space ship doesn't change (it represents an inertial frame of reference), the form (and validity) of the Momentum Principle is unaffected by the motion of the space ship.

However, if your space ship increases its speed, or changes direction, $\Delta \vec{v}_s \neq \vec{0}$, an object's motion relative to you changes without any force acting on it. In that case the Momentum Principle is not valid for the object, because you are not in an inertial frame. Although the Earth is not an inertial frame because it rotates, and goes around the Sun, it is close enough to being an inertial frame for many everyday purposes.

2.11 | *MEASUREMENTS AND UNITS

Using the Momentum Principle requires a consistent way to measure length, time, mass, and force, and a consistent set of units. We state the definitions of the standard Système Internationale (SI) units, and we briefly discuss some subtle issues underlying this choice of units.

Units: Meters, Seconds, Kilograms, Coulombs, and Newtons

Originally the meter was defined as the distance between two scratches on a platinum bar in a vault in Paris, and a second was 1/86,400th of a "mean solar day." Now, however, the second is defined in terms of the frequency of light emitted by a cesium atom, and the meter is defined as the distance light travels in 1/299,792,458th of a second, or about 3.3×10^{-9} seconds (3.3 nanoseconds). The speed of light is defined to be exactly 299,792,458 m/s (very close to 3×10^8 m/s). As a result of these modern redefinitions, it is really speed (of light) and time that are the internationally agreed-upon basic units, not length and time.

By international agreement, one kilogram is the mass of a platinum block kept in that same vault in Paris. As a practical matter, other masses are compared to this standard kilogram by using a balance-beam or spring weighing scale (more about this in a moment). The newton, the unit of force, is defined as that force which acting for 1 second imparts to 1 kilogram a velocity change of 1 m/s. We could make a scale for force by calibrating the amount of stretch of a spring in terms of newtons.

The coulomb, the SI unit of electric charge, is defined in terms of electric currents. The charge of a proton is 1.6×10^{-19} coulomb.

Some Subtle Issues

What we have just said about SI units is sufficient for practical purposes to predict the motion of objects, but here are some questions that might bother you. Is it legitimate to measure the mass that appears in the Momentum Principle by seeing how that mass is affected by gravity on a balance-beam scale? Is it legitimate to use the Momentum Principle to define the units of force, when the concept of force is itself associated with the same law? Is this all circular reasoning, and the Momentum Principle merely a definition with little content? Here is a chain of reasoning that addresses these issues.

Measuring Inertial Mass

When we use balance-beam or spring weighing scales to measure mass, what we're really measuring is the "gravitational mass"— that is, the mass that appears in the law of gravitation and is a measure of how much this object is affected by the gravity of the Earth. In principle, this "gravitational mass" could be different from the so-called "inertial mass"— the mass that appears in the definition of momentum. It is possible to compare the inertial masses of two objects, and we find experimentally that inertial and gravitational mass seem to be entirely equivalent.

Here is a way to compare two inertial masses directly, without involving gravity. Starting from rest, pull on the first object with a spring stretched by some amount s for an amount of time Δt, and measure the increase of speed Δv_1. Then, starting from rest, pull on the second object with the same spring stretched by the same amount s for the same amount of time Δt, and measure the increase of speed Δv_2. We *define* the ratio of the inertial masses as $m_1/m_2 = \Delta v_2/\Delta v_1$. Since one of these masses could be the standard kilogram kept in Paris, we now have a way of measuring inertial mass in kilograms. Having defined inertial mass this way, we find experimentally that the Momentum Principle is obeyed by both of these objects in all situations, not just in the one special experiment we used to compare the two masses.

Moreover, we find to extremely high precision that the inertial mass in kilograms measured by this comparison experiment is exactly the same as the gravitational mass in kilograms obtained by comparing with a standard kilogram on a balance-beam scale (or using a calibrated spring scale), and that it doesn't matter what the objects are made of (wood, copper, glass, etc.). This justifies the convenient use of ordinary weighing scales to determine inertial mass.

Is This Circular Reasoning?

The definitions of force and mass may sound like circular reasoning, and the Momentum Principle may sound like just a kind of definition, with no real content, but there is real power in the Momentum Principle. Forget for a moment the definition of force in newtons and mass in kilograms. The experimental fact remains that any object if subjected to a single force by a spring with constant stretch experiences a change of momentum (and velocity) proportional to the duration of the interaction. Note that it is not a change of *position* proportional to the time (that would be a constant speed), but a change of *velocity*. That's real content. Moreover, we find that the change of velocity is proportional to the amount of stretch of the spring. That too is real content.

Then we find that a different object undergoes a different rate of change of velocity with the same spring stretch, but after we've made one single comparison experiment to determine the mass relative to the standard kilogram, the Momentum Principle works in all situations. That's real content.

Finally come the details of setting standards for measuring force in newtons and mass in kilograms, and we use the Momentum Principle in helping set these standards. Logically, however, this comes after having established the law itself.

SUMMARY

The Momentum Principle

$$\Delta \vec{p} = \vec{F}_{\text{net}} \, \Delta t \text{ (for a short enough time interval } \Delta t)$$

$$\text{Update form: } \vec{p}_f = \vec{p}_i + \vec{F}_{\text{net}} \, \Delta t$$

To analyze the motion of a real-world system, several steps are required:

1. Momentum Principle
   Choose a system, consisting of a portion of the Universe.
   Make a list of objects in the surroundings that exert significant forces on the chosen system, and make a labeled "free-body" diagram showing the external forces exerted by the objects in the surroundings. Apply the Momentum Principle to the chosen system:

$$\vec{p}_f = \vec{p}_i + \vec{F}_{\text{net}} \, \Delta t$$

   For each term in the Momentum Principle, substitute any values you know.

2. Apply the position update formula, if necessary:

$$\vec{r}_f = \vec{r}_i + \vec{v}_{\text{avg}} \, \Delta t$$

3. Check for reasonableness (units, direction, magnitude, etc.).

In many real-world situations the Momentum Principle and position update must be carried out iteratively. If many steps must be taken, this is feasible only by having a computer carry out the iterations. In a computer iteration it is important to choose a time step short enough that the force and velocity do not change very much during each step.

System is a portion of the Universe acted on by the *surroundings*.

Force is a quantitative measure of interactions; units are newtons.

Impulse is the product of force times time $\vec{F}\Delta t$; momentum change equals net impulse (the impulse due to the net force).

Physical models are tractable approximations/idealizations of the real world.

Special case result for average velocity:

$$v_{\text{avg},x} = \frac{(v_{ix} + v_{fx})}{2} \text{ only if } v_x \text{ changes linearly with time}$$

Results are similar for $v_{\text{avg},y}$ and $v_{\text{avg},z}$.

EXERCISES AND PROBLEMS

Section 2.2

2.X.26 An object is moving in the $+x$ direction. Which, if any, of the following statements do you know must be false?

  (a) The net force on the object is in the $+x$ direction.

  (b) The net force on the object is in the $-x$ direction.

  (c) The net force on the object is zero.

2.X.27 A system is acted upon by two forces, $\langle 18, 47, -23 \rangle$ N, and $\langle -20, -13, 41 \rangle$ N. What is the net force acting on the system?

2.X.28 A ball moves in the direction of the arrow labeled c in Figure 2.52. The ball is struck by a stick that briefly exerts a force

on the ball in the direction of the arrow labeled e. Which arrow best describes the direction of $\Delta \vec{p}$, the change in the ball's momentum?

j zero magnitude

Figure 2.52 Exercise 2.X.28

2.X.29 A truck driver slams on the brakes and the momentum of the truck changes from $\langle 65,000, 0, 0 \rangle$ kg·m/s to $\langle 26,000, 0, 0 \rangle$ kg·m/s in 4.1 seconds due to a constant force of the road on the wheels of the truck. As a vector, write the net force exerted on the truck by the surroundings.

2.X.30 An object is moving in the $+y$ direction. Which, if any, of the following statements might be true? Check all that apply.

(a) The net force on the object is zero.

(b) The net force on the object is in the $-y$ direction.

(c) The net force on the object is in the $+y$ direction.

2.X.31 You observe three carts moving to the left. Cart A moves to the left at nearly constant speed. Cart B moves to the left, gradually speeding up. Cart C moves to the left, gradually slowing down. Which cart or carts, if any, experience a net force to the left?

2.X.32 Check all of the following that are correct statements. Read each statement very carefully to make sure that it is exactly correct.

(a) The Momentum Principle can be written $\vec{p}_f = \vec{p}_i + \vec{F}\Delta t$.

(b) The Momentum Principle can be written $\vec{p}_f = \vec{p}_i + \vec{F}_{net}\Delta t$.

(c) At speeds small compared to the speed of light, momentum is approximately equal to $m\vec{v}$.

(d) The Momentum Principle can be written $\Delta \vec{p} = \vec{F}_{net}\Delta t$.

(e) The definition of momentum is $\vec{p} = \dfrac{1}{\sqrt{1 - (|\vec{v}|/c)^2}}m\vec{v}$.

(f) The definition of momentum is $\vec{p} = \dfrac{1}{\sqrt{1 - (|\vec{v}|/c)^2}}mv$.

(g) The relationship among position, velocity, and time is $\vec{r}_f = \vec{r}_i + \vec{v}_{avg}\Delta t$.

(h) The definition of momentum is $\vec{p} = \vec{F}_{net}\Delta t$.

(i) The relationship among position, velocity, and time is $\vec{r}_f = \vec{r}_i + v_{avg}\Delta t$.

(j) The Momentum Principle can be written $\vec{p}_f = \vec{p}_i + F_{net}\Delta t$.

2.X.33 At a certain instant a particle is moving in the $+x$ direction with momentum $+8$ kg·m/s. During the next 0.13 seconds a constant force acts on the particle, with $F_x = -7$ N and $F_y = +5$ N. What is the *magnitude* of the momentum of the particle at the end of this 0.13-second interval?

2.X.34 At $t = 16.0$ seconds an object with mass 4 kg was observed to have a velocity of $\langle 9, 29, -10 \rangle$ m/s. At $t = 16.2$ seconds its velocity was $\langle 18, 20, 25 \rangle$ m/s. What was the average (vector) net force acting on the object?

2.X.35 A proton has mass 1.7×10^{-27} kg. What is the magnitude of the impulse required to increase its speed from $0.990c$ to $0.994c$?

2.X.36 In order to pull a sled across a level field at constant velocity you have to exert a constant force. Doesn't this violate Newton's first and second laws of motion, which imply that no force is required to maintain a constant velocity? Explain this seeming contradiction.

2.P.37 A space shuttle is in a circular orbit near the Earth. An astronaut floats in the middle of the shuttle, not touching the walls. On a diagram, draw and label

(a) the momentum $\vec{p}_1$ of the astronaut at this instant;

(b) all of the forces (if any) acting on the astronaut at this instant;

(c) the momentum $\vec{p}_2$ of the astronaut a short time Δt later;

(d) the momentum change (if any) $\Delta \vec{p}$ in this time interval.

(e) Why does the astronaut seem to "float" in the shuttle?

It is ironic that we say the astronaut is "weightless" despite the fact that the only force acting on the astronaut is the astronaut's weight (that is, the gravitational force of the Earth on the astronaut).

2.P.38 A proton (mass 1.7×10^{-27} kg) interacts electrically with a neutral HCl molecule located at the origin. At a certain time t, the proton's position is $\langle 1.6 \times 10^{-9}, 0, 0 \rangle$ m and the proton's velocity is $\langle 3600, 600, 0 \rangle$ m/s. The force exerted on the proton by the HCl molecule is $\langle -1.12 \times 10^{-11}, 0, 0 \rangle$ N. At a time $t + 3.4 \times 10^{-14}$ s, what is the approximate velocity of the proton? (You may assume that the force was approximately constant during this interval.)

2.P.39 Suppose that you are navigating a spacecraft far from other objects. The mass of the spacecraft is 1.5×10^5 kg (about 150 tons). The rocket engines are shut off, and you're coasting along with a constant velocity of $\langle 0, 20, 0 \rangle$ km/s. As you pass the location $\langle 12, 15, 0 \rangle$ km you briefly fire side thruster rockets, so that your spacecraft experiences a net force of $\langle 6 \times 10^4, 0, 0 \rangle$ N for 3.4 s. The ejected gases have a mass that is small compared to the mass of the spacecraft. You then continue coasting with the rocket engines turned off. Where are you an hour later? Also, what approximations or simplifying assumptions did you have to make in your analysis? Think about the choice of system: what are the surroundings that exert external forces on your system?

2.P.40 SLAC, the Stanford Linear Accelerator Center, located at Stanford University in Palo Alto, California, accelerates electrons through a vacuum tube two miles long (it can be seen from an overpass of the Junipero Serra freeway that goes right over the accelerator). Electrons that are initially at rest are subjected to a continuous force of 2×10^{-12} newton along the entire length of two miles (one mile is 1.6 kilometers) and reach speeds very near the speed of light.

(a) Determine how much time is required to increase the electrons' speed from $0.93c$ to $0.99c$. (That is, the quantity $|\vec{v}|/c$ increases from 0.93 to 0.99.)

(b) Approximately how far does the electron go in this time? What is approximate about your result?

2.P.41 A Ping-Pong ball is acted upon by the Earth, air resistance, and a strong wind. Here are the positions of the ball at several times.

Early time interval:

At $t = 12.35$ s, the position was $\langle 3.17, 2.54, -9.38 \rangle$ m.
At $t = 12.37$ s, the position was $\langle 3.25, 2.50, -9.40 \rangle$ m.

Late time interval:

At $t = 14.35$ s, the position was $\langle 11.25, -1.50, -11.40 \rangle$ m.
At $t = 14.37$ s, the position was $\langle 11.27, -1.86, -11.42 \rangle$ m.

(a) In the early time interval, from $t = 12.35$ s to $t = 12.37$ s, what was the average momentum of the ball? The mass of the Ping-Pong ball is 2.7 grams (2.7×10^{-3} kg). Express your result as a vector.

(b) In the late time interval, from $t = 14.35$ s to $t = 14.37$ s, what was the average momentum of the ball? Express your result as a vector.

(c) In the time interval from $t = 12.35$ s (the start of the early time interval) to $t = 14.35$ s (the start of the late time interval), what was the average net force acting on the ball? Express your result as a vector.

2.P.42 A small space probe, of mass 240 kg, is launched from a spacecraft near Mars. It travels toward the surface of Mars, where it will land. At a time 20.7 seconds after it is launched, the probe is at the location $\langle 4.30 \times 10^3, 8.70 \times 10^2, 0 \rangle$ m, and at this same time its momentum is $\langle 4.40 \times 10^4, -7.60 \times 10^3, 0 \rangle$ kg·m/s. At this instant, the net force on the probe due to the gravitational pull of Mars plus the air resistance acting on the probe is $\langle -7 \times 10^3, -9.2 \times 10^2, 0 \rangle$ N.

  (a) Assuming that the net force on the probe is approximately constant over this time interval, what is the momentum of the probe 20.9 seconds after it is launched?

  (b) What is the location of the probe 20.9 seconds after launch?

2.P.43 In outer space a rock of mass 5 kg is acted on by a constant net force $\langle 29, -15, 40 \rangle$ N during a 4 s time interval. At the end of this time interval the rock has a velocity of $\langle 114, 94, 112 \rangle$ m/s. What was the rock's velocity at the beginning of the time interval?

Section 2.3

2.X.44 You lift a heavy box. We'll consider this process for different choices of system and surroundings.

  (a) Choose the box as the system of interest. What objects in the surroundings exert significant forces on this system?

  (b) Choose you and the box as the system of interest. What objects in the surroundings exert significant forces on this system?

  (c) Choose you, the box, and the Earth as the system of interest. What objects in the surroundings exert significant forces on this system?

2.X.45 For each graph of v_x vs. t numbered 1–6 in Figure 2.53, choose the letter (a–i) corresponding to the appropriate description of motion of a fan cart moving along a track. Not all descriptions will be used. Assume the usual coordinate system ($+x$ to the right, $+y$ up, $+z$ out of the page).

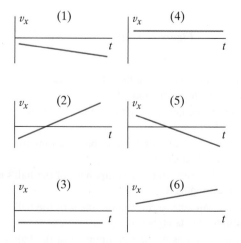

Figure 2.53 Graphs of motion of a fan cart.

  (a) A cart moves to the left, gradually slowing down.
  (b) A cart moves to the right, gradually speeding up.
  (c) A cart moves to the left at constant speed.
  (d) A cart moves to the left, gradually slowing down, stops, and moves to the right, speeding up.
  (e) A cart remains stationary and does not move.
  (f) A cart moves to the right, gradually slowing down.
  (g) A cart moves to the right, gradually slowing down, stops, and moves to the left, speeding up.
  (h) A cart moves to the left, gradually speeding up.
  (i) A cart moves to the right at constant speed.

Section 2.5

2.X.46 The stiffness of a particular spring is 40 N/m. One end of the spring is attached to a wall. When you pull on the other end of the spring with a steady force of 2 N, the spring elongates to a total length of 18 cm. What was the relaxed length of the spring? (Remember to convert to SI units.)

2.X.47 Near the surface of the Earth, what is the magnitude of the gravitational force acting on a 10 kg mass? Acting on a 20 kg mass?

Section 2.5

2.X.48 A comet passes near the Sun. When the comet is closest to the Sun, it is 9×10^{10} m from the Sun. You need to choose a time step to use in predicting the comet's motion. Which of the following would be a reasonable distance for the comet to move in one time step, doing an iterative calculation *by hand*?
(a) 1×10^2 m (b) 1×10^{10} m (c) 1×10^{11} m (d) 1×10^9 m

2.X.49 In a lab experiment you observe that a pendulum swings with a "period" (time for one round trip) of 2 s. In an iterative calculation of the motion, which of the following would NOT be a reasonable choice for Δt, for either hand or computer iterative calculations?
(a) 1 second (b) 0.1 second (c) 0.05 second (d) 0.01 second

Section 2.6

2.X.50 (a) A runner starts from rest and in 3 s reaches a speed of 8 m/s. If we assume that the speed changed at a constant rate (constant net force), what was the average speed during this 3 s interval?
(b) How far did the runner go in this 3 s interval?
(c) The driver of a car traveling at a speed of 18 m/s slams on the brakes and comes to a stop in 4 s. If we assume that the speed changed at a constant rate (constant net force), what was the average speed during this 4 s interval?
(d) How far did the car go in this 4 s interval?

2.X.51 On a straight road with the $+x$ axis chosen to point in the direction of motion, you drive for 3 hours at a constant 30 miles per hour, then in a few seconds you speed up to 60 miles per hour and drive at this speed for 1 hour.

  (a) What was the x component of average velocity for the 4-hour period, using the fundamental definition of average velocity, which is the displacement divided by the time interval?

  (b) Suppose that instead you use the formula $v_{avg,x} = (v_{ix} + v_{fx})/2$. What do you calculate for the x component of average velocity?

(c) Why does the formula used in part (b) give the wrong answer?

2.X.52 You throw a ball. Assume that the origin is on the ground, with the $+y$ axis pointing upward. Just after the ball leaves your hand its position is $\langle 0.06, 1.03, 0 \rangle$ m. The average velocity of the ball over the next 0.7 s is $\langle 17, 4, 6 \rangle$ m/s. At time 0.7 s after the ball leaves your hand, what is the height of the ball above the ground?

2.X.53 A cart rolls with low friction on a track. A fan is mounted on the cart, and when the fan is turned on, there is a constant force acting on the cart. Three different experiments are performed:

(a) Fan off: The cart is originally at rest. You give it a brief push, and it coasts a long distance along the track in the $+x$ direction, slowly coming to a stop.

(b) Fan forward: The fan is turned on, and you hold the cart stationary. You then take your hand away, and the cart moves forward, in the $+x$ direction. After traveling a long distance along the track, you quickly stop and hold the cart.

(c) Fan backward: The fan is turned on facing the "wrong" way, and you hold the cart stationary. You give it a brief push, and the cart moves forward, in the $+x$ direction, slowing down and then turning around, returning to the starting position, where you quickly stop and hold the cart.

Figure 2.54 displays four graphs of p_x (numbered 1–4), the x component of momentum, vs. time. The graphs start when the cart is at rest, and end when the cart is again at rest. Match the experiment with the correct graph.

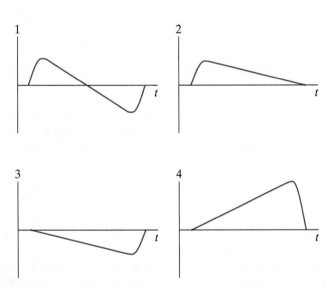

Figure 2.54 Exercise 2.X.53

2.X.54 Consider the three experiments described in Exercise 2.X.53. Figure 2.55 displays four graphs of $F_{net,x}$, the x component of the net force acting on the cart, vs. time. The graphs start when the cart is at rest, and end when the cart is again at rest. Match the experiment with the graph.

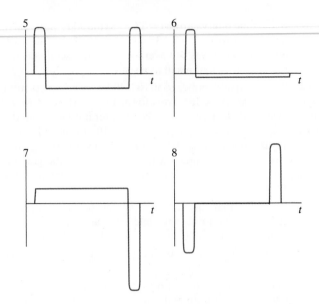

Figure 2.55 Exercise 2.X.54

2.X.55 Consider the three experiments described in Exercise 2.X.53. Figure 2.56 displays four graphs of x, positioned along the track, vs. time. The graphs start when the cart is at rest, and end when the cart is again at rest. Match the experiment with the graph.

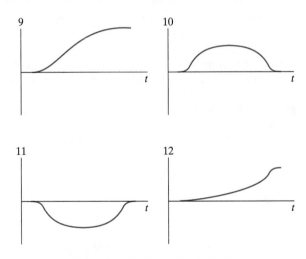

Figure 2.56 Exercise 2.X.55

2.X.56 A ball of mass 0.4 kg flies through the air at low speed, so that air resistance is negligible.

(a) What is the net force acting on the ball while it is in motion?

(b) Which components of the ball's momentum will be changed by this force?

(c) What happens to the x component of the ball's momentum during its flight?

(d) What happens to the y component of the ball's momentum during its flight?

(e) What happens to the z component of the ball's momentum during its flight?

(f) In this situation, why is it legitimate to use the formula for average y component of velocity, $v_{avg,y} = (v_{iy} + v_{fy})/2$, to update the y component of position?

2.P.57 A soccer ball is kicked at an angle of 59 to the horizontal with an initial speed of 20 m/s. Assume for the moment that we can neglect air resistance.

(a) For how much time is the ball in the air?

(b) How far does it go (horizontal distance along the field)?

(c) How high does it go?

2.P.58 A ball is kicked from a location $\langle 9, 0, -6 \rangle$ (on the ground) with initial velocity $\langle -11, 16, -6 \rangle$ m/s. The ball's speed is low enough that air resistance is negligible.

(a) What is the velocity of the ball 0.5 seconds after being kicked? (Use the Momentum Principle!)

(b) In this situation (constant force), which velocity will give the most accurate value for the location of the ball 0.5 seconds after it is kicked: the arithmetic average of the initial and final velocities, the final velocity of the ball, or the initial velocity of the ball?

(c) What is the average velocity of the ball over this time interval (a vector)?

(d) Use the average velocity to find the location of the ball 0.5 seconds after being kicked.

Now consider a different time interval: the interval between the initial kick and the moment when the ball reaches its highest point. We want to find how long it takes for the ball to reach this point, and how high the ball goes.

(e) What is the y-component of the ball's velocity at the instant when the ball reaches its highest point (the end of this time interval)?

(f) Fill in the known quantities in the update form of the Momentum Principle, $mv_{yf} = mv_{yi} + F_{net,y} \, \Delta t$, leaving as symbols anything that is unknown.

(g) How long does it take for the ball to reach its highest point?

(h) Knowing this time, first find the y component of the average velocity during *this* time interval, then use it to find the maximum height attained by the ball.

Now take a moment to reflect on the reasoning used to solve this problem. You should be able to do a similar problem on your own, without prompting. Note that the only equations needed were the Momentum Principle and the expression for the arithmetic average velocity.

2.P.59 A small dense ball with mass 1.5 kg is thrown with initial velocity $\langle 5, 8, 0 \rangle$ m/s at time $t = 0$ at a location we choose to call the origin ($\langle 0, 0, 0 \rangle$ m). Air resistance is negligible.

(a) When the ball reaches its maximum height, what is its velocity (a vector)? It may help to make a simple diagram.

(b) When the ball reaches its maximum height, what is t? You know how v_y depends on t, and you know the initial and final velocities.

(c) Between the launch at $t = 0$ and the time when the ball reaches its maximum height, what is the average velocity (a vector)? You know how to determine average velocity when velocity changes at a constant rate.

(d) When the ball reaches its maximum height, what is its location (a vector)? You know how average velocity and displacement are related.

(e) At a later time the ball's height y has returned to zero, which means that the average value of v_y from $t = 0$ to this time is zero. At this instant, what is the time t?

(f) At the time calculated in part (e), when the ball's height y returns to zero, what is x? (This is called the "range" of the trajectory.)

(g) At the time calculated in part (e), when the ball's height y returns to zero, what is v_y?

(h) What was the angle to the x axis of the initial velocity?

(i) What was the angle to the x axis of the velocity at the time calculated in part (e), when the ball's height y returned to zero?

2.P.60 Determine two different possible ways for a player to make a free throw in basketball. In both cases give the initial speed, initial angle, and initial height of the basketball. The rim of the basket is 10 feet (3.0 m) above the floor. It is 14 feet (4.3 m) along the floor from the free-throw line to a point directly below the center of the basket.

2.P.61 You have probably seen a basketball player throw the ball to a teammate at the other end of the court, 30 m away. Estimate a reasonable initial angle for such a throw, and then determine the corresponding initial speed. For your chosen angle, how long does it take for the basketball to go the length of the court? What is the highest point along the trajectory, relative to the thrower's hand?

2.P.62 You are a detective investigating why someone was hit on the head by a falling flowerpot. One piece of evidence is a home video taken in a 4th-floor apartment, which happens to show the flowerpot falling past a tall window. Inspection of individual frames of the video shows that in a span of 6 frames the flowerpot falls a distance that corresponds to 0.85 of the window height seen in the video. (*Note:* standard video runs at a rate of 30 frames per second.) You visit the apartment and measure the window to be 2.2 m high. What can you conclude? Under what assumptions? Give as much detail as you can.

2.P.63 In a cathode ray tube (CRT) used in older television sets, a beam of electrons is steered to different places on a phosphor screen, which glows at locations hit by electrons. The CRT is evacuated, so there are few gas molecules present for the electrons to run into. Electric forces are used to accelerate electrons of mass m to a speed $v_0 \ll c$, after which they pass between positively and negatively charged metal plates that deflect the electron in the vertical direction (upward in Figure 2.57, or downward if the sign of the charges on the plates is reversed).

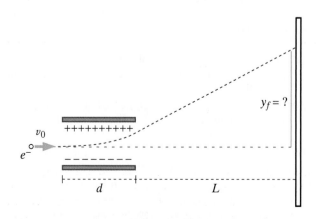

Figure 2.57 Problem 2.P.63: a cathode ray tube.

While an electron is between the plates, it experiences a uniform vertical force F, but when the electron is outside the plates there is negligible force on it. The gravitational force on the electron is negligibly small compared to the electric force in this situation. The length of the metal plates is d, and the phosphor screen is a distance L from the metal plates. Where does the electron hit the screen? (That is, what is y_f?)

2.P.64 Apply the general results obtained in the full analysis of motion under the influence of a constant force in Section 2.6 to answer the following questions. You hold a small metal ball of mass m a height h above the floor. You let go, and the ball falls to the floor. Choose the origin of the coordinate system to be on the floor where the ball hits, with y up as usual. Just after release, what are y_i and v_{iy}? Just before hitting the floor, what is y_f? How much time Δt does it take for the ball to fall? What is v_{fy} just before hitting the floor? Express all results in terms of m, g, and h. How would your results change if the ball had twice the mass?

Section 2.7

2.X.65 This is a review of the discussion of the colliding students. Make the simplifying assumption that the two students have the same mass and the same speed. Check all the statements that are correct:

(a) We're interested in the force exerted on one student, so we apply the Momentum Principle to a system consisting of one student.

(b) A good estimate of the initial speed of one of the running students is 0.1 m/s.

(c) The initial momentum of the one-student system is nonzero.

(d) The Momentum Principle can't be applied in this analysis because we have no way of estimating Δt.

(e) The final momentum of the one-student system is nonzero.

(f) We can estimate Δt from the compression distance in the collision.

2.P.66 A steel safe with mass 2200 kg falls onto concrete. Just before hitting the concrete its speed is 40 m/s, and it smashes without rebounding and ends up being 0.06 m shorter than before. What is the approximate magnitude of the force exerted on the safe by the concrete? How does this compare with the gravitational force of the Earth on the safe? Explain your analysis carefully, and justify your estimates on physical grounds.

2.P.67 In a crash test, a truck with mass 2500 kg traveling at 24 m/s smashes head-on into a concrete wall without rebounding. The front end crumples so much that the truck is 0.72 m shorter than before.

(a) What is the average speed of the truck during the collision (that is, during the interval between first contact with the wall and coming to a stop)?

(b) About how long does the collision last? (That is, how long is the interval between first contact with the wall and coming to a stop)?

(c) What is the magnitude of the average force exerted by the wall on the truck during the collision?

(d) It is interesting to compare this force to the weight of the truck. Calculate the ratio of the force of the wall to the gravitational force mg on the truck. This large ratio shows why a collision is so damaging.

(e) What approximations did you make in your analysis?

2.P.68 A 0.7 kg block of ice is sliding by you on a very slippery floor at 2.5 m/s. As it goes by, you give it a kick perpendicular to its path. Your foot is in contact with the ice block for 0.003 seconds. The block eventually slides at an angle of 22 degrees from its original direction. The overhead view shown in Figure 2.58 is approximately to scale. The arrow represents the average force your toe applies briefly to the block of ice.

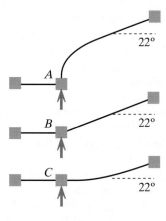

Figure 2.58

(a) Which of the possible paths shown in the diagram corresponds to the correct overhead view of the block's path?

(b) Which components of the block's momentum are changed by the impulse applied by your foot? (Check all that apply. The diagram shows a top view, looking down on the xz plane.)

(c) What is the unit vector $\hat{p}$ in the direction of the block's momentum after the kick?

(d) What is the x component of the block's momentum after the kick?

(e) Remember that $\vec{p} = |\vec{p}|\hat{p}$. What is the magnitude of the block's momentum after the kick?

(f) Use your answers to the preceding questions to find the z-component of the block's momentum after the kick (drawing a diagram is helpful).

(g) What was the magnitude of the average force you applied to the block?

2.P.69 A tennis ball has a mass of 0.057 kg. A professional tennis player hits the ball hard enough to give it a speed of 50 m/s (about 120 miles per hour). The ball hits a wall and bounces back with almost the same speed (50 m/s). As indicated in Figure 2.59, high-speed photography shows that the ball is crushed 2 cm (0.02 m) at the instant when its speed is momentarily zero, before rebounding.

Making the very rough approximation that the large force that the wall exerts on the ball is approximately constant during contact, determine the approximate magnitude of this force. *Hint:* Think about the approximate amount of time it takes for the ball to come momentarily to rest. (For comparison note that the gravitational force on the ball is quite small, only about $(0.057 \text{ kg})(9.8 \text{ N/kg}) \approx 0.6$ N. A force of 5 N is approximately the same as a force of one pound.)

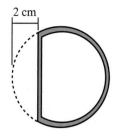

2 cm

Figure 2.59 Problem 2.P.69: A high-speed tennis ball deforms when it hits a wall.

Computational Problems

For guidance on these problems, refer to the discussion of how to structure iterative calculations in a computational model in Section 2.5.

2.P.70 Write a computer program to model and animate the motion of a low-friction fan cart, moving under the influence of a constant force. You will use the working program to explore the effect of varying the initial position, initial momentum, and the force on the cart due to the air. Construct your program so that it produces both an animation of the cart's motion and a graph of x vs. time. A typical fan cart used in instructional physics laboratories has a mass of about 500 grams. The track should be 2 meters long, and oriented along the x axis. It is convenient to position the track so the origin is at the left end.

(a) Zero net force

(1) Start the cart at the origin. Give the cart an initial velocity such that it will travel the full length of the track in 6 seconds.

(2) Use the position update formula with a time step Δt small enough to produce a smooth animation, set the net force to $\vec{0}$, and animate the motion of the cart as it travels from one end of the track to the other.

(3) Check to make sure the displayed animation makes sense, and that the graph of x vs. t matches the motion.

(b) Constant net force in the $+x$ direction

(1) Predict how the motion of the cart and the graph of x vs. t will change if a constant net force to the right is applied to the cart.

(2) Start the cart at the origin with the same initial velocity as in the previous part, but with a nonzero net force of $\langle 0.2, 0, 0 \rangle$ N. Iteratively apply the Momentum Principle, using this force, to animate the motion of the cart and produce the graph of x vs. t. Compare your results to your predictions.

(c) Turning around

(1) Predict what the graph of x vs. t would look like if the cart were to slow down, stop, and head back the other direction.

(2) Is it possible to apply a constant net force to the cart that will make the cart do this? If so, what would be the direction of this force?

(3) In your program start the cart with the same initial velocity as in the previous part. Empirically (by trial and error) determine the constant net force required to make the cart slow down, turn around just before

reaching the end of the track, and head back in the opposite direction.

(d) Motion in the $-x$ direction
What do you need to change in your program in order to start the cart from the right end of the track, heading to the left, and have it slow down, come to a stop, and head back the other way? Try this.

2.P.71 Write a program to model the motion of the vertical mass–spring system discussed in the example in Section 2.5. Use the parameters (mass, spring stiffness, etc.) given in the example, and compute the motion of the system over a total time of 1 second. Animate the motion, and also make a graph of y vs t.

(a) First use a time step of 0.1 s, as is done in the example. Compare your graph to the graph in Figure 2.29.

(b) Next use a time step of 0.01 s. Compare your graph to Figure 2.30.

(c) Try using a time step of 0.001 s. Is this calculation significantly better? How can you tell?

(d) From your graph you can find the period of this oscillatory motion. Use your computer model to determine whether and how the period of the motion is changed if the initial compression of the spring is decreased.

2.P.72 Write a program to model the motion of a horizontal mass–spring system, in which one end of the spring is attached to a wall and the other end to a 25 gram block that slides on a surface with negligible friction. The relaxed length of the spring is 50 cm, and the spring stiffness is 1.3 N/m. Your program should animate the motion of the system as well as producing a graph of x vs. t for the block.

(a) Do you expect the period of the oscillations of the system to depend on whether the spring is initially stretched or compressed? Make a prediction and explain your reasoning.

(b) In your program start the system from rest, with a stretch of 14 cm. How does the system behave?

(c) Now start the system from rest, with a stretch of -14 cm (in other words, the spring is compressed). If the behavior of the system is different, describe the differences.

(d) Start the system with the spring completely relaxed. What do you need to change in your program to make the system oscillate?

(e) Start the system with the spring compressed and with the block already in motion. What do you observe?

2.P.73 Write a program to model the motion of a 20 gram mass attached to a spring hung vertically from a support. The spring stiffness is 0.9 N/m and the relaxed length of the spring is 20 cm. Use the full vector form of the spring force in your computation. Have the moving mass leave a trail, so it is possible to track its motion.

(a) What do you expect to see if you start the system from rest, with the spring stretched 10 cm straight downward? Run your program and compare your observations to your prediction.

(b) What do you expect to see if you start the system from rest, with the spring stretched in both the x and y directions, as in Figure 2.22? Set the initial conditions appropriately, run your program, and compare your observations to your prediction.

(c) What initial conditions would be necessary to get the mass to move in 3 D (not just in a single plane)? Try this, using the trail left by the moving mass to allow you to see the path of the mass over time.

A N S W E R S   T O   E X E R C I S E S

2.X.1 $\langle 0.1, 0.3, 0.4 \rangle$ kg·m/s

2.X.2 $\langle 60, -60, 0 \rangle$ N

2.X.3 1.5 kg·m/s

2.X.4 $\langle -60, -24, 96 \rangle$ N·s; $\langle -60, -24, 96 \rangle$ kg·m/s

2.X.5 $\langle 0, -2, 0 \rangle$ N·s; $\langle 0, -0.67, 0 \rangle$ N

2.X.6 $\langle 10, 0, 13 \rangle$ kg·m/s

2.X.7 (a) $\langle -15000, 0, 3000 \rangle$ kg·m/s; (b) $\langle -15000, 0, 3000 \rangle$ N·s; (c) $\langle -5000, 0, 1000 \rangle$ N

2.X.8 $\langle 9.4, 0.3, 0 \rangle$ kg·m/s

2.X.9 250 km; 50 km/hr; v does not change linearly with time

2.X.10 35 m/s; 350 m

2.X.11 b

2.X.12 0.275 N

2.X.13 (a) 3571 N/m; (b) 71.4 N

2.X.14 (a) $\langle 0, 0.11, 0 \rangle$ m; (b) 0.11 m; (c) $\langle 0, 1, 0 \rangle$; (d) -0.04 m; (e) $\langle 0, 3.8, 0 \rangle$ N

2.X.15 $\langle 0, -49, 0 \rangle$ N

2.X.16 0.1 kg

2.X.17 The initial position and momentum are the final values from step 2.

2.X.18 Momentum *change* is proportional to force, not momentum itself. The block is headed up, and the downward net force is decreasing its upward momentum.

2.X.19 c

2.X.20 a

2.X.21 $\langle 0, 0, -987 \rangle$ m

2.X.22 The final height is 56 m below the initial height (it fell). The y component of the force on a falling object is $-mg$; the vector force is $\langle 0, -mg, 0 \rangle$. g is the magnitude of the gravitational field, and is $+9.8$ N/kg near the Earth's surface.

2.X.23 (a) 0.903 s; (b) 8.85 m/s

2.X.24 (a) $\langle -10, 7.12, -5 \rangle$ m/s; (b) $\langle 3, 6.04, -8 \rangle$ m; (c) 1.33 s; (d) 8.65 m; (e) 2.65 s; (f) $\langle -17.5, 0, -18.3 \rangle$ m

2.X.25 (a) Student on right; (b) $F_{net,y}$ and $F_{net,z}$; (c) $0 = p_{ix} - F\Delta t$; (d) the displacement during the collision; (e) from the Momentum Principle, $F = p_{ix}/\Delta t$; (f) $35mg$

The Fundamental Interactions

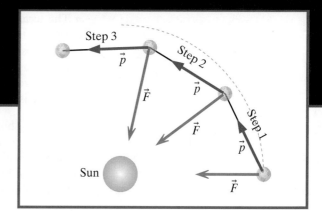

Step 3

Step 2

Step 1

Sun

KEY IDEAS

- **There are only four kinds of fundamental interactions: gravitational, electromagnetic, strong (or "nuclear"), and weak.**
- **Gravitational and electric forces depend on distance and are inversely proportional to $1/r^2$.**
- **Motion of stars and planets can be predicted from gravitational forces in an iterative computation.**
- **Momentum is a conserved quantity.**
- **The Momentum Principle applies to multiparticle systems.**

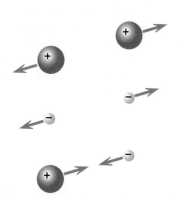

Figure 3.1 The electric interaction: protons repel each other; electrons repel each other; protons and electrons attract each other.

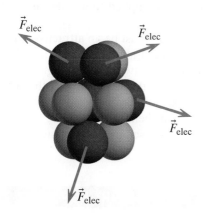

Figure 3.2 The strong interaction: the protons in the nucleus of an atom exert repulsive electric forces on each other, but the strong interaction (which involves neutrons as well as protons) holds the nucleus together despite this electric repulsion.

3.1 THE FUNDAMENTAL INTERACTIONS

In the world we see what seem to be many different kinds of interactions. Planets orbit stars. A falling leaf whirls in the wind. You bend a metal rod. Carbon and oxygen react to form carbon dioxide. In a nuclear power station, uranium nuclei split, making water boil, which drives electric generators. Despite the variety of effects we observe, it became clear in the 20th century that all the changes we see are due to just four different kinds of fundamental interactions: gravitational, electromagnetic, "strong" (also referred to as the nuclear interaction), and "weak."

- The *gravitational* interaction is responsible for an attraction between objects that have mass. For example, the Earth exerts a gravitational force on the Moon, and the Moon exerts a gravitational force on the Earth.
- The *electromagnetic* interaction is responsible for attraction or repulsion between objects that have electric charge. Electric forces are responsible for sparks, static cling, and the behavior of electronic circuits, and magnetic forces are responsible for the operation of motors driven by electric current. Protons repel each other electrically, as do electrons, whereas protons and electrons attract each other (Figure 3.1). Electric forces bind protons and electrons to each other in atoms, and are responsible for the chemical bonds between atoms in molecules. The force of a stretched or compressed spring is due to electric forces between the atoms that make up the spring.
- The *strong* or nuclear interaction occurs between objects made of quarks, such as protons and neutrons, which are held together in the nucleus of an atom despite the large mutual electric repulsion of the protons (Figure 3.2). (The neutrons are not electrically charged and don't exert electric forces.)

■ The *weak* interaction affects all kinds of elementary particles but is much weaker than the strong and electromagnetic interactions. An example of its effects is seen in the instability of a neutron. If a neutron is removed from a nucleus, with an average lifetime of about 15 minutes the neutron decays into a proton, an electron, and a ghostly particle called the antineutrino. This change is brought about by the weak interaction.

In this chapter we will be concerned primarily with gravitational and electric interactions. In later chapters dealing with energy we will encounter situations in which the strong interaction plays an important role. An example of the weak interaction appears in the section on conservation of momentum later in this chapter. The weak interaction is not important in most everyday interactions, so we will not discuss it extensively. We will defer a discussion of magnetic interactions, the other part of the electromagnetic interaction, until later chapters.

Although it continues to be fruitful to classify interactions into four types, it was found in the second half of the 20th century that the electromagnetic interaction and the weak interaction can be considered to be different manifestations of one type of interaction, now called the "electroweak" interaction. Soon after this discovery, it became possible to unify the strong interaction and the electroweak interaction within one powerful theory, the "Standard Model," which also explains the nature of subatomic particles such as the proton and neutron. At present there seem to be really only two fundamental categories of interactions: those explained by the Standard Model and those explained by the gravitational interaction. Physicists are aggressively searching for ways to unify the Standard Model with gravity; this is one of the major scientific quests of the present era.

3.2 | THE GRAVITATIONAL FORCE

The motion of stars and planets is in important ways simpler than other mechanical phenomena, because there is no friction to worry about. These massive bodies interact through the gravitational force, which is always an attractive force. Studying how to predict the motion of stars and planets is one of the most direct ways to understand in general how the Momentum Principle determines the behavior of objects in the real world. The basic ideas used to predict the motion of stars and planets can be applied to a wide range of everyday and atomic phenomena.

In the 1600s Isaac Newton deduced that there must be an attractive force associated with a gravitational interaction between any pair of objects. The gravitational force acts along a line connecting the two objects (Figure 3.3), is proportional to the mass of one object and to the mass of the other object, and is inversely proportional to the square of the distance between the *centers* of the two objects (not the gap between their surfaces). The gravitational force exerted on object 2 by object 1 is expressed by this compact vector equation:

Figure 3.3 The gravitational force exerted on object 2 by object 1.

NEWTON'S GRAVITATIONAL FORCE

$$\vec{F}_{\text{grav on 2 by 1}} = -G\frac{m_1 m_2}{|\vec{r}|^2}\hat{r}$$

$\vec{r} = \vec{r}_2 - \vec{r}_1$ extends from the center of object 1 to the center of object 2.

G is a universal constant: $G = 6.7 \times 10^{-11} \dfrac{\text{N} \cdot \text{m}^2}{\text{kg}^2}$.

This is often called Newton's "universal law of gravitation," reflecting its central importance in the development of human understanding of the Universe. The word "universal" is used because all massive objects in the Universe attract all other massive objects in the Universe.

The gravitational force involves a lot of different symbols and may look pretty intimidating at first. Let's take the expression apart and look at the individual pieces to try to make sense of the formula.

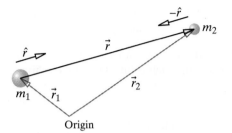

Figure 3.4 $\vec{r} = \vec{r}_2 - \vec{r}_1$, the location of object 2 relative to object 1: "final minus initial." The gravitational force on 2 is in the direction of $-\hat{r}$.

The Relative Position Vector

The relative position vector $\vec{r}$ extends from the center of object 1 to the center of object 2, as shown in Figure 3.4. (This vector can also be written as $\vec{r}_{2-1}$; we will use the notation $\vec{r}$, which is more compact, but requires that you remember that object 1 is the initial location and object 2 is the final location.) The unit vector $\hat{r}$ points in the same direction as $\vec{r}$, but has magnitude 1. The magnitude of $\vec{r}$ is the distance between the centers of the two objects.

The Direction of the Gravitational Force

The direction of the gravitational force on object 2 by object 1 is specified by $-\hat{r}$ (Figure 3.4), which is the unit vector $\hat{r}$, in combination with the minus sign, as indicated in Figure 3.5.

$$\vec{F}_{\text{on 2 by 1}} = -G\frac{m_1 m_2}{|\vec{r}|^2}\hat{r}$$

Figure 3.5 The direction of the gravitational force is opposite to the direction of the unit vector $\hat{r}$, which points from object 1 to object 2.

> **QUESTION** Why is the minus sign necessary?

The minus sign is necessary because the force on object 2 due to object 1 is in the direction opposite to $\hat{r}$ (Figure 3.4).

Mass and Magnitude of the Gravitational Force

As highlighted in Figure 3.6, the gravitational force is proportional to the product of the two masses, $m_1 m_2$. If you double either of these masses, keeping the other one the same, the force will be twice as big.

$$\vec{F}_{\text{on 2 by 1}} = -G\frac{m_1 m_2}{|\vec{r}|^2}\hat{r}$$

Figure 3.6 The magnitude of the gravitational force depends on the masses of both interacting objects.

> **QUESTION** If both masses are doubled, how much larger will the force be?

If you double both of the masses, the force will be four times as big.

Note that since $m_2 m_1 = m_1 m_2$, the magnitude of the force exerted on object 1 by object 2 is exactly the same as the magnitude of the force exerted on object 2 by object 1, but the direction is opposite. This important property of the gravitational force is called "reciprocity" and is discussed in more detail in Section 3.4.

> **3.X.1** A star exerts a gravitational force of magnitude 4×10^{25} N on a planet. What is the magnitude of the gravitational force that the planet exerts on the star?
>
> **3.X.2** A star exerts a gravitational force of magnitude 4×10^{25} N on a planet. If the mass of the planet were twice as large, what would the magnitude of the gravitational force on the planet be?

Distance and Magnitude of the Gravitational Force

The gravitational force is an "inverse square" force. As highlighted in Figure 3.7, the square of the center-to-center distance appears in the denominator. This means that the gravitational force depends very strongly on the distance between the objects. For example, if you double the distance between them, the only thing that changes is the denominator, which gets four times bigger

$$\vec{F}_{\text{on 2 by 1}} = -G\frac{m_1 m_2}{|\vec{r}|^2}\hat{r}$$

Figure 3.7 The gravitational force is an "inverse square" force.

(2 squared is 4), so the force is only 1/4 as big as before:

$$|\vec{F}_d| = \frac{Gm_1m_2}{|\vec{r}|^2}$$

$$|\vec{F}_{2d}| = \frac{Gm_1m_2}{|2\vec{r}|^2} = \frac{1}{4}\frac{Gm_1m_2}{|\vec{r}|^2} = \frac{1}{4}|\vec{F}_d|$$

QUESTION   If you move the masses 10 times farther apart than they were originally, how does the gravitational force change?

The force goes down by a factor of 100. Evidently when two objects are very far apart, the gravitational forces they exert on each other will be vanishingly small: big denominator, small force.

3.X.3 A planet exerts a gravitational force of magnitude 8×10^{20} N on a star. If the distance between the star and planet were three times larger, what would the magnitude of this force be?

3.X.4 Masses M and m attract each other with a gravitational force of magnitude F. Mass m is replaced with a mass $3m$, and it is moved four times farther away. Now what is the magnitude of the force?

The Gravitational Constant G

$$G = 6.7 \times 10^{-11}\frac{\text{N}\cdot\text{m}^2}{\text{kg}^2}$$

We say that the gravitational constant G is "universal" because it is the same for any pair of interacting masses, no matter how big or small they are, or where they are located in the Universe. Because G is universal, it can be measured for a particular pair of objects and then used with other pairs of objects. As we describe in Section 3.15, Cavendish was the first person to make such a measurement.

"Factoring" the Force into a Magnitude and a Direction

A useful way to think about the gravitational force is to factor it into magnitude and direction, like this:

A vector is a magnitude times a direction: $\vec{F}_{grav} = |\vec{F}_{grav}|\hat{F}_{grav}$

- Magnitude: $|\vec{F}_{grav}| = G\frac{m_1m_2}{|\vec{r}|^2}$
- Direction (unit vector): $\hat{F}_{grav} = -\hat{r}$

It is usually simplest to calculate the magnitude and direction separately, then combine them to get the vector force. That way you can focus on one thing at a time rather than getting confused (or intimidated!) by the full complexity of the vector force.

3.X.5 A 3 kg ball and a 5 kg ball are 2 m apart, center to center. What is the magnitude of the gravitational force that the 3 kg ball exerts on the 5 kg ball? What is the magnitude of the gravitational force that the 5 kg ball exerts on the 3 kg ball?

3.X.6 Measurements show that the Earth's gravitational force on a mass of 1 kg near the Earth's surface is 9.8 N. The radius of the Earth is 6400 km (6.4×10^6 m). From these data determine the mass of the Earth.

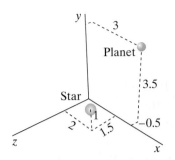

Figure 3.8 A star and a planet interact gravitationally. Distances shown should be multiplied by 1×10^{11} m.

Calculating Gravitational Force

Calculating the gravitational force on an object due to another object requires several steps. Use the formula for gravitational force as a guide that tells you to carry out these steps:

CALCULATING GRAVITATIONAL FORCE

- Calculate $\vec{r} = \vec{r}_2 - \vec{r}_1$, the position of the center of object 2 relative to the center of object 1.
- Calculate $|\vec{r}|$, the center-to-center distance between the objects.
- Calculate $Gm_1m_2/|\vec{r}|^2$, the magnitude of the force.
- Calculate $-\hat{r} = -\vec{r}/|\vec{r}|$, the direction of the force.
- Multiply the magnitude times the direction to get the vector force.

EXAMPLE **The Force on a Planet by a Star**

Figure 3.8 shows a star of mass 4×10^{30} kg located at this moment at position $\langle 2 \times 10^{11}, 1 \times 10^{11}, 1.5 \times 10^{11}\rangle$ m and a planet of mass 3×10^{24} kg located at position $\langle 3 \times 10^{11}, 3.5 \times 10^{11}, -0.5 \times 10^{11}\rangle$ m. (These are typical values for stars and planets. Notice that the mass of the star is much greater than that of the planet.) In Figure 3.8 we show the x, y, and z components of the positions, to be multiplied by 1×10^{11} m. Make sure you understand how the numbers on the diagram correspond to the positions given as vectors.

(a) Calculate the gravitational force exerted on the planet by the star.

(b) Calculate the gravitational force exerted on the star by the planet.

Solution   QUESTION   Which object is object 1, and which is object 2?

In this case **(a)** the planet is object 2 (we want the force on the planet), and the star is object 1.

Relative position vector:

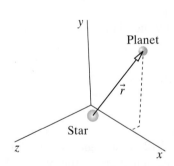

Figure 3.9 The position vector of the planet relative to the star. Initial location: center of star. Final location: center of planet.

$$\vec{r} = \vec{r}_2 - \vec{r}_1$$
$$\vec{r} = \langle 3 \times 10^{11}, 3.5 \times 10^{11}, -0.5 \times 10^{11}\rangle \text{ m} - \langle 2 \times 10^{11}, 1 \times 10^{11}, 1.5 \times 10^{11}\rangle \text{ m}$$
$$\vec{r} = \langle 1 \times 10^{11}, 2.5 \times 10^{11}, -2 \times 10^{11}\rangle \text{ m}$$

Check: Do the signs of the components make sense? Yes, see Figure 3.9. Distance:

$$|\vec{r}| = \sqrt{(1 \times 10^{11})^2 + (2.5 \times 10^{11})^2 + (-2 \times 10^{11})^2} \text{ m}$$
$$|\vec{r}| = 3.35 \times 10^{11} \text{ m}$$

■ This looks like a big force, but it's acting on a planet with a big mass, so it isn't obvious whether this is really a "big" force in terms of what it will do.

Magnitude of force:

$$|\vec{F}_{\text{on } P \text{ by } S}| = G\frac{m_1 m_2}{|\vec{r}|^2}$$
$$|\vec{F}_{\text{on } P \text{ by } S}| = \left(6.7 \times 10^{-11}\frac{\text{N} \cdot \text{m}^2}{\text{kg}^2}\right)\frac{(3 \times 10^{24}\text{ kg})(4 \times 10^{30}\text{ kg})}{(3.35 \times 10^{11}\text{ m})^2}$$
$$|\vec{F}_{\text{on } P \text{ by } S}| = 7.15 \times 10^{21} \text{ N}$$

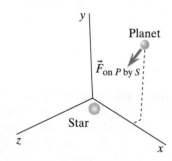

Figure 3.10 Unit vectors.

Direction of force on planet:

$$\hat{F}_{\text{on }P\text{ by }S} = -\hat{r} = -\frac{\vec{r}}{|\vec{r}|}$$

$$\hat{F}_{\text{on }P\text{ by }S} = -\frac{\langle 1 \times 10^{11}, 2.5 \times 10^{11}, -2 \times 10^{11}\rangle \text{ m}}{3.35 \times 10^{11} \text{ m}}$$

$$\hat{F}_{\text{on }P\text{ by }S} = \langle -0.298, -0.745, 0.596\rangle$$

Check: Units? Okay, unit vector has no units. Direction? Figure 3.10 shows that $\hat{F}_{\text{on }P\text{ by }S}$ points from the planet toward the star, which is correct for the direction of the force. Magnitude of the unit vector? $\sqrt{(-0.298)^2 + (-0.745)^2 + (0.596)^2} = 0.995$, okay to 3 significant figures.

Calculate force as a vector:

$$\vec{F}_{\text{on }P\text{ by }S} = |\vec{F}_{\text{on }P\text{ by }S}|\hat{F}_{\text{on }P\text{ by }S}$$

$$\vec{F}_{\text{on }P\text{ by }S} = (7.15 \times 10^{21} \text{ N})\langle -0.298, -0.745, 0.596\rangle$$

$$\vec{F}_{\text{on }P\text{ by }S} = \langle -2.13 \times 10^{21}, -5.33 \times 10^{21}, 4.26 \times 10^{21}\rangle \text{ N}$$

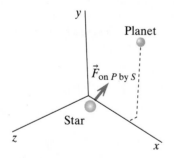

Figure 3.11 Gravitational force on the planet by the star.

Check: Direction? Figure 3.11 shows the force in the correct direction. Magnitude? Do a rough "order of magnitude" check: the distance between star and planet is *very* roughly 1×10^{11} m, the mass of the star is *very* roughly 1×10^{30} kg, the mass of the planet is *very* roughly 1×10^{24} kg, and the gravitational constant is *very* roughly 1×10^{-11} N·m²/kg². Therefore we expect the magnitude of the force to be *very* roughly this:

$$|\vec{F}_{\text{on }P\text{ by }S}| = \left(1 \times 10^{-11}\frac{\text{N·m}^2}{\text{kg}^2}\right)\frac{(1 \times 10^{24} \text{ kg})(1 \times 10^{30} \text{ kg})}{(1 \times 10^{11} \text{ m})^2} = 1 \times 10^{21} \text{ N}$$

The fact that we get within an order of magnitude of the result 7.16×10^{21} N is evidence that we haven't made any huge mistakes.

The magnitude of the force on the star by the planet (part b) will be the same. The only change is in the direction, because $\vec{r}_{S-P} = -\vec{r}_{P-S}$.

$$\vec{F}_{\text{on }S\text{ by }P} = \langle 2.13 \times 10^{21}, 5.33 \times 10^{21}, -4.26 \times 10^{21}\rangle \text{ N}$$

FURTHER DISCUSSION   Note that the signs of the force components for the force on the star are consistent with Figure 3.12. You might be puzzled that the planet pulls just as hard on the star as the star pulls on the planet. We'll discuss this in more detail later in this chapter.

Figure 3.12 The gravitational force exerted on the star by the planet.

3.X.7 At a particular instant Ganymede and Europa, two of the moons of Jupiter, are aligned as shown in Figure 3.13. Coordinate axes are shown in the diagram. In calculating the gravitational force on Ganymede by Europa:

   (a) What is the direction of $\vec{r}$?

   (b) What is the direction of $-\hat{r}$?

   (c) What is the direction of the gravitational force?

In calculating the gravitational force on Europa by Ganymede:

   (d) What is the direction of $\vec{r}$?

   (e) What is the direction of $-\hat{r}$?

   (f) What is the direction of the gravitational force?

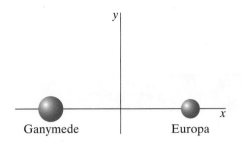

Figure 3.13 Two of Jupiter's moons, aligned along the *x* axis. (Exercise 3.X.7)

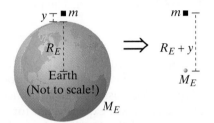

Figure 3.14 Determining the gravitational force by the Earth on an object a height *y* above the surface. The nearly spherical Earth acts as if all its mass were concentrated at its center.

3.X.8 The mass of the Earth is 6×10^{24} kg, and the mass of the Moon is 7×10^{22} kg. At a particular instant the Moon is at location $\langle 2.8 \times 10^8, 0, -2.8 \times 10^8 \rangle$ m, in a coordinate system whose origin is at the center of the Earth.

    (a) What is $\vec{r}_{M-E}$, the relative position vector from the Earth to the Moon?

    (b) What is $|\vec{r}_{M-E}|$?

    (c) What is the unit vector $\hat{r}_{M-E}$?

    (d) What is the gravitational force exerted by the Earth on the Moon? Your answer should be a vector.

3.3 APPROXIMATE GRAVITATIONAL FORCE NEAR THE EARTH'S SURFACE

Earlier we used the expression *mg* to represent the magnitude of the gravitational force on an object near the Earth's surface. This is an approximation to the actual force, but it is a good one. The magnitude of the gravitational force that the Earth exerts on an object of mass *m* near the Earth's surface (Figure 3.14) is

$$F_g = G \frac{M_E\, m}{(R_E + y)^2}$$

where *y* is the distance of the object above the surface of the Earth, R_E is the radius of the Earth, and M_E is the mass of the Earth.

A spherical object of uniform density can be treated as if all its mass were concentrated at its center (see Section 3.14 for a discussion of this). As a result, we can treat the effect of the Earth on the object as though the Earth were a tiny, very dense ball a distance $R_E + y$ away.

Consider the difference in the gravitational force exerted by the Earth on an object when it is one meter above the surface compared with when it is two meters above the surface. The radius of the Earth is about 6.4×10^6 m. Then

$$F_{1\,\text{m}} = G \frac{M_E\, m}{(6.400001 \times 10^6 \text{ m})^2} \quad \text{whereas} \quad F_{2\,\text{m}} = G \frac{M_E\, m}{(6.400002 \times 10^6 \text{ m})^2}$$

For most purposes this difference is not significant. In fact, for all interactions of objects near the surface of the Earth, it makes sense to use the same approximate value, $R_E = 6.4 \times 10^6$ m, for the distance from the center of the Earth to the object. This simplifies calculation of gravitational forces by allowing us to combine all the constants into a single lumped constant, *g*:

$$F_g \approx \left(G \frac{M_E}{R_E^2} \right) m = gm$$

$$g = G \frac{M_E}{R_E^2}$$

QUESTION The mass of the Earth is 6×10^{24} kg and the radius of the Earth is 6.4×10^6 m. What is the the value of *g*?

$$g = \left(6.7 \times 10^{-11} \frac{\text{N} \cdot \text{m}^2}{\text{kg}^2} \right) \frac{(6 \times 10^{24} \text{ kg})}{(6.4 \times 10^6 \text{ m})^2}$$

$$= 9.8 \frac{\text{N}}{\text{kg}}$$

The constant g, called "the magnitude of the gravitational field," has the value $g = +9.8$ newtons/kilogram near the Earth's surface. Note that g is a positive number.

MAGNITUDE OF GRAVITATIONAL FIELD NEAR EARTH'S SURFACE

$$g = G\frac{M_E}{R_E^2}$$

$$g = +9.8\,\frac{\text{N}}{\text{kg}}$$

M_E is the mass of the Earth, and R_E is the radius of the Earth. The units of g are newtons/kg.

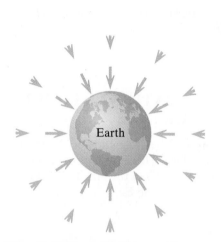

Figure 3.15 The "gravitational field" around the Earth. At a position where the field is $\vec{g}$, an object of mass m experiences a force $m\vec{g}$. The magnitude g of the field decreases with distance away from the Earth.

The "gravitational field" $\vec{g}$ at a location in space is defined to be the vector gravitational force per kg, so the force on a 1 kg mass placed at that location would be $(1\,\text{kg})\vec{g}$, with $\vec{g}$ having units of N/kg. A 2 kg mass would experience a force $(2\,\text{kg})\vec{g}$ at that location. In general, a mass m will experience a force $m\vec{g}$ of magnitude mg. The gravitational field around the Earth is shown in Figure 3.15.

If the Earth consisted of layers of uniform-density spherical shells, the Earth's gravitational field would be perfectly spherically symmetric, and g would be truly constant all over the surface of the Earth. However, the Earth due to its spin bulges out at the equator, so g is smaller at the equator than at the poles because you're farther from the center of the Earth. Also, regions in the Earth of higher or lower density make g above these regions vary slightly from the average value, and geologists use these small variations to find valuable minerals and oil.

If the $+y$ axis points up, away from the center of the Earth, $\vec{g} = \langle 0, -mg, 0 \rangle$. Note that g is a *positive* number, the magnitude of the gravitational field $\vec{g}$.

We will not use the concept of field further until we study electric and magnetic fields in later chapters, other than to use the magnitude of the gravitational field near the Earth's surface ($g = +9.8\,\text{N/kg}$).

EXAMPLE   **A Falling Object**

Drop a rock with mass m near the surface of the Earth. What is its (vector) acceleration?

Solution   Use the Momentum Principle: $\Delta\vec{p} = \vec{F}_{\text{net}}\,\Delta t$. Since the speed is small compared to the speed of light, and the mass isn't changing, we can write $\vec{p} \approx m\vec{v}$. If we neglect air resistance, the only force is the gravitational force, so $\Delta(m\vec{v}) = m\vec{g}\,\Delta t$. Divide through by m and by Δt and we get this:

$$\frac{\Delta\vec{v}}{\Delta t} = \vec{g}$$

However, the acceleration of the rock is the rate of change of the velocity, $\dfrac{\Delta\vec{v}}{\Delta t}$, so we have

$$\vec{a} = \vec{g} = \langle 0, -g, 0 \rangle$$

The rock accelerates toward the ground (the $-y$ direction) at a rate of 9.8 m/s/s. Evidently the units of N/kg are equivalent to the units of m/s/s.

FURTHER DISCUSSION   The reason that the mass dropped out of the calculation is that both the momentum and the gravitational force are proportional to m, which is not true for other kinds of force. This has the interesting effect that if the gravitational force is the only force acting, a heavy rock and a light rock will fall in exactly the same way. However, the air resistance force depends on how wide the rock is, and how blunt it is, and how fast the rock is moving, and is not proportional to m, so if the air resistance force is significant, different masses will fall differently. An obvious example is that if you drop a metal ball and a partially crumpled sheet of paper, the metal ball will hit the ground before the paper does.

> **3.X.9** The mass of the Moon is 7×10^{22} kg, and the radius of the Moon is 1.75×10^6 m. **(a)** What is the value of the constant g at a location on the surface of the Moon? **(b)** What would be the magnitude of the gravitational force exerted by the Moon on a 70 kg astronaut standing on the Moon? **(c)** How does this compare to the gravitational force on the same astronaut when standing on the surface of the Earth?
>
> **3.X.10** The mass of Mars is 6.4×10^{23} kg and its radius is 3.4×10^6 m. What is the value of the constant g on Mars?

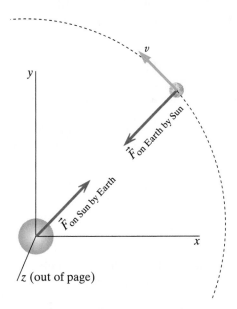

Figure 3.16 The Sun and the Earth exert equal and opposite forces on each other.

3.4   RECIPROCITY

An important aspect of the gravitational interaction is that the force that object 1 exerts on object 2 is equal and opposite to the force that object 2 exerts on object 1 (Figure 3.16). That the magnitudes must be equal is clear from the algebraic form of the gravitational force, because $m_1 m_2 = m_2 m_1$. The directions of the forces are along the line connecting the centers, and in opposite directions.

This property of gravitational interactions is called "reciprocity" or "Newton's third law of motion" and as we will see later in this chapter it is also a property of electric interactions, including the electric interactions between neighboring atoms.

RECIPROCITY
$\vec{F}_{\text{on 1 by 2}} = -\vec{F}_{\text{on 2 by 1}}$
(gravitational and electric forces)

The force that the Earth exerts on the massive Sun in Figure 3.16 is just as big as the force that the Sun exerts on the Earth, so in the same time interval the momentum changes are equal in magnitude and opposite in direction ("equal and opposite" for short):

$$\Delta \vec{p}_1 = \vec{F}_{\text{on 1 by 2}} \, \Delta t = -\Delta \vec{p}_2$$

However, the velocity change $\Delta \vec{v} = \Delta \vec{p}/m$ of the Sun is extremely small compared to the velocity change of the Earth, because the mass of the Sun is enormous in comparison with the mass of the Earth. The mass of our Sun, which is a rather ordinary star, is 2×10^{30} kg. This is enormous compared to the mass of the Earth (6×10^{24} kg) or even to the mass of the largest planet in our Solar System, Jupiter (2×10^{27} kg). Nevertheless, very accurate measurements of small velocity changes of distant stars have been used to infer the presence of unseen planets orbiting those stars.

Magnetic forces do not have the property of reciprocity. Two electrically charged particles that are both moving can interact magnetically as well as

2 grams

electrically, and the magnetic forces that these two particles exert on each other need not be equal in magnitude or opposite in direction. Reciprocity applies to gravitational and electric forces, but not in general to magnetic forces acting between individual moving charges.

As we will see in Chapter 4, the forces associated with contact between two objects are actually electric forces at the microscopic level. Thus contact forces also have the property of reciprocity.

Why Reciprocity?

The algebraic form of the expression for the gravitational force indicates that reciprocity should hold. A diagram may be helpful in explaining *why* the forces behave this way.

Consider two small objects, a 2 gram object made up of two 1 gram balls, and a 3 gram object made up of three 1 gram balls (Figure 3.17). The distance between centers of the two objects is large compared to the size of either object, so the distances between pairs of 1 gram balls are about the same for all pairs.

You can see that in the 2 gram object at the top of Figure 3.17 each ball has three gravitational forces exerted on it by the distant balls in the 3 gram object.

3 grams

Figure 3.17 Reciprocity: there are six forces acting on the 3 gram object and six forces acting on the 2 gram object.

QUESTION How many of these individual forces act on the 2 gram object?

There is a net force of 2 × 3 = 6 times the force associated with one pair of balls.

QUESTION Similarly, consider the forces acting on the 3 gram object at the bottom of Figure 3.17. How many of these individual forces act on the 3 gram object?

There is again a net force of 3 × 2 = 6 times the force associated with one pair of balls.

The effect is that the force exerted by the 3 gram object on the 2 gram object has the same magnitude as the force exerted by the 2 gram object on the 3 gram object. The same reciprocity holds, for similar reasons, for the electric forces between a lithium nucleus containing 3 protons and a helium nucleus containing 2 protons.

3.X.11 A 60 kg person stands on the Earth's surface. **(a)** What is the approximate magnitude of the gravitational force on the person by the Earth? **(b)** What is the magnitude of the gravitational force on the Earth by the person?

3.5 PREDICTING MOTION OF GRAVITATIONALLY INTERACTING OBJECTS

The iterative approach we used in Chapter 2 to predict the motion of a block attached to a spring can be used to predict the motion of objects that interact gravitationally. The only difference is in the force we use: the gravitational force instead of the spring force. The gravitational force changes magnitude and direction as the objects move, so at the beginning of every time step we need to recalculate the gravitational force, both magnitude and direction.

EXAMPLE

Earth and Sun

Let's try to predict approximately the motion of the Earth around the Sun. Choose a coordinate system whose origin is at the center of the Sun, and in which the Earth orbits the Sun in the xy plane, as shown in Figure 3.18. The mass of the Sun is 2×10^{30} kg; the mass of the Earth is 6×10^{24} kg.

Suppose that at a particular instant the Earth has these "initial conditions": position $\langle 1.5 \times 10^{11}, 0, 0 \rangle$ m and velocity $\langle 0, 3 \times 10^4, 0 \rangle$ m/s. Iteratively updating the momentum and position should yield a circular orbit, because the Earth's actual speed in its nearly circular orbit is in fact 3×10^4 m/s, as can be calculated from $2\pi r/T$, where $r = 1.5 \times 10^{11}$ m is the radius of the circle and T is one year, measured in seconds. What does the Momentum Principle predict will be the location of the Earth 3 months later? How well does the prediction match the actual orbit?

Solution **The First Time Step:**

To get an approximate answer and to illustrate the method, we'll divide the 3-month time interval into three intervals each 1 month long. This will give us only a rough approximation to the motion, because the gravitational force acting on the Earth changes quite a lot in one month, which is one-twelfth of the way around the circular orbit.

Figure 3.18 Initial location and velocity of the Earth. The origin of the coordinate system is at the center of the Sun. Sizes of the Sun and Earth are exaggerated.

1. Momentum Principle
   System: Earth
   Surroundings: Sun (ignore all other planets, stars); see the free-body diagram in Figure 3.18.

$$\Delta t = (1 \text{ mo}) \left(30 \frac{\text{days}}{\text{mo}} \right) \left(24 \frac{\text{hr}}{\text{day}} \right) \left(60 \frac{\text{min}}{\text{hr}} \right) \left(60 \frac{\text{sec}}{\text{min}} \right) = 2.6 \times 10^6 \text{ s}$$

$$\vec{p}_i = (6 \times 10^{24} \text{ kg}) \langle 0, 3 \times 10^4, 0 \rangle \text{ m/s} = \langle 0, 1.8 \times 10^{29}, 0 \rangle \text{ kg} \cdot \text{m/s}$$

$$\vec{r} = \vec{r}_E - \vec{r}_S = \langle 1.5 \times 10^{11}, 0, 0 \rangle \text{ m} - \langle 0, 0, 0 \rangle \text{ m} = \langle 1.5 \times 10^{11}, 0, 0 \rangle \text{ m}$$

■ The initial location and the initial momentum of the objects are called the "initial conditions" of the system.

Now that we know $\vec{r}$, we can calculate $|\vec{r}|$, $\hat{r}$, $\hat{F} = -\hat{r}$, $|\vec{F}|$, and $\vec{F} = |\vec{F}|\hat{F}$ as we have done before. The result is

$$\vec{F}_{\text{on } E \text{ by } S} = \langle -3.6 \times 10^{22}, 0, 0 \rangle \text{ N}$$

Now we update the momentum using the Momentum Principle:

■ $\vec{p} \approx m\vec{v}$ because $v \ll c$.

$$\vec{p}_f = \vec{p}_i + \vec{F}_{\text{net}} \, \Delta t$$
$$\vec{p}_f = \langle 0, 1.8 \times 10^{29}, 0 \rangle \text{ kg} \cdot \text{m/s} + \langle -3.6 \times 10^{22}, 0, 0 \rangle \text{ N} \cdot (2.6 \times 10^6 \text{ s})$$
$$\vec{p}_f = \langle -9.3 \times 10^{28}, 1.8 \times 10^{29}, 0 \rangle \text{ kg} \cdot \text{m/s}$$

2. Position update
   We use the final momentum to find an approximate value for the average velocity of the Earth; because the force is changing this is not necessarily a worse approximation than the arithmetic average would be. Details are not shown (you can work this out yourself). The results of the first iteration are shown in Figure 3.19.

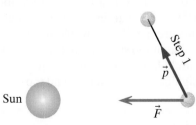

Figure 3.19 Step 1 of 3-step iterative calculation. Gravitational force on Earth at beginning of step and final momentum of Earth are shown.

$$\vec{r}_{E,f} = \langle 1.1 \times 10^{11}, 7.8 \times 10^{10}, 0 \rangle \text{ m}$$

The Second Time Step:

■ It is necessary to recalculate the relative position vector and the force vector at the beginning of every time step, since their directions and magnitudes may have changed.

1. Momentum principle
   Assume that the Sun's motion was negligible. Since $m_{Sun} m_{Earth} = m_{Earth} m_{Sun}$, the magnitude of the force on the Sun was the same as that on the Earth (reciprocity of gravitational forces; see Section 3.4), so the magnitude of the Sun's momentum change was the same as the Earth's. However, because the Sun's mass is very large, the change in the Sun's velocity is very small.

   We need to recalculate the gravitational force acting on the Earth, because the Earth has moved.

$$\vec{r} = \vec{r}_E - \vec{r}_S = \langle 1.1 \times 10^{11}, 7.8 \times 10^{10}, 0 \rangle \text{ m}$$

■ Now the gravitational force on the Earth has a nonzero y component.

After some familiar calculations, which you can check, we obtain the new force:

$$\vec{F}_{\text{on } E \text{ by } S} = \langle -3.6 \times 10^{22}, -2.6 \times 10^{22}, 0 \rangle \text{ N}$$

We update the momentum, using the Momentum Principle:

$$\vec{p}_f = \vec{p}_i + \vec{F}_{\text{net}} \Delta t$$
$$\vec{p}_f = \langle -9.3 \times 10^{28}, 1.8 \times 10^{29}, 0 \rangle \text{ kg} \cdot \text{m/s}$$
$$+ \langle -3.6 \times 10^{22}, -2.6 \times 10^{22}, 0 \rangle \text{ N} \cdot (2.6 \times 10^6 \text{ s})$$
$$\vec{p}_f = \langle -1.9 \times 10^{29}, 1.1 \times 10^{29}, 0 \rangle \text{ kg} \cdot \text{m/s}$$

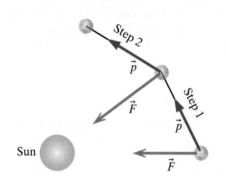

Figure 3.20 Results of first two steps of the iterative calculation. The gravitational force on the Earth is different in the two steps.

2. Position update
   We again use the final momentum to find an approximate value for the average velocity of the Earth. Details are not shown (you can work this out yourself). The results from the second time step are shown in Figure 3.20.

$$\vec{r}_{E,f} = \langle 2.9 \times 10^{10}, 1.3 \times 10^{11}, 0 \rangle \text{ m}$$

The Third Time Step:

In the third time step we must again calculate $\vec{r}$ and $\vec{F}_{\text{on } E \text{ by } S}$. The results of the third iteration are shown in Figure 3.21.

$$\vec{r}_{E,f} = \langle -6.3 \times 10^{10}, 1.2 \times 10^{11}, 0 \rangle \text{ m}$$

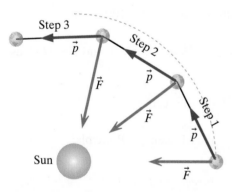

Figure 3.21 The results of all three steps in the iterative calculation of the motion of the Earth under the influence of the Sun. The dashed arc shows the actual orbit of the Earth in three months.

Check: Is the path reasonable? Yes. Although this was a rough calculation, it does predict that the Earth will orbit around the Sun in a roughly circular orbit. If we had used smaller time steps, our predicted path would have come much closer to the actual orbit. One indication of the sizable inaccuracy in our calculation is that in 3 months the Earth should have gone only one-quarter of the way around the circle, but it has gone significantly farther around the circle in Figure 3.21.

FURTHER DISCUSSION As you can see, the procedure outlined here is very tedious if done by hand, which is why the natural way to carry out an iterative prediction of motion is to program a computer to do the repetitive calculations. Computational modeling is now an extremely important tool in all of science and engineering.

In a computer program one would certainly choose a step size much smaller than 1 month, since the Earth's position changes a lot in 1 month, which means that the force varies quite a lot during one time step. When we use the Momentum Principle in the form $\vec{p}_f = \vec{p}_i + \vec{F}_{net}\Delta t$, we're assuming that the force doesn't change much during the interval Δt, which isn't true for a gravitational force if the position changes a lot. In computational modeling one should choose a step size small enough that force doesn't change much during each step.

QUESTION How would you have to change the calculation described above to include in the prediction how the Sun will move?

You would need to update the momentum and position of the Sun each time you update the momentum and position of the Earth. Fortunately, the force acting on the Sun is just the opposite of the force acting on the Earth: $\vec{F}_{on\,S\,by\,E} = -\vec{F}_{on\,E\,by\,S}$ thanks to the reciprocity of gravitational forces (Section 3.4). So it is unnecessary to repeat all of the usual steps in calculating the gravitational force that acts on the Sun; just reverse the force vector.

Because its mass is so large, you would find that the Sun moves very little. Nevertheless, in recent years many planets have been discovered orbiting distant stars by observing the very small wobbles of the stars caused by the gravitational forces of the planets acting on the stars.

EXAMPLE

If the Step Size Is Too Big

By taking a step size of 1 month, we predicted that the Earth would orbit the Sun, though our prediction was rather imprecise, because in 1 month the gravitational force and the momentum change a lot. It is instructive to see what happens if we take an even bigger step size, making the results even more imprecise. Taking a step size of 3 months instead of 1 month, what will we predict the location of the Earth to be 3 months later?

Solution

■ $\vec{p} \approx m\vec{v}$ because $v \ll c$.

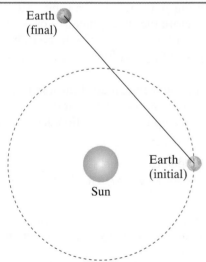

Figure 3.22 The displacement of the Earth predicted by a one-step calculation using the initial value of gravitational force, and the final velocity. The dashed line shows the actual path of the Earth around the Sun.

1. Momentum principle
   System: Earth.
   Surroundings: Sun (ignore all other planets, stars)

$$\Delta t = (3\,\text{mo})\left(30\frac{\text{days}}{\text{mo}}\right)\left(24\frac{\text{hr}}{\text{day}}\right)\left(60\frac{\text{min}}{\text{hr}}\right)\left(60\frac{\text{sec}}{\text{min}}\right) = 7.8 \times 10^6\,\text{s}$$

$$\vec{p}_i = \langle 0, 1.8 \times 10^{29}, 0 \rangle\,\text{kg}\cdot\text{m/s}$$

$$\vec{p}_f = \vec{p}_i + \vec{F}_{net}\,\Delta t$$

$$\vec{p}_f = \langle 0, 1.8 \times 10^{29}, 0 \rangle\,\text{kg}\cdot\text{m/s} + \langle -3.6 \times 10^{22}, 0, 0 \rangle\,\text{N}\cdot(7.8 \times 10^6\,\text{s})$$

$$\vec{p}_f = \langle -2.8 \times 10^{29}, 1.8 \times 10^{29}, 0 \rangle\,\text{kg}\cdot\text{m/s}$$

2. Update position, approximating the average velocity by the final velocity

$$\vec{v}_f \approx \frac{\vec{p}_f}{m} = \frac{\langle -2.8 \times 10^{29}, 1.8 \times 10^{29}, 0 \rangle\,\text{kg}\cdot\text{m/s}}{6 \times 10^{24}\,\text{kg}}$$

$$\vec{v}_f = \langle -4.6 \times 10^4, 3 \times 10^4, 0 \rangle\,\text{m/s}$$

$$\vec{r}_{E,f} = \vec{r}_i + \vec{v}_{avg}\Delta t$$

$$\vec{r}_{E,f} = \langle 1.5 \times 10^{11}, 0, 0 \rangle\,\text{m} + \langle -4.6 \times 10^4, 3 \times 10^4, 0 \rangle\,\text{m/s}\cdot(7.8 \times 10^6\,\text{s})$$

$$\vec{r}_{E,f} = \langle -2e11, 2.3e11, 0 \rangle\,\text{m}$$

3. Check: Units are ok (meters). Is the location reasonable? NO! The path we predicted, from initial location to final location, is shown in Figure 3.22. This disagrees very much with our more accurate calculation using a time

step of 1 month. Even our 1-month calculation was imprecise; in a computer calculation we would choose a much shorter time step than 1 month.

FURTHER DISCUSSION

QUESTION   What went wrong? In this calculation we used the same procedure we used before.

The major error was the tacit assumption that the force on the Earth by the Sun was constant in magnitude and direction over the 3-month interval. The direction and magnitude of the gravitational force between two objects changes when the relative positions of the objects change. An additional error was introduced by using the final velocity as a (poor) approximation to the actual average velocity during this very lengthy time interval.

Elliptical Orbits

The Earth's orbit around the Sun isn't really a circle; it's actually an ellipse that is very nearly circular. Newton himself drew an instructive diagram to show how the launch speed of an object affects the shape of the orbit (Figure 3.23), by imagining throwing a rock horizontally from the top of a mountain so tall as to stick up above the atmosphere (so there will be no air resistance). In this figure you can see that getting a circular orbit requires throwing the rock at just the right speed. Any other speed will give a noncircular orbit, which Newton was able to prove would be an ellipse (or portion of an ellipse if the rock hits the ground).

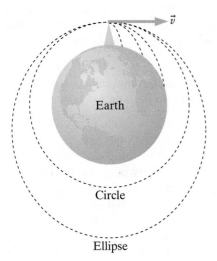

Figure 3.23 Newton imagined throwing a rock horizontally with different speeds from a high mountain resulting in different ellipses. Too small a speed and the rock hits the Earth. High speeds result in full ellipses. A speed that is just right makes a circular orbit.

EXAMPLE

Analyzing an Elliptical Orbit

It is useful to "think iteratively" about a highly elliptical orbit like that of many comets that come near the Sun at one extreme of the orbit and go far beyond Pluto at the other extreme. At some instant the comet is at location A, headed toward the Sun (Figure 3.24). Graphically and qualitatively, carry out one step in the iterative updating of momentum and position, starting with the comet at position A with momentum $\vec{p}_i$ tangent to the orbit. Will the magnitude of the momentum (and therefore the speed) increase or decrease? Can you say why this happens? It is useful to think about the role of the parallel and perpendicular components of the net force.

Figure 3.24 A comet approaches the Sun on a long elliptical orbit.

Solution

(2) Draw impulse
$$\Delta\vec{p} = \vec{F}_{net}\,\Delta t$$
(1) Draw force
$\vec{F}_{net}$

(3) Draw new momentum
$$\vec{p}_f = \vec{p}_i + \Delta\vec{p}$$

$$\Delta\vec{r} = (\vec{p}_f/m)\Delta t$$
(4) Draw displacement

Figure 3.25 Steps in graphical updating of momentum and position.

Consider a short time interval Δt and apply the Momentum Principle to the system of the comet, which experiences a gravitational force exerted by the Sun. Here is a graphical version of momentum update and position update, as illustrated in Figure 3.25:

(1) Draw the net force acting on the comet, due to the Sun.
(2) Draw the impulse $\vec{F}_{net}\,\Delta t$ for a short time interval Δt. The change in momentum $\Delta\vec{p}$ is equal to the impulse.
(3) Draw the new momentum $\vec{p}_f = \vec{p}_i + \Delta\vec{p}$.
(4) Approximate the average velocity by the new velocity $\vec{v}_{avg} \approx \vec{p}_f/m$ and indicate the displacement $\Delta\vec{r} = \vec{r}_f - \vec{r}_i = (\vec{p}_f/m)\Delta t$.

The component of the net force that is parallel to the momentum is in the same direction as the momentum, so it makes the comet speed up. The magnitude of the momentum (and the speed) increase. The perpendicular component of the net force changes the direction of the momentum. See Figure 3.26.

Figure 3.26 Components of the force parallel and perpendicular to the momentum.

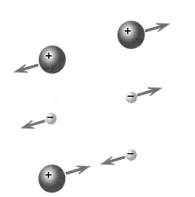

Figure 3.27 The spring force acting on the block is big when the block's momentum is zero. When the spring has its relaxed length the force is zero but the block's momentum is big.

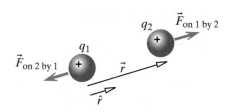

Figure 3.28 Protons repel each other; electrons repel each other; protons and electrons attract each other.

Figure 3.29 The electric force exerted on object 2 by object 1. (The force exerted on object 2 by object 1 has the same magnitude but opposite direction.)

QUESTION A student said, "The reason why the comet speeds up is because the force gets bigger as the comet moves toward the Sun. Bigger force, bigger speed." What's wrong with this analysis?

Force is related to the *change* in momentum, not directly to the momentum itself: $\Delta \vec{p} = \vec{F}_{net} \Delta t$. A striking example of this is the motion of a mass connected to a spring (Figure 3.27). The force is biggest when the spring is stretched or compressed the most, at times when the speed is zero: big force, zero momentum. When the stretch is zero, the force is zero but the mass is traveling with big speed: zero force, big momentum. So momentum is not directly related to force. Rather it is momentum *change* that is related to force (or more precisely, to impulse).

3.X.12 Now consider the situation when the comet is at position *B* in Figure 3.24, heading away from the Sun. Mentally take one step in the iterative update of momentum and position. What happens to the magnitude of the momentum, and the speed? What happens to the direction of the momentum? Explain briefly in terms of the effects of the parallel and perpendicular components of the net force acting on the comet.

3.6 THE ELECTRIC FORCE

There are electric interactions between "charged" particles such as the protons and electrons found in atoms. It is observed that two protons repel each other, as do two electrons, while a proton and an electron attract each other (Figure 3.28). Protons are said to have "positive electric charge" and electrons have "negative electric charge."

The force corresponding to this electric interaction is similar to the gravitational force, and it is known as "Coulomb's law" to honor the French scientist who established this behavior of electric forces in the late 1700s, during the same period when Cavendish measured the gravitational constant *G*. See Figure 3.29.

THE ELECTRIC FORCE (COULOMB'S LAW)

$$\vec{F}_{elec\,on\,2\,by\,1} = \frac{1}{4\pi\varepsilon_0} \frac{q_1 q_2}{|\vec{r}|^2} \hat{r}$$

$\vec{r} = \vec{r}_2 - \vec{r}_1$ is the position of 2 relative to 1.

$\frac{1}{4\pi\varepsilon_0}$ is a universal constant: $\frac{1}{4\pi\varepsilon_0} = 9 \times 10^9 \frac{N \cdot m^2}{C^2}$.

Both the gravitational and electric forces are proportional to the inverse square of the center-to-center distance ("inverse square" forces). The universal electric constant (read as "one over four pi epsilon-zero") is very much larger than the gravitational constant, reflecting the fact that electric interactions are intrinsically much stronger than gravitational interactions. For example, consider a heavy weight that hangs from a thin metal wire. The small number of atoms interacting electrically in the wire have as big an effect on the hanging weight as a very much larger number of atoms in the huge Earth, interacting gravitationally.

The charges q_1 and q_2 must be measured in SI units called "coulombs" and abbreviated "C." The proton has a charge of $+1.6 \times 10^{-19}$ C and the electron has a charge of -1.6×10^{-19} C.

QUESTION The electric force looks very much like the gravitational force. What aspects of these forces are the same?

$$\vec{F}_{\text{electric}} = \frac{1}{4\pi\varepsilon_0}\frac{q_1 q_2}{|\vec{r}|^2}\hat{r}$$

Figure 3.30 Like the gravitational force, the magnitude of the electric force is inversely proportional to the square of the distance between charged objects, and its direction depends on the unit vector $\hat{r}$.

$$\vec{F}_{\text{electric}} = \frac{1}{4\pi\varepsilon_0}\frac{q_1 q_2}{|\vec{r}|^2}\hat{r}$$

Figure 3.31 Like the gravitational force, the electric force involves the product of a property of each object (charge), and a scalar constant $1/(4\pi\epsilon_0)$.

Both forces involve the relative position vector between two objects, and both involve the unit vector associated with the relative position vector. Like the gravitational force, the electric force is an inverse square force, and is proportional to one over the square of the distance between the objects (Figure 3.30).

Both forces also involve the product of a property of each object: mass or charge. In both cases, there is a scalar multiplicative constant: G or $1/(4\pi\varepsilon_0)$; see Figure 3.31.

Because $q_1 q_2 = q_2 q_1$, the electric force has the property of reciprocity, like the gravitational force (Section 3.4). The electric force that charge 1 exerts on charge 2 has the same magnitude as the electric force that charge 2 exerts on charge 1.

QUESTION Study the three cases in Figure 3.28. Why is no minus sign needed in the electric force, unlike the case with the gravitational force?

The difference is that two positively charged particles such as two protons repel each other, whereas masses are always positive but the gravitational force is attractive. Two negatively charged particles such as electrons also repel each other and minus times minus gives plus. Only if the two particles have opposite charges will they attract each other, and then the factor $q_1 q_2$ contributes the necessary minus sign, just as in the gravitational force.

QUESTION Two protons repel each other, due to the electric force. They also attract each other gravitationally. What does this tell you about the relative strengths of the electric and gravitational interactions in this two-proton situation?

Evidently the electric force must be bigger than the gravitational force, since we do observe the protons pushing each other away from each other. The following example makes a quantitative comparison of the two forces.

EXAMPLE **Comparison of Electric and Gravitational Forces**

Two protons are a distance r from each other, center to center. Find the ratio of the magnitude of the electric force one proton exerts on the other to the magnitude of the gravitational force one proton exerts on the other.

Solution

$$F_{\text{elec}} = (9\times10^9\,\text{N}\cdot\text{m}^2/\text{C}^2)\frac{(1.6\times10^{-19}\text{C})^2}{r^2}$$

$$F_{\text{grav}} = (6.7\times10^{-11}\,\text{N}\cdot\text{m}^2/\text{kg}^2)\frac{(1.7\times10^{-27}\,\text{kg})^2}{r^2}$$

$$\frac{F_{\text{elec}}}{F_{\text{grav}}} = 1.2\times10^{36}$$

Notice that r canceled in this ratio, because both the electric and gravitational forces are proportional to the same factor $1/r^2$. The units of newtons also canceled.

FURTHER DISCUSSION This ratio is astoundingly huge. The electric repulsion of two protons is greater than their gravitational attraction by a factor of

1,200,000,000,000,000,000,000,000,000,000,000,000 !!

As a result, in the interactions of subatomic charges with each other you can ignore their extremely tiny gravitational interactions.

If electric interactions are intrinsically so enormously much stronger than gravitational interactions, why is it that gravitational forces seem so large and important in our daily lives? Matter is made of atoms that normally have the same number of negatively charged electrons around the nucleus as there are positively charged protons in the nucleus; we say that an atom is electrically "neutral." The electric effects of the electrons and protons cancel each other out. It is possible to transfer some charge from one object to another by rubbing, leaving one object with an excess of electrons and the other with an excess of protons, and then you see electric effects such as "static cling." We'll explore such effects in depth in later chapters on electric interactions. However, the amount of charge imbalance is extremely tiny; almost all of the atoms in both objects remain neutral. The fact that you see static cling despite very little charge having been transferred is due to the extreme strength of electric interactions compared with the strength of gravitational interactions.

■ Electron charge: $-e = -1.6 \times 10^{-19}$ C
Proton charge: $+e = 1.6 \times 10^{-19}$ C
Neutron charge: 0

3.X.13 A proton and an electron are separated by 1×10^{-10} m, the radius of a typical atom. Calculate the magnitude of the electric force that the proton exerts on the electron, and the magnitude of the electric force that the electron exerts on the proton.

3.X.14 A moving electron passes near the nucleus of a gold atom, which contains 79 protons and 118 neutrons. At a particular moment the electron is a distance of 3×10^{-9} m from the gold nucleus. **(a)** What is the magnitude of the force exerted by the gold nucleus on the electron? **(b)** What is the magnitude of the force exerted by the electron on the gold nucleus?

Interatomic Forces

When two objects touch each other they exert forces on each other. At the microscopic level, these contact forces are due to electric interactions between the protons and electrons in atoms on the surface of one object and the protons and electrons in atoms on the surface of the other object. Although atoms are electrically neutral, atoms that are right next to each other deform each other in such a way that the electric forces do not entirely cancel out. In the next chapter we will examine interatomic forces in more detail.

A table is able to remain standing, defying the pull of the giant Earth, because the interatomic electric forces in the table are intrinsically far stronger than gravitational interactions.

3.7   THE STRONG INTERACTION

Inside nuclei the protons repel each other electrically, yet the protons and neutrons stick together due to the strong interaction (Figure 3.32). The strong (or nuclear) interactions among protons and neutrons are essentially zero unless the particles touch each other, which means that the center-to-center distance must be about equal to the diameter of a proton or neutron (the radius of a proton or neutron is about 1×10^{-15} m, very much smaller than the radius of an atom, which is about 1×10^{-10} m). At these very short distances, the electric force is extremely large, since it is inversely proportional to the tiny distance squared, yet the strong force is even larger in stable nuclei.

There is a delicate balance between the electric repulsion and the strong attraction that determines whether a nucleus can exist at all. For example, a nucleus consisting of just two protons and no neutrons does not exist in nature because the strong force between the two protons is smaller than the electric force. This unstable helium nucleus would be called He-2, meaning the total number of protons and neutrons is 2. He-3 (two protons and one

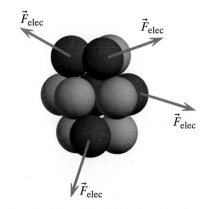

Figure 3.32 The strong force: the protons in the nucleus of an atom exert repulsive electric forces on each other as shown in the diagram. The strong interaction (which involves neutrons as well as protons) holds the nucleus together despite this electric repulsion.

neutron) and He-4 (2 protons and 2 neutrons) are stable "isotopes" of helium (different isotopes of the same element have the same number of protons but different numbers of neutrons). In these stable nuclei the protons experience attractive strong interactions with the neutrons as well as with the other proton, and this is enough to overcome the electric repulsion between the two protons.

At the start of the periodic table you'll find about the same number of neutrons as protons in the nuclei, as in He-4. For example, the most common isotope of oxygen has 8 protons and 8 neutrons (O-16). As you go farther in the periodic table you'll see that more and more neutrons are needed to offset the proton repulsions. The most common isotope of uranium has 92 protons but many more neutrons, 146 in U-238. In uranium not only are there more neutrons for the protons to interact with but also the many neutrons dilute the concentration of protons, so that on average the protons are farther apart, thereby decreasing the electric repulsion.

While there are (relatively) simple formulas for the gravitational force and the electric force, the analysis of strong interactions in the context of the Standard Model is much more complicated. In fact, the concept of force turns out to be not very productive in analyzing nuclei, and energy plays the major role.

In summary, the gravitational force is intrinsically extremely weak despite being very noticeable when you fall, because you're interacting with the entire Earth. The electric force and the strong force are comparable when two protons touch each other, but if the protons are not in contact the strong force is zero.

3.X.15 Look at the periodic table on the inside front cover of this textbook. As is standard practice, each entry gives the "atomic number" (the number of protons in the nucleus; 8 for oxygen, 92 for uranium) and the "atomic mass" in grams per mole, which is equal to the number of protons plus neutrons in a nucleus, averaged over the various isotopes for that element, which contain different numbers of neutrons. How many protons and neutrons are in a typical silicon nucleus (Si)? In a tin nucleus (Sn)? In a gold nucleus (Au)? In a thorium nucleus (Th)? You should see a trend: the farther you go in the periodic table, the more "excess" neutrons with their strong interactions are needed to offset the proton repulsions.

3.8 NEWTON AND EINSTEIN

Newton devised a particular explanatory scheme in which the analysis of motion is divided into two distinct parts:

1. Quantify the interaction in terms of a concept called "force." Specific examples are Newton's gravitational force and Coulomb's electric force.
2. Quantify the change of motion in terms of the change in a quantity called "momentum." The change in the momentum is equal to the force times Δt.

This scheme, called the "Newtonian synthesis," has turned out to be extraordinarily successful in explaining a huge variety of diverse physical phenomena, from the fall of an apple to the orbiting of the Moon. Yet we have no way of asking whether the Universe "really" works this way. It seems unlikely that the Universe actually uses the human concepts of "force" and "momentum" in the unfolding motion of an apple or the Moon.

We refer to the "Newtonian synthesis" both to identify and honor the particular, highly successful analysis scheme introduced by Newton, but also to remind ourselves that this is not the only possible way to view and analyze the Universe.

Einstein's Alternative View

Newton identified the gravitational force and represented it algebraically but could give no explanation for it. He was content with showing that it correctly predicted the motion of the planets, and this was a huge advance, the real beginning of modern science.

Einstein made a further huge advance by giving a deeper explanation for gravity, as a part of his Theory of General Relativity. He realized that it was possible to think of the massive Sun bending space and time (!) in such a way as to make the planets move the way they do. The equations in Einstein's General Theory of Relativity make it possible to calculate the curvature of space and time due to massive objects, and to predict how other objects will move in this altered space and time.

Moreover, Einstein's Theory of General Relativity accurately predicts some tiny effects that Newton's gravitational force does not, such as the slight bending of light as it passes near the Sun. General Relativity also explains some bizarre large-scale phenomena such as black holes and the observed expansion of the space between the galaxies.

Einstein's earlier Theory of Special Relativity established that nothing, not even information, can travel faster than light. Because Newton's gravitational force formula depends only on the distance between objects, not on the time, something's wrong with Newton's formula, since this implies that if an object were suddenly yanked away, its force on another object would vanish instantaneously, thus giving (in principle) a way to send information from one place to another instantaneously. Einstein's Theory of General Relativity doesn't have this problem.

Since the equations of General Relativity are very difficult to work with, and Newton's gravitational force works very well for most purposes, in this textbook we will use Newton's approach to gravity. However, you should be aware that for the most precise calculations one must use the Theory of General Relativity. For example, the highly accurate atomic clocks in the satellites that make up the Global Positioning System (GPS) have to be continually corrected using Einstein's Theory of General Relativity. Otherwise, GPS positions would be wrong by several kilometers after just one day of operation!

3.9   PREDICTING THE FUTURE OF COMPLEX SYSTEMS

We can use the Newtonian method and numerical integration to try to predict the future of a group of objects that interact with each other mainly gravitationally, such as the Solar System consisting of the Sun, planets, moons, asteroids, comets, and so on. There are other, nongravitational interactions present. Radiation pressure from sunlight makes a comet's tail sweep away from the Sun. Streams of charged particles (the "solar wind") from the Sun hit the Earth and contribute to the Northern and Southern Lights (auroras). However, the main interactions within the Solar System are gravitational.

The Solar System as a group is orbiting around the center of our Milky Way galaxy, and our galaxy is interacting gravitationally with other galaxies in a local cluster of galaxies (Figure 3.33), but changes in these speeds and directions take place very slowly compared to those within the Solar System. To a very good approximation we can neglect these interactions while studying the inner workings of the Solar System itself.

We have modeled the motion of planets and stars using the Momentum Principle and the gravitational force. We can use the same approach to model the gravitational interactions of more complex systems involving three, four,

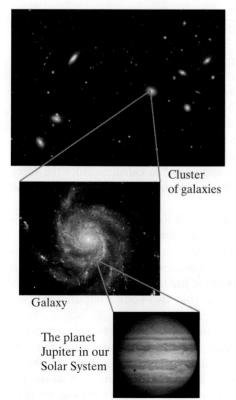

Cluster of galaxies

Galaxy

The planet Jupiter in our Solar System

Figure 3.33 Our Solar System orbits around the center of our galaxy, which interacts with other galaxies.

or any number of gravitationally interacting objects. All that is necessary is to include the forces associated with all pairs of objects.

In principle we could use these techniques to predict the future of our entire Solar System. Of course our prediction would be only approximate, because we take finite time steps that introduce numerical errors, and we are neglecting various small interactions. We would need to investigate whether the prediction of our model is a reasonable approximation to the real motion of our real Solar System.

Our Solar System contains a huge number of objects. Most of the planets have moons, and Jupiter and Saturn have many moons. There are thousands of asteroids and comets, not to mention an uncounted number of tiny specks of dust. This raises a practical limitation: the fastest computers would take a very long time to carry out even one time step for all the pieces of the Solar System.

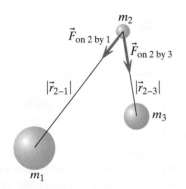

Figure 3.34 In the "three-body problem" each of the three objects interacts with two other objects.

The Three-Body Problem

As we have seen, in order to carry out a numerical integration we need to know the position and velocity of each object at some time t, from which we can calculate the forces on each object at that time. This enables us to calculate the position and momentum for each object at a slightly later time $t + \Delta t$. We need the net force on an object, which we get from vector addition of all the pairwise forces calculated from Newton's gravitational force.

Suppose that there are three gravitating objects, as shown in Figure 3.34. The force on object m_2, which is acted upon gravitationally by objects m_1 and m_3, is a natural extension of what we did earlier for the analysis of a star and a planet. We simply use the superposition principle to include the interaction with m_3 as well as the interaction with m_1:

$$\vec{F}_{\text{net on } m_2} = -G\frac{m_2 m_1}{|\vec{r}_{2-1}|^2}\hat{r}_{2-1} - G\frac{m_2 m_3}{|\vec{r}_{2-3}|^2}\hat{r}_{2-3}$$

We use this force to update the momentum of object m_2. Similar computations apply to the other masses, m_1 and m_3. These calculations are tedious, but let the computer do them! We can carry out a numerical integration for each mass: given the current locations of all the masses we can calculate the vector net force exerted on each mass at a time t, and then we can calculate the new values of momentum and position at a slightly later time $t + \Delta t$.

Reciprocity simplifies these calculations somewhat. For example, for the pair of interacting objects 2 and 3 you only need to calculate fully the force of object 2 on object 3. The force of object 3 on object 2 is just the opposite: $\vec{F}_{3 \text{ on } 2} = -\vec{F}_{2 \text{ on } 3}$.

While there exist analytical (nonnumerical) solutions for two-body gravitational motion, except for some very special cases the "three-body problem" has not been solved analytically. However, in a numerical integration, the computer doesn't mind that there are lots of quantities and a lot of repetitive calculations to be done. You yourself can solve the three-body problem numerically, by adding additional calculations to your two-body computations (see the homework problems 3.P.74 and 3.P.75).

For a single object orbiting a very massive object, or for two objects orbiting each other (with zero total momentum), the possible trajectories are only a circle, ellipse, parabola, hyperbola, or straight line. The motion of a three-body system can be vastly more complex and diverse. Figure 3.35 shows the numerical integration of a complicated three-body trajectory for one particular set of initial conditions. Adding just one more object opens up a vast range of complex behaviors.

Imagine how complex the motion of a galaxy can be!

Figure 3.35 An example of three bodies interacting gravitationally.

3.X.16 Suppose you have four stars with given initial conditions (positions and momenta). The net force on each star is the vector sum of three forces exerted by the other three stars, for a total of 12 forces that must be evaluated in each iterative step. In each step, how many forces would you have to calculate fully, and how many forces could you calculate simply by reversing vectors already calculated?

Figure 3.36 Two different orbits (one red, the other blue), starting from rest but from slightly different initial values of *x* for a low-mass object. The two orbits end with crashes on the two different massive objects. The mass of the red object is twice the mass of the blue object; both of these objects are somehow nailed down and cannot move.

Sensitivity to Initial Conditions

For a two-body system, slight changes in the initial conditions make only slight changes in the orbit, such as changing a circle into an ellipse, or an ellipse into a slightly different ellipse, but you never get anything other than a simple trajectory.

The situation is very different in systems with three or more interacting objects. Consider a low-mass object orbiting two massive objects, which we imagine somehow to be nailed down so that they cannot move. (This could also represent two positive charges that are fixed in position, with a low-mass negative charge orbiting around them.) In Figure 3.36 are two very different orbits (one shown in red, the other in blue), starting from rest with slightly different initial values of *x*. The two orbits end with crashes on the two different massive objects. The trajectories are wildly different!

This sensitivity to initial conditions generally becomes more extreme as you add more objects, and the Solar System contains lots of objects, big and small. Also, we can anticipate that if small errors in specifying the initial conditions can have large effects, so too it is likely that failing to take into account the tiny force exerted by a small asteroid might make a big difference after a long integration time.

The Solar System is actually fairly predictable, because there is one giant mass (the Sun), and the other, much smaller masses are very far apart. This is unlike the example shown here, which deliberately emphasizes the sensitivity observed when there are large masses near each other.

3.10   DETERMINISM

If you know the net force on an object as a function of its position and momentum (or velocity), you can predict the future motion of the object by simple step-by-step calculations based on the Momentum Principle. Newton's demonstration of the power of this approach induced many 17th and 18th century philosophers and scientists to adopt the view already proposed by Descartes, Boyle, and others, that the Universe is a giant clockwork device, whose future is completely determined by the present positions and momenta of all the macroscopic and microscopic objects in it. Just turn the crank, and predict the future! This point of view is called "determinism," and taken to its extreme it raises the question of whether humans actually have any free will, or whether all of our actions are predetermined. Scientific and technological advances in the 20th century, however, have led us to see that although the Newtonian approach can be used to predict the long-term future of a simple system or the short-term future of a complicated system, there are both practical and theoretical limitations on predictions in some systems.

Practical Limitations

One reason we may be unable to predict the long-term future of even a simple system is a practical limitation in our ability to measure its initial conditions with sufficient accuracy. Over time, even small inaccuracies in initial conditions can lead to large cumulative errors in calculations. This is a practical limitation rather than a theoretical one, because in principle if we could measure initial conditions more accurately, we could perform more accurate calculations.

Another practical limitation is our inability to account for all interactions in our model. Every object in the Universe interacts with every other object. In constructing simplified models, we ignore interactions whose magnitude is extremely small. However, over time, even very small interactions can lead to significant effects. Even the tiny effects of sunlight have been shown to affect noticeably the motion of an asteroid over a long time. With larger and faster computers, we can include more and more interactions in our models, but our models can never completely reflect the complexity of the real world.

A branch of current research that focuses on how the detailed interactions of a large number of atoms or molecules lead to the bulk properties of matter is called "molecular dynamics." For example, one way to test our understanding of the nature of the interactions of water molecules is to create a computational model in which the forces between molecules and the initial state of a large number of molecules—several million—are described, and then see whether running the model produces a virtual fluid that actually behaves like water. However, even with an accurate expression for the (electric) forces between atoms or molecules, it is not feasible to compute and track the motions of the 10^{25} molecules in a glass of water. Significant efforts are under way to build faster and faster computers, and to develop more efficient computer algorithms, to permit realistic numerical integration of more complex systems.

Chaos

A second kind of limitation occurs in systems that display an extreme sensitivity to initial conditions such as the situation seen in Figure 3.36, where small changes in initial conditions produce large changes in behavior. In recent years scientists have discovered systems in which *any* change in the initial conditions, no matter how small (infinitesimal), can lead to complete loss of predictability: the difference in the two possible future motions of the system diverges exponentially with time. Such systems are called "chaotic." It is thought that over a long time period the weather may be literally unpredictable in this sense.

An interesting popular science book about this new field of research is *Chaos: Making a New Science*, by James Gleick (Penguin, 1988). The issue of small changes in initial conditions is a perennial concern of time travelers in science fiction stories. Perhaps the most famous such story is Ray Bradbury's "The Sound of Thunder," in which a time traveler steps on a butterfly during the Jurassic era, and returns to an eerily changed present time.

Breakdown of Newton's Laws

There are other situations of high interest in which we cannot usefully apply Newton's laws of motion, because these classical laws do not adequately describe the behavior of physical systems. To model systems composed of very small particles such as protons and electrons, quarks (the constituents of protons and neutrons), and photons, it is necessary to use the laws of quantum mechanics. To model in detail the gravitational interactions between massive objects, it is necessary to apply the principles of General Relativity. Given the principles of quantum mechanics and General Relativity, one might assume that it would be possible to follow a procedure similar to the one we have followed in this chapter, and predict in detail the future of these systems.

However, there appear to be more fundamental limitations on what we can know about the future.

Probability and Uncertainty

Our understanding of the atomic world of quantum mechanics suggests that there are fundamental limits to our ability to predict the future, because at the atomic and subatomic levels the Universe itself is nondeterministic. We cannot know exactly what will happen at a given time, but only the probability that certain events will occur within a given time frame.

A simple example may make this clear. A free neutron (one not bound into a nucleus) is unstable and eventually decays into a positively charged proton, a negatively charged electron, and an electrically neutral antineutrino:

$$n \rightarrow p^+ + e^- + \bar{\nu}$$

The average lifetime of the free neutron is about 15 minutes. Some neutrons survive longer than this, and some last a shorter amount of time. All of our experiments and all of our theory are consistent with the notion that *it is not possible to predict when a particular neutron will decay*. There is only a (known) probability that the neutron will decay in the next microsecond, or the next, or the next.

If this is really how the Universe works, as a great deal of evidence demonstrates, then there is an irreducible lack of predictability and determinism of the Universe itself. As far as our own predictions using the Momentum Principle are concerned, consider the following simple scenario. An electron is traveling with constant velocity through nearly empty space, but there is a free neutron in the vicinity (Figure 3.37).

We might predict that the electron will move in a straight line, and some fraction of the time we will turn out to be right (the electrically uncharged neutron exerts no electric force on the electron, there is a magnetic interaction but it is quite small, and the gravitational interaction is tiny). However, if the neutron happens (probabilistically) to decay just as our electron passes nearby, suddenly our electron is subjected to large electric forces due to the decay proton and electron, and will deviate from a straight path. Moreover, the directions in which the decay products move are also only probabilistic, so we can't even predict whether the proton or the electron will come closest to our electron (Figure 3.38 and Figure 3.39).

This is a simple but dramatic example of how quantum indeterminacy can lead to indeterminacy even in the context of the Momentum Principle.

The Heisenberg Uncertainty Principle

The Heisenberg uncertainty principle states that there are actual theoretical limits to our knowledge of the state of physical systems. One quantitative formulation states that the position and the momentum of a particle cannot both be simultaneously measured exactly:

$$\Delta x \, \Delta p_x \geq h$$

This relation says that the product of the uncertainty in position Δx and the uncertainty in momentum Δp_x is equal to a constant h, called Planck's constant. Planck's constant is tiny ($h = 6.6 \times 10^{-34}$ kg $\cdot$ m^2/s), so this limitation is not noticeable for macroscopic systems, but for objects small enough to require a quantum mechanical description, the uncertainty principle puts fundamental limits on how accurately we can know the initial conditions, and therefore how well we can predict the future.

Figure 3.37 The electron may continue to travel in a straight line, but we cannot be certain that it will, because we do not know when the neutron will decay.

Figure 3.38 One possible path of the electron if the neutron does decay. The arrows represent momentum.

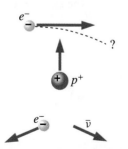

Figure 3.39 Another possible path of the electron if the neutron does decay. The trajectory of the electron depends on the directions of the momenta of the decay products. The arrows represent momentum.

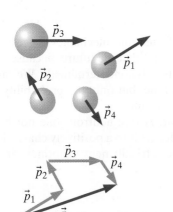

Figure 3.40 The total momentum of the system of four objects is the sum of the individual momenta of each object.

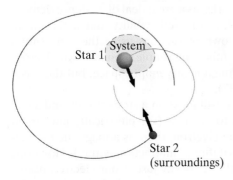

Figure 3.41 A binary star far from other objects. The gray lines show the trajectories of the individual stars. Choose star 1 as the system; star 2 is the surroundings.

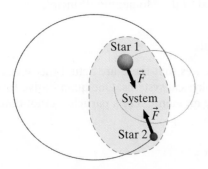

Figure 3.42 A binary star: choose both stars as the system. The momentum of the combined system doesn't change.

3.11   CONSERVATION OF MOMENTUM

The Momentum Principle, $\Delta \vec{p} = \vec{F}_{\text{net}} \Delta t$, predicts how the momentum of a system will change, due to forces exerted on the system by objects in the surroundings of that system. What happens to the momentum of objects in the *surroundings*? For all four kinds of interactions, gravitational, electromagnetic, strong, and weak, it is observed that momentum gained by the system is lost by the surroundings, and we say that momentum is "conserved." That is, you can't create or destroy momentum, you can only move it from one object to another:

> ### CONSERVATION OF MOMENTUM
> $$\Delta \vec{p}_{\text{sys}} + \Delta \vec{p}_{\text{surr}} = \vec{0}$$

The momentum of the system is defined as the (vector) sum of the momenta of all objects in the system (Figure 3.40):

$$\vec{p}_{\text{sys}} = \vec{p}_1 + \vec{p}_2 + \vec{p}_3 + \cdots$$

Similarly, the momentum of the surroundings is the (vector) sum of the momenta of all the objects in the surroundings.

Momentum Can Flow Between System and Surroundings

Consider the case of two stars orbiting each other, a "binary star" far from all other objects (Figure 3.41). If you choose as the system of interest star 1, star 2 is the "surroundings" and exerts a force that changes the first star's momentum. Star 2, the surroundings, experiences a force of equal magnitude and opposite direction (reciprocity of gravitational forces). In an amount of time Δt, the momentum change of the system (star 1) is $\Delta \vec{p}_1 = \vec{F}_{\text{on 1 by 2}} \Delta t$, and the momentum change of the surroundings (star 2) is $\Delta \vec{p}_2 = -\vec{F}_{\text{on 1 by 2}} \Delta t = -\Delta \vec{p}_1$. Whatever momentum the system gains, the surroundings loses: $\Delta \vec{p}_{\text{sys}} + \Delta \vec{p}_{\text{surr}} = \vec{0}$. We say momentum is a "conserved" quantity.

The Momentum of an Isolated System Does Not Change

If on the other hand you choose both stars as your system (Figure 3.42), the surroundings consist of other stars, which may be so far away as to have negligible effects on the binary star. In that case the net force exerted by the surroundings on the system is nearly zero. The momentum of the combined system is $\vec{p}_{\text{sys}} = \vec{p}_1 + \vec{p}_2$. Let's see how the momentum of the system consisting of both stars changes:

$$\begin{aligned}
\Delta \vec{p}_{\text{sys}} &= \Delta \vec{p}_1 + \Delta \vec{p}_2 \\
&= \vec{F}_{\text{on 1 by 2}} \Delta t + \vec{F}_{\text{on 2 by 1}} \Delta t \\
&= \vec{F}_{\text{on 1 by 2}} \Delta t + (-\vec{F}_{\text{on 1 by 2}} \Delta t) \\
&= \vec{0}
\end{aligned}$$

For the two-star system, the Momentum Principle $\vec{p}_f = \vec{p}_i + \vec{F}_{\text{net}} \Delta t$ reduces to $\vec{p}_{\text{sys},f} = \vec{p}_{\text{sys},i}$ and the momentum of the system $\vec{p}_{\text{sys}} = \vec{p}_1 + \vec{p}_2$ remains constant in magnitude and direction. This is an important special case: the momentum of an isolated system, a system with negligible interactions with the surroundings, doesn't change but stays constant. The momentum of each star changes, due to interactions with the other star, but momentum lost by one star is gained by the other star, and the total momentum of the system doesn't change.

Stating the Momentum Principle in this form highlights another aspect of the predictive power of this fundamental principle. It allows us to predict and explain aspects of the behavior of complex systems without knowing the details of the large number of interactions inside the systems. It can be particularly useful in analyzing the behavior of systems of many particles whose interactions with the surroundings are negligible during the time interval of interest.

EXAMPLE **A Balloon Rocket**

Suppose that you blow up a balloon, and hold it closed with your fingers. When you release your grip, air rushes out the opening in the balloon. If the balloon is oriented horizontally, as shown in Figure 3.43, so the outflowing air travels in the $-x$ direction, what must happen to the balloon?

Solution System: Balloon plus all the air initially inside it
Surroundings: Everything else
Initial time: Just before the balloon is released
Final time: Just after the balloon is released
Assumption: Negligible interactions with surroundings during this short Δt
Momentum Principle:

Figure 3.43 When the balloon is released, the total momentum of the system must still be zero.

$$\Delta \vec{p}_{\text{sys}} + \Delta \vec{p}_{\text{surr}} = \vec{0}$$

Let $\vec{p}$ represent $\vec{p}_{\text{sys}}$, the total momentum of the system; that is, it is the sum of the momenta of the air inside the balloon plus the momentum of the rubber balloon itself.

$$\vec{p} = \vec{p}_{\text{air}} + \vec{p}_{\text{rubber}}$$

Initially the balloon and its contents are at rest, so

$$\vec{p}_i = \vec{0}$$

Because the interaction with the surroundings was negligible over this short interval, it follows that

$$\vec{p}_f - \vec{p}_i = \vec{0}$$
$$\vec{p}_f = \vec{0}$$

In the final state the total momentum of the system must still be zero. However, the momentum of the air coming out of the balloon is not zero; it is in the $-x$ direction. Therefore the momentum of the rubber balloon must be nonzero, in the $+x$ direction, so the balloon will move in the $+x$ direction. In this way the x components of the momenta of all of the objects in the system add up to zero.

EXAMPLE **The Weak Interaction: Neutron Decay**

A free neutron, one that is outside a nucleus, is unstable and decays into a positively charged proton, a negatively charged electron, and an uncharged antineutrino:

$$n \rightarrow p^+ + e^- + \bar{\nu}$$

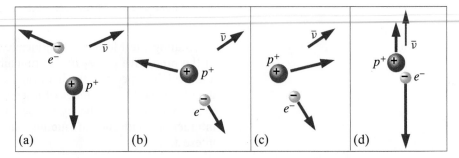

Figure 3.44 Which of these decays of a neutron at rest is impossible? The arrows represent momentum.

This decay of a free neutron is caused by the weak interaction. The precise time that the neutron will decay is governed by quantum mechanical laws of probability and the nature of the weak interaction and is not predictable for any particular neutron, but the average lifetime of a neutron is about 15 minutes. Also probabilistic are the magnitudes and directions of the momenta of the decay products (proton, electron, and antineutrino), and Figure 3.44 shows some decays of an isolated free neutron that was at rest. Which of the decays shown in Figure 3.44 is actually impossible, if the neutron was at rest?

Solution   The decay shown in Figure 3.44c is not actually possible. Conservation of momentum applies to all of the four fundamental interactions, including the weak interaction. The momentum of the isolated neutron was zero, so the momenta of the decay products must also add up to zero. In Figure 3.44c, however, the total momentum of the proton, electron, and antineutrino has a positive x component, which is not possible.

What's interesting about this is that we can rule out the possibility of the decay shown in Figure 3.44c without knowing anything about the details of the weak interaction. In fact, in quantum mechanical calculations at the atomic level concerned with electromagnetism, the strong interaction, or the weak interaction, force isn't even a particularly useful concept, yet conservation of momentum continues to play an important role even in the subatomic world.

3.X.17 A system consists of a 1 kg block moving with velocity $\langle 5, 2, 0 \rangle$ m/s and a 3 kg block moving with velocity $\langle -3, 4, 0 \rangle$ m/s. **(a)** What is the momentum of this two-block system? **(b)** Next, due to interactions between the two blocks, each of their velocities change, but the two-block system is nearly isolated from the surroundings. Now what is the momentum of the two-block system?

The Center of Mass

The momentum of a system such as a binary star can be expressed in terms of the velocity of the "center of mass" of the system. The position of the center of mass of a system of particles is defined as a weighted average of their positions:

$$\vec{r}_{CM} = \frac{m_1 \vec{r}_1 + m_2 \vec{r}_2 + m_3 \vec{r}_3 + \cdots}{M_{total}}$$

$M_{total} = m_1 + m_2 + m_3 + \cdots$ is the total mass of the system. It can be shown that this definition is also valid if object 1 is a large object (not just a small particle) as long as m_1 is the mass of object 1 and $\vec{r}_1$ is the position of the center of mass of object 1. If you apply this definition to a binary star system, you find that the center of mass is between the two stars, closer to the more massive star.

Figure 3.45 A 10 kg ball is located at ⟨2, 0, 0⟩ m and a 2 kg ball is located at ⟨8, 2, 0⟩ m. Where is the center of mass?

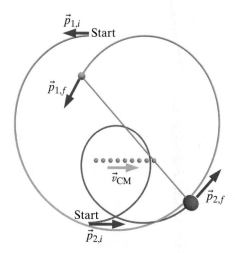

Figure 3.46 A binary star drifts through space. The center of mass moves with constant velocity. A line connecting the two stars passes through the center of mass, which is closer to the more massive star.

Figure 3.47 The total momentum of the two-star system does not change, as can be seen by the graphical sum of the vector momenta.

3.X.18 A 10 kg ball is located at ⟨3, 0, 0⟩ m, and a 2 kg ball is located at ⟨8, 2, 0⟩ m (Figure 3.45). Find the center of mass of the two-ball system. You should find that the center of mass is close to the heavier ball.

The Velocity of the Center of Mass

Taking the derivative with respect to time of the definition of center of mass, and assuming no changes in the masses, we find the following (since $\vec{v} = d\vec{r}/dt$):

$$\vec{v}_{CM} = \frac{m_1\vec{v}_1 + m_2\vec{v}_2 + m_3\vec{v}_3 + \cdots}{M_{total}}$$

Multiplying through by the total mass, we have this:

$$M_{total}\vec{v}_{CM} = m_1\vec{v}_1 + m_2\vec{v}_2 + m_3\vec{v}_3 + \cdots$$

However, if the speeds are small compared to the speed of light ($\gamma \approx 1$), the quantity $m_1\vec{v}_1 + m_2\vec{v}_2 + m_3\vec{v}_3 + \cdots$ is the total momentum of the system, so we can express the momentum of a system in terms of its center-of-mass velocity:

MOMENTUM AND CENTER OF MASS VELOCITY

$$\vec{p}_{sys} = M_{total}\vec{v}_{CM}$$

(If the speeds are small compared to the speed of light.)

If there are negligible interactions with the surroundings of a binary star, the velocity of the center of mass does not change but is constant as the binary star drifts through space (or is zero if the binary star's total momentum is zero). Figure 3.46 shows the motion of the center of mass for a binary star that is drifting through space. The two stars orbit around the (moving) center-of-mass point. A line connecting the two stars passes through the center of mass, which is closer to the more massive star.

The initial momentum of the two-star system is $\vec{p}_{1,i} + \vec{p}_{2,i}$, which is also equal to $(m_1 + m_2)\vec{v}_{CM}$. Later, both momenta of the stars have changed, but the vector sum $\vec{p}_{1,f} + \vec{p}_{2,f}$ is still the same and still equal to $(m_1 + m_2)\vec{v}_{CM}$. Figure 3.47 shows graphical addition of the momenta of the two stars to make the total system momentum, which does not change.

3.X.19 A system consists of a 2 kg block moving with velocity ⟨−3, 4, 0⟩ m/s and a 5 kg block moving with velocity ⟨2, 6, 0⟩ m/s. Calculate the velocity of the center of mass of the two-block system, given that the momentum of the system is $M_{total}\vec{v}_{CM}$.

EXAMPLE   **What Is *Not* Conserved?**

In later chapters we will see that energy and angular momentum are also conserved quantities. However, most quantities, including velocity, are not conserved quantities. Consider a collision between two blocks on ice (so that there is low friction), each with mass m. One block is initially at rest and the other is moving with initial velocity ⟨v_i, 0, 0⟩. In the collision they stick together and move with final velocity ⟨v_f, 0, 0⟩. What is the final velocity? Is velocity a conserved quantity?

Solution   As the blocks slide on the ice there are no changes in momentum in the vertical direction (the y components of momentum are zero at all times). In the

xz plane of the ice there are negligible interactions with the surroundings, so the momentum of the surroundings doesn't change, and $\Delta \vec{p}_{\text{sys}} = \vec{0}$. Therefore $\vec{p}_{\text{sys}} = \vec{p}_1 + \vec{p}_2$ doesn't change. Before the collision the total momentum is

$$\vec{p}_{\text{sys}} = m\langle v_i, 0, 0 \rangle + m\vec{0} = \langle m v_i, 0, 0 \rangle$$

After the collision the total momentum is

$$\vec{p}_{\text{sys}} = m\langle v_f, 0, 0 \rangle + m\langle v_f, 0, 0 \rangle = \langle 2 m v_f, 0, 0 \rangle$$

Since $\vec{p}_{\text{sys}}$ doesn't change, we have

$$\langle 2 m v_f, 0, 0 \rangle = \langle m v_i, 0, 0 \rangle$$

$$v_f = \frac{1}{2} v_i$$

The final speed (and velocity) is half the initial value. Nothing in the surroundings changed, so $\Delta \vec{v}_{\text{system}} + \Delta \vec{v}_{\text{surr}} \neq \vec{0}$. We say that velocity is not a conserved quantity, but momentum is. This is another reason why momentum is such an important physical quantity.

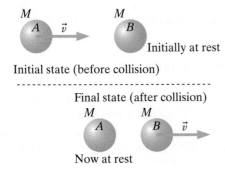

Initial state (before collision)

- -

Final state (after collision)

Now at rest

Figure 3.48 A head-on collision between two identical bowling balls, each of mass M.

3.X.20 Consider the head-on collision of two identical bowling balls (Figure 3.48). Ball A with velocity $\vec{v}$ strikes ball B, which was at rest. Then ball A stops and ball B moves with the same velocity $\vec{v}$ that ball A had initially.

  (a) Choose a system consisting only of ball A. What is the momentum change of the system during the collision? What is the momentum change of the surroundings?

  (b) Choose a system consisting only of ball B. What is the momentum change of the system during the collision? What is the momentum change of the surroundings?

  (c) Choose a system consisting of both balls. What is the momentum change of the system during the collision? What is the momentum change of the surroundings?

3.X.21 You hang from a tree branch, then let go and fall toward the Earth. As you fall, the y component of your momentum, which was originally zero, becomes large and negative.

  (a) Choose yourself as the system. There must be an object in the surroundings whose y momentum must become equally large, and positive. What object is this?

  (b) Choose yourself and the Earth as the system. The y component of your momentum is changing. Does the total momentum of the system change? Why or why not?

Relativistic Momentum Conservation

Particle accelerators produce beams of particles such as electrons, protons, and pions at speeds very close to the speed of light. When these high-speed particles interact with other particles, experiments show that $\Delta \vec{p}_{\text{sys}} + \Delta \vec{p}_{\text{surr}} = \vec{0}$, but only if the momentum of each particle is defined in the way Einstein proposed, $\vec{p} \equiv \gamma m\vec{v}$. When the speed v approaches the speed of light c, the low-speed approximation $\gamma \approx 1$ ($\vec{p} \approx m\vec{v}$) is not valid. The quantity $m\vec{v}$ is not a conserved quantity when the speed v approaches the speed of light c, but momentum $\vec{p} = \gamma m\vec{v}$ is conserved.

THE MULTIPARTICLE MOMENTUM PRINCIPLE

In our applications of the Momentum Principle we have often chosen a single object as the system of interest. However, we have seen that in some situations it can be very useful to choose a system that consists of two or more interacting objects. (Note that we can choose any object or set of objects as the system. Sometimes one choice is more useful than another, but any choice is valid.)

We will call forces "external" if they are exerted on objects in the system by objects in the surroundings. We call forces "internal" if they are forces exerted on objects in the system by other objects that are also in the system. What is remarkable is that the internal forces do not appear in the Momentum Principle for multiparticle systems:

MOMENTUM PRINCIPLE FOR MULTIPARTICLE SYSTEMS

$$\Delta \vec{p}_{\text{sys}} = (\vec{p}_{\text{sys},f} - \vec{p}_{\text{sys},i}) = \vec{F}_{\text{net}} \Delta t$$

$$\vec{p}_{\text{sys}} = \vec{p}_1 + \vec{p}_2 + \vec{p}_3 + \cdots$$
$$\vec{F}_{\text{net}} = \vec{F}_1 + \vec{F}_2 + \vec{F}_3 + \cdots \quad \text{(sum of all } external \text{ forces)}$$

This is an interesting statement of the Momentum Principle: it says that if we know the external forces acting on a multi-object system, forces exerted on objects in the chosen system by objects in the surroundings, we can draw conclusions about the change of momentum of the system over some time interval without worrying about any of the details of the interactions of the objects inside the system with each other. This can greatly simplify the analysis of the motion of some very complex systems. In a way, it should not be surprising that internal forces alone cannot change a system's total momentum. If they could, you could lift yourself off the ground by pulling up on your own feet!

We have implicitly been using this multiparticle version of the Momentum Principle when we have treated macroscopic objects like humans, spacecraft, and planets as if they were single pointlike objects.

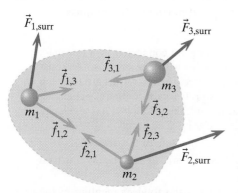

Figure 3.49 External and internal forces acting on a system of three particles.

Internal Forces Cancel

In a three-particle system (Figure 3.49) we show all of the forces acting on each particle, where the lower case $\vec{f}$'s are forces the particles exert on each other (so-called "internal" forces), and the upper case $\vec{F}$'s are forces exerted by objects in the surroundings that are not shown and are not part of our chosen system (these are so-called "external" forces, such as the gravitational attraction of the Earth, or a force that you exert by pulling on one of the particles).

We will use the following shorthand notation: $\vec{f}_{1,3}$ will denote the force exerted on particle 1 by particle 3; $\vec{F}_{2,\text{surr}}$ will denote the force on particle 2 exerted by objects in the surroundings, and so on. We start by applying the Momentum Principle to each of the three particles separately:

$$\Delta \vec{p}_1 = (\vec{F}_{1,\text{surr}} + \vec{f}_{1,2} + \vec{f}_{1,3})\Delta t$$
$$\Delta \vec{p}_2 = (\vec{F}_{2,\text{surr}} + \vec{f}_{2,1} + \vec{f}_{2,3})\Delta t$$
$$\Delta \vec{p}_3 = (\vec{F}_{3,\text{surr}} + \vec{f}_{3,1} + \vec{f}_{3,2})\Delta t$$

Nothing new so far—but now we add up these three equations. That is, we create a new equation by adding up all the terms on the left sides of the three equations, and adding up all the terms on the right sides, and setting them equal to each other:

$$\Delta \vec{p}_1 + \Delta \vec{p}_2 + \Delta \vec{p}_3 = (\vec{F}_{1,\,surr} + \vec{f}_{1,2} + \vec{f}_{1,3} + \vec{F}_{2,\,surr} + \vec{f}_{2,1} + \vec{f}_{2,3}$$
$$+ \vec{F}_{3,\,surr} + \vec{f}_{3,1} + \vec{f}_{3,2})\Delta t$$

Many of these terms cancel. By the principle of reciprocity (see Section 3.4), which is obeyed by gravitational and electric interactions, we have this:

$$\vec{f}_{1,2} = -\vec{f}_{2,1}$$
$$\vec{f}_{1,3} = -\vec{f}_{3,1}$$
$$\vec{f}_{2,3} = -\vec{f}_{3,2}$$

Thanks to reciprocity, all that remains after the cancellations is this:

$$\Delta \vec{p}_1 + \Delta \vec{p}_2 + \Delta \vec{p}_3 = (\vec{F}_{1,\,surr} + \vec{F}_{2,\,surr} + \vec{F}_{3,\,surr})\Delta t$$

The total momentum of the system is $\vec{p}_{sys} = \vec{p}_1 + \vec{p}_2 + \vec{p}_3$, so we have:

$$\Delta \vec{p}_{sys} = \vec{F}_{net,\,surr}\, \Delta t$$

The importance of this equation is that reciprocity has eliminated all of the internal forces (the forces that the particles in the system exert on each other); internal forces cannot affect the motion of the system as a whole. All that matters in determining the change of (total) momentum is the net external force. The equation has the same form as the Momentum Principle for a single particle.

As we have seen, at low speeds compared to the speed of light, the momentum of the system can be expressed as $\vec{p}_{sys} = M_{total}\vec{v}_{CM}$, where the center of mass is a mathematical point obtained from a weighted average of the masses in the system.

Moreover, if an object is a sphere whose density is only a function of radius, the object exerts a gravitational force on other objects as though the sphere were a point particle. We can predict the motion of a star or a planet or an asteroid as though it were a single point particle of large mass.

3.X.22 Two balls of mass 0.3 kg and 0.5 kg are connected by a low-mass spring. This device is thrown through the air with low speed, so air resistance is negligible. The motion is complicated: the balls whirl around each other, and at the same time the system vibrates, with continually changing stretch of the spring. At a particular instant, the 0.3 kg ball has a velocity $\langle 4, -3, 2 \rangle$ m/s and the 0.5 kg ball has a velocity $\langle 2, 1, 4 \rangle$ m/s.

  (a) At this instant, what is the total momentum of the device?
  (b) What is the net gravitational (vector) force exerted by the Earth on the device?
  (c) At a time 0.1 seconds later, what is the total momentum of the device?

Low-Speed Form of the Momentum Principle

In the case of speeds small compared to the speed of light, since $\gamma \approx 1$, we have

$$\vec{p}_{sys} = M_{total}\vec{v}_{CM}$$

If the mass is constant, the change in momentum is then

$$\Delta \vec{p}_{sys} = M_{total}\Delta \vec{v}_{CM} = \vec{F}_{net}\,\Delta t$$

Dividing by Δt, we have this:

$$\frac{\Delta \vec{p}_{sys}}{\Delta t} = M_{total}\frac{\Delta \vec{v}_{CM}}{\Delta t} = \vec{F}_{net}$$

In the limit as $\Delta t \rightarrow 0$, this is

$$\frac{d\vec{p}_{sys}}{dt} = M_{total}\frac{\vec{v}_{CM}}{dt} = \vec{F}_{net}$$

The rate of change of $\vec{v}_{CM}$ is the acceleration of the center of mass, $\vec{a}_{CM}$, so finally we have this alternative, special form of the Momentum Principle, valid when the masses do not change and speeds are small compared to the speed of light:

SPECIAL FORM OF THE MOMENTUM PRINCIPLE

$$M_{total}\vec{a}_{CM} = \vec{F}_{net}$$

(This is valid if the speeds are small compared to the speed of light and the mass is constant.)

EXAMPLE

Accelerating Sports Car

A sports car with a mass of 1500 kg goes from 0 to 60 miles per hour in 4.2 seconds. What is the average net force on the car during this acceleration?

Solution

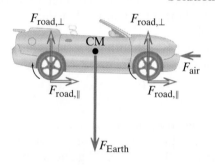

Figure 3.50

Choose the car as the system and use the Momentum Principle. Figure 3.50 shows the forces exerted on the car by the Earth, the road, and the air (air resistance), which are the objects in the surroundings that have significant effects on the car. The forces of the road on the tires are shown split into components parallel and perpendicular to the momentum.

A car is a complicated machine inside. Pistons accelerate up and down due to explosions of gasoline, gears rotate, the wheels go around. Yet there is one aspect of the car's motion that is easy to analyze: how the center of mass moves, which is governed by the Momentum Principle, either in the form $\Delta \vec{p}_{sys} = \vec{F}_{net}\,\Delta t$ or the special form $M_{total}\vec{a}_{CM} = \vec{F}_{net}$.

Whichever form we use, we need to convert the known data to SI units. One mile is about 1600 meters (1.6 km):

$$60\frac{\text{miles}}{\text{hour}} \times \frac{1600\text{ m}}{\text{mile}} \times \frac{1\text{ hour}}{3600\text{ s}} = 26.7\text{ m/s}$$

The average acceleration is $\Delta \vec{v}/\Delta t$, with magnitude (26.7 m/s)/(4.2 s) = 6.3 m/s/s. This is a high acceleration of about two-thirds of a "g" (9.8 m/s/s).

From $M_{total}\vec{a}_{CM} = \vec{F}_{net}$ we obtain the magnitude of the average force to be (1500 kg)(6.3 m/s/s) = 9500 N. Compare this with the weight of the car, which is mg = (1500 kg)(9.8 N/ kg) = 14700 N.

Alternatively, one can obtain the same result starting from $\Delta \vec{p}_{sys} = \vec{F}_{net}\,\Delta t$. The magnitude of the change in momentum is $|\Delta \vec{p}_{sys}| = (1500\text{ kg})(26.7\text{ m/s}) = 40050\text{ kg}\cdot\text{m/s}$, which is equal to $|\vec{F}_{net}|\Delta t$. Dividing by the time interval we again find that the magnitude of the average force is (40050 kg·m/s)/(4.2 s) = 9500 N.

FURTHER DISCUSSION   Since the vertical component of momentum p_y is always zero on a flat road, p_y isn't changing, so the vertical component of the net force must be zero. The vertical components of the forces exerted by the road support the weight of the car. The horizontal component of the net force includes the forward forces exerted by the road and the backward force of air resistance. How does the road "know" to apply a force in the forward direction?

The engine turns the wheels so that the bottom of a tire moves backward as the car moves forward. Atoms in the bottoms of the tires push backward on atoms in the road. Since the interatomic forces are electric and have the property of reciprocity, the atoms in the road push forward on the atoms in the tires with the same magnitude of force that the tires exert on the road.

On glare ice the tires may not be able to apply much of a force to the ice, and there may be little or no forward net force on the tires. The engine spins the wheels, but there is little or no forward acceleration because there is little or no forward force.

> **3.X.23** A 100 kg device consisting of many moving and rotating parts is subject to a net external force of $\langle 300, 500, -200 \rangle$ N. What is the acceleration of the center of mass of this complicated device?

3.13   COLLISIONS: NEGLIGIBLE EXTERNAL FORCES

An event is called a "collision" if it involves an interaction that takes place in a relatively short time and has a large effect on the momenta of the objects compared to the effects of other interactions during that short time. A collision does not necessarily involve actual physical contact between objects, which may be interacting via long-range forces like the gravitational force or the electric force. A collision can involve attractive forces. For example, a spacecraft is deflected as it passes close to Jupiter on its way to Saturn, and Jupiter exerts a large gravitational force for a relatively short time. During that short time the effects of the other planets are negligible.

Often it can be useful to analyze collisions by choosing a system that includes all of the colliding objects, as we will see in the following example.

Figure 3.51 Two lumps of clay traveling through the air, just before colliding.

| EXAMPLE | **Two Lumps of Clay Collide** |

Two lumps of clay travel through the air toward each other, at speeds much less than the speed of light (Figure 3.51), rotating as they move. When the lumps collide they stick together. The mass of lump 1 is 0.2 kg and its initial velocity is $\langle 6, 0, 0 \rangle$ m/s, and the mass of lump 2 is 0.5 kg and its initial velocity is $\langle -5, 4, 0 \rangle$ m/s. What is the final velocity of the center of mass of the stuck-together lumps?

Solution   The Momentum Principle applies to the motion of the center of mass, even though the objects are rotating.

■ $v \ll c$; neglect air resistance and gravitational force (which are negligible compared to the large contact forces during the very short time of the collision).

1. Momentum Principle
   System: both lumps
   Surroundings: air, Earth

$$\vec{p}_{\text{total}, f} = \vec{p}_{\text{total}, i} + \vec{F}_{\text{net}} \, \Delta t$$

$$\vec{p}_{\text{total}, f} \approx m_1 \vec{v}_{1i} + m_2 \vec{v}_{2i} + \langle 0, 0, 0 \rangle \Delta t$$

$$(m_1 + m_2) \vec{v}_f = m_1 \vec{v}_{1i} + m_2 \vec{v}_{2i}$$

$$\vec{v}_f = \frac{m_1 \vec{v}_{1i} + m_2 \vec{v}_{2i}}{m_1 + m_2}$$

$$\vec{v}_f = \frac{(0.2 \text{ kg}) \langle 6, 0, 0 \rangle + (0.5 \text{ kg}) \langle -5, 4, 0 \rangle}{(0.7 \text{ kg})} \text{ m/s}$$

$$\vec{v}_f = \langle -1.86, 2.86, 0 \rangle \text{ m/s}$$

2. Not necessary to update position

3. Check: The units are correct.

vf

m_1

m_2

Figure 3.52 The momentum of the stuck-together lumps of clay just after the collision is equal to the sum of the initial momenta of the two lumps of clay. The object may rotate.

FURTHER DISCUSSION By choosing both lumps as the system, we were able to find the final momentum of the system without needing to know anything about the details of the complex forces that the lumps exerted on each other during the collision.

> QUESTION Why doesn't it matter that the lumps may be rotating after the collision (Figure 3.52)?

The rotation of an object doesn't affect its momentum; the total momentum of the system is still the sum of the individual momenta of the objects within the system and equal to $M_{total}\vec{v}_{CM}$. However, the Momentum Principle does not tell us anything about how fast the stuck-together objects will be rotating; for this, we will have to apply the Angular Momentum Principle, which is discussed in a later chapter.

The initial velocities are the center-of-mass velocities of each lump, and the final velocity is the center-of-mass velocity of the combined system after the two lumps stick together. We can also speak of the velocity of the center of mass just before the collision, which is the velocity of a mathematical point between the two lumps, calculated as the weighted average of their positions.

> QUESTION What was the center-of-mass velocity of the two lumps just before they collided?

It was $\langle -1.86, 2.86, 0 \rangle$ m/s, the same as after the collision, since the total momentum can be calculated as $M_{total}\vec{v}_{CM}$ both before and after the collision.

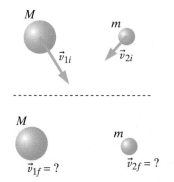

M

$\vec{v}_{1i}$

m

$\vec{v}_{2i}$

M

$\vec{v}_{1f} = ?$

m

$\vec{v}_{2f} = ?$

Figure 3.53 The initial momenta of the colliding objects are known, but the final momenta after the collision are unknown.

Some Problems Require More than One Principle

Because the lumps in the previous example stuck together, we had enough information to solve for the final velocity of the stuck-together lumps—we had one equation with only one unknown (vector) quantity. In more complex situations, we will find that we sometimes do not have enough information to solve for all of the unknown quantities by applying only the Momentum Principle.

For example, consider a collision between two balls that bounce off each other, as shown in Figure 3.53. The Momentum Principle tells us that

$$\vec{p}_{total,f} = \vec{p}_{total,i}$$
$$\vec{p}_{1f} + \vec{p}_{2f} = \vec{p}_{1i} + \vec{p}_{2i}$$

but we are left with two unknown (vector) quantities, $\vec{p}_{1f}$ and $\vec{p}_{2f}$, and only one equation. We will not be able to analyze problems of this kind fully until we can invoke the Energy Principle (Chapter 6) and the Angular Momentum Principle (Chapter 11), along with the Momentum Principle.

> **3.X.24** You and a friend each hold a lump of wet clay. Each lump has a mass of 20 grams. You each toss your lump of clay into the air, where the lumps collide and stick together. Just before the impact, the velocity of one lump was $\langle 5, 2, -3 \rangle$ m/s, and the velocity of the other lump was $\langle -3, 0, -2 \rangle$ m/s. What was the total momentum of the lumps just before impact? What is the momentum of the stuck-together lump just after the impact? What is its velocity?
>
> **3.X.25** In outer space, far from other objects, two rocks collide and stick together. Before the collision their momenta were $\langle -10, 20, -5 \rangle$ kg·m/s and $\langle 8, -6, 12 \rangle$ kg·m/s. What was their total momentum before the collision? What must be the momentum of the combined object after the collision?

– Equivalent:

Figure 3.54 A sphere whose density depends only on distance from the center acts like a point particle with all the mass at the center.

Figure 3.55 The Cavendish experiment.

3.14 | POINTS AND SPHERES

The expression for the gravitational force applies to objects that are "point-like" (very small compared to the center-to-center distance between the objects). In later chapters we will be able to show that any hollow spherical shell with uniform density acts gravitationally on external objects as though all the mass of the shell were concentrated at its center. The density of the Earth is not uniform, because the central iron core has higher density than the outer layers. However, by considering the Earth as layers of hollow spherical shells, like an onion, with each shell of nearly uniform density, we get the result that the Earth can be modeled for most purposes as a point mass located at the center of the Earth (for very accurate calculations one must take into account small irregularities in the Earth's density from place to place). Similar statements can be made about other planets and stars. In Figure 3.54, the gravitational force is correctly calculated using the center-to-center distance $|\vec{r}|$.

This is not an obvious result. After all, in Figure 3.54 some of the atoms are closer and some farther apart than the center-to-center distance $|\vec{r}|$, but the net effect after adding up all the interactions of the individual atoms is as though the two objects had collapsed down to points at their centers. This is a very special property of $1/r^2$ forces, both gravitational and electric, and is not true for forces that have a different dependence on distance.

3.15 | MEASURING THE GRAVITATIONAL CONSTANT G

In order to make quantitative predictions and analyses of physical phenomena involving gravitational interactions, it is necessary to know the universal gravitational constant G. In 1797–1798 Henry Cavendish performed the first experiment to determine a precise value for G (Figure 3.55). In this kind of experiment, a bar with metal spheres at each end is suspended from a thin quartz fiber that constitutes a "torsional" spring. From other measurements, it is known how large a tangential force measured in newtons is required to twist the fiber through a given angle. Large balls are brought near the suspended spheres, and one measures how much the fiber twists due to the gravitational interactions between the large balls and the small spheres.

If the masses are measured in kilograms, the distance in meters, and the force in newtons, the gravitational constant G has been measured in such experiments to be

$$G = 6.7 \times 10^{-11} \frac{\text{N} \cdot \text{m}^2}{\text{kg}^2}$$

This extremely small number reflects the fact that gravitational interactions are inherently very weak compared with electromagnetic interactions. The only reason that gravitational interactions are significant in our daily lives is that objects interact with the entire Earth, which has a huge mass. It takes sensitive measurements such as the Cavendish experiment to observe gravitational interactions between two ordinary-sized objects.

SUMMARY

Four fundamental interactions:
- Gravitational interactions (all objects attract each other gravitationally)
- Electromagnetic interactions (electric and magnetic in-

teractions, closely related to each other); interatomic forces are electric in nature
- "Strong" interactions (inside the nucleus of an atom)
- "Weak" interactions (neutron decay, for example)

The gravitational force (Figure 3.56)

$$\vec{F}_{\text{grav on 2 by 1}} = -G\frac{m_1 m_2}{|\vec{r}|^2}\hat{r}$$

$\vec{r} = \vec{r}_2 - \vec{r}_1$ is the position of 2 relative to 1

G is a universal constant: $G = 6.7 \times 10^{-11}\dfrac{\text{N} \cdot \text{m}^2}{\text{kg}^2}$

Figure 3.56 Near the Earth's surface, $|\vec{F}_{\text{grav}}| = mg$, where $g = +9.8\,\text{N/kg}$.

The electric force (Figure 3.57)

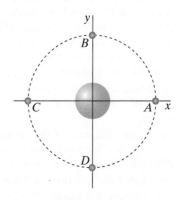

Figure 3.57

$$\vec{F}_{\text{elec on 2 by 1}} = \frac{1}{4\pi\varepsilon_0}\frac{q_1 q_2}{|\vec{r}|^2}\hat{r}$$

$\vec{r} = \vec{r}_2 - \vec{r}_1$ is the position of 2 relative to 1

$\dfrac{1}{4\pi\varepsilon_0}$ is a universal constant: $\dfrac{1}{4\pi\varepsilon_0} = 9 \times 10^9\dfrac{\text{N} \cdot \text{m}^2}{\text{C}^2}$.

Reciprocity
$\vec{F}_{\text{on 1 by 2}} = -\vec{F}_{\text{on 2 by 1}}$ (gravitational and electric forces). This is also called "Newton's third law of motion."

Conservation of momentum

$$\Delta\vec{p}_{\text{sys}} + \Delta\vec{p}_{\text{surr}} = \vec{0}$$

The multiparticle Momentum Principle

$$\Delta\vec{p}_{\text{sys}} = \vec{F}_{\text{net,surr}}$$

Additional results
Uniform-density spheres act as though all the mass were at the center.

Motion with speeds small compared to the speed of light; constant mass

$$\vec{r}_{\text{CM}} = \frac{m_1\vec{r}_1 + m_2\vec{r}_2 + m_3\vec{r}_3 + \cdots}{M_{\text{total}}}$$

$$\vec{v}_{\text{CM}} = \frac{m_1\vec{v}_1 + m_2\vec{v}_2 + m_3\vec{v}_3 + \cdots}{M_{\text{total}}}$$

$$\vec{p}_{\text{sys}} = M_{\text{total}}\vec{v}_{\text{cm}}$$

$$M_{\text{total}}\vec{a}_{\text{CM}} = \vec{F}_{\text{net}}$$

EXERCISES AND PROBLEMS

Section 3.1

3.X.26 Which fundamental interaction (gravitational, electromagnetic, strong, or weak) is responsible for each of these processes?

(a) A neutron outside a nucleus decays into a proton, electron, and antineutrino.

(b) Protons and neutrons attract each other in a nucleus.

(c) The Earth pulls on the Moon.

(d) Protons in a nucleus repel each other.

Section 3.2

3.X.27 At a particular instant the magnitude of the gravitational force exerted by a planet on one of its moons is $3 \times 10^{23}\,\text{N}$. If the mass of the moon were three times as large, what would the magnitude of the force be? If instead the distance between the moon and the planet were three times as large (no change in mass), what would the magnitude of the force be?

3.X.28 A star exerts a gravitational force of magnitude $|\vec{F}|$ on a planet. The distance between the star and the planet is r. If the planet were 3 times farther away (that is, if the distance between the bodies were $3r$), by what factor would the force on the planet due to the star change?

3.X.29 A planet exerts a gravitational force of magnitude $7 \times 10^{22}\,\text{N}$ on a star. If the planet were 4 times closer to the star (that is, if the distance between the star and the planet were 1/4 what it is now), what would be the magnitude of the force on the star due to the planet?

3.X.30 A moon orbits a planet in the xy plane, as shown in Figure 3.58. You want to calculate the force on the moon by the

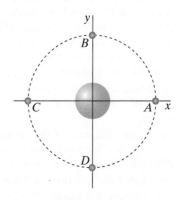

Figure 3.58 Exercise 3.X.30.

planet at each location labeled by a letter (A, B, C, D). At each of these locations, what are: **(a)** the unit vector $\hat{r}$, and **(b)** the unit vector $\hat{F}$ in the direction of the force?

3.X.31 The mass of the Sun is 2×10^{30} kg, and the mass of Mercury is 3.3×10^{23} kg. The distance from the Sun to Mercury is 4.8×10^{10} m.

  (a) Calculate the magnitude of the gravitational force exerted by the Sun on Mercury.

  (b) Calculate the magnitude of the gravitational force exerted by Mercury on the Sun.

3.X.32 Measurements show that Jupiter's gravitational force on a mass of 1 kg near Jupiter's surface would be 24.9 N. The radius of Jupiter is 71500 km (71.5×10^6 m). From these data determine the mass of Jupiter.

3.X.33 A planet of mass 4×10^{24} kg is at location $\langle 5e11, -2e11, 0\rangle$ m. A star of mass 5×10^{30} kg is at location $\langle -2e11, 3e11, 0\rangle$ m. It will be useful to draw a diagram of the situation, including the relevant vectors.

  (a) What is the relative position vector $\vec{r}$ pointing from the planet to the star?

  (b) What is the distance between the planet and the star?

  (c) What is the unit vector $\hat{r}$ in the direction of $\vec{r}$?

  (d) What is the magnitude of the force exerted on the planet by the star?

  (e) What is the magnitude of the force exerted on the star by the planet?

  (f) What is the (vector) force exerted on the planet by the star?

  (g) What is the (vector) force exerted on the star by the planet?

3.X.34 A planet of mass 4×10^{24} kg is at location $\langle -6e11, 3e11, 0\rangle$ m. A star of mass 6×10^{30} kg is at location $\langle 5e11, -3e11, 0\rangle$ m. What is the force exerted on the planet by the star? (It will probably be helpful to draw a diagram, including the relevant vectors.)

3.P.35 Two copies of this textbook are standing right next to each other on a bookshelf. Make a rough estimate of the magnitude of the gravitational force that the books exert on each other. Explicitly list all quantities that you had to estimate, and all simplifications and approximations you had to make to do this calculation. Compare your result to the gravitational force on the book by the Earth.

3.P.36 In June 1997 the NEAR spacecraft ("Near Earth Asteroid Rendezvous"; see http://near.jhuapl.edu/), on its way to photograph the asteroid Eros, passed within 1200 km of asteroid Mathilde at a speed of 10 km/s relative to the asteroid. (See Figure 3.59.) From photos transmitted by the 805 kg spacecraft, Mathilde's size was known to be about 70 km by 50 km by 50 km. The asteroid is presumably made of rock. Rocks on Earth have a density of about 3000 kg/m³ (3 grams/cm³).

  (a) Make a rough diagram to show qualitatively the effect on the spacecraft of this encounter with Mathilde. Explain your reasoning.

  (b) Make a very rough estimate of the change in momentum of the spacecraft that would result from encountering Mathilde. Explain how you made your estimate.

  (c) Using your result from part (b), make a rough estimate of how far off course the spacecraft would be, one day after the encounter.

  (d) From actual observations of the location of the spacecraft one day after encountering Mathilde, scientists concluded that Mathilde is a loose arrangement of rocks, with lots of empty space inside. What was it about the observations that must have led them to this conclusion?

Figure 3.59 Problem 3.P.36.

Experimental background: The position was tracked by very accurate measurements of the time that it takes for a radio signal to go from Earth to the spacecraft followed immediately by a radio response from the spacecraft being sent back to Earth. Radio signals, like light, travel at a speed of 3×10^8 m/s, so the time measurements had to be accurate to a few nanoseconds ($1 \text{ ns} = 1 \times 10^{-9}$ s).

Section 3.3

3.X.37 Why is the value of the constant g different on Earth and on the Moon? Explain in detail.

3.X.38 At what height above the surface of the Earth is there a 1% difference between the approximate magnitude of the gravitational field (9.8 N/kg) and the actual magnitude of the gravitational field at that location? That is, at what height y above the Earth's surface is $GM_E/(R_E + y)^2 = 0.99 \, GM_E/R_E^2$?

3.X.39 Calculate the approximate gravitational force exerted by the Earth on a human standing on the Earth's surface. Compare with the approximate gravitational force of a human on another human at a distance of 3 meters. What approximations or simplifying assumptions must you make? (See data on the inside back cover.)

3.P.40 A steel ball of mass m falls from a height h onto a scale calibrated in newtons. The ball rebounds repeatedly to nearly the same height h. The scale is sluggish in its response to the intermittent hits and displays an *average* force F_{avg}, such that $F_{avg} T = F\Delta t$, where $F\Delta t$ is the brief impulse that the ball imparts to the scale on every hit, and T is the time between hits.

  Calculate this average force in terms of m, h, and physical constants. Compare your result with what the scale reads if the ball merely rests on the scale. Explain your analysis carefully (but briefly).

Section 3.4

3.X.41 You hold a tennis ball above your head, then open your hand and release the ball, which begins to fall. At this moment, which statement about the gravitational forces between the Earth and the ball is correct?

  A. The force on the ball by the Earth is larger than the force on the Earth by the ball.

B. The force on the Earth by the ball is larger than the force on the ball by the Earth.

C. The forces are equal in magnitude.

D. Not enough information is given.

Section 3.5

3.P.42 Suppose that you are going to program a computer to carry out an iterative calculation of motion involving nonconstant forces. Which of the following calculations should be done *before* starting the repetitive "loop" calculations?

A. Define constants such as G.

B. Update the (vector) position of each object.

C. Calculate the (vector) forces acting on the objects.

D. Specify the initial (vector) momentum of each object.

E. Specify an appropriate value for the time step.

F. Specify the mass of each object.

G. Update the (vector) momentum of each object.

H. Specify the initial (vector) position of each object.

Inside the repetitive "loop," assuming that we use the final velocity in each time interval as the approximation to the average velocity in that time interval, what should be the correct sequence of the following calculations?

A. Update the (vector) momentum of each object.

B. Calculate the (vector) forces acting on the objects.

C. Update the (vector) position of each object.

3.P.43 At $t = 532.0$ s after midnight, a spacecraft of mass 1400 kg is located at position $\langle 3 \times 10^5, 7 \times 10^5, -4 \times 10^5 \rangle$ m, and at that time an asteroid whose mass is 7×10^{15} kg is located at position $\langle 9 \times 10^5, -3 \times 10^5, -12 \times 10^5 \rangle$ m. There are no other objects nearby.

(a) Calculate the (vector) force acting on the spacecraft.

(b) At $t = 532.0$ s the spacecraft's momentum was $\vec{p}_i$, and at the later time $t = 538.0$ s its momentum was $\vec{p}_f$. Calculate the (vector) change of momentum $\vec{p}_f - \vec{p}_i$.

3.P.44

(a) In outer space, far from other objects, block 1 of mass 40 kg is at position $\langle 7, 11, 0 \rangle$ m, and block 2 of mass 1000 kg is at position $\langle 18, 11, 0 \rangle$ m. What is the (vector) gravitational force acting on block 2 due to block 1? It helps to make a sketch of the situation.

(b) At 4.6 seconds after noon both blocks were at rest at the positions given above. At 4.7 seconds after noon, what is the (vector) momentum of block 2?

(c) At 4.7 seconds after noon, what is the (vector) momentum of block 1?

(d) At 4.7 seconds after noon, which one of the following statements is true?

A. Block 1 and block 2 have the same speed.

B. Block 2 is moving faster than block 1.

C. Block 1 is moving faster than block 2.

3.P.45 A star of mass 7×10^{30} kg is located at $\langle 5 \times 10^{12}, 2 \times 10^{12}, 0 \rangle$ m. A planet of mass 3×10^{24} kg is located at $\langle 3 \times 10^{12}, 4 \times 10^{12}, 0 \rangle$ m and is moving with a velocity of $\langle 0.3 \times 10^4, 1.5 \times 10^4, 0 \rangle$ m/s.

(a) At a time 1×10^6 seconds later, what is the new velocity of the planet?

(b) Where is the planet at this later time?

(c) Explain briefly why the procedures you followed in parts (a) and (b) were able to produce usable results but wouldn't work if the later time had been 1×10^9 seconds instead of 1×10^6 seconds after the initial time. Explain briefly how you could use a computer to get around this difficulty.

3.P.46 At $t = 0$ a star of mass 4×10^{30} kg has velocity $\langle 7 \times 10^4, 6 \times 10^4, -8 \times 10^4 \rangle$ m/s and is located at $\langle 2.00 \times 10^{12}, -5.00 \times 10^{12}, 4.00 \times 10^{12} \rangle$ m relative to the center of a cluster of stars. There is only one nearby star that exerts a significant force on the first star. The mass of the second star is 3×10^{30} kg, its velocity is $\langle 2 \times 10^4, -1 \times 10^4, 9 \times 10^4 \rangle$ m/s, and this second star is located at $\langle 2.03 \times 10^{12}, -4.94 \times 10^{12}, 3.95 \times 10^{12} \rangle$ m relative to the center of the cluster of stars.

(a) At $t = 1 \times 10^5$ s, what is the approximate momentum of the first star?

(b) Discuss briefly some ways in which your result for (a) is approximate, not exact.

(c) At $t = 1 \times 10^5$ s, what is the approximate position of the first star?

(d) Discuss briefly some ways in which your result for (b) is approximate, not exact.

Section 3.6

3.X.47 Figure 3.60 shows two positively charged objects and one negatively charged object. What is the direction of the net electric force on the negatively charged object? If the net force is zero, state this explicitly.

Figure 3.60 Exercise 3.X.47.

3.X.48 Figure 3.61 shows two negatively charged objects and one positively charged object. What is the direction of the net electric force on the positively charged object? If the net force is zero, state this explicitly.

Figure 3.61 Exercise 3.X.48.

3.X.49 Figure 3.62 shows a proton and an electron. What is the direction of the electric force on the electron by the proton? What is the direction of the electric force on the proton by the electron? How do the magnitudes of these forces compare? Figure 3.63 shows two electrons. What is the direction of the electric force on electron A due to electron B? What is the direction of the electric force on electron B due to electron A?

+ Proton

Figure 3.62 A proton and an electron.

⊖ Electron

⊖ Electron A

⊖ Electron B

Figure 3.63 Two electrons.

3.X.50 An alpha particle contains two protons and two neutrons, and has a net charge of $+2e$. The alpha particle is 1 mm away from a single proton, which has a charge of $+e$. Which statement about the magnitudes of the electric forces between the particles is correct?

A. The force on the proton by the alpha particle is larger than the force on the alpha particle by the proton.

B. The force on the alpha particle by the proton is larger than the force on the proton by the alpha particle.

C. The forces are equal in magnitude.

D. Not enough information is given.

3.P.51 Use data from the inside back cover to calculate the gravitational and electric forces two electrons exert on each other when they are 1×10^{-10} m apart (about one atomic radius). Which interaction between two electrons is stronger, the gravitational attraction or the electric repulsion? If the two electrons are at rest, will they begin to move toward each other or away from each other? Note that since both the gravitational and electric forces depend on the inverse square distance, this comparison holds true at all distances, not just at a distance of 1×10^{-10} m.

3.P.52 Two thin hollow plastic spheres, about the size of a Ping-Pong ball with masses ($m_1 = m_2 = 2 \times 10^{-3}$ kg), have been rubbed with wool. Sphere 1 has a charge $q_1 = -2 \times 10^{-9}$ C and is at location $\langle 50e-2, -20e-2, 0 \rangle$ m. Sphere 2 has a charge $q_2 = -4 \times 10^{-9}$ C and is at location $\langle -40e - 2, 40e - 2, 0 \rangle$ m. It will be useful to draw a diagram of the situation, including the relevant vectors.

(a) What is the relative position vector $\vec{r}$ pointing from q_1 to q_2?

(b) What is the distance between q_1 and q_2?

(c) What is the unit vector $\hat{r}$ in the direction of $\vec{r}$?

(d) What is the magnitude of the gravitational force exerted on q_2 by q_1?

(e) What is the (vector) gravitational force exerted on q_2 by q_1?

(f) What is the magnitude of the electric force exerted on q_2 by q_1?

(g) What is the (vector) electric force exerted on q_2 by q_1?

(h) What is the ratio of the magnitude of the electric force to the magnitude of the gravitational force? (You see that electric forces between two small charged objects are typically very much larger than gravitational forces between those same small objects. It takes the entire mass of the Earth to exert a sizable gravitational force on a small object.)

(i) If the two masses were 4 times farther away (that is, if the distance between the masses were $4\vec{r}$), what would be the ratio of the magnitude of the electric force to the magnitude of the gravitational force now?

3.P.53 A car of mass M moving in the x direction at high speed v strikes a hovering mosquito of mass m, and the mosquito is smashed against the windshield. The interaction between the mosquito and the windshield is an electric interaction between the electrons and protons in the mosquito and those in the windshield.

(a) What is the approximate momentum change of the mosquito? Give magnitude and direction. Explain any approximations you make.

(b) At a particular instant during the impact, when the force exerted on the mosquito by the car is F, what is the magnitude of the force exerted on the car by the mosquito?

(c) What is the approximate momentum change of the car? Give magnitude and direction. Explain any approximations you make.

(d) Qualitatively, why is the collision so much more damaging to the mosquito than to the car?

Sections 3.11–3.12

3.X.54 Two rocks are tied together with a string of negligible mass and thrown into the air. At a particular instant, rock 1, which has a mass of 0.1 kg, is headed straight up with a speed of 5 m/s, and rock 2, which has a mass of 0.25 kg, is moving parallel to the ground, in the $+x$ direction, with a speed of 7.5 m/s. What is the total momentum of the system consisting of both rocks and the string?

Section 3.13

3.X.55 The windshield of a speeding car hits a hovering insect. Consider the time interval from just before the car hits the insect to just after the impact.

(a) For which choice of system is the change of momentum zero?

(b) Is the magnitude of the change of momentum of the bug bigger than, the same as, or smaller than that of the car?

(c) Is the magnitude of the change of velocity of the bug bigger than, the same as, or smaller than that of the car?

3.X.56 When they are far apart, the momentum of a proton is $\langle 3.4 \times 10^{-21}, 0, 0 \rangle$ kg·m/s as it approaches another proton that is initially at rest. The two protons repel each other electrically, without coming close enough to touch. When they are once again far apart, one of the protons now has momentum $\langle 2.4 \times 10^{-21}, 1.55 \times 10^{-21}, 0 \rangle$ kg·m/s. At that instant, what is the momentum of the other proton?

3.P.57 You and a friend each hold a lump of wet clay. Each lump has a mass of 30 grams. You each toss your lump of clay into the air, where the lumps collide and stick together. Just before the impact, the velocity of one lump was $\langle 3, 3, -3 \rangle$ m/s, and the velocity of the other lump was $\langle -3, 0, -3 \rangle$ m/s.

  (a) What was the the total momentum of the lumps just before the impact?

  (b) What is the momentum of the stuck-together lump just after the collision?

  (c) What is the velocity of the stuck-together lump just after the collision?

3.P.58 A car of mass 2800 kg collides with a truck of mass 4700 kg, and just after the collision the car and truck slide along, stuck together. The car's velocity just before the collision was $\langle 40, 0, 0 \rangle$ m/s, and the truck's velocity just before the collision was $\langle -14, 0, 29 \rangle$ m/s.

  (a) What is the velocity of the stuck-together car and truck just after the collision?

  (b) In your analysis in part (a), why can you neglect the effect of the force of the road on the car and truck?

3.X.59 A bullet traveling horizontally at a very high speed embeds itself in a wooden block that is sitting at rest on a very slippery sheet of ice. You want to find the speed of the block just after the bullet embeds itself in the block.

  (a) What should you choose as the system to analyze?

    1. The bullet

    2. The block

    3. The bullet and the block

  (b) Which of the following statements is true?

    1. After the collision, the speed of the block with the bullet stuck in it is the same as the speed of the bullet before the collision.

    2. The momentum of the block with the bullet stuck in it is the same as the momentum of the bullet before the collision.

    3. The momentum of the block with the bullet stuck in it is less than the momentum of the bullet before the collision.

3.X.60 One kind of radioactivity is called "alpha decay." For example, the nucleus of a radium-220 atom can spontaneously split into a radon-216 nucleus plus an alpha particle (a helium nucleus containing two protons and two neutrons).

Consider a radium-220 nucleus that is initially at rest. It spontaneously decays, and the alpha particle travels off in the $+z$ direction. What can you conclude about the motion of the new radon-216 nucleus? Explain your reasoning.

  A. It is also moving in the $+z$ direction.

  B. It remains at rest.

  C. It is moving in the $-z$ direction.

3.X.61 A bowling ball is initially at rest. A Ping-Pong ball moving in the $+z$ direction hits the bowling ball and bounces off it, traveling back in the $-z$ direction. Consider a time interval Δt extending from slightly before to slightly after the collision.

  (a) In this time interval, what is the sign of Δp_z for the system consisting of both balls?

    1. Positive

    2. Negative

    3. Zero—no change in p_z.

  (b) In this time interval, what is the sign of Δp_z for the system consisting of the bowling ball alone?

    1. Positive

    2. Negative

    3. Zero—no change in p_z.

3.P.62 A bullet of mass 0.105 kg traveling horizontally at a speed of 300 m/s embeds itself in a block of mass 2 kg that is sitting at rest on a nearly frictionless surface. What is the speed of the block after the bullet embeds itself in the block?

3.P.63 Object A has mass $m_A = 8$ kg and initial momentum $\vec{p}_{A,i} = \langle 20, -5, 0 \rangle$ kg·m/s, just before it strikes object B, which has mass $m_B = 11$ kg. Just before the collision object B has initial momentum $\vec{p}_{B,i} = \langle 5, 6, 0 \rangle$ kg·m/s.

  (a) Consider a system consisting of both objects A and B. What is the total initial momentum of this system just before the collision?

  (b) The forces that A and B exert on each other are very large but last for a very short time. If we choose a time interval from just before to just after the collision, what is the approximate value of the impulse applied to the two-object system due to forces exerted on the system by objects outside the system?

  (c) Therefore, what does the Momentum Principle predict that the total final momentum of the system will be just after the collision?

  (d) Just after the collision, object A is observed to have momentum $\vec{p}_{A,f} = \langle 18, 5, 0 \rangle$ kg·m/s. What is the momentum of object B just after the collision?

3.P.64 In outer space a small rock with mass 5 kg traveling with velocity $\langle 0, 1800, 0 \rangle$ m/s strikes a stationary large rock head-on and bounces straight back with velocity $\langle 0, -1500, 0 \rangle$ m/s. After the collision, what is the vector momentum of the large rock?

3.P.65 Two rocks collide in outer space. Before the collision, one rock had mass 9 kg and velocity $\langle 4100, -2600, 2800 \rangle$ m/s. The other rock had mass 6 kg and velocity $\langle -450, 1800, 3500 \rangle$ m/s. A 2 kg chunk of the first rock breaks off and sticks to the second rock. After the collision the 7 kg rock has velocity $\langle 1300, 200, 1800 \rangle$ m/s. After the collision, what is the velocity of the other rock, whose mass is 8 kg?

3.P.66 Two rocks collide with each other in outer space, far from all other objects. Rock 1 with mass 5 kg has velocity $\langle 30, 45, -20 \rangle$ m/s before the collision and $\langle -10, 50, -5 \rangle$ m/s after the collision. Rock 2 with mass 8 kg has velocity $\langle -9, 5, 4 \rangle$ m/s before the collision. Calculate the final velocity of rock 2.

3.P.67 In outer space two rocks collide and stick together. Here are the masses and initial velocities of the two rocks:
Rock 1: mass = 15 kg, initial velocity = $\langle 10, -30, 0 \rangle$ m/s
Rock 2: mass = 32 kg, initial velocity = $\langle 15, 12, 0 \rangle$ m/s
What is the velocity of the stuck-together rocks after colliding?

3.P.68 Suppose that all the people of the Earth go to the North Pole and, on a signal, all jump straight up. Estimate the recoil speed of the Earth. The mass of the Earth is 6×10^{24} kg, and there are about 6 billion people (6×10^9).

3.P.69 A bullet of mass m traveling horizontally at a very high speed v embeds itself in a block of mass M that is sitting at rest on a nearly frictionless surface. What is the speed of the block after the bullet embeds itself in the block?

3.P.70 A satellite that is spinning clockwise has four low-mass solar panels sticking out as shown. A tiny meteor traveling at high speed rips through one of the solar panels and continues in the same direction but at reduced speed. Afterward, calculate the v_x and v_y components of the center-of-mass velocity of the satellite. In Figure 3.64 $\vec{v}_1$, and $\vec{v}_2$ are the initial and final velocities of the meteor, and $\vec{v}$ is the initial velocity of the center of mass of the satellite, in the x direction.

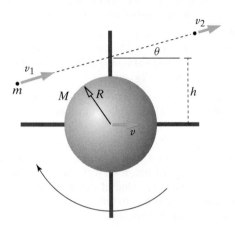

Figure 3.64 Problem 3.P.70.

3.P.71 A ball of mass 0.05 kg moves with a velocity $\langle 17, 0, 0 \rangle$ m/s. It strikes a ball of mass 0.1 kg that is initially at rest. After the collision, the heavier ball moves with a velocity of $\langle 3, 3, 0 \rangle$ m/s.

(a) What is the velocity of the lighter ball after impact?

(b) What is the impulse delivered to the 0.05 kg ball by the heavier ball?

(c) If the time of contact between the balls is 0.03 s, what is the force exerted by the heavier ball on the lighter ball?

3.P.72 A space station has the form of a hoop of radius R, with mass M. Initially its center of mass is not moving, but it is spinning. Then a small package of mass m is thrown by a spring-loaded gun toward a nearby spacecraft as shown in Figure 3.65; the package has a speed v after launch. Calculate the center-of-mass velocity of the space station (v_x and v_y) after the launch.

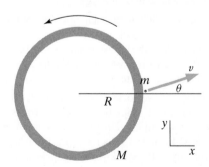

Figure 3.65 Problem 3.P.72.

Computational Problems

3.P.73 Place a spacecraft of mass 15000 kg at a location 10 Earth radii to the left of the center of the Earth and give it an initial velocity as shown, then let it coast, subject only to the gravitational attraction of the Earth, whose mass is 6×10^{24} kg. Neglect the tiny motion of the Earth. Plot a trail behind the spacecraft so that you can see the shape of the trajectory. At each step in the iteration, check for the spacecraft being closer than one Earth radius to the center of the Earth, and stop the program. (See Figure 3.66.)

Figure 3.66 A spacecraft coasts near the Earth.

(a) Predict what will happen if the initial speed of the spacecraft is zero. Then try it.

(b) Try various launch speeds. What is the minimum speed required to go around the Earth without colliding? This orbit can be shown to be an ellipse, as was proved by Newton.

(c) Cut the time step size Δt by a factor of two and make sure that you still get essentially the same trajectory seen in (b). If there is a significant change in the trajectory, keep cutting the step size until you no longer see a change in the trajectory. Repeat part (b) if your earlier step size gave poor accuracy. On the other hand, if the orbit plots very slowly you might have chosen a step size that is much smaller than necessary. Try doubling the step size repeatedly until you start to see significant changes in the trajectory, then reduce the step size somewhat.

(d) Deliberately make Δt large enough that you can easily see straight-line portions of what should be the ellipse seen in (b). You may have to slow down the orbit display, or stop after one orbit, in order to see this clearly. The iterative procedure you're using means that even a seemingly smooth ellipse is actually plotted as very short straight lines, but now the endpoints of those lines are noticeably wrong. What causes this inaccuracy in the computations?

(e) Reset the step size Δt to give good accuracy. Find the launch speed that leads to a circular orbit.

(f) Predict what will happen if the initial speed is somewhat greater than the speed that gives a circular orbit. Then try it.

(g) What is the minimum launch speed for which the spacecraft escapes, never coming back?

(h) Try launching the spacecraft to the left, straight away from the Earth, with the speed that gave a circular orbit in (e). What happens?

(i) With launching to the left, what is the minimum launch speed for which the spacecraft escapes, never coming back? Compare with your result in (f). We will understand this later when we study energy.

3.P.74 This problem is an extension of the previous problem, Problem 3.P.73. When there are only two bodies interacting with each other as in Problem 3.P.73, there are not many different kinds of trajectories possible. If the center of mass of the two-body system doesn't move, the only possible trajectories are straight lines, circles, ellipses, parabolas, and hyperbolas, and slight changes in the initial conditions for an ellipse just yield a slightly different ellipse. In a three-body situation there is a vastly richer set of possible trajectories, and there is high sensitivity to initial conditions; a slight change in the initial conditions may lead to a dramatically different trajectory.

To see these effects, investigate the motion of a spacecraft attracted to the Earth and Moon, which are in their proper relative positions as shown in Figure 3.67 but not free to move, to simplify the situation. Place a spacecraft of mass 15000 kg at a location 10 Earth radii to the left of the center of the Earth and give it an initial velocity as shown, then let it coast, subject only to the gravitational attractions of the stationary Earth and Moon, whose mass is 7×10^{22} kg. Plot a trail behind the spacecraft so that you can see the shape of the trajectory. At each step in the iteration, check for the spacecraft being closer than one Earth radius to the center of the Earth, and stop the program; make similar checks for hitting the Moon.

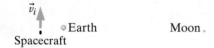

Figure 3.67 A spacecraft coasts near the Earth and Moon.

(a) What is the minimum initial speed required to hit the Moon?

(b) What initial speed leads to a symmetric figure-8 orbit, looping around the Moon and returning?

(c) Find an initial speed such that the spacecraft reaches a turning point (speed equal to or nearly equal to zero) and turns back on its previous path. Cut the step size by a factor of two and make sure that you still get a similar trajectory.

(d) What happens when you make a slight change in the speed found in (c)? Is the trajectory sensitive to initial conditions?

(e) In terms of physics principles, explain briefly why the spacecraft speeds up when it approaches the Earth or Moon.

A more challenging project is to have the Earth and Moon orbit each other, with the spacecraft interacting with these moving bodies. You could even have the Earth and Moon orbit the Sun, in which case you would have a quite accurate model of the real situation.

3.P.75 Carry out a numerical integration of the motion of a three-body gravitational system, with all three objects free to move, and plot the trajectories, leaving trails behind the objects. Calculate all of the forces before using these forces to update the momenta and positions of the objects. Otherwise the calculations of gravitational forces would mix positions corresponding to different times.

Try different initial positions and initial momenta. Find at least one set of initial conditions that produces a long-lasting orbit, one set of initial conditions that results in a collision with a massive object, and one set of initial conditions that allows one of the objects to wander off without returning. Report the masses and initial conditions that you used.

3.P.76 In this problem you will study the motion of a planet around a star. To start with a somewhat familiar situation, you will begin by modeling the motion of our Earth around our Sun. Write answers to questions either as comments in your program or on paper, as specified by your instructor.

Planning

(a) The Earth goes around the Sun in a nearly circular orbit, taking one year to go around. Using data on the inside back cover, what initial speed should you give the Earth in a computer model, so that a circular orbit should result? (If your computer model does produce a circular orbit, you have a strong indication that your program is working properly.)

(b) Estimate an appropriate value for Δt to use in your computer model. Remember that t is in seconds, and consider your answer to part (a) in making this estimate. If your Δt is too small, your calculation will require many tedious steps. Explain briefly how you decided on an appropriate step size.

Circular Orbit

(c) Starting with speed calculated in (a), and with the initial velocity perpendicular to a line connecting the Sun and the Earth, calculate and display the trajectory of the Earth. Display the whole trail, so you can see whether you have a closed orbit (that is, whether the Earth returns to its starting point each time around). Include the Sun's position $\vec{r}_s = \langle x_s, y_s, z_s \rangle$ in your computations, even if you set its coordinates to zero. This will make it easier to modify the computation to let the Sun move in the following problem, on binary stars.

Computational Accuracy

(d) As a check on your computation, is the orbit a circle as expected? Run the calculation until the accumulated time t is the length of a year: does this take the Earth around the orbit once, as expected? What is the largest step size you can use that still gives a circular orbit and the correct length of a year? What happens if you use a much larger step size?

Noncircular Trajectories

(e) Set the initial speed to 1.2 times the Earth's actual speed. Make sure the step size is small enough that the orbit is nearly unaffected if you cut the step size. What kind of orbit do you get? What step size do you need to use to get results you can trust? What happens if you use a much larger step size? Produce at least one other qualitatively different noncircular trajectory for the Earth. What difficulties would humans have in surviving on the Earth if it had a highly noncircular orbit?

Force and Momentum

(f) Choose initial conditions that give a noncircular orbit. Continuously display a vector showing the momentum of the Earth (with its tail at the Earth's position), and a different-colored vector showing the force on the Earth by the Sun (with its tail at the Earth's position). You must scale the vectors appropriately to fit into the scene. A way to figure out what scale factor to use is to print the numerical values of momentum and force, and compare with the scale of the scene. For example, if the width of the scene is W meters, and the force vector has a typical magnitude F (in newtons), you might scale the force vector by a factor $0.1W/F$, which would make the length of the vector one-tenth the width of the scene. Is the force in the same direction as the momentum? How does the momentum depend on the distance from the star?

3.P.77 About a half of the visible "stars" are actually systems consisting of two stars orbiting each other, called "binary stars." In your computer model of the Earth and Sun (Problem 3.P.76), replace the Earth with a star whose mass is half the mass of our Sun, and take into account the gravitational effects that the second star has on the Sun.

(a) Give the second star the speed of the actual Earth, and give the Sun zero initial momentum. What happens? Try a variety of other initial conditions. What kinds of orbits do you find?

(b) One of the things to try is to give the Sun the same magnitude of momentum as the second star, but in the opposite direction. If the initial momentum of the second star is in the $+y$ direction, give the Sun the same magnitude of momentum but in the $-y$ direction, so that the total momentum of the binary star system is initially zero, but the stars are not headed toward each other. What is special about the motion you observe in this case?

3.P.78 Modify your orbit computation to use a different force, such as a force that is proportional to $1/r$ or $1/r^3$, or a constant force, or a force proportional to r^2 (this represents the force of a spring whose relaxed length is nearly zero). How do orbits with these forces differ from the circles and ellipses that result from a $1/r^2$ force? If you want to keep the magnitude of the force roughly the same as before, you will need to adjust the force constant G.

3.P.79 In Problem 3.P.76 you analyzed a simple model of the Earth orbiting the Sun, in which there were no other planets or moons. Venus and the Earth have similar size and mass. At its closest approach to the Earth, Venus is about 40 million kilometers away (4×10^{10} m). The Moon's mass is about 7×10^{22} kg, and the distance from the Earth to the Moon is about 4×10^8 m (400,000 km, center to center).

(a) Calculate the ratio of the gravitational forces on the Earth exerted by Venus and the Sun. Does ignoring the effect of Venus when modeling the motion of the Earth around the Sun give a good approximation?

(b) Calculate the ratio of the gravitational forces on the Earth exerted by the Moon and the Sun. Does ignoring the effect of the Moon when modeling the motion of the Earth around the Sun give a good approximation?

3.P.80 The first U.S. spacecraft to photograph the Moon close up was the unmanned *Ranger 7* photographic mission in 1964. The spacecraft, shown in Figure 3.68 (NASA photograph), contained television cameras that transmitted close-up pictures of the Moon back to Earth as the spacecraft approached the Moon. The spacecraft did not have retro-rockets to slow itself down, and it eventually simply crashed onto the Moon's surface, transmitting its last photos immediately before impact.

Figure 3.68 Problem 3.P.80.

Figure 3.69 is the first image of the Moon taken by a U.S. spacecraft, *Ranger 7,* on July 31, 1964, about 17 minutes before impact on the lunar surface. The large crater at center right is Alphonsus (108 km diameter); above it (and to the right) is Ptolemaeus and below it is Arzachel. The Ranger 7 impact site is off the frame, to the left of the upper left corner. To find out more about the actual *Ranger* lunar missions, see http://nssdc.gsfc.nasa.gov/planetary/lunar/ranger.html.

Figure 3.69 The first *Ranger 7* photo of the Moon (NASA photograph).

To send a spacecraft to the Moon, we put it on top of a large rocket containing lots of rocket fuel and fire it upward. At first the huge ship moves quite slowly, but the speed increases rapidly. When the "first-stage" portion of the rocket has exhausted its fuel and is empty, it is discarded and falls back to Earth. By discarding an empty rocket stage we decrease the amount of mass that must be accelerated to even higher speeds. There may be several stages that operate for a while and then are discarded before the spacecraft has risen above most of Earth's atmosphere (about 50 km, say, above the Earth), and has acquired a high speed. At that point all the fuel available for this mission has been used up, and the spacecraft simply coasts toward the Moon through the vacuum of space.

We will model the *Ranger 7* mission. Starting 50 km above the Earth's surface (5×10^4 m), a spacecraft coasts toward the Moon with an initial speed of about 1×10^4 m/s. Here are the data we will need:

mass of spacecraft = 173 kg; mass of Earth $\approx 6 \times 10^{24}$ kg; mass of Moon $\approx 7 \times 10^{22}$ kg; radius of Moon = 1.75×10^6 m; distance from Earth to Moon $\approx 4 \times 10^8$ m (400,000 km, center to center).

We're going to ignore the Sun in a simplified model even though it exerts a sizable gravitational force. We're expecting the Moon mission to take only a few days, during which time the Earth (and Moon) move in a nearly straight line with respect to the Sun, because it takes 365 days to go all the way around the Sun. We take a reference frame fixed to the Earth as representing (approximately) an inertial frame of reference with respect to which we can use the Momentum Principle.

For a simple model, make the Earth and Moon be fixed in space during the mission. Factors that would certainly influence the path of the spacecraft include the motion of the Moon around the Earth and the motion of the Earth around the Sun. In addition, the Sun and other planets exert gravitational forces on the spacecraft. As a separate project you might like to include some of these additional factors.

(a) Compute the path of the spacecraft, and display it either with a graph or with an animated image. Here, and in remaining parts of the problem, report the step size Δt that gives accurate results (that is, cutting this step size has little effect on the results).

(b) By trying various initial speeds, determine the approximate *minimum* launch speed needed to reach the Moon, to two significant figures (this is the speed that the spacecraft obtains from the multistage rocket, at the time of release above the Earth's atmosphere). What happens if the launch speed is less than this minimum value? (Be sure to check the step size issue.)

(c) Use a launch speed 10% larger than the approximate minimum value found in part (b). How long does it take to go to the Moon, in hours or days? (Be sure to check the step size issue.)

(d) What is the "impact speed" of the spacecraft (its speed just before it hits the Moon's surface)? Make sure that your spacecraft crashes on the surface of the Moon, not at the Moon's center! (Be sure to check the step size issue.)

You may have noticed that you don't actually need to know the mass m of the spacecraft in order to carry out the computation. The gravitational force is proportional to m, and the momentum is also proportional to m, so m cancels. However, nongravitational forces such as electric forces are not proportional to the mass, and there is no cancellation in that case. We kept the mass m in the analysis in order to illustrate a general technique for predicting motion, no matter what kind of force, gravitational or not.

3.P.81 In the previous analysis (the Moon voyage), you used a simplified model in which you neglected among other things the effect of Venus. An important aspect of physical modeling is making estimates of how large the neglected effects might be. Venus and the Earth have similar size and mass. At its closest approach to the Earth, Venus is about 40 million kilometers away $(4 \times 10^{10} \text{ m})$. In the real world, Venus would attract the Earth and the Moon as well as the spacecraft, but to get an idea of the size of the effects, imagine that the Earth, the Moon, and Venus are all fixed in position. (See Figure 3.70)

Moon

Venus

Earth

Figure 3.70 The relative positions of Venus, Earth, and Moon.

If we take Venus into account, make a rough estimate of whether the spacecraft will miss the Moon entirely. How large a sideways deflection of the crash site will there be? Explain your reasoning and approximations.

ANSWERS TO EXERCISES

3.X.1 4×10^{25} N

3.X.2 8×10^{25} N

3.X.3 8.9×10^{19} N

3.X.4 $3F/16 = 0.1875F$

3.X.5 2.5×10^{-10} N; 2.5×10^{-10} N

3.X.6 6×10^{24} kg

3.X.7 **(a)** toward Ganymede, **(b)** toward Europa, **(c)** toward Europa; **(d)** toward Europa, **(e)** toward Ganymede, **(f)** toward Ganymede

3.X.8 **(a)** $\langle 2.8 \times 10^8, 0, -2.8 \times 10^8 \rangle$ m; **(b)** 3.96×10^8 m; **(c)** $\langle 0.707, 0, -0.707 \rangle$; **(d)** $\langle -1.3 \times 10^{20}, 0, 1.3 \times 10^{20} \rangle$ N

3.X.9 **(a)** 1.53 N/kg; **(b)** 107 N; **(c)** 0.15 as much (about 1/6)

3.X.10 3.7 N/kg

3.X.11 **(a)** 588 N; **(b)** 588 N

3.X.12 Magnitude of momentum (and speed) decrease because parallel component of net force is opposite to momentum; perpendicular component of net force changes the direction toward $+y$ direction)

3.X.13 2.3×10^{-8} N; 2.3×10^{-8} N

3.X.14 **(a)** 2.02×10^{-9} N; **(b)** 2.02×10^{-9} N

3.X.15 Si 14 p, 14 n; Sn 50 p, 69 n; Au 79 p, 118 n; Th 90 p, 142 n

3.X.16 6; 6

3.X.17 **(a)** $\langle -4, 14, 0 \rangle$ kg·m/s; **(b)** $\langle -4, 14, 0 \rangle$ kg·m/s

3.X.18 $\langle 3.83, 0.33, 0 \rangle$ m

3.X.19 $\langle 0.57, 5.43, 0 \rangle$ m/s/s

3.X.20 **(a)** $-M\vec{v}, +M\vec{v}$; **(b)** $+M\vec{v}, -M\vec{v}$; **(c)** $\vec{0}, \vec{0}$

3.X.21 **(a)** Earth; **(b)** No, because the combined system is isolated and its momentum doesn't change

3.X.22 **(a)** $\langle 2.2, -.4, 2.6 \rangle$ kg·m/s; **(b)** $\langle 0, -7.84, 0 \rangle$ N; **(c)** $\langle 2.2, -1.184, 2.6 \rangle$ kg·m/s

3.X.23 $\langle 3, 5, -2 \rangle$ m/s/s

3.X.24 $\langle 0.04, 0.04, -0.1 \rangle$ kg·m/s; $\langle 0.04, 0.04, -0.1 \rangle$ kg·m/s; $\langle 1, 1, -2.5 \rangle$ m/s

3.X.25 $\langle -2, 14, 7 \rangle$ kg·m/s; $\langle -2, 14, 7 \rangle$ kg·m/s

Contact Interactions

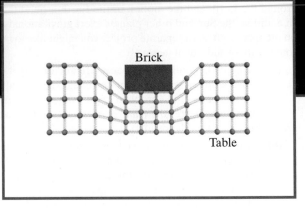

- **A solid object may be modeled as a lattice of balls (atoms) connected by springs (chemical bonds).**
- **Contact forces between solid objects are due to the compression or stretching of springlike interatomic bonds.**
- **Contact forces include**
  - Stretch forces ("tension" forces)
  - Compression forces ("normal" forces)
  - Friction forces
  - Buoyancy forces
- **The Momentum Principle can be written in a derivative (instantaneous) form.**
- **A spring–mass system oscillates sinusoidally.**
- **The speed of sound in a solid can be understood in terms of atomic oscillations.**

Figure 4.1 Tarzan hangs motionless from a vine, which does not break.

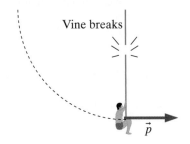

Figure 4.2 The vine breaks midway through Tarzan's swing.

4.1 TARZAN AND THE VINE

Tarzan wants to use a vine to swing across a river. To make sure the vine is strong enough to support him, he tests it by hanging motionless on the vine for several minutes (Figure 4.1). The vine passes this test, so Tarzan grabs the vine and swings out over the river. He is annoyed and perplexed when the vine breaks midway through the swing (Figure 4.2), and he ends up drenched and shivering in the middle of the cold river, to the great amusement of the onlooking apes.

Why did the vine break while Tarzan was swinging on it, but not while he hung motionless from it? To answer this question we will need to understand how objects like vines, wires, and strings exert forces. We will also need to invoke the derivative form of the Momentum Principle, which will be introduced in this chapter.

4.2 A MODEL OF A SOLID: BALLS CONNECTED BY SPRINGS

The great 20th-century American physicist Richard Feynman said:
"If, in some cataclysm, all of scientific knowledge were to be destroyed, and only one sentence passed on to the next generations of creatures, what

Figure 4.3 Two balls connected by a spring. Top: a relaxed spring exerts no forces on the balls. Middle: a stretched spring exerts forces to bring balls closer together. Bottom: a compressed spring exerts forces to move balls apart.

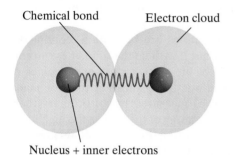

Figure 4.4 The ball–spring model for two atoms connected by a chemical bond (represented by a spring), superimposed on a space-filling model in which the entire electron clouds of the atoms are shown.

Figure 4.5 STM images of two surfaces of a silicon crystal (courtesy of Randall Feenstra, Carnegie Mellon University).

statement would contain the most information in the fewest words? I believe it is the atomic hypothesis (or the atomic fact, or whatever you wish to call it) that all things are made of atoms—little particles that move around in perpetual motion, attracting each other when they are a little distance apart, but repelling upon being squeezed into one another. In that one sentence, you will see, there is an enormous amount of information about the world, if just a little imagination and thinking are applied." (*The Feynman Lectures on Physics* by R. P. Feynman, R. B. Leighton, & M. Sands, 1965; Palo Alto: Addison-Wesley.) In the quote above, Feynman summarized the basic properties of atoms and of interatomic forces. The main properties of atoms that will be important to us in this chapter are these:

- All matter consists of atoms, whose typical radius is about 1×10^{-10} meter.
- Atoms attract each other when they are close to each other but not too close.
- Atoms repel each other when they get too close to each other.
- Atoms in solids, liquids, and gases keep moving even at very low temperatures.

These properties were established during a century of intense study of atoms by physicists and chemists, using a wide variety of experimental techniques.

QUESTION What properties of a block of aluminum, observable without special equipment, support the claims that "atoms attract each other when they are not too close to each other" and "atoms repel each other when they get too close to each other"?

Since the block of metal does not spontaneously fall apart or evaporate, the atoms in the block must be attracting each other. It is very difficult to compress the block, so the atoms must resist an attempt to push them closer together. Evidently the atoms in a block of aluminum are normally at just the right distance from each other, not too close and not too far away. This just-right distance is called the "equilibrium" distance between atoms.

A Chemical Bond Is Like a Spring

Two atoms linked by a chemical bond behave in a manner very similar to two macroscopic balls attached to the ends of a very low mass spring (Figure 4.3). The ball–spring system is a good model for the atomic system. In this model:

- Each ball in the model represents a massive atomic nucleus, surrounded by the inner electrons of the atom. Almost all the mass of an atom is concentrated in the tiny nucleus.
- The spring in the model represents the chemical bond, which is due to the shared outer electrons of both atoms.

The microscopic atomic system behaves much like the macroscopic model system, as long as the stretch or compression is small, as it is in ordinary processes. If the atoms are moved farther apart, they experience forces resisting the separation. If the atoms are pushed closer together, they experience forces resisting the compression. In Figure 4.4 the ball-spring model of two bonded atoms is superimposed on a space-filling representation showing the full electron cloud of each atom.

A Ball–Spring Model for a Solid Object

A solid object contains many atoms, not just two. We know from x-ray studies and from images produced by scanning tunneling microscopes (STM) and atomic force microscopes (AFM) that many solid objects, such as metals, are

Figure 4.6 A simple model of a solid: tiny balls in constant motion, connected by springs. This figure shows only a small section of a solid object, which has many more atoms than are depicted here.

Figure 4.7 A heavy ball hangs motionless at the end of a thin wire.

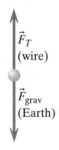

$\vec{F}_T$
(wire)

$\vec{F}_{grav}$
(Earth)

Figure 4.8 Free-body diagram for the ball.

■ Since the ball remains at rest indefinitely, this is true for any value of Δt (other than zero).

---

crystals composed of regular arrays of atoms, as shown in the STM images of silicon surfaces in Figure 4.5. In these images, the spheres indicate the entire electron cloud associated with each atom, not just the nucleus and inner electrons represented by the balls in our model.

These STM images show only one surface of a solid, but inside the solid, atoms are arranged in 3D patterns, like the balls and springs in the model solid shown in Figure 4.6. Solids in which the atoms are arranged in regular "lattices" like those shown in Figure 4.6 are called crystals; these include metals, quartz, diamond, ice, and table salt (NaCl), but not most organic solids such as plastic or wood.

The lattice shown in Figure 4.6 is the simplest kind of crystal, and is called a "cubic" lattice because the atoms in the crystal are located at the corners of adjacent cubes. More complex lattice arrangements are possible; for example, in one common variant called "body centered cubic" there is an additional atom at the center of each cube. In Figure 4.5 the lower image shows a hexagonal arrangement of atoms. Although most crystals are more complex than the simple cubic lattice, we will use this simple cubic ball-and-spring model because it incorporates all the important features we need.

In a solid at room temperature, the atoms are in motion, continually oscillating around their equilibrium positions. If we heat the solid, these atomic oscillations become more vigorous. One of the most important outcomes of research on the atomic nature of matter was the realization that for many materials, the temperature of an object, as measured by an ordinary thermometer, is just an indicator of the average energy of the atoms; the higher the temperature, the more vigorous the atomic motion. Figure 4.6 shows atoms in a model solid frozen in their equilibrium positions.

4.3 TENSION FORCES

By applying the Momentum Principle we can show that an object like a wire, string, or vine can exert a force on an object attached to it. Figure 4.7 shows a heavy iron ball hanging motionless on the end of a wire. The y momentum of the ball is not changing, so the y component of the net force on the ball must be zero. Since the Earth is pulling down on the ball, the wire must be pulling up. The force exerted by an object like a wire or a string is often called a "tension" force, or sometimes just "the tension in the wire," and usually labeled $\vec{F}_T$. A tension force always acts along the wire or string.

> **QUESTION** If the mass of the ball is 1 kg, how large an upward force does the wire exert on the ball?

System: Ball
Surroundings: Earth, wire (see the free-body diagram in Figure 4.8)
Momentum Principle:

$$\Delta p_y = F_{net,y}\Delta t$$

Choose a value for Δt : 10 seconds

$$\Delta p_y = (F_T - mg)(10\,\text{s})$$
$$0 = (F_T - mg)(10\,\text{s})$$
$$F_T = mg = (1\,\text{kg})(9.8\,\text{N/kg}) = 9.8\,\text{N}$$

> **QUESTION** If the mass of the ball were 2 kg, how large an upward force would the wire exert on the ball?

$$F_T = (2\,\text{kg})(9.8\,\text{N/kg}) = 19.6\,\text{N}$$

This simple result has an interesting implication. Evidently the magnitude of the tension force F_T exerted by the wire depends on the mass of the ball.

QUESTION   How does an inanimate object like a wire "know" how large a force to exert in a given situation?

We have already seen that the force exerted by a spring depends on how much the spring is stretched. A wire can be thought of as a very stiff spring. When a weight is hung on the end of a wire, the wire does stretch, although the stretch is usually very small. The ball–spring model of a solid helps us understand what happens when a wire stretches.

Microscopic View: Tension in a Wire

We can model a wire as a chain of balls and springs (atoms connected by springlike chemical bonds). For simplicity, we'll consider a wire that is only one atom thick.

When the wire lies on a table, each of the springs (bonds) is relaxed. When the wire hangs vertically with nothing attached to it, the bonds stretch a tiny amount, just to support the weight of the atoms below them. However, when a massive object is hung from the wire, the springlike bonds between atoms stretch significantly, because each bond must support the weight of everything below it (Figure 4.9). If we make the approximation that the mass of the atoms in the wire is negligible compared to the mass of the hanging object, then we can say that each bond in the wire is stretched by approximately the same amount, or equivalently, that the tension is the same throughout the wire.

QUESTION   How does the wire manage to exert a larger upward force when a more massive object is hung from it?

The more massive object stretches the interatomic bonds more than a less massive object. Of course, there is a limit to how much the wire can stretch. A sufficiently heavy weight will break the wire.

Figure 4.9 The interatomic bonds in a wire stretch when a heavy mass is hung from the wire. In this idealized diagram the wire is only one atom thick, and the stretch of each bond is greatly exaggerated.

4.4   LENGTH OF AN INTERATOMIC BOND

Our goal in the next sections is to determine the stiffness of an interatomic bond, considered as a spring. The first step is to determine the length of an interatomic bond in a particular material. Bond lengths will be slightly different in different materials (for example, aluminum vs. lead), depending on the size of the atoms. We will determine the interatomic bond length in solid copper.

We will define the length of one interatomic bond as the center-to-center distance between two adjacent atoms (Figure 4.10). If we consider the space-filling model of a solid, this distance is equal to twice the radius of an atom, or the diameter of an atom.

The diameter of an atom in a solid is one of the important properties of matter that plays a role in interactions. We can calculate atomic diameters for crystals of particular elements by using the measured density of the material in kilograms per cubic meter and Avogadro's number (the number of atoms in one mole of the material).

QUESTION   Does the density of a block of aluminum depend on the dimensions of the block? Does it depend on the mass of the block?

Density does not depend on the size, shape, or mass of an object. Density is a property of the material itself, and the ratio of mass to volume will always

Bond length d

Atomic radius r Atomic diameter d

Figure 4.10 The length of an interatomic bond is defined as the center-to-center distance between adjacent atoms. This is the same as the diameter of an atom (including the full electron cloud).

be the same for objects made of the same solid material. The density of a wide range of materials has been measured and may easily be looked up in reference materials.

EXAMPLE

Diameter of a Copper Atom (length of a bond in copper)

One mole of copper (6.02×10^{23} atoms) has a mass of 64 grams (see the periodic table on the inside front cover). The density of copper is 8.94 grams/cm^3. What is the approximate diameter, in meters, of a copper atom in solid copper?

Solution

$$\left(8.94 \frac{g}{cm^3}\right) \frac{(1\,kg)}{(1 \times 10^3\,g)} \frac{(1 \times 10^2\,cm)^3}{(1\,m)^3} = 8.94 \times 10^3 \frac{kg}{m^3} \text{ (SI units)}$$

Figure 4.11 shows a cube that contains $5 \times 5 \times 5 = 125$ atoms. The number of atoms on one side of the cube is $\sqrt[3]{125} = 5$.

In a cube that is 1 meter on each side, how many copper atoms are there?

$$(8.94 \times 10^3\,kg)\left(\frac{1\,mol}{0.064\,kg}\right)\frac{(6.02 \times 10^{23}\,atoms)}{(1\,mol)} = 8.41 \times 10^{28} \text{ atoms}$$

Along one edge of the cube, which is 1 m long, there are

$$\sqrt[3]{8.41 \times 10^{28}} = 4.38 \times 10^9 \text{ atoms}$$

A row of 4.38×10^9 atoms is 1 m long, so the diameter of one atom is

$$d = \left(\frac{1\,m}{4.38 \times 10^9\,atoms}\right) = 2.28 \times 10^{-10}\,m$$

In a solid block of copper, the length of a bond between two adjacent copper atoms is 2.28×10^{-10} m.

■ **Unit conversion: density**

Often densities are given in grams per cubic centimeter rather than in kilograms per cubic meter. For example, the density of liquid water is 1 gram/cm^3, the density of aluminum is 2.7 grams/cm^3, and the density of lead is 11.4 grams/cm^3. We often need to convert these densities to SI units, like this:

$$\left(1 \frac{gram}{cm^3}\right)\left(\frac{(100\,cm)^3}{1\,m^3}\right)\left(\frac{1\,kg}{1000\,grams}\right)$$

$$= 1000\,kg/\,m^3$$

Note that since 100 cm = 1 m,

$$(100\,cm)^3 = 1 \times 10^6\,cm^3 = 1\,m^3$$

FURTHER DISCUSSION   An alternative approach to this calculation involves a "micro" view of density. Since density is independent of the amount of an object, the density of one copper atom ought to be the same as the density of a large block of copper. Figure 4.11 shows that the volume of space taken up by each atom is a tiny cube d on a side, where d is the diameter of an atom in the solid and is also the distance from the center of one atom to the center of a neighboring atom. The micro density is the mass of one atom divided by the volume of the tiny cube associated with each atom. This micro density is the same as the macro density, mass per volume:

$$\text{density} = \frac{\text{mass of } 8.41 \times 10^{28} \text{ atoms}}{\text{volume of } 8.41 \times 10^{28} \text{ atoms}} = \frac{\text{mass of 1 atom}}{\text{volume of 1 atom}} = \frac{m_a}{d^3}$$

The mass of one atom can be determined using the mass of one mole and the knowledge that one mole contains 6.02×10^{23} atoms (Avogadro's number):

$$m_a = \frac{\text{mass of one mole}}{6.02 \times 10^{23} \text{ atoms/mole}}$$

Figure 4.11 A simple cubic arrangement of atoms. The volume of space associated with each atom is a tiny cube d by d by d.

Knowing the macroscopic density and the mass of one atom, you can solve for the unknown atomic diameter d. It might be tempting to put these concepts together into a formula of the form "$d = \ldots$," but try to avoid the temptation.

It is much more memorable and much safer to think through these physical relationships each time you make the calculation, to avoid serious mistakes. Just set the macroscopic and microscopic densities equal to each other and solve for d.

Few solid elements actually have cubic lattices, because a cubic lattice is rather unstable. However, assuming that all crystalline solids have cubic lattices greatly simplifies our model, and gives adequately accurate results for our purposes.

> **4.X.1** The density of aluminum is 2.7 grams/cm³. What is the approximate diameter of an aluminum atom (length of a bond) in solid aluminum?
>
> **4.X.2** The density of lead is 11.4 grams/cm³. What is the approximate diameter of a lead atom (length of a bond) in solid lead?

■ The diameters of the smallest and largest atoms differ by only about a factor of 8. Most metallic elements have similar radii, on the order of 1.5×10^{-10} m. It is useful to remember that the radius of an "average" atom is on the order of 1×10^{-10} m.

Figure 4.12 A solid wire considered as many parallel long chains of atoms connected by springs (interatomic bonds). For clarity, the horizontal bonds are not shown; they aren't significant when stretching the wire.

4.5 | THE STIFFNESS OF AN INTERATOMIC BOND

Knowing the length of an interatomic bond in solid copper (and the diameter of a copper atom), we can now use experimental data to find the stiffness of the interatomic bond, considered as a spring. You have probably had some experience with ordinary macroscopic springs. A Slinky, for example, is a soft spring; its spring stiffness is small—around 1 N/m. The stiffness of the spring on a pogo stick is much larger—around 5000 N/m. It is difficult to measure the stiffness of an interatomic bond directly, but we can analyze data from macroscopic (large-scale) experiments to determine this quantity.

The basic idea is to hang heavy masses on a long wire and measure the stretch of the wire; then, by figuring out how many interatomic "springs" there are in the wire, we can determine the spring constant of a single interatomic bond. To do this, we need to be able to relate the stiffness of an object composed of many springs to the stiffness of each individual spring. We will consider the wire to be composed of many parallel long chains of atoms connected by springs (Figure 4.12). We will need to determine how the stiffness of one of the short springs (bonds) is related to the stiffness of an entire chain of springs, and how the stiffness of the entire wire is related to the stiffness of one chain of springs.

Two Springs Linked End-to-End (in Series)

Suppose that we have a long, low-mass spring whose spring stiffness is 100 N/m. If we hang a block whose weight is 100 N (about a 10 kg mass) on the end of the spring, the spring will stretch 1 meter. (This is a big stretch, so the relaxed spring must be several meters long; we are using simple numbers here to make our reasoning easier.)

> **QUESTION** We make a longer spring by linking two identical springs end-to-end (in "series"). Each shorter spring has a stiffness of 100 N/m. What will the stiffness of the longer spring be?

When we hang the 100 N block on the end of the longer spring, this longer spring will stretch 2 meters, because each of the individual springs stretches 1 meter (Figure 4.13). We can use the Momentum Principle to get the stiffness of this longer spring. When the block hangs motionless, its momentum is not changing. Choosing a Δt of 10 seconds, we find this:

System: block
Surroundings: spring, Earth (see the free-body diagram in Figure 4.14)

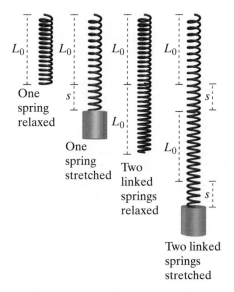

Figure 4.13 Two identical springs linked end to end stretch twice as much as one spring when the same force is applied. The combined spring therefore is only half as stiff as the individual springs.

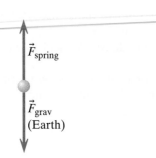

Figure 4.14 Free-body diagram for hanging block.

Momentum Principle:

$$\Delta p_y = F_{\text{net},y}\Delta t$$
$$0 = (k_s s - mg)(10\,\text{s})$$
$$k_s s = mg$$
$$k_s(2\,\text{m}) = 100\,\text{N}$$
$$k_s = 50\,\text{N/m}$$

A long spring made of two identical springs linked end-to-end is only half as stiff as each of the shorter springs.

> **4.X.3** If a chain of 20 identical short springs linked end to end has a stiffness of 40 N/m, what is the stiffness of one short spring?

Figure 4.15 Two springs side by side supporting a block.

Two Springs in Parallel (Side-by-Side)

QUESTION Suppose that we support a 100 N weight with two identical springs side by side (Figure 4.15). If the spring stiffness of each spring is 100 N/m, how much will each spring stretch? What is the effective stiffness of the two-spring system?

Informally, we see that each spring supports only 50 N, so each spring will stretch only 0.5 m. You can work out the formal solution yourself using the Momentum Principle, remembering to include two separate upward spring forces.

We can think of the two springs as a single, wider spring. The effective spring stiffness of this "double-width" spring is:

$$k_{s,\text{effective}} = \frac{100\,\text{N}}{0.5\,\text{m}} = 200\,\text{N/m}$$

Two springs side by side are effectively twice as stiff as a single spring.

> **4.X.4** Nine identical springs are placed side by side (in parallel, as in Figure 4.15), and connected to a large massive block. The stiffness of the nine-spring combination is 2700 N/m. What is the stiffness of one of the individual springs?

Cross-Sectional Area

The cross-sectional area of an object is the area of a flat surface made by slicing through the object (Figure 4.16). A cylindrical object like a round pencil has a circular cross-sectional area (imagine sawing crosswise through a pencil). A cylindrical pencil 10 cm long and 0.5 cm in diameter has a cross-sectional area of

$$A = \pi(0.025\,\text{m})^2 = 1.96 \times 10^{-3}\,\text{m}^2$$

Figure 4.16 The cross-sectional area of a cylinder is the area of a circle. The cross-sectional area of a rectangular solid is the area of a rectangle.

Note that the length of the object is irrelevant.

A long object with four flat sides, like a board, has a cross-sectional area that is rectangular, or perhaps square. A wooden board 7 meters long, 5 cm wide, and 3 cm high, has a cross-sectional area of

$$A = (0.05\,\text{m})(0.03\,\text{m}) = 1.5 \times 10^{-3}\,\text{m}^2$$

EXAMPLE	**Interatomic Bond Stiffness in Copper**

A copper wire is 2 meters long. The wire has a square cross-section (that is, it has four flat sides—it is not round). Each side of the wire is 1 mm in width. Making sure the wire is straight, you hang a 10 kg mass on the end of the wire. Careful measurement shows that the wire is now 1.51 mm longer. From these measurements, determine the stiffness of one interatomic bond in copper.

Solution   **1.** What is the spring stiffness $k_{s,\text{wire}}$ of the entire wire, considered as a single macroscopic (large scale), very stiff spring?
System: mass
Surroundings: Earth, wire
Momentum Principle:

$$\Delta p_y = 0 = (k_{s,\text{wire}}s - mg)\Delta t$$

$$k_{s,\text{wire}} = \frac{mg}{s} = \frac{(10\,\text{kg})(9.8\,\text{N/kg})}{(1.51 \times 10^{-3}\,\text{m})} = 6.49 \times 10^4\,\text{N/m}$$

2. How many side-by-side atomic chains (long springs) are there in this wire? This is the same as the number of atoms on the bottom surface of the copper wire (see Figure 4.17).
Cross-sectional area of wire:

$$A_{\text{wire}} = (1 \times 10^{-3}\,\text{m})^2 = 1 \times 10^{-6}\,\text{m}^2$$

Cross-sectional area of one copper atom (even though an atom itself is spherical, each atom occupies a cubical space in the crystal):

$$A_{1\,\text{atom}} \approx (2.28 \times 10^{-10}\,\text{m})^2 = 5.20 \times 10^{-20}\,\text{m}^2$$

Number of side-by-side atomic chains in the wire:

$$N_{\text{chains}} = \frac{A_{\text{wire}}}{A_{1\,\text{atom}}} = 1.92 \times 10^{13}$$

3. How many interatomic bonds are there in one atomic chain running the length of the wire?

$$N_{\text{bonds in 1 chain}} = \frac{L_{\text{wire}}}{d} = \frac{2\,\text{m}}{2.28 \times 10^{-10}\,\text{m}} = 8.77 \times 10^9$$

4. What is the stiffness $k_{s,i}$ of a single interatomic spring?
Using our results about springs linked end to end and springs arrayed side by side:

$$k_{s,\text{wire}} = \frac{(k_{s,i})(N_{\text{chains}})}{N_{\text{bonds in 1 chain}}} \quad (\text{solve for } k_{s,i})$$

$$k_{s,i} = \frac{(6.49 \times 10^4\,\text{N/m})(8.77 \times 10^9)}{(1.92 \times 10^{13})} = 29.6\,\text{N/m}$$

An interatomic bond in copper is stiffer than a Slinky, but less stiff than a pogo stick. The stiffness of a single interatomic bond is very much smaller than the stiffness of the entire wire (which we found to be around $6 \times 10^4\,\text{N/m}$).

Top of wire

Edge of wire

Bottom of wire

Figure 4.17 A wire modeled as an assembly of side-by-side chains of balls and springs, viewed from the bottom. Colored lines connect the bottom layer of atoms in the wire (only nine ball–spring chains are shown here). The horizontal bonds are not shown; they aren't stretched significantly.

4.X.5 The 2 meter copper wire with square cross section of 1 mm by 1 mm stretched 1.51 mm when it supported a 10 kg mass. Cut a length of this wire 0.2 m long and hang a 10 kg mass from it. How much will this short wire stretch?

4.6 STRESS, STRAIN, AND YOUNG'S MODULUS

QUESTION Suppose that we had used a different copper wire, which was 3 meters long, and had a circular cross section, with a diameter of 0.9 mm. Would our result for the interatomic bond stiffness have been different?

The result should not be different, because the interatomic bond stiffness is a property of the material (solid copper). The macroscopic dimensions of the wire will not change the intrinsic properties of copper.

You will not find tables of interatomic spring stiffness for different solid materials in reference libraries. Instead, what is published is a macroscopic quantity called Young's modulus. Like density, and interatomic spring stiffness, Young's modulus is a property of a particular material (for example, copper), and is independent of the shape or size of a particular object made of that material. Young's modulus is a macroscopic measure of the "stretchability" of a solid material. It relates the fractional change in length of an object to the force per square meter of cross-sectional area applied to the object.

Strain

The longer a wire is, the more atomic bonds it has along its length, and the more it will stretch when a force is applied to it. In order to define a quantity (Young's modulus) that is a property of the material and doesn't depend on the shape of the wire, we need to take into account not only the measured stretch, but also the original length of the wire. If the wire has a length L, we call the amount of stretch of the wire ΔL (a small increment in the length, what we have called s in a spring), and the fractional stretch $\Delta L/L$ is called the "strain."

DEFINITION OF STRAIN
$$\text{strain} \equiv \frac{\Delta L}{L}$$

Stress

As we saw above, all the chains of atoms help hold up the weight attached to the wire. In order to define a quantity that doesn't depend on the thickness of the wire, we need to take into account not only the tension force F_T, but also the cross-sectional area of the wire A (Figure 4.18). The tension force per unit area is called the "stress."

DEFINITION OF STRESS
$$\text{stress} \equiv \frac{F_T}{A}$$

Young's Modulus

As long as the strain isn't too big, the strain is proportional to the stress. At the atomic level, the stress F_T/A can be related to the force that each long chain of

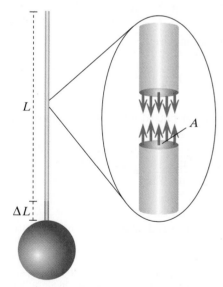

Figure 4.18 The heavy ball stretches the wire an amount ΔL. A is the cross-sectional area—the area of a slice through the wire, perpendicular to its length.

atomic bonds must exert, and the strain $\Delta L/L$ can be related to the stretch of the interatomic bond. The ratio of stress to strain is a property of the material and differs for steel, aluminum, and so on, but doesn't depend on the length or thickness of the wire we use in our measurement of Young's modulus.

DEFINITION OF YOUNG'S MODULUS

$$Y \equiv \frac{\text{stress}}{\text{strain}} = \frac{\left(\dfrac{F_T}{A}\right)}{\left(\dfrac{\Delta L}{L}\right)}$$

Young's modulus is the ratio of stress to strain for a particular material. Young's modulus is a property of the material, and does not depend on the size or shape of an object. The stiffer the material, the larger is Young's modulus.

Using Young's modulus, a quantitative form of the relation between stress and strain in a wire is written as follows:

$$\frac{F_T}{A} = Y\frac{\Delta L}{L}$$

Note the similarity to the spring force, $F_T = k_s s\ (= k_s \Delta L)$.

You may have the opportunity to measure Young's modulus in a laboratory experiment, and to use your measurements to determine the interatomic spring stiffness for a particular material (Problem 4.P.47).

Limit of Applicability of Young's Modulus

If you apply too large a stress, the wire "yields" (suddenly stretches a great deal) or breaks, and the proportionality of stress and strain is no longer even approximately true. Figure 4.19 shows a graph of the strain $\Delta L/L$ that results as a function of the applied stress F_T/A for a particular aluminum alloy.

You see that for moderate stress the resulting strain is proportional to the applied stress: double the stress, double the strain. However, once you reach the yield stress, a further slight increase in the stress leads to a very large increase in the length of the metal. This can be quite dramatic: as you add more weights to the end of a wire the wire gets slightly longer, then suddenly it starts to lengthen a lot, very rapidly, and then the wire breaks. In materials reference manuals, stress is usually plotted against strain, so that the slope of the line is equal to Young's modulus, stress divided by strain. We plot strain vs. stress here to emphasize that stress is the *cause* (the independent variable) and strain the *effect* (the dependent variable).

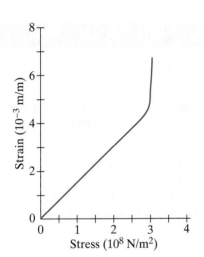

Figure 4.19 Strain vs. stress for a particular aluminum alloy.

> **4.X.6** In Figure 4.19, what is the approximate value of Young's modulus for this aluminum alloy?
>
> **4.X.7** Suppose that we hang a heavy ball with a mass of 10 kg (about 22 pounds) from a steel wire 3 m long that is 3 mm in diameter. Steel is very stiff, and Young's modulus for steel is unusually large, 2×10^{11} N/m^2. Calculate the stretch ΔL of the steel wire. This calculation shows why in many cases it is a very good approximation to pretend that the wire doesn't stretch at all ("ideal nonextensible wire").

Relating Young's Modulus to Interatomic Spring Stiffness

From macroscopic measurements of Young's modulus we can calculate an approximate spring stiffness $k_{s,i}$ for the interatomic bond by again making a "macro-micro" connection.

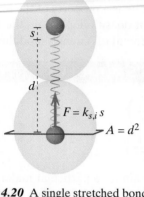

Figure 4.20 A single stretched bond in a solid wire. The effective cross-sectional area of one atom is d^2, and the force acting on that portion is $k_{s,i}s$.

Consider a single stretched interatomic bond in a wire that is being stretched. Let d represent both the relaxed length of an interatomic bond and the diameter of one atom. The cross-sectional area occupied by one atom is d^2 (Figure 4.20) (recall that even though an atom is approximately spherical, it occupies a cube of space in the crystal lattice).

QUESTION What is the magnitude of the stress on one atom in terms of the interatomic force $k_{s,i}s$?

If the stretch of the interatomic bond is s, the atomic stress (force per unit area) is $\dfrac{k_{s,i}s}{d^2}$.

QUESTION In terms of the stretch s of the interatomic "spring," what is the strain for one atom (change of height of the atom, divided by its normal height)?

The strain for one atom is s/d, because its height was originally d, and the change in its height is s. Therefore we can express Young's modulus in terms of atomic quantities:

YOUNG'S MODULUS IN TERMS OF ATOMIC QUANTITIES

$$Y = \frac{(k_{s,i}s/d^2)}{(s/d)} = \frac{k_{s,i}}{d}$$

where $k_{s,i}$ is the stiffness of an interatomic bond in a solid, and d is the length of an interatomic bond (and the diameter of an atom).

If you know Young's modulus for some metal, you can calculate the effective stiffness of the interatomic bond, modeled as a spring: $k_{s,i} = Yd$, where d is the atomic diameter in the metal. This is another example of an important theme, that of relating macroscopic properties to microscopic (atomic-level) properties.

Tarzan and the Vine Revisited

We now know that interatomic bonds in the vine stretch when Tarzan hangs from the vine. It is not yet clear why the vine breaks when Tarzan swings on it, but not when he hangs motionless. Apparently the vine must stretch more when he swings—perhaps it must apply a larger force? In Chapter 5 we will learn how to apply the Momentum Principle to Tarzan's situation.

Figure 4.21 A brick lies on a table. The table pushes up on the brick with a force of magnitude F_{Table}.

Figure 4.22 Forces acting on the brick.

4.7 COMPRESSION (NORMAL) FORCES

A closely related situation is that of a brick lying at rest on a metal table (Figure 4.21). In this situation the table is compressed, while in the previous discussion the wire was stretched.

Take the brick as the system of interest (Figure 4.22). The Earth pulls down on the brick of mass M, and the magnitude of the gravitational force is Mg.

Since the brick remains at rest, the vertical component of the net force must be zero:

$$0 = +F_{\text{Table}} - Mg$$

We conclude that the force F_{Table} exerted by the table is equal to the weight Mg of the brick, an unsurprising result.

How does the metal table exert a force on the brick? If the brick were extremely heavy, we might not be surprised to observe the table sagging under its weight. Even if the table does not sag visibly, the molecules in the top layers of

Figure 4.23 A schematic view of a portion of a brick resting on a portion of a table (scale greatly exaggerated). Interatomic "springs" in the contact region are compressed by the weight of atoms above.

Figure 4.24 When you apply a force on the brick, it runs into atoms of the table, compressing interatomic bonds in front and stretching them behind (scale greatly exaggerated). The force of the table now has a component to the left, a "frictional force."

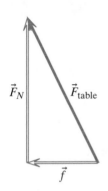

Figure 4.25 If the brick is pulled to the right, the table now exerts a force upward and to the left. It is convenient to express the force $\vec{F}_{\text{table}}$ as the sum of two perpendicular components: the upward "normal" force $\vec{F}_N$, and the sideways "friction" force $\vec{f}$.

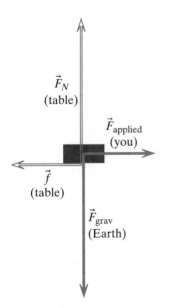

Figure 4.26 The free-body diagram for the brick now includes the components of the force by the table, as well as the force by you and the force by the Earth.

the table are pushed down by molecules in the bottom layer of the brick. The table actually deforms somewhat—its interatomic "springs" are compressed. (Of course, atoms in the brick are compressed somewhat, too.)

Atoms in the table are squeezed closer than their normal equilibrium positions, compressing the springlike interatomic bonds. Figure 4.23 schematically shows the contact region between the brick and the table.

It would be appropriate to call the upward force of the table F_C for "compression force," but it is common practice to label this kind of force as a "normal" force. The word is used in the mathematical sense, meaning "perpendicular to," because the direction of the force is perpendicular to the surface of the table. Usually "normal" forces are labeled as F_N.

Note: Neither "normal" force nor "tension" force is a different kind of fundamental force, like the gravitational force or electric force. These are merely descriptive names for particular kinds of interatomic electric interactions.

4.8 | FRICTION

As you've just seen, a brick sitting on a table sinks into the table, compressing the interatomic bonds, and the table pushes upward on the brick. If you then drag the brick across the table as in Figure 4.24, the brick continually runs into previously uncompressed regions of the table, compressing bonds ahead of the brick, so there is a component of the force of the table that is parallel to the velocity (and opposite to the velocity in this case), which is called a "frictional force," as shown in Figure 4.25. The full free-body diagram for the brick is shown in Figure 4.26. After the brick has passed some point along the table, at that point the bonds relax back to their equilibrium lengths, and the atoms go back to their equilibrium positions.

However, the disturbance of the atoms away from their equilibrium positions leaves the atoms vibrating about their equilibrium positions. As we will discuss in detail in later chapters, this increased vibrational motion or "internal energy" is associated with higher temperature. Dragging the brick across the table makes the temperature of the table and brick rise. The increase in internal energy means that you must continually apply a force and expend energy to do the dragging. If you stop pulling, the brick will slow down and stop as its motional ("kinetic") energy is dissipated away into internal energy in the form of agitation of atoms throughout the objects. We say that friction is a "dissipative" process.

Friction is a complex process, and this picture of friction at the atomic level, based on the ball–spring model of solids, is necessarily simplified. In a later chapter we will discuss friction in more detail, after we have studied energy.

QUESTION To keep the brick moving at a constant speed, you have to keep pulling. It's tempting to conclude that "a constant force is needed to keep an object moving at a constant speed," which sounds like a violation of Newton's first law of motion and of the Momentum Principle. What's going on here?

It is the net force that governs the motion of the brick, not just the force that you apply. Since the net force is zero, the brick moves with constant speed. You are pulling with a force whose magnitude is equal to that of the parallel component of the force of the table, so the net force $\vec{F}_{\text{net}}$ is zero.

Friction Doesn't Always Slow Things Down

It is often the case that the direction of the friction force is opposite to the velocity, tending to slow down the object. However, this is not always true. For example, if you drop a box onto a conveyor belt, bonds at the back of the box are compressed and the friction force pushes the box forward with increasing

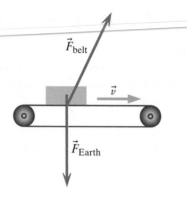

Figure 4.27 Place a box on a conveyor belt, and the horizontal component of $\vec{F}_{\text{belt}}$ accelerates the box.

speed (Figure 4.27). Eventually the box reaches the speed of the moving belt, at which time the friction force drops to zero and the box moves with constant speed, subject to a zero net force.

When a sprinter speeds up, she pushes her feet backwards against the ground, and the horizontal component of the force of the ground on her feet points forward, accelerating her. On a very slick surface with almost no friction she can't accelerate.

Sliding Friction

When one object slides on another, the component of force exerted by one object on the other has a component parallel (or antiparallel) to the motion with the following properties:

THE SLIDING FRICTION FORCE

$$f_{\text{friction}} \approx \mu_k F_N$$

F_N is the "normal force," the perpendicular component of the force that is squeezing the two objects into each other.

μ_k (Greek lower-case mu) is the "coefficient of kinetic friction," which depends on the nature of the two materials in contact, with values typically in the range of 0 to 1.

The sliding friction force does not depend on the speed of the object.

The more you squeeze the two objects together, the deeper their interpenetration and the more difficult it is to slide sideways, which is why the force is proportional to F_N.

It is interesting that the friction force does not depend on the speed. This is in contrast to air resistance, another dissipative force. Air resistance is larger at higher speeds, as you can easily verify by putting your hand out of the window of a moving car and noticing how much greater the air resistance force is on your hand at high speed than it is at low speed.

$f_{\text{friction}} \approx \mu_k F_N$ is only approximate and does not have the fundamental character of the gravitational and electric forces. Nevertheless, it is a good approximation to the effects of friction when one object slides on another.

QUESTION   If you are pulling a brick at a constant speed, what happens if you increase your force to be bigger than $\mu_k F_N$?

If while the brick is sliding you apply a force bigger than $\mu_k F_N$, the net force becomes nonzero, so the momentum of the brick will increase.

Static Friction

QUESTION   What will happen if you exert a horizontal force *less* than $\mu_k F_N$ on a brick lying on a table?

In this case, although the brick doesn't slide, there is a sideways compression of the atomic bonds, so there is a parallel component of the force of the table on the brick, equal to the force you're applying. This situation is called "static friction," where "static" means "not moving." The static friction force can have any value that is less than $\mu_s F_N$, where μ_s is called the "coefficient of static friction." As you apply a bigger and bigger force, but less than $\mu_s F_N$, the bonds are compressed more but the object doesn't slide.

For some pairs of materials it may be necessary to exert a force somewhat bigger than $\mu_k F_N$ in order to get the object to start to move, but once the object has started to slide, you only need a force of $\mu_k F_N$ to keep it moving. In such situations we say that the coefficient of static friction μ_s is larger than the one for sliding friction μ_k ("k" for kinetic, meaning with motion).

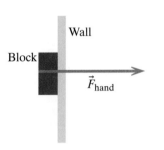

Figure 4.28 You press a block against a wall.

STATIC FRICTION

$$\text{static } f_{\text{friction}} \leq \mu_s F_N$$

μ_s is the "coefficient of static friction."
   The parallel component f_{friction} of the force squeezing the two objects together can be less than $\mu_s F_N$. The maximum static friction force is $\mu_s F_N$.

When μ_s is larger than μ_k, there can be "stick-slip" motion. You pull on a string attached to the brick and apply a large force $\mu_s F_N$ to start the motion, then the friction force decreases to $\mu_k F_N$, the net force becomes nonzero, and the brick accelerates, making the string go slack if you don't accelerate your hand. Now that the string has gone slack, the brick slows down and may stop, and you have to start over, hence the stick-slip motion.

EXAMPLE   **Holding a Block Against a Wall**

You hold a 3 kg metal block against a wall by applying a horizontal force of 40 N, as shown in Figure 4.28. The coefficient of friction for the metal–wall pair of materials is 0.6 for both static friction and sliding friction. Does the block slip down the wall?

Solution   Momentum Principle: $\Delta \vec{p} = \vec{F}_{\text{net}} \Delta t$
System: Block
Surroundings: Earth, wall, hand
Free-body diagram: Figure 4.29

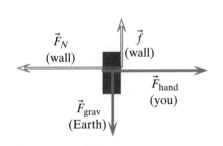

Figure 4.29 You press a block against a wall.

$$x: \Delta p_x = (F_{\text{hand}} - F_N)\Delta t = 0$$
$$y: \Delta p_y = (\mu_k F_N - mg)\Delta t, \text{ assuming it slides}$$

Combining these two equations, we have

$$\Delta p_y = (\mu_k F_{\text{hand}} - mg)\Delta t = (0.6(40\,\text{N}) - (3\,\text{kg})(9.8\,\text{N/kg}))\Delta t$$
$$\Delta p_y = (-5.4\,\text{N})\Delta t$$

Since there is a nonzero impulse in the $-y$ direction, the block will slip downward with increasing speed.

FURTHER DISCUSSION   To prevent the block from sliding downward, you would have to apply a force such that $F_{\text{net},y} = 0$, in which case $\mu_s F_{\text{hand}} = mg$, so $F_{\text{hand}} = (3\,\text{kg})(9.8\,\text{N/kg})/0.6 = 49\,\text{N}$. Note that F_N isn't always equal to mg; here it is equal to the force of your hand.
   If you apply a force greater than 49 N, how big is the static friction force f? If the block doesn't move, the y component of the net force $F_{\text{net},y}$ is zero, so the static friction force will always be mg, no matter how hard you press the block against the wall, with $f_{\text{friction}} = mg < \mu_s F_N$.

EXAMPLE   **Sliding to a Stop**

You take the same 3 kg metal block and slide it along the floor, where the coefficient of friction is only 0.4. You release the block with an initial velocity of $\langle 6, 0, 0 \rangle$ m/s. How long will it take for the block to come to a stop? How far does the block move?

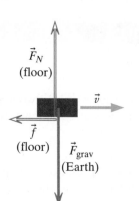

Figure 4.30 F_N is the normal component of the force of the floor and $f = \mu_k F_N$ is the parallel component.

Solution   Momentum Principle: $\Delta \vec{p} = \vec{F}_{net}\,\Delta t$
System: Block
Surroundings: Earth, floor
Free-body diagram: Figure 4.30
Initial state: $v_x = 6$ m/s
Final state: $v_x = 0$

$$x: \Delta p_x = -\mu_k F_N \Delta t$$
$$y: \Delta p_y = (F_N - mg)\Delta t = 0$$

Combining these two equations and writing $p_x = mv_x$, we have

$$\Delta(mv_x) = -\mu_k mg\,\Delta t$$
$$\Delta v_x = -\mu_k g\,\Delta t$$
$$\Delta t = \frac{0 - v_{xi}}{-\mu_k g} = \frac{v_{xi}}{\mu_k g}$$
$$\Delta t = \frac{6\,\text{m/s}}{0.4(9.8\,\text{N/kg})} = 1.53\,\text{s}$$

Since the net force was constant, $v_{x,\text{avg}} = (v_{xi} + v_{xf})/2$, so

$$\Delta x/\Delta t = ((6 + 0)/2)\,\text{m/s} = 3\,\text{m/s}$$
$$\Delta x = (3\,\text{m/s})(1.53\,\text{s}) = 4.5\,\text{m}$$

4.X.8 When you apply a horizontal force of 80 N to a block, the block moves across the floor at a constant speed ($\mu_k = \mu_s$).

(a) What happens if instead you apply a force of 60 N?

(b) When you apply a force of 60 N, what is the magnitude of the horizontal component of the force that the floor exerts on the block?

(c) What happens if instead you exert a force of 100 N?

(d) When you apply a force of 100 N, what is the magnitude of the horizontal component of the force that the floor exerts on the block?

4.X.9 A 5 kg box with initial speed 4 m/s slides across the floor and comes to a stop after 0.7 s.

(a) What is the coefficient of kinetic friction?

(b) How far does the box move?

(c) You put a 3 kg block in the box, so the total mass is now 8 kg, and you launch this heavier box with an initial speed of 4 m/s. How long does it take to stop?

4.9   SPEED OF SOUND IN A SOLID AND INTERATOMIC BOND STIFFNESS

When we have considered objects that exert tension forces, such as vines, strings, and wires, we have made the (very good) approximation that the stretch of the interatomic bonds is nearly the same throughout the length of the object. Likewise, we assume that inside objects exerting compression forces, the interatomic bonds are all compressed nearly the same amount.

The Propagation of Stretch or Compression

The equal stretch or compression of bonds does not, however, occur instantaneously. When Tarzan first starts pulling down on the end of a vine, he very slightly lengthens the interatomic bonds between neighboring atoms in the

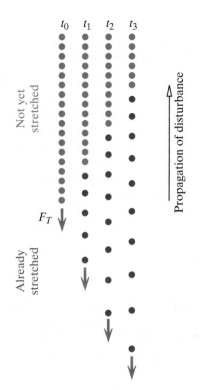

Figure 4.31 At time t_0 Tarzan starts pulling downward on the vine with a force F_T. The "disturbance" in the vine propagates upward. At successive times t_1, t_2, and t_3, more and more of the interatomic bonds in the vine are stretched. Atoms in the stretched region are indicated by colored dots. The amount of stretch is greatly exaggerated here.

Figure 4.32 Clap your hands—sound travels to your ear as a pulse of air compression.

nearby section of the vine (Figure 4.31). As a result of their downward displacement, these atoms stretch the bonds of their neighbors, and very quickly this new interatomic bond-stretching propagates upward, all the way to the other end of the vine, and the whole vine is then in tension.

It is the boundary between the already stretched region of the vine and the newly stretched region that moves upward, along the length of the vine. Individual atoms move extremely short distances, though for clarity in the diagram we have enormously exaggerated the interatomic stretches and the distance any individual atom moves. The process is the same for compression.

This process is fast, but it is not instantaneous. The rate of propagation of the boundary between the stretched and unstretched regions is called the "speed of sound" in the material. There are various ways to measure this speed. One of the simplest methods is to place a microphone at one end of a metal bar, strike the other end of the bar with a hammer, and use a computer interface to measure how long it takes for the disturbance to travel from one end of the bar to the other.

What Does This Have to Do with Sound?

What we call "sound" is a disturbance in a material. We are most familiar with sound traveling through air. When you clap your hands, a pulse of sound propagates through the air to your ear (Figure 4.32). This pulse starts as a momentary compression of the air (an increase in the air density) made by your hands. This compression is passed on to neighboring regions of the air, and the air near the hands relapses to its original density.

Eventually your eardrum is hit by the pulse and moves in response to the sudden increase in density (and pressure). The motion of the eardrum is detected by your inner ear and passed on to the brain, which interprets the signal as meaning that you clapped your hands. Note that no individual air molecule moves from near your hands to your ear. Rather it is the disturbance that moves from one place to another. The propagation of a disturbance is called a wave.

An Iterative Model of Sound Propagation in a Metal

A copper rod is modeled as a single horizontal chain of one hundred copper atoms, each with mass $(64)(1.67 \times 10^{-27}\,\text{kg})$. The initial distance between any two atoms is the length of the interatomic bond in solid copper ($2.4 \times 10^{-10}\,\text{m}$, see Section 4.4), and the stiffness of each interatomic spring is $27\,\text{N/m}$ (see Section 4.5). Initially each atom is at rest. The calculation is organized in a way similar to other iterative calculations we have done; the only difference is that we must update the position and momentum of 100 objects after each time step. An outline of the calculation is this:

Displace the leftmost atom by a small amount
Choose a very short time step ($\Delta t \approx 1 \times 10^{-14}\,\text{s}$)
Set elapsed time $= 0$
Then, repeatedly:

- From the current positions of the atoms, find the stretch of every spring.
- Find the net force on each atom, due to the two springs attached to it.
- Calculate the change in momentum of each atom.
- Calculate the new momentum of each atom.
- Update the position of each atom.
- Add Δt to the elapsed time.

Equilibrium

$t = 0$

$t = \Delta t$

Figure 4.33 Top: a chain of atoms and bonds in their equilibrium position. At $t = 0$ one atom is displaced to the right, starting a wave that moves along the chain.

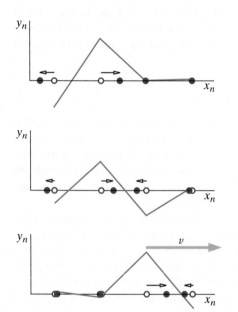

Figure 4.34 Visualizing the propagation by plotting the displacements y_n away from the equilibrium positions x_n.

■ **The derivative of a vector**

The derivative of a vector is itself a vector. We are already familiar with one such derivative, the instantaneous velocity:

$$\vec{v} = \frac{d\vec{r}}{dt} = \left\langle \frac{dx}{dt}, \frac{dy}{dt}, \frac{dz}{dt} \right\rangle$$

Like any other vector, the derivative of a vector has both a magnitude and a direction.

Stop the calculation when the rightmost atom is displaced. Divide the distance traveled (100 atoms)$(2.4 \times 10^{-10}$m/atom) by elapsed time to get the speed at which the disturbance traveled down the chain.

Figure 4.33 shows a visualization of the first step in this process. At the top is a chain of atoms and bonds in their equilibrium positions. At time $t = 0$ atom 1 is displaced to the right, compressing a bond, which then exerts a force on atom 2 (arrows denote the net forces on each atom). After applying the Momentum Principle to all atoms, and updating positions of all atoms over a time step Δt, both atoms 1 and 2 are displaced from equilibrium; the second bond is compressed, and the first is now stretched. Iterative updates continue, involving progressively more atoms after each time step.

Visualizing the propagation of the disturbance can be done in various ways. A particularly useful visualization is to plot the horizontal displacements (y_n) vertically, at the locations of the horizontal equilibrium locations (x_n). An example is shown in Figure 4.34, where you see that a displacement to the left is plotted below the axis, and a displacement to the right is plotted above the axis. For further details, see Problem 4.P.92.

It is clearly desirable to program a computer to do this calculation. When we carry out this calculation for a copper bar, using the values given above, we obtain a value for the speed of sound in copper of about 3530 m/s. Experimentally measured values given in engineering and science reference materials are around 3600 m/s, so our simple model gives a surprisingly good result.

An Algebraic Expression

Although predictions based on iterative calculations can assure us that our understanding of the underlying nature of a phenomenon is correct, it is also useful to have a simple algebraic expression for a central relationship such as that between interatomic bond stiffness and speed of sound. In order to be able to do this, we will need to derive an analytical solution for the oscillations of a single mass and spring. This will give us added insight into the propagation of a disturbance along a chain of masses and springs.

4.10 **DERIVATIVE FORM OF THE MOMENTUM PRINCIPLE**

The forms of the Momentum Principle we have used so far

$$\Delta \vec{p} = \vec{F}_{\text{net}} \Delta t \quad \text{and} \quad \vec{p}_f = \vec{p}_i + \vec{F}_{\text{net}} \Delta t$$

are particularly useful when we know the momentum at a particular time and want to predict what the momentum will be at a later time. We use these forms in repetitive computer calculations to predict future motion.

In solving problems involving objects whose momentum was not changing at all, such as a ball hanging motionless on the end of a wire, we saw that although we had to choose a value for Δt, it didn't matter what value we chose, as long as it was greater than zero. In such problems, we could also have used a different form of the Momentum Principle, which we obtain by dividing by the time interval Δt:

$$\frac{\Delta \vec{p}}{\Delta t} = \vec{F}_{\text{net}}$$

Just as we did in finding the instantaneous velocity of an object in Chapter 1, we can find the instantaneous rate of change of momentum by letting the time interval Δt approach zero (an infinitesimal time interval). The ratio of the infinitesimal momentum change (written as $d\vec{p}$) to the infinitesimal time interval (written as dt) is the instantaneous rate of change of the momentum. As the time interval gets very small, $\Delta t \to 0$, the ratio of the momentum change to the time interval approaches the ratio of the infinitesimal quantities

$$\lim_{\Delta t \to 0} \frac{\Delta \vec{p}}{\Delta t} = \frac{d\vec{p}}{dt}$$

and we obtain the following form:

THE MOMENTUM PRINCIPLE (DERIVATIVE FORM)

$$\frac{d\vec{p}}{dt} = \vec{F}_{net}$$

In words, "the instantaneous time rate of change of the momentum of an object is equal to the net force acting on the object." In calculus terms, since the limit of a ratio of infinitesimal quantities is called a derivative, we can say that "the derivative of the momentum with respect to time is equal to the net force acting on the object."

This form of the Momentum Principle is useful when we know something about the rate of change of the momentum at a particular instant. Knowing the rate of change of momentum, we can use this form of the Momentum Principle to deduce the net force acting on the object, which is numerically equal to the rate of change of momentum. Knowing the net force, we may be able to figure out particular contributions to the net force.

> **4.X.10** At a certain instant the z component of the momentum of an object is changing at a rate of $4 \, kg \cdot m/s$ per second. At that instant, what is the z component of the net force on the object?
>
> **4.X.11** If an object is moving with constant momentum $\langle 10, -12, -8 \rangle$ $kg \cdot m/s$, what is the rate of change of momentum $d\vec{p}/dt$? What is the net force acting on the object?

Acceleration

In previous studies you may have seen a simplified, approximate form of the Momentum Principle, written $F = ma$. Where does this come from, and under what circumstances is it useful?

If we make the following assumptions:

- The mass of the system is constant, and
- The speed of the system is much less than the speed of light,

then if we take the approximate derivative of $\vec{p}$ with respect to time, we get:

$$\frac{d\vec{p}}{dt} \approx \frac{d(m\vec{v})}{dt} \quad \text{(assuming } v \ll c\text{)}$$

$$\frac{d(m\vec{v})}{dt} = \left(\frac{dm}{dt}\right)\vec{v} + m\frac{d\vec{v}}{dt} \quad \text{(product rule for derivatives)}$$

$$\frac{d(m\vec{v})}{dt} = 0 + m\frac{d\vec{v}}{dt} \quad \text{(assuming system mass is constant)}$$

The rate of change of velocity with respect to time is called acceleration, and is usually denoted by the symbol $\vec{a}$.

DEFINITION OF ACCELERATION

$$\vec{a} \equiv \frac{d\vec{v}}{dt}$$

Acceleration is the rate of change of velocity with respect to time. It is a vector quantity.

Using this definition we can write the approximate rate of change of momentum as

$$\frac{d\vec{p}}{dt} \approx m\vec{a}$$

and an approximate form of the Momentum Principle, usually referred to as Newton's second law, as

$$m\vec{a} \approx \vec{F}_{net} \quad \text{(nonrelativistic form; constant mass)}$$

A highly simplified scalar form of this equation, $F = ma$, deals only with magnitudes, involves a single force, and gives no information about directions. Although $F = ma$ is usually referred to as Newton's second law, Newton himself worked with the Momentum Principle in the forms emphasized in this textbook, $d\vec{p}/dt = \vec{F}_{net}$ and $\Delta\vec{p} = \vec{F}_{net}\,\Delta t$.

The definition of acceleration is useful, because it is occasionally important to know the acceleration of an object (the rate at which its velocity is changing). However, we will generally find that the full form of the Momentum Principle, $d\vec{p}/dt = \vec{F}_{net}$, is more appropriate for our analyses.

- $d\vec{p}/dt = \vec{F}_{net}$ is relativistically correct (correct at any speed), and we will sometimes deal with fast-moving particles.
- $d\vec{p}/dt = \vec{F}_{net}$ involves momentum, which is a conserved quantity; we need to work with momentum when we analyze situations such as collisions, as we did in Chapter 3.
- $d\vec{p}/dt = \vec{F}_{net}$ is correct even if the mass of an object is not constant, such as a rocket with exhaust gases ejecting out the back.
- $d\vec{p}/dt = \vec{F}_{net}$ is a vector principle, and contains information about directions. The arrows over the symbols are extremely important; they remind us that there are really three separate component equations, for x, y, and z.
- $d\vec{p}/dt = \vec{F}_{net}$ reminds us that we have to add up all vector forces to give the "net" force, so we write $\vec{F}_{net}$, not just F.

4.X.12 The velocity of a ball changes from $\langle 5, -3, 0 \rangle$ m/s to $\langle 4.98, -3.04, 0 \rangle$ m/s in 0.01 s, due to the gravitational attraction of the Earth and to air resistance. The mass of the ball is 80 grams. What is the acceleration of the ball? What is the rate of change of momentum of the ball? What is the net force acting on the ball?

4.11   ANALYTICAL SOLUTION: SPRING–MASS SYSTEM

In Chapter 2 we applied the finite-time form of the Momentum Principle iteratively to predict the motion of a mass attached to a spring. Now, using the derivative form of the Momentum Principle, we can derive an analytical (algebraic) solution for the motion of an ideal mass–spring system.

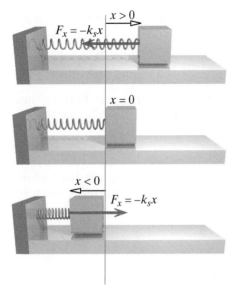

Figure 4.35 A block connected by a spring to a wall slides back and forth with little friction. The x component of the force that the spring exerts on the block is equal to $-k_s x$.

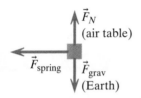

Figure 4.36 Free-body diagram.

■ We know that $dp_y/dt = 0$ because the y momentum of the block is zero and remains zero.

Applying the Momentum Principle to a Spring–Mass System

Consider a block of mass m connected to a spring whose stiffness is k_s (that is, the force exerted by the spring is $k_s s$ when the spring is stretched or compressed an amount s). The other end of the spring is attached to the wall (Figure 4.35). Our goal is to derive an equation for the position x as a function of time t

$$x(t) = \dots$$

which predicts the position of the block at any time in the future.

We will assume that:

■ The block slides with almost no friction on an air table, where it is supported on a cushion of air.
■ The mass of the spring is negligible compared to the mass of the block.
■ Air resistance is negligible.

We can simplify our calculations by choosing to measure the position of the block with respect to an origin chosen to be at the equilibrium position, the location of the block when the spring is relaxed, so that the stretch s of the spring is equal to the position x of the block:

$$L = L_0 + x$$
$$s = L - L_0 = (L_0 + x) - L_0 = x$$

Therefore the x component of the force on the block by the spring is $F_x = -k_s x$, as shown in Figure 4.35.

> QUESTION What would be the equation for the spring force in terms of the position of the block if the origin were located at the wall, and the relaxed length of the spring is L_0?

In that case we would write $F_x = -k_s(x - L_0)$, because what matters in this formula is the stretch (change in length of the spring), which is $(x - L_0)$. A check on this is the fact that when the block is at location $x = L_0$, the stretch $(x - L_0)$ is zero.

System: Block
Surroundings: Spring, Earth, air table (see free-body diagram in Figure 4.36)

Momentum Principle:

$$d\vec{p}/dt = \vec{F}_{net}$$

y component:

$$\frac{dp_y}{dt} = F_N - mg$$
$$0 = F_N - mg$$
$$F_N = mg$$

x component:

$$\frac{dp_y}{dt} = -k_s x$$
$$\frac{d(mv_x)}{dt} = \frac{d}{dt}\left(m\frac{dx}{dt}\right) = -k_s x$$
$$m\frac{d^2 x}{dt^2} = -k_s x$$

■ Since $v_x = dx/dt$, dp_x/dt contains the second derivative of x with respect to t.

If we ignore the constants and signs for a moment, this equation (which is a "differential equation") says that if we know x as a function of time, taking the second derivative of that function will give us back the original function.

QUESTION   What function appears in its own second derivative?

If we differentiate a cosine function twice, we get a cosine again. Further, since the block oscillates back and forth, cosine sounds like a reasonable guess. Let's try this:

$$x = A\cos(\omega t)$$

In this expression, A and ω (lowercase Greek omega) are constants whose values we will need to determine. We will look for values that make this functional relation satisfy the Momentum Principle.

■ Guessing a function isn't as peculiar as it may seem. This is actually one of the standard approaches to solving a differential equation.

We will put this expression for $x(t)$ back into the Momentum Principle equation:

$$m\frac{d^2}{dt^2}(A\cos(\omega t)) = -k_s A\cos(\omega t)$$

$$m\frac{d}{dt}(-A\omega\sin(\omega t)) = -k_s A\cos(\omega t)$$

$$-Am\omega^2\cos(\omega t) = -k_s A\cos(\omega t)$$

$$m\omega^2 = k_s$$

$$\omega = \sqrt{\frac{k_s}{m}}$$

■ Cosine has values between 1 and −1. Since the block may not move 1 meter, we need to multiply by a constant A to change the maximum displacement to a reasonable value.

We have shown that $x = A\cos(\omega t)$ is an analytical solution for the position of the mass as a function of time, if we choose ω to have the value $\sqrt{k_s/m}$.

■ If the block completed one oscillation in 2π seconds, the constant ω (Greek lowercase omega) would be 1. Otherwise, we need ω to be able to adjust the time taken for a round-trip oscillation.

AN ANALYTICAL SOLUTION FOR MASS–SPRING

$$x = A\cos(\omega t), \text{ with } \omega = \sqrt{\frac{k_s}{m}}$$

A (the "amplitude") is a constant that depends on the initial conditions. A is equal to the maximum stretch of the spring during an oscillation. ω is called the "angular frequency" and has units of radians/s.

Angular Frequency, Period, and Frequency

The constant A is called the "amplitude" of the motion. It is equal to the maximum stretch of the spring. If the spring stretches or compresses a maximum of 5 cm, then A will have the value 0.05 m.

The constant ω in $x = A\cos(\omega t)$ is called the "angular frequency" and is measured in radians per second (ω is lowercase Greek omega). When the argument ωt of the cosine increases by 2π radians, the motion repeats (one complete cycle), so if we call the "period" T (the time required for one complete cycle, Figure 4.37), we have $\omega T = 2\pi$.

One other quantity often used to describe oscillating systems is the "frequency" f, which is the number of complete cycles per second: $f = (1\text{ cycle})$ per T seconds. Cycles per second are also called "hertz."

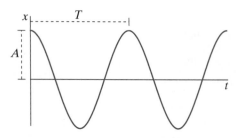

Figure 4.37 Amplitude A and period T, shown on a plot of x vs. t for the mass.

Angular frequency $\omega = \dfrac{2\pi}{T} = \sqrt{\dfrac{k_s}{m}}$, radians per second

Period $T = \dfrac{2\pi}{\omega}$, seconds

Frequency $f = \dfrac{1}{T} = \dfrac{\omega}{2\pi}$, cycles per second or "hertz"

A Vertical Mass–Spring System

■ Because there is no friction in this idealized system, the amplitude of the motion does not decrease over time.

It can be shown that this analytical solution is also valid for a vertical mass–spring system, in which the gravitational force on the mass acts parallel or antiparallel to the force on the mass by the spring. The details of this analysis are given in Section 4.14.

EXAMPLE

Mass–Spring Oscillation (Analytical Solution)

A 50 gram mass is attached to a low-mass horizontal spring whose relaxed length is 25 cm and whose stiffness is 4 N/m. The block slides on a very slippery, low-friction surface. You stretch the spring so it is 30 cm long, and release the block. **(a)** What is the amplitude of the oscillations? **(b)** How long will it take the block to make one full oscillation? **(c)** What would be the amplitude of the oscillations if the initial length of the spring were 35 cm instead of 25 cm? **(d)** How long would it take the block to make one full oscillation if the initial length of the spring were 35 cm instead of 25 cm? **(e)** How many oscillations will the block make in 10 seconds? **(f)** The initial length of the stretched spring is 35 cm. What will the location of the block be 3.15 seconds after you release it? (Remember to use radians, not degrees.)

Solution

System: Mass
Surroundings: Spring, Earth, table
Are the assumptions for an ideal mass–spring system satisfied?

■ Spring mass negligible: yes
■ Friction negligible: yes
■ Air resistance negligible: yes

(a) $A = s_i = 0.3\,\text{m} - 0.25\,\text{m} = 0.05\,\text{m}$

(b) Time for one round trip = T (period)

■ The expression for the period

$$T = \frac{2\pi}{\omega} = \frac{2\pi}{\sqrt{\dfrac{k_s}{m}}}$$

does not contain A, so the period must not depend at all on the amplitude of the oscillations.

$$\omega = \sqrt{\frac{k_s}{m}} = \sqrt{\frac{4\,\text{N/m}}{0.05\,\text{kg}}} = 8.94\,\text{s}^{-1}$$

$$T = \frac{2\pi}{\omega} = \frac{2\pi}{8.94\,\text{s}^{-1}} = 0.70\,\text{s}$$

(c) $A = s_i = 0.35\,\text{m} - 0.25\,\text{m} = 0.10\,\text{m}$

(d) $T = 0.70\,\text{s}$

(e)

$$f = \frac{1}{T} = \frac{1}{0.70\,\text{s}} = 1.43 \;\text{cycles/s}$$

$$\text{number of cycles} = ft = (1.43\,\text{cycles/s})(10\,\text{s}) = 14.3$$

(f)

$$x(t) = A\cos(\omega t)$$

$$x(3.15\,\text{s}) = (0.10\,\text{m})\cos((8.94\,\text{s}^{-1})(3.15\,\text{s})) = -0.0994\,\text{m}$$

Check: The units are ok. Are the results reasonable? Yes, because the origin was put at the end of the relaxed spring, so this means the spring is compressed slightly less than maximum compression (see Figure 4.35).

4.X.13 You have a rubber band whose relaxed length is 8.5 cm. You hang a coffee cup, whose mass is 330 grams, from the rubber band, which stretches to a length of 14 cm. Consider the rubber band to be a single spring. **(a)** What is the stiffness of the rubber band? **(b)** If you start this mass–spring system oscillating, how many round-trip oscillations will the coffee cup make in 5 seconds?

4.X.14 When a particular mass–spring system is started by stretching the spring 4 cm, one complete oscillation takes 2 seconds. **(a)** If instead the initial stretch had been 6 cm, how long would it take for one complete oscillation? **(b)** If the mass were tripled, what would the period of the system be?

Harmonic and Anharmonic Oscillators

The idealized mass–spring system we have analyzed is called a "harmonic oscillator." The special features of a harmonic oscillator are:

- The position of the mass as a function of time is given by a cosine function.
- The period (round-trip time) is independent of the amplitude of the oscillations.

■ The fact that the period of a harmonic oscillator does not depend on the amplitude is surprising. With a larger amplitude, the mass travels farther in one period. However, it also reaches higher speeds, and for an ideal mass–spring system, these two effects exactly cancel.

Not all oscillating systems are harmonic. If you drop a rubber ball and let it bounce, as the bounce amplitude decreases, the period decreases too: the time between bounces gets shorter and shorter as the height of the bounces gets lower and lower. The bouncing ball is an example of an "anharmonic oscillator." "Anharmonic" simply means "not harmonic."

Real Springs Have Mass

Real springs have mass, but we have ignored this fact of the real world. It is not easy to include the effect of the mass of the spring, either in a numerical integration or in an analytical treatment. One way to model a real spring with mass is with a chain of point masses connected by massless springs (Figure 4.38).

Figure 4.38 A massive spring may be modeled as a chain of point masses connected by massless springs.

QUESTION   Under what circumstances is it likely to be a good approximation to neglect the mass of the spring?

In many cases the spring has much less mass than the objects that are connected to it, in which case the effect of the spring's mass is likely to be small. A contrasting example is an oscillating Slinky, in which the only mass is the mass of the spring.

Existence of Analytical Solutions

For the mass–spring system we were able to obtain an analytical solution—a prediction for all time of the motion of our model spring–mass system, not subject to the additional inaccuracies introduced by a finite time step. However, for many systems the differential equation corresponding to the derivative form of the Momentum Principle has a mathematical form that cannot be solved analytically and must be solved numerically. The power of iterative application of the Momentum Principle is that it can be used even in those situations where no analytical solution seems possible.

■ One might feel that analytical solutions are better than numerical ones, but actually these are not as different as one might think. For example, one of the ways to calculate a table of cosine values is to do a numerical integration of the equation that applies to spring–mass systems! In general, a fruitful way to describe a mathematical function is by specifying the differential equation for which it is a solution, because then a numerical integration of that equation yields numerical values of the function.

For example, $x = A \cos(\omega t)$ *cannot* solve the following "nonlinear" differential equation for any value of ω:

$$\frac{dp_x}{dt} = m\frac{d^2 x}{dt^2} = -k_s x^3$$

There may or may not exist an analytical solution of this differential equation in terms of well-known mathematical functions, but *any* differential equation can be solved numerically, just as we did for the spring–mass equation. Given x and p at some instant in time, we can calculate the force $(-k_s x^3)$ and determine the new values of x and p at a time Δt later, then repeat as often as necessary to reach the time of interest.

It is important to keep clearly in mind that either an analytical or a numerical prediction of the future applies only to our *model* of the actual real-world situation. If our model is a good approximation to the real world, our prediction will be a good approximation to what will actually happen, but even an analytical prediction cannot be exact because of the practical impossibility of knowing exactly the initial conditions and the net force due to all the other objects in the Universe.

In our idealized mass–spring system there is no friction or air resistance. In a real system, both of these are often present. In this case, the idealized model will not give an accurate prediction of the position of the mass as a function of time; the amplitude of the real system will decrease, but the amplitude of the idealized model-system will not.

Initial Conditions for a Mass–Spring System

We need to know the initial conditions (position and momentum of the mass) in order to predict the position of the system at some future time. In the previous example, the mass was initially at rest, and the spring was stretched. It is also possible to start an oscillator by placing it at its equilibrium position and giving it a kick—in this case its initial stretch is zero, but its initial momentum is nonzero. In an iterative solution, this is easy to take into account. A discussion at the end of this chapter explains how this is taken into account in the analytical solution.

4.12   ANALYTICAL EXPRESSION FOR SPEED OF SOUND

Under ordinary conditions, the speed of sound in air is about 340 m/s. If your friend stands 340 meters away and yells, you will hear the sound a second later. The speed of sound in a solid object is much faster than it is in air. In aluminum, the speed of sound is about 5000 m/s. In lead, which is a much softer material than aluminum, the speed of sound is only about 1200 m/s. There are two significant differences between lead and aluminum at the atomic level. First, the mass of a lead atom is about eight times larger than the mass of an aluminum atom. Second, the interatomic bond stiffness in aluminum is greater than the interatomic bond stiffness in lead (see Problem 4.P.44). Since the propagation of a disturbance through a metal rod involves displacements of atoms, it seems reasonable that both of these factors should play a role in determining the speed with which the disturbance propagates. Our goal in the following discussion is to use the ball–spring model of a solid to predict this relationship.

Qualitative Guess

Think of a solid metal rod as a network of atoms (balls) connected by springs (interatomic bonds) as in Figure 4.39.

QUESTION   How should the speed of sound depend on m_a, the mass of an atom in a material? (Think of hitting one end of a metal rod with a hammer.)

A given applied force produces the same change of momentum for a massive atom like lead or a lighter atom like aluminum. However, the change in speed of the lead atom will be less than the change in speed of the aluminum atom. This suggests that the lower the mass of the atoms, the faster a sound pulse (disturbance) might propagate through a rod.

Figure 4.39 A model of a metal: atoms (balls) connected by springs (interatomic bonds).

QUESTION How should the speed of sound depend on the stiffness of the interatomic bonds in a material?

The stiffer the bond, the greater the force the "spring" will exert on a neighboring atom in response to the same displacement. Stiffer interatomic bonds should make a sound pulse propagate faster through a rod.

These initial qualitative guesses don't tell us details of the mathematical relationships involved: for example, whether speed of sound is proportional to $1/m_a, 1/m_a^2, 1/\sqrt{m_a}$, or some other factor. We need to apply our understanding of mass–spring systems to refine these guesses.

Dimensional Analysis

We need to construct a mathematical expression that has the correct units, or "dimensions." Speed has units of m/s, so we need an equation involving interatomic bond stiffness $k_{s,i}$ and atomic mass m, ending up with meters in the numerator and seconds in the denominator.

$$v = \ldots$$

The angular frequency ω involves both $k_{s,i}$ and m_a, and has units of s^{-1}, so ω might appear in the numerator of the expression:

$$v = \omega(?)$$

QUESTION Is it reasonable that v should be directly proportional to ω?

Yes. The larger the angular frequency ω is, the faster an atom oscillates back and forth, compressing and stretching intermolecular bonds, which affect neighboring atoms.

QUESTION What other quantity should go in the equation?

Since we need to end up with m/s, we need to multiply by something that has units of meters. ω is related to the rate at which one atom hits its neighbor, propagating a change through a distance of one interatomic spacing. Therefore multiplying by the interatomic bond length d might be appropriate.

$$v = \omega d$$

This informal reasoning is an example of "dimensional analysis," in which we identify meaningful quantities in the situation and combine them to guess a formula for some other quantity of interest. While this can be a fruitful procedure, we shouldn't be surprised if we find that the speed of sound is actually given by $v = 2\omega d$, or $v = \omega d/\pi$, for example. Moreover, note that atomic oscillations in the x direction involve two atomic "springs" on each side of an atom, not one, so we may have omitted a factor of 2. Nevertheless, we can guess that $(\sqrt{k_{s,i}/m_a})d)$ is proportional to the actual speed of sound, with the correct dependence on $k_{s,i}$ and m_a. As it happens, it can be shown that this *is* the correct formula. The full derivation, which is quite difficult, is not given here; it can be found in advanced textbooks on classical mechanics.

SPEED OF SOUND IN A SOLID

$$v = \omega d = \sqrt{\frac{k_{s,i}}{m_a}}\, d$$

v is the speed of sound (m/s), $k_{s,i}$ is the stiffness of the interatomic bond, m_a is the mass of one atom, d is the length of the interatomic bond.

EXAMPLE

Speed of Sound in Copper

We found (Section 4.5) that the stiffness of the interatomic bond in solid copper is about 27 N/m, and that the interatomic bond length in solid copper is about 2.28×10^{-10} m (Section 4.4). There are 64 grams of copper in one mole. Use these data to calculate the speed of sound in solid copper.

Solution

$$m_a = \frac{0.064\,\text{kg}}{6.02 \times 10^{23}\,\text{atoms}} = 1.06 \times 10^{-25}\,\text{kg/atom}$$

$$v = \sqrt{\frac{27\,\text{N/m}}{1.06 \times 10^{-25}\,\text{kg}}}(2.28 \times 10^{-10}\,\text{m}) = 3640\,\text{m/s}$$

This agrees well with experimentally measured values of around 3600 m/s.

Speed of Sound in Different Materials

In aluminum, which is very stiff (large k_s) and has a small atomic mass m_a, the speed of sound is very high, about 5000 m/s. If a wire is half a meter long, it takes about

$$(0.5\,\text{m})/(5000\,\text{m/s}) = 1 \times 10^{-4}\,\text{s} \quad \text{(a tenth of a millisecond)}$$

from the time you pull on one end for the other end to become tense. On the other hand, lead is quite soft (small k_s) and has a large atomic mass m_a, and the speed of sound in lead is only about 1200 m/s, much lower than in aluminum. (The speed of sound in solids is much higher than the speed of sound in air, which is about 340 m/s. Because air molecules are usually far from one another, the speed of sound in air is determined by the average speed of the air molecules, not by interatomic forces.)

4.X.15 A certain metal with atomic mass 2×10^{-25} kg has an interatomic bond with length 2.1×10^{-10} m and stiffness 40 N/m. What is the speed of sound in a rod made of this metal?

4.X.16 A rod made of this metal is 3 m long. If you hit one end of the rod, when would a microphone at the other end first detect a disturbance?

4.13   CONTACT FORCES DUE TO GASES

Until now we have considered only contact forces due to solids. However, gases can also exert contact forces. In everyday circumstances, the most significant of these is the force that the air exerts on objects—even stationary objects. Consider the surrounding air that is in contact with a ball hanging from a wire, as shown in Figure 4.40. You might think that the air has no effect on the ball, especially since the ball is not moving, so we don't need to worry about air resistance. To be certain, let's consider this issue at the microscopic level.

Buoyancy

Air is a gas, composed of about 80% nitrogen and 20% oxygen, with very small amounts of other gases. The molecules in the air are continuously in motion, moving in random directions. Molecules in the air near the ball will frequently bump into the ball, exerting a small force on it. If there is no wind, the motion of the gas molecules is random, so in a short time period there should

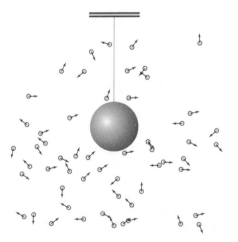

Figure 4.40 Air molecules in motion near a hanging ball. The density of the air beneath the ball is slightly greater than the air density above the ball, so there is a small upward force on the ball exerted by the air.

be about as many molecules colliding with the right-hand side of the ball as with the left-hand side, and there should on average be no sideways force on the ball.

However, the situation is not quite the same for collisions with the top and bottom of the ball. The directions of molecular velocities are still random, but there are more air molecules per cubic meter below the ball than there are above it. The density of the Earth's atmosphere decreases as one moves upward away from the Earth's surface, until finally at a sufficient distance (approximately 50 km) there is essentially no air left. You may have observed this variation yourself if you have ever gone from sea level to a location several thousand meters higher; you may feel light-headed if you exercise at high altitude because the density of oxygen is significantly lower.

This variation in air density occurs over kilometers; can it possibly be important over a distance of a few centimeters? Interestingly enough, the surprising answer is yes! There is actually a "buoyant force" upward on the ball, because the number of air molecules hitting the bottom of the ball per second is slightly greater than the number of molecules hitting the top of the ball per second.

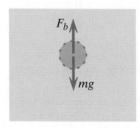

Figure 4.41 A system consisting of a ball-shaped spherical region of fluid, surrounded by more fluid.

Figure 4.42 Forces on the sphere of fluid. Here m is the mass of the fluid (air or water) in the sphere.

A Macroscopic View of Buoyancy

How can we determine the magnitude of the buoyant force on the ball due to the air? Let's consider a macroscopic viewpoint. The following discussion, based on the Momentum Principle, applies equally well to the buoyant force exerted on an object by any fluid (air, water, other liquids, other gases). The result we will obtain is called the Archimedes principle, for the Greek thinker who first understood it.

Imagine a box filled with air (or water). Consider a ball-shaped region that a ball will eventually occupy, at a moment when the ball is not yet there, so the ball-shaped region is filled with the fluid. Let's choose this ball-shaped mass of fluid as our system, and consider what forces must be acting on it (Figure 4.41). This may seem odd, but it is a perfectly legitimate choice of system.

> QUESTION Is the momentum of the indicated sphere of fluid changing? What must the net force on this sphere be?

The momentum of the indicated sphere of fluid is not changing. The Earth exerts a downward gravitational force on it, but the sphere does not sink. Because its momentum isn't changing, the net force on the sphere must be zero, so there must be an upward force that balances the gravitational force.

> QUESTION What exerts this upward force on the sphere?

The only object in contact with this sphere is the rest of the fluid, so it must be that the rest of the fluid exerts an upward force F_b ("buoyant force") whose magnitude is mg, where m is the mass of the sphere of fluid (Figure 4.42).

Now remove the sphere of fluid and replace it with the ball. It must be that initially the rest of the fluid still exerts the same upward force F_b of magnitude mg (Figure 4.43). This force may not be large enough to counteract the downward gravitational force on the ball; in the case of a ball hanging in air it clearly is not, since if the ball were not supported by a wire it would begin accelerating downward.

The upward force is called a "buoyant" force but is simply an interatomic contact force due to fluid molecules striking atoms in the surface of the ball. The key point is that despite the complexity of the interatomic interactions, the net effect is simply an upward force whose magnitude is the mass of an equivalent volume of fluid, times g.

Figure 4.43 A ball hanging from a wire, surrounded by air. The tension force is actually less than the weight of the ball, because the buoyant force F_b due to the air is also in the upward direction.

When Can the Buoyant Force Be Neglected?

Often it is reasonable to neglect buoyant forces in air (though usually not in water). To see why, let's compare the buoyant force to the gravitational force F_g on a ball hanging in air, where we write V for the volume of the ball:

$$\frac{F_b}{F_g} = \frac{m_{air}\,g}{M_{ball}\,g} = \frac{m_{air}}{M_{ball}} = \frac{m_{air}/V}{M_{ball}/V} = \frac{\rho_{air}}{\rho_{ball}}$$

The ratio of the forces is equal to the ratio of ρ_{air} (the density of air) to ρ_{ball} (the density of the ball; ρ is the Greek lowercase letter rho). The density of air at STP ("standard temperature and pressure," 0°C and 1 atmosphere) can be calculated if we remember from chemistry that one mole of a gas at STP occupies 22.4 liters. The molecular mass of N_2 is 28 and that of O_2 is 32, which gives an average molecular mass of about 29 for air (which is about 80% nitrogen and 20% oxygen):

$$\rho_{air} \approx \frac{29 \text{ grams/mole}}{22.4 \times 10^3 \text{ cm}^3/\text{mole}} \approx 1.3 \times 10^{-3} \text{ grams/cm}^3$$

What is the density of the ball? We can estimate it by noting that the density of water is 1 gram/cm^3, and that most solids are more dense than water. Using 1 g/cm^3 as the density of the ball, we find that $F_b/F_g \approx 1.3 \times 10^{-3}/1$.

The buoyant force $m_{air}\,g$ of air on a solid object is therefore very small compared to the gravitational force mg. In many cases we can safely choose to neglect it. Since the upward force exerted by the air on the ball is small compared to the gravitational force or the tension force exerted by the wire, we can choose to neglect it in an analysis of the forces on a hanging ball. However, it is not always appropriate to neglect the buoyant force; in accurate analytical chemistry it is necessary to correct for buoyancy when weighing a sample of liquid or low-density solid on an analytical balance.

A final note: the buoyant force being equal to $m_{air}\,g$ is really a time average, because the buoyant force is due to random collisions with air molecules and therefore has an intermittent character. However, the collision rate is so high that the force seems nearly continuous and constant. If, however, the ball is of microscopic size, it may not be possible to ignore the intermittent nature of the collisions. In a microscope it is possible to observe small particles being jostled about by the random collisions with the water molecules. This effect is called "Brownian motion."

> **4.X.17** Calculate the buoyant force in air on a kilogram of iron (whose density is about 8 grams per cubic centimeter). Compare with the weight mg of this much iron.

Pressure

QUESTION   Why is it difficult to remove a suction cup from a surface?

There is an interesting case in which the effect of the air is very significant, and cannot be neglected. Suppose that an object lies on a table and the surfaces of the table and object are so smooth that upon squeezing them together all air is expelled. Such a situation occurs when a suction cup is pressed down onto a smooth surface. Now there are no longer any air molecules striking the bottom surface of the suction cup, while air molecules continue to strike the top surface. The force of the air on the top surface is quite large, about 1×10^5 newtons on each square meter at sea level!

This large force per unit area is called the "pressure" P of the air. The pressure can be thought of as due to the weight of a one-square-meter column

Figure 4.44 A one-square-meter column of air extending upward through the entire atmosphere.

of the entire atmosphere, extending upward many kilometers (although with diminishing density), as illustrated in Figure 4.44.

DEFINITION OF PRESSURE

Pressure $P = F/A$ (force per unit area)

Atmospheric pressure at sea level is about $1 \times 10^5 \, \text{N}/\text{m}^2$. In older units this is approximately 15 pounds/inch^2.

A Constant-Density Model of the Atmosphere

If the atmosphere had a constant density ρ like that near the surface (1.3 grams per cubic centimeter), we could calculate the height h of the atmosphere in the following way. Choose the column of air as the system. The volume of a column of air of height h and cross-sectional area A is Ah, and its mass is $M_{\text{air}} = \rho(Ah)$. Because the center of mass of the column is at rest, the Momentum Principle tells us that the upward force of the ground on the bottom of the air column must equal the downward gravitational force on the total mass of air in the column. By the reciprocity of interatomic electric forces (air molecules hitting the ground), the force of the air on the ground must also be equal to the weight of the air in the column. Therefore the atmospheric pressure on the ground is the weight $M_{\text{air}}\, g$ divided by the cross-sectional area: $P = M_{\text{air}}\, g/A = \rho g h$.

The height of a constant-density atmosphere, $h = P/(\rho g)$, would be about 8000 meters (about 5 miles), about the height of Mount Everest. In reality air density is not constant but decreases as you go higher, so there is still some low-density air at much higher altitudes than this.

This macroscopic picture, taking the whole column of air as the system, is a useful one, but remember that the actual force exerted on a square meter at sea level is due to air molecules hitting an area, and they strike surfaces at all angles. Every square meter of the walls and ceiling of a room at sea level, as well as the floor, is subjected to the same rate of molecular bombardments. The momentum change per second $\Delta \vec{p}/\Delta t$ of all the molecules colliding with one square meter is equal in magnitude to the force of 1×10^5 newtons.

4.X.18 Do the calculation and verify that the height of a constant-density atmosphere would be about 8000 meters (8 km).

Figure 4.45 The weight of the suction cup is very small compared to the downward force by the air or the normal force by the table.

Suction

A free-body diagram for a suction cup looks like Figure 4.45, where PA represents the force of the air on the area A of the top surface of the suction cup. The normal force exerted upward by the table and the downward force of the air are huge compared to the weight of the suction cup mg.

The normal force $F_N = PA + mg \approx PA$, since the force of the air is so much larger than the weight mg of the suction cup. For example, a rubber suction cup with a diameter of 4 cm has an area of $\pi(0.02\,\text{m})^2 = 1.3 \times 10^{-3}\,\text{m}^2$, so the force of the air on the suction cup is

$$PA = (1 \times 10^5\,\text{N}/\text{m}^2)(1.3 \times 10^{-3}\,\text{m}^2) = 130\,\text{newtons}$$

whereas its own weight mg might be about 0.01 kg (10 grams) times 9.8 N/kg, or only about 0.1 newton. Note that a force of 130 newtons is the weight of an object whose mass is about 13 kg, or about 29 pounds, which is why it isn't easy to pull even a small suction cup off a flat surface.

Every solid object in air, such as a hanging ball, is subjected to large compression forces by the pressure of the surrounding air (Figure 4.46). The

Figure 4.46 A ball hanging in air is subject to large compression forces due to the surrounding air.

upward buoyant force is the tiny difference between these huge forces. A hollow object must be very strong to support this compression, if air is removed from the inside. There is no problem, however, if air can flow into and out of the container, because then it is just the solid walls that are compressed, since the air exerts comparable pressure on the inner and outer surfaces.

All of this is even more relevant under water, because the density of water is about 1000 times greater than the density of air, and the compression forces are about 1000 times as strong. A hollow submersible vehicle that dives very deep must have an extremely strong hull. When a scuba diver descends to greater depth where the water pressure is larger, the breathing apparatus automatically increases the pressure of the air supplied to the diver to equal the increased pressure of the outside water at the new depth, to prevent the lungs from being crushed.

4.X.19 It is hard to imagine that there can be enough air between a book and a table so that there is a net upward (buoyant) force on the book despite the large downward force on the top of the book. About how many air molecules are there between a textbook and a table, if there is an average distance of about 0.01 mm between the uneven surfaces of the book and table?

Figure 4.47 A vertical spring–mass system minimizes sliding friction.

Figure 4.48 Free-body diagram for the hanging block.

4.14   *A VERTICAL SPRING–MASS SYSTEM

To reduce friction, it is easier to experiment with a mass oscillating vertically up and down on a spring (Figure 4.47) rather than sliding on an air table. We will show that we expect the same period whether the spring is vertical or horizontal.

We choose a coordinate system with the positive y axis pointing downward (Figure 4.47). We'll measure y downward from the location at which the end of the relaxed spring would be (the relaxed length of the spring is L).

Neglecting air resistance and gravitational interactions with objects other than the Earth, the physics diagram of the forces acting on the hanging block is that shown in Figure 4.48.

We then have the following:

$$\frac{dp_y}{dt} = -k_s y + mg$$

We choose to analyze only simple vertical motion, with no bouncing side to side, so we know this:

$$p_x = 0 \text{ at all times, so } \frac{dp_x}{dt} = 0$$

We can reduce the p_y equation to a familiar form if we measure a new coordinate, w, from the equilibrium position, when the block is hanging motionless, rather than from the unstretched spring position (Figure 4.49). When the mass hangs motionless from the spring, the spring is stretched an amount $k_s s_0 = mg$, so $s_0 = mg/k_s$. We'll measure w downward from the end of the spring, so

$$y = (s_0 + w) = \frac{mg}{k_s} + w$$

and therefore

$$\frac{dp_y}{dt} = -k_s \left(\frac{mg}{k_s} + w \right) + mg = -k_s w$$

Figure 4.49 Measure w from the place where the block can hang motionless.

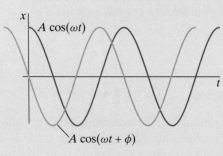

Figure 4.50 Phase shift ϕ.

This says that if we measure from the end of the spring at its equilibrium location, mg/k_s below the bottom of the unstretched spring, the equation for the motion of the mass is the same as in a horizontal spring–mass system, whose motion we already know. This means that the analytical solution we obtained for the horizontal case is also the solution for the vertical case (substituting y for x in the appropriate places).

The moving mass will oscillate up and down around this equilibrium position, as you can observe if you experiment with a vertical mass–spring system.

4.15 *GENERAL SOLUTION FOR THE MASS–SPRING SYSTEM

It is easy to show that the function $x = A\cos(\omega t + \phi)$ is a general solution for a spring–mass system with arbitrary k_s, m, and initial conditions, where $\omega = \sqrt{k_s/m}$, and where the amplitude A and the "phase shift" ϕ (measured in radians and denoted by the Greek letter phi) are constants determined by the initial conditions, as is explained below. The "phase shift" essentially shifts the starting point of the oscillation. For example, if $\phi = \pi/2$, the cosine function actually represents a negative sine function (Figure 4.50):

$$x = A\cos\left(\omega t + \frac{\pi}{2}\right) = -A\sin(\omega t)$$

Determining the Amplitude and Phase Shift

The amplitude A and phase shift ϕ can be determined from the initial values at time t_0 of position x_0 and velocity v_0:

$$x_0 = A\cos\left(\sqrt{\frac{k_s}{m}}t_0 + \phi\right)$$

$$v_0 = -\sqrt{\frac{k_s}{m}}A\sin\left(\sqrt{\frac{k_s}{m}}t_0 + \phi\right)$$

Using the important trigonometric identity $\sin^2\theta + \cos^2\theta = 1$ (which is the Pythagorean theorem applied to a triangle whose hypotenuse has length 1), we can obtain an equation that can be solved for the amplitude A:

$$x_0^2 + \frac{m}{k_s}v_0^2 = A^2, \text{ so we have } A = \sqrt{x_0^2 + \frac{m}{k_s}v_0^2}$$

For example, if the initial velocity is zero, the amplitude is simply equal to the initial stretch.

The value for A can be plugged back into either the x_0 equation or the v_0 equation to obtain the phase shift ϕ. Alternatively, by dividing the x_0 equation by the v_0 equation we obtain an equation that can be solved for the phase shift ϕ:

$$\frac{v_0}{x_0} = -\sqrt{\frac{k_s}{m}}\tan\left(\sqrt{\frac{k_s}{m}}t_0 + \phi\right)$$

$$\sqrt{\frac{k_s}{m}}t_0 + \phi = \arctan\left(-\sqrt{\frac{m}{k_s}}\frac{v_0}{x_0}\right)$$

$$\phi = \arctan\left(-\sqrt{\frac{m}{k_s}}\frac{v_0}{x_0}\right) - \sqrt{\frac{k_s}{m}}t_0$$

4.X.20 Suppose that the system is not oscillating and you strike it with a hammer, applying a large force F for a *very* short time Δt. What is the initial speed v_0? Explain briefly. What will be the amplitude (the maximum displacement away from the equilibrium position)?

4.X.21 In the case you just analyzed (striking with a hammer at time $t_0 = 0$), what is the phase shift ϕ? (Does this make sense?)

Comment: Two Constants for a Second-Order Differential Equation

Note that whether we specify the starting conditions in terms of x_0 and v_0 or in terms of the amplitude A and the phase shift ϕ, two constants are required. You can see why two constants are needed by looking at the start of a numerical integration. You need an initial position x_0 in order to be able to calculate the initial force and impulse, so that you can use the Momentum Principle to update the initial momentum p_0, from which you get the new velocity that lets you update the position; then repeat.

We have been working with what is called a "second-order differential equation." That is, the highest derivative in the equation is a second derivative—the acceleration $dv/dt = d(dx/dt)/dt$. A second-order differential equation normally requires two constants to specify the starting conditions completely.

Linear Systems

We saw that doubling the amplitude of a spring–mass system simply multiplies the solution by a factor of two without changing the time dependence of the motion. This is an extremely important property of systems described by "linear" differential equations (equations in which the variables appear only to the first power, not squared, for example). More formally, in the Momentum Principle for a spring–mass system, let x be replaced by the quantity Rx, where R is a constant:

$$m\frac{d}{dt}\left(\frac{d(Rx)}{dt}\right) = -k_s(Rx)$$

$$Rm\frac{d}{dt}\left(\frac{d(x)}{dt}\right) = -Rk_s(x)$$

The R's cancel, leaving us with the original equation. Therefore starting with a different initial stretch will simply scale up the vertical axis of the graph of x vs. t. Also, since the motion is periodic, a change in the initial velocity corresponds to starting the motion at a different time. For example, instead of starting with an initial stretch and zero velocity, you could start with zero stretch and some initial velocity, and this would correspond to a point on the original graph where the function crosses the time axis.

In contrast, consider a differential equation in which there is an x^3, and try scaling by R:

$$m\frac{d}{dt}\left(\frac{d(Rx)}{dt}\right) = -b(Rx)^3$$

$$Rm\frac{d}{dt}\left(\frac{d(x)}{dt}\right) = -R^3bx^3$$

In this case the R's do *not* cancel. The solutions to this "nonlinear" equation do not scale up in the simple way that solutions to a linear equation do.

S U M M A R Y

Contact forces:

- A solid object may be modeled as a 3D lattice of balls (atoms) connected by springs (interatomic bonds).
- Based on this model, and given the density and atomic weight of an element, one can find the length of an interatomic bond in a solid metal.
- From macroscopic measurements of stress and strain, one can determine the stiffness of an interatomic bond in a solid material.
- Solid objects exert contact forces because the interatomic bonds in the solid are stretched (a tension force) or compressed (a normal force).
- Friction forces arise when one object slides on another.
- The speed of sound in a solid is the speed at which a disturbance propagates through the material.

Definitions:

For a long object like a rod, which is stretched or compressed:

$$\text{strain} \equiv \frac{\Delta L}{L}$$

$$\text{stress} \equiv \frac{F_T}{A}$$

L is the original length of the wire, ΔL is the change in length, F_T is the force applied to stretch or compress the object, and A is the cross-sectional area of the object.

$$\text{Young's modulus: } Y \equiv \frac{\text{stress}}{\text{strain}} = \frac{\left(\dfrac{F_T}{A}\right)}{\left(\dfrac{\Delta L}{L}\right)}$$

Macro/micro connections:

Young's modulus in terms of microscopic quantities is

$$Y = \frac{k_{s,i}}{d}$$

Y is Young's modulus, $k_{s,i}$ is the stiffness of the interatomic bond, and d is the length of the interatomic bond (distance between adjacent nuclei).

$$\text{Speed of sound in a solid: } v = \sqrt{\frac{k_{s,i}}{m_a}}\, d$$

$k_{s,i}$ is the stiffness of the interatomic bond, d is the length of the interatomic bond (distance between adjacent nuclei), m_a is the mass of one atom.

Momentum Principle: Derivative form

$$\frac{d\vec{p}}{dt} = \vec{F}_{\text{net}}$$

The instantaneous time rate of change of the momentum of an object is equal to the net force acting on the object.

A simplified version of the Momentum Principle is $\vec{F}_{\text{net}} = m\vec{a}$, valid for $v \ll c$ and constant mass.

Analytical solution for mass–spring oscillations

$$x = A \cos(\omega t), \text{ with } \omega = \sqrt{\frac{k_s}{m}}$$

assuming that the spring's mass is negligible, and there is no friction and no air resistance.

A (the "amplitude") is a constant equal to the maximum stretch of the spring during an oscillation. ω is the "angular frequency," and its units are radians/s. k_s is the spring stiffness, and m is the mass of the object.

The "period" T is the time for one complete cycle, in seconds (Figure 4.51).

$$T = \frac{2\pi}{\omega}$$

The "frequency" f is the number of complete cycles per second.

$$f = \frac{1}{T} = \frac{\omega}{2\pi}$$

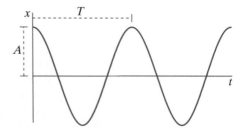

Figure 4.51

Contact forces due to gases

Upward buoyancy force = $m_{\text{fluid}}g$, where m_{fluid} is mass of equivalent volume of fluid.

Pressure is force per unit area; air pressure at sea level is about 1×10^5 N/m^2.

EXERCISES AND PROBLEMS

Sections 4.1–4.4

4.X.22

(a) A climber whose mass is 55 kg hangs motionless from a rope. What is the tension in the rope?

(b) Later, a different climber whose mass is 88 kg hangs from the same rope. Now what is the tension in the rope?

(c) Compare the physical state of the rope when it supports the heavier climber to the state of the rope when it supports the lighter climber. Which statements about the physical state of the rope are true? Check all that apply.

  (1) Because the same rope is used, the tension in the rope must be the same in both cases.

  (2) The interatomic bonds in the rope are stretched more when the rope supports the heavier climber than when the rope supports the lighter climber.

  (3) The rope is slightly longer when it supports the heavier climber than when it supports the lighter climber.

4.X.23 Why are two balls connected by a spring a good model for two atoms connected by a chemical bond?

(a) If the two atoms get farther apart than the equilibrium interatomic distance, they attract each other.

(b) Each spring represents a real microscopic coiled metal wire that connects two adjacent atoms. If the two atoms get closer together than the equilibrium interatomic distance, they repel each other.

(c) The magnitude of the force one atom exerts on another is proportional to the stretch or compression of the bond between them.

4.X.24 Consider a 5 kg block of gallium modeled as a cubic array of balls connected by springs. What is the mass in kg of a single "ball" in the ball–spring model? One mole (6.02×10^{23} atoms) of gallium has a mass of 70 grams, as shown in the periodic table on the inside front cover of the textbook.

4.X.25 Approximately what is the radius of a copper atom? Is it 1×10^{-15} m, 1×10^{-12} m, 1×10^{-10} m, 1×10^{-8} m, or 1×10^{-6} m?

4.X.26 The diameter of a copper atom is approximately 2.28×10^{-10} m. The mass of one mole of copper is 64 grams. Assume that the atoms are arranged in a simple cubic array. Remember to convert to S.I. units.

(a) What is the mass of one copper atom, in kg?

(b) How many copper atoms are there in a cubical block of copper that is 4.6 cm on each side?

(c) What is the mass of the cubical block of copper, in kg?

4.X.27 One mole of tungsten (6.02×10^{23} atoms) has a mass of 184 grams, as shown in the periodic table on the inside front cover of the textbook. The density of tungsten is 19.3 g/cm³. What is the approximate diameter of a tungsten atom (length of a bond) in a solid block of the material? Make the simplifying assumption that the atoms are arranged in a "cubic" array.

Sections 4.5–4.7

4.X.28 You hang a 10 kg mass from a copper wire, and the wire stretches by 8 mm. Then you suspend an identical mass from two copper wires, identical to the first. What happens?

A. Each wire stretches 4 mm.

B. Each wire stretches 8 mm.

C. Each wire stretches 16 mm.

4.X.29 You hang a 10 kg mass from a copper wire, and the wire stretches by 8 mm. Then you hang the same mass from a second copper wire, whose cross-sectional area is half as large (but whose length is the same). What happens?

A. The second wire stretches 4 mm.

B. The second wire stretches 8 mm.

C. The second wire stretches 16 mm.

4.X.30 You hang a 10 kg mass from a copper wire, and the wire stretches by 8 mm. Then you hang the same mass from a second copper wire, which is twice as long, but has the same diameter. What happens?

A. The second wire stretches 4 mm.

B. The second wire stretches 8 mm.

C. The second wire stretches 16 mm.

4.X.31 You hang a mass M from a spring, which stretches an amount s_1. Then you cut the spring in half, and hang a mass M from one half. How much does the half-spring stretch?

A. $2s_1$

B. s_1

C. $s_1/2$

4.X.32 A spring has stiffness k_s. You cut the spring in half. What is the stiffness of the half-spring?

A. $2k_s$

B. k_s

C. $k_s/2$

4.X.33 If a chain of 50 identical short springs linked end to end has a stiffness of 270 N/m, what is the stiffness of one short spring?

4.X.34 A certain spring has stiffness 190 N/m. The spring is then cut into two equal lengths. What is the stiffness of one of these half-length springs?

4.X.35 Forty-five identical springs are placed side by side (in parallel) and connected to a large massive block. The stiffness of the 45-spring combination is 20250 N/m. What is the stiffness of one of the individual springs?

4.X.36 A certain spring has stiffness 140 N/m. The spring is then cut into two equal lengths. What is the stiffness of one of these half-length springs?

4.X.37 Five identical springs, each with stiffness 390 N/m, are attached in parallel (that is, side by side) to hold up a heavy weight. If these springs were replaced by an equivalent single spring, what should be the stiffness of this single spring?

4.X.38 Lead is much softer than aluminum, and can be more easily deformed or pulled into a wire. What difference between the two materials best explains this?

A. Pb and Al atoms have different sizes.

B. Pb and Al atoms have different masses.

C. The stiffness of the interatomic bonds is different in Pb and Al.

4.X.39 Two wires with equal lengths are made of pure copper. The diameter of wire A is twice the diameter of wire B. When

6 kg masses are hung on the wires, wire *B* stretches more than wire *A*. You make careful measurements and compute Young's modulus for both wires. What do you find?

    A. $Y_A > Y_B$

    B. $Y_A = Y_B$

    C. $Y_A < Y_B$

4.X.40 Suppose you attempt to pick up a very heavy object. Before you tried to pick it up, the object was sitting still—its momentum was not changing. You pull very hard, but do not succeed in moving the object. Is this a violation of the Momentum Principle? How can you be exerting a large force on the object without causing a change in its momentum? What does change when you apply this force?

4.X.41 Suppose that you are going to measure Young's modulus for three rods by measuring their stretch when they are suspended vertically and weights are hung from them (Figure 4.52).

Figure 4.52

Rod 1 is 2.7 meters long and cylindrical with radius 4 mm (1 millimeter is 0.001 m). Rod 2 is 3.2 meters long by 12 mm wide by 6 mm deep. Rod 3 is 3 meters long by 6 mm wide by 6 mm deep. The definition of Young's modulus, $Y = (F/A)/(\Delta L/L)$, includes the quantity A, the cross-sectional area.

    (a) What is the cross-sectional area of rod 1?

    (b) What is the cross-sectional area of rod 2?

    (c) What is the cross-sectional area of rod 3?

4.X.42 Two wires are made of the same kind of metal. Wire *A* has a diameter of 2.4 mm and is initially 2.8 m long. You hang a 8 kg mass from wire *A*, measure the amount of stretch, and determine Young's modulus to be $Y_A = 1.37 \times 10^{12}$ N/m^2.

Wire *B*, which is made of the same kind of metal as wire *A*, has the same length as wire *A* but twice the diameter. You hang the same 8 kg mass from wire *B*, measure the amount of stretch, and determine Young's modulus, Y_B.

Which one of the following is true?

    1. $Y_B = Y_A$

    2. $Y_B > Y_A$ or

    3. $Y_B < Y_A$

4.X.43 Steel is very stiff, and Young's modulus for steel is unusually large, 2×10^{11} N/m^2. A cube of steel 28 cm on a side supports a load of 85 kg that has the same horizontal cross section as the steel cube.

    (a) What is the magnitude of the "normal" force that the steel cube exerts on the load?

    (b) What is the compression of the steel cube? That is, what is the small change in height of the steel cube due to the load it supports? Give your answer as a positive number. The compression of a wide, stiff support can be extremely small.

4.P.44 Young's modulus for aluminum is 6.2×10^{10} N/m^2. The density of aluminum is 2.7 grams/cm^3, and the mass of one mole (66.02×10^{23} atoms) is 27 grams. If we model the interactions of neighboring aluminum atoms as though they were connected by springs, determine the approximate spring constant of such a spring. Repeat this analysis for lead: Young's modulus for lead is 1.6×10^{10} N/m^2, the density of lead is 11.4 grams/cm^3, and the mass of one mole is 207 grams. Make a note of these results, which we will use for various purposes later on. Note that aluminum is a rather stiff material, whereas lead is quite soft.

4.P.45 **(a)** In outer space, a rod is pushed to the right by a constant force F (Figure 4.53). Describe the pattern of interatomic distances along the rod. Include a specific comparison of the situation at locations A, B, and C. Explain briefly in terms of fundamental principles.

Figure 4.53

Hint: Consider the motion of an individual atom inside the rod, and various locations along the rod.

    (b) After the rod in part (a) reaches a speed v, the object that had been exerting the force on the rod is removed. Describe the subsequent motion of the rod and the pattern of interatomic distances inside the rod. Include a specific comparison of the situation at locations A, B, and C. Explain briefly.

4.P.46 Two blocks of mass m_1 and m_3, connected by a rod of mass m_2, are sitting on a low-friction surface, and you push to the left on the right block (mass m_1) with a constant force (Figure 4.54).

Figure 4.54

    (a) What is the acceleration dv_x/dt of the blocks?

    (b) What is the compression force in the rod (mass m_2) near its right end? Near its left end?

    (c) How would these results change if you *pull* to the left on the *left* block (mass m_3) with the same force, instead of pushing the right block?

4.P.47 Measure and graph the strain vs. stress curve for a long metal wire by hanging weights from the end of the wire and carefully measuring the stretch as a function of the hanging weight. Determine Young's modulus for this material. If possible, go beyond the "yield" stress and observe the rapid, large increase in the strain. BE CAREFUL! Once you exceed the yield stress, the wire may break and dump the weights on whatever is underneath.

4.P.48 A hanging titanium wire with diameter 2 mm (2×10^{-3} m) is initially 3 m long. When a 5 kg mass is hung from it, the wire stretches an amount 0.4035 mm, and when a 10 kg mass is hung from it, the wire stretches an amount 0.807 mm. A mole of titanium has a mass of 48 grams, and its density is 4.51 grams/cm^3. Find the approximate value of the effective spring stiffness of the interatomic force, and explain your analysis.

4.P.49 One mole of tungsten (6.02×10^{23} atoms) has a mass of 184 grams, and its density is 19.3 grams per cubic centimeter, so the center-to-center distance between atoms is 2.51×10^{-10} m. You have a long thin bar of tungsten, 2.5 m long, with a square cross section, 0.15 cm on a side. You hang the rod vertically and attach a 415 kg mass to the bottom, and you observe that the bar becomes 1.26 cm longer. From these measurements, it is possible to determine the stiffness of one interatomic bond in tungsten.

(a) What is the spring stiffness of the entire wire, considered as a single macroscopic (large scale), very stiff spring?

(b) How many side-by-side atomic chains (long springs) are there in this wire (Figure 4.55)? This is the same as the number of atoms on the bottom surface of the tungsten wire. Note that the cross-sectional area of one tungsten atom is $(2.51 \times 10^{-10})^2$ m^2.

Top of wire

Edge of wire

Bottom of wire

Figure 4.55

(c) How many interatomic bonds are there in one atomic chain running the length of the wire?

(d) What is the stiffness of a single interatomic "spring"?

4.P.50 You hang a heavy ball with a mass of 14 kg from a gold wire 2.5 m long that is 2 mm in diameter. You measure the stretch of the wire, and find that the wire stretched 0.00139 m.

(a) Calculate Young's modulus for the wire.

(b) The atomic mass of gold is 197 g/mole, and the density of gold is 19.3 g/cm^3. Calculate the interatomic spring stiffness for gold.

4.P.51 A hanging copper wire with diameter 1.4 mm (1.4×10^{-3} m) is initially 3.5 m long. When a 36 kg mass is hung from it, the wire stretches an amount 1.83 mm, and when a 72 kg mass is hung from it, the wire stretches an amount 3.66 mm. A mole of copper has a mass of 63 grams, and its density is 9 grams/cm^3. Find the approximate value of the effective spring stiffness of the interatomic force.

4.P.52 A hanging iron wire with diameter 0.08 cm is initially 2.5 m long. When a 52 kg mass is hung from it, the wire stretches an amount 1.27 cm. A mole of iron has a mass of 56 grams, and its density is 7.87 g/cm^3. Find the approximate value of the effective spring stiffness of the interatomic force.

4.P.53 A hanging iron wire with diameter 0.09 cm is initially 2.0 m long. When a 65 kg mass is hung from it, the wire stretches an amount 1 cm. A mole of iron has a mass of 56 grams, and its density is 7.87 g/cm^3.

(a) What is the length of an interatomic bond in iron (diameter of one atom)?

(b) Find the approximate value of the effective spring stiffness of one interatomic bond in iron.

4.P.54 A hanging wire made of an alloy of iron with diameter 0.09 cm is initially 2.2 m long. When a 66 kg mass is hung from it, the wire stretches an amount 1.12 cm. A mole of iron has a mass of 56 grams, and its density is 7.87 g/cm^3.

(a) Based on these experimental measurements, what is Young's modulus for this alloy of iron?

(b) Find the approximate value of the effective spring stiffness of one interatomic bond in iron.

4.P.55 A certain coiled wire with uneven windings has the property that to stretch it an amount s from its relaxed length requires a force that is given by $F = bs^3$, so its behavior is different from a normal spring. You suspend this device vertically, and its unstretched length is 25 cm.

(a) You hang a mass of 18 grams from the device, and you observe that the length is now 29 cm. What is b, including units?

(b) Which of the following were needed in your analysis in part (a)?

A. The Momentum Principle

B. The fact that the gravitational force acting on an object near the Earth's surface is approximately mg

C. The force law for an ordinary spring ($F = k_s s$)

D. The rate of change of momentum being zero

(c) Next you take hold of the hanging 18 gram mass and throw it straight downward, releasing it when the length of the device is 33 cm and the speed of the mass is 5 m/s. After a very short time, 0.0001 s later, what is the stretch of the device, and what was the change in the speed of the mass (including the correct sign of the change) during this short time interval? It helps enormously to draw a diagram showing the forces that act on the mass after it leaves your hand.

Section 4.8

4.X.56 Bob is pushing a box across the floor at a constant speed of 1 m/s, applying a horizontal force whose magnitude is 20 N.

Alice is pushing an identical box across the floor at a constant speed of 2 m/s, applying a horizontal force.

(a) What is the magnitude of the force that Alice is applying to the box?

(b) With the two boxes starting from rest, explain qualitatively what Alice and Bob did to get their boxes moving at different constant speeds.

4.X.57 A 3 kg block measures 5 cm by 10 cm by 20 cm. When it slides on a 10 cm by 20 cm face, it moves with constant speed when pulled horizontally by a force whose magnitude is 3 N. How big a horizontal force must be applied to pull it with constant speed if it slides on a 5 cm by 20 cm face?

4.X.58 A 15 kg box sits on a table. The coefficient of static friction μ_s between table and box is 0.3, and the coefficient of kinetic friction μ_k is 0.2.

(a) What is the force required to start the box moving?

(b) What is the force required to keep it moving at constant speed?

4.X.59 A 20 kg box is being pushed across the floor by a constant force $\langle 90, 0, 0 \rangle$ N. The coefficient of kinetic friction for the table and box is 0.25. At $t = 5$ s the box is at location $\langle 8, 2, -1 \rangle$ m, traveling with velocity $\langle 3, 0, 0 \rangle$ m/s. What is its position and velocity at $t = 5.6$ s?

Sections 4.9–4.12

4.X.60 In a spring–mass oscillator, when is the magnitude of momentum of the mass largest: when the magnitude of the net force acting on the mass is largest, or when the magnitude of the net force acting on the mass is smallest?

4.X.61 For a vertical spring–mass oscillator that is moving up and down, which of the following statements are true? (More than one statement may be true.)

A. At the lowest point in the oscillation, the momentum is zero.

B. At the lowest point in the oscillation, the rate of change of the momentum is zero.

C. At the lowest point in the oscillation, $mg = k_s s$.

D. At the lowest point in the oscillation, $mg > k_s s$.

E. At the lowest point in the oscillation, $mg < k_s s$.

4.X.62 Suppose that the period of a spring–mass oscillator is 1 s. What will the period be if we double the mass?

4.X.63 Suppose that the period of a spring–mass oscillator is 1 s. What will be the period if we replace the spring with a spring that is twice as stiff?

4.X.64 Suppose that the period of a spring–mass oscillator is 1 s. What will be the period if we cut the spring in half and use just one of the pieces?

4.X.65 Suppose that the period of a spring–mass oscillator is 1 s with an amplitude of 5 cm. What will be the period if we increase the amplitude to 10 cm, so that the total distance traveled in one period is twice as large?

4.X.66 Suppose that the period of a spring–mass oscillator is 1 s with an amplitude of 5 cm. What will be the period if we take the oscillator to a massive planet where $g = 19.6$ N/kg?

4.X.67 Write the Momentum Principle for a horizontal spring–mass system with spring stiffness k_s and mass m, and write an analytical solution.

4.X.68 In terms of k_s and m, what is the angular frequency? The frequency? The period? The amplitude?

4.X.69 If you double the amplitude, by what factor does the period change? If you double the mass, by what factor does the period change? If you double the spring stiffness, by what factor does the period change?

4.X.70 How should you start the system going at $t = 0$ in order for the motion to be $A \cos(\omega t)$? How should you start the system going at $t = 0$ in order for the motion to be $A \sin(\omega t)$?

4.X.71 Describe two examples of oscillating systems that are *not* harmonic oscillators.

4.X.72 A ball whose mass is 1.4 kg is suspended from a spring whose stiffness is 4 N/m. The ball oscillates up and down with an amplitude of 14 cm.

(a) What is the angular frequency ω?

(b) What is the frequency?

(c) What is the period?

(d) Suppose this apparatus were taken to the Moon, where the strength of the gravitational field is only 1/6 of that on Earth. What would the period be on the Moon? (Consider carefully how the period depends on properties of the system; look at the equation.)

4.X.73 A mass of 2.2 kg is connected to a horizontal spring whose stiffness is 8 N/m. When the spring is relaxed, $x = 0$. The spring is stretched so that the initial value of $x = +0.18$ m. The mass is released from rest at time $t = 0$. Remember that when the argument of a trigonometric function is in radians, on a calculator you have to switch the calculator to radians or convert the radians to degrees. Predict the position x when $t = 1.15$ s.

4.X.74 Two metal rods are made of different elements. The interatomic spring stiffness of element A is three times larger than the interatomic spring stiffness for element B. The mass of an atom of element A is three times greater than the mass of an atom of element B. The atomic diameters are approximately the same for A and B. How does the speed of sound in rod A compare to the speed of sound in rod B?

A. $v_A > v_B$

B. $v_A = v_B$

C. $v_A < v_B$

4.X.75 Two rods are both made of pure titanium. The diameter of rod A is twice the diameter of rod B, but the lengths of the rods are equal. You tap on one end of each rod with a hammer and measure how long it takes the disturbance to travel to the other end of the rod. In which rod did it take longer?

A. Rod A

B. Rod B

C. The times were equal

4.X.76 A bouncing ball is an example of an anharmonic oscillator. If you quadruple the maximum height, what happens to the period? (Assume that the ball keeps returning almost to the same height.)

4.X.77 A particular spring–mass oscillator oscillates with period T. Write out the general equation for the period of such an oscillator to use as a guide when answering the following questions.

(a) If you double the mass but keep the stiffness the same, by what numerical factor does the period change? (That is, if the original period was T and the new period is bT, what is b?)

(b) If, instead, you double the spring stiffness but keep the mass the same, what is the factor b?

(c) If, instead, you double the mass and also double the spring stiffness, what is the factor b?

(d) If, instead, you double the amplitude (keeping the original mass and spring stiffness), what is the factor b?

4.P.78 You hang a heavy ball with a mass of 41 kg from a silver rod 2.6 m long by 1.5 mm by 3.1 mm. You measure the stretch of the rod, and find that the rod stretched 0.002898 m. Using these experimental data, what value of Young's modulus do you get? The atomic mass of silver is 108 g/mole, and the density of silver is 10.5 g/cm^3. Using this information along with the measured value of Young's modulus, calculate the speed of sound in silver.

4.P.79 Here on Earth you hang a mass from a vertical spring and start it oscillating with amplitude 1.7 cm. You observe that it takes 2.1 s to make one round trip. You construct another vertical oscillator with a mass 6 times as heavy and a spring 10 times as stiff. You take it to a planet where $g_{planet} = 6.8$ N/kg. You start it oscillating with amplitude 3.3 cm. How long does it take for the mass to make one round trip?

4.P.80 You will be given a spring and one or more masses with which to study the motion of a mass hanging from a spring (Figure 4.56).

Figure 4.56

(a) Measure the value of the spring stiffness k_s of the spring in N/m. The magnitude of a spring force is $k_s s$, where s is the stretch of the spring (change from the unstretched length). Explain briefly how you measured k_s. Include in your report the unstretched length of the spring.

(b) Measure the period (the round-trip time) of a mass hanging from the vertical spring. Report the mass that you use, and the amplitude of the oscillation. Amplitude is the maximum displacement, plus or minus, from the equilibrium position (the position where the mass can hang motionless).

(c) With twice the amplitude that you reported in part (b), measure the period again. Since the mass has to move twice as far, one might expect the period to lengthen. Does it?

How to measure the period accurately

There are unavoidable fluctuations in starting and stopping the timing, but you can minimize the error this contributes by timing many complete cycles so that the starting and stopping fluctuations are a small fraction of the total time measured.

It is good practice to count out loud *starting from zero, not one.* To count five cycles, say out loud "Zero, one, two, three, four, five." If on the other hand you say "One, two, three, four, five," you have actually counted only four cycles, not five.

Since the motion is periodic, you can start (say "zero") at any point in the motion. It is best to start and stop when the mass is moving fast past some marker near the equilibrium point, because it is difficult to estimate the exact time when the mass reaches the very top or the very bottom, because it is moving slowly at those turnaround points. Be sure to measure full round-trip cycles, not the half-cycles between returns to the equilibrium point (but going in the opposite direction).

4.P.81 In the approximation that the Earth is a sphere of uniform density, it can be shown that the gravitational force it exerts on a mass m inside the Earth at a distance r from the center is $mg(r/R)$, where R is the radius of the Earth. (Note that at the surface, the force is indeed mg, and at the center it is zero). Suppose that there were a hole drilled along a diameter straight through the Earth, and the air were pumped out of the hole. If an object is released from one end of the hole, how long will it take to reach the other side of the Earth? Include a numerical result.

4.P.82 It was found that a 20 gram mass hanging from a particular spring had an oscillation period of 1.2 seconds.

(a) When two 20 gram masses are hung from this spring, what would you predict for the period in seconds? Explain briefly.

Figure 4.57

(b) When one 20 gram mass is supported by two of these vertical, parallel springs (Figure 4.57), what would you predict for the period in seconds? Explain briefly.

(c) Suppose that you cut one spring into two equal lengths, and you hang one 20 gram mass from this half spring. What would you predict for the period in seconds? Explain briefly.

(d) Suppose that you take a single (full-length) spring and a single 20 gram mass to the Moon and watch the system oscillate vertically there. Will the period you observe on the Moon be longer, shorter, or the same as the period you measured on Earth? (The gravitational field strength on the Moon is about one-sixth that on the Earth.) Explain briefly.

4.P.83 In Problem 4.P.44 you can find the effective spring stiffness corresponding to the interatomic force for aluminum and lead. Let's assume for the moment that, *very* roughly, other atoms have similar values.

(a) What is the (very) approximate frequency f for the vibration of H_2, a hydrogen molecule?

(b) What is the (very) approximate frequency f for the vibration of O_2, an oxygen molecule?

(c) What is the approximate vibration frequency f of the molecule D_2, both of whose atoms are deuterium atoms (that is, each nucleus has one proton and one neutron)?

(d) Explain why the *ratio* of the deuterium frequency to the hydrogen frequency is quite accurate, even though you have estimated each of these quantities very approximately, and the effective spring stiffness is normally expected to be significantly different for different atoms. (*Hint:* What interaction is modeled by the effective "spring"?)

4.P.84 One mole of nickel (6.02×10^{23} atoms) has a mass of 59 grams, and its density is 8.9 grams per cubic centimeter. You have a bar of nickel 2.5 m long, with a square cross section, 2 mm on a side. You hang the rod vertically and attach a 40 kg mass to the bottom, and you observe that the bar becomes 1.2 mm longer. Next you remove the 40 kg mass, place the rod horizontally, and strike one end with a hammer. How much time T will elapse before a microphone at the other end of the bar will detect a disturbance? Be sure to show clearly the steps in your analysis.

4.P.85 Uranium-238 (U^{238}) has three more neutrons than uranium-235 (U^{235}). Compared to the speed of sound in a bar of U^{235}, is the speed of sound in a bar of U^{238} higher, lower, or the same? Explain your choice, including justification for assumptions you make.

Section 4.13

4.X.86 Calculate the buoyant force in air on a kilogram of lead (whose density is about 11 grams per cubic centimeter). The density of air is 1.3×10^{-3} grams per cubic centimeter. Remember that the buoyant force is equal to the weight of a volume of air that is equal to the volume of the object. (Compare with the weight mg of this much lead.)

4.X.87 Air pressure at the surface of a fresh water lake near sea level is about $1 \times 10^5 \, \text{N}/\text{m}^2$. At approximately what depth below the surface does a diver experience a pressure of $3 \times 10^5 \, \text{N}/\text{m}^2$? How would this be different in sea water, which has higher density than fresh water?

4.P.88 Here are two examples of floating objects:

(a) A block of wood 20 cm long by 10 cm wide by 6 cm high has a density of 0.7 grams/cm^3 and floats in water (whose density is 1.0 grams/cm^3). How far below the surface of the water is the bottom of the block? Explain your reasoning.

(b) An advertising blimp consists of a gas bag, the supporting structure for the gas bag, and the gondola hung from the bottom, with its cabin, engines, and propellers. The gas bag is about 30 meters long with a diameter of about 10 meters, and it is filled with helium (density about 4 grams per 22.4 liters under these conditions). Estimate the total mass of the blimp, including the helium.

Computational Problems

4.P.89 A "simple" pendulum (Figure 4.58) consists of a small mass of mass m swinging at the end of a low-mass string of length L (in contrast to a pendulum whose mass is distributed, such as a rod swinging from one end).

(a) Show that the tangential momentum p of the small mass along the arc to the right (increasing θ) obeys the following equation, where s is the arc length $L\theta$:

$$\frac{dp}{dt} = -mg \sin \theta = -mg \sin\left(\frac{s}{L}\right)$$

Figure 4.58

(b) What form does this equation take for small-amplitude swings? (Note that if θ is measured in radians, $\sin \theta \approx \theta$ for small angles. For example, even for an angle as large as $\theta = 30°$, $\theta = \pi/6 = 0.524$ radians, which is close to $\sin(30°) = 0.5$, and the approximation gets better for smaller angles.)

(c) Compare the form of the approximate equation you obtained in part (b) with the form of the Momentum Principle for a spring–mass system. By comparing these equations for different systems, determine the period of a simple pendulum for small-amplitude swings. (Note that any time you can approximate the Momentum Principle for some system by an equation that looks like the equation for a mass on a spring, you know that the system will move approximately sinusoidally.)

(d) Make a simple pendulum and predict its period, then measure the period. Do this for a long pendulum and for a short pendulum. Report your experimental data and results, with comparisons to your theoretical predictions.

(e) For large-amplitude swings, the small-angle approximation is not valid, but you can use numerical integration to calculate the motion. For a mass on a string, the largest possible amplitude is 90°, but if the string is replaced by a lightweight rod, the amplitude can be as large as 180° (standing upside-down, with the mass at the top). It turns out that such a pendulum can be modeled by the same equation as the equation derived in part (a), but the concepts of "torque" and "angular momentum" are required to prove it.

Assume that the equation of part (a) is valid, and use numerical integration to predict the motion of the pendulum. Think of s and p as being like x and p_x so that your program is basically a one-dimensional calculation, with a force $-mg \sin(s/L)$ instead of $-k_s s$. Plot the position s and momentum p as a function of time for a pendulum whose amplitude is nearly 180°.

Note how different these plots are from sines or cosines, because the system is not well approximated by a "spring–mass" equation. Also note that for this anharmonic oscillator the period increases with increasing amplitude, unlike the situation with a harmonic oscillator.

4.P.90 Create a computer model of a long metal bar as many atomic masses connected by interatomic springs, all in a straight line. Give the first atom a sudden push and compute the motions of all the atoms, one time step after another. Calculate all of the forces before using these forces to update the momenta and positions of the objects. Otherwise the calculations of the forces would mix positions corresponding to different times.

After each time step, display the displacements of all the atoms away from their equilibrium positions, and determine the speed of sound (the distance between the first and last atoms, divided by the amount of time it takes for the pulse to reach the last atom). Compare with the prediction that $v = (\sqrt{k_s/m_a})d$.

ANSWERS TO EXERCISES

4.X.1 2.6×10^{-10} m

4.X.2 3.1×10^{-10} m

4.X.3 800 N/m

4.X.4 300 N/m

4.X.5 0.151 mm

4.X.6 $Y \approx 6.5 \times 10^{10}$ N/m^2

4.X.7 $\Delta L \approx 0.2$ mm

4.X.8 **(a)** The box would not move; **(b)** 60 N; **(c)** The box would accelerate; **(d)** 80 N

4.X.9 **(a)** 0.58; **(b)** 1.4 m; **(c)** 0.7 s

4.X.10 $F_z = 4$ N

4.X.11 $\vec{0}$; $\vec{0}$

4.X.12 $\langle -2, -4, 0 \rangle$ m/s^2; $\langle -0.16, -0.32, 0 \rangle$ (kg·m/s)/s; $\langle -0.16, -0.32, 0 \rangle$ N

4.X.13 **(a)** 58.8 N/m; **(b)** 10.6 round trips

4.X.14 **(a)** 2 seconds; **(b)** 3.46 seconds

4.X.15 2970 m/s

4.X.16 1.01 milliseconds later

4.X.17 $F_{\text{buoyant}} = 1.6 \times 10^{-3}$ N; $\dfrac{F_{\text{buoyant}}}{F_{\text{weight of iron}}} = 1.6 \times 10^{-4}$

4.X.19 For a book the size of this book, there are approximately 1×10^{19} air molecules between the book and the table.

4.X.20 $v_0 = \dfrac{F}{m}\Delta t$; $A = v_0\sqrt{\dfrac{m}{k}}$

4.X.21 90° ($\pi/2$ radians), corresponding to a sine rather than a cosine (start at $x = 0$)

Rate of Change of Momentum

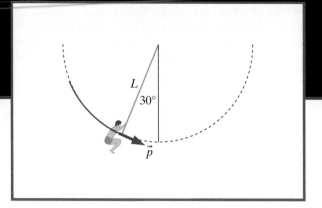

- The magnitude and direction of unknown forces, including contact forces, may be deduced by applying the derivative form of the Momentum Principle.
- Changing direction is caused by a force component perpendicular to momentum.
- Changing speed is caused by a force component parallel to momentum.
- Every force on a system is due to an interaction with an object in the surroundings, either by contact or through a distant (gravitational or electric) interaction.

5.1 IDENTIFYING FORCES ON A SYSTEM

In this chapter we will show how to use the derivative form of the Momentum Principle to understand a wide variety of situations. One of the key problems is to deduce the magnitude and direction of an unknown contact force. How much tension must a particular wire be able to support? Why was the tension in the vine so large that Tarzan fell into the river? What is the force of the seat on an airplane passenger when the plane makes a sudden change of direction, and why is it different from the passenger's weight?

Here is the basic approach we will use to answer questions about unknown contact forces:

- Observe the rate at which a system's momentum changes ($d\vec{p}/dt$).
- List all the objects in the surroundings that exert forces on the system, and evaluate all the known forces that contribute to $\vec{F}_{net}$.
- Use the Momentum Principle ($d\vec{p}/dt = \vec{F}_{net}$) to relate the net force to the observed rate of change of the momentum. It is important to understand that $d\vec{p}/dt$ and $\vec{F}_{net}$ are different quantities, separately evaluated; the Momentum Principle tells us that these two vectors $d\vec{p}/dt$ and $\vec{F}_{net}$ have the same vector components.
- Solve for the unknown forces.

This procedure will work only if we are able to identify *all* of the contributions to the net force. This requires a systematic approach in order to make sure that all forces are counted, that none are counted twice, and that no extra forces are incorrectly included. It is important to be very clear and explicit about the choice of system, and to include all other objects in the surroundings. A list of objects in the surroundings that exert forces on objects in the chosen system is the basis for correctly accounting for all contributions to $\vec{F}_{net}$. We will illustrate these issues with a very simple example for which you already know the answer.

EXAMPLE A Hanging Ball

A 1 kg ball hangs motionless at the end of a wire (Figure 5.1). The top of the wire is attached to an iron support. How large an upward force does the wire exert on the ball?

Solution   Here we use the convenient and widely used convention that F_T means the magnitude of the vector $\vec{F}_T$.

System: Ball
Surroundings: Earth, wire
Free-body diagram: Figure 5.2
Momentum Principle, y-component:

$$dp_y/dt = F_{net,y}$$
$$dp_y/dt = 0 \quad \text{ball's momentum isn't changing}$$
$$F_{net,y} = F_T - mg \quad \text{from the free-body diagram}$$
$$0 = F_T - mg \quad \text{applying the Momentum Principle}$$
$$F_T = mg \quad \text{solve for the unknown}$$
$$F_T = 1\,\text{kg} \cdot 9.8\,\text{N/kg} = 9.8\,\text{N}$$

Figure 5.1 A heavy ball hangs motionless at the end of a thin wire.

Figure 5.2 Free-body diagram.

FURTHER DISCUSSION You almost certainly knew before going through these careful steps that the tension in the wire is equal to the weight of the ball. The reason we show you this analysis without omitting any steps is that this same procedure can be used in all situations, including situations much more complicated than this one.

This approach to analyzing systems is essentially an intelligent form of "plug and chug."

- Choose a system.
- Find the rate of change of the system's momentum $d\vec{p}/dt$.
- Identify the forces exerted on the system by the surroundings and add them to make the net force $\vec{F}_{net}$.
- Use the Momentum Principle to equate $d\vec{p}/dt$ to $\vec{F}_{net}$.
- Solve for the unknown quantities.

Since the Momentum Principle involves vectors, in general you'll have three equations to work with.

QUESTION   Why didn't we include the iron support in the surroundings? Why didn't the force diagram show a force due to the support?

We chose the ball to be the system. Only two kinds of forces can act on the ball: forces involving interaction at a distance, like gravitational and electric forces, and contact forces. To exert a contact force, the atoms of one object must be in contact with the atoms of another object.

The support does not touch the ball, so the interatomic bonds in the ball are not stretched or compressed by the support. The support does exert a very tiny gravitational force on the ball (acting over a distance), but this force is too small to have a noticeable effect.

It is of course important for the iron support to be present, but the support does not interact directly with the ball. The support exerts a force on the

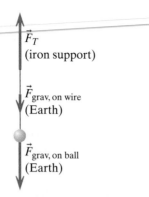

Figure 5.3 The free-body diagram for the system consisting of the ball plus the wire. The gravitational force of the Earth on the wire is exaggerated for clarity.

wire, but the wire is not the system of interest; it does get taken into account indirectly, since the wire would not be under tension if it were not supported by something.

QUESTION   What if we had chosen the ball plus the wire as the system?

In this case the free-body diagram would look like Figure 5.3. What's different here is the list of objects in the surroundings that exert forces on the chosen system: now these objects are the Earth and the iron support, which is in contact with the top of the wire. Also, because the wire is now included in the system, there is a gravitational force exerted by the Earth on the wire. The tension in the iron support must hold up both the ball and the wire. The Momentum Principle for this choice of system leads to

$$0 = F_T - mg - m_{wire}g$$

It is interesting to note that the tension in the top of the wire is a little greater than the tension at the bottom of the wire, because the top of the wire supports not only the ball, but the rest of the wire as well. Of course if the mass of the wire is small compared to the mass of the ball, it is a good approximation to say that the interatomic bonds in the wire are stretched almost as much at the bottom as at the top.

RULES FOR IDENTIFYING FORCES ON A SYSTEM

A free-body diagram is a compact way of identifying all forces on a system and the object in the surroundings responsible for each force. Every force on a system must be due to an interaction with an object in the surroundings.

(a) Choose a system.

(b) *Distance forces:* Show all forces due to objects in the surroundings that interact at a distance with the system (gravitationally or electrically). Label each distance force with the name of the interacting object.

(c) *Contact forces:* Show all forces due to objects in the surroundings that are touching the system. Label each force with the name of the object.

Don't include any other forces on the system—there aren't any! If you can't name an *object* in the surroundings that is responsible for the force, there is no such force on the system.

An interaction involves two objects, so each force on the system must be due to an object in the surroundings. Identifying each force by the name of the interacting object helps avoid counting a force twice, or missing a force.

If you find yourself identifying objects named "tension," "normal," "gravity," or "centripetal," you are in danger of double-counting forces, or including nonexistent forces (Figure 5.4).

Figure 5.4 To avoid missing or double counting forces, or including nonexistent forces, identify each force with the name of the interacting object in the surroundings.

5.X.1 In each of the following cases identify all objects in the surroundings that exert forces on the system.

(a) You hit a baseball with a bat. Choose the baseball as the system. At the instant of contact with the bat, what objects should be included in the surroundings (what objects exert forces on the ball)?

(b) You are playing with a yo-yo. Choose the yo-yo as the system. As the yo-yo moves downward, you pull up on the string. At that moment, what objects should be included in the surroundings (what objects exert forces on the yo-yo?

(c) Tarzan swings on a vine, which is attached to the branch of a tree. Choose Tarzan as the system. What are the objects in the surroundings that exert forces on Tarzan?

(d) You ski down a hill. Choose yourself plus your skis as the system. What objects in the surroundings exert forces on the system?

5.2   MOMENTUM NOT CHANGING (STATICS)

When we say a system is in uniform motion we mean that its momentum is constant, and does not change with time. A special case of uniform motion is a situation in which an object is at rest and remains at rest; this situation is called "static equilibrium," or simply "equilibrium."

Net Force in Equilibrium

QUESTION   If the system never moves, what is $\dfrac{d\vec{p}}{dt}$?

The momentum of the system isn't changing, so the rate of change of the momentum is zero:

$$\frac{d\vec{p}}{dt} = \vec{0} = \langle 0, 0, 0 \rangle \quad \text{(equilibrium)}$$

QUESTION   What does this imply about the net force acting on the system?

From the derivative form of the Momentum Principle, $d\vec{p}/dt = \vec{F}_{\text{net}}$, we deduce that the net force acting on a system that never moves must be zero:

$$\vec{F}_{\text{net}} = \vec{0} = \langle 0, 0, 0 \rangle \quad \text{(equilibrium)}$$

Statics Problems

Problems involving equilibrium are called "statics" problems. If you are studying engineering you may take an entire course on this topic. In a typical problem involving equilibrium situations you are asked to deduce the unknown magnitude of a contact force by applying the Momentum Principle. More difficult statics problems often involve several contact forces, acting in more than one direction, so one must work with x, y, and z components of forces. These problems are not necessarily conceptually more difficult but require more algebra.

Forces Exerted by Strings

Because a string is flexible, the tension force exerted by a string is always in the direction of the string. A string cannot exert a force perpendicular to the string. Also, of course, a string cannot push—it can only pull. It is common to refer to the tension force exerted by a string as "the tension in the string."

EXAMPLE

Equilibrium with Tension Forces

A 3 kg block hangs from string 1 (Figure 5.5). Then you pull the block to the side by pulling horizontally with a second string (string 2). **(a)** What is the tension in string 2 in order for the block to hang motionless with string 1 at an angle of 33° to the vertical, as shown in Figure 5.5? **(b)** In this situation, what is the tension in string 1 (the force on the block by the string)? **(c)** In which situation are the interatomic bonds in string 1 stretched more: block hanging straight down, suspended from string 1, or block hanging at an angle, pulled on by string 1 and string 2?

Solution

System: Block
Surroundings: String 1, string 2, and Earth
Free-body diagram: Figure 5.6
Momentum Principle:

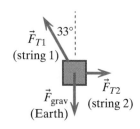

Figure 5.6

$$\frac{d\vec{p}}{dt} = \vec{F}_{net}$$

$$\langle 0, 0, 0 \rangle = \vec{F}_{T1} + \vec{F}_{T2} + \vec{F}_{grav}$$

Write out forces as vectors, using direction cosines (again, F_{T1} means $|\vec{F}_{T1}|$):

$$\hat{F}_{T1} = \langle \cos(90° + 33°), \cos 33°, 0 \rangle$$

$$\vec{F}_{T1} = F_{T1}\hat{F}_{T1} = F_{T1}\langle \cos 123°, \cos 33°, 0 \rangle$$

$$\vec{F}_{T2} = F_{T2}\hat{F}_{T2} = F_{T2}\langle 1, 0, 0 \rangle$$

$$\vec{F}_{grav} = \langle 0, -(3 \text{ kg})(9.8 \text{ N/kg}), 0 \rangle = \langle 0, -29.4, 0 \rangle \text{ N}$$

Separate components:

$$\frac{dp_x}{dt} = 0 = F_{net,x} \quad \text{so} \quad 0 = F_{T1}\cos 123° + F_{T2} + 0$$

$$\frac{dp_y}{dt} = 0 = F_{net,y} \quad \text{so} \quad 0 = F_{T1}\cos 33° + 0 - 29.4 \text{ N}$$

$$\frac{dp_z}{dt} = 0 = F_{net,z} \quad \text{so} \quad 0 = 0 + 0 + 0 \quad \text{(not very informative)}$$

We now have two equations in two unknowns (F_{T1} and F_{T2}). Solve for the unknown force magnitudes:

$$F_{T1} = \frac{29.4 \text{ N}}{\cos 33°} = 35.1 \text{ N}$$

$$F_{T2} = -F_{T1}(\cos 123°) = -(35.1 \text{ N})(-0.545) = 19.1 \text{ N}$$

(c) When the block hangs straight down, the force exerted by string 1 must be equal in magnitude to the force by the Earth on the block, which is 29.4 N. Evidently the tension in string 1 is greater when the block is pulled to the side than when the block hangs straight down. The interatomic bonds in string 1 must be stretched more when the block is pulled to the side.

Check: The units are correct (N). Special cases: Consider other angles.

Call the angle between the string and the vertical θ. If $\theta = 0°$ instead of 33°, the string 1 is vertical, and we expect the tension in this string to be mg. Rewriting the y equation algebraically:

$$F_{T1} = \frac{mg}{\cos\theta}$$

Figure 5.5 A block hangs from a string. Then you pull the block to the side with another string.

■ Why isn't your hand included in the surroundings? Your hand and the ceiling do not touch the block, so they exert only tiny gravitational forces on the system. If we chose the system to include the strings as well as the block, then your hand and the ceiling would exert contact forces on the system.

■ The angle between the $+x$ axis and the string is $(90° + 33°)$. The procedure for finding a unit vector from a known angle using direction cosines is discussed in Chapter 1.

■ Recall that the x component of momentum is affected only by the x component of force; the same for y and z.

■ Use the y equation to solve for F_{T1}.

■ Use the x and y equations to solve for F_{T2}.

Since $\cos(0°) = 1$, we do indeed find $F_{T1} = mg$. Also, $F_{T2} = 0$, as it should be in this case (no pulling).

On the other hand, if $\theta = 90°$, $\cos(90°) = 0$, the strings are horizontal, with $F_{T1} = mg/0$, which is infinitely large and impossible. This makes sense: if both strings are horizontal, there is no y component to support the block.

FURTHER DISCUSSION   Suppose that we replace string 1 with a thin steel wire whose length is 40 cm and thickness is 0.2 mm. How much will the wire stretch while supporting the 3 kg block?

Young's modulus for steel is approximately 2×10^{11} N/m², though different kinds of steel differ in their stiffness.

$$Y = \frac{F/A}{\Delta L/L}$$

$$\Delta L = \frac{F/A}{Y/L} = \frac{(35.1 \text{ N})/[\pi((0.2 \times 10^{-3} \text{ m})/2)^2]}{(2 \times 10^{11} \text{ N/m}^2)/(0.4 \text{ m})}$$

$$\Delta L = 2.2 \times 10^{-3} \text{ m} = 2.2 \text{ mm}$$

QUESTION   The horizontal string applied a 19.1 N force to the right on the block. What was the force that your hand applied to the other end of that string?

Choose the string as the system of interest, so the surroundings consist of the block, the Earth, and your hand. By reciprocity of (interatomic) electric forces, the force that the block exerts to the left on the string is 19.1 N. Since the string would have to have a mass of nearly $m = 2$ kg for mg to be 19.1 N, we can neglect the force exerted by the Earth. Even if the string were accelerating, $d\vec{p}/dt = md\vec{v}/dt$ would be very small, so $\vec{F}_{net} \approx 0$. We conclude that your hand must exert a force of 19.1 N to the right on the string (Figure 5.7).

The forces at each end of a low-mass string in tension must be nearly equal in magnitude, and equal to the tension in the string.

$F_T = 19.1$ N $F_T = 19.1$ N

(block)   (hand)

(Neglect Earth force)

Figure 5.7 If the tension in a low-mass string is F_T, the forces applied to the string at each end have magnitude F_T.

Tarzan and the Vine: Further Insights

In the preceding example we saw that the tension in the vertical string or wire was 29.4 N, but this increased to 35.1 N when the string or wire was not vertical. In different situations the tension force exerted by an object may be different, associated with different amounts of stretch of interatomic bonds. You may now suspect that the tension in the vine is greater when Tarzan swings on the vine than when he hangs motionless; this will turn out to be true, as we will be able to show later in this chapter.

EXAMPLE

Equilibrium with Tension Forces

A 350 kg load is suspended from two aluminum wires of radius 1.2 mm, as shown in Figure 5.8. Young's modulus for this aluminum alloy is 6.9×10^{10} N/m². **(a)** Calculate the tension in each wire (that is, the magnitude of the tension force exerted by each wire). **(b)** What is the fractional stretch (the strain) of each wire?

Solution

(a) System: Load
Surroundings: Rope 1, rope 2, Earth
Free-body diagram: Figure 5.9

Figure 5.8 A load hangs from two aluminum wires. What are the tensions in the wires, and what is the fractional stretch of each wire?

Figure 5.9 Free-body diagram.

■ Recall that the x component of momentum is affected only by the x component of force; the same for y and z.

Momentum Principle:

$$\frac{d\vec{p}}{dt} = \vec{F}_{net}$$

$$\langle 0, 0, 0 \rangle = \vec{F}_{T1} + \vec{F}_{T2} + \vec{F}_{grav}$$

Write out the forces as vectors, using direction cosines:

$$\hat{F}_{T1} = \langle \cos(180° - 55°), \cos(90° - 55°), 0 \rangle$$
$$\vec{F}_{T1} = F_{T1}\hat{F}_{T1} = F_{T1}\langle \cos 125°, \cos 35°, 0 \rangle$$
$$\hat{F}_{T2} = \langle \cos 70°, \cos(90° - 70°), 0 \rangle$$
$$\vec{F}_{T2} = F_{T2}\hat{F}_{T2} = F_{T2}\langle \cos 70°, \cos 20°, 0 \rangle$$
$$\vec{F}_{grav} = \langle 0, -(350\,\text{kg})(9.8\,\text{N/kg}), 0 \rangle = \langle 0, -3430, 0 \rangle\ \text{N}$$

Separate components:

$$\frac{dp_x}{dt} = 0 = F_{net,x} \quad \text{so} \quad 0 = F_{T1}\cos 125° + F_{T2}\cos 70° + 0$$
$$\frac{dp_y}{dt} = 0 = F_{net,y} \quad \text{so} \quad 0 = F_{T1}\cos 35° + F_{T2}\cos 20° - 3430\,\text{N}$$
$$\frac{dp_z}{dt} = 0 = F_{net,z} \quad \text{so} \quad 0 = 0 + 0 + 0 \quad \text{(not very informative)}$$

We now have two equations in two unknowns (F_{T1} and F_{T2}). Solve for the unknown force magnitudes.

$$F_{T1} = -\frac{\cos 70°}{\cos 125°}F_{T2} = 0.596\,F_{T2}$$
$$0 = (0.596\,F_{T2})\cos 35° + F_{T2}\cos 20° - 3430\,\text{N}$$
$$F_{T2} = \frac{3430\,\text{N}}{0.596\cos 35° + \cos 20°} = 2402\,\text{N}$$
$$F_{T1} = 0.596\,F_{T2} = 0.596(2402\,\text{N}) = 1431\,\text{N}$$

Check: The units are correct (N). We can plug these results back into the momentum equations to verify that they satisfy those equations.

(b) Fractional stretch:

$$Y = \frac{F/A}{\Delta L/L}$$
$$A = \pi(1.2 \times 10^{-3}\,\text{m})^2 = 4.52 \times 10^{-6}\,\text{m}^2$$

wire 1: $$\Delta L/L = \frac{\left(\dfrac{1431\,\text{N}}{4.52 \times 10^{-6}\,\text{m}^2}\right)}{6.9 \times 10^{10}\,\text{N/m}^2} = 4.6 \times 10^{-3} \quad \text{or about 0.5\%}$$

wire 2: $$\Delta L/L = \frac{\left(\dfrac{2402\,\text{N}}{4.52 \times 10^{-6}\,\text{m}^2}\right)}{6.9 \times 10^{10}\,\text{N/m}^2} = 7.7 \times 10^{-3} \quad \text{or about 0.8\%}$$

Uniform Motion: Momentum Not Changing

If a system is at rest and remains at rest, it is clear that its momentum is not changing, and therefore that the net force on the system must be zero. Newton's first law reminds us of another situation in which momentum is not changing: uniform motion. If the momentum of a system is not zero, but remains constant, then the net force on the object must be zero, just as if the object were at rest.

| EXAMPLE | **Uniform Motion** |

You place a 4 kg crate on the floor of an elevator that is temporarily stopped. Then the elevator starts moving, briefly speeding up, then moving upward at a constant speed of 2 m/s for many seconds (it is in a very tall building). While the elevator is moving upward at constant speed, what is the magnitude of the force exerted by the elevator floor on the crate?

Solution

System: Crate
Surroundings: Earth, elevator floor
Free-body diagram: Figure 5.10
Momentum Principle:

Figure 5.10

■ During the brief time when the elevator was speeding up, the crate's momentum was increasing, and the net force on the crate was not zero. Now, however, the crate's momentum is constant, and therefore the rate of change of its momentum is zero. In this situation the net force on the crate must now also be zero.

$$\frac{d\vec{p}}{dt} = \vec{F}_{net}$$

$$\langle 0, 0, 0 \rangle = \vec{F}_N + \vec{F}_{grav}$$

$$y : \frac{dp_y}{dt} = 0 = F_N - mg$$

$$F_N = (4\,kg)(9.8\,N/kg) = 39.2\,N$$

Check: The units are ok (N). Is the result reasonable? Yes, it agrees with Newton's first law.

■ Note that the speed of the crate (2 m/s) does not actually appear anywhere in the solution. The crate and elevator could have been traveling at any speed—even in principle a speed close to the speed of light—and this analysis would still be valid.

FURTHER DISCUSSION   While the elevator and the crate move at constant speed, the interatomic bonds in the elevator floor will be compressed exactly as much as they were when there was no motion. This is an example of the principle of relativity—that physical laws (in this case, the Momentum Principle) work in the same way in all reference frames that are in uniform motion (that is, constant velocity) with respect to the cosmic microwave background, which is the stage on which phenomena play themselves out.

| EXAMPLE | **Pulling at an Angle** |

You pull a loaded sled whose mass is 40 kg at constant speed in the x direction with a rope at an angle of 35°. The coefficient of friction between the snow and the sled runners is 0.2. What is the tension in the rope?

Solution

System: Sled
Surroundings: Earth, snow, rope
Free-body diagram: Figure 5.11

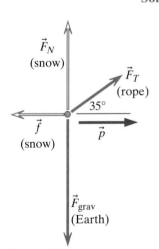

Figure 5.11 The free-body diagram for pulling a sled at constant speed.

$$\frac{d\vec{p}}{dt} = \vec{F}_{net}$$

$$\langle 0, 0, 0 \rangle = \vec{F}_N + \vec{f} + \vec{F}_{grav} + \vec{F}_T$$

$$x: \quad 0 = F_T \cos 35° - \mu_k F_N$$

$$y: \quad 0 = F_N + F_T \cos(90° - 35°) - mg$$

Combine these two equations:

$$F_T \cos 35° - \mu_k(mg - F_T \cos 55°) = 0$$

$$F_T(\cos 35° + \mu_k \cos 55°) = \mu_k mg$$

$$F_T = \frac{\mu_k mg}{\cos 35° + \mu_k \cos 55°}$$

$$F_T = \frac{0.2(40\,kg)(9.8\,N/kg)}{0.819 + 0.2(0.574)} = 84.0\,N$$

FURTHER DISCUSSION   Pulling at an angle rather than horizontally makes the force squeezing the sled and snow together be less than mg, which it would be if you pulled horizontally. Essentially you've partly lifted the sled out of the snow, which reduces the vertical component of the force of the snow, F_N, and therefore reduces the horizontal component of the force of the snow, the friction force $\mu_k F_N$.

5.X.2 An object that is in equilibrium is acted on by three objects in the surroundings. Two of the forces are $\langle 100, 50, -60 \rangle$ N and $\langle -130, 40, -70 \rangle$ N. What is the third force?

5.X.3 If an object is moving with constant momentum $\langle 10, -12, -8 \rangle$ kg · m/s, what is the rate of change of momentum $d\vec{p}/dt$? What is the net force acting on the object?

Preview of the Angular Momentum Principle

In many equilibrium situations, the Momentum Principle is not sufficient for carrying out a full analysis and we also need the Angular Momentum Principle, which we will discuss in a later chapter.

For example, consider two children sitting motionless on a seesaw (Figure 5.12). Choose as the system the two children and the (lightweight) board. For this choice of system, the objects in the surroundings that exert significant (external) forces on the system are the Earth (pulling down) and the support pivot (pushing up). The momentum of the system isn't changing, so the Momentum Principle correctly tells us that the net force must be zero, and the support pivot pushes up with a force equal to the weight of the children (we are neglecting the low-mass board).

However, the Momentum Principle alone doesn't tell us where the children have to sit in order to achieve balance and equilibrium. The force on a child exerts a twist (the technical term is "torque") about the pivot that tends to make the board turn. The torque associated with the forces on the two children have to add up to zero (one tends to twist the board clockwise, the other counterclockwise). Torque is defined as force times lever arm, and a nonzero net *torque* causes changes in a quantity called angular momentum, just as a nonzero net *force* causes changes in the ordinary momentum we have been studying.

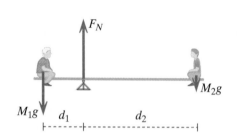

Figure 5.12 Two children sit on a seesaw in equilibrium.

| 5.3 | **FINDING THE RATE OF CHANGE OF MOMENTUM** |

In order to apply the derivative form of the Momentum Principle, it is important to be able to decide whether $d\vec{p}/dt$, the instantaneous rate of change of momentum, is zero or nonzero, and if it is nonzero, what its magnitude and direction are. The procedure for finding $d\vec{p}/dt$ is based on the procedure for finding $\Delta \vec{p}$ that you practiced in Chapter 1, and requires that you draw a simple diagram, as shown in Figure 5.13.

Figure 5.13 Graphical calculation of $\Delta \vec{p}$, using the procedure for graphical subtraction of vectors.

1. Draw two arrows:
   one representing $\vec{p}_i$, the momentum of the object a short time $\Delta t/2$ *before* the instant of interest, and
   a second arrow representing $\vec{p}_f$, the momentum of the object a short time $\Delta t/2$ *after* the instant of interest.
2. Graphically find $\Delta \vec{p} = \vec{p}_f - \vec{p}_i$ by placing the arrows tail to tail, and drawing the resultant arrow starting at the tip of $\vec{p}_i$ and going to the tip of $\vec{p}_f$.

■ You may wish to review exercises in Chapter 1 dealing with finding change of momentum.

3. This arrow indicates the direction of $\Delta \vec{p}$. Since Δt is positive, this is the same direction as $\Delta \vec{p}/\Delta t$, which in the limit as Δt gets very small (goes to zero) is what is meant by the derivative of the momentum with respect to time, $d\vec{p}/dt$. So visualize what the direction of $\Delta \vec{p}$ will be for very small Δt, and you have the direction of the time derivative of the momentum, $d\vec{p}/dt$.

A System Momentarily at Rest

It is clear that if a system remains at rest for a long time, its momentum is not changing. This corresponds to the fact that for a system whose momentum (magnitude and direction) remains constant, the time derivative $d\vec{p}/dt = \vec{0}$. However, situations in which the momentum of an object is only *momentarily* zero can be confusing if you do not use the procedure for finding $d\vec{p}/dt$ described in the section above. Two examples of such situations are these:

- You hang a block on a spring, stretch the spring downward, and let go. The block oscillates straight up and down. At the moment when the block reaches its lowest point, is the rate of change of its momentum zero or nonzero? If nonzero, what is the direction of $d\vec{p}/dt$? The lowest point is called a "turning point," a point where the momentum slows down, comes to a stop, and then speeds up again in the opposite direction.
- Tarzan hangs from a vine, swinging back and forth in a gentle arc. At the moment when he reaches his maximum displacement from the vertical and is momentarily at rest, is the rate of change of Tarzan's momentum zero or nonzero? If nonzero, what is the direction of $d\vec{p}/dt$? This is another situation described as a "turning point," where Tarzan slows down, comes to a stop, and then speeds up again in the opposite direction.

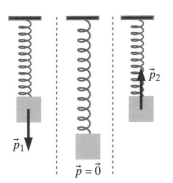

Figure 5.14 The momentum of a block oscillating on a spring, at an instant before it comes to a momentary rest at the bottom of its path (1) and an instant after it starts back upward (2).

| QUESTION | Before reading further, decide for yourself whether $d\vec{p}/dt$ is zero or nonzero, and if nonzero, what the direction of $d\vec{p}/dt$ is, for each of the situations described in the bulleted paragraphs above. Draw diagrams and use the graphical procedure described above to find $\Delta \vec{p}$ and $d\vec{p}/dt$.

EXAMPLE **Block Oscillating on a Spring**

Consider the situation of a block oscillating up and down on a spring, as shown in Figure 5.14. When the block comes to a momentary rest at the bottom of its oscillation, is $d\vec{p}/dt$ zero or nonzero at this turning point? If nonzero, what is its direction?

Solution Use the procedure for finding $d\vec{p}/dt$ given above:

Immediately before reaching bottom, the block is still moving downward, and the momentum of the block is in the $-y$ direction ($\vec{p}_1$ in Figure 5.14).

Immediately after reaching bottom, the block begins moving upward, and its momentum is in the $+y$ direction ($\vec{p}_2$ in Figure 5.14). $\Delta \vec{p} = \vec{p}_2 - \vec{p}_1$ is therefore in the $+y$ direction (Figure 5.15).

Therefore $d\vec{p}/dt$ is nonzero, and its direction is $+y$.

Figure 5.15 The direction of $d\vec{p}/dt$ is the direction of $\Delta \vec{p} = \vec{p}_2 - \vec{p}_1$.

FURTHER DISCUSSION Since $d\vec{p}/dt = \vec{F}_{\text{net}}$, it must be that at this turning point $\vec{F}_{\text{net}}$ is nonzero. This makes physical sense. This is not a case of equilibrium, so the net force cannot be zero. If the net force were zero when the momentum is zero, nothing would make the block pick up speed again.

QUESTION   At the turning point, which is larger in magnitude, the force on the block by the spring or the force on the block by the Earth?

The net force on the block at this instant acts in the direction of $d\vec{p}/dt$, which is upward. Therefore the upward force on the block by the stretched spring is larger than the downward force on the block by the Earth. This upward force is needed to make the block's speed increase from zero, heading upward.

EXAMPLE

Tarzan Swinging on a Vine

Tarzan hangs from a vine, swinging back and forth. At the moment when he reaches his maximum displacement from the vertical, as indicated in Figure 5.16, Tarzan is momentarily at rest (a "turning point"). At this moment, is the rate of change of Tarzan's momentum zero or nonzero? If nonzero, what is the direction of $d\vec{p}/dt$?

Solution   Use the procedure for finding $d\vec{p}/dt$.

Both immediately before this moment and immediately after this moment, Tarzan's momentum is nonzero; this is a turning point. The directions of his momenta at these two instants are shown in Figure 5.17, as is $\Delta\vec{p} = \vec{p}_2 - \vec{p}_1$. Therefore, at the instant Tarzan is motionless, $d\vec{p}/dt$ is nonzero, and its direction is the direction of $\Delta\vec{p}$, as shown in Figure 5.18.

Note that it was important to consider a very short time interval in order to get the direction of $d\vec{p}/dt$. The direction of $\Delta\vec{p}$ would be different for a longer time interval (and not the same as the direction of $d\vec{p}/dt$).

Figure 5.16 At the extreme point of his swing, Tarzan is momentarily at rest.

Applying the Momentum Principle, we conclude that because Tarzan's momentum is changing (he does not remain at rest), the net force on Tarzan (which is the vector sum of the tension force by the vine and the force by the Earth) must be nonzero. The direction of this net force must be the same as the direction of $d\vec{p}/dt$ at this instant.

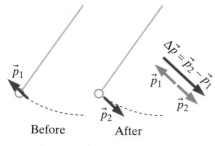

Before   After

Figure 5.17 Tarzan's momentum an instant before and an instant just after he is momentarily at rest, and $\Delta\vec{p}$.

Momentarily at Rest vs. Uniform Motion and Equilibrium

From the previous examples and exercises it is clear that a system that is only momentarily at rest is not in the same situation as a system in equilibrium.

QUESTION   What is the difference in $d\vec{p}/dt$ between a system in equilibrium, a system in uniform motion, and a system that is momentarily at rest?

For a system in equilibrium, $d\vec{p}/dt = \vec{0}$.

For a system in uniform motion, $d\vec{p}/dt = \vec{0}$.

For a system that is only momentarily at rest, $d\vec{p}/dt \neq \vec{0}$.

QUESTION   Is the net force on a system that is only momentarily at rest zero or nonzero?

Since the only way to change the momentum of a system is to apply a nonzero net force, there must be a nonzero net force acting on such a system.

Figure 5.18 The direction of $d\vec{p}/dt$ at the instant Tarzan is motionless. This is also the direction of the net force $\vec{F}_{net}$ due to the Earth and the vine.

5.X.4 Answer these questions about $d\vec{p}/dt$:

(a) A block oscillating up and down on a spring comes to the top of its path, where it is momentarily at rest. At the instant it is at rest, is $d\vec{p}/dt$ zero or nonzero? If nonzero, what is its direction?

(b) You throw a ball straight up into the air. At the instant the ball reaches its highest location, is $d\vec{p}/dt$ zero or nonzero? If nonzero, what is its direction?

(c) A basketball travels straight down, hits the floor, and bounces back straight up. At the moment when it is in contact with the floor and its speed is zero, is the rate of change of its momentum zero or nonzero? If nonzero, what is the direction of $d\vec{p}/dt$?

(d) A tennis ball traveling horizontally to the right hits a wall and bounces straight back with nearly the same speed. While the ball is in contact with the wall, is $d\vec{p}/dt$ zero or nonzero? If nonzero, what is its direction?

Figure 5.19 To make the ball turn somewhat to the left, in what direction should you give it a kick?

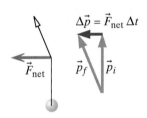

Figure 5.20 A kick to the left makes the ball turn somewhat to the left. The impulse changes the momentum, $\Delta\vec{p} = \vec{F}_{net}\,\Delta t$, and the new momentum is $\vec{p}_f = \vec{p}_i + \vec{F}_{net}\,\Delta t$.

Figure 5.21 To make a 90° turn, you must kick backward at an angle to reduce the forward component of momentum to zero.

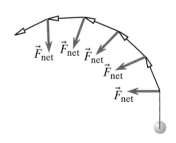

Figure 5.22 If you give the ball repeated big kicks to the left of its current heading, you can make the ball move in a roughly circular path.

5.4 CURVING MOTION

When the path of an object curves, the object's momentum is continually changing. Even if an object moves with constant speed, a change of direction means that $d\vec{p}/dt$ is nonzero. From the Momentum Principle we can conclude that an object moving along a curving path must be subject to a nonzero net force. We have previously applied iterative methods to the prediction of curving motion: in Chapter 2 we predicted the motion of an object under the influence of a constant net force (for example, projectile motion), and in Chapter 3 we predicted orbital motion due to a gravitational force that was constantly changing in both magnitude and direction. In this section we will develop useful tools for applying the instantaneous form of the Momentum Principle to curving motion.

Changing Direction with a Single Impulse

We start with a simple example of a force changing the direction of the momentum. In Figure 5.19 we're looking down on a ball that is rolling straight ahead. You want to change its direction of motion to turn a bit toward the left, by kicking it just once.

QUESTION In what direction should you aim your kick to make the ball turn a bit toward the left? Forward? Backward? To the left? To the right?

Evidently you have to kick to the left, perpendicular to the momentum, in order to get the ball to turn a bit toward the left. (Your kick doesn't have to be exactly perpendicular to the momentum, but it must have a component to the left, perpendicular to the motion.) A kick perpendicular to the motion doesn't change the forward component of the ball's momentum, so the ball turns less than 90°. Your kick to the left gives the ball some sideways momentum, which adds vectorially to the momentum it already had (Figure 5.20). A kick in the forward direction would make the ball move faster but not turn it to the left. A kick in the backward direction would make the ball move slower and might even make it change direction to head in the backward direction (change of direction at a turning point), but a backward kick cannot make the ball turn to the left or right.

Final Direction Depends on Initial Momentum

It is important to realize that because the ball already had some momentum before you kicked it, its new direction will not be in the same direction as your kick. For example, if you really want the ball to turn 90°, you have to kick backward at an angle with a big enough impulse to reduce the forward component of momentum to zero, as shown in Figure 5.21.

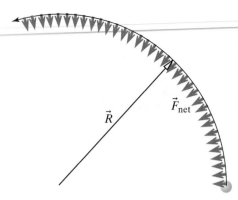

Figure 5.23 With a large number of small sideways kicks you can make the ball move along a path that is nearly indistinguishable from a circular path. Each kick must be aimed perpendicular to the momentum, toward the center of the circle.

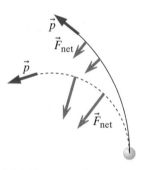

Figure 5.24 If you continually apply a bigger perpendicular force, you make the object turn more sharply, so the circle has a smaller radius.

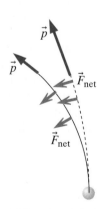

Figure 5.25 If the object's momentum is greater, it's harder to change its direction. If you continually apply the same perpendicular force, the circle has a larger radius.

Applying Many Successive Impulses

After the ball has moved a ways in the new direction shown in Figure 5.20, you give it another sideways kick, perpendicular to the motion, with the same impulse as before, then another and another as shown in Figure 5.22. By repeated perpendicular kicks with the same impulse, and with equal times between kicks, you can make the ball move along a roughly circular path, always turning to the left from its previous direction.

QUESTION Suppose that instead of giving a series of infrequent big kicks you give a large number of small sideways kicks (with equal short times between these small kicks). What would the motion of the ball look like?

The ball would move along a path that is nearly indistinguishable from a circular path, as seen in Figure 5.23. We see that in order to make a ball move along a circular path, we have to continually push inward on the ball. If we always push perpendicular to the motion (and therefore toward the center of the circle), there is no parallel component of the force and so no change in the magnitude of the momentum, in which case the ball moves along a circular path with no change in speed, just a continuous change of direction.

QUESTION Suppose that you kick just as often, but each of your kicks applies a bigger force than before. Would the radius of the circular path be larger or smaller?

Bigger kicks change the direction more, making sharper turns, so the radius of the circular path would be smaller (Figure 5.24). Bigger force, smaller radius; smaller force, bigger radius. In fact, if you apply a very small sideways force, the motion will be nearly a straight line, which corresponds to an extremely large radius of curvature. In the limit of infinitesimal force, the radius will be infinite: the object moves in a straight line in the absence of a sideways force.

QUESTION Suppose that you don't change the force of your kicks, but the ball has a higher speed. Would the radius of the circular path be larger or smaller?

Your sideways kicks make less change in the direction of the higher-speed ball, so the radius of the circular path would be larger. Bigger momentum, bigger radius (Figure 5.25).

The Kissing Circle

By considering smaller and smaller time intervals between impulses, and by taking the ratio of $\Delta \vec{p}/\Delta t$, we approach a limit, in which we can relate the instantaneous rate of change of momentum, $d\vec{p}/dt$, to the net force applied at that instant.

Consider motion along the curving path shown in Figure 5.26. We want to find the rate of change of momentum of a particle traveling along this path, at the instant when the particle is at location A. In Figure 5.26 a dashed circle has been drawn inside the path, tangent to the path at location A. This dashed circle of radius R is called the "osculatory" or "kissing" circle, because it just "kisses" the curving path at location A, fitting into the trajectory as smoothly as possible, with the circle and the trajectory sharing the same tangent ($\vec{p}$ is tangent to the path) and the same radius of curvature R at location A.

From the preceding discussion in which we considered applying a sequence of impulses, we see that both the perpendicular component of the net force, $\vec{F}_{\text{net}\perp}$, and the component of $d\vec{p}/dt$ perpendicular to the instantaneous momentum point toward the center of the kissing circle. If this force were bigger, the radius of the kissing circle would be smaller (a sharper turn). If on the other hand the momentum were bigger, the force would make a smaller

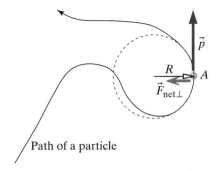

Figure 5.26 The "kissing circle" fits the trajectory at A as smoothly as possible.

change in direction in the same amount of time, and the kissing circle would be bigger.

> **5.X.5** Which of the dashed circles with radii R_1, R_2, or R_3 in Figure 5.27 best represents the "kissing circle" tangent to the path of the particle, with the same radius of curvature as that of the path at the location marked by "x"? At the location marked "x," what is the direction of $\vec{F}_{net\perp}$ for the particle?

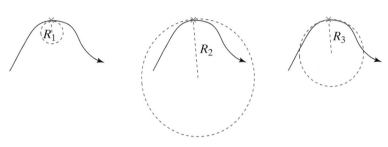

Figure 5.27

Qualitative Aspects of Curving Motion

Here is a summary of what we've learned about making an object follow a curving path:

- The component of the net force that is perpendicular to the momentum, $\vec{F}_{net\perp}$, turns the momentum in a new direction. To make an object turn to the left, apply a force to the left.
- The perpendicular component of the net force, $\vec{F}_{net\perp}$, points toward the center of the kissing circle.
- Bigger force, smaller radius of curvature.
- Bigger momentum, bigger radius of curvature.
- The component of the net force that is parallel to the momentum, $\vec{F}_{net\parallel}$, makes the momentum increase or decrease, and at a turning point it makes the direction reverse, but it cannot make the momentum turn to the left or the right.

We will make quantitative the relationships among force, momentum, and the radius of curvature along a curving path.

EXAMPLE   **A Turning Car**

Suppose that you're riding as the passenger in a convertible with left-hand drive (American or continental European), so you are sitting on the right (Figure 5.28). Assume that you have foolishly failed to fasten your seat belt. The driver makes a sudden right turn. Before studying physics you would probably have said that you are "thrown to the left," and in fact you do end up closer to the driver. But there's something wrong with this "thrown to the left" idea. What object in the surroundings exerted a force to the left to make you move to the left?

Solution   Don't confuse your kinesthetic perceptions with the physical analysis. There is no object that exerts an outward (leftward) force, so there is no such force! To understand what's happening, watch the turning car from a fixed vantage point above the convertible, and observe carefully what happens to the passenger. In the absence of forces to change the passenger's direction of motion, the passenger keeps moving ahead, in a straight line at the same speed as before (Figure 5.29). It is the car that is yanked to the right, out from under the passenger. The driver moves closer to the passenger; it isn't that the passenger moves toward the driver.

The driver may also feel "thrown to the left" against the door. However, it is the door that runs into the driver, forcing the driver to the *right*, not to the left.

Figure 5.28 You are a passenger in a car that makes a sudden right turn.

FURTHER DISCUSSION Sometimes people invent fictitious forces such as the so-called "centrifugal" force to account for the passenger and driver being "thrown to the left." However, this just adds a second confusion to the first. The passenger and driver experience an impulse to the right in a right turn, not to the left, and there are forces to the right that make them go to the right (for example, the force of the door on the driver's left shoulder).

Parallel and Perpendicular Components of $d\vec{p}/dt$

By taking components of each side of the Momentum Principle $d\vec{p}/dt = \vec{F}_{net}$, in the direction of the momentum and in a direction perpendicular to the momentum, we have the following, where $\vec{F}_{net\parallel}$ is parallel to the momentum and $\vec{F}_{net\perp}$ is perpendicular to the momentum (Figure 5.30):

Driver runs into door

Passenger now closer to driver

Figure 5.29 The passenger moves straight ahead and is now closer to the driver, who was thrown to the right by the door pushing the driver to the right.

PARALLEL AND PERPENDICULAR COMPONENTS OF THE MOMENTUM PRINCIPLE

$$\left(\frac{d\vec{p}}{dt}\right)_{\parallel} = \vec{F}_{net\parallel}$$

$$\left(\frac{d\vec{p}}{dt}\right)_{\perp} = \vec{F}_{net\perp}$$

The quantity $(d\vec{p}/dt)_{\parallel}$ is associated with the rate of change of the magnitude of the momentum (near a turning point, it gives the rate of the decrease just before reaching the turning point and the rate of increase just after).

The quantity $(d\vec{p}/dt)_{\perp}$ gives the rate of change of the direction of the momentum, which is numerically equal to the sideways (perpendicular) component of the net force. There is a simple expression for the magnitude of $(d\vec{p}/dt)_{\perp}$:

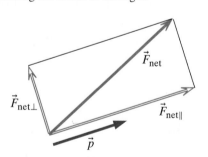

Figure 5.30 Split the net force into components parallel to the momentum (which changes the magnitude of the momentum) and perpendicular to the momentum (which changes the direction of the momentum).

RATE OF CHANGE OF DIRECTION OF $d\vec{p}/dt$

We will derive the following result:

$$\left|\left(\frac{d\vec{p}}{dt}\right)_{\perp}\right| = |\vec{p}|\frac{|\vec{v}|}{R}$$

R is the radius of the kissing circle.
This result is valid even if v is comparable to c.

It is important to note that $|(d\vec{p}/dt)_{\perp}| = |\vec{p}|(|\vec{v}|/R)$ does not contain anything about the forces that cause this change.

EXAMPLE   **Calculating $(d\vec{p}/dt)_{\perp}$**

Suppose that a 3 kg object is moving with a speed of 4 m/s and at a certain instant the radius of curvature of the kissing circle (the best fit to the curving motion at that point) is 2 m. What are the magnitude and direction of $(d\vec{p}/dt)_{\perp}$?

Solution

$$\left|\left(\frac{d\vec{p}}{dt}\right)_{\perp}\right| = |\vec{p}|\frac{|\vec{v}|}{R} = (3\,\text{kg} \cdot 4\,\text{m/s})\frac{4\,\text{m/s}}{2\,\text{m}} = 24\,\text{kg}\cdot\text{m/s}^2$$

The direction of $(d\vec{p}/dt)_{\perp}$ is toward the center of the kissing circle.

FURTHER DISCUSSION We didn't need to know anything about the force in order to be able to calculate the perpendicular component of the rate of change of the momentum. All we needed to know was the mass, speed, and radius of curvature of the kissing circle.

QUESTION However, what does the Momentum Principle let us conclude about the net force that is causing this curving motion?

Since $(d\vec{p}/dt)_{\perp} = \vec{F}_{net\perp}$, and we know $|(d\vec{p}/dt)_{\perp}| = 24\,\text{kg}\cdot\text{m/s}^2$, we can conclude that the magnitude of the perpendicular component of the net force, $|\vec{F}_{net\perp}|$, must be equal to 24 N, and this component of the net force must point toward the center of the kissing circle.

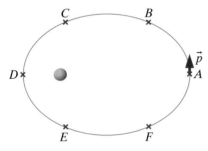

Figure 5.31 The elliptical orbit of a comet around a star. An arrow represents the momentum of the comet when it is at location *A*.

5.X.6 A comet orbits a star in an elliptical orbit, as shown in Figure 5.31. The momentum of the comet at location *A* is shown in the diagram. At the instant the comet passes each location labeled *A*, *B*, *C*, *D*, *E*, and *F*, answer the following questions:

(a) Draw an arrow representing the gravitational force on the comet by the star. Remember that the magnitude of the gravitational force is proportional to $1/|\vec{r}|^2$.

(b) Is the parallel component of the net force on the comet zero or nonzero?

(c) Is the perpendicular component of the net force on the comet zero or nonzero?

(d) Is the magnitude of the rate of change of the comet's momentum positive, negative, or zero?

(e) Is the direction of the comet's momentum changing or not changing?

5.5 | RATE OF CHANGE OF DIRECTION

We will show that $|(d\vec{p}/dt)_{\perp}|$, the magnitude of the perpendicular component of $d\vec{p}/dt$, is equal to $|\vec{p}|(|\vec{v}|/R)$. This component gives the rate of change of the direction of momentum. We start by "factoring" momentum into the product of the magnitude of the momentum (a scalar) and a unit vector in the direction of the momentum, and then taking the derivative, by applying the product rule. The resulting expression has two pieces, which correspond to the two ways in which momentum can change, in magnitude and in direction.

$$\vec{p} = |\vec{p}|\hat{p}$$

$$\frac{d\vec{p}}{dt} = \frac{d|\vec{p}|}{dt}\hat{p} + |\vec{p}|\frac{d\hat{p}}{dt}$$

This may look unfamiliar, but we can make sense of it by considering each term separately.

QUESTION Which term in the expression above involves a change in magnitude of momentum (change in speed)? Which term involves

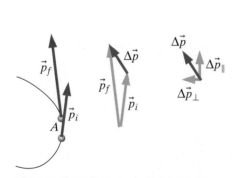

$\vec{p}$ $\vec{p}$

B✕ A✕

Ball falling   Ball rising

Figure 5.32 Falling ball (left): Magnitude of momentum is increasing, direction of $(d|\vec{p}|/dt)\hat{p}$ is $\hat{p}$. Rising ball (right): Magnitude of momentum is decreasing; direction of $(d|\vec{p}|/dt)\hat{p}$ is $-\hat{p}$.

Figure 5.33 An object moves in a circle at constant speed (top view). At location A, the direction but not the magnitude of momentum is changing.

a change in direction? How are these terms related to $\vec{F}_{net\,\|}$ and $\vec{F}_{net\,\perp}$?

$\dfrac{d|\vec{p}|}{dt}\hat{p}$ is nonzero if speed is changing, and equal to $\vec{F}_{net\,\|}$

$|\vec{p}|\dfrac{d\hat{p}}{dt}$ is nonzero if direction is changing, and equal to $\vec{F}_{net\,\perp}$

Rate of Change of Magnitude of Momentum: $\dfrac{d|\vec{p}|}{dt}\hat{p}$

The rate of change of magnitude, the scalar quantity $d|\vec{p}|/dt$, should be familiar to you—it is just the instantaneous slope of a plot of $|\vec{p}|$ vs. time. Physically, if only this term is nonzero it indicates that the speed of the system is changing, but the line along which it is moving is not (Figure 5.32). The only possible change of direction related to this term is at a turning point, where the motion flips 180° but stays on the same line.

The direction of $(d|\vec{p}|/dt)\hat{p}$ is either parallel or antiparallel to $\hat{p}$. If an object is speeding up, the direction of the quantity $(d|\vec{p}|/dt)\hat{p}$ is $\hat{p}$, and if the object is slowing down, the direction of the quantity $(d|\vec{p}|/dt)\hat{p}$ is $-\hat{p}$. This is because the quantity $d|\vec{p}|/dt$ is positive if $|\vec{p}|$ is increasing, and negative if $|\vec{p}|$ is decreasing.

Rate of Change of Direction of Momentum: $|\vec{p}|\dfrac{d\hat{p}}{dt}$

The rate of change of direction of momentum, $|\vec{p}|d\hat{p}/dt$, is itself a vector. Physically, if only this term is nonzero, it indicates that the speed of a system is constant, but the direction of motion is changing as in Figure 5.33, where an object moves at constant speed in a circle.

The direction of $|\vec{p}|(d\hat{p}/dt)$ is perpendicular to the direction of momentum at any particular instant, toward the center of the kissing circle. $d\hat{p}/dt$ cannot have a component in the direction of $\hat{p}$. If it did, the magnitude of $\hat{p}$ would be changing. This is impossible: the magnitude of a unit vector is always 1. We can also see this graphically by applying the standard procedure for determining the change of a vector, as shown in Figure 5.34. Note that $d\hat{p}/dt$ does not necessarily have magnitude 1.

Figure 5.34 Left: Find values of $\hat{p}$ slightly before and slightly after the instant of interest. Center: determine $\Delta\hat{p}$. Right: $d\hat{p}/dt$ is perpendicular to $\hat{p}$.

Speed and Direction Changing

If both the speed and direction of an object's motion are changing, there are both parallel and perpendicular components of $d\vec{p}/dt$, as seen in Figure 5.35.

Figure 5.35 If both the magnitude and the direction of the momentum change, then $\Delta\vec{p}$ has a component parallel or antiparallel to the motion as well as a component perpendicular to the motion.

Figure 5.36 A person rides on a Ferris wheel at constant speed (Exercise 5.X.7). When the person is at the location shown, what is the direction of the net force acting on the person?

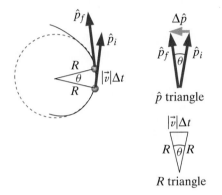

Figure 5.37 Find $\Delta \hat{p} = \hat{p}_f - \hat{p}_i$ graphically, by putting the vectors tail to tail. The two triangles are similar, with the same acute angle θ, because the sides of one triangle were originally perpendicular to the sides of the other.

5.X.7 A person rides on a Ferris wheel at constant speed (Figure 5.36). When the person is at the location shown, what is the direction of $(d|\vec{p}|/dt)\hat{p}$ for the person? What is the direction of $|\vec{p}|(d\hat{p}/dt)$ for the person? What must be the direction of the net force acting on the person? What objects in the surroundings exert the forces that contribute to the net force on the person?

A Geometrical Derivation

We can use geometry to calculate the rate at which the direction of momentum changes, $d\hat{p}/dt$. In Figure 5.37 we show $\hat{p}$ slightly before and slightly after the object passes a point, and we show the kissing circle at that point. In order to determine the change in $\hat{p}$, we place $\hat{p}_1$ and $\hat{p}_2$ tail to tail as usual and draw an arrow representing the change, $\Delta \hat{p}$. Under the $\hat{p}$ triangle we redraw the circular sector whose angle is θ, rotated 90°, where we approximate the arc of that circular sector by a straight line, with length $|\vec{v}|\Delta t$, which is the distance the object moves along the arc with speed $|\vec{v}|$ in a short time Δt.

The $\hat{p}$ triangle and the R triangle are similar to each other, with the same small acute angle θ, because the radius vectors in the R triangle and the momentum vectors in the $\hat{p}$ triangle are at right angles to each other. To say it another way, you can rotate one of the triangles through 90° and the small acute angles will clearly be the same. Since $\hat{p}$ is a unit vector, $|\hat{p}_f| = |\hat{p}_i| = 1$.

Since the two triangles are similar, the ratio of the side opposite the acute angle to the side adjacent to the acute angle must be the same for both triangles:

$$\frac{|\Delta \hat{p}|}{|\hat{p}|} = \frac{|\vec{v}|\Delta t}{R}$$

We are interested in the rate of change of $\hat{p}$, so rearrange the formula:

$$\frac{|\Delta \hat{p}|}{\Delta t} = \frac{|\vec{v}|}{R} \text{ since } |\hat{p}| = 1$$

Since the time is increasing, Δt is always positive and therefore the same as $|\Delta t|$, so we can express the magnitude of the rate of change of the momentum like this:

$$\left| \frac{\Delta \hat{p}}{\Delta t} \right| = \frac{|\vec{v}|}{R}$$

In the limit as Δt becomes infinitesimally small, we have the derivative:

$$\lim_{\Delta t \to 0} \left| \frac{\Delta \hat{p}}{\Delta t} \right| = \left| \frac{d\hat{p}}{dt} \right| = \frac{|\vec{v}|}{R} \text{ on a curving path}$$

Check Units

A simple check on the results of a calculation or derivation is a units check.

QUESTION   Do the units on both sides of the preceding equation match?

Like all unit vectors, $\hat{p}$ is dimensionless, so $|d\hat{p}/dt|$ has units of $1/s$. Since $|\vec{v}|/R$ also has units of $1/s$, the units match. If they had not matched, we would know the equation was not correct, and we would have to look for errors in our derivation.

RATE OF CHANGE OF DIRECTION

$$\left|\frac{d\hat{p}}{dt}\right| = \frac{|\vec{v}|}{R}$$

R is the radius of the kissing circle. The direction of $d\hat{p}/dt$ is perpendicular to the particle's momentum, toward the center of the kissing circle.

Although we've proved this important result for the rate of change of $\hat{p}$, if you look over the derivation you'll see that it also applies to the rate of change of any unit vector that changes its direction at the same rate as the momentum unit vector, such as $\hat{v}$. Therefore we have the following important results for the perpendicular component of $d\vec{p}/dt$ (where we use the fact that $\vec{p} = \gamma m\vec{v}$) and the perpendicular component of the acceleration $\vec{a} = d\vec{v}/dt$:

PERPENDICULAR COMPONENT OF $d\vec{p}/dt$ AND $\vec{a} = d\vec{v}/dt$

$$\left|\left(\frac{d\vec{p}}{dt}\right)_{\perp}\right| = |\vec{p}|\left|\frac{d\hat{p}}{dt}\right| = |\vec{p}|\frac{|\vec{v}|}{R} = \frac{\gamma m v^2}{R}$$

$$|\vec{a}_{\perp}| = \left|\left(\frac{d\vec{v}}{dt}\right)_{\perp}\right| = |\vec{v}|\left|\frac{d\hat{v}}{dt}\right| = |\vec{v}|\frac{|\vec{v}|}{R} = \frac{v^2}{R}$$

An interesting way to remember these formulas is to note that because $|\vec{v}|/R$ has units of 1/seconds, multiplying a quantity by this factor gives a result that is a rate—something per second. For example, multiplying $|\vec{p}|$ times $|\vec{v}|/R$ gives a result that has the units of kg · m/s per second, which are the correct units for $|(d\vec{p}/dt)_{\perp}|$. Multiplying $|\vec{v}|$ times $|\vec{v}|/R$ gives a result that has the units of m/s per second, which are the correct units for $|\vec{a}_{\perp}|$. The factor $|\vec{v}|/R$ converts a rotating vector such as $|(d\vec{p}/dt)_{\perp}|$ or $|\vec{v}|$ into its derivative with respect to time. Even multiplying $|\vec{r}|$ times the factor $|\vec{v}|/R$ gives $R|\vec{v}|/R = |\vec{v}|$, with units of meters per second, which are the correct units for $|d\vec{r}/dt|$.

In nearly straight line motion the radius R of the kissing circle is very large, and the rate of change of momentum is nearly zero, which makes sense, since there is almost no change of direction if the motion is nearly in a straight line. In the limit of truly straight-line motion, the radius R of the kissing circle approaches infinity. You can think of straight-line motion as a special kind of curving motion in which the perpendicular component of the momentum change is zero, the radius of curvature of the kissing circle is infinite, and the perpendicular component of the net force is zero, all of which makes physical sense.

Qualitative Conclusions

In Section 5.4 we showed in a qualitative way that in curving motion a bigger force leads to a smaller radius of curvature, and a bigger momentum leads to a bigger radius of curvature. We can understand these effects quantitatively using the new results. The perpendicular component of the Momentum Principle says that $\gamma m v^2/R = \vec{F}_{net\perp}$. If we keep the same v but make $\vec{F}_{net\perp}$ bigger, R gets smaller, because it is inversely proportional to $\vec{F}_{net\perp}$. If we make the momentum bigger, R must get bigger if we keep $\vec{F}_{net\perp}$ constant. If $\gamma \approx 1$, if we double the speed but don't change the force that's causing the curving motion, the radius of curvature will quadruple since there is a factor of v^2 in the numerator.

> **5.X.8** A car makes a turn with a radius of 200 m while moving with a constant speed of 30 m/s. What is the magnitude of the acceleration of the car? Compare with the magnitude of the acceleration of a falling object.
>
> **5.X.9** A particle moving at nearly the speed of light ($v \approx c$) passes through a region where it is subjected to a magnetic force of constant magnitude that is always perpendicular to the momentum and has a magnitude of 2×10^{-10} N. As a result, the particle moves along a circular arc with a radius of 8 m. What is the magnitude of the momentum of this particle?

$\vec{F}_{net}$ and $d\vec{p}/dt$ Are NOT the Same!

It will be extremely important in what follows to distinguish clearly between $d\vec{p}/dt$ and $\vec{F}_{net}$. Although the Momentum Principle tells us that these two vectors are equal to each other, the power of the Momentum Principle comes from the fact that it links two very different kinds of things. If you know the net force, you know how the momentum will change, and you can update the momentum.

If on the other hand you observe the motion of an object and measure how its momentum is changing, you know what net force must be acting on the object. In particular, if you see that an object of mass m is moving with speed v and curving with a radius of curvature R, you can calculate how big must be the perpendicular component of the net force.

One way to summarize this relationship is to say that momentum change is *caused* by the net force. We might usefully write the Momentum Principle with a leftward going arrow to emphasize the causal nature of this principle:

$$\frac{d\vec{p}}{dt} \leftarrow \vec{F}_{net}$$

Applications Involving Curving Motion

EXAMPLE **A Child on a Merry-Go-Round**

A child whose mass is 30 kg sits on a merry-go-round at a distance of 3 m from the center. The merry-go-round makes one revolution every 8 s. What are the magnitude and direction of the net force acting on the child?

Solution System: Child
Surroundings: Earth, seat
Momentum Principle: $d\vec{p}/dt = \vec{F}_{net}$
Parallel component (tangential):

Left-hand side of equation: $\dfrac{d|\vec{p}|}{dt} = 0$ because speed is not changing

Therefore, right-hand side is $F_{net\parallel} = 0$

Perpendicular component (toward center):

Left-hand side of equation: $\left| \left(\dfrac{d\vec{p}}{dt} \right)_{\perp} \right| = \gamma m v \dfrac{v}{R} \approx \dfrac{mv^2}{R}$

Therefore, right-hand side is $F_{net\perp} = \dfrac{mv^2}{R}$

What is the speed v? The child goes a distance of the circumference $2\pi R$ in a time of 8 s, so the speed is $v = 2\pi (3\,\text{m})/(8\,\text{s}) = 2.36$ m/s. The only nonzero component of the net force is $F_{net\perp}$, so the net force is directed toward the center with magnitude

$$F_{net} = \frac{(30\,\text{kg})(2.36\,\text{m/s})^2}{3\,\text{m}} = 55.5\,\text{N}$$

EXAMPLE   **Earth Satellite**

How long does it take a satellite in a circular orbit of radius R to make one complete revolution about the Earth?

Solution   System: Satellite
Surroundings: Earth
Free-body diagram: Figure 5.38
Momentum Principle: $d\vec{p}/dt = \vec{F}_{net}$
Parallel component (tangential):

Right-hand side of equation: $F_{net\|} = 0$

Therefore, left-hand side is $\dfrac{d|\vec{p}|}{dt} = 0$ (constant speed)

Perpendicular component (toward center):

Left-hand side of equation: $\left|\left(\dfrac{d\vec{p}}{dt}\right)_{\perp}\right| = \gamma m v \dfrac{v}{R} \approx \dfrac{m v^2}{R}$

Right-hand side of equation: $F_{net\perp} = \dfrac{GMm}{R^2}$

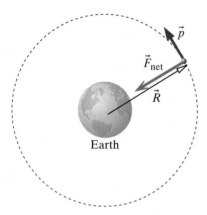

Figure 5.38 A satellite in a circular orbit around the Earth.

These two very different things, one the rate of change of the satellite's momentum and the other the gravitational attraction between the Earth and the satellite, are related to each other by the Momentum Principle to be numerically equal:

$$\frac{m v^2}{R} = \frac{GMm}{R^2}$$
$$v^2 = \frac{GM}{R}$$
$$v = \sqrt{\frac{GM}{R}}$$

The satellite goes a distance of the circumference $2\pi R$ in a time T, called the "period" of the orbit (and we say that the repetitive motion is "periodic"). So we have

$$\frac{2\pi R}{T} = \sqrt{\frac{GM}{R}}$$
$$T = \frac{2\pi R}{\sqrt{\dfrac{GM}{R}}} = \frac{2\pi R^{3/2}}{\sqrt{GM}}$$

FURTHER DISCUSSION   Since the period is proportional to $R^{3/2}$, the smaller the radius of the orbit, the shorter the period T. The shortest period for an Earth satellite is for one in which the satellite orbits just high enough that air resistance isn't significant. This minimum height is about 50 kilometers (5×10^4 m), so the radius of the orbit of a "near-Earth" satellite is $R_{Earth} = 6.4 \times 10^6$ m plus 5×10^4 m, which is not very different from the radius of the Earth. The period for a near-Earth satellite is therefore

$$T = \frac{2\pi (5 \times 10^4 \text{ m})^{3/2}}{\sqrt{(6.7 \times 10^{-11} \text{N} \cdot \text{m}^2 / \text{kg}^2)(6 \times 10^{24} \text{ kg})}} = 5074 \text{ s} = 85 \text{ minutes}$$

The speed of a near-Earth satellite is $v = 2\pi R / T = 2\pi (6.4 \times 10^6 \text{ m})/(5074 \text{ s}) = 7925 \text{ m/s} \approx 8 \text{ km/s}$. A chemical rocket must accelerate from rest on the launch pad to 8 km/s in order to go into a minimal near-Earth orbit.

QUESTION   Were we justified in approximating $\gamma \approx 1$?

Yes. v/c is only $8 \times 10^3 / 3 \times 10^8 = 2.7 \times 10^{-5} \ll 1$.

EXAMPLE

Swinging a Bucket in a Vertical Plane

You swing a bucket of water in a vertical plane at the end of a rope. The center of mass of the bucket and water moves approximately in a circle of radius 1.2 m. The speed of the bucket is faster at the bottom and slower at the top. The bucket with the water has a mass of 3 kg. At the instant when the rope is horizontal, the speed of the bucket is 9 m/s. What is the tension in the rope when the rope is horizontal? What is the rate of change of the speed at this moment?

Solution

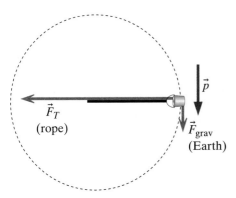

Figure 5.39 Swinging a bucket of water in a vertical plane.

System: Bucket with water
Surroundings: Earth, rope; see free-body diagram in Figure 5.39
Momentum Principle: $d\vec{p}/dt = \vec{F}_{net}$
Parallel component (tangential):

Right-hand side: $F_{net\|} = mg$

Therefore, left-hand side is $\dfrac{d|\vec{p}|}{dt} = m\dfrac{dv}{dt} = ma_\| = mg$

$a_\| = g = 9.8 \text{ N/ kg} = 9.8 \text{ m/s/s} = 9.8 \text{ m/s}^2$

Perpendicular component (toward center)

Left-hand side of equation: $\left| \left(\dfrac{d\vec{p}}{dt} \right)_\perp \right| = \gamma m v \dfrac{v}{R} \approx \dfrac{mv^2}{R}$

Therefore, right-hand side is $F_T = \dfrac{mv^2}{R}$

$F_{rope} = (3 \text{ kg})(9 \text{ m/s})^2 / (1.2 \text{ m}) = 202.5 \text{ N}$

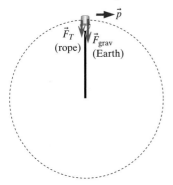

Figure 5.40 The bucket just barely makes it over the top with very little tension in the rope.

FURTHER DISCUSSION   If the bucket simply hangs motionless from the rope, the tension in the rope would be only $mg = (3 \text{ kg})(9.8 \text{ N/ kg}) = 29.4 \text{ N}$, much less than 202.5 N. The tension can be much larger during the circular motion. Here is a hint at what happened to Tarzan: during curving motion the tension in the vine might be greater than mg.

There is a minimum speed necessary for the bucket to make it over the top without the rope going slack. In Figure 5.40 we see that as long as the rope is not slack there are two downward forces acting on the bucket, due to the rope and the Earth. If the bucket just barely makes it over the top of the circular loop with the rope just barely taut, the force of the rope is negligible compared to the force of the Earth, so we have $mv^2/R = mg$, or $v = \sqrt{gR} = \sqrt{(9.8 \text{ N/ kg})(1.2 \text{ m})} = 3.4 \text{ m/s}$. Any speed at the top that is less than 3.4 m/s will mean that the rope has gone slack, and the motion won't be a circle any more.

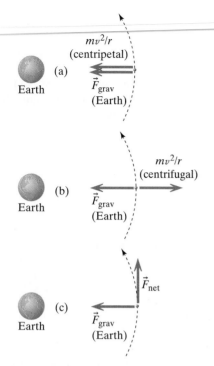

Figure 5.41 Incorrect free-body diagrams for a satellite in a circular orbit around the Earth.

Notice that as long as the speed at the top is at least 3.4 m/s the water will not fall out of the bucket, because its motion in a circle is perfectly compatible with the Momentum Principle. Of course the water actually is falling, moving downward and to the right, but you keep pulling inward on the bucket by pulling on the rope, and you're continually "catching" the water as it falls.

> **5.X.10** A child of mass 40 kg sits on a wooden horse on a carousel. The wooden horse is 5 meters from the center of the carousel, which completes one revolution every 90 seconds. What is $(d|\vec{p}|/dt)\hat{p}$ for the child, both magnitude and direction? What is $|\vec{p}|(d\hat{p}/dt)$ for the child? What is the net force acting on the child? What objects in the surroundings exert this force?
>
> **5.X.11** Figure 5.41 shows three incorrect free-body diagrams for a satellite in a circular orbit around the Earth. In each case (**a**, **b**, and **c**), state what mistakes have been made.

5.6 | WHY DOES THE VINE BREAK?

We now have all the tools we need to answer the question about Tarzan and the vine posed at the beginning of Chapter 4. In the following problem, we can calculate $d\vec{p}/dt$, but the magnitude of the tension force on Tarzan by the vine is unknown. We can find the unknown $|\vec{F}_{\text{vine}}|$ by applying the Momentum Principle.

EXAMPLE   **Tarzan and the Broken Vine**

Tarzan, whose mass is 90 kg, wants to use a vine to swing across a river. To make sure the 8 m long vine is strong enough to support him, he tests it by hanging motionless on the vine for several minutes. The vine passes this test, so Tarzan grabs the vine and swings out over the river. He is annoyed and perplexed when the vine breaks midway through the swing, when his speed is 12 m/s (Figure 5.42), and he ends up drenched and shivering in the middle of the cold river, to the great amusement of the onlooking apes. Why did the vine break while Tarzan was swinging on it, but not while he hung motionless from it?

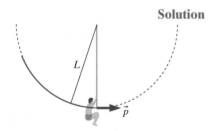

Figure 5.42 Tarzan at the bottom of his swing, the instant before the vine breaks. The colored line indicates his path up to this moment. The radius of the kissing circle (dashed line) is L, the length of the vine.

■ At this moment we know that the net force has no parallel component. The tension force due to the vine points straight up, along the vine. The gravitational force due to the Earth points straight down.

Solution   System: Tarzan
Surroundings: Vine, Earth (neglect air)
Free-body diagram: Figure 5.43
Momentum Principle: $d\vec{p}/dt = \vec{F}_{\text{net}}$
Parallel component (horizontal):

Right-hand side of equation: $F_{\text{net}\parallel} = 0$

Therefore, left-hand side is $\dfrac{d|\vec{p}|}{dt} = 0$

Speed is increasing just before this instant, and decreasing just after; $|\vec{p}|$ is going through a maximum.

Perpendicular component (vertical):

Direction of $F_{\text{net}\perp}$ is up, toward center of kissing circle

Left-hand side of equation: $\left|\left(\dfrac{d\vec{p}}{dt}\right)_{\perp}\right| = \dfrac{mv^2}{L}$

Right-hand side of equation: $F_{\text{net}\perp} = F_T - mg$

Figure 5.43

These two different quantities, the rate at which the momentum is changing, and the net force consisting of the vector sum of the upward force of the vine and the downward force of the Earth, are related to each other by the Momentum Principle to be numerically equal:

$$\frac{mv^2}{L} = F_T - mg$$

$$\frac{mv^2}{L} = \frac{(90\,\text{kg})(12\,\text{m/s})^2}{(8\,\text{m})} = 1620\,\frac{\text{kg}\cdot\text{m}}{\text{s}^2}$$

$$1620\text{ N} = F_T - mg$$

$$F_T = 1620\text{ N} + (90\,\text{kg})(9.8\text{ N/kg}) = 2502\text{ N}$$

■ 1620 N is the net force on Tarzan at this instant, and is due to both the vine and the Earth. The tension force exerted by the vine must be larger than the gravitational force exerted by the Earth.

QUESTION When Tarzan hung motionless on the vine, what was the tension in the vine?

$$\frac{dp_y}{dt} = F_T - mg$$

$$0 = F_T - (90\,\text{kg})(9.8\text{ N/kg})$$

$$F_T = 882\text{ N}$$

The tension force exerted by the vine was nearly three times as great when Tarzan was midway through his swing as it was when he hung motionless. The interatomic bonds in the vine were stretched three times as much while he swung—evidently too much!

Check: Is the result reasonable? The result shows that the vine has to exert an upward force that is greater than Tarzan's weight. That makes sense. If Tarzan were hanging motionless (equilibrium), the vine would only have to exert an upward force of *mg* (net force would be zero). However, to make Tarzan's momentum turn from horizontal to upward requires that the net force be upward.

FURTHER DISCUSSION Why must the vine exert a greater force when Tarzan's momentum is changing than when he hangs motionless? It is not simply that Tarzan is moving—zero net force would be required if Tarzan's momentum were constant. The "extra" tension in the vine is necessary to change the direction of Tarzan's momentum. To swing upward, increasing his p_y, Tarzan has to pull down harder on the vine, and therefore the vine stretches more, pulling up more on Tarzan.

It may seem odd that at the instant in question, the *magnitude* of Tarzan's momentum is not changing. We know that his speed increased as he swung downward, and his speed would have decreased as he swung upward if the vine had not broken. It is important to keep in mind that it is the instantaneous rate of change of magnitude of momentum (and speed) that is zero. At this instant only the direction of momentum is changing, as the net force is straight up. A nanosecond later, both the momentum and the net force would have changed, and there would be a nonzero parallel force component, so the magnitude of momentum would then be changing as well.

It is important to keep in mind that $\vec{F}_{\text{net}}$ and $d\vec{p}/dt$ are not the same thing. The equal sign in the Momentum Principle relates cause ($\vec{F}_{\text{net}}$) to effect ($d\vec{p}/dt$). In the example above, we calculated the effect, and from this deduced the cause.

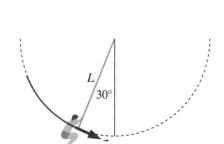

Figure 5.44 Tarzan swinging downward, with a speed of 11.1 m/s.

| EXAMPLE | **Magnitude and Direction of Momentum Changing** |

Consider the instant shown in Figure 5.44, before the vine breaks, when the angle between the vine and the vertical is 30°. At this instant Tarzan's speed is 11.1 m/s, and both the magnitude and direction of his momentum are changing. As before, Tarzan's mass is 90 kg, and the vine is 8 m long. **(a)** What is the tension in the vine at this moment? **(b)** What is the rate of change of the magnitude of Tarzan's momentum?

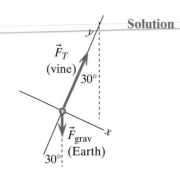

Figure 5.45 Free-body diagram.

■ It is convenient to choose axes aligned along directions parallel and perpendicular to $\hat{p}$, as shown in the free-body diagram.

■ The tension in the vine is greater than when Tarzan was hanging motionless, but not as large as it will be at the bottom of the swing.

Solution   System: Tarzan
Surroundings: Earth, vine (ignore air); see free-body diagram in Figure 5.45
Momentum Principle: $d\vec{p}/dt = \vec{F}_{net}$
(a) Perpendicular (y) component:

Direction of $F_{net\perp}$ is toward center of kissing circle, along y

Left-hand side of equation is $\left| \left(\dfrac{d\vec{p}}{dt} \right)_{\perp} \right| = \dfrac{mv^2}{L}$

Right-hand side of equation is $F_{net\perp} = F_T + mg\cos(180° - 30°)$

The Momentum Principle says that these two things are numerically equal:

$$\frac{mv^2}{L} = F_T + mg\cos 150°$$

$$F_T = \frac{mv^2}{L} - mg\cos 150°$$

$$F_T = \frac{(90\,\text{kg})(11.1\,\text{m/s})^2}{(8\,\text{m})} - (90\,\text{kg})(9.8\,\text{N/kg})\cos 150°$$

$$F_T = 2150\ \text{N}$$

(b) Parallel (x) component:

Right-hand side of equation:

$$F_{\parallel} = mg\cos(90 - 30)°$$

$$F_{\parallel} = (90\,\text{kg})(9.8\,\text{N/kg})\cos 60° = 441\ \text{kg}\cdot\text{m/s}^2$$

Therefore, left-hand side is $\dfrac{d|\vec{p}|}{dt} = 441\ \text{kg}\cdot\text{m/s}^2$

FURTHER DISCUSSION   In this example, both the magnitude and direction of Tarzan's momentum were changing. The net force on Tarzan had nonzero parallel and perpendicular components.

EXAMPLE   **Compression Force**

A passenger of mass M rides in an airplane flying along a curved path at constant speed v as shown in Figure 5.46, where the radius of the kissing circle at the top of the path is R. At the instant shown, when the airplane is at the top of its arc, how large is the compression force that the seat bottom exerts on the passenger? Compare this force to the force the seat exerts on the passenger when the airplane moves in a straight line at constant speed.

Solution   System: Passenger
Surroundings: Seat, Earth; see free-body diagram in Figure 5.47
Momentum Principle: $d\vec{p}/dt = \vec{F}_{net}$
    Parallel component (along x axis):
        Right-hand side of equation: $F_{net\parallel} = 0$

■ At this moment we know that the net force has no parallel component. The compression (normal) force due to the seat points straight up. The gravitational force due to the Earth points straight down.

        Therefore, left-hand side is $\dfrac{d|\vec{p}|}{dt} = 0$
    Speed is decreasing just before this instant, and increasing just after; $|\vec{p}|$ is going through a minimum.

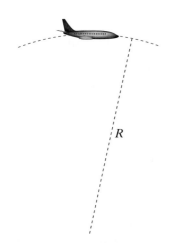

Figure 5.46 An airplane goes over the top of its path with speed v. The kissing circle has radius R at the top of the path.

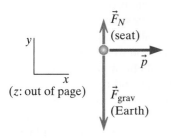

Figure 5.47 Free-body diagram.

Perpendicular component (along y axis): $\dfrac{dp_y}{dt} = F_{net,y}$

Direction is toward center of kissing circle, in the $-y$ direction, so $\left(\dfrac{d\vec{p}}{dt}\right)_{\perp,y}$ is negative

Left-hand side of equation (y component) is $\left(\dfrac{d\vec{p}}{dt}\right)_{\perp,y} = -\dfrac{Mv^2}{R}$

Right-hand side of equation (y component) is $F_{net,y} = F_N - Mg$

The Momentum Principle relates these two things:

$$-\frac{Mv^2}{R} = F_N - Mg$$

$$F_N = Mg - \frac{Mv^2}{R}$$

Alternatively, we could have equated the magnitude of the rate of change of the perpendicular component of momentum, $\left|(d\vec{p}/dt)_{\perp}\right|$, to the magnitude of the perpendicular component of the net force, $F_{net\perp}$, which is $Mg - F_N$, since the force points in the $-y$ direction. This is another way to deal with the signs properly. It of course leads to the same result.

Check: units are okay (newtons)

FURTHER DISCUSSION Physical significance: The minus sign in the result is important. The upward force of the seat F_N on the passenger's bottom is *less* than the passenger's weight Mg. The interatomic bonds in the seat are compressed less than they would be if the airplane's momentum were constant. The seat is being pulled down from under the passenger.

QUESTION The net force points downward. Does that make sense?

Yes, because the net force should point toward the center of the kissing circle, which is below the airplane.

QUESTION But since $F_N = Mg - Mv^2/R$, if the airplane's speed v is big enough, the force exerted by the seat could be zero. Does that make any sense?

Yes, that's the case where the airplane changes direction so quickly that the seat is essentially jerked out from under the passenger, and there is no longer contact between the seat and the passenger. No contact, no force of the seat on the passenger.

The Feeling of Weight and of Weightlessness

In the preceding example, what does the situation look and feel like to the passenger? From the passenger's point of view it is natural to think that the passenger has been thrown upward, away from the seat. Actually, though, the seat has been yanked out from under the passenger, in which case the passenger is falling toward the Earth (being no longer supported by the seat), but is changing velocity less rapidly than the airplane and so the ceiling comes closer to the passenger.

In the extreme case, if the airplane's maneuver is very fast, it may seem as though the passenger is "thrown up against the cabin ceiling." In actual fact, however, it is the cabin ceiling that is yanked down and hits the passenger!

You can see the value of wearing a seat belt, to prevent losing contact with the seat. When the airplane yanks the seat downward, the seat belt yanks the passenger downward. This may be uncomfortable, but it is a lot better than having the cabin ceiling hit you hard in the head.

QUESTION How would this analysis change if the airplane goes through the bottom of a curve rather than the top (Figure 5.48)?

Figure 5.48 An airplane goes through the bottom of its path.

Now the center of the kissing circle is above the airplane, so $d\vec{p}/dt$ points upward. That means that the net force must also point upward.

> QUESTION   Which is larger, the force of the seat on the passenger, or the gravitational force of the Earth on the passenger?

The force of the seat is greater than Mg, since the net force is upward. The passenger sinks deeper into the seat. Or to put it another way, the airplane yanks the seat up harder against the passenger's bottom, squeezing the seat against the passenger.

We Perceive Contact Forces

As you sit reading this text, you feel a sensation that you associate with weight. If someone suddenly yanks the chair out from under you, you feel "weightless," which feels funny and odd. It also feels scary, because you know from experience that whenever you lose the support of objects under you (chair, airplane seat, floor, mountain path), something bad is about to happen.

In reality, what we perceive as "weight" is actually not the gravitational force at all but rather the forces of atoms in the chair (or airplane seat or floor or mountain path) on atoms in your skin. You have nerve endings that sense the compression of interatomic bonds in your skin, and you interpret this as evidence for gravity acting on you. However, if you were placed in a spaceship that was accelerating, going faster and faster, its floor would squeeze against you in a way that would fool your nerve endings and brain into thinking you were subjected to gravity, even if there are no stars or planets nearby.

If you lose contact with the seat in an airplane that is going rapidly over the top of a curving path, your nerve endings no longer feel any contact forces, and you feel "weightless." Yet this is a moment when the only force acting on you is in fact the Earth's gravitational force. Weightlessness near the Earth paradoxically is associated with being subject only to your weight Mg.

■ Einstein came to the theory of General Relativity through his realization that the effect of a gravitational force in an inertial reference frame is indistinguishable from the effect of being in an accelerated reference frame without any massive objects nearby.

Contact Forces Internal to the Body

It isn't just nerve endings in your skin that give you the illusion of "weight." As you sit here reading, your internal organs press upward on other organs above them, making the net force zero. You feel these internal contact forces. If you are suddenly "weightless," a main reason for feeling funny is that the forces one organ exerts on another inside your body are suddenly gone.

To train astronauts, NASA has a cargo plane that is deliberately flown in a curving motion over the top, so that people in the padded interior lose contact with the floor and seem to float freely (in actual fact, they are accelerating toward the Earth due to the Mg forces acting on them, but the plane is also deliberately accelerating toward the Earth rather than flying level, so the people don't touch the walls and appear to be floating). The brain is crying out for the usual comforting signals from the nerve endings and not getting them. This airplane is sometimes called the "vomit comet" in honor of the effects it can have on trainees.

5.7   PROBLEM SOLVING

Forward Reasoning

A common technique for solving problems is to start by considering what you want to know (or what you have been asked to determine), and to work backward from that to the solution. Physics often approaches the analysis of

Figure 5.49 A circular pendulum, moving in the xz plane.

physical systems in the opposite way. We begin by organizing our knowledge about the system according to a fundamental principle—in this case the Momentum Principle—and systematically recording our knowledge in equations and diagrams. After we have recorded all the information we know, whatever is missing is what we need to figure out. Figuring out the unknown information may be straightforward, or it may involve several steps.

This way of working problems, sometimes called "forward reasoning," is typical of the way expert physicists solve problems of this kind—starting each new problem from the fundamental principles. This approach works! It is a powerful method for solving complex or unfamiliar problems when you feel as though you have no idea even how to get started. Approaching problems in this way may require a bit of faith—it is not necessarily obvious how the desired answer will emerge from the analysis. However, if you practice approaching problems in this way, you will develop confidence in the approach and will gain important experience in explaining complex phenomena starting from fundamental principles.

| EXAMPLE | **Circular Pendulum** |

A ball is suspended from a string, and after being given a push, moves along a horizontal circular path, as shown in Figure 5.49. You measure the length of the string L and the angle θ, shown in Figure 5.50. You also time the motion, and find that it takes T seconds for the ball to make one complete circular trip. From these measurements, determine the gravitational constant g.

| Solution | System: Ball |

Surroundings: String, Earth (neglect air resistance)

Free-body diagram: Figure 5.50, which shows a snapshot of a side view of the circular pendulum, at an instant when the string (of length L to the center of the ball) lies in the xy plane (at an angle θ to the vertical), and the ball is at the leftmost point in its circular path of radius $R = L\cos(90° - \theta) = L\sin\theta$. At this instant the ball's momentum is in the $+z$ direction (out of the page, toward you).

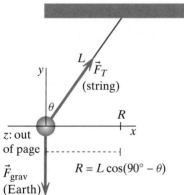

Figure 5.50 The circular pendulum at an instant when the string is in the xy plane, and the ball is moving in the $+z$ direction.

Momentum Principle: $d\vec{p}/dt = \vec{F}_{net}$

$$d\vec{p}/dt = \vec{F}_{net}$$

$$\hat{F}_T = \langle \cos(90° - \theta), \cos\theta, 0 \rangle$$

$$\vec{F}_T = F_T \langle \cos(90° - \theta), \cos\theta, 0 \rangle$$

$$\vec{F}_{net} = \langle F_T \cos(90° - \theta), (F_T \cos\theta - mg), 0 \rangle$$

Parallel component $(+z)$:

$$\frac{dp_z}{dt} = F_{net,z} = 0$$

- Unit vector is in direction of string force.

- We know that there are no forces in the z direction, so we deduce that $dp_z/dt = 0$.

- We know that $dp_y/dt = 0$ because the ball stays at the same horizontal level; p_y is zero and remains zero.

Perpendicular components:

$$y : \frac{dp_y}{dt} = F_T \cos\theta - mg \text{ but } \frac{dp_y}{dt} = 0$$

$$F_T = \frac{mg}{\cos\theta}$$

$$x : \frac{dp_x}{dt} = F_T \cos(90° - \theta)$$

Direction: toward center of circle $\langle 1, 0, 0 \rangle$

$$\frac{mv^2}{R} = F_T \sin\theta$$

$$\frac{mv^2}{L\sin\theta} = \left(\frac{mg}{\cos\theta}\right)\sin\theta$$

$$g = \frac{v^2 \cos\theta}{L(\sin\theta)^2}$$

■ Here the fact that $\cos(90° - \theta) = \sin\theta$ is used to simplify the final expression.

Measuring the speed v, length L, and angle θ determines g. v can be found by measuring T, the time to go around once, since $v = (2\pi R)/T$, where $R = L\sin\theta$.

Check: Units (N/kg). Special cases: If $\theta = 0$ the string hangs vertically and there is no motion, so the tension in the string ought to be mg, which is consistent with the equation for F_T that we found above.

FURTHER DISCUSSION   A large circular pendulum was used by Newton to measure g, the magnitude of the gravitational field, with considerable accuracy. You could observe and measure the motion of a circular pendulum yourself, and your analysis of the motion would let you deduce a value for g, which you could compare with the accepted value of 9.8 N/kg.

Note that $\vec{F}_{net}$ and $d\vec{p}/dt$ are not the same thing. In the example above, we knew the x and y components of $d\vec{p}/dt$ (the effect), and from this deduced the x and y components of $\vec{F}_{net}$ (the cause). In the z direction, it was the other way around: we knew that there were no forces in this direction, so we deduced that the z component of momentum was not changing.

Because we systematically identified each force with an object in the surroundings, we did *not* make the mistake of introducing a radially outward "centrifugal force" (Figure 5.51). The only forces on the ball are due to interactions with real objects (string, Earth). If there were such a force, it would make the net force zero, and the ball would have to move in a straight line, not a circle.

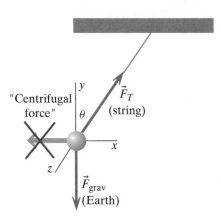

Figure 5.51 There is no "centrifugal force." There is no object in the surroundings that can exert this fictitious force. Moreover, if the net force were zero, the ball would have to move in a straight line!

Constrained Motion

Note that there could in principle be many possible motions of this system. The ball could swing back and forth in a plane; this is the motion of a "simple pendulum." The ball could go around in a path something like an ellipse, in which case the ball's height would vary, just as with simple pendulum motion. The ball could move so violently that the string goes slack for part of the time, with abrupt changes of momentum every time the string suddenly goes taut. If the string can stretch noticeably, there can be a sizable oscillation superimposed on the swinging motion.

All of these motions are difficult to analyze. In principle, we could use a computer to predict the general motion of the ball hanging from the string if we had a formula for the tension force in the string, as a function of its length. Essentially, we would need the effective spring stiffness for this stiff "spring." However, the string may be so stiff that even a tiny stretch implies a huge tension (a nearly "inextensible" string). The observable length of the string is nearly constant, but the tension force that the string applies to the ball can vary a great deal. This makes it challenging to do a computer numerical integration, because a tiny error in position makes a huge change in the force.

| EXAMPLE | **An Amusement Park Ride** |

There is an amusement park ride that some people love and others hate in which a bunch of people stand against the wall of a cylindrical room of radius R, and the room starts to rotate at higher and higher speed (Figure 5.52). The surface of the wall is designed to maximize friction between the person and the wall, so it is fuzzy or sticky, not slick. When a certain critical speed is reached, the floor drops away, leaving the people stuck against the wall as they whirl around at constant speed. Why do the people stick to the wall without falling down? Include a carefully labeled force diagram of a person, and discuss how the person's momentum changes, and why.

Solution

System: Person
Surroundings: Wall, Earth (neglect air resistance)
Free-body diagram: Figure 5.53

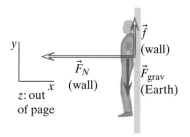

Figure 5.52 An amusement park ride.

Momentum Principle: $d\vec{p}/dt = \vec{F}_{net}$

At this instant the person is stuck to the right wall, with momentum directed into the page $(-z)$. The force due to the wall has been separated into two parts: a normal part $\vec{F}_N$ that turns the direction of the person's momentum and a frictional part $\vec{f}$ that keeps the person from sliding downward.

Parallel component (z): No change in speed, so $\dfrac{d|\vec{p}|}{dt} = 0$

Perpendicular components:

$x : \dfrac{dp_x}{dt} = -F_N$; direction: $\langle -1, 0, 0 \rangle$, toward center

$$-\frac{mv^2}{R} = -F_N$$

$$F_N = \frac{mv^2}{R}$$

$$y : \frac{dp_y}{dt} = 0 = f - mg$$

friction force $f = mg$

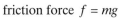

Figure 5.53 Free-body diagram.

The y component of momentum is not changing, since the person remains at the same horizontal level, and does not start to move up or down.

Check: Units okay. Also, the faster the ride, the bigger the inward force of the wall, which sounds right.

FURTHER DISCUSSION We deduce that the wall exerts an unknown force that must have a y component f (because the person isn't falling) and an x component $-F_N$ normal to the wall (because the person's momentum is changing direction). The only significant objects in the surroundings are the Earth and the wall; therefore only these objects interact with the person. There is a momentum change inward; if the net force were zero, the person would move in a straight line.

The vertical component f of the wall force is a frictional force. If the wall has friction that is too low, the person won't be supported; $f \le \mu_s F_N$, and the "coefficient of static friction" μ_s has a value for many materials between about 0.1 and 1.0. The more strongly two objects are squeezed together, the harder it is to slide one along the other.

The speed of the ride has to be great enough that $f \leq \mu_s F_N$, so we have

$$mg \leq \mu_s \left(\frac{mv^2}{R} \right)$$

$$\mu_s \geq \frac{gR}{mv^2}$$

This predicts that the lower the speed v, the greater the coefficient of friction must be to hold the person up on the wall, which sounds right. Conversely, if the ride spins very fast, the wall doesn't have to have a very large coefficient of friction to hold the people against the wall.

SUMMARY

Momentum Principle: Derivative form

$$\frac{d\vec{p}}{dt} = \vec{F}_{net}$$

The instantaneous time rate of change of the momentum of an object is equal to the net force acting on the object.

Finding the rate of change of momentum

1. Draw two arrows (Figure 5.54):
   (a) the first arrow represents $\vec{p}_i$, the momentum of the object a short time before the instant of interest.
   (b) the second arrow represents $\vec{p}_f$, the momentum of the object a short time after the instant of interest.
2. Graphically find $\Delta\vec{p} = \vec{p}_f - \vec{p}_i$ by placing the arrows tail to tail, and drawing the resultant arrow starting at the tip of $\vec{p}_i$ and going to the tip of $\vec{p}_f$.
3. This arrow indicates the direction of $\Delta\vec{p}$, which is the same as the direction of $d\vec{p}/dt$, provided the time interval involved is sufficiently small.

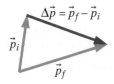

Figure 5.54

Momentarily at rest vs. uniform motion and equilibrium

For a system in equilibrium $d\vec{p}/dt = \vec{0}$.
For a system in uniform motion $d\vec{p}/dt = \vec{0}$.
For a system that is momentarily at rest, $d\vec{p}/dt \neq \vec{0}$.

The Momentum Principle: Parallel and perpendicular components

$$\frac{d|\vec{p}|}{dt}\hat{p} = \vec{F}_{\parallel} \quad \text{and} \quad |\vec{p}|\frac{d\hat{p}}{dt} = \vec{F}_{\perp}$$

Valid only for a moving object, because $\hat{p}$ and $d\hat{p}/dt$ are undefined when $\vec{p} = \vec{0}$.

Component of net force parallel to momentum changes magnitude of momentum.

Component of net force perpendicular to momentum changes direction of momentum.

Effect of perpendicular component of net force on a particle moving along a curving path

$$|\vec{p}|\left|\frac{d\hat{p}}{dt}\right| = |\vec{p}|\frac{|\vec{v}|}{R} = |\vec{F}_{net\perp}|$$

R is the radius of the kissing circle. The component of net force perpendicular to the particle's momentum changes its direction but not the magnitude of its momentum (speed).

EXERCISES AND PROBLEMS

Sections 5.1–5.2

5.X.12 The velocity of a 90 gram ball changes from $\langle 10, -6, 0 \rangle$ m/s to $\langle 10.1, -6.6, 0 \rangle$ m/s in 0.07 s, due to the gravitational attraction of the Earth and to air resistance.

   (a) What is the acceleration of the ball?
   (b) What is the rate of change of momentum of the ball?
   (c) What is the net force acting on the ball?

5.X.13 A rope is attached to a block, as shown in Figure 5.55. The rope pulls on the block with a force of 210 N, at an angle of

$\theta = 23°$ to the horizontal (this force is equal to the tension in the rope).

Figure 5.55

(a) What is the *x* component of the force on the block due to the rope?

(b) What is the *y* component of the force on the block due to the rope?

5.P.14 A box of mass 40 kg hangs motionless from two ropes, as shown in Figure 5.56. The angle is 38 degrees. Choose the box as the system. The *x* axis runs to the right, the *y* axis runs up, and the *z* axis is out of the page.

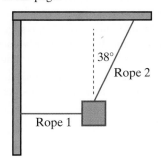

Figure 5.56

(a) Is $d\vec{p}/dt$ of the box zero or nonzero?

(b) Which of the free-body diagrams in Figure 5.57 correctly describes the forces acting on the box?

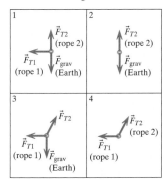

Figure 5.57

(c) What is the *y* component of the gravitational force acting on the block? (A component can be positive or negative).

(d) What is the *y* component of the force on the block due to rope 2?

(e) What is the magnitude of $\vec{F}_2$?

(f) What is the *x* component of the force on the block due to rope 2?

(g) What is the *x* component of the force on the block due to rope 1?

5.P.15 You pull with a force of 255 N on a rope that is attached to a block of mass 30 kg, and the block slides across the floor at a constant speed of 1.1 m/s. The rope makes an angle $\theta = 40°$ with the horizontal. Both the force and the velocity of the block are in the *xy* plane. See Figure 5.58.

Figure 5.58

(a) What is the direction of $d\vec{p}/dt$ of the block?

(b) What is the net force on the block?

(c) Which objects exert forces on the block with nonzero *x* components?

(d) What is the *x* component of the tension force exerted by the rope on the block? (A component may be positive or negative.)

(e) What is the *x* component of the force exerted by the floor on the block (the friction force)?

(f) Which objects exert significant forces on the block with nonzero *y*-components?

(g) What is the *y* component of the contact force exerted by the rope on the block?

(h) What is the *y* component of the gravitational force exerted by the Earth on the block?

(i) What is the *y* component of the contact force exerted by the floor on the block?

5.P.16 A ball of mass 450 g hangs from a spring whose stiffness is 110 newtons per meter. A string is attached to the ball and you are pulling the string to the right, so that the ball hangs motionless, as shown in Figure 5.59. In this situation the spring is stretched, and its length is 15 cm. What would be the relaxed length of the spring, if it were detached from the ball and laid on a table?

Figure 5.59

5.P.17 A cylindrical steel rod with diameter 0.7 mm is 2 m long when lying horizontally (Figure 5.60). The density of this particular steel is 7.85 grams/cm³, and Young's modulus is 2×10^{11} N/m². **(a)** You now hang the rod vertically from a support and attach a 60 kg mass to the bottom end. How much does the rod stretch? **(b)** Next you remove the 60 kg mass so the rod hangs under its own weight. Describe qualitatively how the stretch of the interatomic bonds depends on position along the rod. **(c)** How much does the rod stretch? This shows why it is often appropriate to neglect how much a rod stretches under its own weight. (This is a challenging question and requires the use of integral calculus, because the tension and strain vary along the rod.) **(d)** If the rod were twice as thick, what would be the answer to part **(c)**?

Figure 5.60

5.P.18 An 800 kg load is suspended as shown in Figure 5.61. **(a)** Calculate the tension in all three wires (that is, the magnitude of the tension force exerted by each of these wires). **(b)** These wires are made of a material whose value for Young's modulus is 1.3×10^{11} N/m². The diameter of the wires is 1.1 mm. What is the strain (fractional stretch) in each wire?

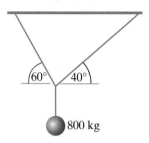

Figure 5.61

5.P.19 A helicopter flies to the right (in the $+x$ direction) at a constant speed of 12 m/s, parallel to the surface of the ocean. A 900 kg package of supplies is suspended below the helicopter by a cable as shown in Figure 5.62: the package is also traveling to the right in a straight line, at a constant speed of 12 m/s. The pilot is concerned about whether or not the cable, whose breaking strength is listed at 9300 N, is strong enough to support this package under these circumstances.

Figure 5.62

(a) Choose the package as the system. What is the rate of change of the momentum of the system?

(b) List all the objects that are exerting forces on the package. (Note that "gravity" is not an object.)

(c) Represent the package as a dot. Draw a careful force diagram ("free-body diagram") showing all the forces acting on the package. You should have the same number of arrows as you had objects in your list in part (b), and the tail of each arrow should be on the package. Label each force with the name of the object exerting the force.

(d) What is the magnitude of the tension in the cable supporting the package? Carefully show all steps in your work.

(e) Write the force exerted on the package by the cable as a vector.

(f) What is the magnitude of the force exerted by the air on the package? Carefully show all steps in your work.

(g) Write the force on the package by the air as a vector.

(h) Is the cable in danger of breaking?

Sections 5.3–5.4

5.X.20 A review of the basics of circular motion: The radius of a merry-go-round is 11 meters, and it takes 12 seconds to go around once. We want to calculate the speed of an atom in the outer rim.

(a) First, how far does the atom go in 12 s? (This distance is the circumference of the circle.)

(b) Therefore, what is the speed of the atom?

5.X.21 If the radius of a merry-go-round is 5 meters, and it takes 14 seconds to go around once, what is the speed of an atom at the outer rim? What is the direction of the velocity of this atom: toward the center, away from the center, or tangential?

5.X.22 In Figure 5.63, locations A and B indicate the position of the Earth at two different times, as it orbits the Sun in a nearly circular orbit, at nearly constant speed.

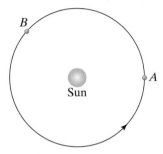

Figure 5.63 Locations A and B indicate the position of the Earth at two different times.

(a) At both A and B, draw a vector representing the momentum of the Earth, with its tail on the Earth.

(b) At both A and B, draw a vector representing the gravitational force acting on the Earth, with its tail on the Earth.

(c) At both A and B, draw a short vector representing the change $\Delta \vec{p}$ in the momentum of the Earth in the next hour. Place the tail of the short $\Delta \vec{p}$ vector at the tip of the momentum vector.

(d) At both A and B, draw a vector representing the new momentum of the Earth after the hour. Place the tail of the vector on the Earth.

5.X.23 You throw a ball straight up in the air.

(a) At the instant the ball reaches its highest location, what is the direction of the momentum of the ball?

(b) At the instant the ball reaches its highest location, what is the direction of the rate of change of momentum $d\vec{p}/dt$ of the ball?

(c) At the instant the ball reaches its highest location, what is the direction of the net force $\vec{F}_{net}$ acting on the ball?

(d) You launch a fan cart to the right along a track, with the fan blowing against the motion. At the instant the cart reaches its farthest location to the right (just before moving back to the left), what is the direction of the momentum $\vec{p}$ of the cart?

(e) At the instant the cart reaches its farthest location to the right, what is the direction of the rate of change of momentum $d\vec{p}/dt$ of the cart?

(f) At the instant the cart reaches its farthest location to the right, what is the direction of the net force $\vec{F}_{net}$ acting on the cart?

(g) A block oscillates up and down at the end of a vertical spring. At the instant the block reaches its lowest position, what is the direction of the momentum $\vec{p}$ of the block?

(h) At the instant the block reaches its lowest position, what is the direction of the rate of change of momentum $d\vec{p}/dt$ of the block?

(i) At the instant the block reaches its lowest position, what is the direction of the net force $\vec{F}_{net}$ acting on the block?

5.X.24 The radius of a merry-go-round is 7 meters, and it takes 12 seconds to make a complete revolution.

(a) What is the speed of an atom on the outer rim?

(b) Which statement below (i, ii, or iii) best describes the direction of the momentum of this atom?

(c) Which statement below (i, ii, or iii) best describes the direction of the rate of change of the momentum of this atom?

(i) Inward, toward the center
(ii) Outward, away from the center
(iii) Tangential

5.X.25 Figure 5.64 shows the orbit of the Moon around the Earth. At the instant when the Moon is at each of the locations marked by a number (1–5):

(a) which arrow (a–m) in Figure 5.65 best describes the direction of the Moon's momentum?

(b) which arrow (a–m) best describes the direction of the rate of change of the Moon's momentum?

(c) which arrow (a–m) best describes the direction of the net force on the Moon?

Figure 5.64

Directions:

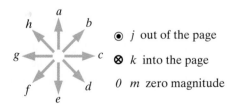

Figure 5.65

5.X.26 A comet orbits a star in an elliptical orbit, as shown in Figure 5.66. The momentum of the comet at location A is also shown.

Figure 5.66

Figure 5.67

(a) At location D, what is the direction (a–j in Figure 5.67) of the gravitational force exerted on the comet by the star?

(b) At location D, is the parallel component of the net force on the comet zero or nonzero?

(c) At location F, is the perpendicular component of the net force on the comet zero or nonzero?

(d) At location D, is $d|\vec{p}|/dt$, the rate of change of the magnitude of the comet's momentum, positive, negative, or zero?

(e) At location D, is $\hat{p}$, the direction of the comet's momentum, changing or not changing?

(f) At location C, what is the direction (a–j) of the gravitational force exerted on the comet by the star?

(g) At location C, is the parallel component of the net force on the comet zero or nonzero?

(h) At location C, is the perpendicular component of the net force on the comet zero or nonzero?

(i) At location C, is $d|\vec{p}|/dt$, the rate of change of the magnitude of the comet's momentum, positive, negative, or zero?

(j) At location C, is $\hat{p}$, the direction of the comet's momentum, changing or not changing?

(k) At location F, what is the direction (a–j) of the gravitational force exerted on the comet by the star?

(l) At location F, is the parallel component of the net force on the comet zero or nonzero?

(m) At location F, is the perpendicular component of the net force on the comet zero or nonzero?

(n) At location F, is $d|\vec{p}|/dt$, the rate of change of the magnitude of the comet's momentum, positive, negative, or zero?

(o) At location F, is $\hat{p}$, the direction of the comet's momentum, changing or not changing?

5.X.27 Figure 5.68 shows the path of a particle. Assuming that the particle's speed does not change as it travels along this path, at each location labeled "×" draw an arrow showing the direction of $d\vec{p}/dt$ for the particle as it passes that location.

Figure 5.68 Path of a particle traveling at constant speed (Exercise 5.X.27).

5.X.28 Which of the following statements are true about the motion of an object?

A. When an object moves at constant speed along a curving path, the net force on the object must act straight outward from the center of the kissing circle.

B. A force perpendicular to the momentum changes the direction of the momentum but not its magnitude.

C. To change the magnitude of the momentum you need to apply a force parallel to the momentum (either in the same direction as the momentum or the opposite direction).

D. To make an object turn to the left, something has to exert a force to the right on the object.

5.X.29 You are riding in the passenger seat of an American car, sitting on the right side of the front seat. The car makes a sharp left turn. You feel yourself "thrown to the right" and your right side hits the right door. Is there a force that pushes you to the right? What object exerts that force? What really happens? Draw a diagram to illustrate and clarify your analysis.

Sections 5.5–5.7

5.X.30 A student said, "When the Moon goes around the Earth, there is an inward force due to the Moon and an outward force due to centrifugal force, so the net force on the Moon is zero." Give two or more physics reasons why this is wrong.

5.X.31 An object moving at a constant speed of 23 m/s is making a turn with a radius of curvature of 4 m (this is the radius of the "kissing circle"). The object's momentum has a magnitude of 78 kg·m/s. What is the magnitude of the rate of change of the momentum? What is the magnitude of the net force?

5.X.32 A proton follows the curving path shown in red. The dashed circle is the kissing circle tangent to the path when the proton is at location A. The proton is traveling at a constant speed of 7.0×10^5 m/s, and the radius of the kissing circle is 0.08 m. The mass of a proton is 1.7×10^{-27} kg. Refer to the directional arrows shown at the right in Figure 5.69 when answering the questions below.

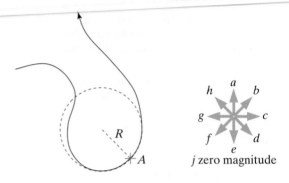

Figure 5.69

(a) When the proton is at location A, what are the magnitude and direction of $(d|\vec{p}|/dt)\hat{p}$, the parallel component of $d\vec{p}/dt$?

(b) When the proton is at location A, what are the magnitude and direction of $|\vec{p}|\hat{p}/dt$, the perpendicular component of $d\vec{p}/dt$?

5.X.33 A proton follows the curving path shown in red, traveling at constant speed in the direction shown. The dashed circle is the kissing circle tangent to the path when the proton is at location A. Refer to the directional arrows shown at the right in Figure 5.70 when answering the questions below.

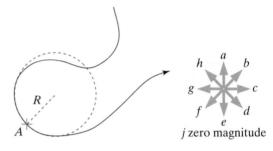

Figure 5.70

(a) When the proton is at location A, what is the direction of the proton's momentum?

(b) When the proton is at location A, what is the direction of $d\vec{p}/dt$?

(c) The mass of a proton is 1.7×10^{-27} kg. The proton is traveling at a constant speed of 6.0×10^5 m/s, and the radius of the kissing circle is 0.07 m. What is the magnitude of $d\vec{p}/dt$ of the proton?

5.X.34 A child of mass 35 kg sits on a wooden horse on a carousel. The wooden horse is 3.3 m from the center of the carousel, which rotates at a constant rate and completes one revolution every 5.2 seconds.

(a) What are the magnitude and direction (tangential in direction of velocity, tangential in the opposite direction of the velocity, radial outward, radial inward) of $(d|\vec{p}|/dt)\hat{p}$, the parallel component of $d\vec{p}/dt$ for the child?

(b) What are the magnitude and direction of $|\vec{p}|d\hat{p}/dt$, the perpendicular component of $d\vec{p}/dt$ for the child?

(c) What are the magnitude and direction of the net force acting on the child?

(d) What objects in the surroundings contribute to this horizontal net force acting on the child? (There are also vertical forces, but these cancel each other if the horse doesn't move up and down.)

5.X.35 A 30 kg child rides on a playground merry-go-round, 1.4 m from the center. The merry-go-round makes one complete revolution every 5 seconds. How large is the net force on the child? In what direction does the net force act?

5.X.36 The orbit of the Earth around the Sun is approximately circular, and takes one year to complete. The Earth's mass is 6×10^{24} kg, and the distance from the Earth to the Sun is 1.5×10^{11} m. What is $(d|\vec{p}|/dt)\hat{p}$ of the Earth? What is $|\vec{p}|(d\hat{p}/dt)$ of the Earth? What is the magnitude of the gravitational force the Sun (mass 2×10^{30} kg) exerts on the Earth? What is the direction of this force?

5.X.37 A circular pendulum of length 1.1 m goes around at an angle of 28 degrees to the vertical. Predict the speed of the mass at the end of the string. Also predict the period, the time it takes to go around once. Remember that the radius of the circle is the length of the string times the sine of the angle that the string makes to the vertical.

5.X.38 Tarzan swings from a vine. When he is at the bottom of his swing, as shown in Figure 5.71:

(a) Which arrow (a–m) in Figure 5.72 best describes the direction of the rate of change of Tarzan's momentum?

(b) Which is larger in magnitude: the force by the Earth on Tarzan, the force by the vine (a tension force) on Tarzan, or neither (same magnitude)? Explain how you know this.

Figure 5.71

Directions:

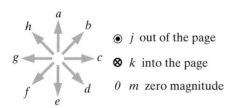

Figure 5.72

5.X.39 Tarzan swings back and forth on a vine. At the microscopic level, why is the tension force on Tarzan by the vine greater than it would be if he were hanging motionless?

A. The interatomic bonds in the vine are stretched less than they would be if Tarzan's momentum were not changing.

B. The interatomic bonds in the vine are stretched more than they would be if Tarzan's momentum were not changing.

C. The interatomic bonds are not stretched at all.

5.X.40 An airplane flies at constant speed along a curve, as shown in Figure 5.73. At the top of the arc, what is the direction of $d\vec{p}/dt$ on a passenger riding in the plane? What is the direction of $d\hat{p}/dt$?

Figure 5.73

5.X.41 At the moment shown in Figure 5.73, the magnitude of the upward (normal) force exerted by the seat on a passenger in the airplane is not zero, but it is less than it would be if the plane and passenger were sitting still on the ground. Which of the following statements correctly explains this at a microscopic level?

A. The interatomic bonds in the seat are compressed less than they would be if the passenger's momentum weren't changing.

B. The interatomic bonds in the seat are compressed more than they would be if the passenger's momentum weren't changing.

C. The interatomic bonds are not compressed at all.

5.X.42 At a particular instant an airplane is flying in the $-x$ direction. At this instant the magnitude of its momentum is changing, but the direction is not changing. Which of the following quantities are not zero? If a quantity is not zero, give a possible direction for that quantity.

A. $d|\vec{p}|/dt$

B. $d\hat{p}/dt$

C. $\vec{F}_{\parallel}$

D. $\vec{F}_{\perp}$

5.X.43 At a particular instant an airplane is flying in the $-z$ direction. At this instant the direction of its momentum is changing, but the magnitude is not changing. Which of the following quantities are not zero? If a quantity is not zero, give a possible direction for that quantity.

A. $d|\vec{p}|/dt$

B. $d\hat{p}/dt$

C. $\vec{F}_{\parallel}$

D. $\vec{F}_{\perp}$

5.P.44 The Ferris wheel in Figure 5.74 is a vertical, circular amusement ride with radius 10 m. Riders sit on seats that swivel to remain horizontal. The Ferris wheel rotates at a constant rate, going around once in 10.5 s. Consider a rider whose mass is 56 kg.

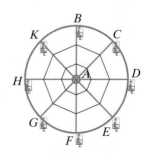

Figure 5.74

(a) At the bottom of the ride, what is the rate of change of the rider's momentum?

(b) At the bottom of the ride, what is the vector gravitational force exerted by the Earth on the rider?

(c) At the bottom of the ride, what is the vector force exerted by the seat on the rider?

(d) Next consider the situation at the top of the ride. At the top of the ride, what is the rate of change of the rider's momentum?

(e) At the top of the ride, what is the vector gravitational force exerted by the Earth on the rider?

(f) At the top of the ride, what is the vector force exerted by the seat on the rider?

A rider *feels* heavier if the electric, interatomic contact force of the seat on the rider is larger than the rider's weight mg (and the rider sinks more deeply into the seat cushion). A rider *feels* lighter if the contact force of the seat is smaller than the rider's weight (and the rider does not sink as far into the seat cushion).

(g) Does a rider *feel* heavier or lighter at the bottom of a Ferris wheel ride?

(h) Does a rider *feel* heavier or lighter at the top of a Ferris wheel ride?

5.P.45 In the dark in outer space, you observe a glowing ball of known mass 2 kg moving in the xy plane at constant speed in a circle of radius 6.5 m, with the center of the circle at the origin $(\langle 0, 0, 0 \rangle$ m). You can't see what's making it move in a circle. At time $t = 0$ the ball is at location $\langle -6.5, 0, 0 \rangle$ m and has velocity $\langle 0, 40, 0 \rangle$ m/s.

On your own paper draw a diagram of the situation showing the circle and showing the position and velocity of the ball at time $t = 0$. The diagram will help you analyze the situation. Use letters a–j (Figure 5.75) to answer questions about directions ($+x$ to the right, $+y$ up).

Figure 5.75

(a) At time $t = 0$, what is the direction of the vector $\vec{p}$?

(b) At time $t = 0$, what are the magnitude and direction of $(d|\vec{p}|/dt)\hat{p}$, the parallel component of $d\vec{p}/dt$?

(c) At time $t = 0$, what are the magnitude and direction of $|\vec{p}|d\hat{p}/dt$, the perpendicular component of $d\vec{p}/dt$?

(d) At time $t = 0$, even though you can't see what's causing the motion, what can you conclude must be the direction of the vector $\vec{F}_{net}$?

(e) At time $t = 0$, even though you can't see what's causing the motion, what can you conclude must be the vector $\vec{F}_{net}$?

(f) You learn that at time $t = 0$, two forces act on the ball, and that at this instant one of these forces is $\vec{F}_1 = \langle 196, -369, 0 \rangle$ N. What is the other force?

5.P.46 You're driving a vehicle of mass 1350 kg and you need to make a turn on a flat road. The radius of curvature of the turn is 76 m. The coefficient of static friction and the coefficient of kinetic friction are both 0.25.

(a) What is the fastest speed you can drive and still make it around the turn? Invent symbols for the various quantities and solve algebraically before plugging in numbers.

(b) Which of the following statements are true about this situation?

(1) The net force is nonzero and points away from the center of the kissing circle.

(2) The rate of change of the momentum is nonzero and points away from the center of the kissing circle.

(3) The rate of change of the momentum is nonzero and points toward the center of the kissing circle.

(4) The momentum points toward the center of the kissing circle.

(5) The centrifugal force balances the force of the road, so the net force is zero.

(6) The net force is nonzero and points toward the center of the kissing circle.

(c) Look at your algebraic analysis and answer the following question. Suppose that your vehicle had a mass 5 times as big (6750 kg). Now what is the fastest speed you can drive and still make it around the turn?

(d) Look at your algebraic analysis and answer the following question. Suppose that you have the original 1350 kg vehicle but the turn has a radius twice as large (152 m). What is the fastest speed you can drive and still make it around the turn?

This problem shows why high-speed curves on freeways have very large radii of curvature, but low-speed entrance and exit ramps can have smaller radii of curvature.

5.P.47 In outer space a rock of mass 4 kg is attached to a long spring and swung at constant speed in a circle of radius 9 m. The spring exerts a force of constant magnitude 760 N.

(a) What is the speed of the rock?

(b) What is the direction of the spring force?

(1) In the direction of motion (tangential to the circle)

(2) Toward the center of the circle (radially inward)

(3) Opposite the direction of motion (tangential to the circle)

(4) Away from the center of the circle (radially outward)

(c) The relaxed length of the spring is 8.7 m. What is the stiffness of this spring?

5.P.48 A child of mass 26 kg swings at the end of an elastic cord. At the bottom of the swing, the child's velocity is horizontal, and the speed is 12 m/s. At this instant the cord is 4.30 m long.

(a) At this instant, what is the parallel component of the rate of change of the child's momentum?

(b) At this instant, what is the perpendicular component of the rate of change of the child's momentum?

(c) At this instant, what is the *net* force acting on the child?

(d) What is the magnitude of the force that the elastic cord exerts on the child? (It helps to draw a diagram of the forces.)

(e) The relaxed length of the elastic cord is 4.22 m. What is the stiffness of the cord?

5.P.49 A child of mass 24 kg swings at the end of an elastic cord. At the bottom of the swing, the child's velocity is horizontal, and the speed is 8 m/s. At this instant the cord is 3.10 m long.

(a) At this instant, what is the magnitude of the rate of change of the child's momentum?

(b) At this instant, what is the (vector) *net* force acting on the child?

(c) What is the magnitude of the force that the elastic cord exerts on the child? (It helps to draw a diagram of the forces.)

(d) The relaxed length of the elastic cord is 3.06 m. What is the stiffness of the cord?

5.P.50 In outer space a rock of mass 7 kg is attached to a long spring and swung at constant speed in a circle of radius 7.5 m. The spring exerts a force of constant magnitude 720 N.

(a) How long does it take for the rock to go once around the circle?

(b) What is the speed of the rock?

(c) What is the direction of the spring force?

  (1) In the direction of motion (tangential to the circle)

  (2) Opposite the direction of motion (tangential to the circle)

  (3) Away from the center of the circle (radially outward)

  (4) Toward the center of the circle (radially inward)

5.P.51 What is the minimum speed v that a roller coaster car must have in order to make it around an inside loop and just barely lose contact with the track at the top of the loop (see Figure 5.76)? The center of the car moves along a circular arc of radius R. Include a carefully labeled force diagram. State briefly what approximations you make. Design a plausible roller coaster loop, including numerical values for v and R.

Figure 5.76

5.P.52 Use a circular pendulum to determine g. You can increase the accuracy of the time it takes to go around once by timing N revolutions and then dividing by N. This minimizes errors contributed by inaccuracies in starting and stopping the clock. It is wise to start counting from zero (0, 1, 2, 3, 4, 5) rather than starting from 1 (1, 2, 3, 4, 5 represents only 4 revolutions, not 5). It also improves accuracy if you start and stop timing at a well-defined event, such as when the mass crosses in front of an easily visible mark.

This was the method used by Newton to get an accurate value of g. Newton was not only a brilliant theorist but also an excellent experimentalist. For a circular pendulum he built a large triangular wooden frame mounted on a vertical shaft, and he pushed this around and around while making sure that the string of the circular pendulum stayed parallel to the slanting side of the triangle.

5.P.53 An engineer whose mass is 70 kg holds onto the outer rim of a rotating space station whose radius is 14 m and which takes 30 s to make one complete rotation. What is the magnitude of the force the engineer has to exert in order to hold on? What is the magnitude of the net force acting on the engineer?

5.P.54 After the NEAR spacecraft passed Mathilde, on several occasions rocket propellant was expelled to adjust the spacecraft's momentum in order to follow a path that would approach the asteroid Eros, the final destination for the mission. After getting close to Eros, further small adjustments made the momentum just right to give a circular orbit of radius 45 km (45×10^3 m) around the asteroid. So much propellant had been used that the final mass of the spacecraft while in circular orbit around Eros was only 500 kg. The spacecraft took 1.04 days to make one complete circular orbit around Eros. Calculate what the mass of Eros must be.

5.P.55 A Ferris wheel is a vertical, circular amusement ride. Riders sit on seats that swivel to remain horizontal as the wheel turns. The wheel has a radius R and rotates at a constant rate, going around once in a time T. At the bottom of the ride, what are the magnitude and direction of the force exerted by the seat on a rider of mass m? Include a diagram of the forces on the rider.

5.P.56 By "weight" we usually mean the gravitational force exerted on an object by the Earth. However, when you sit in a chair your own perception of your own "weight" is based on the contact force the chair exerts upward on your rear end rather than on the gravitational force. The smaller this contact force is, the less "weight" you perceive, and if the contact force is zero, you feel peculiar and "weightless" (an odd word to describe a situation when the only force acting on you is the gravitational force exerted by the Earth!). Also, in this condition your internal organs no longer press on each other, which presumably contributes to the odd sensation in your stomach.

(a) How fast must a roller coaster car go over the top of a circular arc for you to feel "weightless"? The center of the car moves along a circular arc of radius R (see Figure 5.77). Include a carefully labeled force diagram.

Figure 5.77

(b) How fast must a roller coaster car go through a circular dip for you to feel three times as "heavy" as usual, due to the upward force of the seat on your bottom being three times as large as usual? The center of the car moves along a circular arc of radius R (see Figure 5.78). Include a carefully labeled force diagram.

Figure 5.78

5.P.57 A ball of unknown mass m is attached to a spring. In outer space, far from other objects, you hold the other end of the spring and swing the ball around in a circle of radius 1.5 m at constant speed.

(a) You time the motion and observe that going around 10 times takes 6.88 seconds. What is the speed of the ball?

(b) Is the momentum of the ball changing or not? How can you tell?

(c) If the momentum is changing, what interaction is causing it to change? If the momentum is not changing, why isn't it?

(d) The relaxed length of the spring is 1.2 m, and its stiffness is 1000 N/m. While you are swinging the ball, since the radius of the circle is 1.5 m, the length of the spring is also 1.5 m. What is the magnitude of the force that the spring exerts on the ball?

(e) What is the mass m of the ball?

5.P.58 A sports car (and its occupants) of mass M is moving over the rounded top of a hill of radius R. At the instant when the car is at the very top of the hill, the car has a speed v. You can safely neglect air resistance.

(a) Taking the sports car as the system of interest, what object(s) exert non-negligible forces on this system?

(b) At the instant when the car is at the very top of the hill, draw a diagram showing the system as a dot, with force vectors whose tails are at the location of the dot. Label the force vectors (that is, give them algebraic names). Try to make the lengths of the force vectors be proportional to the magnitudes of the forces.

(c) Starting from the Momentum Principle, calculate the force exerted by the road on the car.

(d) Under what conditions will the force exerted by the road on the car be zero? Explain.

5.P.59 A small block of mass m is attached to a spring with stiffness k_s and relaxed length L. The other end of the spring is fastened to a fixed point on a low-friction table. The block slides on the table in a circular path of radius $R > L$. How long does it take for the block to go around once?

5.P.60 The planets in our Solar System have orbits around the Sun that are nearly circular, and $v \ll c$. Calculate the period T (a "year"—the time required to go around the Sun once) for a planet whose orbit radius is r. This is the relationship discovered by Kepler and explained by Newton. (It can be shown by advanced techniques that this result also applies to elliptical orbits if you replace r by the "semimajor axis," which is half the longer, "major" axis of the ellipse.) Use this analytical solution for circular motion to predict the Earth's orbital speed, using the data for Sun and Earth on the inside back cover of the textbook.

5.P.61 A person of mass 70 kg rides on a Ferris wheel whose radius is 4 m. The person's speed is constant at 0.3 m/s. The person's location is shown by a dot in Figure 5.79.

(a) What is the magnitude of the rate of change of the momentum of the person at the instant shown?

Figure 5.79

(b) What is the direction of the rate of change of momentum of the person at the instant shown?

(c) What is the magnitude of the *net* force acting on the person at the instant shown? Draw the *net* force vector on the diagram at this instant, with the tail of the vector on the person.

5.P.62 When a particle with electric charge q moves with speed v in a plane perpendicular to a magnetic field B, there is a magnetic force at right angles to the motion with magnitude qvB, and the particle moves in a circle of radius r (see Figure 5.80). This formula for the magnetic force is correct even if the speed is comparable to the speed of light. Show that

$$p = \frac{mv}{\sqrt{1 - (|\vec{v}|/c)^2}} = qBr$$

even if v is comparable to c.

This result is used to measure relativistic momentum: if the charge q is known, we can determine the momentum of a particle by observing the radius of a circular trajectory in a known magnetic field.

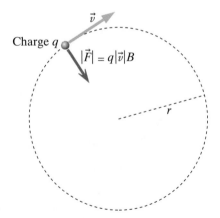

Charge q $\vec{v}$ $|\vec{F}| = q|\vec{v}|B$ r

Figure 5.80

5.P.63 In the 1970s the astronomer Vera Rubin made observations of distant galaxies that she interpreted as indicating that perhaps 90% of the mass in a galaxy is invisible to us ("dark matter"). She measured the speed with which stars orbit the center of a galaxy, as a function of the distance of the stars from the center. The orbital speed was determined by measuring the "Doppler shift" of the light from the stars, an effect that makes light shift toward the red end of the spectrum ("red shift") if the star has a velocity component away from us, and makes light shift toward the blue end of the spectrum if the star has a velocity component toward us. She found that for stars farther out from the center of the galaxy, the orbital speed of the star hardly changes with distance from the center of the galaxy, as is indicated in Figure 5.81. The visible components of the galaxy (stars, and illuminated clouds of dust) are most dense at the center of the galaxy and thin out rapidly as you move away from the center, so most of the visible mass is near the center.

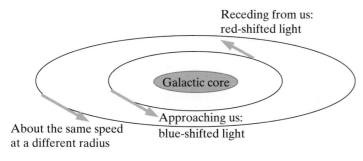

Figure 5.81

(a) Predict the speed v of a star going around the center of a galaxy in a circular orbit, as a function of the star's distance r from the center of the galaxy, assuming that almost all of the galaxy's mass M is concentrated at the center.

(b) Construct a logical argument as to why Rubin concluded that much of the mass of a galaxy is not visible to us. Reason from principles discussed in this chapter, and your analysis of part (a). Explain your reasoning. You need to address the following issues:

Rubin's observations are not consistent with your prediction in (a).

Most of the *visible* matter is in the center of the galaxy.

Your prediction in (a) assumed that most of the mass is at the center.

This issue has not yet been resolved, and is still a current topic of astrophysics research. Here is a discussion by Rubin of her work: "Dark Matter in Spiral Galaxies" by Vera C. Rubin, *Scientific American*, June 1983 (96–108). You can find several graphs of the rotation curves for spiral galaxies on page 101 of this article.

5.P.64

(a) Many communication satellites are placed in a circular orbit around the Earth at a radius where the period (the time to go around the Earth once) is 24 hours. If the satellite is above some point on the equator, it stays above that point as the Earth rotates, so that as viewed from the rotating Earth the satellite appears to be motionless. That is why you see dish antennas pointing at a "fixed" point in space. Calculate the radius of the orbit of such a "synchronous" satellite. Explain your calculation in detail.

(b) Electromagnetic radiation including light and radio waves travels at a speed of 3×10^8 m/s. If a phone call is routed through a synchronous satellite to someone not very far from you on the ground, what is the minimum delay between saying something and getting a response? Explain. Include in your explanation a diagram of the situation.

(c) Some human-made satellites are placed in "near-Earth" orbit, just high enough to be above almost all of the atmosphere. Calculate how long it takes for such a satellite to go around the Earth once, and explain any approximations you make.

(d) Calculate the orbital speed for a near-Earth orbit, which must be provided by the launch rocket. (The advantages of near-Earth communications satellites include making the signal delay unnoticeable, but with the disadvantage of having to track the satellites actively and having to use many satellites to ensure that at least one is always visible over a particular region.)

(e) When the first two astronauts landed on the Moon, a third astronaut remained in an orbiter in circular orbit near the Moon's surface. During half of every complete orbit, the orbiter was behind the Moon and out of radio contact with the Earth. On each orbit, how long was the time when radio contact was lost?

5.P.65 There is no general analytical solution for the motion of a three-body gravitational system. However, there do exist analytical solutions for very special initial conditions. Figure 5.82 shows three stars, each of mass m, which move in the plane of the page along a circle of radius r. Calculate how long this system takes to make one complete revolution. (In many cases three-body orbits are not stable: any slight perturbation leads to a breakup of the orbit.)

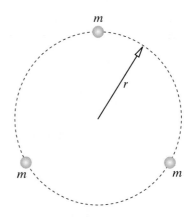

Figure 5.82

5.P.66 Remarkable data indicate the presence of a massive black hole at the center of our Milky Way galaxy. The W. M. Keck 10 meter diameter telescopes in Hawaii were used by Andrea Ghez and her colleagues to observe infrared light coming directly through the dust surrounding the central region of our galaxy (visible light is multiply-scattered by the dust, blocking a direct view). Stars were observed for several consecutive years to determine their orbits near a motionless center that is completely invisible ("black") in the infrared but whose precise location is known due to its strong output of radio waves, which are observed by radio telescopes. The data were used to show that the object at the center must have a mass that is huge compared to the mass of our own Sun, whose mass is a "mere" 2×10^{30} kg.

Figure 5.83 shows positions from 1995 to 2008 of one of the stars, called S0-20, orbiting around the galactic center. The orbit appears nearly circular with the radius shown.

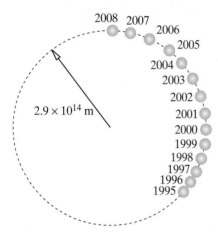

Figure 5.83 Positions on the sky of the star S0-20 near the center of our galaxy.

(a) Using the positions and times shown in Figure 5.83, what is the approximate speed of this star in m/s? Also express the speed as a fraction of the speed of light.

(b) This is an extraordinarily high speed for a macroscopic object. Is it reasonable to approximate the star's momentum as mv? (Some other stars near the galactic center with highly elliptical orbits move even faster when they are closest to the center.)

(c) Based on these data, estimate the mass of the massive black hole about which this star is orbiting. (You can see in Figure 5.83 that the speed appears greater in 2007–2008 than in 1995–1996. You're looking at a projection of an elliptical orbit, which only happens to look circular as viewed from Earth, and the black hole is not at the center of this circle. Therefore your result will be approximate and will differ from that obtained by Ghez and colleagues, who carefully made a consistent fit of the position data for seven different stars.)

(d) How many of our Suns does this represent?

It is now thought that most galaxies have such a black hole at their centers, as a result of long periods of mass accumulation. When many bodies orbit each other, sometimes in an interaction an object happens to acquire enough speed to escape from the group, and the remaining objects are left closer together. Some simulations show that over time, as much as half the mass may be ejected, with agglomeration of the remaining mass. This could be part of the mechanism for forming massive black holes.

For more information, see

http://www.astro.ucla.edu/ghezgroup/gc

You may see the term "arc seconds," which is an angular measure of how far apart objects appear in the sky, and to "parsecs," which is a distance equal to 3.3 light-years (a light-year is the distance light goes in one year).

5.P.67 You put a 10 kg object on a bathroom scale at the North Pole, and the scale reads exactly 10 kg (actually, it measures the force F_N that the scale exerts on the object, but displays a reading in kg). At the North Pole you are 6357 km from the center of the Earth. At the equator, the scale reads a different value due to two effects: (1) The Earth bulges out at the equator (due to its rotation), and you are 6378 km from the center of the Earth. (2) You are moving in a circular path due to the rotation of the Earth (one rotation every 24 hours). Taking into account **both** of these effects, what does the scale read at the equator?

Computational Problems

5.P.68 In Chapter 3 you wrote a program to model the motion of a spacecraft around the Earth. Modify this program to display continuously the parallel and perpendicular components of the net force on the spacecraft as it orbits the Earth, or the Earth and the Moon.

One way to find these components is to use the Momentum Principle to deduce $\vec{F}_\parallel$. You know that

$$\frac{d|\vec{p}|}{dt}\hat{p} = \vec{F}_\parallel \quad \text{and} \quad \frac{d|\vec{p}|}{dt} \approx \frac{\Delta|\vec{p}|}{\Delta t}$$
$$\text{Then } \vec{F}_\perp = \vec{F}_{net} - \vec{F}_\parallel$$

A different way to find $\vec{F}_\parallel$ is to take the vector dot product of $\vec{F}_{net}$ and $\hat{p}$. The dot product is introduced in the next chapter.

ANSWERS TO EXERCISES

5.X.1 (a) Bat, Earth, air; (b) Earth, string, air; (c) Earth, vine, air; (d) Earth, snow, air

5.X.2 $\langle 30, -90, 130 \rangle$ N

5.X.3 $0; \vec{0}$

5.X.4 (a) nonzero, downward; (b) nonzero, downward; (c) nonzero, upward; (d) nonzero, to the left

5.X.5 number 3; on the right

5.X.6 (a) Figure 5.84 shows the forces:

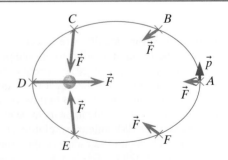

Figure 5.84

A: **(b)** zero **(c)** nonzero **(d)** zero **(e)** changing
B: **(b)** nonzero **(c)** nonzero **(d)** positive **(e)** changing
C: **(b)** nonzero **(c)** nonzero **(d)** positive **(e)** changing
D: **(b)** zero **(c)** nonzero **(d)** zero **(e)** changing
E: **(b)** nonzero **(c)** nonzero **(d)** negative **(e)** changing
F: **(b)** nonzero **(c)** nonzero **(d)** negative **(e)** changing
5.X.7 $(d|\vec{p}|/dt)\hat{p} = \vec{0}$, $|\vec{p}|(d\hat{p}/dt)$ and $\vec{F}_{net}$ point toward the center of the wheel, $\vec{F}_{net}$ is due to the Earth and the seat.
5.X.8 $4.5\,\text{m/s}^2$; about half the magnitude of acceleration of a falling object

5.X.9 $5.3 \times 10^{-18}\,\text{kg} \cdot \text{m/s}$
5.X.10 $\vec{0}$ (speed not changing); $0.97\,\text{kg} \cdot \text{m/s}^2$ toward center; 0.97 N toward center; horse (Earth pulls down, air resistance is opposite velocity)
5.X.11 **(a)** There is no object in the surroundings named "centripetal," so the force labeled mv^2/r is not a real force. **(b)** There is no object in the surroundings named "centrifugal," so the force labeled mv^2/r is not a real force. **(c)** There is no object in the surroundings responsible for the force labeled $\vec{F}_{net}$, so this force is not a real force.

The Energy Principle

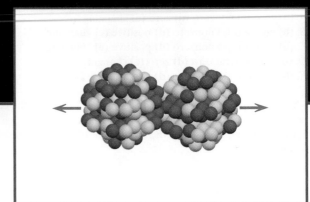

- **Energy is conserved: The change in energy of a system plus the change in energy of the surroundings is zero.**
- **Energy is a scalar quantity.**
- **A single particle is the simplest possible system.**
  Particle energy has two parts:
  - rest energy (energy an object has even when at rest)
  - kinetic energy (energy associated with motion)
- **Multiparticle system energy has three parts:**
  - rest energy
  - kinetic energy
  - potential energy (the interaction energy of pairs of particles within a multiparticle system)
- **The energy of a system can be changed only by external inputs.**
- **Work is mechanical energy transfer into or out of a system.**
  - Work involves an external force acting through a distance.
  - Work can be positive or negative.

6.1   THE ENERGY PRINCIPLE

You know a lot about the flow of energy from daily life. You ingest chemical energy in the form of food, and you use up this chemical energy when you are active. You put chemical energy in the form of gasoline into a car, and the car uses up this chemical energy to run the engine. You put electrical energy into a toaster, and the toaster raises the temperature of the bread. You compress a spring, and the mechanical energy stored in the spring can launch a ball.

The Energy Principle is a fundamental principle, just as the Momentum Principle is. The validity of the Energy Principle has been verified through a very wide variety of observations and experiments, involving large and small objects, moving slowly or at speeds near the speed of light, and even undergoing nuclear reactions that change the identity of the objects. It is a summary of the way energy flows in the real world.

THE ENERGY PRINCIPLE

$$\Delta E_{\text{system}} = W_{\text{surr}} + \text{other energy transfers}$$

Change in energy of a system (E_{system}) is equal to the work done by the surroundings (W_{surr}), and to other kinds of energy transfers between system and surroundings to be studied in Chapter 7.

■ The Energy Principle is a fundamental principle because:

■ It applies to every possible system, no matter how large or small it is (from clusters of galaxies to subatomic particles) or how fast it is moving.

■ It is true for any kind of interaction (gravitational, electromagnetic, strong, weak).

■ It relates an effect (change in energy of a system) to a cause (an interaction with the surroundings).

ΔE is a *change* in energy, which is the final energy minus the initial energy, $E_{\text{final}} - E_{\text{initial}}$. The current chapter will consider only situations in which work is the only kind of energy transfer between system and surroundings; other kinds of energy transfers will be discussed in Chapter 7. Work, the result of a force acting through a displacement, will be discussed in Section 6.3.

As with the Momentum Principle, the left-hand side of the Energy Principle equation describes an effect on a system; the right-hand side describes the causes, which involve interactions of the system with its surroundings.

Conservation of Energy

Energy cannot be created or destroyed, but it can change form. The only way for a system to gain or lose energy is if the surroundings lose or gain the same amount of energy.

CONSERVATION OF ENERGY

$$\Delta E_{\text{system}} + \Delta E_{\text{surroundings}} = 0$$

6.2 THE SIMPLEST SYSTEM: A SINGLE PARTICLE

The simplest possible system consists of a single particle. "A particle" could refer to a proton or an electron, but it could also refer to a baseball or even a planet if during the process of interest there are no significant changes internal to the "particle" such as changes of shape or rotation or vibration or temperature.

In 1905 Einstein discovered that to be consistent with his Special Theory of Relativity and with existing measures of energy, including an already existing definition of mechanical energy transfer to a system, called "work," the energy of a single particle is given by this expression:

DEFINITION: ENERGY OF A SINGLE PARTICLE

$$E_{\text{particle}} \equiv \gamma m c^2$$

where $\gamma = \dfrac{1}{\sqrt{1 - \left(\dfrac{v}{c}\right)^2}}$. The unit of energy is the joule, abbreviated "J."

One joule is equal to one $\text{kg} \cdot (\text{m/s})^2$.

The speed of light $c = 3 \times 10^8$ m/s. Later in this chapter we'll see that this definition of particle energy is consistent with the definition of a particle's momentum, $\vec{p} = \gamma m \vec{v}$.

Similarities between particle energy and particle momentum are that

■ Both contain the same factor γ.

■ Both are proportional to the mass m.

Differences between particle energy and particle momentum are as follows:

- Particle energy is a scalar.
- Momentum is a vector.

The units of energy are $\text{kg} \cdot (\text{m/s})^2$. This combination of units is given the name "joule" to honor the British physicist James Joule, who in the 1840s did some of the earliest experiments to establish the nature of energy. The abbreviation for joule is J.

Rest Energy

QUESTION   If a particle is at rest, what is its energy?

If $v = 0$ then $\gamma = 1$, and we find the famous equation $E = mc^2$. This has the striking interpretation that a particle has energy even when it is sitting still, which was a revolutionary idea when Einstein proposed it. We call mc^2 the "rest energy," the amount of energy a particle has when it is sitting still.

REST ENERGY

$$E_{\text{rest}} = mc^2$$

The "rest energy" is the energy of a particle at rest.

Even more strikingly, Einstein realized that this means that the mass of an object is its energy content divided by a constant: $m = E_{\text{rest}}/c^2$. For example, a hot object, with more internal energy, has very slightly more mass than a cold object. In a real sense, mass and energy are the same thing, although for historical reasons we use different units for them: mass in kilograms, energy in joules, with a constant factor of c^2 difference.

Note another difference between energy and momentum: A particle at rest has zero momentum but it has nonzero energy, its rest energy.

Kinetic Energy

QUESTION   If a particle is moving with speed v, is its energy greater than mc^2?

If $v > 0$, $\sqrt{1 - (v/c)^2}$ is less than 1, so $\gamma = 1/\sqrt{1 - (v/c)^2}$ is greater than 1, and the particle energy γmc^2 increases with increasing speed. As a consequence of its motion, the particle has additional, "kinetic" energy K:

KINETIC ENERGY K OF A PARTICLE

$$K = \gamma mc^2 - mc^2$$

The kinetic energy K of a particle is the energy a moving particle has in addition to its rest energy. It is often useful to turn this equation around:

$$E_{\text{particle}} = mc^2 + K$$

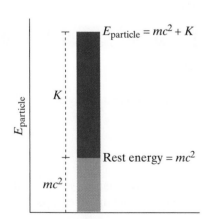

Figure 6.1 Particle energy is equal to rest energy plus kinetic energy K. This particle has high speed, and its kinetic energy is large compared to its rest energy.

If a particle is at rest, its particle energy is just its rest energy mc^2. If the particle moves, it has not only its rest energy mc^2 but also an additional amount of energy, which we call the kinetic (motional) energy K (Figure 6.1).

EXAMPLE

Energy of a Fast-Moving Proton

A proton in a particle accelerator has a speed of 2.91×10^8 m/s. **(a)** What is the energy of the proton? **(b)** What would the energy of the proton be if it were sitting still? **(c)** What is the kinetic energy of the moving proton?

Solution

(a) Moving proton:

$$\gamma = \frac{1}{\sqrt{1 - \left(\dfrac{2.91 \times 10^8}{3 \times 10^8}\right)^2}} = 4.11$$

$$E = (4.11)(1.7 \times 10^{-27}\, \text{kg})(3 \times 10^8\, \text{m/s})^2$$

$$E = 6.288 \times 10^{-10}\, \text{J}$$

(b) Proton at rest:

$$E = mc^2$$

$$E = (1.7 \times 10^{-27}\, \text{kg})(3 \times 10^8\, \text{m/s})^2$$

$$E = 1.53 \times 10^{-10}\, \text{J}$$

■ In this example, the kinetic energy of the fast-moving proton turns out to be about 3 times as large as the rest energy of the proton—about the ratio shown in Figure 6.1.

(c) Kinetic energy of proton:

$$K = E - mc^2$$

$$K = 6.288 \times 10^{-10}\, \text{J} - 1.53 \times 10^{-10}\, \text{J} = 4.76 \times 10^{-10}\, \text{J}$$

A Universal Speed Limit

Delivering more and more energy to a particle increases its energy more and more, and there is no theoretical limit to how much energy a particle can acquire. As the speed approaches the speed of light $c = 3 \times 10^8$ m/s, the denominator $\sqrt{1 - (v/c)^2}$ becomes very small, and the energy becomes very large.

However, even though you can increase the energy, it becomes very difficult to increase the speed v, because as you near the speed of light a tiny speed increase would require a huge energy increase. There is effectively a universal speed limit: a particle can't reach the speed of light c because an infinite amount of energy input would be required to achieve this.

The Low-Speed Approximation for Kinetic Energy

Next consider the case of a particle moving with a speed v that is small compared to the speed of light ($v \ll c$), the low-speed or "nonrelativistic" approximation. The situation is shown in Figure 6.2, where the kinetic energy is very small compared to the rest energy, so the particle energy is only slightly larger than the rest energy. An approximate formula for the energy at low speeds can be obtained through use of the mathematics of the "binomial expansion," as is shown in Section 6.21 at the end of this chapter:

$$\frac{mc^2}{\sqrt{1 - (v/c)^2}} = mc^2\left[1 + \frac{1}{2}\left(\frac{v}{c}\right)^2 + \frac{3}{8}\left(\frac{v}{c}\right)^4 + \frac{5}{16}\left(\frac{v}{c}\right)^6 + \cdots\right]$$

$$\frac{mc^2}{\sqrt{1 - (v/c)^2}} \approx mc^2 + \frac{1}{2}mv^2 \quad \text{if } v/c \ll 1$$

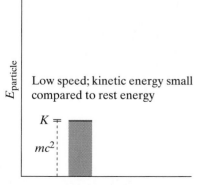

Figure 6.2 At low speeds the kinetic energy K is small compared to the rest energy.

Since we define kinetic energy through the relation $E_{\text{particle}} = mc^2 + K$, at low speeds the kinetic energy is approximately the following:

LOW-SPEED KINETIC ENERGY K OF A PARTICLE

$$K \approx \frac{1}{2}mv^2 \quad \text{(only if } v \ll c\text{)}$$

Approximate kinetic energy can also be written in terms of approximate magnitude of momentum.

$$K \approx \frac{1}{2}mv^2 \approx \frac{1}{2}m\left(\frac{p}{m}\right)^2 = \frac{p^2}{2m} \quad \text{(if } v \ll c\text{)}$$

EXAMPLE **Approximate Kinetic Energy**

The solar wind consists of charged particles streaming away from the Sun. The speed of a proton in the solar wind can be as high as 1×10^6 m/s. What is the kinetic energy of such a proton?

QUESTION   Is it appropriate to use the approximate formula for kinetic energy in this case?

Solution

$$\left(\frac{v}{c}\right)^2 = \left(\frac{1 \times 10^6 \text{ m/s}}{3 \times 10^8 \text{ m/s}}\right)^2 = (3.33 \times 10^{-3})^2 = 1.11 \times 10^{-5}$$

so $\gamma \approx 1$.

■ The kinetic energy of this relatively slow-moving proton is much less (nearly a million times less) than its rest energy, which was calculated in the previous example.

It is appropriate to use the approximate formula for K.

$$K \approx \frac{1}{2}mv^2 = \frac{1}{2}(1.7 \times 10^{-27} \text{ kg})(1 \times 10^6 \text{ m/s})^2 = 8.5 \times 10^{-16} \text{ J}$$

6.X.1 An electron has mass 9×10^{-31} kg. If the electron's speed is $0.988c$ (that is, $v/c = 0.988$), what is its particle energy? What is its rest energy? What is its kinetic energy?

6.X.2 A commercial airliner flew 20,000 km nonstop from Seattle to Kuala Lumpur at an average speed of 900 km/hr. If its average mass was 140,000 kg, what was its average kinetic energy? (We take an average because the mass of an airplane changes as it exhausts burnt fuel.)

EXAMPLE **Pull a Block**

If you pull on a 3 kg block and change its speed from 4 m/s to 5 m/s, what is the minimum change of chemical (food) energy in you? Give magnitude, sign, and units. (Your temperature rises when you do this, involving an increase in thermal energy, which we'll study in the next chapter, so there is more actual chemical cost than the minimum amount you calculate.)

Solution The speed is tiny compared to c, so the change in kinetic energy of the block is

$$\Delta K = \frac{1}{2}mv_f^2 - \frac{1}{2}mv_i^2 = \frac{1}{2}(3 \text{ kg})[(5 \text{ m/s})^2 - (4 \text{ m/s})^2]$$

$$\Delta K = 13.5 \text{ J}$$

Because you increased the kinetic energy of the block by 13.5 J, and your temperature also rose a bit, the change of your store of chemical (food) energy is at least -13.5 J.

6.X.3 An automobile traveling on a highway has an average kinetic energy of 1.1×10^5 J. Its mass is 1.5×10^3 kg; what is its average speed? Convert your answer to miles per hour to see whether it makes sense.

6.X.4 If you could use all of the mc^2 rest energy of some amount of fuel to provide the car in the previous exercise with its kinetic energy of 1.1×10^5 J, what mass of fuel would you need?

Relativistic Energy and Momentum

Because both relativistic energy and momentum involve the factor

$$\gamma = \frac{1}{\sqrt{1 - \left(\frac{v}{c}\right)^2}}$$

it seems likely that they can be related in some way. As you can show in Problem 6.P.108, the relationship turns out to be the following, where p represents $|\vec{p}|$:

RELATIVISTIC ENERGY AND MOMENTUM OF A PARTICLE

$$E^2 - (pc)^2 = (mc^2)^2$$

where $E = \gamma mc^2$; $m =$ rest mass.

This relation is true in all reference frames, for all particle speeds. Both energy and momentum will change if a phenomenon is viewed in a different reference frame, but this relation will still be true. The quantity $E^2 - (pc)^2$ is therefore called an "invariant" quantity—a quantity that does not vary when the reference frame is changed.

This relation turns out to be $E^2 - (pc)^2 = 0$ for those particles such as photons that always travel at the speed of light. Such particles have zero rest mass, and can never be at rest. The relation for so-called "massless" particles reduces to $E^2 = (pc)^2$, or $E = pc$. This relation is also approximately true for any particle that is traveling at very high speed, because in that case mc^2 is small compared to E. Neutrinos have extremely small mass and travel at nearly the speed of light.

The relation $E^2 - (pc)^2 = (mc^2)^2$ can be used to find another formula for kinetic energy, valid for both high and low speeds:

■ The difference of squares $E^2 - (mc^2)^2$ factors into $(E + mc^2)(E - mc^2)$. Note that $K = E - mc^2$.

$$E^2 - (mc^2)^2 = (pc)^2$$

$$(E + mc^2)(E - mc^2) = p^2c^2$$

$$(\gamma mc^2 + mc^2)(K) = p^2c^2$$

$$K = \frac{p^2}{(\gamma + 1)m}$$

At low speeds, $\gamma \approx 1$, and we find as before that $K \approx p^2/(2m)$.

EXAMPLE

Proton Momentum and Energy

A proton has a mass of 1.7×10^{-27} kg. If its total energy is 3 times its rest energy, what is the magnitude of its momentum?

Solution

$$mc^2 = (1.7 \times 10^{-27} \text{ kg})(3 \times 10^8 \text{ m/s})^2 = 1.5 \times 10^{-10} \text{ J}$$

$$E_{\text{particle}} = mc^2 + K = 3mc^2 \text{ and } (pc)^2 = (E_{\text{particle}})^2 - (mc^2)^2$$

$$pc = \sqrt{(E_{\text{particle}})^2 - (mc^2)^2} = \sqrt{(3mc^2)^2 - (mc^2)^2}$$

$$= \sqrt{8(mc^2)^2} = \sqrt{8}(mc^2)$$

$$p = \sqrt{8}(mc) = \sqrt{8}(1.7 \times 10^{-27} \text{ kg})(3 \times 10^8 \text{ m/s}) = 1.4 \times 10^{-18} \text{ kg} \cdot \text{m/s}$$

6.3 | WORK: MECHANICAL ENERGY TRANSFER

We are interested in changes in the energy of a system that are due to the interaction of the system with objects in its surroundings. Our goal is to figure out how to describe such changes quantitatively.

Displacement $\Delta \vec{r}$ Instead of Time Interval Δt

The Momentum Principle relates a change in momentum (effect) to the net impulse applied to a system. Impulse is the product of force and time.

$$\Delta \vec{p} = (\text{net force})\Delta t$$

We might ask what property of a system changes due to forces acting not for a time but through a distance, something like this:

$$\Delta ? = (\text{net force})(\text{distance})$$

The quantity that changes turns out to be the energy of the system, γmc^2 in the case of a single particle. The quantity involving force times distance is called "work."

In Section 6.6 we will show that this is correct. For now we'll assume this to be true, and after explaining how to calculate work we'll use the Energy Principle for a particle to analyze several kinds of processes.

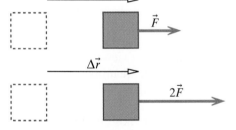

Figure 6.3 Pulling a block over a nearly frictionless surface with a force twice as big (through the same distance) means you expend twice as much chemical energy.

Magnitude of Applied Force

Suppose that you apply a force with constant magnitude F in a constant direction to a block that was initially at rest on a nearly frictionless surface, pulling it through a displacement Δr (Figure 6.3). The block moves faster and faster, acquiring a certain amount of kinetic energy. If you were to pull twice as hard ($2F$) through the same displacement Δr, the block would presumably acquire twice as much kinetic energy (and you would expend twice as much chemical energy). Note that the rest energy mc^2 of the block doesn't change, just the kinetic energy.

Distance Through Which Force Acts (Displacement)

Similarly, if you lift a heavy box to a height h above the ground, you will have expended a certain amount of chemical energy (Figure 6.4). If you lift the box twice as high (a height $2h$ above the ground), you will presumably have had to expend twice as much energy.

Figure 6.4 Lifting a box twice as high (with the same force) means you expend twice as much chemical energy.

The Product of Force Times Displacement

Because the amount of energy expended appears to be proportional both to the magnitude of the force exerted and to the distance through which an object is moved, we'll tentatively define the amount of energy transfer between you and the object as the product of force times displacement, $F\Delta r$. This type of mechanical energy transfer from the surroundings into the system is called "work." Here we use the word "work" as a technical term in the context of physics. This technical meaning is only loosely connected with everyday uses of the word.

The unit of work is newton · meter. From the Momentum Principle we know that a newton is the same as a kg · (m/s)/s, so a newton · meter is the same as a kg · m^2/s^2, which is a joule (same units as mc^2). Work has the same units as energy, as it should if we expect the change in energy of a system to be equal to the work.

Distance vs. Time

<u>QUESTION</u>   Is doubling the displacement the same as doubling the time?

No, it isn't. If you pull a block through twice the displacement, you do twice as much work, and the block acquires twice as much kinetic energy. However, because the block is moving faster in the second half of the doubled displacement, the time interval is less than doubled (and the impulse is less than doubled), so the momentum change is not doubled.

Note carefully: work and kinetic energy change are proportional to displacement, whereas impulse and momentum change are proportional to time interval. In symbols: work $F\Delta r$ is not the same as impulse $F\Delta t$.

Work and Displacement in 3D

We need to define work carefully. To see why, suppose you and a co-worker are maneuvering a heavy box on a space station (Figure 6.5). You pull in the x direction with a force F_x through a displacement Δx, so you do work

$$W_{\text{by you}} = F_x \Delta x$$

while at the same time your colleague is pulling in the y direction with a force F_y through a displacement Δy, doing work

$$W_{\text{by co-worker}} = F_y \Delta y$$

The total work done on the box by the two of you is

$$W = F_x \Delta x + F_y \Delta y$$

and this will be equal to the change in the energy of the box.

In general, the work done by a constant force $\vec{F} = \langle F_x, F_y, F_z \rangle$ acting through a displacement $\Delta \vec{r} = \langle \Delta x, \Delta y, \Delta z \rangle$ is defined as follows:

WORK DONE BY A CONSTANT FORCE

$$W = F_x \Delta x + F_y \Delta y + F_z \Delta z$$

for a constant force $\vec{F} = \langle F_x, F_y, F_z \rangle$. Work is a scalar quantity.

This is a special kind of product of two vectors, resulting in a scalar result. It is called the "scalar product" of the two vectors $\vec{F}$ and $\Delta \vec{r}$. A compact notation for this kind of product involves a dot between the two vectors. For that reason it is also called the "dot product":

Your co-worker pulls in the +y direction

You pull in the +x direction

Figure 6.5 You and a co-worker maneuver a box on a space station.

$$\vec{F} \bullet \Delta\vec{r} = F_x \Delta x + F_y \Delta y + F_z \Delta z$$

There's no new information in the notation $\vec{F} \bullet \Delta\vec{r}$. It's just shorthand for $F_x \Delta x + F_y \Delta y + F_z \Delta z$. Using this shorthand we can say that the work done by a force acting through a displacement is $\vec{F} \bullet \Delta\vec{r}$.

As we'll discuss in more detail in a moment, work can be positive or negative. If you push or pull in the direction of motion, you do positive work and you increase the energy of the system. If you push or pull in the direction opposite to the motion, you do negative work and you decrease the energy of the system.

■ You and your co-worker pulled on the box for the same time interval but through different displacements in the directions of the forces. Thus the calculation of work is of a different kind than the calculation of impulse, where Δt is the same for both of you.

EXAMPLE

Pull a Block

You pull a block 1.5 m across a table in the $-x$ direction while exerting a force of $\langle -0.3, 0.25, 0 \rangle$ N. How much work do you do on the block?

Solution

$$W = \vec{F} \bullet \Delta\vec{r} = \langle -0.3, 0.25, 0 \rangle \, \text{N} \bullet \langle -1.5, 0, 0 \rangle \, \text{m}$$
$$W = (-0.3 \, \text{N})(-1.5 \, \text{m}) + (0.25 \, \text{N})(0) + 0 \cdot 0 = 0.45 \, \text{J}$$

6.X.5 A paper airplane flies from position $\langle 6, 10, -3 \rangle$ m to $\langle -12, 2, -9 \rangle$ m. The net force acting on it during this flight, due to the Earth and the air, is nearly constant at $\langle -0.03, -0.04, -0.09 \rangle$ N. What is the total work done on the paper airplane by the Earth and the air?

6.X.6 A 2 kg ball rolls off a 30-m-high cliff, and lands 25 m from the base of the cliff. Express the displacement and the gravitational force in terms of vectors and calculate the work done by the gravitational force. Note that the gravitational force is $\langle 0, -mg, 0 \rangle$, where g is a *positive* number (+9.8 N/kg).

Positive and Negative Work

The Energy Principle states that the change in energy of a system (in this case, a single particle) is equal to the energy input from work done by the surroundings.

$$\Delta E_{\text{sys}} = W_{\text{surr}}$$

Since the kinetic energy of a particle might increase or decrease, it must be the case that the work done by the surroundings can be positive or negative.

QUESTION What is the meaning of negative work?

Suppose that you want to slow down a moving object. You push in a direction opposite to the object's motion, and although the object keeps moving in the original direction, it gradually slows down. The object's kinetic energy decreased, so you must have done negative work on the object. Evidently a force acting in a direction opposite to the displacement of a system does negative work.

It is important to determine the correct sign of the work done on a system, because increasing energy is associated with positive work, and decreasing energy is associated with negative work.

Consider four simple situations that bring out the main issues. In Figure 6.6 a box is shown moving 2 meters to the right or left, acted on by a force of magnitude 3 N to the right or left. Since the forces and displacements are all in the $+x$ or $-x$ direction, in each case $W = F_x \Delta x + F_y \Delta y + F_z \Delta z$ reduces to $W = F_x \Delta x$.

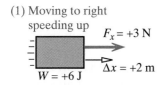

(1) Moving to right
speeding up

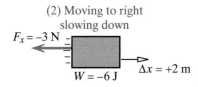

(2) Moving to right
slowing down

(3) Moving to left
speeding up

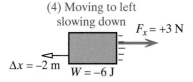

(4) Moving to left
slowing down

Figure 6.6 Sign of work in four cases: force to the right or left, while moving to the right or the left.

Case 1: The box is speeding up so its energy is increasing. We expect the work done on the box to be positive, and we can check the sign:

$$W = F_x \Delta x = (+3\,\text{N})(+2\,\text{m}) = +6\,\text{J}$$

Case 2: The box is slowing down so its energy is decreasing. We expect the work done on the box to be negative, and we can check the sign:

$$W = F_x \Delta x = (-3\,\text{N})(+2\,\text{m}) = -6\,\text{J}$$

Case 3: The box is speeding up so its energy is increasing. We expect the work done on the box to be positive, and we can check the sign:

$$W = F_x \Delta x = (-3\,\text{N})(-2\,\text{m}) = +6\,\text{J}$$

Case 4: The box is slowing down so its energy is decreasing. We expect the work done on the box to be negative, and we can check the sign:

$$W = F_x \Delta x = (+3\,\text{N})(-2\,\text{m}) = -6\,\text{J}$$

It is important to look at the physical situation first, to see whether the energy is increasing or decreasing, which tells you the sign of the work. Then check that the definition $W = F_x \Delta x + F_y \Delta y + F_z \Delta z$ does in fact give the same sign.

Another way to think about the sign of the work is this:

- If the force acts in the direction of the motion, it makes the object speed up, and the work is positive (cases 1 and 3).
- If the force acts opposite to the direction of the motion, it makes the object slow down, and the work is negative (cases 2 and 4).

It's not whether the force points left or right, or up or down, but in which direction it points *relative to the motion*.

A Common Mistake

A common mistake that is easy to make is to try to get the sign of the work merely by looking at the direction of the force, without paying attention to the direction of the displacement. For example, in case 3 the force is to the left but the work is positive, because the displacement is also to the left. Always ask yourself whether the force tends to increase or decrease the energy, and that will tell you what the sign of the work should be.

Do the following exercises and check the answers at the end of the chapter to make sure that you can reliably determine the correct sign of the work, which is an essential skill.

6.X.7

(a) A ball is moving upward, acted on by a downward gravitational force. Is the kinetic energy of the ball increasing or decreasing? Is the work done by the gravitational force positive or negative?

(b) A ball is falling downward, acted on by a downward gravitational force. Is the kinetic energy of the ball increasing or decreasing? Is the work done by the gravitational force positive or negative?

(c) A car is moving to the left, and Superman pushes on it to the right to slow it down, backing up to the left as he pushes to the right. Is the kinetic energy of the car increasing or decreasing? Is the work Superman does positive or negative?

(d) You throw a ball downward. Is the kinetic energy of the ball increasing or decreasing? Do you do positive work or negative work?

6.X.8 A boat is coasting toward a dock you're standing on, and as it comes toward you, you push back on it with a force of 300 N. As you do this, you back up a distance of 2 m. How much work do you do on the boat?

Figure 6.7 A hockey puck slides to the right with low friction on ice, sped up by a hockey stick pushing on the puck.

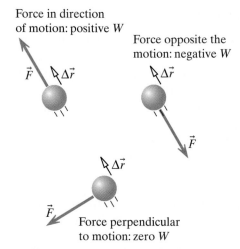

Figure 6.8 Force in direction of motion, opposite the motion, and perpendicular to motion.

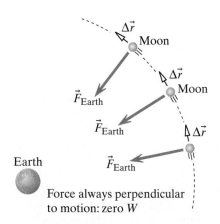

Figure 6.9 The gravitational force of the Earth on the Moon in a (nearly) circular orbit is always (nearly) perpendicular to the motion, so the work is zero.

Zero Work

In Figure 6.7 a puck is sliding with low friction to the right on ice, being sped up by a hockey stick pushing on it. There are three different objects in the surroundings that are interacting with the puck. The hockey stick pushes horizontally, doing positive work to speed up the puck, the Earth pulls down, and the ice pushes up, supporting the puck.

QUESTION How much work is done by the force exerted downward by the Earth? Does this force speed up or slow down the puck? Is the work done by the Earth positive or negative?

The work done by the Earth's gravitational force is zero. It doesn't make the puck speed up or slow down. We can check this:

$$W = F_x\Delta x + F_y\Delta y + F_z\Delta z = (0)\Delta x + (-mg)(0) + (0)(0) = 0$$

A force perpendicular to the motion does zero work.

QUESTION How much work is done by the force exerted upward by the ice? Does this force speed up or slow down the puck? Is the work done by the ice positive or negative?

Again, the work done by the force exerted by the ice is zero. It doesn't make the puck speed up or slow down (remember that we're neglecting any horizontal, frictional component of the force that the ice exerts on the puck).

Sign of Work Depends on Direction of Force and Direction of Motion

There are three important special cases for the sign of work (Figure 6.8):

- A force in direction of motion does positive work.
- A force opposite to the direction of motion does negative work.
- A force perpendicular to the direction of motion does zero work.

A particularly striking case of zero work is circular motion at constant speed, such as the Moon going around the Earth in a (nearly) circular orbit under the influence of the Earth's gravitational force, which always acts perpendicular to the motion if the orbit is a circle (Figure 6.9).

QUESTION Is the Moon speeding up or slowing down? Therefore what can you say about the work done on the Moon by the Earth's gravitational attraction? Is this consistent with the definition of work?

The Moon's speed and kinetic energy aren't changing (and its rest energy isn't changing), so its energy isn't changing. Therefore the work done should be zero. This makes sense, because the force is always perpendicular to the motion.

QUESTION Then what does the Earth's gravitational force do, if it doesn't do any work?

It changes the momentum of the Moon. The momentum is a vector quantity, and change of direction represents a change of the vector momentum. A force is required to change the momentum of the Moon, otherwise it would move in a straight line, not a circle.

Again we see important differences between energy and momentum, and between work and impulse. There is no change in the Moon's energy, since that depends on the *magnitude* of the velocity, but there is a change of the Moon's momentum, since that depends on the *direction* of the velocity.

Another Case of Zero Work

There is another important case of a force doing zero work. Suppose a person crouches down and jumps straight up, and consider the jumper as the system (Figure 6.10). The surroundings are the Earth, the floor, and the air (but we'll neglect the small air resistance force). As the jumper rises, the Earth pulls down, doing negative work. How much work is done by the force of the floor, acting on the jumper's feet? Zero! The force of the floor acts through zero distance and does zero work. The force of the floor contributes to the net force that determines the change in the jumper's momentum, but the floor force does no work. The only significant work done by the surroundings is the negative work done by the force of the Earth (correspondingly, the chemical energy of the jumper decreases).

A similar example is an ice skater who pushes off from a wall. The wall exerts a force on the hands, which changes the skater's momentum, but that force does no work, because at the point of contact there is no displacement. We'll discuss such situations in more detail in Chapter 9.

Figure 6.10 The floor exerts an upward force through zero distance and does zero work.

Work Done by Parallel and Perpendicular Forces

Figure 6.11 shows the force exerted by the Sun on a passing comet, and the components of that force parallel and perpendicular to the comet's momentum. The parallel component of the force does work on the comet, changing the kinetic energy (and changing the magnitude of the momentum). The perpendicular component of the force does zero work but changes the direction of the momentum.

Work Done by a Force at an Angle to the Displacement

Consider the motion of a block sliding to the right, pulled through a small displacement Δx by a string held at an angle θ to the horizontal, as shown in Figure 6.12. We resolve the string force into vertical and horizontal components, as shown in Figure 6.13. We pull at an angle such that the block does not move in the vertical direction.

> QUESTION In calculating the amount of work done to pull the block, should we use the magnitude of the force F, or just the horizontal component of the force $F_x = F \cos \theta$?

Let's use the definition of work:

$$W = F_x \Delta x + F_y \Delta y + F_z \Delta z = F_x \Delta x + F_y(0) + (0)(0) = F_x \Delta x$$

or alternatively

$$W = (F \cos \theta) \Delta x$$

Only the x component of the force, the component in the direction of the motion, does work. Because only the component of a force parallel to the direction of motion does any work, the work done is given by the product of the component of the force parallel to the displacement times the displacement:

$$\text{work} = F_\| \Delta r$$

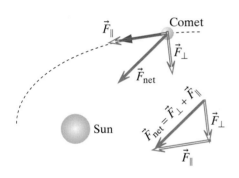

Figure 6.11 The Sun exerts a force on a comet. The parallel component of the force does work, changing the kinetic energy (and the magnitude of the momentum). The perpendicular component does zero work but changes the direction of the momentum.

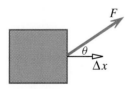

Figure 6.12 Exerting a force of magnitude F at an angle to the motion.

The Dot Product Using Angles

As we saw above, the component $F_\|$ of the force vector $\vec{F}$ in the direction of the displacement $\Delta \vec{r}$ is equal to $F \cos \theta$, where θ is the angle between the two vectors $\vec{F}$ and $\Delta \vec{r}$. The work done by this force is therefore $(F \cos \theta) \Delta r$, or

Figure 6.13 x and y components of the force.

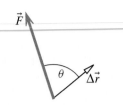

Figure 6.14 θ is the angle between the two vectors.

■ The alternative forms of the dot product offer a way to determine the angle between any two given vectors $\vec{a} = \langle a, b, c \rangle$ and $\vec{q} = \langle q, r, s \rangle$.
$\vec{a} \bullet \vec{q} = (aq + br + cs)$ and also
$\vec{a} \bullet \vec{q} = |\vec{a}||\vec{q}| \cos \theta$
So $\cos \theta = \dfrac{(aq + br + cs)}{|\vec{a}||\vec{q}|}$, from which θ can be determined.

$F\Delta r \cos \theta$. This means that we have another way of evaluating the dot product of the two vectors, as shown in Figure 6.14.

DOT PRODUCT: FORM USING ANGLE θ

$$W = \vec{F} \bullet \Delta \vec{r} = |\vec{F}||\Delta \vec{r}| \cos \theta \quad \text{(a scalar quantity)}$$

Remember that the result of the dot product is a *scalar* quantity, a single number, and not a vector.

QUESTION   Using the definition of the dot product, what is the work done by a force of magnitude F acting through a displacement of magnitude Δr if the angle is $0°$? $90°$? $180°$?

If the force is in the same direction as the displacement ($0°$), the work is positive: $F\Delta r \cos(0°) = F\Delta r$. If the force is perpendicular to the displacement ($90°$), the work is zero: $F\Delta r \cos(90°) = 0$. If the force is opposite to the displacement ($180°$), the work is negative: $F\Delta r \cos(180°) = -F\Delta r$.

Here is a summary of the various equivalent ways to calculate work. Sometimes one of these is more convenient than the others, depending on the situation.

SUMMARY: CALCULATING WORK

$$W = \vec{F} \bullet \Delta \vec{r}$$
$$= F_x \Delta x + F_y \Delta y + F_z \Delta z$$
$$= F_\parallel \Delta r$$
$$= |\vec{F}||\Delta \vec{r}| \cos \theta$$

EXAMPLE   **Throw a Ball Straight Up**

You throw a 250 gram ball straight up, and it rises 8 meters into the air (Figure 6.15). Neglecting air resistance, is the work positive or negative? How much work is done by the gravitational force?

Solution   System: Ball
Surroundings: Earth

Physical reasoning: The ball slows down on its way up, so its kinetic energy is decreasing. Therefore the work must be negative.

Calculate work:

$$F_y = -mg = -(0.25 \, \text{kg})(+9.8 \, \text{N/kg}) = -2.45 \, \text{N}$$
$$W = F_y \Delta y = (-2.45 \, \text{N})(8 \, \text{m}) = -19.6 \, \text{J}$$

Negative work is done on the ball, and the ball slows down, losing some kinetic energy.

Figure 6.15 A ball travels straight up.

6.X.9 A jar of honey with a mass of 0.5 kg is nudged off the kitchen counter and falls 1 meter to the floor. What force acts on the jar during its fall? How much work is done by this force?

6.X.10 (a) About how much mechanical work is done by the Sun on the Earth during six months? (A separate kind of energy transfer is solar radiation, which warms up the Earth.) **(b)** Contrast work and impulse. If the magnitude of the Earth's momentum is P, what is the magnitude of the impulse that the Sun applies to the Earth during six months?

6.4 **UPDATE FORM OF THE ENERGY PRINCIPLE**

■ Remember that in this chapter we're considering situations in which the only kind of energy transfer from the surroundings to the system is mechanical work W.

The Energy Principle

$$\Delta E_{sys} = W_{surr}$$

involves a difference in the energy of a system in its final state and the energy in its initial state. We will refer to this as the "difference form" of the Energy Principle. Since $\Delta E_{sys} = E_{sys,f} - E_{sys,i}$, we can also write the Energy Principle this way:

THE ENERGY PRINCIPLE (UPDATE FORM)

$$E_{sys,f} = E_{sys,i} + W_{surr} \text{ (update form)}$$

Sometimes it's easier to think about the situation using the difference form of the Energy Principle, and sometimes the update form is more appropriate. Use whichever form you're comfortable with, but make sure you understand both forms.

EXAMPLE **Pushing a Box in Space**

You're outside a spacecraft, pushing on a heavy box whose mass is $m = 2000$ kg. You exert a force $\vec{F} = \langle 300, 500, 0 \rangle$ N while the box moves through a displacement $\Delta \vec{r} = \langle 0.1, -0.3, 0.2 \rangle$ m. Initially the box had a speed $|\vec{v}_i| = 0.7$ m/s. **(a)** How much work do you do? **(b)** What is the final kinetic energy of the box? **(c)** What is the final speed of the box? **(d)** What is the direction of the final velocity of the box?

Solution System: Box (considered to be a particle)
Surroundings: You

■ We use the Energy Principle instead of the Momentum Principle because we know the displacement $\Delta \vec{r}$, not the time interval Δt.
In this case the rest energy of the box mc^2 is the same in the initial and final states, so the mc^2 terms cancel. We know this is so because the identity of the box did not change (for example, it did not turn into solid gold).
You did negative work, pushing against the box to slow it down, so the kinetic energy of the box decreased.

Initial state: When you start pushing the box
Final state: When the box has been displaced the amount $\Delta \vec{r}$

Energy Principle:

$$E_f = E_i + W$$
$$(\cancel{mc^2} + K_f) = (\cancel{mc^2} + K_i) + W$$
$$K_f = K_i + W$$
$$K_f = K_i + (F_x \Delta x + F_y \Delta y + F_z \Delta z)$$
$$K_f = K_i + (300\,\text{N})(0.1\,\text{m}) + (500\,\text{N})(-0.3\,\text{m}) + (0\,\text{N})(0.2\,\text{m})$$
$$K_f = K_i + (-120\,\text{J})$$
$$K_f \approx \left(\frac{1}{2}mv_i^2\right) + (-120\,\text{J})$$
$$K_f \approx \frac{1}{2}(2000\,\text{kg})(0.7\,\text{m/s})^2 + (-120\,\text{J})$$
$$K_f \approx (490\,\text{J}) + (-120\,\text{J}) = 370\,\text{J}$$

Final speed:

$$\frac{1}{2}mv_f^2 = K_f$$
$$v_f = \sqrt{\frac{2K_f}{m}}$$
$$v_f = \sqrt{\frac{2(370\,\text{J})}{2000\,\text{kg}}} = 0.608\,\text{m/s}$$

(a) $W = -120\,\text{J}$

(b) $K_f \approx 370\,\text{J}$

(c) $v_f = 0.608\,\text{m/s}$

(d) We don't have enough information to answer this.

Check: The final speed is smaller than the initial speed, corresponding to the work being negative. We slowed down the box. The units also check.

■ It is interesting that we can find the final *speed* of the box without knowing the exact trajectory of the box, or how long it took the box to move from its initial location to its final location. However, because energy is a scalar quantity, we don't learn about the final *direction* of motion.

FURTHER DISCUSSION We answered all the questions except for **(d)**, which asked for the direction of the final velocity of the box. We can't answer that question! The Energy Principle is a scalar principle, and only the magnitude of the velocity appears in the particle energy, not the direction. In order to figure out the direction of the final velocity we would have to use the Momentum Principle, which is a vector principle. However, note that it was not stated how long the process took, so we can't apply the Momentum Principle directly. We would have to determine the approximate Δt from the known displacement and initial speed. (The actual trajectory is a curve; we've approximated the small displacement by a straight line.)

EXAMPLE **Throwing a Ball**

You hold a ball of mass 0.5 kg at rest in your hand, then throw it forward, underhand, so that as it leaves your hand its speed is 20 m/s. How much work did you do on the ball? Note that you do not know how big a force you applied, nor do you know the distance through which you applied this force.

Solution
System: Ball
Surroundings: You (neglect the gravitational effect of the Earth in this short process)

Initial state: At the start of the throw, when the ball's speed is zero
Final state: When the ball just leaves your hand at 20 m/s

Energy Principle:

■ Simple and powerful problem-solving procedure:
■ Write out all terms in the Energy Principle.
■ Identify terms that don't change and therefore cancel.
■ Substitute in all numerical values you know.
■ Solve for the unknown quantity.

$$E_f = E_i + W$$
$$(mc^2 + K_f) = (mc^2 + K_i) + W$$
$$K_f = K_i + W$$
$$\frac{1}{2}mv_f^2 = \frac{1}{2}mv_i^2 + W$$
$$\frac{1}{2}(0.5\,\text{kg})(20\,\text{m/s})^2 = \frac{1}{2}m(0)^2 + W$$
$$100\,\text{J} = W$$

■ No change of identity, so no change in rest energy. mc^2 terms cancel.

Check: Work is positive; this makes sense because the final speed is larger than the initial speed. The units are correct.

■ The energy of the ball increased; it sped up from 0 to 20 m/s. You did positive work, pushing the ball to speed it up.

FURTHER DISCUSSION We were able to calculate the work you did on the ball even though we didn't know how large a force you applied or how large was the displacement of your hand in the throwing motion. This is one aspect of the power of the Energy Principle—that simply by determining the change in energy of a system you know how much energy input there was to that system.

Energy is a conserved quantity: $\Delta E_{system} + \Delta E_{surroundings} = 0$. The energy of the system (the ball) increased by 100 J.

■ Actually, you need to expend more than 100 J of chemical energy (an energy decrease), because in addition to increasing the kinetic energy of the ball by 100 J, you also raised your body temperature a bit (thermal energy increase). Your muscles aren't 100% efficient; it takes more than 100 J of chemical energy to give the ball 100 J of kinetic energy.
The net energy change of the surroundings is −100 J, which could consist (say) of −180 J of chemical energy change and +80 J of thermal energy rise in your body.

QUESTION What was the change in the energy of the surroundings? Can you say what kind of energy it was in the surroundings that changed?

Evidently the change in energy of the surroundings was −100 J. This decrease of energy in the surroundings corresponds to a decrease in the chemical energy stored in your body, coming from food you consumed. In throwing the ball you converted some stored chemical energy into kinetic energy of the ball.

6.X.11 In the example above, suppose that you wanted the final speed of the ball to be twice as big (40 m/s). How much work would you have to do?

EXAMPLE **The Energy Principle at High Speed**

An electron (mass 9×10^{-31} kg) in a particle accelerator is acted on by a constant electric force $\langle 5 \times 10^{-13}, 0, 0 \rangle$ N (this is much greater than the force of gravity, which is only $mg \approx 1 \times 10^{-29}$ N, so we can neglect the effect of the Earth on the electron). The initial velocity of the electron is $\langle 0.995c, 0, 0 \rangle$, where c as usual is the speed of light. The electron moves through a displacement of $\langle 5, 0, 0 \rangle$ m. What is its final speed, expressed as a fraction of the speed of light?

Solution System: Electron
Surroundings: Accelerator (which applies the electric force); neglect the gravitational effect of the Earth in this short process.

Initial state: Beginning of given displacement
Final state: End of given displacement

The unknown quantity, final speed, is part of γ_f. It's easiest to find γ_f first, then solve for v_f.

■ Because the electron's speed is near the speed of light, we can't use an approximate expression for K. We need to use the exact, relativistic equation for particle energy.

To solve for v_f it is necessary to invert both sides of the equation.

Energy Principle:

$$E_f = E_i + W$$
$$\gamma_f mc^2 = \gamma_i mc^2 + W$$
$$\gamma_f mc^2 = \gamma_i mc^2 + (F_x \Delta x + F_y \Delta y + F_z \Delta z)$$
$$\gamma_f mc^2 = \gamma_i mc^2 + (5 \times 10^{-13} \text{N})(5 \text{m}) + (0 \text{N})(0 \text{m}) + (0 \text{N})(0 \text{m})$$
$$\gamma_f mc^2 = \gamma_i mc^2 + (2.5 \times 10^{-12} \text{J})$$
$$\gamma_f mc^2 = \frac{1}{\sqrt{1 - (0.995c/c)^2}} mc^2 + (2.5 \times 10^{-12} \text{J})$$
$$\gamma_f mc^2 = 10.0 mc^2 + (2.5 \times 10^{-12} \text{J})$$
$$\gamma_f = \frac{10.0 mc^2 + (2.5 \times 10^{-12} \text{J})}{mc^2}$$
$$\gamma_f = 10.0 + \frac{2.5 \times 10^{-12} \text{N}}{(9 \times 10^{-31} \text{kg})(3 \times 10^8 \text{m/s})^2} = 40.9$$
$$\sqrt{1 - (v_f/c)^2} = \frac{1}{40.9} = 2.44 \times 10^{-2}$$

$$1 - (v_f/c)^2 = (2.44 \times 10^{-2})^2 = 5.98 \times 10^{-4}$$
$$(v_f/c)^2 = 1 - 5.98 \times 10^{-4} = 0.9994$$
$$v_f/c = \sqrt{0.9994} = 0.9997$$
$$v_f = 0.9997c$$

Check: The final speed is greater than the initial speed, consistent with positive work. The units look right; in particular, the values for γ were dimensionless, as they should be.

FURTHER DISCUSSION   The speed increased only slightly, from $0.995c$ to $0.9997c$, even though the energy changed a lot, from $10.0mc^2$ to $40.9mc^2$. The electron's energy increased by about a factor of 4, but the electron's speed increased very little. When a particle is traveling at nearly the speed of light, adding a lot of energy to the particle doesn't change its speed very much.

> **6.X.12** In the preceding example, at the final speed, $0.9997c$, what was the particle energy as a multiple of the rest energy mc^2? (That is, if it was twice mc^2, write $2mc^2$.) What was the kinetic energy as a multiple of mc^2? Was the kinetic energy large or small compared to the rest energy? At low speeds, is the kinetic energy large or small compared to the rest energy?

6.5   CHANGE OF REST ENERGY

As we saw in the preceding examples, if a particle doesn't change its identity, both the initial and the final energies of the system include mc^2 rest-energy terms and these cancel out in the energy equation:

$$mc^2 + K_f = mc^2 + K_i + W$$
$$K_f = K_i + W$$

This is an important and common special case, but it is not the only case. In some processes the rest mass of a particle can change because the identity of the particle changes.

Neutron decay is a good example to illustrate the issues. A free neutron (one not bound into a nucleus) is unstable, with an average lifetime of about 15 minutes. As shown in Figure 6.16, it decays into a proton, an electron, and a nearly massless antineutrino, which travels at nearly the speed of light. All three of these particles have kinetic energy. That is, they have energy above and beyond their rest energy; the energy of the antineutrino is nearly all kinetic, because it has almost no rest energy. This is a situation with a change of particle identity, and a change in particle rest energy.

$$n \rightarrow p^+ + e^- + \overline{\nu}$$
$$m_n c^2 = (m_p c^2 + K_p) + (m_e c^2 + K_e) + E_\nu$$

Figure 6.16 A stationary neutron decays into a proton, an electron, and a nearly massless antineutrino. This is an example of a process in which there is a change of identity and therefore a change in rest energy.

An Electron Volt (eV) Is a Unit of Energy

The mass of a neutron is 1.6749×10^{-27} kg, so its rest energy mc^2 is about 1.51×10^{-10} J. Because the rest energy of a single particle is such a small number of joules, it is usually expressed in different units of a million electron volts, abbreviated "MeV." One electron volt is the amount of energy an electron acquires when moving through an electric potential difference of 1 volt. (Electric potential difference is introduced in a later chapter; an ordinary flashlight battery has an electric potential difference across it of 1.5 volts.)

DEFINITION OF ELECTRON VOLT AND MeV

$$1 \text{ eV} = 1.6 \times 10^{-19} \text{ J}$$

"eV" is the abbreviation for electron volt. $1 \text{ MeV} = 1 \times 10^6 \text{ eV}$

The rest energy of a neutron is 939.6 MeV, and the rest energy of a proton is 938.3 MeV—slightly less than that of a neutron. An electron is much less massive; its rest energy is 0.511 MeV.

Scientists studying elementary particles often use the unit MeV/c^2 as a unit of mass, because dividing energy by c^2 gives mass. In these units the rest mass of a proton is $938.3 \text{ MeV}/c^2$. The rest mass of a neutron is $939.6 \text{ MeV}/c^2$. The rest mass of the electron is $0.511 \text{ MeV}/c^2$. A neutrino or antineutrino is nearly massless—almost all of its energy is kinetic energy.

EXAMPLE

Neutron Decay

In the decay of a free neutron at rest (Figure 6.16), how much kinetic energy (in MeV) do the decay products (proton, electron, and antineutrino) share? Why can't a proton decay into a neutron?

Solution

System: All the particles, both before and after the decay
Surroundings: Nothing

■ As long as we include all particles in the system, it is okay if some of them change identity during the process.

Initial state: Neutron at rest
Final state: Proton (p), electron (e), and antineutrino ($\bar{v}$) far away from each other

Energy Principle:

■ The Earth can be neglected, since the process is so fast that work done by gravitational forces is negligible.

$$m_v c^2 \approx 0$$

Here the neutron does undergo a change of identity, so there is a change in rest energy, and the mc^2 terms do not cancel.
We found the total kinetic energy of all the particles in the final state. We don't know the separate kinetic energies of the individual particles.

$$E_f = E_i + W$$
$$(m_P c^2 + K_P) + (m_e c^2 + K_e) + (K_{\bar{v}}) = (m_N c^2 + K_N) + W$$
$$(938.3 \text{ MeV} + K_P) + (0.511 \text{ MeV} + K_e) + (K_{\bar{v}}) = (939.6 \text{ MeV} + 0) + 0$$
$$(K_P + K_e + K_{\bar{v}}) + 938.8 \text{ MeV} = 939.6 \text{ MeV}$$
$$(K_P + K_e + K_{\bar{v}}) = (939.6 - 938.8) \text{ MeV}$$
$$(K_P + K_e + K_{\bar{v}}) = 0.8 \text{ MeV}$$

A proton can't decay into a neutron, because the neutron has more rest energy than a proton, and the kinetic energy of the decay products would turn out to be a negative number, which is impossible. Kinetic energy is always positive.

Check: The kinetic energy of the decay products is positive, which it must be for this reaction to be possible. The units check.

FURTHER DISCUSSION   In this example we couldn't cancel the rest energy terms because there was a change of identity. The neutron turned into a lower-mass proton, an electron, and an antineutrino. Because the sum of the rest energies of these three decay particles is less than the rest energy of the parent neutron, there is energy left over that shows up as kinetic energy shared among the three decay particles. How this kinetic energy is divided up among the three particles varies from one neutron decay to the next, governed by probabilities that can be calculated using the science of quantum mechanics and the properties of the "weak interaction" that is responsible for neutron decay.

In a chemical reaction such as carbon combining with oxygen in the fire of a coal-burning power plant, the resulting kinetic energy (which is responsible for heating water to make steam) is on the order of 1 eV per molecule. Nuclear reactions produce vastly more kinetic energy, in the case of neutron decay almost a million times more per nucleus (1 MeV/1 eV is one million). This is why some power plants use nuclear rather than chemical reactions.

> **6.X.13** In the preceding example, what fraction of the original neutron's rest energy was converted into kinetic energy?

Intelligent "Plug and Chug"

Look back at the last four worked-out examples, and you'll see that the same procedure was used to solve each one. Applying the Energy Principle is basically an intelligent sort of "plug and chug" procedure.

- Specify a system.
- Specify surroundings: objects that exert significant forces on the system.
- Specify the initial state.
- Specify the final state.
- Write out the Energy Principle in detail for this system.
- Use the given information to evaluate all the terms you can.
- Solve for the unknown quantity, which may be inside one of the remaining terms.
- Check the units, and check consistency of signs.

The four examples were quite different, but the same procedure worked in each case. The central idea is to let the terms in the Energy Principle tell you what to calculate, and then go ahead and calculate them.

Often students have the impression that physics consists of "knowing what formula to use" for a particular type of problem, with a different formula for every problem. But look again at the last four examples. There weren't four "formulas" giving the four answers. Rather there was one logical procedure that worked in every case, based on the fundamental Energy Principle that applies to all situations.

In following this procedure, you still need to think, but let your thinking be organized by the principle itself. You need to decide whether to try using the Momentum Principle or the Energy Principle, you need to choose a system and initial and final states intelligently, and so on. However, the basic framework never changes.

■ QUESTION How do you decide whether to use the Momentum Principle or the Energy Principle?
Think about what quantities are given, and what you must find:
Work involves a distance (energy), and impulse involves a time (momentum).
Momentum is a vector, and involves information about directions; energy is a scalar, and has no directional information.

■ You should never find yourself saying, "I don't have any idea even where to start on this problem!"
If you do find yourself saying this, take a deep breath, choose a system, identify the objects in the surroundings that exert forces on the system, write down the Energy Principle, choose initial and final states, and flesh out the various terms in the Energy Principle for the particular situation at hand, then solve for the unknown quantities. Physics can be easier than most people think, if you go with the flow. If on the other hand you try to invent a new and different approach for every new problem you encounter, you'll waste huge amounts of time and get very frustrated.

6.6   PROOF OF THE ENERGY PRINCIPLE FOR A PARTICLE

We have asserted that the following is true for a particle:

$$\Delta E = W_{\text{surr}}$$

To prove this, we need to show that $\Delta E = W_{\text{surr}}$ is true given our definitions of particle energy and work. Consider a single pointlike particle of mass m. For convenience in the reasoning we are about to do, choose coordinate axes so that the x axis is in the direction of the motion at this instant, so the displacement in the next short time interval is simply Δx (Figure 6.17). There is a net force $\vec{F}$ (magnitude F) acting at an angle θ to the displacement. The physical results we will obtain do not depend on our choice of coordinate axes; we are free to choose axes that simplify our calculations.

The component of the net force that is parallel to the displacement, F_x, affects the speed and hence the kinetic energy of the particle. The perpendicular component F_y merely changes the direction of the momentum.

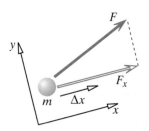

Figure 6.17 A particle of mass m. We choose the x axis to be in the direction of the motion. The net force acting on the particle has a component F_x in the direction of motion.

In the very short displacement Δx, the energy of the particle E_{particle} changes by a small amount, and this is supposed to be equal to the work done on the particle:

$$\Delta E = F_x \Delta x$$

According to the Momentum Principle $F_x = \dfrac{\Delta p_x}{\Delta t}$, so

$$\Delta E = \left(\frac{\Delta p_x}{\Delta t}\right) \Delta x$$

Dividing by Δx, we have this:

$$\frac{\Delta E}{\Delta x} = \frac{\Delta p_x}{\Delta t}$$

In the limit as $\Delta t \to 0$, this equation relates the spatial rate of change of energy to the time rate of change of momentum:

$$\frac{dE}{dx} = \frac{dp_x}{dt}$$

■ This equation reflects the fact that a change in position (Δx or dx) is associated with a change in energy, whereas a change in time (Δt or dt) is associated with a change in momentum.

In words, this says that the change in energy of the particle per unit distance (in the direction of the motion) is equal to the change in the particle's momentum in that direction per unit time. If we can show that this equation is true if $E = \gamma mc^2$, we will have proven that $\Delta E = W$ is correct.

Proof for Low Speeds

In the special case that the speed of a particle is small compared to the speed of light, the proof is simple. We need to show that if particle energy is given by the approximate equation:

$$E \approx mc^2 + \frac{1}{2}mv^2 \quad (v \ll c)$$

then the equation derived above:

$$\frac{dE}{dx} = \frac{dp_x}{dt}$$

is satisfied.

We chose the x axis to be in the direction of the motion, so the particle energy is:

$$E \approx mc^2 + \frac{1}{2}mv_x^2 \quad (v \ll c)$$

Left-hand side of equation:

$$\frac{dE}{dx} = \frac{d}{dx}\left(\frac{1}{2}mv_x^2\right)$$

$$\frac{dE}{dx} = \frac{1}{2}m\frac{d}{dx}(v_x^2) = mv_x\frac{dv_x}{dx} = m\frac{dx}{dt}\frac{dv_x}{dx} = m\frac{dv_x}{dt}$$

Right-hand side of equation:

■ Since mc^2 is constant, $\dfrac{d}{dx}(mc^2) = 0$.

$$\frac{dp_x}{dt} = \frac{d}{dt}(mv_x) = m\frac{dv_x}{dt} \quad (v \ll c)$$

So

$$\frac{dE}{dx} = \frac{dp_x}{dt}$$

is indeed true with our definitions of energy and work. We have proved that at least at low speed the Energy Principle for a particle $\Delta E = W$ is correct.

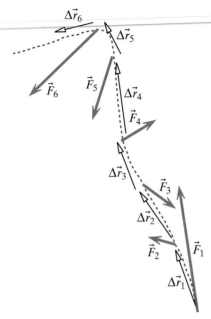

Figure 6.18 A varying force is exerted on an object moving along a curved path.

The proof that the relativistically correct expression $E = \gamma mc^2$ satisfies the relation

$$\frac{dE}{dx} = \frac{dp_x}{dt}$$

is given in Section 6.22, at the end of the chapter. It follows the same reasoning as was used to prove the low-speed case but is algebraically more complex.

6.7   WORK DONE BY A NONCONSTANT FORCE

If a force changes in magnitude or direction during a process, we can't calculate the work simply by multiplying a constant force times the net displacement. The nonconstant force acts on the object along a path, and we split the path into small increments $\Delta \vec{r}$ (Figure 6.18). This is similar to the way in which we took small time steps Δt when computing the trajectories of objects.

If our increments along the path are sufficiently small, the force is approximately constant in magnitude and direction within one increment. We can write the total work done as a sum along the path:

$$W = \vec{F}_1 \bullet \Delta \vec{r}_1 + \vec{F}_2 \bullet \Delta \vec{r}_2 + \vec{F}_3 \bullet \Delta \vec{r}_3 + \cdots \text{(small steps)}$$

More compactly, we can use the symbol Σ (Greek capital sigma) to mean "sum":

$$W = \Sigma \vec{F} \bullet \Delta \vec{r}$$

> **6.X.14** You push a heavy crate out of a carpeted room and down a hallway with a waxed linoleum floor. While pushing the crate 2.3 m out of the room you exert a force of 30 N; while pushing it 8 m down the hallway you exert a force of 15 N. How much work do you do in all?
>
> **6.X.15** An electron traveling through a curving wire in an electric circuit experiences a constant force of 5×10^{-19} N, always in the direction of its motion through the wire. How much work is done on the electron by this force as it travels through 0.5 m of the wire?

We have refined our definition of work to allow us to calculate the work done by varying forces. This more complete definition of work allows us to apply the Energy Principle in situations where forces vary.

Work as an Integral

If we go to the limit where the increments are infinitesimal $(\Delta \vec{r} \to d\vec{r})$, the sum used for calculating work turns into a sum of an infinite number of infinitesimal contributions. Such a summation is called a "definite integral." It is written as a distorted "S" for "sum," with an indication of the initial position "i" and the final position "f":

$$W = \int_i^f \vec{F} \bullet d\vec{r}$$

If you have already studied integration in a calculus course, you already know some techniques for evaluating definite integrals such as this; if not, you will learn them soon in calculus. However, in many real-world cases it is not possible to find an analytical form for evaluating an integral, in which case one approximates the infinite sum by a finite sum, with finite increments $\Delta \vec{r}$.

This procedure is called "numerical integration." The iterative calculations you have done with the Momentum Principle, whether by hand or on a computer, have been numerical integration calculations.

GENERAL DEFINITION OF WORK

$$W \equiv \int_i^f \vec{F} \bullet d\vec{r} \approx \Sigma \vec{F} \bullet \Delta \vec{r}$$

This general definition applies to any force, whether constant or not.

EXAMPLE   **Work Done by a Spring**

A ball traveling horizontally runs into a horizontal spring, whose stiffness is 100 N/m (Figure 6.19). As the spring is compressed, the ball slows down. If the relaxed length of the spring is 30 cm, and the spring is compressed until its length is 10 cm, how much work does the spring do on the ball?

Solution   Initial state: $s = 0$

Final state: $s = (0.10 - 0.30)\,\mathrm{m} = -0.2\,\mathrm{m}$

Choose the origin at the end of the relaxed spring, so $F_x = -k_s x$.

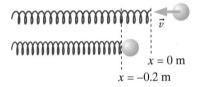

$x = 0$ m
$x = -0.2$ m

Figure 6.19 A ball runs into a horizontal spring.

$$W = \int_0^{-0.2} (-100\,\mathrm{N/m})x\, dx$$

$$W = (-100\,\mathrm{N/m})\frac{1}{2}x^2 \Big|_{0\,\mathrm{m}}^{-0.2\,\mathrm{m}}$$

$$W = (-100\,\mathrm{N/m})\left(\frac{1}{2}(-0.2)^2 - 0\right) = -2\,\mathrm{N} \cdot \mathrm{m}$$

The spring does -2 J of work on the ball.

Check: Sign ok—negative work decreases kinetic energy of ball. Units ok ($\mathrm{N} \cdot \mathrm{m} = \mathrm{J}$).

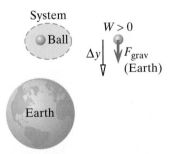

System
Ball
$W > 0$
Δy F_{grav} (Earth)
Earth

Figure 6.20 If the ball alone is chosen as the system, the Earth is part of the surroundings and does positive work on the ball, increasing the kinetic energy of the ball.

6.8   POTENTIAL ENERGY IN MULTIPARTICLE SYSTEMS

Up to this point, our discussions of energy have been confined to systems consisting of a single particle. However, we saw in the case of momentum that it was often productive to choose a system containing two or more objects. We'll look at the effect of choice of system on the energy analysis of a simple situation: a ball falls from rest.

System: Ball Only

If the ball alone is the system, the Earth is part of the surroundings (Figure 6.20). The kinetic energy of the system (ball) increases, due to positive work done by the Earth on the system. The gravitational force acts in a direction in the same direction as the displacement of the ball, so the work done by the Earth on the ball is positive.

Ball at rest

y_i

y_f $\vec{v}$

Initial state   Final state

Figure 6.21 A ball falls from rest, and at a lower location its kinetic energy has increased.

| EXAMPLE | **Ball And Earth** |

In Figure 6.21 a ball of mass 100 g is released from rest 7 m above the ground. When the ball has fallen to 4 m above the ground, what is its kinetic energy K_f?

Solution

System: Ball
Surroundings: Earth

Energy Principle:

$$\Delta(m_{ball}c^2) + \Delta K_{ball} = W_{by\ Earth}$$
$$W_{by\ Earth} = F_y \Delta y = -mg\Delta y$$
$$W_{by\ Earth} = -(0.1\,kg)(9.8\,N/kg)(-3\,m) = 2.94\,J$$
$$K_f - 0 = W_{by\ Earth}$$
$$K_f = 2.94\,J$$

FURTHER DISCUSSION   The Earth did an amount of work +2.94 J on the ball, and the kinetic energy of the ball increased from zero to 2.94 J.

System: Ball + Earth

If the ball plus the Earth are the system, there is nothing significant in the surroundings (Figure 6.22). As before, we see that the kinetic energy of the system increases. However, there is a problem: the surroundings did no work on the system! The change in kinetic energy of the system is not equal to the work done by the surroundings.

Before deciding that the Energy Principle has been violated, we should consider the possibility that we have overlooked a kind of energy that is present in systems containing more than one interacting object. In fact, this is the case.

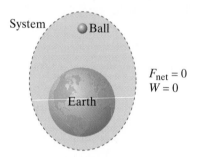

Figure 6.22 If the ball + the Earth are chosen as the system, there are no significant objects in the surroundings. No work is done on the system, but the kinetic energy of the system increases anyway. The increase in kinetic energy is accompanied by a decrease in the interaction energy (potential energy) of the ball + Earth system.

Potential Energy Belongs to Pairs of Interacting Objects

There is a kind of energy associated with pairs of interacting objects inside a system, such as the ball and the Earth. This interaction energy is called "potential energy." A change in potential energy is associated with a change in the separation of interacting objects. As the ball and the Earth get closer together, the kinetic energy of the system increases, but we will see that the potential energy (interaction energy) decreases. The net change in the energy of the system is in fact zero, consistent with the fact that no work was done on the system by the surroundings.

In any system containing two or more interacting particles, such as compressed or stretched springs, galaxies of stars interacting gravitationally, or atoms in which the protons and electrons interact electrically, there is energy associated with the interactions between pairs of particles inside the system. This interaction energy is not the same as the rest energies or kinetic energies of the individual particles. Changes in interaction energy are associated with changes in the shape of the multiparticle system, such as the ball and Earth moving closer together, or stretching or compressing a spring, or moving an electron away from a proton.

When we take the ball + Earth as the system, the force acting on the ball is exerted by another object inside the system. We call such a force an "internal force," which does "internal work." A simple rearrangement of the Energy Principle for the system of the ball alone leads to a useful way to account for

■ By the reciprocity of gravitational forces, the ball exerts an upward force mg on the Earth. In the same time interval Δt the Earth gets as much magnitude of momentum p as the ball, but its speed $v = p/M_{Earth}$ is extremely small, so $\Delta y = v\Delta t$ for the Earth is extremely small. Therefore the work done on the Earth, $W = F_y\Delta y$, is extremely small, and the Earth gets very little kinetic energy.

Another way to see this is to write $K = p^2/(2M)$; the same p but much larger M means that the Earth gets negligible kinetic energy.

energy in the ball + Earth system:

$$\Delta K_{ball} = W_{by\ Earth}$$

Rewrite this by moving the work to the left side of the equation:

$$\Delta K_{ball} + (-W_{by\ Earth}) = 0$$

We define "the change of potential energy ΔU" to be the negative of the internal work: $\Delta U = -W_{by\ Earth}$. For the system of the ball + Earth, $\Delta U = -(F_y \Delta y) = -(-mg)\Delta y = \Delta(mgy)$. Actually, because we approximated the gravitational force by a constant mg, this is an approximate result valid only for height changes near the Earth's surface that are small compared to the radius of the Earth.

ΔU for Object + Earth Near Surface of Earth

$$\Delta U_{grav} \approx \Delta(mgy)$$

for object + Earth near the surface of the Earth

With this definition we can write the Energy Principle for the ball + Earth system as $\Delta K_{object} + \Delta U = 0$. This has the advantage of putting on the left side of the equation quantities associated with the system (kinetic energy of the ball, interaction energy of the ball + Earth), and leaving on the right side of the equation only those quantities associated with the surroundings (in this case, nothing).

EXAMPLE   **Ball and Earth**

In Figure 6.21 a ball of mass 100 g is 7 m above the ground, initially at rest ($K_i = 0$). When the ball is 4 m above the ground, what is its kinetic energy K_f? Choose the ball + Earth as the system.

Solution   System: Ball + Earth
Surroundings: Nothing significant

Energy Principle:

$$\Delta(m_{Earth}c^2) + \Delta K_{Earth} + \Delta(m_{ball}c^2) + \Delta K_{ball} + \Delta U = 0$$
$$(K_f - 0) + \Delta(mgy) = 0$$
$$(K_f + 0.1\,kg)(9.8\,N/kg)(-3\,m) = 0$$
$$K_f + (-2.94\,J) = 0$$
$$K_f = 2.94\,J$$

FURTHER DISCUSSION   When we chose the ball alone as the system, the Earth was in the surroundings and did +2.94 J of work on the ball. When we choose the ball + Earth as the system, no work is done on the system, but the potential energy of the ball–Earth pair of interacting objects decreases; $\Delta U = -2.94$ J. We of course get the same result for the final kinetic energy of the ball no matter which system we choose, but the viewpoint is rather different in the two cases.

Interaction Energy in a Multiparticle System

Having illustrated the basic idea of potential energy in a situation where the internal force was constant (mg), we now offer a more general analysis. Consider a system of three particles that exert forces on each other, and that are

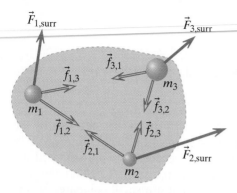

Figure 6.23 A system consisting of three particles. They exert "internal" forces on each other (lower case f's), and additional "external" forces (black arrows, upper case F's) are exerted by objects in the surroundings, which are not shown.

also acted upon by objects in the surroundings, outside this system, as shown in Figure 6.23. For example, these could be three charged particles that exert electric forces on each other but also have springs connecting them to objects in the surroundings. The forces that are exerted by other particles in the system are called "internal" forces, and are denoted here by lowercase "f's." The forces exerted by objects in the surroundings are called "external" forces, and are denoted here by uppercase "F's."

The work done by these internal and external forces on each particle changes the particle energy $E_{particle}$ of that particle. Using the symbol W_{int} for work done by internal forces, and W_{surr} for work done by external forces, we can write the Energy Principle for each particle:

$$\Delta E_1 = (\vec{f}_{1,2} + \vec{f}_{1,3} + \vec{F}_{1,surr}) \bullet \Delta \vec{r}_1 = W_{1,int} + W_{1,surr}$$
$$\Delta E_2 = (\vec{f}_{2,1} + \vec{f}_{2,3} + \vec{F}_{2,surr}) \bullet \Delta \vec{r}_2 = W_{2,int} + W_{2,surr}$$
$$\Delta E_3 = (\vec{f}_{3,1} + \vec{f}_{3,2} + \vec{F}_{3,surr}) \bullet \Delta \vec{r}_3 = W_{3,int} + W_{3,surr}$$

Adding up the changes in the particle energy for each of the three particles (the left sides of the equations) and the work done on each particle (the right sides of the equations), we can write this:

$$\Delta(E_1 + E_2 + E_3) = W_{int} + W_{surr}$$

where W_{int} is the sum of the work done by internal forces on each of the particles, and W_{surr} is the sum of the work done by external forces on each of the particles. Added together, these two terms equal the sum of the changes in the individual particle energies. Note that the work done by each individual external force involves the displacement of that force (equivalently, the displacement of the particle on which that particular force acts).

Separating Internal and External Energy Terms

As we saw with the falling ball, it can be useful to have all terms relating to energy changes internal to the system on one side of the equation and terms relating to objects in the surroundings on the other side. As we did with the falling ball, we can do this by making a simple algebraic rearrangement, moving the work done on the particles by the internal forces, W_{int}, to the other side of the equation:

$$\Delta(E_1 + E_2 + E_3) + (-W_{int}) = W_{surr}$$

We relabel the term $(-W_{int})$ as a change of potential energy U, energy associated with the interactions of the particles with each other.

DEFINITION OF CHANGE IN POTENTIAL ENERGY

$$\Delta U \equiv -W_{int}$$

The change in potential energy (ΔU) of a system is defined to be the negative of the work done by forces internal to the system.

The left side of the Energy Principle equation for a multiparticle system now represents all changes in the energy of the system, and the right side represents energy inputs from the surroundings (work). It is customary to drop the subscript and write W_{surr} simply as W, with the understanding that the work W is mechanical energy transfer from the surroundings to the system, with a resulting change in the energy of the system.

Because the "internal work" term involves the forces exerted on each particle by each other particle in the system, the potential energy U is really the sum of the interaction energies of all pairs of particles in the system:

$$U = U_{12} + U_{23} + U_{31}$$

THE MULTIPARTICLE ENERGY PRINCIPLE

$$\Delta(E_1 + E_2 + E_3 + \ldots) + \Delta(U_{12} + U_{13} + U_{23} + \ldots) = W$$

$E_1 = m_1c^2 + K_1$ is the particle energy of object 1, etc.

Multiparticle Systems vs. Single-Particle Systems

The Energy Principle is the same for a multiparticle system as it is for a single-particle system. To the left of the equal sign are quantities that are internal to the system, while to the right of the equal sign are the external effects on the system due to objects in the surroundings. The interpretation is, "if the surroundings do work on a system, the energy of the system changes" (Figure 6.24).

$$\Delta E_{\text{sys}} = W$$

The difference is in the kinds of energy found inside each kind of system:

- A single-particle system has only particle energy (rest energy and kinetic energy).
- A multiparticle system has particle energy and potential energy (interaction energy).

Work done by the surroundings can affect both particle energy and potential energy inside a multiparticle system.

You get to choose which system you want to analyze, but once you have specified the portion of the Universe you're considering to be the system of interest, you must be consistent about what energy flows into or out of the system (work done by the surroundings), versus what energy changes occur inside your system (possible changes in particle energies or potential energy).

The Experimental Basis for the Energy Principle for Complex Systems

We were able to make a mathematical proof of the Energy Principle for a single particle, starting from the Momentum Principle. For a complex, multiparticle system, the more general Energy Principle ultimately is based on over a hundred years of experiments starting in the early 1800s. Slowly scientists learned how to identify flows of energy in many different kinds of systems, and once the energy was accounted for, they found that conservation of energy, $\Delta E_{\text{sys}} + \Delta E_{\text{surr}} = 0$, was always true, not only in physics but also in chemistry and biology.

Invisible Particles: Neutrinos

A striking example of this occurred in the early part of the 20th century, when scientists found that apparently energy was not conserved in nuclear reactions called "beta decays," in which certain radioactive nuclei spontaneously emit

Figure 6.24 Change of system energy is equal to work. Work can be either positive or negative.

■ Recently neutrinos have been in the scientific news again. For many years it was thought that neutrinos had only kinetic energy, with zero rest energy (and zero rest mass). However, recent careful experiments have verified theoretical predictions that neutrinos should have a very tiny rest mass.

an electron (early in the history of atomic physics, electrons were called "beta rays"). At the same time a neutron in the nucleus changes into a proton (similar to the decay of a free neutron outside a nucleus). It was observed that the emitted electrons had a wide range of energies, when only one specific energy was expected, corresponding to the change in mc^2 of the nucleus. The Swiss physicist Wolfgang Pauli made the bold proposal that an unseen particle, later named the "neutrino," was also emitted, with energy shared between it and the electron.

Although at the time some scientists wondered whether perhaps energy conservation didn't apply to the new science of nuclear physics, Pauli was so convinced of the universal correctness of energy conservation that he was willing to postulate an unseen particle rather than discard energy conservation. It took many years before experiments were sensitive enough to detect the elusive neutrinos, but eventually Pauli's analysis was confirmed.

> **6.X.16** Consider an isolated system, by which we mean a system free of external forces. If the sum of the particle energies in the system increases by 50 joules, what must have been the change ΔU in the system's potential energy?
>
> **6.X.17** If a system contains four particles, how many potential energy pairs U_{12} etc. are there? List them.

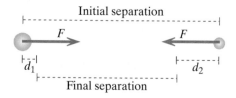

Figure 6.25 Two stars move toward each other during a short time interval. The internal work for the two-star system is F times the change in the separation distance of this pair of objects. The displacements are shown overly large for clarity.

6.9   GRAVITATIONAL POTENTIAL ENERGY

We found that for a system of a ball falling near the surface of the Earth, where the force of the Earth was a constant mg, the change of gravitational potential energy for the ball + Earth system was $\Delta U \approx \Delta(mgy)$. Next we will find a formula for gravitational potential energy in the more general case where the force depends on distance, $\vec{F}_{grav} = -(GMm/r^2)\hat{r}$.

A big star and a small star are initially at rest (Figure 6.25). They start to move toward each other along the x axis due to the gravitational forces they exert on each other, with the same magnitude F (reciprocity). In a short time the big star moves a distance d_1 and the small star moves a distance d_2 (the displacements are shown overly large for clarity). The increase in the magnitude of the momentum of each star is the same, because the magnitude of the impulse $F\Delta t$ is the same, so the small star reaches a higher speed ($v \approx p/m$) and moves farther: $d_2 > d_1$, as indicated in Figure 6.25. Therefore more work is done on the small star.

The impulse (and momentum change) have the same magnitude for both stars because both forces act for the same time interval Δt. However, the work done is different because the two forces act through different distances. This is an important difference between impulse and work, and therefore between the Momentum Principle and the Energy Principle.

Consider the system consisting of both stars. The internal work W_{int} is the work done by forces internal to the system (not exerted by objects in the surroundings):

$$W_{int} = Fd_1 + Fd_2 = F(d_1 + d_2)$$

Change of potential energy is defined as the negative of the internal work:

$$\Delta U = -W_{int} = -F(d_1 + d_2)$$

Since ΔU for the two stars is negative, the potential energy decreased, which would imply that the kinetic energy increased, which is indeed what happened.

Initial separation

Final separation

Figure 6.26 We can calculate internal work as though one of the stars were stationary and the other moved.

Figure 6.27 Two stars whose relative positions are changing (in this case, the stars move closer together).

■ According to our definition of $\vec{r}$,

$$\vec{r}_{2i} - \vec{r}_{1i} = \vec{r}_i$$

$$\vec{r}_{2f} - \vec{r}_{1f} = \vec{r}_f$$

■ For every pair of interacting particles in a system there is a potential energy term (U_{12}, U_{23}, etc.). Change of potential energy is associated with change of distance between pairs of particles.

Change in Separation of Two Objects

The stars got closer to each other; their separation decreased by an amount $(d_1 + d_2)$. We can calculate internal work by multiplying the force one object exerts on another by the change in the separation between the two objects. The individual positions of the objects don't matter. The way the algebra works out, the internal work is calculated as though a single force F acted through a distance equal to the change in the separation of the pair, as in Figure 6.26.

Next we'll prove this more generally. Again consider a system of two interacting stars. We define the relative position vector $\vec{r}$ to be the difference between the positions of two stars, as shown in Figure 6.27. When the stars move a little closer together, star 1 does work on star 2, and star 2 does work on star 1. Since both stars are inside the system, the change in gravitational potential energy of the system is the negative of the sum of these work terms. Assume that the changes in position are small enough that we can consider the forces to be constant:

$$\Delta U = -W_{\text{int}} = -(\vec{f}_{1,2} \bullet \Delta \vec{r}_1 + \vec{f}_{2,1} \bullet \Delta \vec{r}_2)$$

$$\vec{f}_{1,2} = -\vec{f}_{2,1}, \text{ so } \Delta U = -\vec{f}_{2,1} \bullet (\Delta \vec{r}_2 - \Delta \vec{r}_1)$$

The term $(\Delta \vec{r}_2 - \Delta \vec{r}_1)$ involves the change of position of each star (Figure 6.27). We can rewrite this in terms of $\Delta \vec{r}$, the change in relative position of the two stars:

$$\Delta \vec{r}_2 - \Delta \vec{r}_1 = (\vec{r}_{2f} - \vec{r}_{2i}) - (\vec{r}_{1f} - \vec{r}_{1i})$$
$$= (\vec{r}_{2f} - \vec{r}_{1f}) - (\vec{r}_{2i} - \vec{r}_{1i})$$
$$= \vec{r}_f - \vec{r}_i$$
$$= \Delta \vec{r}$$

Using this result we can rewrite the change in potential energy:

$$\Delta U = -\vec{f}_{2,1} \bullet \Delta \vec{r}$$

$\vec{r}$ is the position of star 2 relative to star 1. Therefore we can calculate the amount of change of potential energy, ΔU, in terms of the *relative displacements* or separation between *pairs of particles*, and any change of potential energy is associated with a change of "shape," with pairs of particles getting closer together or farther apart. The algebra works out in such a way that the force we use in calculating ΔU is just one of the pairs of forces, not both. The result is the same if the interaction is an electric interaction instead of a gravitational interaction, since the reasoning above depended only on reciprocity of forces, and not on any other details of the force.

We write U_{12} for the potential energy of the pair consisting of particles 1 and 2, U_{23} for pair 2 and 3, and so on. If a system consists of two particles, there are three energy terms: two particle energies E_1 and E_2, and one potential energy term, U_{12}, since there is only one pair of particles. (Of course the particle energies can be further subdivided into rest energy plus kinetic energy.)

If a system consists of three particles, there are six energy terms: three particle energies E_1, E_2, and E_3, and three potential energy terms, U_{12}, U_{23}, and U_{31}, since there are three pairs of particles.

Force Is the Negative Gradient of U

Since it is the relative displacement that matters, even if both objects move we can pretend that object 1 remains stationary, and just calculate the work done on object 2. The force is in a line with $\vec{r}$, so we can evaluate the dot product and write the following, where F_r is the component of the gravitational force

Figure 6.28 The r component of the gravitational force, F_r.

Gradient = 0.07 = 7%

7 m

100 m

Figure 6.29 The gradient is the rate of change of y with respect to x.

on object 2 in the direction of $\vec{r}$, and r is the distance from object 1 to object 2 (Figure 6.28):

$$dU = -F_r\,dr$$

Therefore if we already knew the formula for potential energy, we could calculate the associated force like this: $F_r = -dU/dr$. A rate of change of a quantity with respect to position such as this is called a "gradient." For example, if a hill is said to have "a 7% grade," this means that it rises 7 m vertically for every 100 meters horizontally—the tangent of the angle of the hill (the slope) is 0.07 (Figure 6.29).

FORCE IS THE NEGATIVE GRADIENT OF POTENTIAL ENERGY

$$F_r = -\frac{dU}{dr}$$

All that remains to be done to find a formula for gravitational energy is to think of an expression whose (negative) derivative is the component of the gravitational force acting on m_2 in the direction of $\vec{r}$. From Figure 6.28 we see that this component of the gravitational force is negative: $F_r = -G(m_1 m_2/r^2)$, where r is the center-to-center distance. The minus sign reflects the fact that the attractive gravitational force on object 2 points back toward object 1.

QUESTION   Can you think of a function of r whose (negative) derivative with respect to r has this value of F_r? (If necessary, review the appendix on basic calculus.)

The derivative of r^n with respect to r is nr^{n-1}. We have a r^{-2} factor in the gravitational force, so $n = -1$, and the "antiderivative" of r^{-2} (the function whose derivative gives r^{-2}) must be $-r^{-1}$.

QUESTION   Therefore, what is the formula for gravitational energy?

Being careful about the signs, we find that $U = -Gm_1 m_2/r$.

QUESTION   Check that the negative gradient of this gravitational energy is indeed the force, by differentiating with respect to r.

What about the minus sign on U? Is it correct? Yes, because the potential energy decreases as the particles get closer together. When stars fall toward each other, their kinetic energies increase and the pair-wise potential energies must decrease to more and more negative values. In summary,

---

GRAVITATIONAL POTENTIAL ENERGY

$$U = -G\frac{m_1 m_2}{r}$$

r is the center-to-center separation of m_1 and m_2.

QUESTION   Could there be an additional constant in this equation?

It is true that the gradient of $U = (-Gm_1 m_2/r + \text{constant})$ would also give us the correct gravitational force, because the derivative of a constant is zero. However, as we will soon discuss in more detail, when the objects are very far apart the total energy of the system must be equal just to the sum of the particle energies, and the potential energy must be zero.

QUESTION When the particle separation r is very large, what is the value of $-Gm_1m_2/r$?

When r is very large, $-Gm_1m_2/r$ is nearly zero. Therefore, for the potential energy to be zero when the separation is very large, the value of the constant in $U = (-Gm_1m_2/r + \text{constant})$ must be zero, in order for U to be zero at large separations. The pair-wise gravitational potential energy must be simply $U = -Gm_1m_2/r$, with no added constant.

> **6.X.18** Calculate the gravitational potential energy of the interacting pair of the Earth and a 1 kg block sitting on the surface of the Earth. You would need to supply the absolute value of this result to move the block to a location very far from the Earth (actually, you would need to use even more energy than this due to the gravitational potential energy associated with the Sun–block interacting pair).

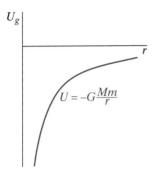

Figure 6.30 Gravitational potential energy is negative. It increases with increasing separation r.

$U = -G\dfrac{Mm}{r}$

The Minus Sign

It may seem odd that the expression for gravitational potential energy has a minus sign in it. Figure 6.30 shows that although U is negative, it does increase with increasing separation r, which means that you have to do work to move two gravitationally attracting objects farther apart.

Here are two little exercises that will help you see why the minus sign is appropriate:

> **6.X.19** If a meteor falls toward the Earth, the kinetic energy increases and the gravitational potential energy (which is proportional to $-1/r$) must decrease. Suppose that the separation r decreases from 1×10^8 m to 1×10^7 m. What is $(-1/r_f) - (-1/r_i)$? Is this change negative, as expected?
>
> **6.X.20** If a spacecraft coasts away from the Earth, the kinetic energy decreases and the gravitational potential energy (which is proportional to $-1/r$) must increase. Suppose that the separation r increases from 1×10^7 m to 1×10^8 m. What is $(-1/r_f) - (-1/r_i)$? Is this change positive, as expected?

EXAMPLE

Figure 6.31 Initial and final states.

A Robot Spacecraft Leaves an Asteroid

A robot spacecraft lands on an asteroid, picks up a sample, and blasts off to return to Earth; its total mass is 1500 kg. When it is 200 km (2×10^5 m) from the center of the asteroid, its speed is 5.0 m/s, and the rockets are turned off. At the moment when it has coasted to a distance 500 km (5×10^5 m) from the center of the asteroid, its speed has decreased to 4.1 m/s. You can use these data to determine the mass of the asteroid. Before starting your analysis, draw a diagram of the initial and final states you will consider in your analysis. Label your diagram to show the distances involved. This will help avoid mistakes in your calculation.

(a) Draw and label the diagram.

(b) Calculate the mass of the asteroid. Make an accurate calculation, not a rough, approximate calculation.

Solution **(a)** Figure 6.31 shows the situation. It is important to keep in mind that r is measured center to center in $U = -Gm_1m_2/r$.

System: Asteroid and spacecraft
Surroundings: Nothing significant
Initial state: Rockets off, 5.0 m/s, 200 km from center of asteroid

■ By choosing both the asteroid and the spacecraft to be in the system, we won't have to calculate work, since there are no significant external forces.

No change in rest energies. No work by surroundings (no interacting objects in surroundings).

As discussed below, the kinetic energy of the massive asteroid changes very little: $(K_{M,f} \approx K_{m,i})$.

$$v \ll c \text{ so } K \approx \frac{1}{2}mv^2$$

In many cases it is useful to solve the problem algebraically before putting in numbers, as is done here, because it reduces the number of calculations and makes it easier to track down algebra errors. Note that m canceled; we didn't need to know or use the value of 1500 kg for the mass.

Final state: Speed 4.1 m/s, 500 km from center of asteroid

Energy Principle: $E_f = E_i + W$

$$\cancel{Mc^2} + K_{M,f} + \cancel{mc^2} + K_{m,f} + U_f = \cancel{Mc^2} + K_{M,i} + \cancel{mc^2} + K_{m,i} + U_i + \cancel{W}$$

$$\cancel{K_{M,f}} + K_{m,f} + U_f = \cancel{K_{M,i}} + K_{m,i} + U_i$$

$$\frac{1}{2}\cancel{m}v_f^2 + \left(-G\frac{M\cancel{m}}{r_f}\right) = \frac{1}{2}\cancel{m}v_i^2 + \left(-G\frac{M\cancel{m}}{r_i}\right)$$

$$\frac{1}{2}v_f^2 + \left(-G\frac{M}{r_f}\right) = \frac{1}{2}v_i^2 + \left(-G\frac{M}{r_i}\right)$$

$$GM\left(\frac{1}{r_i} - \frac{1}{r_f}\right) = \frac{1}{2}(v_i^2 - v_f^2)$$

$$M = \frac{\frac{1}{2}(v_i^2 - v_f^2)}{G(1/r_i - 1/r_f)}$$

$$M = \frac{\frac{1}{2}(5.0^2 - 4.1^2)\,\text{m/s}^2}{(6.7 \times 10^{-11}\,\text{N} \cdot \text{m}^2/\text{kg}^2)\left(\dfrac{1}{2 \times 10^5\,\text{m}} - \dfrac{1}{5 \times 10^5\,\text{m}}\right)} = 2 \times 10^{16}\,\text{kg}$$

FURTHER DISCUSSION We could not have used the Momentum Principle to carry out an accurate analysis without writing an iterative computer program because we didn't know how much time the process took. The velocity was changing at a nonconstant rate, so knowing the distance wouldn't give us the time, since we didn't know the average velocity. We were able to use the Energy Principle because we knew something about distances.

6.X.21 What was the advantage of choosing an "inclusive" system (including both the asteroid and the spacecraft)?

6.X.22 Concerning the separation r in $U = -Gm_1m_2/r$: Is this the distance from the spacecraft to the surface of the asteroid, or from the spacecraft to the center of the asteroid?

Multiparticle Systems: Same Time Interval, Different Displacements

In the previous example the spacecraft pulls just as hard on the asteroid as the asteroid pulls on the spacecraft (reciprocity of gravitational forces). Yet we claimed that the kinetic energy of the asteroid hardly changed compared to the change in the spacecraft's kinetic energy. We can show why by using the Momentum Principle and the Energy Principle together.

Apply the Momentum Principle

First consider the Momentum Principle. The two different forces act on the asteroid and the spacecraft for the same amount of time Δt. Since the magnitude of the two forces is the same, the magnitude of the impulse is the same, and the magnitude of the change of momentum is the same. Since $v = p/M$ at low speeds, the change in speed of the asteroid is extremely small compared to the change in speed of the spacecraft (ratio of m/M), and if the asteroid was initially at rest it will move a very small distance compared to the displacement of the spacecraft.

Apply the Energy Principle

Next consider the Energy Principle. The work done on the asteroid by the spacecraft is extremely small, because the displacement of the asteroid is extremely small. The two objects exert the same magnitude of force on each other, but through enormously different distances. In the example above, the spacecraft moved 300 km, whereas the asteroid moved roughly (m/M) of that distance, or about $(1500/2 \times 10^{16})(3 \times 10^5 \, \text{m}) \approx 2 \times 10^{-8} \, \text{m}$, or about 100 atomic diameters!

This illustrates yet another important difference in the role played by forces acting in and on a multiparticle system in the Momentum Principle (all forces act for the same time interval, and the net force is all that matters) and in the Energy Principle (different forces may act through very different distances, and change of shape matters).

Another way of seeing why the asteroid gets very little kinetic energy is to recall that kinetic energy can be expressed in terms of momentum:

$$\frac{1}{2}mv^2 = \frac{1}{2}m\left(\frac{p}{m}\right)^2 = \frac{1}{2}\frac{p^2}{m}$$

The asteroid and spacecraft experience the same change in the magnitude of momentum, but the p^2 term is divided by a huge mass in the case of the asteroid.

When a ball falls toward the Earth, the magnitude mg of the force that the Earth exerts on the ball is the same as that which the ball exerts on the Earth, and both experience the same change in the magnitude of momentum in the same time interval. However, the massive Earth acquires an extremely small velocity, so the Earth moves an extremely small distance while the ball moves a large distance. Same force mg, very different distances, very different amounts of work, and the Earth gets very little kinetic energy.

6.10 GENERAL PROPERTIES OF POTENTIAL ENERGY

There are some general properties of potential energy that are true not only for gravitational interactions but also for other kinds of interactions as well, including electric interactions. Here is a list of important properties:

- Potential energy depends on the separation between pairs of particles, not on their individual positions. (As we saw with gravitational interactions, this depends on the reciprocity of gravitational and electric forces.)
- Potential energy must approach zero as the separation between particles becomes very large, as we will soon show.
- If an interaction is attractive, potential energy becomes negative as the distance between particles decreases. (We saw this in the case of gravitational interactions.)
- If an interaction is repulsive, potential energy becomes positive as the distance between particles decreases.

Let's look in a bit more detail at the consequences of the fact that potential energy depends on the separation between the interacting objects.

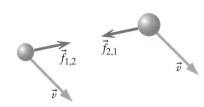

Figure 6.32 The two particles move together, with the same velocity.

QUESTION Suppose that two particles move together, with the same velocity (Figure 6.32). In that case, what is $\Delta \vec{r}_{21}$ for this two-particle system? What is ΔU_{12}?

In this case the relative positions don't change, $\Delta \vec{r}_{21}$ is zero, and there is no change in the pair-wise potential energy.

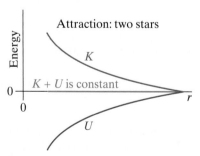

Figure 6.33 The two particles rotate around each other.

QUESTION Suppose instead that the two particles rotate around each other (Figure 6.33). What can you say about ΔU_{12}?

This is a bit trickier, because if there is rotation, $\Delta \vec{r}_{21}$ isn't zero, because $\vec{r}_{21}$ is now a rotating vector. In rotation, however, $\Delta \vec{r}_{21}$ is perpendicular to $\vec{r}_{21}$ and to the force $\vec{f}_{2,1}$, so the dot product $\vec{f}_{2,1} \bullet \Delta \vec{r}_{21}$ is zero. So again there is no change in the pair-wise potential energy.

QUESTION What can you conclude about potential energy for a rigid system?

Evidently U is constant for a rigid system, one whose shape doesn't change. For the potential energy to change, there must be a change of shape or size. Conversely, a change of shape or size is an indicator that U may have changed.

$U \rightarrow 0$ When the Particles Are Very Far Apart

Consider a system consisting of two identical stars, which attract each other gravitationally. If the stars are at rest, very, very far apart—almost infinitely far apart, so that the force one exerts on the other is nearly zero—then the total energy of the two-star system must be simply the sum of their rest energies:

$$E_{\text{sys}} = mc^2 + mc^2$$

This implies that the potential energy associated with the interaction of these two stars must be zero if the stars are "infinitely" far apart.

More generally, the total energy of a multiparticle system is given by the expression $E_{\text{system}} = (E_1 + E_2 + \cdots) + U$, where $E_1 = \gamma m_1 c^2$ is the particle energy of particle 1, and so on. When the particles are very far apart, we must define U to be zero, so that the total energy of these far-apart particles is simply the sum of the particle energies:

POTENTIAL ENERGY OF A PAIR OF PARTICLES THAT ARE VERY FAR APART
$U \rightarrow 0 \quad \text{as} \quad r \rightarrow \infty$

Figure 6.34 Initially two stars are very far apart, at rest, but they attract each other.

Attractive Interaction: U Becomes Negative as r Decreases

If the particles start to come closer together, we find that U starts to become negative if the interactions are attractive, while U starts to become positive if the interactions are repulsive. To see this, first consider an isolated system consisting of two stars that are very far apart, initially at rest, so the initial energy of the system is just the sum of the rest energies of the two stars. At first slowly, then more and more rapidly, their kinetic energies increase due to their mutual gravitational attractions (Figure 6.34).

QUESTION U was initially zero, so what must be the sign of U when the stars have acquired significant kinetic energies?

If the system is isolated, there are no external forces and hence no external work, so E_{sys} does not change. If the individual particle energies increase (so the total kinetic energy increases), the pair-wise potential energy must decrease. Because U was defined to be zero initially, U must become negative, as we found for gravitational potential energy. A graph of the energies is shown in Figure 6.35, omitting the very large rest energies of the stars. Since the rest energies don't change, $K + U$ is a constant.

Does this negative sign make sense in terms of the definition of potential energy change, $\Delta U = -W_{\text{int}}$? The attractive gravitational forces that the stars

Figure 6.35 The energies of two stars (with rest energies omitted) as a function of separation r. U must be zero at large r. As r decreases, K increases and U decreases. The total energy of the two-star system is constant, and $K + U$ is constant.

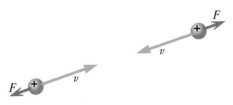

Figure 6.36 Initially two protons are very far apart, heading inward. The mutual repulsions slow the protons down.

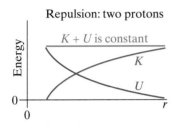

Repulsion: two protons

Figure 6.37 The energies of two protons as a function of separation (with rest energies omitted). U must be zero at large r. As r decreases, K decreases and U increases. The total energy of the two-proton system is constant, and $K + U$ is constant.

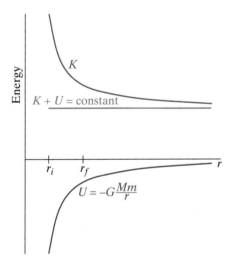

Figure 6.38 Energy vs. the separation distance between the asteroid and the spacecraft. (We omit the rest energies, which aren't changing.)

exert on each other certainly do a positive amount of work on each star, so the potential energy change must indeed be negative.

Repulsive Interaction: U Becomes Positive as r Decreases

Next consider an isolated system consisting of two protons that are very far apart, and suppose that initially they have high speeds, heading inward (Figure 6.36). Their mutual electric repulsions make the protons slow down (their electric repulsion is enormously larger than their gravitational attraction).

QUESTION U was initially zero, so what must be the sign of U when the protons have slowed down somewhat?

Again, there is no external work done on this isolated multiparticle system, so E_{sys} does not change. The particle energies decrease, so the potential energy must increase; U becomes positive (Figure 6.37). Also, the repulsive electric forces do negative work on each other, so $\Delta U = -W_{\text{int}}$ is positive. The graph in Figure 6.37 omits the rest energies of the protons. Since the rest energies don't change, $K + U$ is a constant.

Note that the change in potential energy for the two stars or the two protons was associated with a change in shape (change in configuration) of the two-particle system. If there is no change in shape or size, there is no change in potential energy.

6.11 PLOTTING ENERGY vs. SEPARATION

A graphical representation of energy can be exceptionally helpful in reasoning about a process. In Figure 6.38 we show an energy graph for the spacecraft leaving the asteroid, the process analyzed in Section 6.9 (the graph continues beyond the separation $r_f = 500\,\text{km}$). We plot the pair-wise gravitational potential energy U, the spacecraft's kinetic energy K, and the sum $K + U$, as a function of the separation distance r between the center of the asteroid and the spacecraft. We omit the rest energies, which are not changing during the motion.

Important features of the graph are these: The gravitational potential energy U is negative because the interaction is attractive, not repulsive, and the potential energy increases with increasing separation r. As the potential energy increases, the kinetic energy K of the spacecraft decreases (the gravitational force exerted on the spacecraft by the asteroid is slowing it down). Because there is negligible work done on the combined system of asteroid plus spacecraft, $K_f + U_f = K_i + U_i$, so $K + U$ has a constant value at all times, and the graph of $K + U$ is a straight horizontal line on the graph, representing a constant positive value.

As the spacecraft gets farther and farther from the asteroid, you can see from the graph that its kinetic energy K will continue to fall, and the graph of K will approach the $K + U$ line at large separation. When the two objects are very far apart, U is nearly zero, and $K + U$ is just K.

This is an example of an "unbound" system, because the separation between spacecraft and asteroid will increase without bound. An unbound system has $K + U$ greater than zero.

As an example of a "bound" system, again consider the spacecraft and the asteroid but this time give the spacecraft a smaller initial velocity, headed away from the asteroid. As a function of the center-to-center asteroid-to-spacecraft separation r, we again plot a graph of the potential energy $-GMm/r$ (Figure 6.39), and a graph of kinetic energy K. As the spacecraft moves away from the asteroid, its kinetic energy K falls as the pair-wise potential energy U rises. You can see on the graph that K falls to zero, and it cannot go negative, because kinetic energy is always a positive quantity. What happens next? The

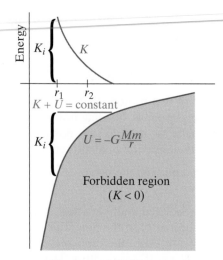

Figure 6.39 Energy vs. the separation distance between the asteroid and the spacecraft. (We omit the rest energies, which aren't changing.) At r_1 the potential energy is U_1 and the kinetic energy is K_1.

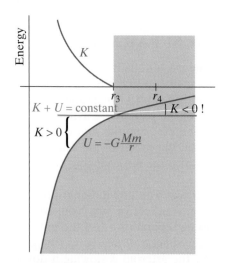

Figure 6.40 The asteroid and spacecraft cannot get farther away from each other than r_3; at r_4 the kinetic energy would be negative. (We omit the rest masses, which aren't changing.)

spacecraft momentarily comes to a stop (a turning point), then turns around and speeds up toward the asteroid; eventually it will crash onto the asteroid's surface.

We add to the graph a horizontal line whose height above or below the axis represents the total kinetic plus potential energy of the two-object system. The line on the graph for $K + U$ is horizontal because no external work is done on this isolated system, and there is no change of the rest masses (no change of identity), so $K + U$ is constant during the motion.

> **QUESTION**   Is $K + U$ positive or negative for the system shown in Figure 6.39?

The horizontal line representing the constant value of $K + U$ is below the axis representing $U = 0$, so $K + U$ is negative. This corresponds to a "bound" state of the spacecraft and asteroid; the spacecraft and asteroid cannot get completely away from each other.

The vertical distance from the U curve up to the $K + U$ line is equal to K, the total kinetic energy of the spacecraft and asteroid. At all separations between the spacecraft and the asteroid, the height of the $K + U$ line above the U curve tells you the total kinetic energy. Notice that when K falls to zero, $K + U = U$ and the right end of the $K + U$ line touches the U graph.

The shaded area in Figure 6.39 represents a forbidden region, where the kinetic energy K would be negative, no matter what the value of $K + U$ (more about this in the following discussion).

> **6.X.23** The asteroid–spacecraft separation increases from r_1 to r_2 (Figure 6.39). Does the kinetic energy of the asteroid–spacecraft system increase, decrease, or remain constant?

Limits on Possible Motion

The energy graph shows at a glance the limits on the motion for a given energy. In Figure 6.40, the asteroid and spacecraft with $K + U < 0$ cannot get farther away from each other than r_3, a bound that is set by the potential-energy curve. At a larger separation r_4, the kinetic energy would have to be negative, which is "classically forbidden" (that is, impossible according to the laws of classical, prequantum mechanics). Kinetic energy cannot be negative. At low speeds, $K = \frac{1}{2}mv^2$, which is always positive. At high speeds, $K = \gamma mc^2 - mc^2$, which is always positive, because γ is always greater than 1.

If the system has the total $K + U$ shown ($K + U < 0$), the separation of the asteroid and spacecraft can never be larger than r_3, and we say that this is a "bound" system. An energy graph shows at a glance the range of possible positions achievable during the motion, for a given total $K + U$. (We should mention, however, that in the world of atoms where a full analysis requires quantum mechanics, in some cases a system can "tunnel through" a region that is classically forbidden!)

An energy graph can show at a glance whether a system is "bound" or "unbound." An example of a bound system is a planet in circular or elliptical orbit around a star: the planet cannot escape. Another example is an electron bound to an atom.

An example of an unbound system is the Earth plus a spacecraft whose initial speed is great enough that the spacecraft will get away from the Earth and never come back. For another example, if an amount of energy greater than or equal to the ionization energy is supplied to an atom, an electron can become unbound and escape from the atom. An unbound system has total $K + U \geq 0$, which is easy to see on an energy graph, where it is clear that arbitrarily large separations are possible. Only when $K + U \geq 0$ can the system

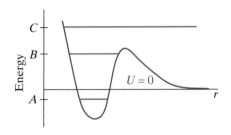

Figure 6.41 An energy graph for some system. Which states are bound? (Exercise 6.X.24.)

become separated by large distances, because K is never negative, and U goes to zero at large separations.

> **6.X.24** In this energy graph for some system (Figure 6.41), consider the various energy states indicated. Which of these values for the energy $K+U$ (A, B, or C) represents a bound state? Which represents an unbound state (the particle can escape)? Which represents a trapped state with enough energy to be unbound but with a barrier that (classically) prevents escape?

Drawing Energy Graphs

It is important to be able to draw (create) energy graphs as well as to be able to read them. Here is a practical scheme:

- Draw U vs. r for the particular interaction (gravitational, electric).
- At some r where you happen to know K, plot the point (r, K).
- Add that value of K to the value U has at the same separation r.
- Plot $K + U$ at that r, then draw a horizontal line through the point.
- At some other r, find a K that when added to U at that r gives $K + U$.
- Now you have two points on the K graph; sketch the behavior of K vs. r.

EXAMPLE   **Making an Energy Graph**

Let's follow through these steps for the energy graph for the spacecraft leaving the asteroid, which is shown again in Figure 6.42.

Solution
- We know the shape of U for gravity; it is $-GMm/r$. If r is small, U is a large negative number. If r is large, U is nearly zero. Plot two such points, then sketch the behavior of U vs. r.
- We know that K is nonzero at large r, because the spacecraft escapes, so at large r, plot some positive K.
- At that distant location, U is nearly zero, so the K you just plotted is nearly the same as $K + U$. Draw a horizontal line representing $K + U$ for all separations.
- At the initial separation r_i, determine graphically a value of K that when added to the value of U at r_i yields the known value of $K + U$. The value of K plus the value of U must be the same as $K + U$.
- Sketch the behavior of K vs. r. Note that its shape is the mirror image of U vs. r.

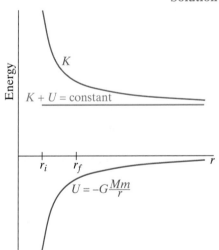

Figure 6.42 Energy vs. the separation distance between the asteroid and the spacecraft. (We omit the rest energies, which aren't changing.)

6.12

APPLYING GRAVITATIONAL POTENTIAL ENERGY

In the next sections we apply the Energy Principle to the analysis of a variety of situations involving gravitational potential energy. We will frequently use energy diagrams as a powerful tool.

Application: Energy Graphs and Types of Orbits

In Figure 6.43 are plotted graphs of potential energy, and potential plus kinetic energy, for three different orbits of a planet and star. The following exercise asks you to interpret the meaning of these graphs.

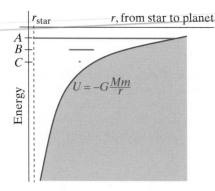

Figure 6.43 What kind of motion is represented by each of these situations? (Exercise 6.X.25).

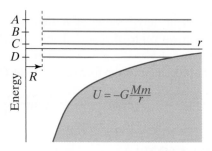

Figure 6.44 A spacecraft is launched with momentum p_i from near an airless planet's surface; it has momentum p_f when it has traveled far away.

Figure 6.45 What is the minimum K required to escape?

6.X.25 In Figure 6.43, what kind of motion is represented by the situation with $K + U = A$? B? C? Think about the range of r in each situation. For example, C represents a circular orbit (constant r).

Application: Escape Speed

An analysis based on the Energy Principle will allow us to answer the following question: With what minimum speed v must a spacecraft leave the surface of an airless planet (no air resistance) of mass M and radius R in order that it can coast away without ever coming back (Figure 6.44)? Assume that there are no other objects nearby.

If you did the Moon voyage problem in Chapter 3 you found that there was a minimum initial speed required to reach the Moon. With an initial speed smaller than this, the spacecraft fell back to Earth without reaching the Moon. Your determination of the minimum initial speed was done by trial and error. However, we can now use energy relationships to calculate such minimum speeds directly.

As is often the case, an energy diagram can be very helpful. Consider Figure 6.45. Motions with $K + U = A$, B, C, or D all start from the surface of an airless planet, with a separation R between the center of the planet and the spacecraft. (Remember that the gravitational force outside a uniform sphere is exactly the same as it would be if all the mass collapsed to the center of the sphere.)

> **QUESTION**   Which of these motions starts with the largest kinetic energy K? Which starts with the least K?

Each of the horizontal lines represents motion with $K + U$ constant. Evidently A is the highest value of $K + U$, and therefore involves the largest initial kinetic energy. D is the lowest value of $K + U$, and starts with the least kinetic energy.

> **QUESTION**   For which of these motions does the spacecraft "escape" and never come back?

For motions with $K + U = A$, B, or C, no matter how far from the planet the spacecraft gets, there is still some kinetic energy, so the spacecraft will never return. This is another example of an unbound state being associated with a positive value of $K + U$.

In contrast, with motion $K + U = D$ the spacecraft will reach a maximum separation from the planet, at which point its kinetic energy has fallen to zero. The spacecraft will fall back to the planet.

> **QUESTION**   How would you describe the motion that requires the least possible initial kinetic energy to achieve escape?

Evidently the least costly escape is to have $K = 0$ when the spacecraft has reached a distance very far from the planet. At a large separation, $U = 0$, so $K + U = 0$. This corresponds on the diagram to a horizontal line lying on the axis. Because $K + U$ doesn't change (no external work on the planet–spacecraft system), it must also be true that $K + U = 0$ at the start of the motion:

MINIMAL CONDITION FOR ESCAPE
$K + U = 0$

The minimal initial kinetic plus potential energy of the system composed of planet plus spacecraft is this, assuming that the kinetic energy of the planet is negligible:

$$K_i + U_i = \frac{1}{2}mv_{\text{esc}}^2 + \left(-G\frac{Mm}{R}\right) = 0 \quad (\text{for } v \ll c)$$

This lets us calculate the minimum speed, which is called the "escape speed" v_{esc}, for a planet with mass M and radius R. Though we speak of "escape speed," it is wrong to say that the spacecraft has "escaped from gravity." As long as the spacecraft is a finite distance from the planet, it feels a finite gravitational attraction to the planet.

If you give the spacecraft more than the minimum kinetic energy, it not only escapes but also has nonzero kinetic energy when it is far away (for example, see motions A and B in Figure 6.45).

It is interesting to note that the initial direction of motion doesn't matter. An energy analysis is insensitive to direction.

Bound vs. Unbound States

If the initial velocity points directly away from the planet, the path is a straight line. If the initial velocity points in some other direction, the path can be shown to be a parabola if the initial speed is exactly equal to escape speed and a hyperbola for higher speeds.

If the speed is below escape speed, there is no escape. The various possible orbits (Figure 6.46) can be classified on the basis of their total kinetic plus potential energy, with initial speed v_i:

$$\frac{1}{2}mv_i^2 + \left(-G\frac{Mm}{R}\right) > 0: \quad \text{escape with } v_\infty > 0; \text{ straight line or hyperbola}$$

$$\frac{1}{2}mv_i^2 + \left(-G\frac{Mm}{R}\right) = 0: \quad \text{escape with } v_\infty = 0; \text{ straight line or parabola}$$

$$\frac{1}{2}mv_i^2 + \left(-G\frac{Mm}{R}\right) < 0: \quad \text{no escape; straight line, circular, or elliptical}$$

This is a specific example of an important general principle. If the kinetic plus potential energy is negative, the system is "bound": the objects cannot become widely separated, because that would require that the kinetic energy become negative, which is impossible (classically). If the kinetic plus potential energy is positive, the system is "unbound" or "free": the objects can become widely separated, with net kinetic energy (and zero potential energy).

BOUND AND UNBOUND STATES
If $K + U < 0$, the system is in a bound state.
If $K + U \geq 0$, the system is unbound (free).

Straight line

Hyperbola
$K + U > 0$

Circle
or
ellipse
$K + U < 0$

Parabola
$K + U = 0$

Figure 6.46 Possible orbital trajectories.

EXAMPLE **Escape Speed**

What is escape speed from Earth?

Solution System: Earth + spacecraft
Surroundings: Nothing significant (ignore Sun and Moon)

Initial state: At surface of Earth, with speed
Final state: Spacecraft extremely far away from Earth, nearly at rest

Energy principle:

$$K_{s,f} + \cancel{K_{E,f}} + U_f = K_{s,i} + \cancel{K_{E,i}} + U_i + \cancel{W}$$

$$0 + 0 = \frac{1}{2}mv_i^2 + \left(-G\frac{Mm}{r_i}\right)$$

$$v_i = \sqrt{\frac{2GM}{r_i}}$$

$$v_{esc} = \sqrt{\frac{2(6.7 \times 10^{-11}\text{N}\cdot\text{m}^2/\text{kg}^2)(6 \times 10^{24}\text{ kg})}{(6.4 \times 10^6 \text{ m})}} = 1.12 \times 10^4 \text{ m/s}$$

■ When we know there is no change in rest energies it is okay to omit these terms from the Energy Principle equation.
Negligible change in kinetic energy of Earth.
No work by surroundings.

6.X.26 This escape speed is slightly larger than the launch speed that was found necessary to reach the Moon in the problem in Chapter 3 on the Ranger 7 mission to the Moon. Explain why.

6.X.27 Turn the argument around. If an object falls to Earth starting from rest a great distance away, what is the speed with which it will hit the upper atmosphere? (Actually, a comet or asteroid coming from a long distance away might well have an even larger speed, due to its interaction with the Sun.) Small objects vaporize as they plunge through the atmosphere, but a very large object can penetrate and hit the ground at very high speed. Such a massive impact is thought to have killed off the dinosaurs (see *T. Rex and the Crater of Doom*, Walter Alvarez, Princeton University Press, 1997).

Application: The Moon Voyage

We have been applying energy considerations to two-particle systems. A homework problem in Chapter 3 involves predicting the motion of the Ranger 7 spacecraft as it travels from Earth to the Moon (where it crash lands). In our simple model for this voyage there are three "particles": the spacecraft with mass m, the Earth with mass M_{Earth}, and the Moon with mass M_{Moon}.

QUESTION How many interaction pairs are there in this three-particle system?

In a three-particle system there are three interaction pairs: U_{12}, U_{13}, and U_{23}. We can write the gravitational potential energy like this:

$$U = \left[-G\frac{M_{Earth}m}{r_{\text{to Earth}}}\right] + \left[-G\frac{M_{Moon}m}{r_{\text{to Moon}}}\right] + \left[-G\frac{M_{Earth}M_{Moon}}{r_{\text{Earth to Moon}}}\right]$$

In our simple model, the Earth and Moon were fixed in space, so the Earth–Moon term of this expression doesn't change. Also, there is no change of identity of the three particles, so their rest masses do not change. Therefore the energy equation for the system can be written like this, where v is the speed of the spacecraft, the only moving particle in the model system (and $v \ll c$):

$$\Delta\left(\frac{1}{2}mv^2\right) + \Delta\left[-G\frac{M_{Earth}m}{r_{\text{to Earth}}}\right] + \Delta\left[-G\frac{M_{Moon}m}{r_{\text{to Moon}}}\right] = 0$$

The change in the energy of the system is zero because no external work is done on the system (we're ignoring the effects of other objects, including the Sun). We can rearrange the terms and write the energy equation in terms of initial and final energies, which must be equal:

$$\frac{1}{2}mv_f^2 + \left[-G\frac{M_{Earth}m}{r_{\text{to Earth},f}}\right] + \left[-G\frac{M_{Moon}m}{r_{\text{to Moon},f}}\right] =$$
$$\frac{1}{2}mv_i^2 + \left[-G\frac{M_{Earth}m}{r_{\text{to Earth},i}}\right] + \left[-G\frac{M_{Moon}m}{r_{\text{to Moon},i}}\right]$$

That is, the final (kinetic plus potential) energy of the system is equal to the initial (kinetic plus potential) energy of the system, because no external work is done on the system (and the rest masses don't change; no change of identity).

6.13   GRAVITATIONAL POTENTIAL ENERGY NEAR THE EARTH'S SURFACE

An important special case of gravitational potential energy arises in projectile motion near the surface of the Earth. Earlier we found that for motion near the surface of the Earth $\Delta U \approx \Delta(mgy)$.

There is no minus sign in this (approximate) expression for the change in gravitational potential energy. As you go up (farther from the center of the Earth), the change $\Delta(mgy)$ is positive as expected; if you go down, it is negative as expected. In the full formula there is a $(-1/r)$ because moving to larger distance r means an increase (gravitational potential energy *less negative*), as can be seen in Figure 6.47.

In the formula $\Delta U = \Delta(mgy)$, Δy must be small compared to the radius of the Earth (6400 kilometers), because the gravitational force gets smaller as you go farther away from the Earth, inversely proportional to the square of the distance from the center of the Earth. However, if y doesn't change much, neither does gravity. If you go up 0.1 km (100 meters, the length of a football field), the gravitational force decreases by a tiny factor:

$$\left(\frac{6400 \, \text{km}}{6400.1 \, \text{km}} \right)^2 \approx 0.99997$$

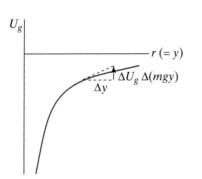

Figure 6.47 A small increase Δy in height y corresponds to a small increase in the (negative) gravitational energy.

| EXAMPLE | A Falling Ball |

At a certain instant a ball is falling with a speed of 6 meters/second, and its position is $y = 35$ meters above the Earth's surface. How fast is the ball falling when it has fallen to a position $y = 20$ meters above the Earth's surface, assuming that we can neglect air resistance?

Solution   System: Ball + Earth
Surroundings: Nothing significant (neglect air resistance)

Initial state: Speed 6 m/s, 35 m above ground
Final state: 20 m above ground

Energy Principle:

■ The *change* in potential energy near the surface of the Earth is $\Delta U \approx \Delta(mgy)$, and the potential energy at the surface is $-GMm/R$, a constant that you calculated in Exercise 6.X.18 in Section 6.9. Therefore the actual potential energy near the Earth is $-GMm/R + mgy$. Since the term $-GMm/R$ doesn't change during an interaction, we simply omit it and write $U_i = mgy_i$ and $U_f = mgy_f$.

$$K_f + U_f = K_i + U_i + \cancel{W}$$
$$\frac{1}{2}mv_f^2 + mgy_f = \frac{1}{2}mv_i^2 + mgy_i$$
$$\frac{1}{2}mv_f^2 = \frac{1}{2}mv_i^2 - (mgy_f - mgy_i)$$
$$v_f^2 = v_i^2 - 2g(y_f - y_i)$$
$$v_f = \sqrt{v_i^2 - 2g(y_f - y_i)}$$
$$v_f = \sqrt{(6 \, \text{m/s})^2 - 2(9.8 \, \text{N/kg})[(20 \, \text{m}) - (35 \, \text{m})]} = 18.2 \, \text{m/s}$$

FURTHER DISCUSSION There is nothing in our equations to specify whether the ball is initially heading upward or downward. In fact, it doesn't matter. In either case, if we neglect air resistance, the speed will be 18.2 m/s when the ball reaches a height of 20 m above the ground. (If it was initially headed upward, it will again have a speed of 6 m/s when it returns to a height of 35 m.)

6.X.28 Suppose that in this situation we measure y from 25 meters above the surface, so that $y_{initial} = +10$ meters, and $y_{final} = -5$ meters. Why do we get the same value for the final speed?

6.X.29 Suppose that a pitcher can throw a ball straight up at 100 miles per hour (about 45 m/s). Use energy conservation to calculate how high the baseball goes. (See the Answers section for a hint, if necessary.) Explain your work. Actually, a pitcher can't attain this high a speed when throwing straight up, so your result will be an overestimate of what a human can do; air resistance also reduces the achievable height.

Including Horizontal Motion

Moving vertically up or down involves a change in gravitational energy approximately equal to $\Delta(mgy)$. What about sideways horizontal movements? There is negligible change in the gravitational energy of the Universe when an object is moved a short distance horizontally (that is, tangent to the Earth's surface). Consider moving a block along a surface with it rolling very easily on low-friction wheels, or a hockey puck sliding easily along the ice. You hardly have to use any energy at all to move the block or puck horizontally, as long as there is little friction.

What if a block is moved horizontally, but at a constant height of 3 meters above the surface? You could support the block with a tall cart, and again there would be little effort involved in moving the block horizontally. Or you could pile up dirt 3 meters deep and place a slick surface on top, and then you could easily move the block horizontally. We conclude that horizontal position doesn't affect the gravitational energy. (Again, we're considering movements over distances small compared with the radius of the Earth, so we can ignore the Earth's curvature.)

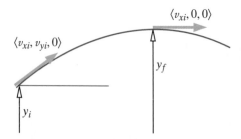

Figure 6.48 Path of a ball thrown at an angle to the horizontal.

6.X.30 Suppose that you throw a ball at an angle to the horizontal, and just after it leaves your hand at a height y_i its velocity is $\langle v_{xi}, v_{yi}, 0 \rangle$ (Figure 6.48). Assuming that we can neglect air resistance, at the top of its trajectory, when it is momentarily traveling horizontally, its velocity is $\langle v_{xi}, 0, 0 \rangle$. What is the height y_f at the top of the trajectory, in terms of the other known quantities? Use the Energy Principle.

| 6.14 | **ELECTRIC POTENTIAL ENERGY** |

In this section we will study the potential energy associated with the electric interactions of charged particles. The electric force law for two charged particles is similar to the gravitational force law:

Figure 6.49 Electric potential energy is positive when like charged objects repel each other.

THE ELECTRIC FORCE
$$F_{elec} = \frac{1}{4\pi\varepsilon_0}\frac{|q_1 q_2|}{r^2}$$
where $\dfrac{1}{4\pi\varepsilon_0} = 9 \times 10^9 \dfrac{\text{N·m}^2}{\text{coulomb}^2}$.

The constant $1/(4\pi\varepsilon_0)$ is read as "one over four pi epsilon-zero." The quantities q_1 and q_2 represent the amount of charge on the two particles (measured in coulombs), and r is the distance between the two particles. Because

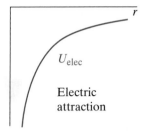

Figure 6.50 Electric potential energy is negative (like gravitational potential energy) when objects attract.

this is an inverse square force, like gravitation, the electric potential energy has a formula very similar to that for gravitational potential energy:

ELECTRIC POTENTIAL ENERGY U_{elec}

$$U_{elec} = \frac{1}{4\pi\varepsilon_0}\frac{q_1 q_2}{r}$$

There is no minus sign in the formula for electric potential energy, because when the electric interaction is attractive, like gravitation, q_1 and q_2 have different signs, and the minus sign shows up automatically. However, if the two charges have the same sign, the interaction is repulsive, and there is a plus sign in the formula for the electric energy (Figure 6.49). If the two charges have opposite signs, the interaction is attractive, and there is a minus sign in the formula for electric energy (Figure 6.50).

A certain amount of energy is required to remove an outer electron from a neutral atom, leaving behind a positive "ion" that is missing one electron and whose charge is $+e$, where $e = 1.6 \times 10^{-19}$ coulomb. (The electron's charge is of course $-e$.) The minimum energy to remove one outer electron is called the "ionization" energy.

EXAMPLE   **Ionization Energy**

A typical atom has a radius of about 1×10^{-10} m. Calculate a typical ionization energy in joules. Then convert your result to "electron volts," a unit of energy equal to 1.6×10^{-19} joules. This small energy unit is frequently used in discussions of atomic phenomena.

Solution   System: Two particles, the outer electron with charge $-e$, and the remaining part of the atom, the ion with charge $+e$.

Surroundings: An object that exerts an external force to remove the electron from the atom.

Initial state: The electron is bound to the atom, at a distance r_{atom} from the center.

Final state: The two particles very far from each other, at rest.
Energy Principle:

■ We don't know the initial K of the electron, so assume it is zero (this is a rough approximation).

$$E_f = E_i + W$$

$$\cancel{K_{ion,f}} + \cancel{K_{e,f}} + \cancel{m_i c^2} + \cancel{m_e c^2} + U_f = \cancel{K_{ion,i}} + \cancel{K_{e,i}} + \cancel{m_i c^2} + \cancel{m_e c^2} + U_i + W$$

$$U_f = U_i + W$$

$$W = U_f - U_i$$

$$W = \frac{1}{4\pi\varepsilon_0}\frac{q_1 q_2}{r_\infty} - \frac{1}{4\pi\varepsilon_0}\frac{q_1 q_2}{r_{atom}}$$

$$W = -\frac{1}{4\pi\varepsilon_0}\frac{(+e)(-e)}{r_{atom}}$$

$$W = \left(9 \times 10^9 \frac{\text{N·m}^2}{\text{C}^2}\right)\frac{(1.6 \times 10^{-19}\,\text{C})^2}{(1 \times 10^{-10}\,\text{m})}$$

$$W = 2.3 \times 10^{-18}\,\text{J}$$

$$W = (2.3 \times 10^{-18}\,\text{J})\left(\frac{1\,\text{eV}}{1.6 \times 10^{-19}\,\text{J}}\right) = 14\,\text{eV}$$

Check: The units are correct.

FURTHER DISCUSSION   This is a rough estimate because we don't know the electron's initial kinetic energy K_i, but typical ionization energies are indeed a few electron volts.

> **6.X.31** Two protons are hurled straight at each other, each with a kinetic energy of 0.1 MeV, where 1 mega electron volt is equal to $1 \times 10^6 \times (1.6 \times 10^{-19})$ joules. What is the separation of the protons from each other when they momentarily come to a stop (a turning point)?

6.15 | THE MASS OF A MULTIPARTICLE SYSTEM

A surprising prediction of Einstein's Special Theory of Relativity (1905) was that mass and energy are actually the same thing. This prediction has been confirmed in many different kinds of experiments. The most dramatic example of mass–energy equivalence is the annihilation reaction that occurs between matter and antimatter, as in the case of the annihilation of an electron and a positron (a positive electron) into two photons (high-energy quanta of light denoted by lowercase Greek gamma, γ):

$$e^- + e^+ \rightarrow \gamma + \gamma$$

Before the reaction, an electron and a positron can be almost at rest far from each other, a situation we would naturally describe as one in which there is mass but seemingly no energy. After the reaction there is light (in the form of two photons), a situation that we would naturally describe as pure energy but seemingly no mass. However, from the point of view of Special Relativity, the mass of the electron–positron system *is* energy, and the energy of the two-photon system *is* mass.

The equivalence of mass and energy is of course visible in Einstein's formula for the energy of a particle:

$$E = \gamma mc^2 = mc^2 + K \quad (\textit{definition of kinetic energy } K)$$

In this equation m is called the rest mass of the particle and is a constant. The particle has "rest energy" mc^2 even if it is not moving. If it is moving, it has additional, "kinetic" energy K and additional mass. A valid way of thinking about a fast-moving particle is that it has more mass, because mass is equivalent to energy, so the mass is E/c^2.

What is the mass of a multiparticle system consisting of interacting particles? Consider the simple case where the multiparticle system is overall "at rest." The individual particles are moving around, but the system as a whole is not moving (technically, the total momentum of the system is zero, and its center of mass is at rest). An example would be a box containing air molecules that are moving around rapidly in random directions, but the box isn't going anywhere. Another example is a diatomic molecule such as O_2 that is vibrating but has no overall translational motion.

It is natural to suppose that the total rest mass of such a system is just the sum of the rest masses of the particles: $M = m_1 + m_2 + \cdots$. However, this is not quite right. Rather, as predicted by the theory of relativity, the rest energy of the multiparticle system is Mc^2, and therefore its rest mass is this:

$$M = \frac{E_{\text{system}}}{c^2} = \frac{m_1 c^2 + K_1 + m_2 c^2 + K_2 + \cdots + U}{c^2}$$

$$M = (m_1 + m_2 + \cdots) + \left(\frac{K_1 + K_2 + \cdots + U}{c^2} \right)$$

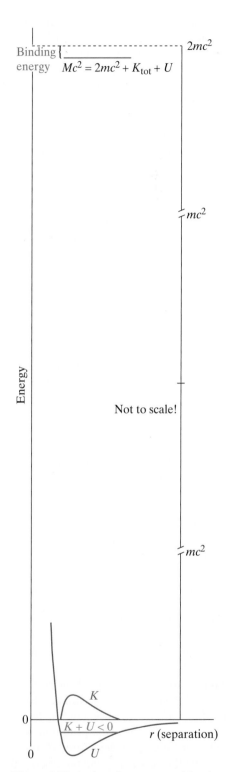

Figure 6.51 A plot of rest energy, kinetic energy, potential energy, and total energy (Mc^2) for O_2. Note that the total energy is less than the sum of the rest energies of the two individual atoms.

The rest mass M of the system may be greater or less than the sum of the individual particle rest masses, because the final term in this equation may be positive or negative.

In most everyday situations the kinetic energies of the individual particles are very small compared to their rest energies (that is, the speeds of the particles are small compared to the speed of light), and the interaction energy U (the sum of all the pair-wise interactions) is also small compared to the rest energies. In that case, the mass of the multiparticle system is very nearly equal to the sum of the individual rest masses. However, this is really only an approximation, and it is often not a good approximation in nuclear and particle physics.

The Total Energy of a Multiparticle System

Consider an oxygen molecule (O_2) at rest. Instead of plotting only the kinetic and potential energy for this two-atom system, let's include the rest energy of the atoms, too, as we attempt to show in Figure 6.51. We can't draw the full vertical scale, because the rest energies are enormous compared to the kinetic and potential energies. In order to show the rest energy to the same scale as K and U, we would need a page stretching from the Earth to the Moon!

The total energy of the system (indicated by the green horizontal line near the top of the graph) is the sum of the rest energies of the two atoms, the potential energy U (which is negative), and K_{tot}, the total kinetic energy of the two atoms. For the bound state drawn on the graph you can see that the rest energy of the two-atom O_2 system is very slightly less than the sum of the individual atomic rest energies, because $K + U < 0$.

Binding Energy

When the diatomic molecule O_2 forms, about $5\,eV$ of energy are given off (for example, by emitting photons, which carry energy). Correspondingly, it takes an input of about $5\,eV$ of energy to break O_2 apart into two oxygen atoms. This is called the "binding energy" of the molecule and is shown at the top of Figure 6.51.

Two Ways of Thinking About the Energy of the System

Given these aspects of particle energy, there are two rather different ways to think about the energy of a multiparticle system:

- The energy of a multiparticle system consists of the individual particle energies of the particles that make up the system, plus their pair-wise interaction energies.
- A multiparticle system itself has energy, rather like a particle, and if the system is at rest (its center of mass is not moving; its net momentum is zero) its energy E is simply Mc^2, where M is the mass of the system.

The Mass of the System

The mass M may be greater or smaller than the sum of the rest masses of the individual particles, depending on whether the total $K + U$ of the particles is positive or negative. The graph in Figure 6.51 indicates that in this case the quantity Mc^2 (where M is the mass of the whole molecule) is less than $2mc^2$, and the difference is the binding energy of the molecule. From the graph we can see that:

$$Mc^2 = 2mc^2 + K + U < 2mc^2$$

Mass of a Hot Object

This way of thinking about mass leads us to conclude that a hot block has a greater mass than a cold block. As we'll discuss in more detail in the next chapter, a hot metal block has more thermal energy than a cold metal block, with more kinetic energy of the atoms and more potential energy in the interactions with neighboring atoms (represented as springs in our simple model of a solid). Although the effect is much too small to measure directly, the equivalence of mass and energy embodied in the special theory of relativity, $M = E/c^2$, leads to the conclusion that a hot block is actually more massive than a cold block! The same impulse $\vec{F}\Delta t$ applied to the two blocks should change the momentum of the cold block very slightly more than the momentum of the hot block.

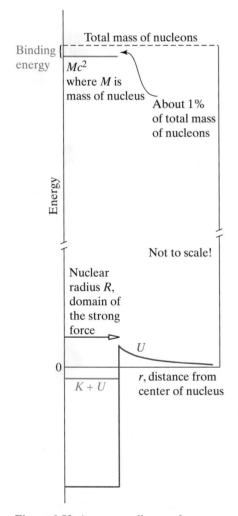

Figure 6.52 An energy diagram for a proton in an iron nucleus, not to scale. The mass of the nucleus is about 1% less than the sum of the masses of the individual nucleons (protons and neutrons).

6.X.32 Calculate the approximate ratio of the binding energy of O_2 (about 5 eV) to the rest energy of O_2. Most oxygen nuclei contain 8 protons and 8 neutrons, and the rest energy of a proton or neutron is about 940 MeV. Do you think you could use a laboratory scale to detect the difference in mass between a mole of molecular oxygen (O_2) and two moles of atomic oxygen?

Application: Nuclear Binding

In chemical reactions it is nearly impossible to measure the change of rest mass directly, but in nuclear interactions the changes are easily measurable. Consider the nuclear binding energy of an iron nucleus, which contains 26 protons and 30 neutrons, a total of 56 nucleons. When the protons and neutrons bind to each other through the "strong" (nuclear) force, the resulting iron nucleus has a mass that is about 1% smaller than the total mass of its individual constituents when separated. There is a sizable negative potential energy U associated with the strong force, which makes the multiparticle system's rest energy measurably smaller than the sum of the individual rest energies.

The average binding energy per nucleon is largest in iron. Lighter nuclei (hydrogen through manganese) are less tightly bound and can in principle fuse together to form the more tightly bound iron, with the release of energy. In fact, fusion reactions in stars build up from hydrogen to heavier and heavier nuclei, stopping at iron (heavier elements are made in supernova explosions).

Heavier nuclei (cobalt through uranium and beyond) are also less tightly bound than iron and can in principle fission to form the more tightly bound iron. However, only a few nuclei such as uranium and plutonium do actually spontaneously fission.

Figure 6.52 shows an energy diagram (not to scale) for one of the protons in an iron nucleus, subjected to the forces of all the other protons and neutrons. There is a deep potential well corresponding to the strong or nuclear force, which is essentially the same for protons and neutrons. This potential well has a very steep side corresponding to the very strong nuclear force (force is negative gradient of potential energy). The well is so steep it can be described approximately as a "square well." This deep well extends only out to the radius R of the nucleus, because the nuclear force has a very short range. A proton (but not a neutron) in addition experiences the repulsive electric force of the other protons (the electric U curve outside the nucleus). The total rest mass of the nucleus is about 1% less than the sum of the rest masses of the individual nucleons (protons and neutrons).

EXAMPLE

Nuclear Binding Energy

The rest energy of an iron nucleus (26 protons and 30 neutrons) is 52,107 MeV. Show that the average binding energy per nucleon is about 1% of the rest energy of a nucleon, which is easily measurable. The rest energy of a proton is 938.3 MeV; of a neutron, 939.6 MeV.

Solution

As shown in Figure 6.52, the binding energy is essentially the difference between the rest energy of the multiparticle system (the nucleus) and the sum of the rest energies of its constituent particles.

$$\text{binding energy} = ((26)(938.3)\,\text{MeV} + (30)(939.6)\,\text{MeV}) - 52107\,\text{MeV}$$
$$= 494.8\,\text{MeV}$$

$$\text{energy per nucleon} = \frac{494.8\,\text{MeV}}{26+30} = 8.8\,\text{MeV}$$

This is about 1% of the rest energy of one nucleon.

6.X.33 The deuteron, the nucleus of the deuterium atom ("heavy" hydrogen), consists of a proton and a neutron. It is observed experimentally that a high-energy photon ("gamma ray") with a minimum energy of 2.2 MeV can break up the deuteron into a free proton and a free neutron; this process is called "photo-dissociation." About what fraction of the deuteron rest energy corresponds to its binding energy? The result shows that the deuteron is very lightly bound compared to the iron nucleus.

Application: Nuclear Fission

The nuclear fission of uranium leads to an easily measured change in rest masses. The average binding energy per nucleon in uranium is about 7.8 MeV, so the uranium nucleus is less tightly bound than an iron nucleus, whose binding is 8.8 MeV per nucleon. This suggests the possibility of obtaining energy from the fission of a uranium nucleus, corresponding to roughly 0.1% (0.7 MeV per nucleon) of its rest energy (about 940 MeV per nucleon).

In other words, the "mass deficiency" of a medium-mass nucleus is about 1%, with a variation of about a tenth of that. Even though there is only a small difference in average binding energy between iron and uranium, the energy involved is vast compared to the energy of a chemical reaction, which only involves the outer electrons of an atom, not the nucleus.

A uranium-235 nucleus contains 92 protons and 143 neutrons. The 92 protons exert strong electric repulsions on each other, but the large number of neutrons spreads the protons throughout a larger volume and reduces the electric repulsions. The neutrons and the protons also attract neighboring protons with strong nuclear forces. (Low-mass nuclei don't have an overabundance of neutrons; for example, ordinary carbon has 6 protons and 6 neutrons, and iron has only 30 neutrons with its 26 protons.)

Uranium-235 has an unusual property that few other nuclei have. If you hit a U-235 nucleus with a slowly moving neutron, the nucleus may deform into a dumbbell-like form and then split apart because the electric repulsion of the protons can now overcome the short-range attraction of the nucleons for each other (Figure 6.53). (In addition to the two main fission fragments there are typically one or more free neutrons in the final state.)

The fission fragments quickly acquire very high speed, and they leave the original atom's 92 electrons behind. These highly charged nuclei storm through the block of uranium metal, tearing up the material and violently heating the metal. In a nuclear power reactor the reaction is controlled in such a

Figure 6.53 A U-235 nucleus in the process of fissioning.

way that the metal merely gets hot enough to boil water. With many uranium-235 nuclei fissioning, the metal can boil water in a nuclear power plant to produce steam and power a turbine to drive an electric generator.

The rest masses of the two fission-fragment nuclei total less than the rest mass of the uranium-235 nucleus. The rest mass difference Δm is only about one-tenth of one percent of the original rest mass, but this rest mass change is easily measurable, and c^2 times this rest mass change is indeed found to be equal to the observed kinetic energy of the fission fragments. This 0.1% change is about ten million times as big an effect as the rest mass changes that occur in chemical reactions, which only involve rearrangements of the outer electrons in the atoms, and do not affect the nucleus.

Nuclear Waste

The problem of radioactive waste from nuclear power plants comes from the fact that the fission products are nuclei with a large overabundance of neutrons, and they emit energetic radiations. For example, uranium can fission into two palladium nuclei, each with 46 protons and 71 or 72 neutrons, whereas ordinary stable palladium has only 60 neutrons (see the inside front cover of this book). A neutron in such a neutron-rich nucleus can change into a proton (Figure 6.54) with the emission of an energetic electron and an antineutrino (the antineutrinos almost never interact with matter, so they aren't dangerous). This process is called "beta decay" (energetic electrons were initially called "beta rays"). After a beta decay the new nucleus, in which a neutron has changed into a proton, may be in an "excited state" and emit a gamma ray (a high-energy photon, see Figure 6.55). Heavy elements such as plutonium also emit alpha particles (the nuclei of helium atoms) and there may be heavy elements mixed in with the rest of the nuclear waste.

When energetic electrons or gamma rays or alpha particles from radioactive nuclei go through matter, energy is absorbed and the matter gets hot. If the radiation goes through living tissue there can be damage to cells, and to the genetic information in the DNA in cells. The range of the radiation is relatively short and it can be stopped by a layer of concrete or other shielding. The big concern is that radioactive wastes might leak out of storage facilities, spread through the ground water, and get into the water supply. Drinking contaminated water would bring radioactive nuclei into the body, close to cells, where damage could occur.

A great deal of engineering and scientific study is under way to try to find a secure storage scheme that would prevent leakage for hundreds or even thousands of years, because some of the radioactive nuclei have very long "half-lives" (the time it takes for half of the nuclei to decay).

$$n \rightarrow p^+ + e^- + \overline{\nu}$$

Figure 6.54 A neutron in a neutron-rich nucleus can change into a proton with emission of an electron and an antineutrino.

$$N^* \rightarrow N + \gamma$$

Figure 6.55 A nucleus in an excited state (called N^*) can drop to a lower energy state and emit a gamma ray (a high-energy photon).

6.16   REFLECTION: WHY ENERGY?

It is appropriate at this point to reflect on why we introduced the topic of energy. Energy methods often make it possible to analyze some aspect of a phenomenon with much less effort than is required by direct application of the Momentum Principle. A striking example is the ability to calculate escape speed, or the impact speed of the *Ranger* spacecraft on the Moon, without carrying out a lengthy numerical integration.

Limits on the Possible

Another powerful capability of energy analyses is that they can easily predict whether some process can occur or not. For example, if there is insufficient energy, the rocket can't escape from the planet, or the atom cannot be ionized, or the biochemical reaction cannot proceed, or a proton can't decay into a neutron.

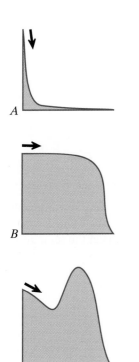

Figure 6.56 The final speed of the snow-boarder will be the same at the bottom of each hill.

Less Detail

However, energy analyses give less detailed information than is given by a step-by-step numerical integration of the Momentum Principle. In particular, energy analyses tell us nothing about how fast a process will proceed; time does not appear in the statement of conservation of energy. Also, since the Energy Principle is a scalar principle, it doesn't tell us about change in the direction of motion.

Here is an example that illustrates the strengths and weaknesses of energy analyses. Consider snowboarding down the three hills shown in Figure 6.56, starting from rest. If we can neglect friction, the final speed at the bottom should be the same in all three cases, because the final kinetic energy should be equal to mgh, the amount by which the gravitational energy decreases.

QUESTION   But how will the amount of *time* compare for the ride down the hill?

On the first hill you get going very fast very quickly, and the trip is short. On the second hill you move very slowly for a long time before dropping to the bottom, and the total trip takes a long time. On the third hill you never get to the bottom, despite the fact that the change in potential energy would be negative, because you would have to pass through a region that is energetically forbidden. So energy methods are powerful, but at the cost of losing some information, including the amount of time required for a process.

6.17   IDENTIFYING INITIAL AND FINAL STATES

Some processes involve several identifiable states, not just two. Any two states can be chosen as initial and final states; the rationale for making the choice depends on what unknown quantity you are trying to determine. In this section we offer two examples of situations involving more than two states, one simple and one complex.

EXAMPLE

Throwing a Ball: Three States

Throw a metal ball with mass 0.3 kg straight up with initial speed 10 m/s; air resistance is negligible. Consider these three states (Figure 6.57):

(1) $y_1 = 7$ m, $v_{y1} = 10$ m/s
(2) $y_2 = ?$, $v_{y2} = 0$ (maximum height; a turning point in the motion)
(3) $y_3 = 4$ m, $v_{y3} = ?$ (after coming back down, to a lower level)

Calculate the maximum height y_2 and the final speed, $|v_{y3}|$.

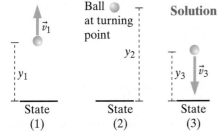

Figure 6.57 Throw a ball straight up. Find the maximum height and the final y component of the velocity.

Solution   System: Ball + Earth
Initial state 1, final state 2
Energy Principle: $E_2 + U_2 = E_1 + U_1 + \cancel{W}$

$$\cancel{K_2} + mgy_2 = K_1 + mgy_1$$

$$y_2 = y_1 + \frac{\frac{1}{2}mv_{y1}^2}{mg}$$

$$y_2 = 7\,\text{m} + \frac{\frac{1}{2}(10\,\text{m/s})^2}{9.8\,\text{N/kg}}$$

$$y_2 = 12.1\,\text{m}$$

Initial state 2, final state 3
Energy Principle: $E_3 + U_3 = E_2 + U_2 + \cancel{W}$

$$K_3 + mgy_3 = \cancel{K_2} + mgy_2$$

$$\frac{1}{2}mv_{y3}^2 = mg(y_2 - y_3)$$

$$v_{y3} = \pm\sqrt{2g(y_2 - y_3)}$$

$$|v_{y3}| = \sqrt{2(9.8\,\text{N/kg})(12.1 - 4)\,\text{m}}$$

$$|v_{y3}| = 12.6\,\text{m/s}$$

(See below for a discussion of the $\pm$ issue.) If we were not interested in the maximum height, we could have just considered states 1 and 3; it's not necessary to go through state 2 in the analysis. We should get the same result for the final speed:

Initial state 1, final state 3
Energy Principle: $E_3 + U_3 = E_1 + U_1 + \cancel{W}$

$$K_3 + mgy_3 = K_1 + mgy_1$$

$$\frac{1}{2}mv_{y3}^2 = \frac{1}{2}mv_{y1}^2 + mg(y_1 - y_3)$$

$$v_{y3} = \pm\sqrt{v_{y1}^2 + 2g(y_1 - y_3)}$$

$$|v_{y3}| = \sqrt{(10\,\text{m/s})^2 + 2(9.8\,\text{N/kg})(7 - 4)\,\text{m}}$$

$$|v_{y3}| = 12.6\,\text{m/s}$$

Figure 6.58 State 1: Alpha and C far apart, moving.

Figure 6.59 State 2: Alpha and C touch.

Figure 6.60 State 3: Excited O* nucleus.

Figure 6.61 State 4: O nucleus and photon.

FURTHER DISCUSSION When using the Energy Principle, you get to choose which two states to use, just as you get to choose the system. Of course you should always try to choose a system and two states that give you the most direct and simplest route to the desired result. For example, in this case it would be a lot of extra work to calculate the maximum height if all you really wanted was the final speed.

Unlike the Momentum Principle, the Energy Principle doesn't give us information about the direction of the velocity. In taking the square root above, v_{y3} could be either +12.6 m/s or −12.6 m/s. Which is correct? Both!

On the way down, $v_{y3} = -12.6$ m, which corresponds to the situation we analyzed. However, the mathematics claims that another possible value is $v_{y3} = +12.6$ m. This solution corresponds to the past, before the time of state (1). If the ball was thrown upward from the ground, v_y was +12.6 m/s as it rose past the height we've called $y_3 = 4$ m. It kept on rising, and v_y was +10 m/s as it rose past the height we've called $y_1 = 7$ m. Whenever you take a square root in a physics equation, you need to use physical sense to decide which root is the one you want.

The next example involves a complex process with several states of interest. If an energetic alpha particle (a helium nucleus: two protons and two neutrons) collides with a carbon nucleus, a nuclear reaction can occur, forming an oxygen nucleus whose rest energy is higher than an ordinary oxygen nucleus (an "excited state"; this concept is discussed in Chapter 8). After a short time, the excited oxygen nucleus can emit its excess energy as a photon (a particle of light), reducing its rest energy to that of an ordinary oxygen nucleus.

In this process there are four identifiable states:

State 1: Alpha particle and carbon nucleus moving toward each other (Figure 6.58)

State 2: Alpha particle and carbon nucleus just touch (Figure 6.59)

State 3: Excited oxygen nucleus at rest (extra rest energy) (Figure 6.60)

State 4: Oxygen nucleus and photon, heading in opposite directions (Figure 6.61)

Selecting different pairs of initial and final states allows us to answer different questions about the process.

| EXAMPLE | **Reaction Between an Alpha Particle and a Carbon Nucleus** |

An alpha particle (the nucleus of a helium atom, containing two protons and two neutrons) moving with magnitude of momentum p_1 is shot toward a carbon-12 nucleus (containing six protons and six neutrons) that is moving with the same magnitude of momentum p_1 toward the alpha particle. The speeds are nonrelativistic.

There is a head-on collision of the alpha particle and the carbon-12 nucleus, and an excited state of an oxygen-16 nucleus is formed at rest. A very short time later the oxygen nucleus drops to its ground state, emitting a photon (called a "gamma ray") whose energy is 10.352 MeV(1 MeV = 1×10^6 eV). The oxygen nucleus recoils with momentum equal and opposite to the photon's momentum, but negligible kinetic energy.

(a) What is the minimum (nonrelativistic) magnitude of momentum p_1 that the alpha particle must have in order for this sequence of phenomena to occur? Masses of the ground (not excited) states of selected nuclei, where $1 \, u = 1.6605 \times 10^{-27}$ kg :

$$He\text{-}4 \, (2p + 2n) = 4.00040868 \, u$$
$$C\text{-}12 \, (6p + 6n) = 11.99670852 \, u$$
$$O\text{-}16 \, (8p + 8n) = 15.99052636 \, u$$

(b) What is the kinetic energy in MeV of an alpha particle with the magnitude of momentum p_1 found in part (a)?

(c) Because the alpha particle and carbon nucleus both have positive electric charges, they repel each other. It could be the case that these particles need to start with extra kinetic energy in order to get close enough to touch and react. Find the minimum initial momentum p_{min} each particle must have to get close enough to touch. Compare this to your answer to part (a) to determine whether this reaction can actually occur.

A proton or neutron has a radius of roughly 1×10^{-15} m, and a nucleus is a tightly packed collection of nucleons. Experiments show that the radius of a nucleus containing N nucleons is approximately $(1.3 \times 10^{-15} \, m) \times N^{1/3}$.

| Solution | **(a)** Find p_1: |

System: All particles in existence (alpha + C, or excited O, or O + photon)
Surroundings: No external objects

Initial state: Very far apart, particles moving toward each other (State 1, Figure 6.58)
Final state: O nucleus and photon moving apart (State 4, Figure 6.61)

Energy Principle

■ $U_f = 0$ because there are no electrical or gravitational interactions (photon has no charge).
$U_i = 0$ because particles are far apart.
$K_O \approx 0$ (negligible recoil kinetic energy).
Photon (γ) has no rest energy, only kinetic energy.

$$E_f = E_i + \cancel{W}$$

$$(m_O c^2 + K_O) + K_\gamma + \cancel{U_f} = (m_\alpha c^2 + K_\alpha) + (m_C c^2 + K_C) + \cancel{U_i}$$

$$m_O c^2 + K_\gamma = m_\alpha c^2 + m_C c^2 + \frac{p_1^2}{2m_\alpha} + \frac{p_1^2}{2m_C}$$

$$p_1 = \sqrt{\frac{(m_O c^2 + K_\gamma - m_\alpha c^2 - m_C c^2)}{\dfrac{1}{2m_\alpha} + \dfrac{1}{2m_C}}}$$

In the numerator we need many significant figures to capture small rest mass changes:

$$(m_O - m_\alpha - m_C) = (15.99052636 - 4.00040868 - 11.99670852)\,u$$
$$= -0.00659084\,u$$

$$(-0.00659084\,u)(1.6605 \times 10^{-27}\,kg/u)(3 \times 10^8\,m/s)^2$$
$$= -9.8497 \times 10^{-13}\,J$$

$$K_\gamma = (10.352 \times 10^6\,eV)(1.60 \times 10^{-19}\,eV/J) = 16.5632 \times 10^{-13}\,J$$

$$p_1 = \sqrt{\frac{(16.5632 \times 10^{-13} - 9.8497 \times 10^{-13})\,J}{\frac{1}{2}\left(\frac{1}{4.000} + \frac{1}{11.997}\right)\left(\frac{1}{1.6605 \times 10^{-27}\,kg}\right)}}$$
$$= 8.18 \times 10^{-20}\,kg \cdot m/s$$

(b) $K_\alpha = \dfrac{p_1^2}{2m} = \dfrac{(8.18 \times 10^{-20}\,kg \cdot m/s)^2}{2(4 \times 1.66 \times 10^{-27}\,kg)}\left(\dfrac{1 \times 10^{-6}\,MeV}{1.6 \times 10^{-19}\,J}\right) = 3.15\,MeV$

Check: The units for momentum are correct.

(c) Find p_{min} for alpha and carbon to touch.

System: alpha particle and carbon nucleus.
Surroundings: No external objects.

Initial state: Very far apart, particles moving toward each other (State 1, Figure 6.58)
Final state: Particles at rest, touching each other (Figure 6.59). No change of identity yet

■ No change of rest energy yet.
Final state: particles at rest, so $K_f = 0$.
Initial state: $U_i = 0$ because particles far apart.
Approximate K is okay because $v \ll c$.

Energy Principle:

$$E_f = E_i + W$$
$$E_f = (m_a c^2 + \cancel{K_{a,f}}) + (m_C c^2 + \cancel{K_{C,f}}) + U_f$$
$$E_i = (m_a c^2 + K_{a,i}) + (m_C c^2 + K_{C,i}) + \cancel{U_i}$$
$$U_f = K_{a,i} + K_{C,i}$$
$$\frac{1}{4\pi\varepsilon_0}\frac{(2e)(6e)}{r} = \frac{p_{min}^2}{2m_a} + \frac{p_{min}^2}{2m_C}$$

■ $2e$ is charge of alpha particle. $6e$ is charge of carbon nucleus.

$$p_{min} = \sqrt{\frac{\frac{1}{4\pi\varepsilon_0}\frac{12e^2}{r}}{\frac{1}{2}\left(\frac{1}{m_a} + \frac{1}{m_C}\right)}}$$

final separation $r = (1.3 \times 10^{-15}\,m)(4^{1/3} + 12^{1/3}) = 5.04 \times 10^{-15}\,m$

$$p_{min} = \sqrt{\frac{\left(9 \times 10^9 \dfrac{N \cdot m^2}{C^2}\right)\dfrac{12(1.6 \times 10^{-19}\,C)^2}{(5.04 \times 10^{-15}\,m)}}{\frac{1}{2}\left(\frac{1}{4} + \frac{1}{12}\right)\left(\dfrac{1}{1.66 \times 10^{-27}\,kg}\right)}} = 7.39 \times 10^{-20}\,kg \cdot m/s$$

$$K_\alpha = \frac{p_{min}^2}{2m} = \frac{(7.39 \times 10^{-20}\,kg \cdot m/s)^2}{2(4 \times 1.66 \times 10^{-27}\,kg)}\left(\frac{1 \times 10^{-6}\,MeV}{1.6 \times 10^{-19}\,J}\right) = 2.57\,MeV$$

Check: The units for momentum and energy are okay.

Since the momentum required to touch is less than the momentum required to initiate the reaction (answer to part **(a)**), the reaction can in fact occur.

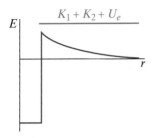

$$K_1 + K_2 + U_e$$

Figure 6.62 The initial kinetic energies are a bit greater than what is needed to overcome the electric potential energy barrier.

FURTHER DISCUSSION The potential energy graph for this process is shown in Figure 6.62. The electric potential energy rises (due to a repulsive interaction) as the particles get closer together, up to the point at which they touch. After this, the attractive strong (nuclear) force dominates the interaction, and the electric repulsion between the particles is insignificant.

This kind of potential energy curve is sometimes described as having a "Coulomb barrier." This means that $K + U$ total must be larger than the maximum value of U_{elec}.

We found that the incoming momentum p_{min} required to create the excited oxygen nucleus is greater than the momentum p_1 required just to overcome the electric potential energy barrier and bring the alpha particle and carbon nucleus in contact. It is necessary to bring nuclei into contact for a reaction to proceed, since the nuclear force has a very short range. However, in this reaction just making contact is insufficient. A bit more energy is needed in order to produce the larger-mass oxygen nucleus.

We never did use State 3 in our analysis. However, if we needed to find the mass of the excited O* nucleus, we would use this state as either the final or initial state.

6.18 *A PUZZLE

The principle of conservation of energy states that when a system's energy increases, the energy of the surroundings decreases by the same amount. For example, take a block as the system of interest. If you push the block away from you and raise its kinetic energy, your store of chemical energy decreases by the same amount. The flow of energy from you (the "surroundings") to the block takes the form of mechanical energy transfer we call "work." We can demonstrate that there is a puzzle concerning the energy in the surroundings of a star whose kinetic energy increases due to an interaction with another star.

Consider two stars of equal mass initially at rest, far apart (Figure 6.63). The two stars do work on each other:

$$\Delta E_1 = W_{on\,1} \quad \text{star 1 as system}$$
$$\Delta E_2 = W_{on\,2} \quad \text{star 2 as system}$$

Figure 6.63 Two stars with equal mass attract each other, starting from rest.

If we add these two energy equations for the two stars, we have this:

$$\Delta E_1 + \Delta E_2 = W_{on\,1} + W_{on\,2} \quad \text{(sum of two energy equations)}$$

Moving $(W_{on\,1} + W_{on\,2})$ to the left side of the equation and renaming the negative of this quantity ΔU, we have this familiar result:

$$\Delta E_1 + \Delta E_2 + \Delta U = 0 \quad \text{(energy equation for two stars)}$$

Our interpretation of this equation is that the gravitational potential energy U of the two-star system (calculated from the work on the individual particles) decreases as the particle energies increase. This point of view is valid, yields physically correct results, and is the view that we will take in this introductory textbook.

However, it is not difficult to show that there is a puzzle concerning energy flow between one of the stars and its surroundings. Consider star 1 as the system. Since it is a single object, there is no potential energy U, no pairwise interaction energy. The energy of star 1 increases due to work done on it: $\Delta E_1 = E_{on\,1}$. There are very strong reasons to believe in energy conservation, so we expect that when star 1 gains an amount of energy ΔE_1 the surroundings of star 1 should lose that much energy.

QUESTION Part of the surroundings is star 2. Does star 2 lose energy?

No! Star 2 doesn't lose energy, it *gains* energy, and of the same amount $\Delta E_2 = \Delta E_1$, due to the symmetry of the situation, with the two stars having equal mass.

What else is there in the surroundings that could experience a change in energy? The only other matter is star 2, but its energy change is inconsistent with the principle of energy conservation. Evidently our model of the world is incomplete. To be able to analyze this situation fully, we need to introduce the abstract idea of a "field." Fields, which are the subject of later chapters of this textbook, are associated with objects having mass (gravitational fields) or charge (electric and magnetic fields), and fields extend throughout all space. It is energy stored in the gravitational field that accounts for the energy in the surroundings of star 1. Fields can also have momentum, as we will see in the chapters that deal with electric and magnetic interactions.

Despite the fact that we haven't yet defined what a "field" is, we can determine indirectly how much the field energy must change. Since the energy in the surroundings of star 1 changes by an amount $-\Delta E_1$, and the energy of star 2 increases by an amount $+\Delta E_1$, it must be that the energy stored in the gravitational field changes by an amount $-2\Delta E_1$. It is an indication of the power of the principle of conservation of energy that we can calculate the energy change of an entity that we know nothing about! This energy change is precisely the amount of change in the potential energy of the two-star system: $\Delta U = -(\Delta E_1 + \Delta E_2) = -2\Delta E_1$.

Because the potential energy U correctly accounts for the energy changes, we don't have to consider field energy in our calculations. Our analyses in terms of U that focus on particles and their interactions, not their surroundings, are completely correct. However, the difficulty in accounting for the energy in the surroundings of a star suggests that eventually we will need to include fields in our models of the world.

Steep negative slope, large force to right

Shallow negative slope, small force to right

Steep positive slope, large force to left

Zero slope, equilibrium position

Figure 6.64 The negative of the slope (gradient) of a graph of potential energy vs. separation gives the force.

6.19 *GRADIENT OF POTENTIAL ENERGY

Because the (negative) gradient of U is equal to the force, the force is the (negative) slope of a graph of potential energy. This means that you can mentally read the force right off a potential energy graph (Figure 6.64). Where the curve is steep, the force is large (large gradient), and where the curve is shallow, the force is small. Where the slope of the curve is horizontal, the force is zero, and this represents a possible equilibrium position for the system. The component of the force is given by the negative gradient, so a negative slope corresponds to a force to the right, while a positive slope corresponds to a force to the left.

Actually, in determining the x component of the force, we differentiate with respect to x while holding y and z constant. This kind of derivative is called a "partial derivative" and has a special symbol to emphasize that everything else is held constant:

$$F_x = -\frac{\partial U}{\partial x}$$

To get all three components of the force we take partial derivatives of the potential energy with respect to x, y, and z:

FORCE IS NEGATIVE GRADIENT OF POTENTIAL ENERGY

$$F_x = -\frac{\partial U}{\partial x}, \; F_y = -\frac{\partial U}{\partial y}, \; F_z = -\frac{\partial U}{\partial z}$$

$$\vec{F} = \left\langle -\frac{\partial U}{\partial x}, -\frac{\partial U}{\partial y}, -\frac{\partial U}{\partial z} \right\rangle$$

(Another standard mathematical notation for this relation is $\vec{F} = -\vec{\nabla}U$.)

6.X.34 With y pointing upward from the center of the Earth, find the y component of the gravitational force from $-\partial U/\partial y$, where $U = -GMm/y$. Find the x component of the force from $-\partial U/\partial x$.

6.20 *INTEGRALS AND ANTIDERIVATIVES

In determining the formula for gravitational potential energy, we started from these complementary formulas:

$$\Delta U_g = -F_r \Delta r \quad \text{and} \quad F_r = -\frac{dU_g}{dr}$$

We calculated the change in potential energy from the negative of the internal work:

$$U_{s,f} - U_{s,i} = \Delta \left(-G\frac{m_1 m_2}{r} \right) = -\Sigma F_r \Delta r$$

However, if we had let Δr approach zero, the finite sum would have turned into an integral:

$$\Delta \left(-G\frac{m_1 m_2}{r} \right) = -\int_i^f F_r \, dr = \int_i^f G\frac{m_1 m_2}{r^2} \, dr$$

You can see that a definite integral (the infinitesimal version of a finite sum) can be evaluated in terms of the antiderivative. The antiderivative of $Gm_1 m_2/r^2$ is $-Gm_1 m_2/r$, and the change in the value of the antiderivative turns out to be equal to the definite integral (the infinite sum of infinitesimal elements).

The reason why this works this way can be traced back to the fact that the force is the derivative of the energy, so the energy must be the antiderivative of the force, yet the energy is also the sum of lots of $F_r \Delta r$'s. Hence the sum is given by the change in the antiderivative.

When the quantity to be integrated (here, the force) is a simple function, it may be easy to find the antiderivative. However, in many practical cases there is no known simple function that is the antiderivative, in which case an integral must be evaluated by numerical integration, as a finite sum.

6.21 *APPROXIMATION FOR KINETIC ENERGY

An approximation for the kinetic energy of a particle at low speeds can be obtained through use of the mathematics of the "binomial expansion":

$$(1 + \varepsilon)^n = 1 + n\varepsilon + \frac{n(n-1)}{2}\varepsilon^2 + \frac{n(n-1)(n-2)}{3 \cdot 2}\varepsilon^3 + \cdots$$

This formula comes from working through what happens when you multiply the quantity $(1 + \varepsilon)$ times itself n times. The first term is 1 multiplied by itself n times (1^n); the next term is made up of all those products, n of them, in which ε appears only once $[n\varepsilon]$; and so on.

Now for the important point that makes the binomial expansion useful in obtaining approximate results. If ε is a number that is very small compared to 1, each term in the expansion is much smaller than the preceding term:

$$n\varepsilon \text{ is much smaller than } 1, \frac{n(n-1)}{2}\varepsilon^2 \text{ is much smaller than } n\varepsilon, \text{ etc.}$$

Therefore we can get a good approximation if we keep just the first few terms and ignore the rest.

Although we have developed the binomial expansion for the case where n is an integer, it can be shown that the formula is valid even for noninteger values of n. In particular,

$$\frac{1}{\sqrt{1-\varepsilon}} = (1-\varepsilon)^{-1/2}$$

$$= 1 + \left(-\frac{1}{2}\right)(-\varepsilon) + \frac{(-\frac{1}{2})(-\frac{3}{2})}{2}(-\varepsilon)^2 + \frac{(-\frac{1}{2})(-\frac{3}{2})(-\frac{5}{2})}{6}(-\varepsilon)^3 + \cdots$$

If v/c is small compared to 1, then $(v/c)^2$ is even smaller compared to 1. For example, 1/10 is small compared to one, but $(1/10)^2 = 1/100$ is very small indeed. Therefore we can write this for low speeds:

$$\frac{mc^2}{\sqrt{1-(v/c)^2}} = mc^2\left[1 + \frac{1}{2}\left(\frac{v}{c}\right)^2 + \frac{3}{8}\left(\frac{v}{c}\right)^4 + \frac{5}{16}\left(\frac{v}{c}\right)^6 + \cdots\right]$$

$$\frac{mc^2}{\sqrt{1-(v/c)^2}} \approx mc^2 + \frac{1}{2}mv^2$$

The last term can be rewritten in terms of momentum by inserting $v \approx p/m$ into $K \approx \frac{1}{2}mv^2$:

$$K \approx \frac{1}{2}mv^2 \approx \frac{1}{2}m\left(\frac{p}{m}\right)^2 = \frac{p^2}{2m}$$

Figure 6.65 Choose the x axis to be in the direction of the motion. The net force acting on the particle has a component F_x in the direction of the motion.

6.22   *FINDING THE FORMULA FOR PARTICLE ENERGY

If we choose our axes so that x is in the direction of the motion at this instant (Figure 6.65) we have

$$\Delta E = F_x \Delta x = \left(\frac{\Delta p_x}{\Delta t}\right)\Delta x$$

$$\text{So } \frac{\Delta E}{\Delta x} = \frac{\Delta p_x}{\Delta t}$$

for the change in energy of a point particle acted on by a net force. In the limit of small displacement and short time interval, the energy E should satisfy the following relationship between two derivatives:

$$\frac{dE}{dx} = \frac{dp_x}{dt}$$

We will show that if we choose the following function for E, the relationship between the two derivatives will be satisfied:

$$E = \frac{mc^2}{\sqrt{1 - (v/c)^2}}$$

The proof proceeds by calculating dE/dx, calculating dp_x/dt, and showing that the two derivatives are equal to each other. First, here is dE/dx:

$$\frac{dE}{dx} = \frac{d}{dx}\left(\frac{mc^2}{\sqrt{1 - (v/c)^2}}\right)$$

$$= \frac{mc^2}{(1 - (v/c)^2)^{3/2}}\left(-\frac{1}{2}\right)\left(-\frac{2v}{c^2}\right)\frac{dv}{dx}$$

$$\frac{dE}{dx} = \frac{mv}{(1 - (v/c)^2)^{3/2}}\frac{dv}{dx}$$

Next we calculate dp_x/dt, keeping in mind that $v_x = v$, because we chose our axes so that x is in the direction of the motion at this instant:

$$\frac{dp_x}{dt} = \frac{d}{dt}\left(\frac{mv}{\sqrt{1 - (v/c)^2}}\right)$$

$$= \frac{m}{\sqrt{1 - (v/c)^2}}\frac{dv}{dt} + \frac{mv}{(1 - (v/c)^2)^{3/2}}\left(-\frac{1}{2}\right)\left(-\frac{2v}{c^2}\right)\frac{dv}{dt}$$

$$\frac{dp_x}{dt} = \frac{m}{(1 - (v/c)^2)^{3/2}}\left[\left(1 - \frac{v^2}{c^2}\right) + \frac{v^2}{c^2}\right]\frac{dv}{dt}$$

$$= \frac{m}{(1 - (v/c)^2)^{3/2}}\frac{dv}{dt}$$

This is almost in the same form as the derivative dE/dx. Now note this:

$$\frac{dv}{dt} = \left(\frac{dv}{dt}\right)\left(\frac{dx}{dx}\right)$$

$$= \left(\frac{dv}{dx}\right)\left(\frac{dx}{dt}\right) = \frac{dv}{dx}v, \quad \text{because } v = \frac{dx}{dt}$$

Substituting this result for dv/dt into the derivative dp_x/dt, we find that this derivative is indeed equal to the derivative dE/dx if E is defined as

$$E = \frac{mc^2}{\sqrt{1 - (v/c)^2}}$$

Hence the relativistic formula for energy does follow from the relativistic formula for momentum.

It might seem that we could define the particle energy with an additive constant, which would drop out when we took the derivative dE/dx. However, when one considers a variety of particle reactions, it is found that to account for energy in a consistent way requires that the additive constant be zero.

This approach to solving a "differential equation" (an equation containing derivatives), by proposing a solution and showing that it satisfies the equation, is a legitimate technique for solving differential equations, and in fact is used quite frequently.

SUMMARY

The unit of energy is the joule, abbreviated "J."

Energy of a single particle

The energy of a particle with rest mass m is:

$$E_{\text{particle}} \equiv \gamma m c^2 \quad \text{where } \gamma = \frac{1}{\sqrt{1 - (v/c)^2}}$$

Particle energy has two parts: rest energy and kinetic energy:

$$E_{\text{particle}} = mc^2 + K$$

The "rest energy" is the energy of a particle at rest.

$$\text{rest energy} = mc^2$$

The kinetic energy K of a particle is the energy a moving particle has in addition to its rest energy.

$$K = \gamma mc^2 - mc^2$$

For a particle whose speed is low compared to the speed of light, kinetic energy is approximately:

$$K \approx \frac{1}{2}mv^2 = \frac{p^2}{2m} \quad \text{(if } v \ll c\text{)}$$

For a particle with any speed:

$$E^2 - (pc)^2 = (mc^2)^2$$

for a massless particle this reduces to

$$E = pc$$

The Energy Principle

$$\Delta E_{\text{sys}} = W_{\text{surr}}$$

The change in energy of a system is equal to the work done by the surroundings.

Work

Work W_F done by a single constant force $\vec{F}$:

$$W_F = \vec{F} \bullet \Delta \vec{r} = F_x \Delta r_x + F_y \Delta r_y + F_z \Delta r_z$$
$$= F \Delta r \cos \theta$$

Work done by a varying force:

$$W_F = \Sigma \vec{F} \bullet \Delta \vec{r} \text{ or } W = \int_i^f \vec{F} \bullet d\vec{r}$$

Work can be positive, negative, or zero. The sign of work depends on the relative directions of the force and the displacement of the point of application of the force.

Conservation of energy

$$\Delta E_{\text{sys}} + \Delta E_{\text{surr}} = 0$$

Potential energy

Multiparticle system energy consists of particle energy and potential energy:

$$E_{\text{sys}} \equiv (K_1 + m_1 c^2) + (K_2 + m_2 c^2) + \cdots$$
$$+ (U_{12} + \cdots)$$

Potential energy U is the energy of interaction of a pair of objects, and change in U is the negative of work done by internal forces:

$$\Delta U = -W_{\text{int}}$$

Force is negative gradient of potential energy:

$$F_x = -\frac{dU}{dx}$$

Potential energy U depends on the separation r between two objects.
Potential energy U goes to zero as objects get very far apart:

$$U \to 0 \text{ as } r \to \infty$$

For an attractive interaction, U becomes negative as separation decreases.
For a repulsive interaction U becomes positive as separation decreases.

Gravitational potential energy

$$U = -G\frac{m_1 m_2}{r}$$
$$G = 6.7 \times 10^{-11} \text{N} \cdot \text{m}^2 / \text{kg}^2$$

m_1 and m_2 are the masses of the two interacting objects, and r is the center-to-center distance between them.

$$\Delta U \approx \Delta (mgy) \text{ near Earth's surface}$$

Electric potential energy

$$U_{\text{elec}} = \frac{1}{4\pi \varepsilon_0} \frac{q_1 q_2}{r}$$

$$\frac{1}{4\pi \varepsilon_0} = 9 \times 10^9 \text{N} \cdot \text{m}^2 / \text{coulomb}^2$$

q_1 and q_2 are the charges of the two interacting particles, in coulombs, and r is the center-to-center distance between them.

Nuclear binding:

Heavy nuclei can fission, light nuclei can fuse, with energy released in both cases.

$$\text{Electron volt: } 1 \text{ eV} = 1.6 \times 10^{-19} \text{ J}$$

EXERCISES AND PROBLEMS

Sections 6.1–6.2

6.X.35 Show that the units of $p^2/(2m)$ and mc^2 are indeed joules. (Note that $1\,\text{N}$ is $1\,\text{kg} \cdot (\text{m/s})^2$.)

6.X.36 It is not very difficult to accelerate an electron to a speed that is 99% of the speed of light, because the electron has such a very small mass. What is the ratio of the kinetic energy K to the rest energy mc^2 in this case? In the definition of what we mean by kinetic energy K, $E = mc^2 + K$, you must use the full relativistic formula for E, because v/c is not small compared to 1.

6.X.37 A pitcher can throw a baseball at about 100 miles/hour (about 44 m/s). What is the ratio of the kinetic energy to the rest energy mc^2? (Can you use $K \approx \frac{1}{2}mv^2$?)

6.X.38 Approximately what is the kinetic energy of a baseball pitcher's fastball? Of a basketball passed from one player to another?

6.X.39 What is the speed of an electron whose total energy is equal to the total energy of a proton that is at rest? What is the kinetic energy of this electron?

6.X.40 The point of this question is to compare rest energy and kinetic energy at low speeds. A baseball is moving at a speed of 17 m/s. Its mass is 145 grams (0.145 kg).

(a) What is its rest energy?

(b) Is it okay to calculate its kinetic energy using the formula $\frac{1}{2}mv^2$?

(c) What is its kinetic energy?

(d) Which is true?

A. The kinetic energy is approximately equal to the rest energy.

B. The kinetic energy is much bigger than the rest energy.

C. The kinetic energy is much smaller than the rest energy.

6.X.41 The point of this question is to compare rest energy and kinetic energy at high speeds. An alpha particle (a helium nucleus) is moving at a speed of 0.9993 times the speed of light. Its mass is 6.40×10^{-27} kg.

(a) What is its rest energy?

(b) Is it okay to calculate its kinetic energy using the formula $\frac{1}{2}mv^2$?

(c) What is its kinetic energy?

(d) Which is true?

A. The kinetic energy is approximately equal to the rest energy.

B. The kinetic energy is much bigger than the rest energy.

C. The kinetic energy is much smaller than the rest energy.

6.X.42 A runner whose mass is 60 kg runs in the $+x$ direction at a speed of 7 m/s.

(a) What is the kinetic energy of the runner?

(b) The runner turns around and runs in the $-x$ direction at the same speed. Now what is the kinetic energy of the runner?

6.X.43 A baseball of mass 144 g has a velocity of $\langle 22, 23, -11 \rangle$ m/s. What is the kinetic energy of the baseball?

6.X.44 One mole of helium atoms has a mass of 4 grams. If a helium atom in a balloon has a kinetic energy of 1.437×10^{-21} J, what is the speed of the helium atom? (The speed is much lower than the speed of light.)

6.X.45 You throw a ball of mass 160 g upward (Figure 6.66). When the ball is 2 m above the ground, headed upward (the initial state), its speed is 19 m/s. Later, when the ball is again 2 m above the ground, this time headed downward (the final state), its speed is 19 m/s. What is the change in the kinetic energy of the ball from initial to final state?

Figure 6.66

6.X.46 A fan cart of mass 0.8 kg initially has a velocity of $\langle 0.9, 0, 0 \rangle$ m/s. Then the fan is turned on, and the air exerts a constant force of $\langle -0.3, 0, 0 \rangle$ N on the cart for 1.5 seconds.

(a) What is the change in momentum of the fan cart over this 1.5 second interval?

(b) What is the change in kinetic energy of the fan cart over this 1.5-second interval?

Sections 6.3–6.4

6.X.47 You push a crate 3 m across the floor with a 40 N force whose direction is 30° below the horizontal. How much work do you do?

6.X.48 You pull your little sister across a flat snowy field on a sled. Your sister plus the sled have a mass of 20 kg. The rope is at an angle of 35 degrees to the ground. You pull a distance of 50 m with a force of 30 N. How much work do you do?

6.X.49 Give brief explanations for your answers to each of the following questions:

(a) You hold a 1 kg book in your hand for 1 minute. How much work do you do on the book?

(b) In a circular pendulum, how much work is done by the string on the mass in one revolution?

(c) For a mass oscillating horizontally on a spring, how much work is done by the spring on the mass in one complete cycle? In a half cycle?

6.X.50 An object that is originally at location $\langle -17, 0, 0 \rangle$ m moves to location $\langle -25, 0, 0 \rangle$ m as shown in Figure 6.67. While it is moving it is acted on by a constant force of $\langle 22, 0, 0 \rangle$ N.

Final   Initial
location   location

Figure 6.67

(a) How much work is done on the object by this force?

(b) Does the kinetic energy of the object increase or decrease?

Figure 6.68

(c) A different object moves from location $\langle -25, 0, 0 \rangle$ m to location $\langle -17, 0, 0 \rangle$ m, as shown in Figure 6.68. While it is moving it is acted on by a constant force of $\langle 22, 0, 0 \rangle$ N. How much work is done on the second object by this force?

(d) Does the kinetic energy of the object increase or decrease?

6.X.51 A ball of mass 0.7 kg falls downward, as shown in Figure 6.69. Initially you observe it to be 4.5 m above the ground. After a short time it is just about to hit the ground.

Figure 6.69

(a) During this interval how much work was done on the ball by the gravitational force?

(b) Does the kinetic energy of the ball increase or decrease?

Figure 6.70

(c) The ball hits the ground and bounces back upward, as shown in Figure 6.70. After a short time it is 4.5 m above the ground again. During this second interval (between leaving the ground and reaching a height of 4.5 m) how much work was done on the ball by the gravitational force?

(d) Does the kinetic energy of the ball increase or decrease?

6.X.52 A lithium nucleus has mass 5.1×10^{-27} kg. If its speed is $0.984c$ (that is, $|\vec{v}|/c = 0.984$), what are the values of the particle energy, the rest energy, and the kinetic energy?

Next an electric force acts on the lithium nucleus and does 4.7×10^{-9} J of work on the particle. Now what are the values of the particle energy, the rest energy, and the kinetic energy?

6.X.53 Check all of the following that are correct statements, where $E = \gamma mc$. Read each statement very carefully to make sure that it is exactly correct.

A. The definition of work is $W = \langle F_x \Delta x, F_y \Delta y, F_z \Delta z \rangle$.
B. The definition $K = E - mc^2$ is valid even for speeds near the speed of light.
C. The definition of work is $W = |\langle F_x \Delta x, F_y \Delta y, F_z \Delta z \rangle|$.
D. At speeds close to the speed of light, kinetic energy is approximately equal to $\frac{1}{2}mv^2$.
E. The definition of work is $W = F_x \Delta x + F_y \Delta y + F_z \Delta z$.
F. The energy principle can be written $E_{\text{sys}, f} = E_{\text{sys},i} + W_{\text{surr}}$.
G. The energy principle can be written $\Delta E_{\text{sys}} = W_{\text{surr}}$.

6.P.54 An object with mass 120 kg moved in outer space from location $\langle 11, -24, -5 \rangle$ to location $\langle 17, -28, -10 \rangle$ m. A single constant force $\langle 250, 440, -220 \rangle$ N acted on the object while the object moved. Its final speed was 7.9 m/s. What was the speed of the object at the initial location?

6.P.55 An object with mass 120 kg moved in outer space. When it was at location $\langle 7, -20, -8 \rangle$ its speed was 12 m/s. A single constant force $\langle 250, 490, -160 \rangle$ N acted on the object while the object moved to location $\langle 10, -29, -13 \rangle$ m. What is the speed of the object at this final location?

6.X.56 You pull a block of mass m across a frictionless table with a constant force. You also pull with an equal constant force a block of larger mass M. The blocks are initially at rest. If you pull the blocks through the same distance, which block has the greater kinetic energy, and which block has the greater momentum? If instead you pull the blocks for the same amount of time, which block has the greater kinetic energy, and which block has the greater momentum?

6.P.57 Jack and Jill are maneuvering a 3000 kg boat near a dock. Initially the boat's position is $\langle 2, 0, 3 \rangle$ m and its speed is 1.3 m/s. As the boat moves to position $\langle 4, 0, 2 \rangle$ m, Jack exerts a force $\langle -400, 0, 200 \rangle$ N and Jill exerts a force $\langle 150, 0, 300 \rangle$ N.

(a) How much work does Jack do?

(b) How much work does Jill do?

(c) Without doing any calculations, say what is the angle between the (vector) force that Jill exerts and the (vector) velocity of the boat. Explain briefly how you know this.

(d) Assuming that we can neglect the work done by the water on the boat, what is the final speed of the boat?

6.P.58 An electron traveling at a speed $0.99c$ encounters a region where there is a constant electric force directed opposite to its momentum. After traveling 3 m in this region, the electron's speed was observed to decrease to $0.93c$. What was the magnitude of the electric force acting on the electron?

6.P.59 A mass of 0.12 kg hangs from a vertical spring in the lab room. You grab hold of the mass and throw it vertically downward. The speed of the mass just after leaving your hand is 3.40 m/s.

(a) While the mass moved downward a distance of 0.07 m, how much work was done on the mass by the Earth?

(b) The speed of the mass has decreased to 2.85 m/s. How much work was done on the mass by the spring?

6.P.60 An electron (mass 9×10^{-31} kg) is traveling at a speed of $0.91c$ in an electron accelerator. An electric force of 1.6×10^{-13} N is applied in the direction of motion while the electron travels a distance of 2 m. You need to find the new speed of the electron. Which of the following steps must be included in your solution to this problem?

   A. Calculate the initial particle energy $\gamma_i mc^2$ of the electron.

   B. Calculate the final particle energy $\gamma_f mc^2$ of the electron.

   C. Determine how much time it takes to move this distance.

   D. Use the formula $\frac{1}{2}m|\vec{v}|^2$ to find the kinetic energy of the electron.

   E. Calculate the net work done on the electron.

   F. Use the final energy of the electron to find its final speed.

What is the new speed of the electron as a fraction of c?

6.P.61 SLAC, the Stanford Linear Accelerator Center, located at Stanford University in Palo Alto, California, accelerates electrons through a vacuum tube two miles long (it can be seen from an overpass of the Junipero Serra freeway that goes right over the accelerator). Electrons that are initially at rest are subjected to a continuous force of 2×10^{-12} N along the entire length of two miles (one mile is 1.6 kilometers) and reach speeds very near the speed of light. A similar analysis in a previous chapter required numerical integration, but with the new techniques of this chapter you can analyze the motion analytically.

   (a) Calculate the final energy, momentum, and speed of the electron.

   (b) Calculate the approximate time required to go the two-mile distance.

Sections 6.5–6.7

6.X.62 One end of a spring whose spring constant is 20 N/m is attached to the wall, and you pull on the other end, stretching it from its equilibrium length of 0.2 m to a length of 0.3 m. Estimate the work done by dividing the stretching process into two stages and using the average force you exert to calculate work done during each stage.

6.X.63 You bring a boat toward the dock by pulling on a rope with a force of 130 newtons through a distance of 6 meters.

   (a) How much work do you do? (Include the appropriate sign.)

   (b) Then you slow the boat down by pushing against it with a force of 40 newtons, opposite to the boat's movement of 5 meters. How much work do you do? (Include the appropriate sign.)

   (c) What is the total amount of work that you do?

6.X.64 A space probe in outer space has a mass of 111 kg, and it is traveling at a speed of 29 m/s. When it is at location $\langle 445, 535, -350 \rangle$ m it begins firing two booster rockets. The rockets exert constant forces of $\langle 90, 150, 195 \rangle$ N, and $\langle 90, -90, -585 \rangle$ N, respectively. The rockets fire until the spacecraft reaches location $\langle 449, 541, -354 \rangle$ m. Now what is its speed? There is negligible mass loss due to the rocket exhaust.

6.P.65 An object with mass 100 kg moved in outer space. When it was at location $\langle 9, -24, -4 \rangle$ its speed was 3.5 m/s. A single constant force $\langle 250, 400, -170 \rangle$ N acted on the object while the object moved from location $\langle 9, -24, -4 \rangle$ m to location $\langle 15, -17, -8 \rangle$ m. Then a different single constant force $\langle 140, 250, 150 \rangle$ N acted on the object while the object moved from location $\langle 15,$

$-17, -8 \rangle$ m to location $\langle 19, -24, -3 \rangle$ m. What is the speed of the object at this final location?

6.P.66 Outside the space shuttle, you and a friend pull on two ropes to dock a satellite whose mass is 700 kg. The satellite is initially at position $\langle 3.5, -1, 2.4 \rangle$ m and has a speed of 4 m/s. You exert a force $\langle -400, 310, -250 \rangle$ N. When the satellite reaches the position $\langle 7.1, 3.2, 1.2 \rangle$ m, its speed is 4.01 m/s. How much work did your friend do?

6.P.67

   (a) You bring a boat toward the dock by pulling on a rope with a force of 240 newtons through a distance of 7 meters. How much work do you do?

   (b) Next you slow the boat down by pushing against it with a force of 60 newtons, opposite to the boat's movement of 3 meters. How much work do you do?

   (c) What is the total amount of work that you do?

6.P.68 You push a box out of a carpeted room and along a hallway with a waxed linoleum floor. While pushing the crate 2 m out of the room you exert a force of 34 N; while pushing it 6 m along the hallway you exert a force of 13 N. To slow it down you exert a force of 40 N through a distance of 2 m, opposite to the motion. How much work do you do in all?

6.P.69 A crate with a mass of 100 kg glides through a space station with a speed of 3.5 m/s. An astronaut speeds it up by pushing on it from behind with a force of 180 N, continually pushing with this force through a distance of 6 m. The astronaut moves around to the front of the crate and slows the crate down by pushing backward with a force of 170 N, backing up through a distance of 4 m. After these two maneuvers, what is the speed of the crate?

6.X.70 In positron-emission tomography (PET) used in medical research and diagnosis, compounds containing unstable nuclei that emit positrons are introduced into the brain, destined for a site of interest in the brain. When a positron is emitted, it goes only a short distance before coming nearly to rest. It forms a bound state with an electron, called "positronium," which is rather similar to a hydrogen atom. The binding energy of positronium is very small compared to the rest energy of an electron. After a short time the positron and electron annihilate. In the annihilation, the positron and the electron disappear, and all of their rest energy goes into two photons (particles of light) that have zero mass; all their energy is kinetic energy. These high-energy photons, called "gamma rays," are emitted at nearly 180° to each other. What energy of gamma ray (in MeV, million electron volts) should each of the detectors be made sensitive to? (The mass of an electron or positron is 9×10^{-31} kg.)

6.P.71 A nucleus whose mass is 3.499612×10^{-25} kg undergoes spontaneous "alpha" decay. The original nucleus disappears and there appear two new particles: a He-4 nucleus of mass 6.640678×10^{-27} kg (an "alpha particle" consisting of two protons and two neutrons) and a new nucleus of mass 3.433132×10^{-25} kg (note that the new nucleus has less mass than the original nucleus, and it has two fewer protons and two fewer neutrons).

   (a) When the alpha particle has moved far away from the new nucleus (so the electric interactions are negligible), what is the combined kinetic energy of the alpha particle and new nucleus?

   (b) How many electron volts is this?

In contrast to this nuclear reaction, chemical reactions typically involve only a few eV.

6.P.72 In a location in outer space far from all other objects, a nucleus whose mass is 3.894028×10^{-25} kg and that is initially at rest undergoes spontaneous "alpha" decay. The original nucleus disappears, and two new particles appear: a He-4 nucleus of mass 6.640678×10^{-27} kg (an "alpha particle" consisting of two protons and two neutrons) and a new nucleus of mass 3.827555×10^{-25} kg. These new particles move far away from each other, because they repel each other electrically (both are positively charged).

Because the calculations involve the small difference of (comparatively) large numbers, you need to keep 7 significant figures in your calculations, and you need to use the more accurate value for the speed of light, 2.99792e8 m/s.

Choose all particles as the system. Initial state: Original nucleus, at rest. Final state: Alpha particle + new nucleus, far from each other.

(a) What is the rest energy of the original nucleus? *Give 7 significant figures.*

(b) What is the sum of the rest energies of the alpha particle and the new nucleus? *Give 7 significant figures.*

(c) Did the portion of the total energy of the system contributed by rest energy increase or decrease?

(d) What is the sum of the kinetic energies of the alpha particle and the new nucleus?

Sections 6.8–6.10

6.X.73 You throw a ball straight up, and it reaches a height of 20 meters above your hand before falling back down. What was the speed of the ball just after it left your hand?

6.X.74 A 1 kg block rests on the Earth's surface. How much energy is required to move the block very far from the Earth, ending up at rest again?

6.X.75 The gravitational force inside the Earth is given approximately by $(GMm/R^3)r = gr/R$, where r is the distance from the center and R is the radius of the Earth; see Problem 6.P.95 for additional discussion. Draw the potential energy as a function of r, from $r = 0$ to very large r ($\gg R$). Note that the potential energy curve must be continuous, with no breaks.

6.X.76 One often hears the statement, "Nuclear energy production is fundamentally different from chemical energy production (such as burning of coal), because the nuclear case involves a change of mass." Critique this statement. Discuss the similarities and differences of the two kinds of energy production.

6.X.77 Figure 6.71 shows the path of a comet orbiting a star.

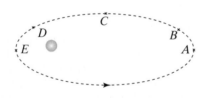

Figure 6.71

(a) Rank-order the locations on the path in terms of the magnitude of the comet's momentum at each location, starting with the location at which the magnitude of the momentum is the largest.

(b) Rank-order the locations on the path in terms of the comet's kinetic energy at each location, starting with the location at which the kinetic energy is the largest.

(c) Consider the system of the comet plus the star. Which of the following statements are correct?

A. External work must be done on the system to speed up the comet.

B. As the comet slows down, energy is lost from the system.

C. As the comet's kinetic energy increases, the gravitational potential energy of the system also increases.

D. As the comet slows down, the kinetic energy of the system decreases.

E. As the kinetic energy of the system increases, the gravitational potential energy of the system decreases.

(d) Still considering the system of the comet plus the star, which of the following statements are correct?

A. The sum of the kinetic energy of the system plus the gravitational potential energy of the system is a positive number.

B. At every location along the comet's path the gravitational potential energy of the system is negative.

C. The gravitational potential energy of the system is inversely proportional to the square of the distance between the comet and star.

D. The sum of the kinetic energy of the system plus the gravitational potential energy of the system is the same at every location along this path.

E. Along this path the gravitational potential energy of the system is never zero.

(e) Rank-order the locations on the path in terms of the potential energy of the system at each location, largest first. (Remember that $-3 > -5$).

6.X.78 Figure 6.72 is a graph of the energy of a system of a planet interacting with a star. The gravitational potential energy U_g is shown as the thick curve, and plotted along the vertical axis are various values of $K + U_g$.

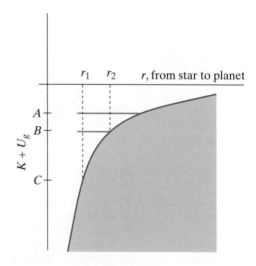

Figure 6.72

Suppose that $K + U_g$ of the system is A. Which of the following statements are true?

- A. The potential energy of the system decreases as the planet moves from r_1 to r_2.
- B. When the separation between the two bodies is r_2, the kinetic energy of the system is $(A - B)$.
- C. The system is a bound system; the planet can never escape.
- D. The planet will escape.
- E. When the separation between the two bodies is r_2, the kinetic energy of the system is $(B - C)$.
- F. The kinetic energy of the system is greater when the distance between the star and planet is r_1 than when the distance between the two bodies is r_2.

Suppose instead that $K + U_g$ of the system is B. Which of the following statements are true?

- A. When the separation between the planet and star is r_2, the kinetic energy of the system is zero.
- B. The planet and star cannot get farther apart than r_2.
- C. This is not a bound system; the planet can escape.
- D. When the separation between the planet and star is r_2, the potential energy of the system is zero.

6.X.79 Under certain conditions the interaction between a "polar" molecule such as HCl located at the origin and an ion located along the x axis can be described by a potential energy $U = -b/x^2$, where b is a constant. What is F_x, the x component of the force on the ion? What is F_y, the y component of the force on the ion?

Sections 6.10–6.11

6.X.80 The radius of the Moon is 1750 km, and its mass is 7×10^{22} kg. What would be escape speed from an isolated Moon? Why was a small rocket adequate to lift the lunar astronauts back up from the surface of the Moon?

6.P.81 The radius of Mars (from the center to just above the atmosphere) is 3400 km (3400×10^3 m), and its mass is 0.6×10^{24} kg. An object is launched straight up from just above the atmosphere of Mars.

- (a) What initial speed is needed so that when the object is far from Mars its final speed is 1000 m/s?
- (b) What initial speed is needed so that when the object is far from Mars its final speed is 0 m/s? (This is called the "escape speed.")

6.P.82 The radius of an airless planet is 2000 km (2×10^6 m), and its mass is 1.2×10^{23} kg. An object is launched straight up from the surface.

- (a) What initial speed is needed so that when the object is far from the planet its final speed is 900 m/s?
- (b) What initial speed is needed so that when the object is far from the planet its final speed is 0 m/s? (This is called the "escape speed.")

6.P.83 The escape speed from an asteroid whose radius is 10 km is only 10 m/s. If you throw a rock away from the asteroid at a speed of 20 m/s, what will be its final speed?

6.P.84 The escape speed from a very small asteroid is only 24 m/s. If you throw a rock away from the asteroid at a speed of 35 m/s, what will be its final speed?

6.P.85 Use energy conservation to calculate analytically (that is, without doing a numerical integration) the final speed of the spacecraft just before it hits the Moon. Include the gravitational effect of the Moon. Use a launch speed of 1.3×10^4 m/s. Modify your program for the Chapter 3 problem on the Ranger 7 mission to the Moon to print out the speed of the spacecraft when it hits the surface of the Moon, and compare this value to your analytical result. What questions that could be addressed in the numerical integration are you *not* able to answer by doing this energy calculation?

6.P.86 A spacecraft is coasting toward Mars. The mass of Mars is 6.4×10^{23} kg and its radius is 3400 km (3.4×10^6 m). When the spacecraft is 7000 km (7×10^6 m) from the center of Mars, the spacecraft's speed is 3000 m/s. Later, when the spacecraft is 4000 km (4×10^6 m) from the center of Mars, what is its speed? Assume that the effects of Mars's two tiny moons, the other planets, and the Sun are negligible. Precision is required to land on Mars, so make an accurate calculation, not a rough, approximate calculation.

6.P.87 Calculate the speed of a satellite in a circular orbit near the Earth (just above the atmosphere). If the mass of the satellite is 200 kg, what is the minimum energy required to move the satellite from this near-Earth orbit to very far away from the Earth?

6.P.88 A comet is in an elliptical orbit around the Sun. Its closest approach to the Sun is a distance of 4×10^{10} m (inside the orbit of Mercury), at which point its speed is 8.17×10^4 m/s. Its farthest distance from the Sun is far beyond the orbit of Pluto. What is its speed when it is 6×10^{12} m from the Sun? (This is the approximate distance of Pluto from the Sun.)

6.P.89 An electron is traveling at a speed of $0.95c$ in an electron accelerator. An electric force of 1.6×10^{-13} N is applied in the direction of motion while the electron travels a distance of 2 m. What is the new speed of the electron?

6.P.90 You stand on a spherical asteroid of uniform density whose mass is 2×10^{16} kg and whose radius is 10 km (10^4 m). These are typical values for small asteroids, although some asteroids have been found to have much lower average density and are thought to be loose agglomerations of shattered rocks.

- (a) How fast do you have to throw the rock so that it never comes back to the asteroid and ends up traveling at a speed of 3 m/s when it is very far away?
- (b) Sketch graphs of the kinetic energy of the rock, the gravitational potential energy of the rock plus asteroid, and their sum, as a function of separation (distance from center of asteroid to rock). Label the graphs clearly.

6.P.91 This problem is closely related to the spectacular impact of the comet Shoemaker-Levy with Jupiter in July 1994:

http://www.jpl.nasa.gov/sl9/sl9.html

A rock far outside our Solar System is initially moving very slowly relative to the Sun, in the plane of Jupiter's orbit around the Sun. The rock falls toward the Sun, but on its way to the Sun it collides with Jupiter. Calculate the rock's speed just before colliding with Jupiter. Explain your calculation and any approximations that you make.

$M_{Sun} = 2 \times 10^{30}$ kg, $M_{Jupiter} = 2 \times 10^{27}$ kg

Distance, Sun to Jupiter = 8×10^{11} m

Radius of Jupiter = 1.4×10^8 m

Section 6.12

6.X.92 Use energy conservation to find the approximate final speed of a basketball dropped from a height of 2 meters (roughly the height of a professional basketball player). Why don't you need to know the mass of the basketball?

6.X.93 **(a)** A 0.5 kg teddy bear is nudged off a window sill and falls 2 m to the ground. What is its kinetic energy at the instant it hits the ground? What is its speed? What assumptions or approximations did you make in this calculation? **(b)** A 1.0 kg flowerpot is nudged off a window sill and falls 2 m to the ground. What is its kinetic energy at the instant it hits the ground? What is its speed? How do the speed and kinetic energy compare to that of the teddy bear in part (a)?

6.X.94 You throw a ball of mass 1.2 kg straight up. You observe that it takes 3.1 s to go up and down, returning to your hand. Assuming we can neglect air resistance, the time it takes to go up to the top is half the total time, 1.55 s. Note that at the top the momentum is momentarily zero, as it changes from heading upward to heading downward.

  (a) Use the Momentum Principle to determine the speed that the ball had just after it left your hand.

  (b) Use the Energy Principle to determine the maximum height above your hand reached by the ball.

6.P.95 In the rough approximation that the density of the Earth is uniform throughout its interior, the gravitational field strength (force per unit mass) inside the Earth at a distance r from the center is gr/R, where R is the radius of the Earth. (In actual fact, the outer layers of rock have lower density than the inner core of molten iron.) Using the uniform-density approximation, calculate the amount of energy required to move a mass m from the center of the Earth to the surface. Compare with the amount of energy required to move the mass from the surface of the Earth to a great distance away.

Section 6.13

6.X.96 Two protons are a distance 4×10^{-9} m apart. What is the electric potential energy of the system consisting of the two protons? If the two protons move closer together, will the electric potential energy of the system increase, decrease, or remain the same?

A proton and an electron are a distance 4×10^{-9} m apart. What is the electric potential energy of the system consisting of the proton and the electron? If the proton and the electron move closer together, will the electric potential energy of the system increase, decrease, or remain the same?

Which of the following statements are true?

  A. In some situations charged particles released from rest would move in a direction that increases electric potential energy, but not in other situations.

  B. If released from rest two protons would move closer together, increasing the potential energy of the system.

  C. If any two charged particles are released from rest, they will spontaneously move in the direction in which the potential energy of the system will be decreased.

6.X.97 Which of the diagrams in Figure 6.73 corresponds to a system of two electrons that start out far apart, moving toward each other (that is, their initial velocities are nonzero and they are heading straight at each other)?

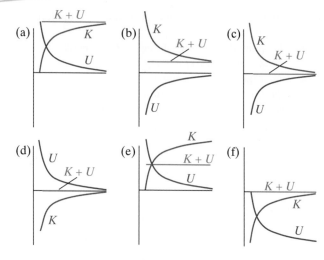

Figure 6.73

Which of the diagrams corresponds to a system of a proton and an electron that start out far apart, moving toward each other (that is, their initial velocities are nonzero and they are heading straight at each other)?

6.P.98 Four protons, each with mass M and charge $+e$, are initially held at the corners of a square that is d on a side. They are then released from rest. What is the speed of each proton when the protons are very far apart?

Section 6.14

6.P.99

  (a) A particle with mass M and charge $+e$ and its antiparticle (same mass M, charge $-e$) are initially at rest, far from each other. They attract each other and move toward each other. Make a graph of energy terms vs. separation distance r between the two particles. Label the various energies involved in this process. Include the rest energy of the particles, assuming that the other energy terms are comparable to the rest energy.

  (b) When the particle and antiparticle collide, they annihilate and produce a different particle with rest mass m (much smaller than M) and charge $+e$ and its antiparticle (same rest mass m, charge $-e$). When these two particles have moved far away from each other, how fast are they going? Is this speed large or small compared to c?

  (c) Now take the specific case of a proton and antiproton colliding to form a positive and negative pion. Each pion has a rest mass of 2.5×10^{-28} kg. When the pions have moved far away, how fast are they going?

  (d) How far apart must the two pions be (in meters) for their electric potential energy to be negligible compared to their kinetic energy? Be explicit and quantitative about your criterion and your result.

6.P.100 For some isotopes of some very heavy nuclei, including nuclei of thorium, uranium, and plutonium, the nucleus will fission (split apart) when it absorbs a slow-moving neutron.

For example, plutonium-241, with 94 protons and 147 neutrons, can fission when it absorbs a neutron and becomes plutonium-242. The two fission fragments, called "daughter" nuclei, can be almost any two nuclei whose charges Q_1 and Q_2 add up to $94e$ (where e is the charge on a proton), and whose nucleons add up to 242 protons and neutrons (Pu-242, formed from Pu-241 plus a neutron).

One of the possible fission modes involves nearly equal fragments, silver nuclei (Ag-121) each with electric charge $Q_1 = Q_2 = 47e$. The rest masses of the two daughter nuclei add up to less than the rest mass of the original parent nucleus. (In addition to the two main fission fragments there are typically one or more free neutrons in the final state; in your analysis make the simplifying assumption that there are no free neutrons, just two daughter nuclei.)

Objects involved in the reaction, where the rest mass is given in atomic mass units, and 1 amu = 1.66054×10^{-27} kg:

Nucleus	Protons	Neutrons	Charge	Rest mass
Pu-242	94	148	+94e	242.007
Ag-121	47	74	+47e	120.894

Although in most problems you solve in this course it is adequate to use values of constants rounded to 2 or 3 significant figures, in this problem you must keep 6 significant figures throughout your calculation. Problems involving mass changes require many significant figures because the changes in mass are small compared to the total mass. In this problem you must use the following values of constants, accurate to 6 significant figures:

Constant	Value to 6 significant figures
c (speed of light)	2.99792×10^8 m/s
e (charge of a proton)	1.60218×10^{-19} coulomb
$\frac{1}{4\pi\varepsilon_0}$	8.98755×10^{09} N·m²/C²

There are three states you should consider in your analysis: (1) The initial state of the Pu-242 nucleus, before it fissions. (2) The state just after fission, when the two silver nuclei are close together, and momentarily at rest. (3) The state when the silver nuclei are very far away from each other, traveling at high speed.

A: The final speed of the fission products
Your first task is to determine the final speed of each of the daughter nuclei in state (3), when they are far from each other.

Figure 6.74 shows three important states in the process:

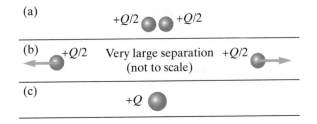

Figure 6.74

Which diagram (a, b, or c) depicts the state of the original plutonium nucleus? We'll take this state to be the initial state.

Which diagram depicts the state when the two daughter nuclei have reached their final speeds? We'll take this state to be the final state.

Compare the initial and final states of the system. Which quantities have changed: potential energy, rest energy, kinetic energy?

What will be the total kinetic energy of the two daughter nuclei when they are very far apart?

Assume for the moment that when the two silver nuclei are very far apart that the speed is small compared to the speed of light. In that case, what is the final speed of one of the nuclei?

(Although this is a very high speed, it is still a rather small fraction of the speed of light, so the low-speed approximation for the kinetic energy is good enough for our purposes.)

B: Fission as an energy source
The kinetic energy of the fast-moving daughter nuclei is eventually absorbed in the surrounding material and raises the temperature of that material. In a fission power plant, this thermal energy is used to boil water to drive a steam turbine and generate electricity. If a mole of plutonium undergoes this fission reaction, how much kinetic energy is generated?

For comparison, only around 1×10^6 joules are obtained from burning a mole of gasoline, which is why energy from fission is of great interest, if some way can be found to deal with the waste products. These silver fission nuclei are not the same as the silver nuclei found in nature; the fission-product isotopes have excess neutrons and are "radioactive," which means that they emit electrons, alpha particles (helium nuclei), and gamma rays (high-energy photons). These radiations can be easily blocked by burying the wastes, but if radioactive nuclei leak into the ground water and are consumed by humans, the emissions can damage tissues inside the body and cause cancer and genetic damage.

C: Testing the simple model of fission
A simple model of fission is that when the original plutonium nucleus splits, the two daughter nuclei are "born" at rest, right next to each other. We can test this model by determining what the Energy Principle tells us about the center-to-center distance of these two nuclei just after the plutonium nucleus splits.

In Figure 6.74, which two states do NOT involve any change in rest energy? It is convenient to take these two states to be the initial and final states.

Compare these initial and final states of the system. Which quantities have changed: potential energy, rest energy, kinetic energy?

Using energy considerations, calculate the distance between centers of the silver nuclei just after fission, when they are momentarily at rest. Keep at least 6 significant figures in your calculations.

A proton or neutron has a radius r of roughly 1×10^{-15} m, and a nucleus is a tightly packed collection of nucleons of radius R. Therefore the volume of the nucleus, $\frac{4}{3}\pi R^3$, is approximately equal to the volume of one nucleon, $\frac{4}{3}\pi r^3$, times the number N of nucleons in the nucleus: $\frac{4}{3}\pi R^3 = N\frac{4}{3}\pi r^3$. So the radius R of a nucleus is about $N^{1/3}$ times the radius r of one nucleon. More precisely, experiments show that the radius of a nucleus containing N nucleons is $(1.3 \times 10^{-15}$ m$)N^{1/3}$. What is the radius of a silver-121 nucleus?

On paper, make a careful scale drawing of the two silver nuclei just after fission, and label the drawing with the distances that you just calculated. If the two silver nuclei are nearly touching, this would be consistent with our model of fission, in which the Pu-242 nucleus fissions into two pieces that are initially nearly at rest. How big is the gap between the surfaces of the two nuclei? (If you have done the calculations correctly, you will indeed find that the gap is a rather small fraction of the center-to-center distance, which means that our simple model for the fission process is a pretty good model.)

Note: One way to check all the calculations in this problem is to apply the Energy Principle to a different choice of initial and final states than those that you considered above. You should of course get the same results.

6.P.101 In a fusion reaction, the nuclei of two atoms join to form a single atom of a different element. In such a reaction, a fraction of the rest energy of the original atoms is converted to kinetic energy of the reaction products. A fusion reaction that occurs in the Sun converts hydrogen to helium. Since electrons are not involved in the reaction, we focus on the nuclei.

Hydrogen and deuterium (heavy hydrogen) can react to form helium plus a high-energy photon called a gamma ray:

$$^1H + \ ^2H \rightarrow \ ^3He + \gamma$$

Objects involved in the reaction, where the rest mass is given in atomic mass units, and 1 amu = 1.66054×10^{-27} kg are:

Particle	Protons	Neutrons	Charge	Rest mass
1H (proton)	1	0	+e	1.0073
2H (deuterium)	1	1	+e	2.0136
3He (helium)	2	1	+2e	3.0155
gamma ray	0	0	0	0

Although in most problems you solve in this course you should use values of constants rounded to 2 or 3 significant figures, in this problem you must keep at least 6 significant figures throughout your calculation. Problems involving mass changes require many significant figures because the changes in mass are small compared to the total mass. In this problem you must use the following values of constants, accurate to 6 significant figures:

Constant	Value to 6 significant figures
c (speed of light)	2.99792×10^8 m/s
e (charge of a proton)	1.60218×10^{-19} coulomb
atomic mass unit	1.66054×10^{-27} kg
$\dfrac{1}{4\pi\varepsilon_0}$	8.98755×10^{09} N·m²/C²

Figure 6.75

A proton (1H nucleus) and a deuteron (2H nucleus) start out far apart. An experimental apparatus shoots them toward each other (with equal and opposite momenta). If they get close enough to make actual contact with each other, they can react to form a helium-3 nucleus and a gamma ray (a high energy photon, which has kinetic energy but zero rest energy). Consider the system containing all particles. Work out the answers to the following questions on paper, using symbols (algebra), before plugging numbers into your calculator.

A: Bringing the particles together
Which diagram in Figure 6.75 depicts the initial state in the process of bringing the particles together?

Which diagram depicts the final state in the process of bringing the particles together?

Compare the initial state and final states of the system. Which quantities have changed: potential energy, rest energy, kinetic energy?

The deuterium nucleus starts out with a kinetic energy of 2.07×10^{-13} joules, and the proton starts out with a kinetic energy of 4.14×10^{-13} joules. The radius of a proton is 0.9×10^{-15} m; assume that if the particles touch, the distance between their centers will be twice that. What will be the total kinetic energy of both particles an instant before they touch?

B: Reaction to make helium
Now that the proton and the deuterium nucleus are touching, the reaction can occur. Take the final state from the previous process to be the initial state of the system for this new process. Which diagram depicts the final state?

Compare the initial state and final states of the system. Which quantities have changed: potential energy, rest energy, kinetic energy?

What is the kinetic energy of the reaction products (helium nucleus plus photon)?

C: Gain of kinetic energy
What was the gain of kinetic energy in this reaction? (The products have more kinetic energy than the original particles did when they were far apart. How much more?)

D: Fusion as energy source
Kinetic energy can be used to drive motors and do other useful things. If a mole of hydrogen and a mole of deuterium underwent this fusion reaction, how much kinetic energy would be generated? (For comparison, around 1×10^6 joules are obtained from burning a mole of gasoline.)

6.P.102 Many heavy nuclei undergo spontaneous "alpha decay," in which the original nucleus emits an alpha particle (a helium nucleus containing two protons and two neutrons), leaving behind a "daughter" nucleus that has two fewer protons and two fewer neutrons than the original nucleus. Consider a radium-220 nucleus that is at rest before it decays to radon-216 by alpha decay.

The mass of the radium-220 nucleus is 219.96274 u (unified atomic mass units) where 1 u = 1.6603×10^{-27} kg (approximately the mass of one nucleon).

The mass of a radon-216 nucleus is 215.95308 u, and the mass of an alpha particle is 4.00151 u. Radium has 88 protons, radon 86, and an alpha particle 2.

(a) Make a diagram of the final state of the radon-216 nucleus and the alpha particle when they are far apart, showing the momenta of each particle to the same relative scale. Explain why you drew the lengths of the momentum vectors the way you did.

(b) Calculate the final kinetic energy of the alpha particle. For the moment, assume that its speed is small compared to the speed of light.

(c) Calculate the final kinetic energy of the radon-216 nucleus.

(d) Show that the nonrelativistic approximation was reasonable.

6.P.103 A proton (1.6726×10^{-27} kg) and a neutron (1.6749×10^{-27} kg) at rest combine to form a deuteron, the nucleus of deuterium or "heavy hydrogen." In this process, a gamma ray (high-energy photon) is emitted, and its energy is measured to be 2.2 MeV (2.2×10^6 eV).

(a) Keeping all five significant figures, what is the mass of the deuteron? Assume that you can neglect the small kinetic energy of the recoiling deuteron.

(b) Momentum must be conserved, so the deuteron must recoil with momentum equal and opposite to the momentum of the gamma ray. Calculate approximately the kinetic energy of the recoiling deuteron and show that it is indeed small compared to the energy of the gamma ray.

6.P.104 Uranium-235 fissions when it absorbs a slow-moving neutron. The two fission fragments can be almost any two nuclei whose charges Q_1 and Q_2 add up to $92e$ (where e is the charge on a proton, $e = 1.6 \times 10^{-19}$ coulomb), and whose nucleons add up to 236 protons and neutrons (U-236; U-235 plus a neutron). One of the possible fission modes involves nearly equal fragments, palladium nuclei with $Q_1 = Q_2 = 46e$. The rest masses of the two palladium nuclei add up to less than the rest mass of the original nucleus. (In addition to the two main fission fragments there are typically one or more free neutrons in the final state; in your analysis make the simplifying assumption that there are no free neutrons, just two palladium nuclei.) The rest mass of the U-236 nucleus is 235.996 u (unified atomic mass units), and the rest mass of each Pd-118 nucleus is 117.894 u, where 1 u $= 1.7 \times 10^{-27}$ kg (approximately the mass of one nucleon).

(a) Calculate the final speed v, when the palladium nuclei have moved far apart (due to their mutual electric repulsion). Is this speed small enough that $p^2/(2m)$ is an adequate approximation for the kinetic energy of one of the palladium nuclei? (It is all right to go ahead and make the nonrelativistic assumption first, but you then must check that the calculated v is indeed small compared to c.)

(b) Using energy considerations, calculate the distance between centers of the palladium nuclei just after fission, when they are starting from rest.

(c) A proton or neutron has a radius of roughly 1×10^{-15} m, and a nucleus is a tightly packed collection of nucleons. Experiments show that the radius of a nucleus containing N nucleons is approximately $(1.3 \times 10^{-15}$ m$) \times N^{1/3}$. What is the approximate radius of a palladium nucleus? Draw a sketch of the two palladium nuclei in part (b), and label the distances you calculated in parts (b) and (c).

6.P.105 One of the thermonuclear or fusion reactions that takes place inside a star such as our Sun is the production of helium-3 (^{3}He, with two protons and one neutron) and a gamma ray (high-energy photon) in a collision between a proton (^{1}H) and a deuteron (^{2}H, the nucleus of "heavy" hydrogen, consisting of a proton and a neutron):

$$^1\text{H} + {}^2\text{H} \rightarrow {}^3\text{He} + \gamma$$

The rest mass of the proton is 1.0073 u (unified atomic mass unit, 1.7×10^{-27} kg), the rest mass of the deuteron is 2.0136 u, the rest mass of the helium-3 nucleus is 3.0155 u, and the gamma ray is massless.

(a) The strong interaction has a very short range and is essentially a contact interaction. For this fusion reaction to take place, the proton and deuteron have to come close enough together to touch. The approximate radius of a proton or neutron is about 1×10^{-15} m. What is the approximate initial total kinetic energy of the proton and deuteron required for the fusion reaction to proceed, in joules and electron volts (1 eV $= 1.6 \times 10^{-19}$ joule)?

(b) Given the initial conditions found in part (a), what is the kinetic energy of the ^{3}He plus the energy of the gamma ray, in joules and in electron-volts?

(c) The net energy released is the kinetic energy of the ^{3}He plus the energy of the gamma ray found in part (b), minus the energy input that you calculated in part (a). What is the net energy release, in joules and in electron volts? Note that you do get back the energy investment made in part (a).

(d) Which of the following potential energy curves (1–4) in Figure 6.76 is a reasonable representation of the interaction in this fusion reaction? Why?

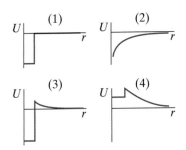

Figure 6.76 Potential energy curves.

As we will study later, the average kinetic energy of a gas molecule is $\frac{3}{2}k_B T$, where k is the "Boltzmann constant," 1.4×10^{-23} joule/Kelvin (J/K), and T is the absolute or Kelvin temperature, measured from absolute zero (so that the freezing point of water is 273 K). The approximate temperature required for the fusion reaction to proceed is very high. This high temperature, required because of the electric repulsion barrier to the reaction, is the main reason why it has been so difficult to make progress toward thermonuclear power generation. Sufficiently high temperatures are found in the interior of the Sun, where fusion reactions take place.

6.P.106 A nucleus whose mass is 3.917268×10^{-25} kg undergoes spontaneous "alpha" decay. The original nucleus disappears and there appear two new particles: a He-4 nucleus of mass 6.640678×10^{-27} kg (an "alpha particle" consisting of two protons and two neutrons) and a new nucleus of mass 3.850768×10^{-25} kg. (Note that the new nucleus has less mass than the original nucleus, and it has two fewer protons and two fewer neutrons.) When the alpha particle has moved far away from the new nucleus (so that the electric potential energy is negligible), what is the combined kinetic energy of the alpha particle and new nucleus?

6.P.107 A pendulum (see Figure 6.77) consists of a very light but stiff rod of length L hanging from a nearly frictionless axle, with a mass m at the end of the rod.

(a) Calculate the gravitational potential energy as a function of the angle θ, measured from the vertical.

(b) Sketch the potential energy as a function of the angle θ, for angles from $-210°$ to $+210°$.

(c) Let $s = L\theta =$ the arc length away from the bottom of the arc. Calculate the tangential component of the force on the mass by taking the (negative) gradient of the energy with respect to s. Does your result make sense?

(d) Suppose that you hit the stationary hanging mass so it has an initial speed v_i. What is the minimum initial speed needed for the pendulum to go over the top ($\theta = 180°$)? On your sketch of the potential energy (part b), draw and label energy levels for the case in which the initial speed is less than, equal to, or greater than this critical initial speed.

Figure 6.77

6.P.108 Show the validity of the relation $E^2_{\text{particle}} - (pc)^2 = (mc^2)^2$ when $m \neq 0$, by making these substitutions:

$$E_{\text{particle}} = \frac{mc^2}{\sqrt{1 - (v/c)^2}} \quad \text{and} \quad p = \frac{mv}{\sqrt{1 - (v/c)^2}}$$

Computational Problems

6.P.109 If you have already done the problem in Chapter 3 on the Ranger 7 mission to the Moon, you can use the program you wrote for that problem, and go directly to part (b). Otherwise:

(a) Write a program to model the journey of a spacecraft coasting from the Earth to the Moon. Start the spacecraft at a height of 50 km above the Earth's surface with a speed of 1.3×10^4 m/s. This is approximately the speed it would have after all the rockets have fired, and is high enough to be above most of the Earth's atmosphere. Include the spacecraft's interactions with both the Earth and the Moon. Use a dt of 5 seconds. Make sure you stop the program when the spacecraft reaches the Moon's surface (not its center!). The data you need may be found on the inside back cover of this book; the mass of the spacecraft was 173 kg. If you are interested in more of the details of this trip, see the problem in Chapter 3 on the Ranger 7 mission to the Moon.

(b) Add a calculation of the work done by the gravitational forces of the Earth and the Moon to your analysis of sending a spacecraft to the Moon. You need to approximate the work by adding up the amount of work done by gravitational forces along each step of the path:

$$W = \Sigma \vec{F} \bullet \Delta \vec{r} = \vec{F}_1 \bullet \Delta \vec{r}_1 + \vec{F}_2 \bullet \Delta \vec{r}_2 + \vec{F}_3 \bullet \Delta \vec{r}_3 + \cdots$$

Compare the numerical value of the work with the change in the kinetic energy (final kinetic energy just before crashing on the Moon, minus initial kinetic energy when released above the Earth's atmosphere). Try shorter time steps to make sure that the errors introduced by a finite time step are not significant. Report your results.

6.P.110 Energy conservation is a powerful check on the accuracy of a numerical integration. Modify the program for the Chapter 3 problem on the Ranger 7 mission to the Moon to plot graphs of kinetic energy, of gravitational potential energy, and of the sum of the kinetic energy and the gravitational potential energy, vs. position. Does the kinetic plus potential energy remain constant? What if you vary the step size (which varies the accuracy of the numerical integration)? Vary the launch speed, and explain the effect that this has on your graphs.

ANSWERS TO EXERCISES

6.X.1 particle energy = 5.2×10^{-13} J, rest energy = 8.1×10^{-14} J; kinetic energy = 4.4×10^{-13} J

6.X.2 4.4×10^9 J

6.X.3 12 m/s; 27 miles per hour

6.X.4 1.2×10^{-12} kg (an extremely small amount of mass!)

6.X.5 1.4 J

6.X.6 $\langle 25, -30, 0 \rangle$ m, $\langle 0, -19.6, 0 \rangle$ N, 588 J

6.X.7 (a) decreasing, negative; (b) increasing, positive; (c) decreasing, negative; (d) increasing, positive

6.X.8 −600 J

6.X.9 4.9 N, downward; 4.9 J

6.X.10 (a) zero; (b) $2P$

6.X.11 400 J

6.X.12 40.9; 39.9; very large; very small

6.X.13 8.5×10^{-4}, approximately 0.1%

6.X.14 189 J

6.X.15 2.5×10^{-19} J

6.X.16 $\Delta U = -50$ J

6.X.17 6 pairs: $U_{12} + U_{13} + U_{14} + U_{23} + U_{24} + U_{34}$

6.X.18 -6.3×10^7 J

6.X.19 − change, because more negative

6.X.20 + change, because less negative

6.X.21 Work is zero, and need not be calculated.

6.X.22 center to center

6.X.23 decreases

6.X.24 A is bound; C is unbound; B is trapped. In a quantum system in state B there is some probability of "tunneling" through the barrier and getting out of the trap!

6.X.25 *A*: go out in a straight line and stop, then fall back; *B*: elliptical (motion with varying separation); *C*: nearly circular (nearly constant separation)

6.X.26 You're going only as far as the Moon, not to infinity. Moreover, the Moon attracts the spacecraft.

6.X.27 1.12×10^4 m/s

6.X.28 same ΔU

6.X.29 *Hint:* At top of trajectory the ball's speed is momentarily zero. Answer is given below.

6.X.30 $\frac{1}{2}m(v_{xi}^2) - \frac{1}{2}m(v_{xi}^2 + v_{yi}^2) + mg(y_f - y_i) = 0$, so $y_f = y_i + v_{yi}^2/(2g)$

6.X.31 7.2×10^{-15} m; the radius of the proton is about 1×10^{-15} m.

6.X.32 About 1.7×10^{-10}; accuracy of laboratory scale is far below what would be needed to detect this.

6.X.33 about 0.1%

6.X.34 $-GMm/y^2$; 0

6.X.29 103 m

Internal Energy

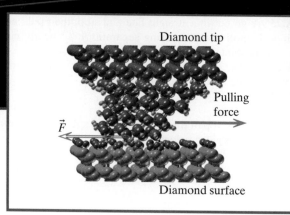

Diamond tip

Pulling force

$\vec{F}$

Diamond surface

KEY IDEAS

- **Macroscopic systems are composed of many interacting particles.**
- **The potential energy associated with stretching or compressing interatomic bonds is similar to the potential energy of macroscopic springs.**
- **The potential energy difference between two states does not depend on how a system gets from the initial to the final state (path independence).**
- **Internal energy includes deformation, rotation, vibration, and thermal energy.**
- **Thermal energy is random kinetic and potential energy of atoms and interatomic bonds within an object.**
  - Temperature is a measure of thermal energy.
  - A temperature difference between system and surroundings leads to energy transfer (Q).
  - Specific heat capacity (C) is the amount of energy required to raise the temperature of a given amount of a substance by 1 kelvin.
- **The choice of system affects the form of the Energy Principle.**
  - If there are energy transfers between the system and the surroundings, the system is an "open" system; otherwise it is a "closed" system.
  - A system is in a "steady state" if there are energy inputs and outputs but no net change in the energy of the system.
- **Energy can be dissipated throughout a system and the surroundings in a form that cannot be fully retrieved.**

7.1 POTENTIAL ENERGY OF MACROSCOPIC SPRINGS

In Chapter 4 we used the behavior of idealized, macroscopic springs as a model for the dynamic behavior of the springlike interatomic bonds in solid objects. This was a productive model, because it allowed us to understand the nature of contact forces (tension and compression forces) and because using this model we were able to predict the speed of sound in different materials.

This model continues to be useful in understanding the flow of energy in macroscopic systems (objects composed of many atoms, such as wires, tables, vines, and humans). To apply it, we need to find an equation for the potential energy of a mass–spring system.

An Ideal Spring

A real spring has a nonzero mass and can be damaged in various ways: it can be deformed or broken, the metal can fatigue, and so on. There is a limit to how much a real spring can be stretched (it will eventually break) or compressed (the coils will eventually touch each other, turning the spring into a stiff hollow rod). An idealized spring has none of these problems. There are no limits on the stretch s in the ideal spring force equation (here we're calling the axis of the spring the "s" axis):

$$F_s = -k_s\, s$$

Despite its limitations, an ideal spring proves to be a useful model for both real macroscopic springs and for springlike interatomic bonds.

To determine the potential energy associated with the spring force, recall that the (negative) gradient of potential energy is equal to the associated force. So to find the spring potential energy U_s we need a solution to this equation:

$$-\frac{dU_s}{ds} = -k_s\, s$$

which simplifies to

$$\frac{dU_s}{ds} = k_s\, s$$

> QUESTION Can you think of a function of s whose derivative with respect to s is $k_s\, s$?

The derivative with respect to s of $\frac{1}{2}k_s s^2$ is $k_s s$, and the negative gradient of $\frac{1}{2}k_s s^2$ is the force $-k_s s$.

IDEAL SPRING POTENTIAL ENERGY

$$U_s = \frac{1}{2}k_s s^2$$

s is stretch, measured from the equilibrium point.

The function $\frac{1}{2}k_s s^2$ is an upturned parabola (Figure 7.1). Note that the stretch s appears squared in the expression for spring energy. This reflects the fact that either a lengthening (positive stretch) or a compression (negative stretch) of the spring involves increased potential energy.

Peculiar Features of the Ideal Spring Potential Energy

This ideal spring potential energy curve has some peculiar features:

- As the absolute value of stretch becomes large, U becomes infinite, instead of approaching zero. (This curve is sometimes called an "infinite potential energy well.")
- For a system of a mass and ideal spring, $K + U$ is always greater than zero, so apparently there can't be any bound states of the system.

As a result of these peculiarities, this ideal potential energy function doesn't fully represent the nature of a macroscopic spring you can hold in your hand, or of an interatomic bond.

One possible modification to the equation would be to subtract a constant E_S. Such a function would still satisfy the gradient equation above, but it would allow for bound states of the system, as in Figure 7.2.

$$U_s = \frac{1}{2}k_s s^2 - E_S$$

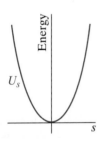

Figure 7.1 Potential energy as a function of stretch s for an ideal spring.

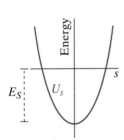

Figure 7.2 Subtracting a constant E_S shifts the ideal spring potential energy curve downward.

For the ideal spring, the constant is conventionally set to zero, but the negative offset may be needed when analyzing real systems.

Real Springs

We are primarily interested in helical (spiral) springs. A modest force applied to a helical spring gives a substantial change in the length of the spring, even though the total length of the coiled wire hardly changes. A large stretch of the spring with a small force implies a much smaller spring stiffness than the wire would have when straightened out.

If you stretch a helical spring far enough, it straightens out into a long straight wire, and it suddenly becomes very difficult to lengthen the wire any further. If you continue to stretch the straight wire, eventually it yields and then breaks. If you compress a helical spring so far that its coils run into each other, it suddenly gets extremely hard to compress the spring any further. Therefore, the linear relation between stretch and force, $F_s = -k_s s$, is valid only over a limited region of stretch, although this linear range is much wider than for the microscopic "spring."

In many situations the entire phenomenon of interest takes place within the valid range of the linear force approximation. One can speak of an "infinite" potential energy well as shown as the dashed curve in Figure 7.3, which is an idealization of the real situation. Within the linear range where $F_s = -k_s s$ is the negative gradient of $\frac{1}{2}k_s s^2$, we approximate the real curve by a parabola. The actual bottom of the parabola is at a negative value of the potential energy. Work would be required to stretch the wire out straight. However, in our calculations with macroscopic springs we're almost always interested only in *changes* in the potential energy, so we shift the origin to the bottom of the well and measure U_s from there.

In Figure 7.3 (which is still an idealization of the real situation) there are very steep sides of the potential energy well, corresponding to the large forces required to compress a spring whose coils are in contact and the large forces required to stretch the spring after it has straightened out into a straight wire. Since force is the negative gradient of the potential energy, a steep slope corresponds to a large force.

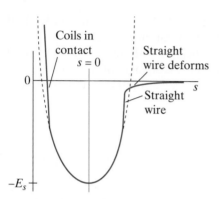

Figure 7.3 The dashed curve represents the spring potential energy

$$U_s = \frac{1}{2}k_s s^2 - E_S$$

within the linear range where $F_s = -k_s s$. The colored solid curve represents a real macroscopic spring.

Figure 7.4 As the compressed spring expands the kinetic energy of the block increases, and spring potential energy decreases.

Figure 7.5 As the stretched spring contracts, the kinetic energy of the block increases and spring potential energy decreases.

Energy in an Oscillating Spring–Mass System

In an oscillating spring–mass system such as the one discussed in Chapter 4, the energy of the system is continually changing from kinetic energy to potential energy, and back again.

Consider a block on a low friction surface, connected by a helical spring to a wall. If you compress the spring and then let go, the block acquires kinetic energy (Figure 7.4), with a loss of spring potential energy. If you stretch the spring and then let go, the block also acquires kinetic energy, again with a loss of spring potential energy (Figure 7.5).

QUESTION At the moment that the spring is released, what is the kinetic energy of the spring–mass system? What is the total energy of the system?

Since the block is released from rest, its kinetic energy is zero. The energy of the system is equal to the energy stored in the spring, $\frac{1}{2}k_s s^2$ (omitting the constant rest energy and constant $-E_S$).

QUESTION At what point in the oscillation is the kinetic energy of the system highest? At that moment, what is the potential energy of the system?

When the contracting spring reaches its equilibrium length, the energy stored in the spring is a minimum, $\frac{1}{2}k_s s^2 = 0$. The maximum possible amount of spring potential energy has now been converted to kinetic energy, so this is the instant when the kinetic energy of the system is highest.

QUESTION At what point in the oscillation will the kinetic energy of the system be lowest?

At either turning point of the oscillation—spring fully compressed or spring fully extended—the instantaneous speed of the mass is zero. The kinetic energy of the system is lowest here (zero in fact), and all the energy has been momentarily converted back into spring potential energy. Since the absolute value of the stretch is highest at these locations, the potential energy $\frac{1}{2}k_s s^2$ is also highest.

Does the Wall Do Work on the System?

For the system of spring + block, the surroundings include the Earth, the low friction table, and the wall. Both the table and the wall touch the system and therefore exert contact forces on it.

The table and Earth exert forces perpendicular to the block's motion, so they do zero work on the system (assuming negligible friction).

QUESTION The force by the wall on the spring is parallel to the block's motion. Does it do nonzero work on the system?

This force does zero work, because the point at which the force acts (the left end of the spring) undergoes zero displacement. So

$$F_{\text{wall},x}\,\Delta x = 0 \quad \text{because } \Delta x = 0 \text{ for the left end of the spring}$$

This force does have an effect on the momentum of the system, however, because the time interval over which it acts is not zero. The momentum of the mass–spring system is changed by the force on the spring by the wall. (Think about what would happen if the wall were not there.) This is another example of the difference between the Momentum Principle (all forces act for the same time interval) and the Energy Principle (different forces may act over different distances).

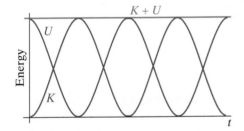

Figure 7.6 Kinetic energy (red), potential energy (purple), and the sum of kinetic + potential energy (green) for an oscillating spring–mass system.

Flow of Energy in the Mass–Spring System

Since the surroundings do no work on the system, the total energy of the system must be constant (in the absence of friction), and energy flows from potential to kinetic and back again, as shown in Figure 7.6.

EXAMPLE **Energy in a Spring–Mass System**

A mass of 0.2 kg is attached to a horizontal spring whose stiffness is 12 N/m. Friction is negligible. At $t = 0$ the spring has a stretch of 3 cm and the mass has a speed of 0.5 m/s. **(a)** What is the amplitude (maximum stretch) of the oscillation? **(b)** What is the maximum speed of the block?

Solution System: Mass and spring
   Surroundings: Earth, table, wall (neglect friction, air)
   Initial state: $s_i = 3$ cm, $v_i = 0.5$ m/s
   Final state: $v_f = 0$ (maximum stretch)

■ From the discussion in the preceding section, we know that no work is done by the surroundings, because the point where the wall force is applied does not move. We also know that when the stretch is maximum, the block is momentarily at rest (a turning point).

■ We could use any known state as the initial state in part (b), because the energy of the system is constant.

(a) Find the maximum stretch:
Energy Principle:

$$E_f = E_i + W$$

$$\cancel{K_f} + U_f = K_i + U_i + \cancel{W}$$

$$0 + \frac{1}{2}k_s s_f^2 = \frac{1}{2}mv_i^2 + \frac{1}{2}k_s s_i^2$$

$$\frac{1}{2}k_s s_f^2 = \frac{1}{2}(0.2 \text{ kg})(0.5 \text{ m/s})^2 + \frac{1}{2}(12 \text{ N/m})(0.03 \text{ m})^2$$

$$\frac{1}{2}k_s s_f^2 = 0.03 \text{ J}$$

$$s_f = \sqrt{2(0.03 \text{ J})/(12 \text{ N/m})} = 0.07 \text{ m}$$

$$s_f = 0.07 \text{ m}$$

(b) Find the maximum speed:
Initial state: $s_i = 3$ cm, $v_i = 0.5$ m/s
Final state: Maximum speed ($s_f = 0$)

$$K_f + 0 = 0.03 \text{ J}$$

$$\frac{1}{2}mv_f^2 = 0.03 \text{ J}$$

$$v_{max} = \sqrt{2(0.03 \text{ J})/(0.2 \text{ kg})} = 0.55 \text{ m/s}$$

FURTHER DISCUSSION   Assuming negligible energy dissipation by friction or air resistance, the energy of the system never changes. If we can find $K + U$ for one state, we can always use this as the initial state in any calculation.

> **7.X.1** How many joules of energy can you store in a spring whose stiffness is 0.6 N/m, by starting from a relaxed spring and stretching it 20 cm?
>
> **7.X.2** A horizontal spring with stiffness 0.5 N/m has a relaxed length of 15 cm. A mass of 20 grams is attached and you stretch the spring to a total length of 25 cm. The mass is then released from rest and moves with little friction. What is the speed of the mass at the moment when the spring returns to its relaxed length of 15 cm?

Nonhelical Macroscopic Springs

Figure 7.7 A bar of metal can be treated as a spring. For a limited range, double the force gives double the stretch.

We saw in Chapter 4 that a bar of metal can be treated as a spring, since stretching or compressing the bar stretches or compresses the interatomic bonds (Figure 7.7). Double the force produces double the change in length, as long as we're in the range where the spring approximation to the interatomic force is adequate. This is usually described for a bar in terms of Young's modulus Y, which we studied in Chapter 4, where the tension force F_T (per unit area) is proportional to the stretch ΔL (per unit length):

$$\frac{F_T}{A} = Y\frac{\Delta L}{L}$$

Since ΔL, like s, represents a change in length, and F_T is a springlike force (tension force), this equation can be rewritten to parallel the spring force equation:

$$F_T = \left(\frac{YA}{L}\right)|\Delta L| \quad \text{is like} \quad F = (k_s)|s|$$

The effective "spring stiffness" of the metal bar is YA/L.
If you stretch the bar too much, two things can make the bar stop behaving in a springlike manner. You might exceed the interatomic stretch for which $k_s s$

Figure 7.8 The bending of a diving board stretches and compresses interatomic bonds.

is a good approximation to the magnitude of the interatomic force. Also, the regular array of atoms may be disrupted by dislocations of the crystal structure, leading to large-scale slippage of crystal planes. Suddenly the bar "yields" and grows very much longer with little applied force.

A straight object is not usually used as a longitudinal spring but may be used in a bending mode. For example, when a diving board bends under a load, atomic bonds in the upper part of the board are stretched, and atomic bonds in the lower part are compressed (Figure 7.8). The net effect is a springlike behavior of the diving board, with the amount of bend proportional to the applied force.

7.2 POTENTIAL ENERGY OF A PAIR OF NEUTRAL ATOMS

In Chapter 6 we found an equation for electric potential energy for pairs of charged particles. In Chapter 4 we studied the springlike forces between electrically *neutral* atoms (a neutral object has equal amounts of positive and negative charge, so it has zero net charge). These interatomic forces are the superposition of electric attractions and repulsions among the positively charged protons and negatively charged electrons of which the atoms are made.

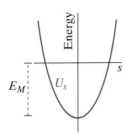

Figure 7.9 Two neutral atoms interact very little when far apart, attract each other at intermediate distances, and repel each other at very short distances.

When two electrically neutral atoms are far from each other (Figure 7.9), they exert almost no force on each other because the attractions between unlike charges (electrons in atom one with protons in atom two, and vice versa) are nearly equal to the repulsions between like charges. However, when the atoms come quite near each other, the electron clouds distort in such a way that the two atoms attract each other. Molecules and solid objects made of two or more atoms are bound together by these electric forces.

However, if you try to push two atoms even closer together, you eventually encounter a rapidly increasing repulsion. The attraction at a small distance and the repulsion when very close are due ultimately to the superposition of the electric forces of the various protons and electrons in the atoms, with the probable locations of these particles governed by the laws of quantum mechanics.

Potential Energy for Springlike Interatomic Bonds

Since an interatomic bond behaves like a spring as long as it is not stretched or compressed excessively, we can guess that the equation for interatomic potential energy has the same form as the potential energy for an ideal spring (Figure 7.10):

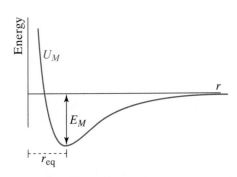

Figure 7.10 The interatomic potential energy is approximately springlike if the bond is not stretched or compressed too much.

Figure 7.11 The Morse potential energy function U_M, representing the potential energy of two interacting neutral atoms as a function of separation.

APPROXIMATE INTERATOMIC POTENTIAL ENERGY

$$U = \frac{1}{2}k_s s^2 - E_M$$

What About Large Separations?

Of course, the potential energy of a real interatomic bond can't become infinitely large at large separation—it needs to approach zero. A more realistic potential energy function for the interatomic bond is called the Morse potential energy and is shown in Figure 7.11. The Morse potential energy is described by this equation:

$$U_M = E_M[1 - e^{-\alpha(r-r_{eq})}]^2 - E_M$$

The Morse approximation is designed to model the basic behavior of the interatomic force and is in approximate agreement with experimental measurements of the interatomic force.

The parameter E_M gives the depth of the potential energy at its minimum, and the parameter r_{eq} (on the order of 1×10^{-10} m) is the distance at which the potential energy curve is a minimum. The parameter α is adjusted to give the curve the right width. When the interatomic separation r is very large, $U_M = 0$, as it should.

The depth of the potential energy well, E_M, is typically a few electron volts (recall that 1 eV is equal to 1.6×10^{-19} joule). For example, the potential energy well for oxygen, O_2, is about 5 eV deep. (As you saw in Chapter 6, a typical ionization energy is also in the range of a few electron volts.) However, for the noble gases such as helium and argon the well is so shallow that these atoms don't form stable diatomic molecules at room temperature, because collisions easily provide enough energy to break the molecules apart.

> **QUESTION** What is the physical significance of the minimum of the curve?

At the minimum of the potential energy curve, the slope is zero, so the force between the two atoms is zero ($F = -dU/dr$). This is the position of stable equilibrium.

> **QUESTION** What is the direction of the force on an atom at a distance smaller than the equilibrium distance?

The left side of the curve has a large negative slope, so the force $F = -dU/dr$ will be large and to the right (in the $+r$ direction), away from the other atom. This makes sense; when the atoms are too close together they push each other apart.

Likewise, if the atom moves to the right (farther away) there should be an attractive restoring force to the left. The slope of this region of the curve is positive, so the sign of the force $F = -dU/dr$ is negative, meaning its direction is to the left, as expected.

As indicated by the dashed curve in Figure 7.12, the bottom of the interatomic potential energy curve can be approximated by the springlike potential energy function:

$$U = \frac{1}{2}k_s s^2 - E_M$$

Because we can make this approximate fit, we know that for ordinary small oscillations around the equilibrium point the two atoms will act as though they are connected by a spring with a $-k_s s$ force. This is the rationale for modeling a solid as a collection of balls connected by springs.

Bound and Unbound States with Interatomic Potential Energy

In Figure 7.13, a horizontal line represents motion in which the sum of the kinetic energies and the potential energy of a two-atom system does not change. A horizontal line corresponds to a possible state of the system when it is isolated from external forces.

> **7.X.3** At separation r_1 in Figure 7.13, what is the physical significance of the quantity A? Of the quantity B? Of the quantity C?
>
> **7.X.4** In Figure 7.14, which of the states are bound states of a two-atom system? Which are unbound states?

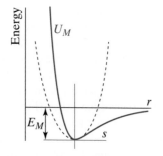

Figure 7.12 The bottom of the interatomic potential energy curve may be approximated by a parabola.

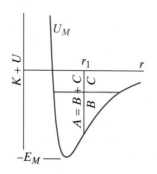

Figure 7.13 A horizontal line represents constant kinetic plus potential energy.

Figure 7.14 Which are bound states? Which are unbound states?

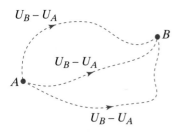

Figure 7.15 Potential energy difference is independent of the path between the initial and final locations.

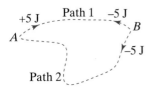

Figure 7.16 A particle moves from A along a path and returns to A.

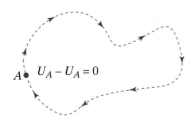

Figure 7.17 Forward and reversed paths, and a round trip.

Figure 7.18 A ball goes up and down with negligible air resistance.

7.3   PATH INDEPENDENCE OF POTENTIAL ENERGY

In a system of several interacting particles, potential energy may be associated with an interaction that depends solely on position. In contrast, in a later section we will consider interactions such as air resistance that depend on speed and do not have an associated potential energy. For example, spring energy depends on the stretch s, and gravitational and electric energy depend on the separation r. It is easy to show that the following must be true:

PATH INDEPENDENCE OF POTENTIAL ENERGY

Change in potential energy ΔU doesn't depend on the path taken.
For a round trip, the change in ΔU is zero.

Here is the reasoning: In Figure 7.15, change in potential energy U of the system when one of its particles moves along some path from location A to location B depends solely on the initial and final values of r or s, so it doesn't matter what path the particle follows to get from A to B. Similarly, in Figure 7.16 if one of the particles in a multiparticle system starts at point A, moves around, and comes back to the starting point, the change in potential energy U of the multiparticle system must be zero, since the initial and final potential energy depends on pair-wise distances r or s, and these return to their original values in a round trip.

There is a connection between the fact that change of potential energy is independent of path, and the fact that ΔU is zero over a round trip.

QUESTION   In Figure 7.17, suppose that along path 1 from A to B the change in potential energy is +5 J. What will be the change in potential energy along the path 1 in the opposite direction, from B to A?

Going backwards from B to A along this same path the change would be -5 J. Since the change is independent of the path, you could take *any* path to return from B to A, such as path 2, with a change of -5 J. Any round trip from A to B and back to A will yield a change of zero.

A simple example is projectile motion with negligible air resistance (Figure 7.18). You throw a ball up in the air with speed v, and when it returns to your hand it has regained the same speed v that you gave it (the gravitational potential energy change of the ball + Earth system is zero, so the change in the kinetic energy is zero). Another example is the motion of a mass on a spring in the absence of friction: every time the mass passes a particular position, it has the same speed (same s implies same v). Note that the direction of the velocity may change, but neither the kinetic energy nor the potential energy depend on direction.

7.X.5 During one complete oscillation of a mass on a spring (one period), what is the change in potential energy of the mass + spring system, in the absence of friction?

EXAMPLE   A Rebounding Block

A metal block of mass 3 kg is moving downward with speed 2 m/s when the bottom of the block is 0.8 m above the floor (Figure 7.19). When the bottom of the block is 0.4 m above the floor, it strikes the top of a relaxed vertical spring 0.4 m in length. The stiffness of the spring is 2000 N/m.

(a) The block continues downward, compressing the spring. When the bottom of the block is 0.3 m above the floor, what is its speed?

(b) The block eventually heads back upward, loses contact with the spring, and continues upward. What is the maximum height reached by the bottom of the block above the floor?

(c) What approximations did you make?

Solution **(a)** Find the speed of block when it is 0.3 m above the floor.
System: Earth, block, spring
Surroundings: Nothing significant
Initial state: Block 0.8 m above floor, moving downward, $v_i = 2$ m/s, spring relaxed
Final state: Block 0.3 m above floor, spring compressed
Energy Principle:

$$E_f = E_i + W$$

$$K_f + U_{s,f} + U_{g,f} = K_i + \cancel{U_{s,i}} + U_{g,i} + \cancel{W}$$

$$\frac{1}{2}mv_f^2 + mgy_f + \frac{1}{2}k_s s_f^2 = mgy_i + \frac{1}{2}mv_i^2$$

$$v_f^2 = v_i^2 + 2g(y_i - y_f) - \frac{k_s}{m}s_f^2$$

$$v_f = \sqrt{(2 \text{ m/s})^2 + 2(9.8 \text{ N/kg})(0.5 \text{ m}) - \frac{2000 \text{ N/m}}{3 \text{ kg}}(0.1 \text{ m})^2}$$

$$v_f = 2.7 \text{ m/s}$$

Figure 7.19 A block falls onto a vertical spring, compresses the spring, then rebounds upward.

■ Both Earth and spring are part of the system, so there are two potential energy terms, one for (block + spring) and one for (block + Earth). Since the mass of the spring is negligible, there is not a (spring + Earth) potential energy term.

■ We could equally well have used the final state from part (a) as the initial state in part (b).

(b) Find the maximum height.
Initial state: Same as in part (a)
Final state: Block at highest point, $v_f = 0$, spring relaxed
Energy Principle:

$$E_f = E_i + W$$

$$\cancel{K_f} + \cancel{U_{s,f}} + U_{g,f} = K_i + \cancel{U_{s,i}} + U_{g,i} + \cancel{W}$$

$$mgy_f = mgy_i + \frac{1}{2}mv_i^2$$

$$y_f = y_i + \frac{v_i^2}{2g} = (0.8 \text{ m}) + \frac{(2 \text{ m/s})^2}{2(9.8 \text{ N/kg})} = 1.0 \text{ m}$$

(c) Approximations: Air resistance and dissipation in the spring are negligible. $U_g \approx mgy$ near Earth's surface. ΔK_{Earth} is negligible.

7.4 INTERNAL ENERGY AND THERMAL ENERGY

Figure 7.20 The object with the compressed spring has more "internal energy." It is kept compressed by a string of negligible mass.

In Figure 7.20 are two identical objects each consisting of two balls connected by identical springs, one compressed (and kept compressed by a string with negligible mass tied around it) and the other relaxed. If these two objects move at the same speed v, they have the same kinetic energy $K = \frac{1}{2}Mv^2$, where M is the total mass, but the object with the compressed spring has more "internal energy."

It is often useful to split the total internal energy into various categories of interest to us in a particular situation. For example, if an object rotates about its center of mass, this internal energy is called "rotational energy" (Figure 7.21). If two balls connected by a spring oscillate relative to the center of mass of the system, this internal energy is called "vibrational energy" (Figure 7.22). If an object is stretched or compressed, there is an associated internal energy

Figure 7.21 More rotation means more internal energy in the form of "rotational energy."

Figure 7.22 More vibration means more internal energy in the form of "vibrational energy."

change, as in Figure 7.20. When you eat, you increase your internal energy in the form of more "chemical energy." If the temperature of an object increases, there is an increase in the internal energy that is called "thermal energy."

From a fundamental perspective, all of these internal energy increases can be thought of as contributing to an increase in the mass of the system: $M_{sys} = E_{sys}/c^2$. However, for many practical macroscopic calculations it makes sense to focus just on the changes in the thermal energy, or vibrational energy, or rotational energy, or deformational energy, or chemical energy, because these changes are extremely small compared to the huge total rest energy Mc^2.

INTERNAL ENERGY

Internal energy = $E_{\text{thermal}} + E_{\text{rotational}} + E_{\text{vibrational}} + E_{\text{chemical}} + \cdots$

In this section we will focus specifically on thermal energy, which is an important category of internal energy. In our idealized models of spring–mass systems, the sum of kinetic plus spring potential energy is constant. In real spring–mass systems, however, the oscillations die away ("damp down") with time, which means that the kinetic plus spring potential energy must be decreasing. The decrease in the amplitude of the oscillations is mainly due to air resistance or to friction with a solid, and some energy is transferred to the surroundings, to the air or to a solid along which the mass slides. On the microscopic level, if we could look inside the solid, we would see increased atomic motion—the atoms in the solid have gained energy, and we call this form of internal energy thermal energy. A macroscopic indication of this increased atomic motion is an increase in the temperature of the object.

The most important difference between thermal energy and vibrational or rotational internal energy is that thermal energy is random in character, not organized into macroscopically recognizable motion or deformation. When two objects collide and there is a loss of some of the original kinetic energy, the increased internal energy can be seen in the form of organized rotation or vibration, or visible deformation, but there may also be random motions at the atomic level, and it is this random component of the internal energy that we call thermal energy and whose change is associated with a change in temperature. Merely increasing the rotational or vibrational internal energy of a system doesn't affect its temperature.

We can measure macroscopic spring potential energy and kinetic energy with simple measurements that don't require a microscope. However, when the temperature of a solid object increases, the increased energy inside the solid is not visible to the naked eye, and we face the problem of how to evaluate the increase in the thermal energy.

Figure 7.23 In our model solid we must account for the kinetic energy of every ball and the spring energy of every spring.

Microscopic Kinetic and Spring Potential Energy

The increased microscopic energy in a solid is in two forms. First, the atoms on average are moving around faster, with increased kinetic energy $\frac{1}{2}mv^2$. Second, there is increased spring potential energy $\frac{1}{2}k_s s^2$ in the interatomic bonds ("springs" in our model of solids), where k_s is the stiffness of an interatomic bond modeled as a spring (Figure 7.23).

In principle the microscopic, thermal energy in a solid could be evaluated by simultaneously measuring at some instant the momenta of *all* of the atoms, and the stretches or compressions of *all* of the springs (that is, changes in atomic positions away from their equilibrium positions, together with knowledge of the spring stiffness). Since there are about 1×10^{23} atoms (and even more "springs") in a macroscopic object, it is in practice impossible to evaluate the energy this way.

We might try to make microscopic measurements of just one atom (and its attached "springs"), assume that this is the average energy of each atom, and then multiply by the number of atoms there are in the object to get the total energy. That doesn't work either, because at a given instant we may choose an atom that is momentarily sitting motionless at its equilibrium position (zero energy), but an instant later will be displaced away from its equilibrium position and moving rapidly. Energy keeps getting passed back and forth among the many atoms and springs, and measuring the energy of just one atom and its attached springs at one instant doesn't give us the correct average energy we would need to determine the total energy of the object.

Temperature

In the 1800s it was realized that for many systems the temperature is essentially a measure of the average energy of the atoms in the system. Therefore we can use a thermometer to measure thermal energy. A familiar example is the mercury thermometer (silver colored) or the alcohol thermometer (tinted red for visibility). In these thermometers, a very thin column of liquid expands more than the glass when heated. By tradition, marks on the thermometer are placed to represent "degrees of temperature," but we could just as well place marks to indicate the average energy of the molecules in the thermometer, in joules (Figure 7.24). One kelvin (or one Celsius degree of temperature) is equivalent to an average molecular energy of about 1×10^{-23} joules.

This is not the whole story. In a later chapter we will make more precise the relationship between temperature and thermal energy, and we will see that temperature is more directly related to a quantity called "entropy" than it is to energy. Nevertheless, for most ordinary systems at room temperature it is true that temperature is a good measure of the average thermal energy of the atoms.

How does a thermometer work? When two objects are in good contact with each other, the average kinetic energies of the molecules in the two objects slowly come to be equal. At a location where two objects touch each other, molecules collide with each other, and the faster molecules on average lose kinetic energy to the slower molecules in a collision. Once the average kinetic energies of molecules in both objects have come to be equal, additional collisions on average make no further change.

To measure the thermal energy of an object, we place a thermometer in contact with the object. We wait for the temperatures (average molecular kinetic energies) of the object and thermometer to equilibrate, then we read the thermometer. A useful thermometer must have a relatively small mass compared to the object of interest, so that attaching the thermometer doesn't add or subtract much energy to or from the object. Another advantage of a small thermometer is that it will reach thermal equilibrium quicker than a large thermometer will.

How was the calibration between one kelvin (or one Celsius degree) and energy established? In one of a series of classic experiments performed by Joule in the 1840s, a paddle wheel in water was turned by a falling weight. The work done by the Earth turned the paddle wheel, which stirred the water. When the paddle wheel stopped, the increase in the internal energy of the water was equal to the work done by the Earth, assuming little rise in internal energy of the paddle wheel and container, and little energy transfer to the surroundings.

It was found for water that the energy required to raise the temperature of one gram of water by one kelvin (1 K) is 4.2 joules. The "heat capacity" of an object is the amount of energy required to raise its temperature one kelvin. The "specific heat capacity" is the heat capacity on a per-gram or per-mole or per-atom basis, and is a property of the material. The specific heat capacity of water is 4.2 joules per kelvin per gram (Figure 7.25). Other materials have

Degrees? Joules?

Figure 7.24 A thermometer might be calibrated in joules instead of in degrees.

4.2 J input — 1 K rise

1 gram of water

Figure 7.25 Energy input of 4.2 joules into a gram of water raises the temperature by 1 K. We say that the specific heat capacity of water is 4.2 J/K/gram.

Figure 7.26 For larger oscillations of two bound atoms the average separation increases.

different specific heat capacities. We will study an atomic theory of specific heat capacity in a later chapter.

Volume and Temperature

A liquid thermometer works because an ordinary liquid has a larger volume at higher temperature. Why does this happen?

For large oscillations of atoms whose interaction is described by a potential energy such as that shown in Figure 7.26, the average stretch increases. Because the actual potential energy curve is not really symmetric around the equilibrium point, with increasing energy the average interatomic bond length gets slightly longer. This is why an object typically expands at higher temperature. Higher temperature implies larger amplitude of oscillations and higher energy, and there is a slight shift in the center of the oscillations at higher energy.

In the interior of a solid the potential energy curve for an atom is symmetric, not asymmetric, because the potential energy is associated with forces exerted by atoms to the left and to the right. However, since the bond lengthening can start at the surface and propagate into the interior, a solid ultimately expands at higher temperatures.

Other Kinds of Thermometers

There are many other kinds of thermometers. All materials change in some way when they get hotter or colder, and some of these effects are the basis for useful temperature indicators. Liquid crystals are increasingly used in thermometers. In a liquid crystal there is some molecular order over long distances in the liquid, unlike a true liquid in which there is hardly any long-range order. In one form of liquid crystal thermometer, helical molecules remain roughly parallel to each other, and the distance between coils in the helix changes with temperature. Because the distance between coils is comparable to the wavelength of visible light, there is an effect on the way light is reflected by or transmitted through the liquid crystal. The result is that a change in temperature affects the appearance of the material.

Calculations Using Specific Heat Capacity

Joule found that it takes 4.2 J of energy input to a gram of water to raise its temperature 1 K. If such experiments are done on other materials, the temperature rise is found to be different. For example, the specific heat capacity of ethanol is found to be 2.4 J/K/gram, and the specific heat capacity of copper is only 0.4 J/K/gram. One of the many unusual and useful properties of water is that it has a very large specific heat capacity on a per-gram basis, which means that it is difficult to change its temperature. Here is the meaning of specific heat capacity on a per-gram basis, where $\Delta E_{thermal}$ is the rise in the internal energy of the system, in the form of increased atomic kinetic and potential energy:

SPECIFIC HEAT CAPACITY C ON A PER-GRAM BASIS
$$C = \frac{\Delta E_{thermal}}{m\Delta T} \quad (m \text{ in grams})$$

In principle, it should be possible to predict the specific heat capacity of a material if something is known about its atomic structure, since a temperature rise is a measure of increased energy at the atomic level. Comparisons of calculations with experimental values of the specific heat capacity data are a good test of our understanding of atomic models of solids, liquids, and gases. In later chapters you will learn how to predict the specific heat capacity of a solid (based on the ball-and-spring model of a solid) and of a gas.

If the specific heat capacity of a material is known, the amount of energy transfer into an object can be determined by measuring the temperature rise.

EXAMPLE   **Energy Input Raises the Temperature**

You stir 12 kg of water vigorously, doing 36,000 joules of work. If the container is well insulated (so that all of your energy input goes into increasing the energy of the water), what temperature rise would you expect?

Solution

$$\Delta E_{sys} = W$$
$$mC\Delta T = W$$
$$\Delta T = \frac{W}{mC} = \left(\frac{36,000 \text{ J}}{12,000 \text{ g}}\right)\left(\frac{1}{4.2 \text{ J/K/g}}\right) = 0.7 \text{ K}$$

Including units helps avoid making calculational mistakes. Note that a lot of work is required to achieve a small temperature change.

EXAMPLE   **Thermal Equilibrium**

A 300 gram block of aluminum at temperature 500 K is placed on a 650 gram block of iron at temperature 350 K in an insulated enclosure. At these temperatures the specific heat capacity of aluminum is approximately 1.0 J/K/gram, and the specific heat capacity of iron is approximately 0.42 J/K/gram. Within a few minutes the two metal blocks reach the same common temperature T_f. Calculate T_f.

Solution   System: The two blocks
Surroundings: Due to the insulation, no objects exchange energy with the chosen system
Initial state: Different temperatures, not in contact
Final state: Blocks have come to thermal equilibrium, same T_f
Energy Principle: $\Delta E_{Al} + \Delta E_{Fe} = 0$ (The total energy of the two blocks does not change, because there is no energy transfer from or to the surroundings.)

$$m_1 C_1 (T_f - T_{1i}) + m_2 C_2 (T_f - T_{2i}) = 0$$

$$(300 \text{ g})(1.0 \text{ J/K/g})(T_f - 500) + (650 \text{ g})(0.42 \text{ J/K/g})(T_f - 350) = 0$$

Solving for the final temperature, we find $T_f = 429$ K.
In words, what happens in this process is that the aluminum temperature falls from 500 K to 429 K, and the iron temperature rises from 350 K to 429 K. The thermal energy decrease in the aluminum is equal to the thermal energy increase in the iron.

7.X.6 300 grams of water whose temperature is 25° Celsius are added to a thin glass containing 800 grams of water at 20° Celsius (about room temperature). What is the final temperature of the water? What simplifying assumptions did you have to make in order to determine your approximate result?

7.X.7 Niagara Falls is about 50 meters high. What is the temperature rise in kelvins of the water from just before to just after it hits the rocks at the bottom of the falls, assuming negligible air resistance during the fall and that the water doesn't rebound but just splats onto the rock? It is helpful (but not essential) to consider a 1 gram drop of water.

7.5 ENERGY TRANSFER DUE TO A TEMPERATURE DIFFERENCE

Figure 7.27 A hot block in contact with a cold block.

Work is mechanical energy transfer into or out of a system involving forces acting through macroscopic displacements. When a hot object is placed in contact with a cold object, energy is transferred from the hot object to the cold one, but there are no macroscopic forces or displacements, so we don't refer to this kind of energy transfer as work.

Instead, we speak of "energy transfer Q due to a temperature difference." At the microscopic level there is actual work; when a hot block is placed in contact with a cold block (Figure 7.27), at the interface the atoms in the two blocks collide with each other, and do work on each other. The atoms in the hot block have greater average kinetic energy than the atoms in the cold block, so in an individual collision it is likely that a fast-moving atom in the hot block loses energy to a slow-moving atom in the cold block.

It can happen that the atom in the hot block happens to be moving slowly and gains energy when it is hit by an atom in the cold object that happens to be moving fast, but this is less likely. On average there is energy flow ("microscopic work") from the hot block to the cold block. These energy changes will diffuse throughout the blocks. If left to themselves, the two blocks will eventually come to the same temperature, intermediate between their two initial temperatures.

Q = ENERGY TRANSFER DUE TO A TEMPERATURE DIFFERENCE

The symbol Q represents an amount of energy that flows from the surroundings into a system, due to a temperature difference between the system and the surroundings. One can also call this process "microscopic work." Q can be positive or negative.

If we choose the cold block as our system of interest, this system gains energy (at the expense of the surroundings, the hot block). At the atomic level, this energy increase is the result of work done on the atoms at the surface of the cold block. Usually we are unable to observe these atomic-level interactions directly. We can infer the amount of energy transfer into the system by observing the temperature rise of that system, if we know the specific heat capacity of the material.

It is common practice to denote energy transfer due to a temperature difference (microscopic work) by the letter Q and work (macroscopic, mechanical energy transfer) by the letter W. The energy equation for an "open" (not isolated) system is then

$$\Delta E_{\text{sys}} = Q + W$$

That is, energy transfer due to a temperature difference and mechanical energy transfers into or out of the system produce a change in the energy of the system. The energy of the system includes particle energy and potential energy, at the microscopic or macroscopic level.

Figure 7.28 If energy flows from the surroundings into the system, Q is positive.

Figure 7.29 If energy flows from the system into the surroundings, Q is negative.

Sign of Q

Like W, Q can be negative. If there is energy transfer out of a system, we give Q a negative sign, because the change in the energy of the system is a decrease. See Figure 7.28 and Figure 7.29. Typically such transfer would be due to the system having a higher temperature than the surroundings. Similarly, if the system does work on its surroundings rather than the other way round, the sign of W is negative. (Some older textbooks on thermal physics reverse the sign of W, counting energy outputs from the system as positive.)

It is thought that for the Universe as a whole Q and W are zero, and the energy of the Universe does not change. However, even a small system may be effectively closed, with its energy unchanging, if it is thermally well insulated (to prevent the flow of energy due to a temperature difference) and if there are no other energy transfers from the surroundings.

Terminology: The Word *Heat*

The technical meaning of the word "heat" in science is not the same as its everyday meaning, just as the scientific meaning of the word "work" is different from its everyday meaning. In science the word "heat" is reserved to refer to energy transfer across a system boundary due to a temperature difference (microscopic work); this quantity is represented by the symbol Q. One cannot say "there is heat in the object;" instead, one says there is thermal energy $E_{thermal}$ in the object. The amount of energy in the object might increase due to "heat" Q (microscopic work). However, even professional scientists sometimes slip in their usage of the word "heat"!

To avoid confusion, we will avoid the use of the word "heat" as a noun. Instead, we will speak of "energy transfer Q due to a temperature difference" between system and surroundings, and change of "thermal energy" inside the system $\Delta E_{thermal}$ (in the form of increased atomic kinetic energy and potential energy).

> **7.X.8** Suppose that you warm up 500 grams of water (half a liter, or about a pint) on a stove while doing 5×10^4 J of work on the water with an electric beater. The temperature of the water is observed to rise from 20° C to 80° C. What was the change in the thermal energy of the water? Taking the water as the system, how much transfer of energy Q due to a temperature difference was there across the system boundary? What was the energy change of the surroundings?

Other Kinds of Energy Transfers

Work W and energy transfer Q due to a temperature difference are very common kinds of energy transfers into or out of an open system, but there are several other kinds of energy transfer. Here is a list, together with examples of each type.

- W, mechanical work $\vec{F} \bullet \Delta \vec{r}$; a force acts through a macroscopic distance: Throw a ball, increasing the ball's kinetic energy; compress a macroscopic spring.
- Q, energy transfer due to a temperature difference between system and surroundings: Boil water over a hot flame; a warm house loses energy to the cold outside air.
- Matter transfer: Put gasoline into a car; natural gas is piped into the home.
- Mechanical waves: Sound enters a microphone and moves a plate to generate electric signals; ocean waves drive an electric generator.
- Electricity: Electrical current enters the house and is used to run appliances; electrical current stores energy in the batteries of an electric car.

- Electromagnetic radiation, including visible light, infrared light, gamma rays, x-rays, and ultraviolet light: Sunlight absorbed by the Earth raises the temperature (and the thermal energy) of the daylight side of the Earth. We will usually account for this kind of energy transfer in terms of photons (packets of electromagnetic energy) crossing the system boundary, to be discussed in a later chapter.

A general form of the Energy Principle is this:

THE ENERGY PRINCIPLE

$$\Delta E_{sys} = W + Q + \text{other energy transfers}$$

Other energy transfers into a system may include matter transfer, mechanical waves, electric current, and electromagnetic radiation.

7.X.9 An electric hot plate raises its own internal energy and the internal energy of a cup of water by 8000 J, and there is at the same time 1000 J transferred to the cooler air (that is, $Q = -1000$ J). How much energy was transferred to the hot plate in the form of electricity?

The Steady State

A system is in a *steady state* if its energy does not change despite the fact that there are energy transfers between the system and the surroundings. In the steady state, the rates of energy flow into and out of the system must be equal.

EXAMPLE   **An Electric Heater**

An electric heater receives an energy input of 5000 J of electric energy. During this time the heating element is maintained at a constant high temperature. What is $\Delta E_{thermal}$ for the heater, and what is Q, the energy transfer between the heater's hot heating element and the cooler air?

Solution   System: Heater
Surroundings: Electric wires, air
Energy Principle: $\Delta E_{sys} = W + Q + \text{other energy transfers}$
      Since there is no internal energy change, $\Delta E_{thermal} = \Delta E_{sys} = 0$. This is a steady-state situation.

$$\Delta E_{sys} = W + Q + \text{electric energy input}$$
$$0 = Q + 5000\,\text{J}$$
$$Q = -5000\,\text{J}$$

FURTHER DISCUSSION   In this situation the sign of Q is negative, reflecting the fact that energy is transferred from the heating element to the air. We describe this situation as a steady state, not equilibrium, because there are energy transfers into and out of the system. However, these energy transfers, plus and minus, add up to zero, so there is no change internally.

If the heater were in an insulating enclosure and no energy could flow into the air, as electric energy continually flowed into the heater the temperature of the heating element would keep increasing until the element melted.

Q Is Not the Same as $\Delta E_{thermal}$

It is very important to see the difference between an energy transfer Q between system and surroundings due to a temperature difference, and a change in the thermal energy inside a system, associated with a change in the temperature of the system.

In the previous example of the electric heater, the thermal or internal energy of the heating element did not change ($\Delta E_{sys} = 0$, steady state), but $Q = -5000\,\text{J}$.

> **7.X.10** In a certain time interval, natural gas with energy content of 10000 J was piped into a house during a winter day. In the same time interval sunshine coming through the windows delivered 1000 J of energy into the house. The temperature of the house didn't change. What was $\Delta E_{thermal}$ of the house, and what was Q, the energy transfer between the house and the outside air?

7.6 REFLECTION: FORMS OF ENERGY

There are three fundamental kinds of energy in a multiparticle system:

- rest energy
- kinetic energy
- potential energy

All other forms of energy are examples of these basic forms. For example, chemicals in a system can react with each other and raise the temperature (and thermal energy) of the system, so we speak of "chemical energy," of which food is an example, with the molecule ATP providing energy storage in the body. The "chemical energy" is the kinetic energy and electric potential energy of molecules. Change of shape (configuration) is associated with change of potential energy, and there may also be changes of the kinetic energy of the molecules.

Nuclear energy is similar. Nucleons can be more or less tightly bound to particular nuclei, associated with different amounts of nuclear potential energy (change of shape) and different amounts of kinetic energy of the nucleons. Also, a change of identity in a nuclear reaction can be associated with a (huge) kinetic energy change associated with a change of rest mass.

7.7 POWER: ENERGY PER UNIT TIME

In technical usage, the word "power" is defined to mean "energy per unit time." From the point of view of energy usage, it makes no difference whether you take a minute or a month to climb a flight of stairs, yet from a practical point of view you certainly do notice the *rate* of energy usage, called power. The units of power are joules per second, or watts (honoring James Watt, the developer of the first efficient steam engine).

POWER
Energy per unit time (J/s or watts)

We can construct an equation for the instantaneous power associated with the work done by a force, through the following reasoning. If a force $\vec{F}$ does an amount of work $\vec{F} \bullet \Delta \vec{r}$ in a time Δt, then:

$$\text{power} = \frac{\vec{F} \bullet \Delta \vec{r}}{\Delta t}$$

so instantaneously

$$\text{power} = \frac{d\vec{F} \bullet d\vec{r}}{dt} = \vec{F} \bullet \frac{d\vec{r}}{dt} = \vec{F} \bullet \vec{v}$$

INSTANTANEOUS POWER

$$\text{power} = \frac{\vec{F} \bullet d\vec{r}}{dt} = \vec{F} \bullet \vec{v}$$

EXAMPLE   **Light Bulb Power**

How much energy is required to run a 100 watt light bulb for an hour?

Solution $(100 \text{ J/s})(1 \text{ hr}) \left(60 \frac{\text{min}}{\text{hr}}\right) \left(60 \frac{\text{s}}{\text{min}}\right) = 3.6 \times 10^5 \text{ J}$

7.X.11 In the Niagara Falls hydroelectric generating plant, the energy of falling water is converted into electricity. The height of the falls is about 50 meters. Assuming that the energy conversion is highly efficient, approximately how much energy is obtained from one kilogram of falling water? Therefore, approximately how many kilograms of water must go through the generators every second to produce a megawatt of power (1×10^6 watts)?

7.X.12 A vehicle with a mass of 1000 kg has an engine whose maximum power output is 50 kilowatts (about 67 horsepower; one horsepower is 746 watts). At a speed of 20 m/s (about 45 miles/hour) and maximum power, determine the maximum acceleration by calculating the force that is acting.

7.8   OPEN AND CLOSED SYSTEMS

We have already had some experience with the difference between a closed system and an open system. Because of the importance of this distinction, in this section we will go more deeply into the issues. Figure 7.30 shows transfers of money into and out of a bank during a particular time period, and the corresponding change in accounts inside the bank.

QUESTION   We have shown a situation in which more money is transferred into the bank than is transferred out. During these transactions, does the total amount of money inside the bank remain unchanged?

In Figure 7.30 $5000 came in and $3000 went out, for a net gain of $2000. We say that the bank is an "open system"—a portion of the world open to transfers in and out, and therefore subject to changes in its internal amount of money. During times when the bank does not permit transfers, the bank is temporarily a "closed system" and its total internal amount of money is unchanged, although there may be changes in the form of money inside the bank such as transfers between checking accounts and savings accounts.

Now consider Figure 7.31, which shows energy transfers during a particular time period in the winter, into and out of a house that is heated by gas, and the corresponding change in the energy inside the house. In this situation more energy comes into the house than goes out, and as a result the temperature inside the house rises.

QUESTION   During these energy transactions, the total amount of energy in the whole Universe is unchanged, but is the amount of energy inside the house unchanged?

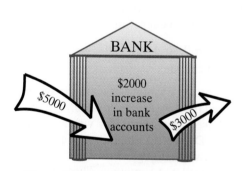

Figure 7.30 A bank is an open system.

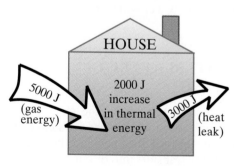

Figure 7.31 A house is an open system.

Figure 7.32 The change of energy of the system is equal to energy transferred into the system across the system boundary minus energy transferred out of the system across the system boundary.

The amount of energy inside the house increased by 2000 joules, and the energy of the surroundings decreased by 2000 joules. We say that the house is an "open system"—a portion of the Universe open to energy transfers, and therefore subject to changes in its internal amount of energy. If the house could be perfectly insulated against heat leakage (and solar radiation), and the gas (and electricity) turned off, the house would be a "closed system" with respect to energy and its total internal amount of energy would be unchanged, although there may be changes in the form of energy inside the house, such as the family dog converting chemical energy into kinetic energy when chasing its tail.

These considerations lead to the following scheme for keeping track of energy. Choose some portion of the Universe and mentally surround it by a dashed line marking the boundary (Figure 7.32). Then the energy transfer into the system across the system boundary minus the energy transfer out of the system across the system boundary is equal to the change of the energy inside the system.

The change in energy *inside* the system, ΔE_{sys}, can be either a positive or a negative amount of energy. For example, take an electric car as the system of interest:

> Charge the battery: ΔE_{sys} is positive (more energy stored in the battery).
> Do work on the car by pushing it: ΔE_{sys} is positive (more kinetic energy).
> Run the headlights (radiate electromagnetic energy): ΔE_{sys} is negative (less energy stored in the battery).

The car is an open system whose total amount of energy changes due to energy transfers into or out of the system: the energy of an *open* system is *not* constant.

In contrast, the total energy of the Universe remains constant, because there are compensating energy changes in the surroundings of the car: the electric company lost a store of chemical energy by running its generators to charge the car, you used up some chemical energy to push the car, and the light from the headlights is absorbed by the surroundings and leads to a rise in temperature of the surroundings. The measurements that we are able to make confirm the premise that the Universe is a closed system whose total amount of energy never changes, although the form of the energy may change.

> For any closed system, *inflow = outflow = 0*, so $\Delta E_{sys} = 0$.
> The energy of a closed system does not change.

The Universe as a whole is the most important example of a closed system, but it is often the case that a portion of the Universe can be considered to be a closed system, at least approximately. For example, put hot water and ice cubes into a very well-insulated container. During the short time that the ice melts and the water gets somewhat cooler, we can neglect the small amount of energy leakage through the walls of the insulated container. There is an increase in the energy of the ice cubes (which changed from solid to liquid), and a decrease in the energy of the hot water, but negligible net change in the energy inside the container, which is approximately a closed system whose total energy is (approximately) unchanged.

7.9   THE CHOICE OF SYSTEM AFFECTS ENERGY ACCOUNTING

What you choose to include in your chosen system affects the form of the energy equation for that system. For example, when we analyze a system consisting of an airless Earth plus a ball moving upward, the energy of the system does not change, because there are no energy transfers into or out of our chosen system (Figure 7.33).

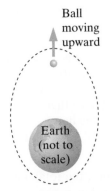

Ball moving upward

Earth (not to scale)

Figure 7.33 System: Earth + ball.

System: Earth + Ball

System: Earth + ball
Surroundings: Nothing significant
Initial state: Ball moving upward
Final state: Ball has risen a distance h, and is momentarily at rest
Energy Principle:

$$E_f = E_i + W$$

$$K_f + U_f = K_i + \cancel{U_i} + \cancel{W}$$

$$0 + mgh = \frac{1}{2}mv_i^2$$

Since only the difference in U matters, we can set $U_i = 0$.

System: Ball

On the other hand, when we analyze a system consisting of just the ball (Figure 7.34), the Earth (which is now outside our chosen system) does work on the ball, and the energy of the ball increases:
System: Ball
Surroundings: Earth
Energy Principle:

Ball
moving
upward

mg

Earth
(not to
scale)

Figure 7.34 System: ball only.

$$E_f = E_i + W$$

$$K_f = K_i + W$$

$$0 = \frac{1}{2}mv_i^2 - mgh$$

There are no potential energy terms for a single-particle system. The Earth does negative work because the force is opposite to the displacement.

Energy terms move back and forth across the equal sign in the energy equation, depending on our choice of the system of interest. It is extremely important to be clear about the choice of system, because this affects the form of the energy equation and helps avoid serious mistakes.

Warning: Avoid Double Counting!

QUESTION   A student said, "The Earth exerts a downward force mg as the ball rises a distance h and does an amount of work $-mgh$, and there is also an increase in gravitational potential energy $+mgh$, so the kinetic energy decreases by an amount $-2mgh$." What is wrong with the student's statement?

The problem here is double-counting due to a lack of clarity about the choice of system. If the Earth is part of the system, the force mg is internal to the system and does no external work on the system; rather, there is work done by internal forces on both the Earth and the ball, and the negative of this work by internal forces is the change in potential energy. On the other hand, if the ball alone is the system, there is no shape change of the ball and no change of gravitational potential energy, but there is an external force mg that does an amount of work $-mgh$.

In the next section we will illustrate the issue of choice of system by analyzing the same problem three times, using three different choices of system. We will find that the form of the Energy Principle will be different in each case, but we obtain the same physical results each time.

The Woman, the Barbell, and the Earth

To help avoid double-counting mistakes, it is useful practice to write the energy equation for various choices of system in a more complicated situation,

Figure 7.35 A woman lifts a barbell a distance h, exerting a force F. In the final state the barbell is moving upward with speed v.

Figure 7.36 System: woman + barbell + Earth.

Figure 7.37 System: barbell only.

■ Warning! It is tempting, but wrong, to include a term for the change in the gravitational energy of the barbell and to write the energy equation for the barbell alone like this:

$$\frac{1}{2}mv^2 + mgh = Fh - mgh$$

The error here is in assigning some gravitational energy to the barbell alone. Gravitational potential energy is associated with a change in the shape of a multi-object system; in particular, to changes in the separations of pairs of interacting objects. The barbell alone is not a pair of objects and does not change shape.

that of a woman applying a constant force F to lift a barbell of mass m from rest through a distance h, at which point the barbell is not only higher above the Earth but also has acquired some speed v (Figure 7.35). We're considering a time when the barbell is still headed upward, before the woman has brought the barbell to a stop above her.

For convenience in the following discussion, let E_w represent the following energy terms:

$$E_w = \text{chemical energy of woman} +$$
$$\text{kinetic energy of woman (moving arm)} +$$
$$\text{gravitational energy of woman and Earth} +$$
$$\text{thermal energy of woman}$$

System: Woman + Barbell + Earth

System: Woman + barbell + Earth (Figure 7.36)
Surroundings: Nothing significant
Initial state: Barbell at rest
Final state: Barbell has moved upward a distance h, and has speed v
Energy Principle:

$$E_f = E_i + W$$
$$K_f + U_f + E_{W,f} = \cancel{K_i} + \cancel{U_i} + E_{W,i} + \cancel{W}$$
$$\frac{1}{2}mv^2 + mgh + E_{W,f} = E_{W,i}$$

If we choose the system to consist of the woman, the barbell, and the Earth, there are no energy transfers into or out of our system, so energy does not change, and the energy equation looks like this as the woman lifts the barbell a distance h, starting from rest (we neglect the tiny kinetic energy of the recoiling Earth). This allows us to calculate the change in the woman's energy from our final equation:

$$\Delta E_w = -\left(\frac{1}{2}mv^2 + mgh\right)$$

As can be seen from this result, the net effect is a decrease in her energy, because the barbell and the Earth are farther apart, and the barbell goes faster, both of which represent energy increases in other parts of the system.

The energy term ΔE_w represents changes in energy associated with the woman, mainly a decrease in her stored chemical energy, because she must perform chemical reactions with food (or with chemicals stored in body tissues after eating food) in order to be able to move her arm. The woman has to burn more energy than the rest of the system gets, since she not only has to supply energy to the rest of the system but also has to supply her own energy increases: increased kinetic energy and gravitational potential energy associated with her moving, raised arm, and increased thermal energy, because her temperature rises a bit when she exerts herself.

System: Barbell

System: Barbell (Figure 7.37)
Surroundings: Woman and Earth
Initial state: Barbell at rest
Final state: Barbell has moved upward a distance h, and has speed v
Energy Principle:

$$E_f = E_i + W$$
$$K_f = \cancel{K_i} + W_{\text{Woman}} + W_{\text{Earth}}$$
$$\frac{1}{2}mv^2 = Fh - mgh$$

If we choose the system to consist just of the barbell, the woman does (positive) work $+Fh$ on the system, and the Earth does (negative) work $-mgh$ on the system (because mg is down and the displacement h is up). Compare this equation to the equation we obtained when considering the system of woman + barbell + Earth. We see that energy terms have moved to the other side of the equation as a result of our different choice of system.

QUESTION Which is bigger, F or mg? Why?

Since the barbell's upward momentum increases, F must be larger than mg, since the net force in the upward direction is $F - mg$.

QUESTION Compare the equations for system 1 (woman + barbell + Earth) and system 2 (barbell only). In terms of F, calculate ΔE_w.

$$\Delta E_w = -\left(\frac{1}{2}mv^2 + mgh\right) \text{ and } Fh = \frac{1}{2}mv^2 + mgh$$

By comparing the two equations, we see that the change in the energy associated with the woman must be $-Fh$. The sign makes sense: we had concluded earlier that her energy change must be negative.

System: Barbell + Earth

System: Barbell + Earth
Surroundings: Woman
Initial state: Barbell at rest
Final state: Barbell has moved upward a distance h, and has speed v
Energy Principle:

$$E_f = E_i + W$$
$$K_f + U_f = \cancel{K_i} + \cancel{U_i} + W_{\text{Woman}}$$
$$\frac{1}{2}mv^2 + mgh = Fh$$

Figure 7.38 System: barbell + Earth.

If we choose the system to consist of the barbell plus the Earth (Figure 7.38), this system changes shape as the woman pushes the two pieces of the system apart. It is almost as though she is stretching an invisible spring connecting the two objects.

QUESTION Does the woman do any work *on the Earth*? Why or why not?

The woman's feet push down on the Earth but do negligible work, because there is essentially no displacement there (the Earth hardly moves). Work done by a force is calculated by taking into account the displacement of the point where the force is applied. In this case, the point of application hardly moves, so the force does hardly any work.

Further Discussion

Looking back over the equations obtained for different choices of system, you can see how energy terms that represent transfers across the system boundary for one choice of system become changes of energy *inside* the system for a different choice of system. Comparing equations for different choices of system can be useful in determining an unknown quantity (such as the energy change in the woman, $-Fh$, in the example you just worked through). Also, analyzing a process for more than one choice of system is a good check on your calculations and your understanding.

7.X.13 Consider a harmonic oscillator (mass on a spring without friction). Taking the mass alone to be the system, how much work is done on the system as the spring of stiffness k_s contracts from its maximum stretch A to its relaxed length? What is the change in kinetic energy of the system during this motion? For what choice of system does energy remain constant during this motion?

7.10	**ENERGY DISSIPATION**

The total energy of the Universe does not change, but useful energy is often "dissipated" into forms less useful to us. Push a chair across the floor, and some of the work that you do goes into raising the temperature of the floor and the chair, rather than into increasing the macroscopic kinetic energy of the chair. Throw a ball up into the air, and some of the initial energy is dissipated into increased microscopic energy of the air. Sliding friction, air resistance, viscous friction—all of these phenomena are examples of energy dissipation discussed in the rest of the chapter.

Figure 7.39 A sequence of video frames of a falling metal ball, which travels farther in each frame, reflecting the increase in its speed.

Air Resistance

Air resistance isn't a major factor when you drop a metal ball a short distance. A video sequence of a falling metal ball shows that the ball moves faster and faster as it falls (an effect that is hard to observe by eye alone). The time between adjacent frames in Figure 7.39 is 1/15 second, and the increasing distances between heights of the ball in adjacent frames show that the speed is getting faster and faster. (Also note the increasing blur due to faster motion of the ball.) The gravitational attraction of the Earth acting on the ball makes the momentum of the ball continually increase. A curve is drawn along the tops of the ball images. The major visible marks on the vertical meter sticks are 10 cm apart. In 7/15 second the ball falls about 1 meter.

In contrast, if you drop a bowl-shaped paper coffee filter, a video sequence of the falling filter shows that the filter's speed does increase at first, but instead of continuing to gain speed it quickly reaches a constant speed despite the gravitational force acting on it (Figure 7.40):

Figure 7.40 A sequence of video frames of a falling coffee filter, which speeds up briefly, then falls at constant speed.

In the video sequence shown above, the time between adjacent frames is again 1/15 second. In 16/15 second the filter falls about 1 meter (the dense ball took only 7/15 second to fall that far). A curve is drawn through the centers of the coffee filter images. The nearly straight line in the later part of the motion indicates motion at constant speed.

Dependence of Air Resistance on Speed

The restraining effect of the air is called "air resistance" or "drag." You have probably observed that it is harder to move something quickly through a fluid such as water or air than to move it slowly. This suggests that air resistance acting on a falling object might depend on the speed of the object.

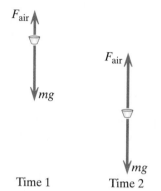

Figure 7.41 Forces acting on a falling coffee filter at Time 1 when the filter's speed is increasing and at Time 2 when a constant speed has been reached.

> **QUESTION** Consider a falling coffee filter at two different times: (1) while the filter's speed is still increasing, and (2) after a constant speed has been reached. How do you know the air resistance is larger at the second instant than at the first instant?

While the filter's speed is increasing there must be a nonzero net force, which means that the air resistance is smaller than the gravitational force (Figure 7.41). Later, however, the net force is zero, because the speed is no longer increasing, which means that the force of the air has grown to be as large as the gravitational force (which is nearly constant at mg and in fact increases very slightly as the coffee filter gets closer to the Earth). Since the change in the density of the air is small over the distance the filter falls, and the change in the gravitational force is very small, we can conclude that the change in the air resistance is due to a speed dependence of that force.

Experiments show that the magnitude of the air resistance force is approximately proportional to the square of an object's speed. You can do a simple experiment with coffee filters to determine the form of the speed dependence, at least for low speeds (see Problem 7.P.60).

Terminal Speed

A sky diver initially speeds up due to the gravitational force acting downward, but the sky diver's speed doesn't keep getting bigger and bigger, because the upward air resistance force increases with speed. Eventually the sky diver reaches a "terminal speed" and falls thereafter at constant speed despite the gravitational force. This terminal speed for falling humans is so high (about 60 m/s, or about 135 miles per hour) that hitting the ground without a parachute is normally fatal, although there have been cases of people hitting deep snow at terminal speed and surviving.

> **7.X.14** A sky diver whose mass is 90 kg is falling at a terminal speed of 60 m/s. What is the magnitude of the force of the air on the sky diver?

Dependence of Air Resistance on Cross-Sectional Area

It is easy to observe that the effect of air resistance increases as the cross-sectional area of an object increases—a sky diver falls much more slowly with an open parachute than with a closed parachute. A low-density paper coffee filter has a large cross-sectional area, so air resistance has a large effect on a falling coffee filter. In contrast, a high-density metal ball has a small cross-sectional area, but a large gravitational force acts on it, and air resistance may be negligible in comparison with gravity.

Dependence of Air Resistance on Shape

QUESTION From your own experience, would you expect larger air resistance (at a particular speed) for a pointed object or a blunt object?

As you might expect, the air resistance force depends not only on density, cross-sectional area, and speed, but also on the shape of the object. There is a larger air resistance force on a blunt object such as a coffee filter than on a ball with the same cross-sectional area. Streamlined objects such as arrows have reduced air resistance.

Dependence of Air Resistance on Air Density

QUESTION If you could double the density of the air through which a coffee filter falls, what change in the air resistance force would you expect?

You would probably expect a larger air resistance force in denser air. In fact, measurements of a variety of ordinary-sized objects moving through air or other fluids show that air resistance is proportional to the density ρ of the air. For example, there is less air resistance at higher altitudes, where the air is less dense.

An Empirical Equation for Air Resistance

An equation describing the air resistance force on a moving object must incorporate all the effects we have listed above: dependence on speed, cross-sectional area, shape of the object, and density of the air. The shape effect is usually taken into account by measuring the air resistance force (for example, by measuring the terminal speed) and determining an experimental "drag coefficient" C such that the following equation expresses the experimental results for the friction force:

APPROXIMATE AIR RESISTANCE FORCE (EMPIRICAL)

For blunt objects at ordinary speeds:

$$F_{\text{air}} \approx \frac{1}{2} C \rho A v^2$$

where $0.3 \leq C \leq 1.0$; blunter objects have higher values of C (ρ is the density of the air). The direction is opposite to the velocity:

$$\vec{F}_{\text{air}} \approx -\frac{1}{2} C \rho A v^2 \hat{v}$$

How important is the effect of air resistance in the everyday world? In Chapter 2 we analyzed the motion of a ball thrown through the air. However, we neglected air resistance, which can have a sizable effect. In Problem 7.P.68 you are asked to include the air resistance force given above. You find that a baseball thrown at high speed by a professional baseball pitcher goes only about half as far in air as it would go in a vacuum.

Even when we include air resistance in our calculations, we still have ignored a force that can have a major effect on the trajectory. If a ball has spin, there is an effect of fluid flow around the ball that raises the air pressure on the side where the rotational motion is in the same direction as the ball's velocity, and lowers the air pressure on the other side, where the rotational motion is

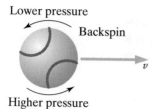

Figure 7.42 A spinning ball experiences additional forces.

in the opposite direction to the velocity. If the force points to the side, the ball curves; this is an important effect in baseball pitching. If the force points upward (due to "backspin," as in Figure 7.42), it lifts the ball and extends the range. If the force points downward (due to "topspin") it decreases the range. The calculations involved in modeling this effect are quite complex, and involve "fluid dynamics" (the study of fluid flow).

It is important to keep in mind that air resistance is *not* a fundamental force like the gravitational or electric force. Air resistance is the average result of a huge number of momentary contact electric interactions of air molecules hitting atoms on the surfaces of the falling object, and the average net air resistance is very difficult to calculate from fundamental principles at the molecular level.

> **7.X.15** As a ball falls, it has to push aside the air underneath it, and the air molecules acquire additional kinetic energy K_{air}. Therefore, will the final speed of the ball be greater or less than you would predict from the constancy of kinetic energy plus gravitational energy? Why?
>
> **7.X.16** A coffee filter of mass 1.4 grams dropped from a height of 2 m reaches the ground with a speed of 0.8 m/s. How much kinetic energy K_{air} did the air molecules gain from the falling coffee filter?

Mechanism of Air Resistance

We would like to understand the details of the interaction of the air with a falling object. In particular, we'd like to find a reason that the air resistance would depend on the speed of the object, the cross-sectional area of the object, and the density of the air.

Let's consider the interactions of the air and object from a microscopic viewpoint. When you hold a coffee filter stationary, air molecules hit it nearly equally from above and below, as seen in Figure 7.43. (Actually there are slightly more collisions per second on the lower surface than on the upper surface, leading to a small upward buoyant force that is very small compared to air resistance.) However, when the filter is moving downward due to the gravitational attraction of the Earth, the bottom side of the filter is running into the air molecules, while the top side is moving away from the air molecules.

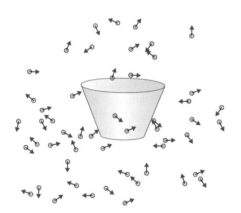

Figure 7.43 Air molecules collide with a falling coffee filter.

On average, the bottom side will have a significantly increased number of collisions per second with air molecules, and with greater impact. On the top side there will be a reduced number of collisions per second, and with less impact. There is a net upward push on the filter. This air resistance force increases with increasing downward speed of the filter, because higher speed increases the rate and impact of collisions on the bottom, and decreases the rate and impact of collisions on the top.

Eventually, when the downward speed of the coffee filter has become big enough, the net upward push by the air molecules becomes as large as the downward gravitational pull of the Earth. From then on the filter falls at a constant speed (terminal speed).

While this model captures the qualitative aspects of the situation, a simple quantitative calculation along these lines (using techniques discussed in a later chapter on gases) predicts a terminal speed that is *a thousand times smaller* than is observed! This is a serious failure of our simple model. The difficulty turns out to be that the falling coffee filter establishes macroscopic "wind"-like motions in the air, and the random motions of the air molecules at any particular location are relative to the bulk wind motion at that location. The techniques of "fluid dynamics" are capable of analyzing such phenomena, but these techniques are mathematically quite difficult and well beyond the scope of this introductory textbook. (Our later study of gases will address simpler situations, where there is no large-scale wind motion.)

7.X.17 When a falling object reaches terminal speed, its kinetic energy reaches a constant value. However, the gravitational energy of the system consisting of object plus Earth continues to decrease. Does this violate the principle of conservation of energy? Explain why or why not.

7.X.18 If you let a mass at the end of a string start swinging, at first the maximum swing decreases rather quickly, but once the swing has become small it takes a long time for further significant decrease to occur. Try it! Explain this simple observation.

Viscous Friction

Very small particles such as dust or droplets of fog falling slowly through air, or small objects moving through a thick liquid such as honey, experience a friction force that is proportional to v rather than v^2. This is called "viscous" friction.

The dependence of the friction force on v (for viscous friction) or v^2 (for air resistance) is approximately valid only in specific situations. The expressions for friction forces do not have the wide applicability of those for the gravitational and electric forces. However, even approximate expressions for friction forces allow us to make much better predictions than we could make if we ignored friction completely.

Viscous Friction in a Mass–Spring System

One of the most obvious things about the motion of a real macroscopic spring–mass system is that the amplitude gets smaller and smaller with time. Our model of a spring–mass system that we developed in Chapter 4 is too simple: we should include the effects of friction. See Problem 7.P.67 on numerical integration of a spring–mass system with friction.

In the case of "viscous" friction, where the friction force is $\vec{f} = -c\vec{v}$, it is possible with elaborate math (or by guessing the solution and plugging it into the Momentum Principle for the system) to find an analytical solution for x position as a function of time.

$$x(t) = Ae^{-(c/2m)t}\cos(\omega t)$$

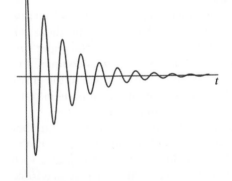

Figure 7.44 Position vs. time for an oscillator with viscous friction.

The amplitude dies away exponentially with time (Figure 7.44). The angular frequency ω of this "damped" oscillator is

$$\omega = \sqrt{\frac{k_s}{m} - \left(\frac{c}{2m}\right)^2}$$

which is approximately $\sqrt{k_s/m}$ for low friction. There is no such simple analytical solution if the friction force is independent of the speed (sliding friction) or proportional to v^2 (air resistance for large blunt objects), but such frictional forces are easily handled in a computational model.

Dissipation with Sliding Friction

In Chapter 4 we saw that when a block slides on a table, the table exerts a force on the block associated with compression of the interatomic bonds in the table and block. This force has two components: $\vec{F}_N$ normal to the surface, and $\vec{f}$ parallel to the surface, called the frictional force. Atoms of the block run into atoms in the table, and the resulting interactions stretch and compress interatomic bonds in the objects.

The frictional force is nearly independent of the speed of the block, and $f \approx \mu_k F_N$, where μ_k is the kinetic (motional) coefficient of friction.

Diamond tip

Pulling force

$\vec{F}$

Diamond surface

Figure 7.45 A computational model of sliding friction between a diamond tip and diamond surface. Image courtesy of Judith A. Harrison, U.S. Naval Academy. The vertical forces are not shown.

Figure 7.46 Forces in the *x* direction on a block pulled with constant speed across a table. (Forces in the *y* direction are not shown.)

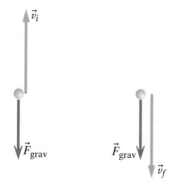

Figure 7.47 In the absence of air resistance, work done on the ball by the Earth is negative going up and positive coming down, and the total work done on the ball is zero. When the ball has returned to its initial position, its kinetic energy has not changed.

If the block does not slide across the table, the frictional force is just big enough to make the net force be zero. In this case, $f \le \mu_s F_N$, where μ_s is called the coefficient of static friction.

Figure 7.45 provides an atomic-level picture of friction. This image is a single frame from a molecular dynamics computer simulation of sliding friction between two objects, created by Judith A. Harrison and coworkers at the United States Naval Academy in Annapolis, Maryland. The computations are similar to the numerical integrations you have done, but they involve a large number of objects (atoms). The forces between the atoms are modeled by springlike forces.

In this image, a diamond tip is dragged to the right across a diamond surface. The large spheres represent carbon atoms, and the small ones are hydrogen atoms on the carbon surfaces. Note that the projecting tip has caught on the lower surface, causing interatomic bonds in the two objects to deform. The stretching and compression of interatomic bonds is responsible for the horizontal component of the force exerted on the moving object by the lower object.

A Paradox Involving Friction

If we are careful in our energy bookkeeping for a system involving motion with friction, we encounter a paradox that cannot easily be resolved. Consider the situation shown in Figure 7.46.

You exert a constant force *F*, and pull a block a distance *d* across a table. Because there is friction between the block and the table, the block travels at constant speed. Since the momentum of the block does not change, we conclude correctly that the net force on the block is zero, and that therefore the magnitude of the friction force must also be *F*.

In terms of energy inputs, you have done an amount of work *Fd* on the block. The friction force, acting in the opposite direction, has apparently done an amount of work $-Fd$. The net work done on the block therefore would seem to be zero. This is consistent with the fact that the kinetic energy of the block did not seem to change.

It is not, however, consistent with the fact that both the block and the table get hotter as they rub together! If the net work done on the block was zero, where did that increased internal energy come from?

This is not a "trick question," but a genuine puzzle. We will return to this issue in a later chapter, when we will deal with it in enough detail to resolve the paradox.

> **7.X.19** You move a block slowly to the right a distance of 0.2 m on a table. You apply a constant force of 20 N. How much work do you do? You then move slowly back to the starting point; how much work do you do in this second move? What change has there been in the thermal energy of the block and table?

7.11 | POTENTIAL ENERGY AND "CONSERVATIVE" FORCES

In Section 7.3 we saw that, in the absence of air resistance, a ball thrown upward will return to your hand with exactly the same kinetic energy it had when it left your hand. Taking the ball as the system, the work done on the ball by the Earth is negative as the ball ascends and positive as the ball descends; the total work done by the Earth during the round trip is zero, and the energy of the system is unchanged (Figure 7.47).

However, if we include air resistance, we find that the total work done by the air resistance force is not zero, and therefore the final kinetic energy

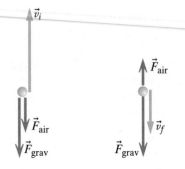

Figure 7.48 Taking into account air resistance, work done on the ball by the air is negative going up and also negative coming down, and the total work done on the ball is negative (the total work done by the Earth is zero). When the ball has returned to its initial position, its kinetic energy has decreased.

of the system (the ball) is less than its initial energy. The work done by the air resistance force is negative as the ball ascends, and it is also negative as the ball descends, because the direction of the air resistance force has changed (Figure 7.48). The kinetic energy of the ball decreases during both parts of the trip.

The decrease in kinetic energy of the ball is accompanied by an increase in the internal energy of the air and the ball, and the temperatures of the air and ball rise. The kinetic energy lost by the ball has been randomly and irreversibly distributed among the molecules of the air and the ball.

If a force depends only on the position of two interacting objects, it is possible to define a potential energy associated with this force. Gravitational and electric forces are examples of such forces, as is the force exerted by a spring. Such forces are called "conservative forces." A conservative force is equal to the negative gradient of the associated potential energy.

Dissipative forces such as friction forces are called "nonconservative" forces. It isn't possible to define a potential energy for a dissipative force. In defining potential energy, we consider two interacting objects to be pointlike, in the sense that their internal states do not change. This is not the case for dissipative forces; the air is a complicated multiparticle system whose internal energy does change. The flow of energy from the ball to the air is irreversible; the energy distributed among the air molecules cannot be recovered and converted back to kinetic energy of the ball.

Irreversibility has far-reaching consequences for the statistical or probabilistic behavior of macroscopic objects consisting of large numbers of atoms. We will consider issues of reversibility and irreversibility in more detail in a later chapter.

7.12   *RESONANCE

Imagine pushing somebody in a swing. If you time your pushes to match the natural frequency of the swing, you can build up an increasingly large amplitude of the swing. However, larger amplitude means higher speed and a higher rate of energy dissipation.

> QUESTION You keep pushing with constant (small) amplitude, and the swing eventually reaches a "steady state" with constant (large) amplitude. What determines the size of this steady-state amplitude?

The steady state is established when the energy dissipated through air resistance on each cycle has grown to be equal to the energy input you make on each cycle. The large steady-state oscillation needs only little pushes at the right times from you to make up for energy dissipation, and the amplitude of the oscillation can be much larger than the amplitude of the pushes (Figure 7.49).

On the other hand, if you deliberately push at the wrong times, not matching the natural frequency of the swing, the amplitude of the swing won't continually build up (Figure 7.50). To get a large response from an oscillating system, inputs should be made at the natural, free-oscillation frequency of that system. This important feature of driven oscillating systems is called "resonance."

Figure 7.49 Small pushes at the right times lead to a big response.

Figure 7.50 Pushes at the wrong times lead to a small response.

*Analytical Treatment of Resonance

Problem 7.P.71 is a numerical integration of a sinusoidally driven spring–mass oscillator, as shown in Figure 7.51.

This problem lets you study the buildup of the steady state as well as the steady state itself. For the steady-state portion of the motion, it is possible to

Figure 7.51 A sinusoidally driven oscillator.

Figure 7.52 Amplitude as a function of driving frequency. The damping in the lower curve is five times greater than in the upper curve.

obtain an analytical solution, if the energy dissipation is due to viscous friction. Here is the Momentum Principle for the driven oscillator in the diagram, with viscous friction force $-cv$:

$$\frac{dp}{dt} = -k_s[x - D\sin(\omega_D t)] - cv$$

where x is the position of the mass, D is the amplitude of the sinusoidal motion, and ω_D is the variable angular frequency of the driven end of the spring. For more details on how this equation was obtained, see the discussion in Problem 7.P.71.

In the steady state x must be a sinusoidal function, with amplitude A and phase shift ϕ depending strongly on the driving angular frequency ω_D:

$$x = A\sin(\omega_D t + \phi)$$

By taking derivatives of this expression to obtain $v = dx/dt$ and dv/dt as a function of time t and plugging these derivatives into the Momentum Principle, it can be shown that this expression for x as a function of time is a solution in the steady state if the amplitude A and phase shift ϕ have values determined by the following expressions:

$$A = \frac{\omega_F^2}{\sqrt{(\omega_F^2 - \omega_D^2)^2 + \left(\frac{c}{m}\omega_D\right)^2}}D$$

$$\cos\phi = \frac{(\omega_F^2 - \omega_D^2)}{\sqrt{(\omega_F^2 - \omega_D^2)^2 + \left(\frac{c}{m}\omega_D\right)^2}}$$

The detailed proof is rather complicated, so we omit it. Instead, in the exercises that follow we give you the opportunity to see that these expressions for amplitude and phase shift are in agreement with your experimental observations and the computer modeling of Problem 7.P.71. Figure 7.52 is a graph of the amplitude A as a function of the driving angular frequency ω_D for two different values of the viscous friction parameter c.

7.X.20 Using the equation for the amplitude A, show that if the viscous friction is small, the amplitude is large when ω_D is approximately equal to ω_F. Using the equation involving the phase shift ϕ, show that the phase shift is approximately 0° for very low driving frequency ω_D, approximately 180° for very high driving frequency ω_D, and 90° at resonance, consistent with your experiment.

7.X.21 Show that with small viscous friction, the amplitude A drops to $1/\sqrt{2}$ of the peak amplitude when the driving angular frequency differs from resonance by this amount:

$$|\omega_F - \omega_D| \approx \frac{c}{2m}\omega_F$$

(Hint: Note that near resonance $\omega_D \approx \omega_F$, so $\omega_F + \omega_D \approx 2\omega_F$.)

Given these results, how does the width of the resonance peak depend on the amount of friction? What would the resonance curve look like if there were very little friction?

Resonance in Other Systems

We have discussed resonance in the context of a particular system—a mass on a spring driven by a motor. We study resonance because it is an important phenomenon in a great variety of situations.

When you choose a radio or television station, you adjust the parameters of an electrically oscillating circuit so that the circuit has a narrow resonance at the chosen station's main frequency, and other stations with significantly different frequencies have little effect on the circuit.

Magnetic resonance imaging (MRI) is based on the phenomenon of nuclear magnetic resonance (NMR) in which nuclei acting like toy tops precess in a strong magnetic field. The precessing nuclei are significantly affected by radio waves only at the precession frequency. Because the precession frequency depends slightly on what kind of atoms are nearby, resonance occurs for different radio frequencies, which makes it possible to identify different kinds of tissue and produce remarkably detailed images.

SUMMARY

Potential energy of an ideal spring–mass system:

$$U_s = \frac{1}{2}k_s s^2$$

Approximate interatomic potential energy:

$$U = \frac{1}{2}k_s s^2 - E_M$$

A better approximation to interatomic potential energy:

$$U_M = E_M[1 - e^{-\alpha(r-r_{eq})}]^2 - E_M$$

(Morse potential energy, Figure 7.53)

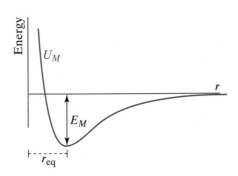

Figure 7.53

Thermal energy: Related to the temperature of the object

Power = $\dfrac{\text{energy}}{\text{time}}$ (J/s = watts); also calculable as $\vec{F} \bullet \vec{v}$

Energy transfer due to a temperature difference, Q

Other energy transfers: Work W, matter transfer, mechanical waves, electricity, electromagnetic radiation

Specific heat capacity = energy per gram to raise temperature 1 K

$$\Delta E_{\text{thermal}} = mC\Delta T$$

$$C = \frac{\Delta E_{\text{thermal}}}{m\Delta T} \quad (C \text{ in J/K per gram, } m \text{ in grams})$$

Open system: energy flows into and/or out of system.
Closed system: no energy flows into or out of system.
The choice of system (open or closed) affects energy accounting.
Dissipation of energy
Terminal speed
Air resistance: collisions between air molecules and moving object
Sliding friction force due to forming and breaking of bonds between two solid objects
Path independence of potential energy
Impossibility of describing friction in terms of potential energy
Resonance: sensitivity of driven oscillator to driving frequency

Air resistance is approximately proportional to speed squared, and in a direction opposite to the velocity:

$$\vec{F}_{\text{air}} \approx -\frac{1}{2}C\rho A v^2 \hat{v}$$

For small particles ("viscous" friction), fluid friction is approximately proportional to v.

*Analytical solution for driven spring–mass system: $x = A\sin(\omega_D t + \phi)$, where $\omega_F = \sqrt{k_s/m}$; viscous friction cv, driven at ω_D:

$$A = \frac{\omega_F^2}{\sqrt{(\omega_F^2 - \omega_D^2)^2 + \left(\dfrac{c}{m}\omega_D\right)^2}}D$$

$$\cos\phi = \frac{(\omega_F^2 - \omega_D^2)}{\sqrt{(\omega_F^2 - \omega_D^2)^2 + \left(\dfrac{c}{m}\omega_D\right)^2}}$$

EXERCISES AND PROBLEMS

Sections 7.1–7.3

7.X.22 A spring has a relaxed length of 6 cm and a stiffness of 100 N/m. How much work must you do to change its length from 5 cm to 9 cm?

7.X.23 A constant force $\langle 23, -12, 32 \rangle$ N acts through a displacement $\langle 0.12, 0.31, -0.24 \rangle$ m. How much work does this force do?

7.X.24 An object with mass 7 kg moves from a location $\langle 22, 43, -41 \rangle$ m near the Earth's surface to location $\langle -27, 11, 46 \rangle$ m. What is the change in the potential energy of the system consisting of the object plus the Earth?

7.X.25 A spring whose stiffness is 800 N/m has a relaxed length of 0.66 m. If the length of the spring changes from 0.55 m to 0.96 m, what is the change in the potential energy of the spring?

7.X.26 A particle moves inside a circular glass tube under the influence of a tangential force of constant magnitude F (Figure 7.54). Explain why we cannot associate a potential energy with this force (which is therefore a "nonconservative" force). How is this situation different from the case of a block on the end of a string, which is swung in a circle?

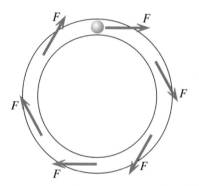

Figure 7.54 A particle moves in a circular glass tube.

7.X.27 There are three different ways to get from location A to location E in Figure 7.55. Along path 1, you take an elevator directly from location A straight up to location E. Along path 2, you walk from A to B, climb a rope from B to C, then walk from C to E. Along path 3, you walk from location A to D, then climb a ramp up to location E. The following questions focus on the work done on a 80 kg person by the Earth while following each of these paths.

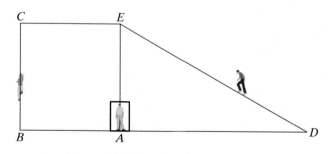

Figure 7.55 Following different paths from A to E.

Taking the origin at location A, the coordinates of the other locations are: $B \langle 0, -3, 0 \rangle$; $C \langle -3, 13, 0 \rangle$; $D \langle 33, 0, 0 \rangle$; $E \langle 0, 13, 0 \rangle$.

(a) Path 1 (A to E): What is the displacement vector $\Delta \vec{r}$ for the path from A to E on the elevator?

(b) What is the gravitational force acting on the person along path 1, as a vector?

(c) What is the work done by the gravitational force of the Earth on the 77 kg person during the elevator ride from A to E?

(d) Path 2 (A to B to C to E): What is the work done by the gravitational force of the Earth on the person as he walks from A to B?

(e) What is the work done by the gravitational force of the Earth on the person as he climbs from B to C?

(f) What is the work done by the gravitational force of the Earth on the person as he walks from C to E?

(g) What is the total work done by the gravitational force of the Earth on the person as the person goes from A to E along path 2?

(h) Path 3 (A to D to E): What is the work done by the gravitational force of the Earth on the person as he walks from A to D?

(i) What is the work done by the gravitational force of the Earth on the person as he climbs from D to E?

(j) What is the total work done by the gravitational force of the Earth on the person as the person goes from A to E along path 3?

(k) You calculated the work done by the gravitational force of the Earth on the person along three different paths. How do these quantities compare with each other?

7.X.28 Refer to Figure 7.56. Calculate the change in electric energy along two different paths in moving charge q away from charge Q from A to B along a radial path, then to C along a circle centered on Q, then to D along a radial path. Also calculate the change in energy in going directly from A to D along a circle centered on Q. Specifically, what are $U_B - U_A$, $U_C - U_B$, $U_D - U_C$, and their sum? What is $U_D - U_A$? Also, calculate the round-trip difference in the electric energy when moving charge q along the path from A to B to C to D to A.

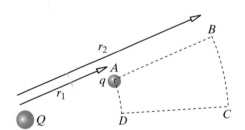

Figure 7.56 Change in electric energy along two different paths.

7.P.29 A horizontal spring–mass system has low friction, spring stiffness 200 N/m, and mass 0.4 kg. The system is released with an initial compression of the spring of 10 cm and an initial speed of the mass of 3 m/s.

(a) What is the maximum stretch during the motion?

(b) What is the maximum speed during the motion?

(c) Now suppose that there is energy dissipation of 0.01 J per cycle of the spring–mass system. What is the average power input in watts required to maintain a steady oscillation?

7.P.30 Write an equation for the total energy of a system consisting of a mass suspended vertically from a spring, and include the Earth in the system. Place the origin for gravitational energy at the equilibrium position of the mass, and show that the changes in energy of a vertical spring–mass system are the same as the changes in energy of a horizontal spring–mass system.

7.P.31 A spring with stiffness k_s and relaxed length L stands vertically on a table. You hold a mass M just barely touching the top of the spring.

(a) You *very slowly* let the mass down onto the spring a certain distance, and when you let go, the mass doesn't move. How much did the spring compress? How much work did *you* do?

(b) Now you again hold the mass just barely touching the top of the spring, and then let go. What is the maximum compression of the spring? State what approximations and simplifying assumptions you made.

(c) Next you push the mass down on the spring so that the spring is compressed an amount s, then let go, and the mass starts moving upward and goes quite high. When the mass is a height of $2L$ above the table, what is its speed?

7.P.32 Figure 7.57 is a potential energy curve for the interaction of two neutral atoms. The two-atom system is in a vibrational state indicated by the blue horizontal line.

Figure 7.57 Potential energy diagram.

(a) At $r = r_1$, what are the approximate values of the kinetic energy K, the potential energy U, and the quantity $K + U$?

(b) What minimum energy must be supplied to cause these two atoms to separate?

(c) In some cases, when r is large, the interatomic potential energy can be expressed approximately as $U = -a/r^6$. For large r, what is the algebraic form of the magnitude of the force the two atoms exert on each other in this case?

7.P.33 Design a "bungee jump" apparatus for adults. A bungee jumper falls from a high platform with two elastic cords tied to the ankles. The jumper falls freely for a while, with the cords slack. Then the jumper falls an additional distance with the cords increasingly tense. Assume that you have cords that are 10 m long, and that the cords stretch in the jump an additional 24 m for a jumper whose mass is 80 kg, the heaviest adult you will allow to use your bungee jump (heavier customers would hit the ground).

(a) It will help you a great deal in your analysis to make a series of 5 simple diagrams, like a comic strip, showing the platform, the jumper, and the two cords at the following times in the fall and the rebound:
- while cords are slack
- when the two cords are just starting to stretch
- when the two cords are half stretched
- when the two cords are fully stretched
- when the two cords are again half stretched, on the way up

On each diagram, draw and label vectors representing the forces acting on the jumper, and the jumper's velocity. Make the relative lengths of the vectors reflect their relative magnitudes.

(b) At what instant is there the greatest tension in the cords? (How do you know?)

(c) What is the jumper's speed at this instant, when the tension is greatest in the cords?

(d) Is the jumper's momentum changing at this instant or not? (That is, is dp_y/dt nonzero or zero?) Give a valid physics reason for your answer.
Check to make sure that the magnitudes of the velocity and force vectors shown in your diagram number 4 are consistent with your analysis of parts (c) and (d).

(e) Focus on this instant of greatest tension and, starting from a fundamental principle, determine the spring stiffness k_s for each of the two cords.

(f) What is the maximum tension that each one of the two cords must support without breaking? (This tells you what kind of cords you need to buy.)

(g) What is the maximum acceleration $|a_y| = |dv_y/dt|$ (in "g's"; acceleration divided by 9.8 m/s^2) that the jumper experiences? (Note that $|dp_y/dt| = m|dv_y/dt|$ if v is small compared to c.)

(h) What is the direction of this maximum acceleration?

(i) What approximations or simplifying assumptions did you have to make in your analysis that might not be adequately valid?

7.P.34 A package of mass m sits on an airless asteroid of mass M and radius R. We want to launch the package in such a way that it will never come back, and when it is very far from the asteroid it will be traveling with speed v. We have a powerful spring whose stiffness is k_s. How much must we compress the spring?

7.P.35 A package of mass 9 kg sits on an airless asteroid of mass 8.0×10^{20} kg and radius 8.7×10^5 m. We want to launch the package in such a way that it will never come back, and when it is very far from the asteroid it will be traveling with speed 226 m/s. We have a large and powerful spring whose stiffness is 2.8×10^5 N/m. How much must we compress the spring?

7.P.36 A mass of 0.3 kg hangs motionless from a vertical spring whose length is 0.8 m and whose unstretched length is 0.65 m. Next the mass is pulled down so the spring has a length of 0.9 m and is given an initial speed upward of 1.2 m/s. What is the maximum length of the spring during the following motion? What approximations or simplifying assumptions did you make?

Sections 7.4–7.9

7.X.37 Substance A has a large specific heat capacity (on a per gram basis), while substance B has a smaller specific heat capacity. If the same amount of energy is put into a 100 gram block of each substance, and if both blocks were initially at the same temperature, which one will now have the higher temperature?

7.X.38 An oil company included in its advertising the following phrase: "Energy—not just a force, it's power!" In technical usage, what are the differences among the terms energy, force, and power?

7.X.39 Electricity is billed in kilowatt-hours. Is this energy or power? How much is one kilowatt-hour in standard physics units? (The typical cost of one kilowatt-hour is 5 to 10 cents.)

7.X.40 State which of the following are open systems with respect to energy, and which are closed: a car; a person; an insulated picnic chest; the Universe; the Earth. Explain why.

7.X.41 A certain motor is capable of doing 3000 joules of work in 11 seconds. What is the power output of this motor?

7.X.42 Starting from rest, a woman lifts a barbell with a constant force F through a distance h, at which point she is still lifting, and the barbell has acquired a speed v. Let E_{woman} stand for the following energy terms associated with the woman:

$$E_{woman} = E_{chemical,woman} + K_{woman \text{ (moving arms etc.)}} +$$
$$U_{grav,woman+Earth} + E_{thermal,woman}$$

The change in the kinetic energy of the barbell is $\frac{1}{2}mv^2 - 0 = \frac{1}{2}mv^2$.

The Energy Principle here is $\Delta E_{sys} = W$ (no other energy transfers). We'll consider terms on the left side of the equation (the ΔE_{sys} side, changes in the energy inside the system) and terms on the right side (the W side, energy inputs from the surroundings).

I. System: Woman + barbell + Earth
For the system consisting of the woman, the barbell, and the Earth, which terms belong on the left side of the energy equation (the ΔE_{sys} side)?
For the system consisting of the woman, the barbell, and the Earth, which terms belong on the right side of the energy equation (the W side)?

II. System: Barbell only
For the system consisting of the barbell only, which terms belong on the left side of the energy equation (the ΔE_{sys} side)?
For the system consisting of the barbell only, which terms belong on the right side of the energy equation (the W side)?

III. System: Barbell + Earth
For the system consisting of the barbell and the Earth, which terms belong on the left side of the energy equation (the ΔE_{sys} side)?
For the system consisting of the barbell and the Earth, which terms belong on the right side of the energy equation (the W side)?

7.X.43 A block of mass m is projected straight upward by a strong spring whose stiffness is k_s. When the block is a height y_1 above the floor, it is traveling upward at speed v_1, and the spring is compressed an amount s_1. A short time later the block is at height y_2, traveling upward at speed v_2, and the spring is compressed an amount s_2. In this discussion we'll use the Energy Principle in the form $E_f = E_i + W$, where we assume that thermal transfer of energy (microscopic work) Q between the block and the air is negligible. We can ignore the rest energies, which don't change, and the kinetic energy of the Earth, which hardly changes.

I. System: Universe (block + spring + Earth)
For the system consisting of the block, the spring, and the Earth, write the Energy Principle in the update form $E_f = E_i + W$.

II. System: Block + spring
For the system consisting of the block plus spring, write the Energy Principle in the update form $E_f = E_i + W$.

III. System: Block alone
For the system consisting of the block alone, write the Energy Principle in the update form $E_f = E_i + W$.

7.X.44 A horse whose mass is M gallops at constant speed v up a long hill whose vertical height is h, taking an amount of time t to reach the top. The horse's hooves do not slip on the rocky ground, so the work done by the force of the ground on the hooves is zero (no displacement of the force). When the horse started running, its temperature rose quickly to a point at which from then on, heat transferred from the horse to the air keeps the horse's temperature constant.

(a) First consider *the horse* as the system of interest. In the initial state the horse is already moving at speed v. In the final state the horse is at the top of the hill, still moving at speed v. Write out the Energy Principle $\Delta E_{sys} = W + Q$ for the system of the horse alone. The terms on the left-hand side should include only energy changes for the system, while the terms on the right-hand side should relate to the surroundings (everything else).
Which of the terms are equal to 0? Which of the terms should go on the *system* (left) side (ΔE_{sys})? Which of the terms should go on the *surroundings* (right) side?

(b) Next consider *the Universe* as the system of interest. Write out the energy principle for this system. Remember that terms on the left-hand side should include all energy changes for the system, while the terms on the right-hand side should relate to the surroundings.
Which of the terms are equal to 0? Which of the terms should go on the *system* (left) side (ΔE_{sys})? Which of the terms should go on the surroundings (right) side?

7.P.45 400 grams of boiling water (temperature $100°$ C, specific heat capacity 4.2 J/K/gram) are poured into an aluminum pan whose mass is 600 grams and initial temperature $20°$ C (the specific heat capacity of aluminum is 0.9 J/K/gram). After a short time, what is the temperature of the water? Explain. What simplifying assumptions did you have to make?

7.P.46 180 grams of boiling water (temperature $100°$ C, heat capacity 4.2 J/K/gram) are poured into an aluminum pan whose mass is 1050 grams and initial temperature $26°$ C (the heat capacity of aluminum is 0.9 J/K/gram).

(a) After a short time, what is the temperature of the water?

(b) What simplifying assumptions did you have to make?

A. The thermal energy of the water doesn't change.

B. The thermal energy of the aluminum doesn't change.

C. Energy transfer between the system (water plus pan) and the surroundings was negligible during this time.

D. The heat capacities for both water and aluminum hardly change with temperature in this temperature range.

(c) Next you place the pan on a hot electric stove. While the stove is heating the pan, you use a beater to stir the water, doing 29,541 J of work, and the temperature of the water and pan increases to 86.9° C. How much energy transfer due to a temperature difference was there from the stove into the system consisting of the water plus the pan?

7.P.47 Here are questions dealing with human power.

(a) If you follow a diet of 2000 food calories per day (2000 kilocalories), what is your average power consumption in watts? (A food or "large" calorie is a unit of energy equal to 2.4×10^3 J; a regular or "small" calorie is equal to 4.2 J.) Compare with the power consumption in a table lamp.

(b) You can produce much higher power for short periods. Make appropriate measurements as you run up some stairs, and report your measurements. Use these measurements to estimate your power output (this is in addition to your basal metabolism—the power needed when resting). Compare with a horsepower (which is about 750 watts) or a toaster (which is about 1000 watts).

7.P.48 Here are questions about human diet.

(a) A typical candy bar provides 280 calories (one "food" or "large" calorie is equal to 4.2×10^3 joules). How many candy bars would you have to eat to replace the chemical energy you expend doing 100 sit-ups? Explain your work, including any approximations or assumptions you make. (In a sit-up, you go from lying on your back to sitting up.)

(b) How many days of a diet of 2000 large calories are equivalent to the gravitational energy difference for you between sea level and the top of Mount Everest, 8848 m above sea level? (However, the body is not anywhere near 100% efficient in converting chemical energy into change in altitude. Also note that this is in addition to your basal metabolism.)

7.P.49 Humans have about 60 milliliters (60 cm³) of blood per kilogram of body mass, and blood makes a complete circuit in about 20 seconds, to keep tissues supplied with oxygen. Make a crude estimate of the *additional* power output of your heart (in watts) when you are standing compared with when you are lying down. Note that we're not asking you to estimate the power output of your heart when you are lying down, just the change in power when you are standing up. You will have to estimate the values of some of the relevant parameters. Because we're only looking for a crude estimate, try to construct a model that is as simple as possible. Describe the approximations and estimates you made.

7.P.50 During three hours one winter afternoon, when the outside temperature was 0° C (32° F), a house heated by electricity was kept at 20° C (68° F) with the expenditure of 45 kwh (kilowatt·hours) of electric energy. What was the average energy leakage in joules per second through the walls of the house to the environment (the outside air and ground)?

The rate at which energy is transferred between two systems is often proportional to their temperature difference. Assuming this to hold in this case, if the house temperature had been kept at 25° C (77° F), how many kwh of electricity would have been consumed?

7.P.51 You observe someone pulling a block of mass 43 kg across a low-friction surface. While they pull a distance of 3 m in the direction of motion, the speed of the block changes from 5 m/s to 7 m/s. Calculate the magnitude of the force exerted by the person on the block. What was the change in internal energy (chemical energy plus thermal energy) of the person pulling the block?

Sections 7.10–7.14

7.X.52 Describe a situation in which it would be appropriate to neglect the effects of air resistance.

7.X.53 Describe a situation in which neglecting the effects of air resistance would lead to significantly wrong predictions.

7.X.54 List all the approximations that are made in order to find a simple analytical solution for the motion of a projectile near the Earth's surface.

7.X.55 You are standing at the top of a 50 m cliff. You throw a rock in the horizontal direction with speed 10 m/s. If you neglect air resistance, where would you predict it would hit on the flat plain below? Is your prediction too large or too small as a result of neglecting air resistance?

7.X.56 A coffee filter of mass 1.8 grams dropped from a height of 4 m reaches the ground with a speed of 0.8 m/s. How much kinetic energy K_{air} did the air molecules gain from the falling coffee filter? Start from the Energy Principle, and choose as the system the coffee filter, the Earth, and the air.

7.X.57 A box with its contents has a total mass of 36 kg. It is dropped from a very high building.

(a) After reaching terminal speed, what is the magnitude of the air resistance force acting upward on the falling box?

(b) The box survived the fall and is returned to the top of the building. More objects are put into the box, and the box with its contents now has a total mass of 71 kg. The box is dropped, and it reaches a higher terminal speed than before. After reaching terminal speed, what is the magnitude of the air resistance force acting upward on the falling box? (The fact that the heavier object reaches a higher terminal speed shows that the air resistance force increases with increasing speed.)

7.X.58 You drag a block across a table while a friend presses down on the block. The coefficient of friction between the table and the block is 0.6. The vertical component of the force exerted by the table on the block is 190 N. How big is the horizontal component of the force exerted by the table on the block?

7.X.59 You drag a block with constant speed v across a table with friction. Explain in detail what you have to do in order to change to a constant speed of $2v$ on the same surface. (That is, the puzzle is to explain how it is possible to drag a block with sliding friction at different constant speeds.)

7.P.60 A paper coffee filter in the shape of a truncated cone falls in a stable way and quickly reaches terminal speed because it has a large area (big air resistance) but low mass (small gravitational force). Use a stopwatch to time the drop of coffee filters from as high a starting position as you can conveniently manage. If a stairwell is available, drop the coffee filters there to time a longer fall. This experiment is easier to do with a partner.

(a) By stacking coffee filters you can change the mass of a falling object without changing the shape. With this scheme you can explore how the terminal speed depends on the mass. Time the fall for different numbers of stacked filters, taking some care that the shape of the bottom filter is always the same (the filters tend to flatten out when removed from a stack). Start the measurement after the filters have fallen somewhat, to allow them to reach terminal speed. Average the results of several repeated measurements of time and height. Can you think of a simple experiment you can do to verify that the coffee filters do in fact reach terminal speed before you start timing?

(b) We want to know quantitatively how air resistance depends on speed. Plot your data for the air resistance force vs. the terminal speed. (The air resistance force is equal to the gravitational force when terminal speed has been reached, so it is proportional to the number of filters in a stack). How does the air resistance force depend on v?

There is an important constraint on the graph of speed dependence: Should the curve of the air resistance force vs. terminal speed go through the origin (force and terminal speed both very small), or not?

Analysis hint: If you suspect that the force is proportional to v^3, you might plot force vs. v^3 and see whether a straight line fits the data.

7.P.61 Figure 7.58 is a portion of a graph of energy terms vs. time for a mass on a spring, subject to air resistance. Identify and label the three curves as to what kind of energy each represents. Explain briefly how you determined which curve represented which kind of energy.

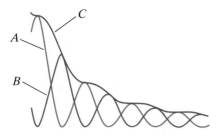

Figure 7.58 Which curve represents which kind of energy?

7.P.62 You drop a single coffee filter of mass 1.7 grams from a very tall building, and it takes 52 seconds to reach the ground. In a small fraction of that time the coffee filter reached terminal speed.

(a) What was the upward force of the air resistance while the coffee filter was falling at terminal speed?

(b) Next you drop a stack of five of these coffee filters. What was the upward force of the air resistance while this stack of coffee filters was falling at terminal speed?

(c) Again assuming that the stack reaches terminal speed very quickly, about how long will the stack of coffee filters take to hit the ground? (*Hint:* Consider the relation between speed and the force of air resistance.)

7.P.63 Design and carry out an experiment to determine the coefficient of sliding friction for a block on a flat plate.

7.P.64 For this problem you will need measurements of the position vs. time of a block sliding on a table, starting with some initial velocity, slowing down, and coming to rest. If you do not have an appropriate laboratory setup for making these measure-

ments, your instructor will provide you with such data. Analyze these data to see how well they support the assertion that the force of sliding friction is essentially independent of the speed of sliding. (Recall that a reason was given for this assertion—the friction force is determined by the strength of temporary interatomic bonds that must be broken.)

7.P.65 You can observe the main effects of resonance with very simple experiments. Hold a spring vertically with a mass suspended at the other end, and observe the frequency of "free" oscillations with your hand kept still. Then stop the oscillations, and move your hand *extremely* slowly up and down in a kind of slow sinusoidal motion. You will see that the mass moves up and down with the same very low frequency.

(a) How does the amplitude (plus or minus displacement from the center location) of the mass compare with the amplitude of your hand? (Notice that the phase shift of the oscillation is 0°; the mass moves up when your hand moves up.)

(b) Next move your hand up and down at a significantly higher frequency than the free-oscillation frequency. How does the amplitude of the mass compare to the amplitude of your hand? (Notice that the phase shift of the oscillation is 180°; the mass moves down when your hand moves up.)

(c) Finally, move your hand up and down at the free-oscillation frequency. How does the amplitude of the mass compare with the amplitude of your hand? (It is hard to observe, but the phase shift of the oscillation is 90°; the mass is at the midpoint of its travel when your hand is at its maximum height.)

(d) Change the system in some way so as to increase the air resistance significantly. For example, attach a piece of paper to increase drag. At the free-oscillation frequency, how does this affect the size of the response?

A strong dependence of the amplitude and phase shift of the system to the driving frequency is called resonance.

Computational Problems

7.P.66 Energy calculations provide a powerful check on the accuracy of a numerical integration. Modify your numerical integration of the motion of a mass on a spring that you did earlier. In addition to plotting graphs of position versus time, also plot graphs of kinetic energy, of spring energy, and of the sum of the kinetic energy and the spring energy, versus time during the motion. It is up to you to choose some value for the potential energy U_0 at the bottom of the well. Label the energy graphs with numerical values and units. Make sure that the energy values are reasonable for your system.

(a) Does the sum of kinetic and potential energy remain constant as a function of time? What if you vary the step size (which varies the accuracy of the numerical integration)?

(b) Increase the energy by increasing the initial speed. How does the amplitude A change? How does the angular frequency ω change?

7.P.67 Modify your numerical integration of the spring–mass system to include the effects of friction. Such a system is called a "damped oscillator."

Simulate the effect of all three kinds of friction: sliding friction (independent of v), viscous friction (proportional to v), and air resistance for large blunt objects (proportional to v^2). Choose friction to be small enough that you get at least two full cycles of oscillation, but large enough that you see an effect. If the friction is so large that the graph never crosses the axis, the system is said to be "over-damped." For each kind of friction, display the following:

(1) Display an animation of the motion of the system.

(2) Plot a graph of position vs. time.

(3) Plot graphs of K, U, and $(K + U)$ vs. time.

On the energy graphs, when is the energy loss rate large and small, and why?

7.P.68 Use a computer to carry out the following step-by-step quantitative calculations for a baseball. Remember that the direction of the air resistance force on an object is opposite to the direction of the object's velocity.

A baseball has a mass of 155 grams and a diameter of 7 cm. The drag coefficient C for a baseball is about 0.35, and the density of air at sea level is about 1.3×10^{-3} grams/cm^3, or 1.3 kg/m^3.

(a) Including the effect of air resistance, compute and plot the trajectory of a baseball thrown or hit at an initial speed of 100 miles per hour, at an angle of 45° to the horizontal. How far does the ball go? Is this a reasonable distance?

(A baseball field is about 400 feet from home plate to the fence in center field. An outfielder cannot throw a baseball in the air from the fence to home plate.)

(b) Plot a graph of $(K + U_g)$ vs. time.

(c) Temporarily neglect air resistance (set $C = 0$ in your computations) and determine the range of the baseball. Check your computations by using the analytical solution for motion of a projectile without air resistance. Compare your result with the range found in part (a), where the effect of air resistance was taken into account. Is the effect of air resistance significant? (The effect is surprisingly large—about a factor of 2!)

(d) Neglecting air resistance, plot a graph of $(K + U_g)$ vs. time.

(e) In Denver, a mile above sea level, the air is about 83% as dense as the air at sea level. Including the effect of air resistance, use your computer model to predict the trajectory and range of a baseball thrown in Denver. How does the range compare with the range at sea level?

7.P.69 Terminal speed for a falling sky diver has been measured to be about 60 m/s. Using your computer model from Problem 7.P.68, determine how far (in meters) and how long (in seconds) a sky diver falls before reaching terminal speed. (You will need to think about the meaning of "terminal speed" and how to test for it.)

Plot graphs of speed vs. time and position vs. time.

7.P.70 In Problem 7.P.60 you experimented with falling coffee filters. It was important to let the coffee filters fall for a while to reach terminal speed before starting to time their fall, in order to measure the terminal speed. Model the fall of the coffee filters computationally, starting from rest, using your computer model from Problem 7.P.68.

7.P.71 You can study resonance in a driven oscillator in detail by modifying your computation for the spring–mass system with friction. Let one end of the spring be moved back and forth sinusoidally by a motor, with a motion given by $D \sin(\omega_D t)$ (see Figure 7.59). Here D is the amplitude of the motor motion and ω_D is the angular frequency of the motor, which can be varied. (The free-oscillation angular frequency $\omega_F = \sqrt{k_s/m}$ of the spring–mass system has a fixed value, determined by the spring stiffness k_s and the mass m.)

We need an expression for the stretch s of the spring in order to be able to calculate the force $-k_s s$ of the spring on the mass. We have to do a bit of geometry to figure out the length of the spring when the mass is displaced a distance x from the equilibrium position, and the motor has moved the other end of the spring by an amount $D \sin(\omega_D t)$.

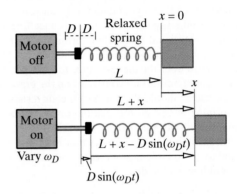

Figure 7.59 A driven spring–mass system.

In the figure, the spring gets longer when the mass moves to the right ($+x$), but the spring gets shorter when the motor moves to the right ($-D \sin(\omega_D t)$). The new length of the spring is $L + x - D \sin(\omega_D t)$, and the net stretch of the spring is this quantity minus the unstretched length L, yielding ($s = x - D \sin(\omega_D t)$). A check that this is the correct expression for the net stretch of the spring is that if the motor moves to the right the same distance as the mass moves to the right, the spring will have zero stretch. Replace x in your computer computation with the quantity $[x - D \sin(\omega_D t)]$.

(a) Use viscous damping (friction proportional to v). The damping should be small. That is, without the motor driving the system ($D = 0$), the mass should oscillate for many cycles. Set ω_D to $0.9\omega_F$ (that is, 0.9 times the free-oscillation angular frequency, $\omega_F = \sqrt{k_s/m}$), and let x_0 be 0. Graph position x vs. time for enough cycles to show that there is a transient buildup to a "steady state." In the steady state the energy dissipation per cycle has grown to exactly equal the energy input per cycle.

Vary ω_D in the range from 0 to $2.0\omega_F$, with closely spaced values in the neighborhood of $1.0\omega_F$. Have the computer plot the position of the motor end as well as the position of the mass as a function of time. For each of these driving frequencies, record for later use the steady-state amplitude and the phase shift (the difference in angle between the sinusoids for the motor and for the mass). If a harmonic oscillator is lightly damped, it has a large response only for driving frequencies near its own free-oscillation frequency.

(b) Is the steady-state angular frequency equal to ω_F or ω_D for these various values of ω_D? (Note that during the transient buildup to the steady state the frequency is not well-defined, because the motion of the mass isn't a simple sinusoid.)

(c) Sketch graphs of the steady-state amplitude and phase shift vs. ω_D. Note that when $\omega_D = \omega_F$ the amplitude of the mass can be much larger than D, just as you observed with your hand-driven spring–mass system. Also note the interesting variation of the phase shift as you go from low to high driving frequencies—does this agree with the phase shifts you observed with your hand-driven spring–mass system?

(d) Repeat part (c) with viscous friction twice as large. What happens to the resonance curve (the graph of steady-state amplitude vs. angular frequency ω_D)?

ANSWERS TO EXERCISES

7.X.1 0.012 J

7.X.2 0.5 m/s

7.X.3 A is $|U|$; B is K; C is $|K + U|$

7.X.4 1–4 are bound; 5 is unbound.

7.X.5 No change; same stretch.

7.X.6 21.36° C; neglect the material of the thin glass

7.X.7 0.12 K

7.X.8 1.26×10^5 J; 7.6×10^4 J; -1.26×10^5 J

7.X.9 9000 J

7.X.10 0, since no temperature change; $-11{,}000$ J

7.X.11 About 500 J; about 2000 kg per second

7.X.12 2.5 m/s^2

7.X.13 $\frac{1}{2}k_s A^2$; $\frac{1}{2}k_s A^2$; spring plus mass

7.X.14 $(90\,\text{kg})(9.8\,\text{N}/\,\text{kg}) = 882$ N; the net force must be zero.

7.X.15 Less, because some of the original energy is dissipated into the air.

7.X.16 2.7×10^{-2} J

7.X.17 No, because the decreasing gravitational potential energy is matched by increasing kinetic energy of the air molecules.

7.X.18 Air resistance is proportional to v^2, so the energy dissipation rate is higher at higher speeds.

7.X.19 +4 J; another +4 J; +8 J

7.X.20 Denominator gets very small when $\omega_D = \omega_F$.
Low frequency:
$$\cos\phi \approx \frac{(\omega_F^2)}{\sqrt{(\omega_F^2)^2 + \left(\frac{c}{m}\omega_D\right)^2}} \approx 1 \text{ so } \phi \approx 0°$$
High frequency:
$$\cos\phi \approx \frac{(-\omega_D^2)}{\sqrt{(-\omega_D^2)^2 + \left(\frac{c}{m}\omega_D\right)^2}} \approx -1, \phi \approx 180°$$
Resonance frequency, $\cos\phi = 0$, so $\phi \approx 90°$

7.X.21 When $\omega_D = \omega_F$, denominator $= \frac{c}{m}\omega_D$.
Denominator $= \sqrt{2}\frac{c}{m}\omega_D$ when $|(\omega_F^2 - \omega_D^2)| = \frac{c}{m}\omega_D$.
However, since near resonance $\omega_D \approx \omega_F$, we have
$(\omega_F^2 - \omega_D^2) = (\omega_F + \omega_D)(\omega_F - \omega_D)$
$\approx 2\omega_D(\omega_F - \omega_D)$
Hence when $|\omega_F - \omega_D| \approx \frac{c}{2m}\omega_D \approx \frac{c}{2m}\omega_F$, the amplitude is down by a factor of $1/\sqrt{2}$. Larger friction, wider resonance peak. If friction is very small, the resonance peak is a very narrow spike.

Energy Quantization

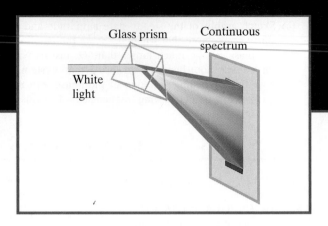

- **A photon is a particle of light with a certain amount of kinetic energy and zero rest energy.**
- **The internal energy of microscopic systems is quantized; that is, the internal energy of the system can have only certain specific values ("energy levels"). The system is never observed to have energies between these levels.**
- **An atom in a higher energy level can drop to a lower energy level, and emit a photon whose energy is the difference in the energy of the two energy levels.**
- **An atom in the lowest-energy state (called the ground state) can absorb a photon and jump to an excited state, if the photon has just the right energy (the difference in the two energy levels).**
- **Collisions with energetic electrons can excite atoms to higher energy states, leaving the electron with less kinetic energy than it had initially.**
- **The energy of an atom plus its electrons, the vibrational and rotational energies of molecules, the energy of a nucleus, and the energy of the system of quarks making up a proton or neutron, are all quantized.**

8.1 PHOTONS

A photon is a particle that has only kinetic energy—its rest energy is zero. A photon can never be at rest; it may be thought of as a packet of energy, traveling at the speed of light. A microscopic object such as an atom can gain energy by absorbing a photon; if this happens, all of the photon's energy is transferred to the object. Conversely, an object can lose energy by emitting a photon with a particular amount of energy. (Photons also have momentum, and the Momentum Principle also applies to absorption and emission of photons.) The general term applied to photons is "electromagnetic radiation" (or simply "light"), and in later chapters on electricity and magnetism we will go more deeply into the electric and magnetic aspects of electromagnetic radiation, and we will also discuss its wavelike aspects.

Historically, photons of different energies have been given different names, usually because the instruments needed to detect and measure electromagnetic radiation depend on the photon energies involved. The entire range of possible photon energies is called the electromagnetic spectrum (Figure 8.1). Only photons whose energies fall between about 1.8 eV (red light) and 3.1 eV (violet light) can be detected by the human eye (Figure 8.2), so photons in this energy range are referred to as visible light. Colloquially we

Increasing energy

10^6 eV — Gamma ray
10^4 eV — X-ray
10^2 eV — Ultra violet
1 eV — Visible
10^{-2} eV — Infrared
— Microwave
10^{-6} eV — Radio
10^{-10} eV

Figure 8.1 The electromagnetic spectrum.

3.1 eV — Violet

1.8 eV — Red

Figure 8.2 The visible portion of the electromagnetic spectrum.

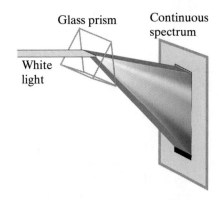

Figure 8.3 A prism separates a narrow beam of white light into a continuous rainbow of colors.

often refer to all photons as "light," whether they are high-energy gamma rays or low-energy radio waves.

Although earlier scientists, including Isaac Newton, had speculated that light might be made up of discrete particles, it was not until early in the 20th century that clear evidence for the particle nature of light was obtained. As in the case of atoms, the noncontinuous, lumpy nature of light was not noticed for a long time because the amount of energy carried by one photon in an ordinary beam of light is extremely small compared to the total energy of the beam. It has now become routine to detect individual photons in a very weak beam of light and to measure the energy of an individual photon. Observations of the interaction of photons and ordinary matter have provided key insights into the internal structure of molecules, atoms, and nuclei.

One can separate a beam of visible light into its component energies by passing a narrow beam of light through a prism. You have probably seen the continuous rainbow spectrum produced when ordinary white light goes through a prism, as in Figure 8.3. Your eye reacts differently to photons of different energies, which is why you see a colorful spectrum.

> **8.X.1** The photon energy for green light lies between the values for red and violet light. What is the approximate energy of the photons in green light?
>
> **8.X.2** The intensity of sunlight above the Earth's atmosphere is about 1400 watts (J/s) per square meter. That is, when sunlight hits perpendicular to a square meter of area, about 1400 watts of energy can be absorbed. Using the average photon energy found in the previous exercise, about how many photons per second strike an area of one square meter? (This is why the lumpiness of light was not noticed for so long.)

Absorption and Emission of Photons

A microscopic object such as an atom or a molecule can gain energy by absorbing a photon. When a photon is absorbed, nothing is left over—the photon no longer exists, and the object that absorbed the photon simply has more energy. We can describe the absorption of a photon by a hydrogen atom in this compact way (the asterisk is used to denote an atom with higher internal energy):

$$H + photon \rightarrow H^*$$

Applying the Energy Principle to this process, and choosing all particles as the system, we can write this equation:

$$E_{H^*,f} = E_{H,i} + K_{photon}$$

Likewise, an atom, molecule, or solid can lose energy by emitting a photon, which comes into existence specifically to carry away a particular amount of energy. For a hydrogen atom, emission of a photon can be described this way:

$$H^* \rightarrow H + photon$$
$$E_{H,f} + K_{photon} = E_{H^*,i}$$

8.2 ELECTRONIC ENERGY LEVELS

One of the most exciting discoveries of the early 20th century was that the internal energy of a small system such as an atom can have only certain discrete values, called energy "levels." One piece of evidence leading to this realization was the fact that a particular atom or molecule emitted only photons with

■ The term "quantized" is often used colloquially as a synonym for "discrete." We will use it this way, although the technical meanings of the two words are not identical.

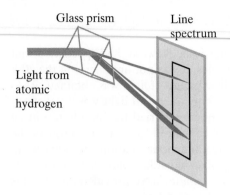

Figure 8.4 The light emitted by excited atomic hydrogen contains only a few separate colors. In a spectrum, light of a particular color and energy is called a "spectral line" due to the linelike shape of the light coming through a slit as seen here.

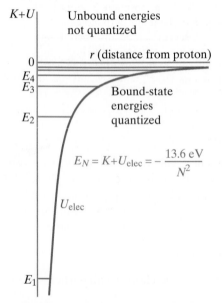

Figure 8.5 Electronic energy levels for a hydrogen atom. The potential energy curve describes the potential energy of the electron–proton system as a function of separation.

particular kinetic energies. For example, in contrast to the continuous spectrum of photon energies in white light (Figure 8.3), the visible light emitted by hydrogen atoms contains photons with only a few specific energies (Figure 8.4). In a spectrum, light of a particular color and energy is called a "spectral line" due to the linelike shape of the light coming through a slit, as seen in Figure 8.4.

An example of discrete energy levels is found in the "electronic" energy associated with the motion of electrons around the nucleus of an atom. You probably know from your study of chemistry that electronic energy levels in an atom are discrete—only certain values of $K + U$ are possible for the system of nucleus plus electrons. The electronic energy levels in atomic hydrogen are shown in Figure 8.5 (atomic hydrogen is a single H atom, composed of one proton and one electron, not the usual diatomic molecule H_2).

The diagram shows the familiar electric potential energy of the proton–electron system. The horizontal lines represent the possible bound energy states for the hydrogen atom; bound states with other values of $(K + U)$ are not observed. Unbound states $(K + U \geq 0)$, however, are not quantized.

The electric potential energy of an electron and a proton bound together in a hydrogen atom is negative for all separations:

$$U_{elec} = \frac{1}{4\pi\varepsilon_0}\frac{q_1 q_2}{r} = \frac{1}{4\pi\varepsilon_0}\frac{(+e)(-e)}{r} = -\frac{1}{4\pi\varepsilon_0}\frac{e^2}{r}$$

A quantum mechanical calculation of the energy levels of a system involves solving a differential equation called the Schrödinger equation, a form of the Energy Principle consistent with the nature of microscopic systems. This equation includes both a potential energy term and a kinetic energy term. Solving this equation for the hydrogen atom predicts that the discrete energy levels for bound states of hydrogen (ignoring the rest energies of the proton and the electron) will be these:

ELECTRONIC ENERGY LEVELS OF A HYDROGEN ATOM
$$E_N = K + U_{elec} = -\frac{13.6\,\text{eV}}{N^2}, \quad N = 1, 2, 3, \text{etc.}$$

This arrangement of hydrogen energy levels has been verified by observations of the interaction of light and hydrogen, as we will see.

These energy levels denote different bound states of the electron + proton system. Recall that a bound state has a negative value of $K + U$, since a potential energy of zero is associated with a very large separation between the electron and the proton. We have seen that bound gravitational orbits are also associated with negative values of $K + U$. As you can see from the energy diagram (Figure 8.5), in higher-energy states the average location of the electron is farther from the proton. Note that there is a lowest-energy state, called the "ground state"; a hydrogen atom cannot have $K + U$ less than $-13.6\,\text{eV}$.

Before the development of the fully quantum mechanical Schrödinger equation, Niels Bohr made a simple model for the hydrogen atom that also yielded the energy formula given above. We will study the Bohr model in a later chapter.

EXAMPLE Raising the Energy of a Hydrogen Atom

A hydrogen atom is originally in its ground (lowest) electronic energy state ($N = 1$). It absorbs a photon, which raises the hydrogen atom's energy to the next energy level ($N = 2$). What was the energy of the photon, in eV?

Solution   System: All particles
Surroundings: Nothing significant
Energy Principle:

$$E_f = E_i + K_{photon}$$

$$\left(-\frac{13.6\,eV}{2^2}\right) = -\frac{13.6\,eV}{1^2} + K_{photon}$$

$$K_{photon} = 10.2\,eV$$

It is not possible to add less than 10.2 eV to a hydrogen atom in its lowest energy level; any smaller amount fails to be absorbed.

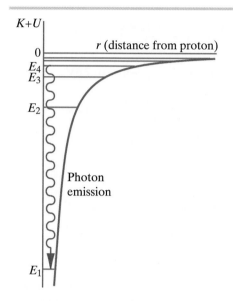

Figure 8.6 An atom can drop to a lower energy level by emitting a photon whose energy corresponds to the difference between the energies of the levels. By convention a squiggly line with an arrowhead indicates a photon.

8.X.3 How much energy in electron volts is required to ionize a hydrogen atom (that is, remove the electron from the proton), if initially the atom is in its lowest energy level?

Emission Spectra

When a hydrogen atom drops from a higher energy level E_4 to a lower energy level E_1 it emits a photon with the associated energy $E_4 - E_1$ (Figure 8.6). A collection of hydrogen atoms can emit photons whose energies correspond to the differences in energy between any two electronic energy levels of the hydrogen atom. The light emitted by excited atomic hydrogen is called the "hydrogen emission spectrum," and the bright lines of light of particular energies in the spectrum are called "spectral lines." The observed energies of light emitted by excited atomic hydrogen are consistent with the energy level differences predicted by quantum mechanics. Other atoms have different quantized energy levels, and a line spectrum can be used to identify the emitting atoms by the pattern of emitted photon energies.

If an isolated hydrogen atom is initially in a high energy level, over time it will spontaneously emit photons and drop to lower and lower states, eventually ending up in the lowest energy level, which is called the "ground state." Once an atom has dropped to the ground state, it cannot lose any more energy, or emit photons, because there is no lower energy level than the ground state.

EXAMPLE	**Emission from an Atom**

Different atoms have different quantized energy levels. Consider a particular atom (not hydrogen) whose first excited state is 6.0 eV above the ground state, and the second excited state is 8.5 eV above the ground state (Figure 8.7). What is the energy of a photon emitted when such an atom drops from the second excited state to the first excited state?

Solution   System: atom and photon
Principle: Energy

$$E_f = E_i + \cancel{W}$$
$$(E_0 + 6.0\ eV) + K_{photon} = (E_0 + 8.5\ eV)$$
$$K_{photon} = 2.5\ eV$$

$E_0 + 8.5\ eV$ ⎯⎯ 2nd excited state
$E_0 + 6.0\ eV$ ⎯⎯ 1st excited state

E_0 ⎯⎯ Ground state

Figure 8.7 Emission from an atom.

FURTHER DISCUSSION   The Energy Principle is valid even for atomic systems with quantized energy. Writing down the Energy Principle explicitly helps to avoid the common mistake of confusing the energy levels of the atom with the energy of the photon. Note that in the example above the photon energy (2.5 eV) is the *difference* between the energies of the second and third energy levels.

| EXAMPLE | **Emission from Hydrogen Atoms** |

Consider a collection of many hydrogen atoms. Initially each individual atom is in one of the lowest three energy states ($N = 1, 2, 3$). What will be the energies of photons emitted by this collection of hydrogen atoms?

Solution There are three possibilities: transitions from $N = 3$ to $N = 2$, from $N = 2$ to $N = 1$, or from $N = 3$ to $N = 1$. The photon energies for the three transitions are these:

$$N = 3 \text{ to } N = 2 : \quad \left(-\frac{13.6\,\text{eV}}{3^2}\right) - \left(-\frac{13.6\,\text{eV}}{2^2}\right) = 1.9\,\text{eV}$$

$$N = 2 \text{ to } N = 1 : \quad \left(-\frac{13.6\,\text{eV}}{2^2}\right) - \left(-\frac{13.6\,\text{eV}}{1^2}\right) = 10.2\,\text{eV}$$

$$N = 3 \text{ to } N = 1 : \quad \left(-\frac{13.6\,\text{eV}}{3^2}\right) - \left(-\frac{13.6\,\text{eV}}{1^2}\right) = 12.1\,\text{eV}$$

8.X.4 How many different photon energies would emerge from a collection of hydrogen atoms that occupy the lowest four energy states ($N = 1, 2, 3, 4$)? (You need not calculate the energies of these transitions.)

Absorption Spectra

An atom can drop from a higher energy level to a lower energy level and emit a photon. An atom can also absorb a photon of an appropriate energy and jump from a lower energy level to a higher energy level (Figure 8.8). If an atom is in its ground state, only photon absorption is possible; photon emission cannot occur. Energy absorption can occur if the photon energy corresponds exactly to the difference between two energy levels of the atom. If we shine white light (containing a mixture of all photon energies in the visible spectrum) through a collection of atoms (such as a container of hydrogen gas), only photons whose energy is the difference of atomic energy levels will be strongly absorbed, while other photons can pass right through the material. In this case, the spectral lines are dark lines (missing energies) on a light background, instead of the bright lines on a dark background observed in an emission spectrum. This is called an "absorption spectrum" (Figure 8.9).

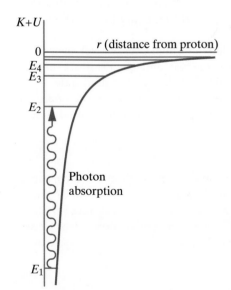

Figure 8.8 An atom can gain energy by absorbing a photon whose energy corresponds to the difference between two energy levels.

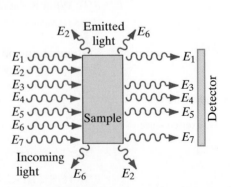

Figure 8.9 Measuring an absorption spectrum. A sample of a material is placed in a transparent container. Some energies of incoming light are absorbed. The outgoing light is depleted of the absorption energies (in this case E_2 and E_6).

Photon absorption may be followed almost immediately by emission from the excited state to a lower state, so at any instant there are few atoms in excited states that could absorb photons. The result is that in dark-line absorption spectra only absorption from the ground state is observed, not from higher-level states.

The atoms that have jumped to higher energy levels through photon absorption will eventually drop to lower energy levels with the emission of photons. However, these photons are emitted in all directions, not just in the direction of the original beam of light, so the light that has passed through the material will have little intensity at photon energies corresponding to the strongly absorbed energies.

EXAMPLE

Why Is Hydrogen Gas Transparent?

At room temperature in a gas of atomic hydrogen almost all of the atoms are in the ground state. Why might you expect the gas to be completely transparent to visible light?

Solution

As we saw in an earlier example, the energy required to go from the ground state of atomic hydrogen to the first excited state is 10.2 eV, but visible light consists of photons whose energies lie between 1.8 eV and 3.1 eV. So visible light cannot be absorbed by ground-state hydrogen atoms. (In a collision between a photon of visible light and an atom, the atom can acquire some kinetic energy, but no internal energy change is possible.)

■ The energy absorbed by the sample may be quickly emitted as photons with energies E_2 and E_6. However, these emitted photons go in all directions in 3D, so almost none of them reach the detector.

8.X.5 Suppose you had a collection of hypothetical quantum objects whose individual energy levels were -4.0 eV, -2.3 eV, and -1.6 eV. If nearly all of the individual objects were in the ground state, what would be the energies of dark spectral lines in an absorption spectrum if visible white light (1.8 to 3.1 eV) passes through the material?

Electron Excitation

In order for an atom to emit a photon, it first has to absorb enough energy to raise the atom to an excited state, above the ground state. As we will see in a later section, collisions with other atoms do not provide enough energy unless the collection of atoms is at a very high temperature (for example, on the surface of the Sun).

A common nonthermal way to excite atoms in a gas is to produce a beam of fast-moving electrons and send the beam through the gas. Because electrons are (negatively) charged electrically, they interact strongly with atoms, and an electron that passes near an atom may excite the atom from the ground state to an excited state whose internal energy is ΔE_{atom} higher, with a corresponding loss of kinetic energy of the electron. Unlike photon absorption, which almost always occurs only if the photon energy exactly matches the difference between energy levels, an electron can give up some kinetic energy and have some kinetic energy left over. The process is illustrated in Figure 8.10 for an electron colliding with a mercury atom, whose first excited state is 4.9 eV above the ground state.

Figure 8.10 An electron with initial kinetic energy of more than 4.9 eV can excite a mercury atom (Hg) from the ground state to the first excited state. The electron loses 4.9 eV of kinetic energy.

QUESTION For example, suppose an electron with kinetic energy of 6 eV collides with a mercury atom that is in the ground state. What can happen?

The first excited state of mercury is 4.9 eV above the ground state. This 6 eV electron has enough kinetic energy to excite the mercury atom to the first excited state, giving up 4.9 eV of energy. The electron moves away with 1.1 eV of kinetic energy remaining. The mercury atom will quickly drop back down to the ground state, emitting a 4.9 eV photon. If a beam containing a large number of energetic electrons passes through a gas of mercury vapor, there will be continual electron excitation and continual photon emission. However, as soon as you turn off the electron beam, the mercury will stop emitting photons.

> QUESTION What happens if you send a beam of electrons with kinetic energy of 4 eV through the mercury vapor?

4 eV isn't enough energy to go from the ground state to the first excited state in mercury, so the electrons can pass right through the gas without changing the electronic energy states of the atoms. (An electron may collide with an atom, but because the mass of an electron is several thousand times smaller than the mass of a mercury atom, the electron will bounce off the atom like a Ping-Pong ball bouncing off a bowling ball—the kinetic energy of the atom will not change significantly.)

The German physicists James Franck and Gustav Hertz performed an electron excitation experiment in 1914 that provided early dramatic evidence for quantized energies in atoms. Electrons with varying amounts of initial kinetic energy were sent through a gas of mercury vapor, and the number of electrons per second making it through the gas was measured. When the initial kinetic energy of the electrons was less than 4.9 eV, many electrons reached the detector. However, when the initial kinetic energy reached 4.9 eV, few electrons reached the detector. The Franck–Hertz experiment was correctly interpreted to mean that the first excited state of a mercury atom is 4.9 eV above the ground state, and an electron with kinetic energy of 4.9 eV can excite the atom, leaving the electron with almost no kinetic energy, so the number of electrons that made it through the gas became very small.

> QUESTION What happens if the initial kinetic energy of the electron is 9.8 eV?

In this case the electron could excite one mercury atom and still have 4.9 eV of kinetic energy available to excite a second mercury atom. In the Franck–Hertz experiment it was observed that the number of electrons reaching the detector again dropped sharply when the initial kinetic energy of the electrons was increased past 9.8 eV.

Every element has a different set of quantized energies. The first excited state in mercury is 4.9 eV above the ground state, whereas the first excited state in hydrogen is 10.2 eV above the ground state. These differences make it possible to identify elements from emission or absorption spectra, or measurements of the Franck–Hertz type.

> QUESTION What happens if you send a beam of electrons with kinetic energy of 25 eV through a gas of atomic hydrogen?

Because the ground state of hydrogen has $K + U = -13.6$ eV, only 13.6 eV is needed to raise a hydrogen atom to any of its excited states, or even to break the atom apart into a proton and an electron. Therefore a beam containing lots of 25 eV electrons passing through the atomic hydrogen gas will keep many excited states populated, and you'll observe many different photon energies being emitted as atoms drop from excited states to lower excited states or to the ground state.

8.X.6 The first excited state of a mercury atom is 4.9 eV above the ground state. A moving electron collides with a mercury atom, and excites the mercury atom to its first excited state. Immediately after the collision the kinetic energy of the electron is 0.3 eV. What was the kinetic energy of the electron just before the collision?

Summary: Three Processes

- Electron excitation: A fast moving electron can transfer all or part of its kinetic energy to an atom through a collision, if $K_{electron}$ is greater than or equal to the energy difference between two atomic states. Therefore, a beam containing many electrons with high kinetic energy can maintain a gas of atoms in many excited states.

- Photon emission: Atoms in excited states emit photons as the atoms drop from higher-energy states to lower-energy states.

- Photon absorption: Atoms in cold material, in the absence of an electron beam, can absorb photons and go from the ground state to an excited state, but only if the photon's energy exactly matches the energy difference between the excited state and the ground state. Absorption leads to a "dark-line" spectrum.

EXAMPLE

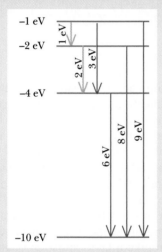

Figure 8.11 A possible energy-level scheme that fits the emission data. Lighter arrows indicate photons not detectable by this detector.

Observed Photon Emission

You observe photon emissions from a collection of quantum objects, each of which is known to have just four quantized energy levels. The collection is continually bombarded by an electron beam, and you detect emitted photons using a detector sensitive to photons in the energy range from 2.5 eV to 30 eV. With this detector, you observe photons emitted with energies of 3 eV, 6 eV, 8 eV, and 9 eV, but no other energies.

(a) It is known that the ground-state energy of each of these objects is -10 eV. Propose two possible arrangements of energy levels that are consistent with the experimental observations. Explain in detail, using diagrams.

(b) You obtain a second detector that is sensitive to photon energies in the energy range from 0.1 eV to 2.5 eV. What additional photon energies do you observe to be emitted? Explain briefly.

(c) You turn off the electron beam so that essentially all the objects are in the ground state. Then you send a beam of photons with a wide range of energies through the material. Using both detectors to determine the absorption spectrum, what are the photon energies of the dark lines? How can this information be used to choose between your two proposed energy level schemes? Explain briefly.

Solution

Fundamental principle: The Energy Principle
System: Everything—some unknown object and a photon
Surroundings: Nothing significant
Initial state: Object in excited state (no photon)
Final state: Object in lower energy state, plus photon
Energy Principle: $E_{obj,f} + K_{photon} = E_{obj,i}$

The energy of any emitted photon must be equal to the difference in energy between an excited state and a lower-energy state.

(a) Figure 8.11 shows a possible energy-level scheme that fits the emission data. The highest photon energy (9 eV) tells us the difference between the highest and lowest states. The rest is trial and error, checking that the correct photon energies would be produced. The arrows are labeled with the energies of photons emitted in transitions between the two states indicated. The 1 eV and 2 eV emissions would not be detected by a detector that is sensitive to the photon energy range 2.5 eV to 30 eV.

Figure 8.12 shows another possible scheme that fits the emission data. This is somewhat less likely, because it is more common to find the energy level spacing decreasing at higher energies, as shown in Figure 8.11.

(b) With this detector we observe the 1 eV and 2 eV photon emissions.

(c) With the electron beam turned off, almost all of the objects are in the ground state. If the energy levels are as shown in Figure 8.11, the dark lines will be at 6 eV, 8 eV, and 9 eV, corresponding to absorption in the ground state. If the energy levels are as shown in Figure 8.12, the dark lines will be at 1 eV, 3 eV, and 9 eV. Therefore the absorption spectrum allows us to distinguish between the two possible energy level schemes.

■ Note that since this was an energy analysis, we didn't need to know what kind of object this was, or whether the energy levels were electronic, vibrational, or rotational energy levels.

FURTHER DISCUSSION Figure 8.12 is an example of an "energy level" diagram, with photon emissions indicated by arrows that start on the upper level and point downward, with arrowheads on the lower level. Photon absorptions are indicated by arrows pointing up from the ground state to an upper level.

8.X.7 The energy levels of a particular quantum object are −8 eV, −3 eV, and −2 eV. If a collection of these objects is bombarded by an electron beam so that there are some objects in each excited state, what are the energies of the photons that will be emitted?

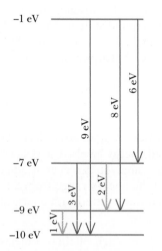

Figure 8.12 Another possible energy level scheme that fits the emission data.

Selection Rules

For simplicity, we've been assuming that photon emissions can occur between any discrete energy level to a lower level. Real atomic and molecular systems may be more complex, and some photon emissions are "forbidden." In a later chapter we will study angular momentum, a quantity that is related to rotational motion and is a conserved quantity, like momentum and energy. Photon emission and absorption, like all other physical processes, must conserve angular momentum, and as a result transitions between two particular energy levels may be forbidden. Such constraints are called "selection rules" and govern photon emission and absorption (electron excitation is not constrained by selection rules). Selection rules are beyond the scope of this introductory course, but you will learn about them if you take a later course in quantum mechanics. For the purposes of this introduction to quantized energy, we will ignore selection rules and assume that all energetically possible transitions are allowed.

We also ignore questions of how probable an allowed transition is, which requires full quantum mechanics to calculate. Suppose that an object is in the second excited state, from which it could drop either to the first excited state or directly to the ground state. One of these two transitions could be much more probable than the other one, even if the low-probability transition isn't forbidden by a selection rule.

8.3 THE EFFECT OF TEMPERATURE

When we observe atoms we are usually observing a collection of atoms, not an isolated single atom. Each atom in the collection (the "sample") interacts with other atoms and perhaps with a container. If an atom absorbs radiation coming from elsewhere, or absorbs energy from being jostled by a neighboring atom, it may jump to a higher energy level, and eventually emit photons and drop

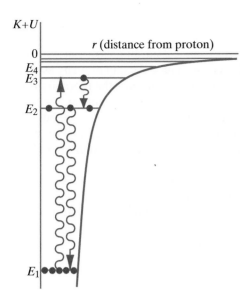

Figure 8.13 A dot represents the energy state of one hydrogen atom in a group of atoms. Energy is exchanged with the surroundings through photon emission and absorption.

Figure 8.14 At a low temperature (system on the left) all atoms are in low-energy states. At a high temperature (system on the right) some atoms are in higher-energy states, though the largest fraction will still be in the ground state.

again to a lower level (Figure 8.13 shows emissions as downward transitions and absorptions as upward transitions for a collection of hydrogen atoms). In thermal equilibrium with its surroundings, an atom's energy continually goes up and down, but with an average energy determined by the temperature of the collection of atoms.

In Figures 8.13 and 8.14 each dot represents one hydrogen atom in a container of atomic hydrogen. If three dots are on a line, this means three atoms are in this particular state. Each hydrogen atom has the same allowed energy levels, and at any particular instant some of the atoms will be in the ground state, some in the first excited state, and so on.

For a particular temperature, at any instant there is a certain probability that one particular atom will be in state 1, 2, 3, and so on. At low temperature almost all atoms are in the ground state (Figure 8.14). At high temperature the various atoms are in various states, from the ground state up to a high energy level, though the largest fraction will still be in the ground state. As we will see in a later chapter on statistical mechanics, in a collection of atoms the fraction of the atoms in a state whose energy is E above the ground state is proportional to the "Boltzmann factor" $e^{-E/k_B T}$, where k_B is the Boltzmann constant (1.38×10^{-23} J/K) and T is the absolute temperature in kelvins (K), where $0°$ Celsius is 273.15 K.

QUESTION Can you observe an emission spectrum if the material is very cold? Can you observe an absorption spectrum if the material is very cold?

Very cold matter will normally be in the ground state and will not emit photons. However, an absorption spectrum will show gaps (dark lines) corresponding to absorption of photons of energies appropriate to lift an atom from its ground state to an excited state. (In order to observe an absorption spectrum you would of course need to use a beam of photons whose energies are sufficiently high to raise the system energy at least to the first excited state above the ground state.)

Cold Samples

In a sample at room temperature (about 300 K), almost all atoms are in their electronic ground state. For example, if the first excited state is 1 eV above the ground state, the Boltzmann factor is

$$\exp(-E/k_B T) = \exp\left(-\frac{1.6 \times 10^{-19} \text{ J}}{(1.4 \times 10^{-23} \text{ J/K})(300 \text{ K})}\right) = \exp(-38) = 3 \times 10^{-17}$$

This extremely tiny number represents the fact that at room temperature an insignificant fraction of the atoms will be excited thermally to 1 eV above the ground state. Therefore, it is reasonable to assume that essentially all the atoms in an ordinary room-temperature absorption measurement start out in the electronic ground state.

On the other hand, the surface temperature of the Sun is about 5000 K, and with $E = 1$ eV the Boltzmann factor is $\exp(-E/k_B T) = \exp(-2.3) = 0.1$, so a sizable fraction of atoms can be in an excited state 1 eV above the ground state, due to thermal excitation.

8.4 VIBRATIONAL ENERGY LEVELS

For concreteness we have described energy quantization for the electronic states of atomic hydrogen. Many other kinds of microscopic systems have quantized energy. We will next discuss vibrational energy quantization in atomic oscillators.

Figure 8.15 Model of a solid as a network of balls and springs.

Figure 8.16 A classical harmonic oscillator (spring–mass system) can vibrate with any amplitude, and hence can have any energy. $K + U$ of the system could have any value in the shaded region.

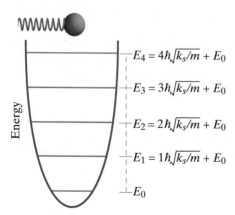

Figure 8.17 A quantum harmonic oscillator can have only certain energies, indicated by the horizontal lines on the graph. This idealized potential energy well represents the situation near the bottom of a real well (which is at a negative energy).

Vibrational Energy in a Classical Oscillator

We have modeled a solid as a network of classical harmonic oscillators, balls and springs (Figure 8.15). Although this classical model explains many aspects of the behavior of solids, the predictions it makes regarding the thermal properties of solids do not correspond to experimental measurements at low temperatures. The classical model also fails to explain some aspects of the interaction of light with solid matter. To explain these phenomena we need to make our model of a solid more sophisticated, by incorporating the quantum nature of these atomic oscillators into our model.

The vibrational energy of a classical (nonquantum) spring–mass system can have *any* value within the colored region in Figure 8.16. Equivalently, the amplitude A (maximum stretch) can have *any* value within the allowed range. We can express the energy as $E = \frac{1}{2}k_s A^2$, because at the maximum stretch A the speed is zero, so the kinetic energy is zero, and the energy $K + U$ is just the potential energy at that instant. Classically, the amplitude A can have continuous rather than quantized values.

Vibrational Energy in a Quantum Mechanical Oscillator

Historically, the first hint that energy might be quantized came from an attempt by the German physicist Max Planck in 1900 to explain some puzzling features of the spectrum of light emitted through a small hole in a hot furnace (so-called "blackbody radiation"). Planck found that he could correctly predict the properties of blackbody radiation if he assumed that transfers of energy in the form of electromagnetic radiation were quantized—that is, an energy transfer could take place only in fixed amounts, rather than having a continuous range of possible values.

Later, with the full development of quantum mechanics, it was recognized that one could model an atom in the solid wall of the furnace as a tiny spring–mass system whose energy levels are discrete (Figure 8.17). Classically, the energy of a mass on a spring is related to the amplitude of its oscillation ($E = \frac{1}{2}k_s A^2$), so saying that a spring–mass oscillator can only have certain discrete amounts of energy is equivalent to saying that only certain amplitudes are allowed. To see this, note that we have

$$K + U = \frac{1}{2}\frac{p^2}{m} + \frac{1}{2}k_s s^2 = \frac{1}{2}k_s A^2$$

since when the stretch s is equal to the amplitude A (the maximum stretch), the momentum is zero.

By solving the quantum mechanical Schrödinger equation for a system involving an atom and a chemical bond, one can show that the spacing ΔE between the allowed, quantized energies has the following value:

$$\Delta E = \hbar\sqrt{\frac{k_s}{m}} \quad \text{or} \quad \Delta E = \hbar\omega_0, \quad \text{where } \omega_0 = \sqrt{\frac{k_s}{m}}$$

In this case, the quantity $k_s = k_{s,i}$ is the interatomic spring stiffness, which we estimated in Chapter 4, $m = m_a$ is the mass of one atom, and $\omega_0 = \sqrt{k_s/m}$ is the angular frequency of the oscillator (in radians per second), as defined in Chapter 4. The quantity $\hbar$, pronounced "h-bar," is equal to h, "Planck's constant," divided by 2π. Planck's constant h has an astoundingly small value:

$$h = 6.6 \times 10^{-34} \text{ joule} \cdot \text{second}$$

The quantized energies for an atomic harmonic oscillator are these:

QUANTIZED VIBRATIONAL ENERGY LEVELS

$$E_N = N\hbar\omega_0 + E_0, \quad N = 0, 1, 2, \ldots, \quad \text{where } \omega_0 = \sqrt{k_s/m}$$

In other words, the energy can have only the following values:

$$E = E_0, \ 1\hbar\omega_0 + E_0, \ 2\hbar\omega_0 + E_0, \ 3\hbar\omega_0 + E_0, \ldots.$$

For an atom and a springlike interatomic bond: $k_s = k_{s,i}$ (interatomic spring stiffness) and $m = m_a$ (mass of an atom)

$$\hbar = \frac{h}{2\pi} = 1.05 \times 10^{-34} \text{ joule} \cdot \text{second (“h-bar”)}$$

It is an unusual and special feature of the quantized harmonic oscillator that the spacing between adjacent energy levels is the same for all levels ("uniform spacing"). This is quite different from the uneven spacing of energy levels in the hydrogen atom. These energy levels are predicted by quantum mechanics, and experiments confirm this prediction.

It is important to understand that in both the quantum and the classical picture, the angular frequency $\omega_0 = \sqrt{k_s/m}$ of a particular oscillator is fixed and is the same whether the energy of the oscillator is large or small. A useful way to think about the situation is to say that the frequency is determined by the parameters k_s and m, while the energy is determined by the initial conditions (or equivalently by the amplitude A). In the quantum world, it is as though the amplitude can have only certain specific values.

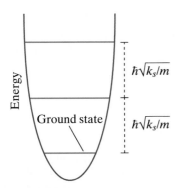

Figure 8.18 A quantum oscillator with a large k_s has widely spaced energy levels.

> **8.X.8** Show that $(\hbar\sqrt{k_s/m})$ has units of joules.
>
> **8.X.9** Suppose that a collection of quantum harmonic oscillators occupies the lowest four energy levels, and the spacing between levels is 0.4 eV. What is the complete emission spectrum for this system? That is, what photon energies will appear in the emissions? Include all energies, whether or not they fall in the visible region of the electromagnetic spectrum.

Spring Stiffness k_s

The energy quantum depends on the stiffness k_s of the "spring." A stiffer spring (larger value of k_s, Figure 8.18) means a steeper, narrower potential curve and larger spacing between the allowed energies. A smaller value of k_s (Figure 8.19) means a shallower, wider potential curve and smaller spacing between the allowed energies.

> **8.X.10** You may have measured the properties of a simple spring–mass system in the lab. Suppose that you found $k_s = 0.7$ N/m and $m = 0.02$ kg, and you observed an oscillation with an amplitude of 0.2 m. What is the approximate value of N, the "quantum number" for this oscillator? (That is, how many levels above the ground state is this oscillator?)
>
> **8.X.11** What is the energy of the photon emitted by a harmonic oscillator with stiffness k_s and mass m when it drops from energy level 5 to energy level 2?

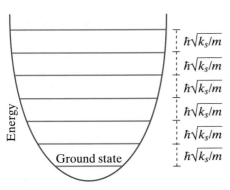

Figure 8.19 A quantum oscillator with a small k_s has closely spaced energy levels.

The Ground State of a Quantum Oscillator

The lowest possible quantized energy E_0 is not zero relative to the bottom of the potential curve, but is $\frac{1}{2}(\hbar\sqrt{k_s/m})$ above the bottom of the well. This is closely related to the "Heisenberg uncertainty principle." When applied to

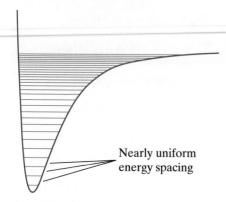

Figure 8.20 Energy levels for the interatomic potential energy, which describes the interaction of two neighboring atoms. This diagram is not to scale; the number of vibrational energy levels in the uniform region may be very large.

Nearly uniform energy spacing

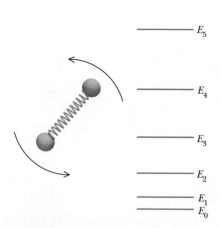

Figure 8.21 The vibrational energy of a diatomic molecule is quantized.

the harmonic oscillator, this principle implies that if you knew that the mass on the spring were exactly at the center of the potential, its speed would be completely undetermined, whereas if you knew that the speed of the mass were exactly zero, the position of the mass could be literally anywhere! The minimum energy $\frac{1}{2}(\hbar\sqrt{k_s/m})$ above the bottom of the well can be viewed as a kind of a compromise, with neither the speed nor the position fixed at zero. The minimum energy is called the ground state of the system.

In our work with the quantized oscillator we will measure energy from the ground state rather than from the bottom of the potential well, so the offset from the bottom of the well won't come into our calculations.

Energy Levels for the Interatomic Potential Energy

If we model the atoms in a solid as simple harmonic oscillators, whose potential energy curve is a parabola, then the energy levels of these systems are uniformly spaced. However, if we use a more realistic interatomic potential energy such as is shown in Figure 8.20, the energy levels of the system follow a more complex pattern. The lowest energy levels are nearly evenly spaced, corresponding to the fact that for small amplitude the bottom of the potential energy curve can be fit rather well by a curve of the form $\frac{1}{2}k_s s^2$. However, for large energies the energy spacing is quite small. Of course there are no bound states for energies larger than the binding energy. The spacing of the vibrational energy levels of a diatomic molecule (Figure 8.21) typically corresponds to photons in the infrared region of the spectrum (see Problem 8.P.35).

Energy Spacing in Various Systems

Most quantum objects have energy-level spacings that decrease as you go to higher levels. The harmonic oscillator potential energy is unusual in leading to uniform energy-level spacing. Even less common is the situation where the energy-level spacing actually increases with higher energy. One such exception is the so-called "square well" with very steep sides and a flat bottom. The energy levels in a square well actually get farther and farther apart for higher energies.

8.5 ROTATIONAL ENERGY LEVELS

A molecule in the gas phase has not only electronic energy and vibrational energy but rotational energy as well. Like electronic and vibrational energy levels, rotational energy levels are also discrete—only certain rotational energies are permitted. As we will see in a later chapter, the angular momentum of a molecule is quantized, and this quantization gives rise to discrete rotational energy levels. In this case all the energy of rotation is kinetic energy—there is no potential energy involved. Rotational energy levels for a diatomic molecule such as N_2 are shown in Figure 8.22. Note that the spacing between levels actually increases with increasing energy, which is the opposite of the trend seen in many atomic systems. Rotational energy levels are typically close together; transitions between these levels usually involve photons in the microwave region of the electromagnetic spectrum. Microwave ovens excite rotational energies of water molecules in food. These excited molecules then give up their energy to the food, making the food hotter.

Case Study: A Diatomic Molecule

Analyzing a diatomic molecule such as N_2 provides a good review of three important types of quantized energy levels: electronic, vibrational, and rotational.

E_5

E_4

E_3

E_2

E_1
E_0

Figure 8.22 Rotational kinetic energy levels of a diatomic molecule.

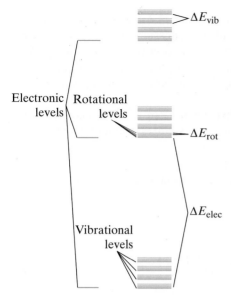

Figure 8.23 The energy bands of a diatomic molecule (not to scale). This energy level structure gives rise to a band spectrum of emitted photons.

Only certain configurations of the electron clouds in a diatomic molecule are possible, corresponding to discrete electronic energy levels. The spacings of the low-lying molecular electronic energy levels are typically in the range of 1 eV or more, like the spacings of the low-lying energy levels in the hydrogen atom.

For each major configuration of the electron clouds, many different vibrational energy states are possible, with energy spacings of the order of 1×10^{-2} eV (see Problem 8.P.35).

For each major configuration of the electron clouds and each vibrational energy state, many different rotational kinetic energy states are possible, with energy spacings of the order of 1×10^{-4} eV, as we will see in a later chapter.

These three different families of energy levels, with their very different energy level spacings, give rise to energy "bands" and a distinctive "band" spectrum of emitted photons. See Figure 8.23, which is not to scale due to the large differences in level spacing for the three kinds of quantized energy levels. An emitted photon can have the energy of an electronic transition plus a smaller-energy vibrational transition plus an even smaller-energy rotational transition. The various possible energy differences give rise to "bands" in the observed emission spectrum.

8.6   OTHER ENERGY LEVELS

It is not only electronic, vibrational, and rotational energies in atoms and molecules that are quantized. Small objects such as nuclei and hadrons (particles made of quarks) also have discrete energy levels.

Nuclear Energy Levels

Just as atoms have discrete states corresponding to allowed configurations of the electron cloud, so nuclei have discrete states corresponding to allowed configurations of the nucleons (protons and neutrons). An excited nucleus can drop to a lower state with the emission of a photon. The energy spacing of nuclear levels is very large, of the order of 1×10^6 eV (1 MeV), so the emitted photons have energies in the range of millions of electron volts. Such energetic photons are called "gamma rays."

> **8.X.12** A gamma ray with energy greater than or equal to 2.2 MeV can dissociate a deuteron (the nucleus of deuterium, heavy hydrogen) into its constituents, a proton and a neutron. What is the nuclear binding energy of the deuteron?

Hadronic Energy Levels

Atoms have energy levels associated with the configurations of electrons in the atom, and nuclei have energy levels associated with the configurations of nucleons in the nucleus. Hadrons are particles made of quarks. In a simple model we represent baryons (particles like protons) as a combination of three quarks, and mesons (particles like pions) as a combination of a quark and an antiquark. Different configurations of quarks correspond to different energy levels of the multiquark system. These different configurations are the different hadrons we observe, and the corresponding energy levels are the observed rest energies of these particles. The energy spacing is of the order of hundreds of MeV, or about 1×10^8 eV. (Leptons, which include electrons, muons, and neutrinos, are not made of quarks.)

An example of a hadron is the Δ^+ particle, which has a rest energy of 1232 MeV. It can be considered to be an excited state of the three-quark system whose ground state is the proton, whose rest energy is 938 MeV. The Δ^+ particle can decay into a proton with the emission of a very high energy photon. However, this "electromagnetic" decay occurs in only about 0.5% of the decays. In most cases the Δ^+ decays into a proton plus a pion, a strong-interaction decay that has a much higher rate, corresponding to the much stronger interaction.

An excited electronic state of a hydrogen atom is still called hydrogen, but for historical reasons we give excited quark states distinctive names, such as Δ^+.

> **8.X.13** What is the approximate energy of the photon emitted when the Δ^+ decays electromagnetically?

Type of state	Typical energy level spacing
hadronic	10^8 eV
nuclear	10^6 eV
electronic (atoms, molecules)	1 eV
vibrational (molecules)	10^{-2} eV
rotational (molecules)	10^{-4} eV

Figure 8.24 Typical energy level spacings in various quantum objects.

8.7 COMPARISON OF ENERGY LEVEL SPACINGS

Figure 8.24 is a summary of typical energy scales for various quantum objects. It is important to distinguish among hadronic, nuclear, electronic, vibrational, rotational, and other kinds of quantized energy situations. We can generalize the main issues to any kind of generic quantum object. The key features of all these different kinds of objects are these:

- discrete energy levels
- discrete emissions whose energies equal differences in energy levels

Some of the homework problems at the end of the chapter ask you to analyze generic objects, independent of the details of what gave rise to the particular discrete energy levels.

8.8 *CASE STUDY: HOW A LASER WORKS

A laser offers an interesting case study of the interaction of light and atomic matter. We will give a basic introduction to lasers, but we can only scratch the surface of a vast topic that continues to be an active field of research and development. Laser action depends on a process called "stimulated emission" and on creating an "inverted population" of quantum states, which can lead to a chain reaction and produce a special kind of light, called "coherent" light.

Stimulated Emission

If an atom can be placed into an excited state—that is, a state of higher energy than the ground state—normally it drops quickly to a lower-energy state and emits a photon in the process. For example, in a neon sign high-energy electrons collide with ground-state neon atoms and raise them to an excited state. An excited neon atom quickly drops back to the ground state, emitting a photon whose energy is the difference in energy between the two quantized energy levels, and you see the characteristic orange color corresponding to these photons.

We call this emission process "spontaneous emission." Spontaneous emission happens at an unpredictable time. All that quantum mechanics can predict is the *probability* that the excited atom will drop to the ground state in the next small time interval. It is truly not possible to predict the exact time of the emission, just as it is impossible to predict exactly when a free neutron will decay into a proton, electron, and antineutrino.

A collection of excited atoms such as those in a neon sign emit spontaneously, at random times. In the full quantum-mechanical description of light, light has both the properties of a particle (the photon picture) and the properties of a wave; see Section 8.9. Emissions at random times have random phases in a wave description: sine, cosine, and all phases between a sine and a cosine. We say that such light is "incoherent."

"Stimulated emission" of light is a different process that produces coherent light. Consider an atom in an energy level E above the ground state. A photon of this same energy E can interact with the atom in a remarkable way, causing the atom to drop immediately to the ground state with the emission of a second, "clone" photon. Where there had been one photon of energy E, there are now two photons of exactly the same energy E, and in a wave description they have exactly the same phase (Figure 8.25). That is, the two waves are both sines, or both cosines, or both with exactly the same phase somewhere between a sine and a cosine. Such light is called coherent light, and it has valuable properties for such applications as making holograms.

Figure 8.25 Stimulated emission produces two identical photons from one.

Chain Reaction

The phenomenon of stimulated emission can be exploited to create a chain reaction. Start with one photon of the right energy E to match the energy difference between the ground state and an excited state of the atoms in the system. If that photon causes stimulated emission from an excited atom, there are now two identical photons, both capable of triggering stimulated emission of other excited atoms, which yields four identical photons, then 8, 16, 32, 64, 128, 256, 512, 1024, and so on.

However, the chain reaction won't proceed if all of the photons escape from the system, so some kind of containment mechanism is required. In a common type of gas laser the gas is in a long tube with mirrors at both ends (Figure 8.26). Photons that happen to be going in the direction of the tube are continually reflected back and forth through the tube, which increases the probability that they will interact with excited atoms and cause stimulated emission. Since the stimulated-emission photons are clones, they too are headed in the direction of the tube and they too are likely to interact. One of the mirrors is deliberately made to be an imperfect reflector, so a fraction of the photons leaks out and provides a coherent beam of light.

Figure 8.26 A gas laser with mirrors at the ends. Photons leak through one of the mirrors.

Unfortunately, the stimulated-emission photons also have just the right energy E to be absorbed by an atom in the ground state, in which case they can no longer contribute to the chain reaction. For this reason we need an "inverted population," with few atoms in the ground state to absorb the photons and stop the chain reaction. We discuss this issue next.

Inverted Population

As we have discussed earlier in this chapter, in a normal collection of atoms more atoms are in the ground state than in any other state, and the number of atoms decreases in each successively higher state. In fact, for a collection of atoms in thermal equilibrium the fraction of atoms in a particular energy state decreases exponentially with the energy above the ground state, as we will see in a later chapter.

By clever means it is possible to invert this population scheme, so that there are more atoms in an excited state than in the ground state (this is not a state of thermal equilibrium). An inverted population is crucial to laser action, so that stimulated emission with cloning of photons will dominate over simple absorption of photons by atoms in the ground state.

Figure 8.27 A three-state scheme for sustaining an inverted population.

There are many different schemes for creating an inverted population. For example, the scheme used in the common helium–neon laser is to "pump" atoms to a high energy level (Figure 8.27). There are diverse energy input schemes, including electric discharges and powerful flashes of ordinary light. Atoms in this high energy level rapidly drop to an intermediate energy level, higher than the ground state. This intermediate state happens to have a rather long lifetime for spontaneous emission, so that it is possible to sustain an inverted population, with more atoms in the intermediate state than in the ground state. Stimulated emission drives these excited atoms down to the ground state, from which they are again pumped up to the high-energy state. Spontaneous emission from the high-energy state to the intermediate state maintains the inverted population.

8.9 *WAVELENGTH OF LIGHT

In this chapter we have emphasized the particle model of light, in terms of photons and their energy. However, as we have hinted earlier, some aspects of the behavior of light (of any energy) can best be understood by considering light to be a wave. For completeness we will mention a connection between the particle model and the wave model of light. Light consisting of photons whose energy is E can also be treated as a wave, and the energy E and wavelength λ are related as follows:

ENERGY AND WAVELENGTH OF LIGHT
$$E_{photon} = \frac{hc}{\lambda_{light}}$$

where h is Planck's constant.

Figure 8.28 A wave packet is a traveling wave of finite spatial extent.

A model for light that incorporates both the features of a wave and a localized particle is a "wave packet," a wave whose magnitude is nonzero only in a small region of space (Figure 8.28). We will return to this topic in a later chapter on waves and particles.

SUMMARY

Light consists of particles of light called photons with energy and momentum but no rest mass, so $E = K$. Their speed is always 3×10^8 m/s.

At the microscopic level the energy of bound states is discrete, not continuous. (Unbound states have a continuous range of energies.)

Quantized systems can have only certain energies. They can emit or absorb photons when they change energy levels.

The temperature is a measure of the average energy of a collection of objects, so high temperature is associated with significant probability of finding particles in any of many states, from the ground state up to a high energy level. Very low temperature is associated with little probability of the particles being in any state other than the ground state.

Discrete hydrogen atom energy levels
$E_N = -(13.6\,\text{eV})/N^2$, where N is a nonzero positive integer $(1, 2, 3 \ldots)$

Discrete harmonic oscillator energy levels
$$E_N = N\hbar\omega_0 + E_0, \quad N = 0, 1, 2, \ldots, \quad \text{where}$$
$\omega_0 = \sqrt{k_s/m}$

Planck's constant $h = 6.6 \times 10^{-34}$ joule·second
$$\hbar = \frac{h}{2\pi} = 1.05 \times 10^{-34} \text{ joule·second}$$

$1\,\text{eV} = 1.6 \times 10^{-19}\,\text{J}$

Photon energy and wavelength: $E_{photon} = \dfrac{hc}{\lambda_{light}}$

EXERCISES AND PROBLEMS

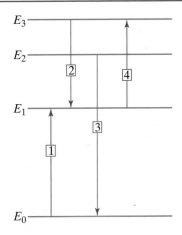

Figure 8.30

Sections 8.1–8.3

8.X.14 How much energy in electron volts is required to ionize a hydrogen atom (that is, remove the electron from the proton), if initially the atom is in the state $N = 2$? (Remember that $N = 1$ if the atom is in the lowest energy level.)

8.X.15 A certain laser outputs pure red light (photon energy 1.8 eV) with power 700 milliwatts (0.7 watts). How many photons per second does this laser emit?

8.X.16 $N = 1$ is the lowest electronic energy state for a hydrogen atom.

(a) If a hydrogen atom is in state $N = 4$, what is $K + U$ for this atom (in eV)?

(b) The hydrogen atom makes a transition to state $N = 2$. Now what is $K + U$ in electron volts for this atom?

(c) What is the energy (in eV) of the photon emitted in the transition from level $N = 4$ to $N = 2$?

(d) Which of the arrows in Figure 8.29 represents this transition?

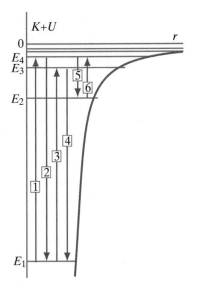

Figure 8.29

8.X.17 Match the description of a process with the corresponding arrow in Figure 8.30:

(a) Absorption of a photon whose energy is $E_1 - E_0$

(b) Absorption from an excited state (a rare event at low temperatures)

(c) Emission of a photon whose energy is $E_3 - E_1$

(d) Emission of a photon whose energy is $E_2 - E_0$

(e) In drawing arrows to represent energy transitions, which of the following statements are correct?

  (1) It doesn't matter in which direction you draw the arrow as long as it connects the initial and final states.

  (2) For emission, the arrow points down.

  (3) For absorption, the arrow points up.

  (4) The tail of the arrow is drawn on the initial state.

  (5) The head of the arrow is drawn on the final state.

  (6) It is not necessary to draw an arrowhead.

8.X.18 When starlight passes through a cold cloud of hydrogen gas, some hydrogen atoms absorb energy, then reradiate it in all directions. As a result, a spectrum of the star shows dark absorption lines at the energies for which less energy from the star reaches us. How does the spectrum of dark absorption lines for very cold hydrogen differ from the spectrum of bright emission lines from very hot hydrogen?

8.X.19 The Franck–Hertz experiment involved shooting electrons into a low-density gas of mercury atoms and observing discrete amounts of kinetic energy loss by the electrons. Suppose that instead a similar experiment is done with a very cold gas of atomic hydrogen, so that all of the hydrogen atoms are initially in the ground state. If the kinetic energy of an electron is 11.6 eV just before it collides with a hydrogen atom, how much kinetic energy will the electron have just after it collides with and excites the hydrogen atom?

8.P.20 Hydrogen atoms:

(a) What is the minimum kinetic energy in electron volts that an electron must have to be able to ionize a hydrogen atom that is in its ground state (that is, remove the electron from being bound to the proton)?

(b) If electrons of energy 12.8 eV are incident on a gas of hydrogen atoms in their ground state, what are the energies of the photons that can be emitted by the excited gas?

(c) If instead of electrons, photons of all energies between 0 and 12.8 eV are incident on a gas of hydrogen atoms in the ground state, what are the energies at which the photons are absorbed?

8.P.21 Suppose we have reason to suspect that a certain quantum object has only three quantum states. When we excite such an object we observe that it emits electromagnetic radiation of three different energies: 2.48 eV (green), 1.91 eV (orange), and 0.57 eV (infrared).

(a) Propose two possible energy-level schemes for this system.

(b) Explain how to use an absorption measurement to distinguish between the two proposed schemes.

8.P.22 The eye is sensitive to photons with energies in the range from about 1.8 eV, corresponding to red light, to about 3.1 eV, corresponding to violet light. White light is a mixture of all the energies in the visible region. If you shine white light through a slit onto a glass prism, you can produce a rainbow spectrum on a screen, because the prism bends different colors of light by different amounts (Figure 8.31).

Figure 8.31

If you replace the source of white light with an electric-discharge lamp containing excited atomic hydrogen, you will see only a few lines in the spectrum, rather than a continuous rainbow (Figure 8.32).

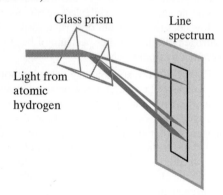

Figure 8.32

Predict how many lines will be seen in the visible spectrum of atomic hydrogen. Specify the atomic transitions that are responsible for these lines, given that the energies of the quantized states in atomic hydrogen are given by $E_N = -(13.6\,\text{eV})/N^2$, where N is a nonzero positive integer (1, 2, 3, ...).

8.P.23 Assume that a hypothetical object has just four quantum states, with the energies shown in Figure 8.33.

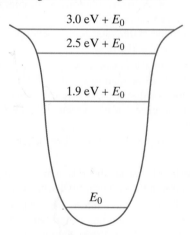

$3.0\,\text{eV} + E_0$

$2.5\,\text{eV} + E_0$

$1.9\,\text{eV} + E_0$

E_0

Figure 8.33

(a) Suppose that the temperature is high enough that in a material containing many such objects, at any instant some objects are found in all of these states. What are all the energies of photons that could be strongly emitted by the material? (In actual quantum objects there are often "selection rules" that forbid certain emissions even though there is enough energy; assume that there are no such restrictions here.)

(b) If the temperature is very low and electromagnetic radiation with a wide range of energies is passed through the material, what will be the energies of photons corresponding to missing ("dark") lines in the spectrum? (Assume that the detector is sensitive to a wide range of photon energies, not just energies in the visible region.)

8.P.24 A certain material is kept at very low temperature. It is observed that when photons with energies between 0.2 and 0.9 eV strike the material, only photons of 0.4 eV and 0.7 eV are absorbed. Next the material is warmed up so that it starts to emit photons. When it has been warmed up enough that 0.7 eV photons begin to be emitted, what other photon energies are also observed to be emitted by the material? Explain briefly.

8.P.25 Suppose we have reason to suspect that a certain quantum object has only three quantum states. When we excite a collection of such objects we observe that they emit electromagnetic radiation of three different energies: 0.3 eV (infrared), 2.0 eV (visible), and 2.3 eV (visible).

(a) Draw a possible energy-level diagram for one of the quantum objects, which has three bound states. On the diagram, indicate the transitions corresponding to the emitted photons, and check that the possible transitions produce the observed photons and no others. Label the energies of each level ($K + U$, which is *negative*). The energy $K + U$ of the ground state is -4 eV.

(b) The material is now cooled down to a very low temperature, and the photon detector stops detecting photon emissions. Next a beam of light with a continuous range of energies from infrared through ultraviolet shines on the material, and the photon detector observes the beam of light after it passes through the material. What photon energies in this beam of light are observed to be significantly reduced in intensity ("dark absorption lines")?

(c) There exists another possible set of energy levels for these objects that produces the same photon emission spectrum. On an alternative energy-level diagram, *different from the one you drew in part (a),* indicate the transitions corresponding to the emitted photons, and check that the possible transitions produce the observed photons and no others. Label the energies of each level ($K + U$, which is *negative*).

(d) For your second proposed energy-level scheme, what photon energies would be observed to be significantly reduced in intensity in an absorption experiment ("dark absorption lines")? (Given the differences from part (b), you can see that an absorption measurement can be used to tell which of your two energy-level schemes is correct.)

8.P.26 Assume that a hypothetical object has just four quantum states, with the following energies:

-1.0 eV (third excited state)

-1.8 eV (second excited state)

−2.9 eV (first excited state)

−4.8 eV (ground state)

(a) Suppose that material containing many such objects is hit with a beam of energetic electrons, which ensures that there are always some objects in all of these states. What are the six energies of photons that could be strongly emitted by the material? (In actual quantum objects there are often "selection rules" that forbid certain emissions even though there is enough energy; assume that there are no such restrictions here.) List the photon emission energies.

(b) Next, suppose that the beam of electrons is shut off so that all of the objects are in the ground state almost all the time. If electromagnetic radiation with a wide range of energies is passed through the material, what will be the three energies of photons corresponding to missing ("dark") lines in the spectrum? Remember that there is hardly any absorption from excited states, because emission from an excited state happens very quickly, so there is never a significant number of objects in an excited state. Assume that the detector is sensitive to a wide range of photon energies, not just energies in the visible region. List the dark-line energies.

8.P.27 Energy graphs:

(a) Figure 8.34 shows a graph of potential energy vs. interatomic distance for a particular molecule. What is the direction of the associated force at location A? At location B? At location C? Rank the magnitude of the force at locations A, B, and C. (That is, which is greatest, which is smallest, and are any of these equal to each other?) For the energy level shown on the graph, draw a line whose height is the kinetic energy when the system is at location D.

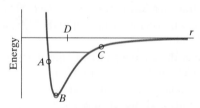

Figure 8.34

(b) Figure 8.35 shows all of the quantized energies (bound states) for one of these molecules. The energy for each state is given on the graph, in electron volts ($1 \text{ eV} = 1.6 \times 10^{-19} \text{ J}$). How much energy is required to break a molecule apart, if it is initially in the ground state? (Note that the final state must be an unbound state; the unbound states are not quantized.)

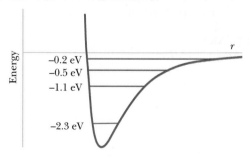

Figure 8.35

(c) At high enough temperatures, in a collection of these molecules there will be at all times some molecules in each of these states, and light will be emitted. What are the energies in electron volts of the emitted light?

(d) The "inertial" mass of the molecule is the mass that appears in Newton's second law, and it determines how much acceleration will result from applying a given force. Compare the inertial mass of a molecule in the top energy state and the inertial mass of a molecule in the ground state. If there is a difference, briefly explain why and calculate the difference. If there isn't a difference, briefly explain why not.

8.P.28 A bottle contains a gas with atoms whose lowest four energy levels are −12 eV, −6 eV, −3 eV, and −2 eV. Electrons run through the bottle and excite the atoms so that at all times there are large numbers of atoms in each of these four energy levels, but there are no atoms in higher energy levels. List the energies of the photons that will be emitted by the gas.

Next the electron beam is turned off, and all the atoms are in the ground state. Light containing a continuous spectrum of photon energies from 0.5 eV to 15 eV shines through the bottle. A photon detector on the other side of the bottle shows that some photon energies are depleted in the spectrum ("dark lines"). What are the energies of the missing photons?

8.P.29 Some material consisting of a collection of microscopic objects is kept at a high temperature. A photon detector capable of detecting photon energies from infrared through ultraviolet observes photons emitted with energies of 0.3 eV, 0.5 eV, 0.8 eV, 2.0 eV, 2.5 eV, and 2.8 eV. These are the only photon energies observed.

(a) Draw and label a possible energy-level diagram for one of the microscopic objects, which has 4 bound states. On the diagram, indicate the transitions corresponding to the emitted photons. Explain briefly.

(b) Would a spring–mass model be a good model for these microscopic objects? Why or why not?

(c) The material is now cooled down to a very low temperature, and the photon detector stops detecting photon emissions. Next a beam of light with a continuous range of energies from infrared through ultraviolet shines on the material, and the photon detector observes the beam of light after it passes through the material. What photon energies in this beam of light are observed to be significantly reduced in intensity ("dark absorption lines")? Explain briefly.

Sections 8.4–8.8

8.X.30 If you try to increase the amplitude of a quantum harmonic oscillator by adding an amount of energy $\frac{1}{2}\hbar\sqrt{k_s/m}$, the amplitude doesn't increase. Why not?

8.X.31 If you double the amplitude, what happens to the frequency in a classical (nonquantum) harmonic oscillator? In a quantum harmonic oscillator?

8.X.32 Summarize the differences and similarities between different energy levels in a quantum oscillator. Specifically, for the first two levels in Figure 8.18, compare the angular frequency $\sqrt{k_s/m}$, the amplitude A, and the kinetic energy K at the same value of s. (In a full quantum-mechanical analysis the concepts of angular frequency and amplitude require reinterpretation. Nevertheless there remain elements of the classical picture. For example, larger amplitude corresponds to a higher probability of observing a large stretch.)

8.X.33 Which energy diagram in Figure 8.36 is appropriate for each of the following situations?

- **(a)** Hadronic (such as Δ^+)
- **(b)** Vibrational states of a diatomic molecule such as O_2
- **(c)** Idealized quantized spring–mass oscillator
- **(d)** Nuclear (such as the nucleus of a carbon atom)
- **(e)** Electronic, vibrational, and rotational states of a diatomic molecule such as O_2
- **(f)** Rotational states of a diatomic molecule such as O_2
- **(g)** Electronic states of a single atom such as hydrogen

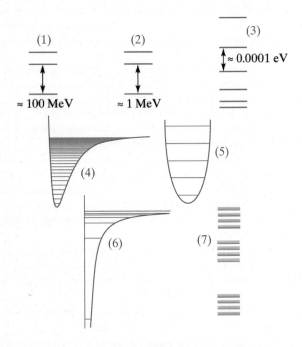

Figure 8.36

8.P.34 Consider a microscopic spring–mass system whose spring stiffness is 50 N/m, and the mass is 4×10^{-26} kg.

- **(a)** What is the smallest amount of vibrational energy that can be added to this system?
- **(b)** What is the difference in mass (if any) of the microscopic oscillator between being in the ground state and being in the first excited state?
- **(c)** In a collection of these microscopic oscillators, the temperature is high enough that the ground state and the first three excited states are occupied. What are possible energies of photons emitted by these oscillators?

8.P.35 Molecular vibrational energy levels:

- **(a)** A HCl molecule can be considered to be a quantized harmonic oscillator, with quantized vibrational energy levels that are evenly spaced. Make a rough estimate of this uniform energy spacing in electron volts (where $1\,\text{eV} = 1.6 \times 10^{-19}$ J). You will need to make some rough estimates of atomic properties based on prior work. For comparison with the spacing of these vibrational energy states, note that the spacing between quantized energy levels for "electronic" states such as in atomic hydrogen is of the order of several electron volts.
- **(b)** List several photon energies that would be emitted if a number of these vibrational energy levels were occupied due to electron excitation. To what region of the spectrum (x-ray, visible, microwave, etc.) do these photons belong? (See Figure 8.1 at the beginning of the chapter.)

8.P.36 A hot bar of iron glows a dull red. Using our simple model of a solid (see Figure 8.15), answer the following questions, explaining in detail the processes involved. The mass of one mole of iron is 56 grams.

- **(a)** What is the energy of the lowest-energy spectral emission line? Give a numerical value.
- **(b)** What is the approximate energy of the highest-energy spectral emission line? Give a numerical value.
- **(c)** What is the quantum number of the highest-energy occupied state?
- **(d)** Predict the energies of two other lines in the emission spectrum of the glowing iron bar.

(*Note*: Our simple model is too simple—the actual spectrum is more complicated. However, this simple analysis gets at some important aspects of the phenomenon.)

ANSWERS TO EXERCISES

8.X.1 About 2.5 eV

8.X.2 About 3.5×10^{21} per second per square meter

8.X.3 13.6 eV

8.X.4 Six different photon energies, corresponding to transitions from 4 to 3, 4 to 2, 4 to 1, 3 to 2, 3 to 1, and 2 to 1

8.X.5 Just one dark line at 2.4 eV

8.X.6 5.2 eV

8.X.7 1 eV, 5 eV, and 6 eV

8.X.9 0.4 eV (three transitions), 0.8 eV (two transitions), and 1.2 eV (one transition)

8.X.10 $N \approx 2 \times 10^{31}$. A system in a very high quantum state behaves like a classical system.

8.X.11 $3\hbar\sqrt{k_s/m}$

8.X.12 2.2 MeV

8.X.13 About 294 MeV

Multiparticle Systems

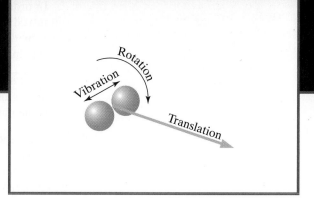

- **The kinetic energy of a multiparticle system has two parts:**
  - Translational kinetic energy (due to motion of center of mass)
  - Kinetic energy relative to the center of mass (rotation + vibration)

    Rotational kinetic energy can be calculated in terms of moment of inertia I and angular speed ω.
- **Treating a multiparticle system as if it were a point particle lets us calculate changes in translational kinetic energy.**
- **Analyzing the real system lets us calculate changes in the other energy terms (rotational, vibrational, potential).**

9.1 | THE MOTION OF THE CENTER OF MASS

Recall that in Chapter 3 we derived an equation that predicts the overall motion of a complex multiparticle system, in terms of the rate of change of the total momentum of the system $\vec{p}_{sys}$:

THE MOMENTUM PRINCIPLE FOR MULTIPARTICLE SYSTEMS

$$\frac{d\vec{p}_{sys}}{dt} = \vec{F}_{net}$$

$\vec{p}_{sys} \approx M_{total}\vec{v}_{CM}$ (if $v \ll c$ and the mass is constant).

There are some unusual aspects of this equation. It correctly predicts that when a person jumps upward from a crouch, the floor pushes up on the feet with a larger force than the downward gravitational force, with the result that the jumper's momentum $M\vec{v}_{CM}$ increases upward (Figure 9.1), and the jumper's center of mass moves upward.

Yet in the Energy Principle the force of the floor does no work on the person, because the displacement of the point of application of this force is zero (or very small, assuming that the floor bends very little). The work done on the jumper is actually negative, because the gravitational force of the Earth pulls down as the jumper moves up, yet v_{CM} increases. Evidently the internal energy of the jumper must decrease in order to pay for the motion of the center of mass, since the work done by objects in the surroundings actually does negative work on the system.

Figure 9.1 The floor does no work on the jumper, yet K increases.

In order to deal effectively with situations of this kind, in this chapter we will introduce methods for separating the motion *of* the center of mass (governed by the net force) from motion *relative to* the center of mass (associated with the internal energy of the system).

> QUESTION   A high diver jumps outward, then tucks into a tight ball and spins rapidly, then straightens out before hitting the water. What is the path of the diver's center of mass?

Although the diver's overall motion is very complex, the center of mass of the diver must move like that of a simple projectile! We do not yet have enough tools to analyze the diver's complicated rotation and stretching relative to the center of mass, yet we can analyze the motion of the center of mass by applying the Momentum Principle.

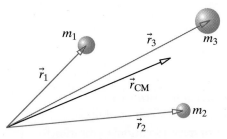

Figure 9.2 Location of the center of mass of the three-particle system.

Center of Mass

We briefly review what was established in Chapter 3 about the concept of the center of mass of a multiparticle system. It is defined as a weighted average of the positions of all the atoms that make up the chosen system (Figure 9.2), with $M_{total} = m_1 + m_2 + m_3 + \cdots$:

CENTER OF MASS
$$\vec{r}_{CM} \equiv \frac{m_1\vec{r}_1 + m_2\vec{r}_2 + m_3\vec{r}_3 + \cdots}{M_{total}}$$

Rearrange the definition above as follows:

$$M_{total}\vec{r}_{CM} = m_1\vec{r}_1 + m_2\vec{r}_2 + m_3\vec{r}_3 + \cdots$$

Differentiate this equation with respect to time ($d\vec{r}/dt = \vec{v}$), in the case where M_{total} is not changing:

$$M_{total}\vec{v}_{CM} = m_1\vec{v}_1 + m_2\vec{v}_2 + m_3\vec{v}_3 + \cdots$$

However, the right side of this equation is the sum of the momenta of all the atoms in the system, as long as $v \ll c$ (so $\gamma \approx 1$ for all atoms), which is what we call the total (nonrelativistic) momentum of the system, $\vec{p}_{sys}$. Therefore we have this important result:

MOMENTUM AND CENTER OF MASS VELOCITY
$$\vec{p}_{sys} = M_{total}\vec{v}_{CM}$$
for speeds small compared to the speed of light, and constant mass.

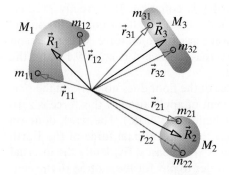

Figure 9.3 Finding the center of mass of a system consisting of several large objects.

Center of Mass of Several Large Objects

Often we need to know the location of the center of mass of a system consisting of several large objects, each of whose center of mass is known. In Figure 9.3, $\vec{R}_1$ is the location (relative to the origin) of the center of mass of the object whose total mass is M_1, and so on.

For each of the large objects we have an equation for its center of mass of the form $M_1\vec{R}_1 = \vec{r}_{11} + m_{12}\vec{r}_{12} + \cdots$, $M_2\vec{R}_2 = m_{21}\vec{r}_{21} + m_{22}\vec{r}_{22} + \cdots$, and so on.

We group terms in the calculation of the location of the center of mass of the total system:

$$\vec{r}_{CM} \equiv \frac{(m_{11}\vec{r}_{11} + m_{12}\vec{r}_{12}) + \cdots + (m_{21}\vec{r}_{21} + m_{22}\vec{r}_{22}) + \cdots}{(m_{11} + m_{12} + \cdots) + (m_{21} + m_{22} + \cdots) + \cdots}$$

$$\vec{r}_{CM} = \frac{M_1\vec{R}_1 + M_2\vec{R}_2 + M_3\vec{R}_3 + \cdots}{M_1 + M_2 + M_3 + \cdots}$$

This looks just like the equation for finding the center of mass of a bunch of atoms, except that the $\vec{R}$ vectors refer to the centers of mass of large objects.

> **9.X.1** Three uniform-density spheres are positioned as follows:
> - A 3 kg sphere is centered at $\langle 10, 20, -5 \rangle$ m.
> - A 5 kg sphere is centered at $\langle 4, -15, 8 \rangle$ m.
> - A 6 kg sphere is centered at $\langle -7, 10, 9 \rangle$ m.
>
> What is the location of the center of mass of this three-sphere system?
>
> **9.X.2** Relative to an origin at the center of the Earth, where is the center of mass of the Earth–Moon system? The mass of the Earth is 6×10^{24} kg, the mass of the Moon is 7×10^{22} kg, and the distance from the center of the Earth to the center of the Moon is 4×10^8 m. The radius of the Earth is 6400 km. One can show that the Earth and Moon orbit each other around this center of mass.

Application: An Ice Skater Pushes Off from a Wall

The Momentum Principle is always valid, but applying it can produce some nonintuitive predictions. Consider a woman on ice skates who pushes on a wall with a nearly constant contact force to the left of magnitude F_N (normal to the wall in Figure 9.4). By the reciprocity of electric interatomic forces, the wall exerts a force of magnitude F_N to the right on the skater. As a result, she moves backward with increasing speed. Her center of mass is marked on each frame of the following sequence:

Figure 9.4

What would we predict for the motion of the skater's center of mass? If M is her total mass, and if we can assume that the frictional force of the ice on the skates is negligibly small, we predict this from the free-body diagram (Figure 9.5):

$$\frac{dp_{sys}}{dt} = F_N$$

The intriguing aspect of this simple result is that during the time that the wall exerts a force F_N on the skater's hand, her hand does not move! (Note that the work done by the force exerted by the wall is zero; more about this later.) The place where the force F_N is applied is very far away from the mathematical point (the center of mass) whose rate of change of speed we predict. It is only

Figure 9.5 Free-body diagram for the skater.

because the electric interactions between neighboring atoms in the skater's body obey the principle of reciprocity (Newton's third law) that these internal forces cancel and lead to a simple prediction for the motion of the center of mass:

$$\frac{d\vec{p}_{sys}}{dt} = \frac{d}{dt}(M_{total}\vec{v}_{CM}) = \vec{F}_{net}$$

Note that as the skater straightens her arms, the location in her body of the center of mass shifts toward her hands a little; it is not a point fixed at some place in her body. The motion that we predict is the motion of this mathematical center of mass point.

Application: Pull on Two Hockey Pucks

We'll look at another odd consequence of this multiparticle version of the Momentum Principle. Tie a string to the center of a hockey puck, and wrap another string around the outside of a second hockey puck, as shown in Figure 9.6. Then two people pull on the two strings so that both strings have the same tension F_T. As shown in the figure, in a time Δt hand 2 pulling the bottom puck has traveled farther, because the string has unrolled from the puck a bit.

The pucks have the same mass M. Assume that friction with the ice rink is negligible. It would be reasonable to suppose that it would make a difference where you attach the string to the puck.

Figure 9.6 One puck is pulled by a string attached to its center. The other puck is pulled by a string wrapped around its edge, which unrolls as the puck is pulled.

| QUESTION | What will be the velocity (direction and magnitude) of the center of the first puck after a short time Δt? Of the center of the second puck? How do you justify your predictions?

Both pucks are subjected to exactly the same external force, so their centers of mass must move in exactly the same way. After a short time Δt, the momentum of the center of mass must be $F_T\Delta t$, in the direction of the string, for both pucks. Despite the fact that one puck is pulled from the side, the change in momentum of its center of mass must be in the direction of the net external force, so it cannot acquire a momentum component at right angles to that force. Doesn't this seem a bit odd? For example, one might have expected the second puck to have some kind of sideways motion.

9.X.3 Is there any difference at all in the motion of the two pucks? Think about the motion of your hands at the other ends of the strings. Which hand does more work? Which puck has more energy? In what forms? Does the Momentum Principle for the motion of the center of mass predict all the details of the motion of both pucks?

9.X.4 A meter stick whose mass is 300 grams lies on ice (Figure 9.7). You pull at one end of the meter stick, at right angles to the stick, with a force of 6 newtons. The ensuing motion of the meter stick is quite complicated, but what are the initial magnitude and direction of the rate of change of the momentum of the stick, dp_{sys}/dt, when you first apply the force? What is the magnitude of the initial acceleration of the center of the stick?

6 N

Figure 9.7. Pull one end of a meter stick that is lying on ice.

| 9.2 | ## SEPARATION OF MULTIPARTICLE SYSTEM ENERGY

We have seen that we can predict the motion of a multiparticle object by treating it as if all its mass were concentrated at the center of mass, acted upon by a force that is the (vector) sum of all the external forces acting on the object. In many situations we find that the energy of a multiparticle system can also be analyzed simply, by dividing the kinetic energy of the system into two parts: the translational kinetic energy, K_{trans}, and K_{rel}, the kinetic energy relative to the center of mass. K_{rel} includes the kinetic energy associated with rotation of the object and the kinetic energy associated with vibration of the object.

MULTIPARTICLE KINETIC ENERGY

$$K_{tot} = K_{trans} + K_{rel}$$

where $K_{rel} = K_{vib} + K_{rot}$.

For example, the kinetic energy of a rotating, vibrating diatomic molecule such as oxygen (O_2) is the sum of the kinetic energy that the moving molecule would have if it were not rotating or vibrating, plus the additional kinetic energy of rotation around the center of mass and kinetic energy of vibration relative to the center of mass (Figure 9.8).

This split of the total kinetic energy into different parts sounds plausible. A formal derivation of this result is given in Section 9.8 at the end of this chapter.

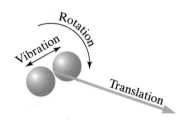

Figure 9.8 A diatomic molecule such as O_2 can vibrate and rotate while also translating.

Translational Kinetic Energy

In physics, the word "translate" means "move from one location to another." Translational motion is the ordinary kind of motion we have studied most so far. The translational motion of a macroscopic object is described by $\vec{v}_{CM}$, the velocity of the object's center of mass. We use the term "translate" here in order to distinguish between translational motion, in which the center of mass of a system moves from one location to another, and other kinds of motion such as rotation and vibration, in which parts of a system move relative to the center of mass.

The translational kinetic energy of a system is kinetic energy associated with translational motion, and does not include kinetic energy associated with rotation or vibration.

TRANSLATIONAL KINETIC ENERGY

$$K_{trans} = \frac{1}{2} M_{total} v_{CM}^2 = \frac{p_{sys}^2}{2 M_{total}}$$

for an object whose speed is small compared to the speed of light ($v \ll c$).

EXAMPLE

Translational Kinetic Energy of a Frisbee™

What is the translational kinetic energy of a spinning 140 gram Frisbee that is traveling at a speed of 15 m/s? Is this the total kinetic energy of the Frisbee?

Solution

$$K_{trans} = \frac{1}{2}(0.14 \, \text{kg})(15 \, \text{m/s})^2 = 15.8 \, \text{J}$$

This does not include the kinetic energy associated with the rotation of the disk.

Vibrational Energy

One form of energy that is internal to a system is vibrational energy, both elastic and kinetic. We already know how to calculate vibrational energy. Consider

Figure 9.9 A vibrating oxygen molecule whose center of mass is at rest.

Figure 9.10 A translating, vibrating oxygen molecule.

an oxygen molecule (O_2) that has no translational motion (that is, the center of mass is stationary), but is vibrating (Figure 9.9), with elastic energy and kinetic energy continually interchanging, but with the sum of the two energies remaining constant, like the mass and spring you modeled in earlier chapters.

VIBRATIONAL ENERGY

$$E_{vib} = K_{vib} + U_{spring}$$

If in addition the oxygen molecule is translating (Figure 9.10), the total energy of the molecule is the sum of the translational kinetic energy, the vibrational kinetic energy (in terms of velocities of the two atoms relative to the center of mass), and the elastic energy of the "spring" holding them together, and the rest energies of the constituent atoms:

$$E_{tot} = \frac{1}{2}(2m)v_{CM}^2 + K_{vib} + \frac{1}{2}k_s s^2 + 2mc^2$$

Note that the elastic energy $\frac{1}{2}k_s s^2$ is clearly unaffected by the translational motion of the center of mass, because it depends only on the stretch of the distance between the two atoms.

The same reasoning applies to the motion of a hot block of metal. In addition to its translational kinetic energy $\frac{1}{2}Mv_{CM}^2$ there is vibrational kinetic and elastic energy of the atoms around their equilibrium positions, which we call thermal energy, with more thermal energy if the block's temperature is higher.

Rotational Kinetic Energy

Just as a vibrating object has kinetic energy associated with vibration, even if its center of mass is at rest, so a rotating object has kinetic energy associated with rotation, even if its center of mass is at rest.

Figure 9.11 A bicycle wheel spinning on its axle, which is at rest. Atoms in the rim have a speed v_{rel} relative to the center of mass.

As an example, suppose that you spin a bicycle wheel on its axis, and hold the axle stationary, as in Figure 9.11. The spinning wheel has some kinetic energy K_{rel}, which in this case we will call K_{rot}, because the atoms are moving in rotational motion relative to the center of mass. Almost all of the mass M of the wheel is in the rim, so if the rim is traveling at a (tangential) speed v_{rel}, the kinetic energy of the wheel is approximately $K_{rot} \approx \frac{1}{2}Mv_{rel}^2$. (This isn't exact, because there is some mass in the spokes, and the atoms in the spokes are moving at speeds smaller than v_{rel}.)

In Section 9.3 we will see how to calculate the rotational kinetic energy of other kinds of rotating objects.

Rotation, Vibration, and Translation

The most general motion of an oxygen molecule involves rotation, vibration, and translation. It is easy to show that the internal (nontranslational) kinetic energy separates cleanly into rotational and vibrational contributions. Consider one atom with momentum $\vec{p}$ relative to the center of mass, and resolve its momentum into "tangential" and "radial" components (Figure 9.12). By "radial" component we mean a component in the direction of the line from the center of mass to the particle. By "tangential" component we mean a component perpendicular to that line. If the particle is in circular motion around the center of mass, the radial component is zero, so tangential momentum is associated with rotation. Radial momentum is associated with vibration.

Figure 9.12 Radial and tangential components of momentum of a single atom, relative to the center of mass of an object.

Find the kinetic energy in terms of these two momentum components:

$$\frac{p^2}{2m} = \frac{(p_{tangential}^2 + p_{radial}^2)}{2m} = K_{rot} + K_{vib}$$

Therefore the kinetic energy relative to the center of mass splits into two terms, rotational energy and vibrational energy:

$$K_{\text{rel}} = K_{\text{rot}} + K_{\text{vib}}$$

The total energy of a translating, rotating, vibrating oxygen molecule can be written like this:

$$E_{\text{tot}} = K_{\text{trans}} + K_{\text{rot}} + K_{\text{vib}} + \frac{1}{2}k_s s^2 + 2mc^2$$

where $K_{\text{trans}} = \dfrac{p_{\text{sys}}^2}{2(2m)} = \dfrac{1}{2}(2m)v_{\text{CM}}^2$.

If the object can fly apart into separate pieces, we might call the kinetic energy relative to the center of mass $K_{\text{explosion}}$! For example, a fireworks rocket has translational kinetic that doesn't change from just before to just after its explosion, and the pieces move at high speeds relative to the center of mass of the rocket. These large kinetic energies relative to the center of mass come from the chemical energy used up in the explosion.

Figure 9.13 The velocity of an atom in the rim is the vector sum of the center-of-mass velocity and the velocity relative to the center of mass.

Why Separating Kinetic Energy Is Simpler

If we do not separate kinetic energy into $K_{\text{trans}} + K_{\text{rel}}$, adding up the kinetic energies of all the particles in a system may be a very complex task. Consider the bicycle wheel shown in Figure 9.13. The fact that this wheel is both translating to the right, and also spinning on its axis, means that particles at different locations in the wheel have quite different speeds. For example, an atom at the bottom of the wheel is actually moving at a speed less than v_{CM}, while an atom at the top of the wheel is moving at a speed greater than v_{CM}. In order to calculate the kinetic energy of the whole system, we would have to figure out the instantaneous speed of each constituent particle!

Gravitational Energy of a Multiparticle System

In previous chapters we have calculated the gravitational potential energy of a system consisting of the Earth and a multiparticle object simply by treating the object as if all its mass were concentrated at a single point (its center of mass). Why is this correct?

Consider the gravitational energy of a block of mass M and the Earth, near the Earth's surface. The gravitational energy associated with the interaction of the Earth with the i-th atom in the block is $m_i g y_i$ (Figure 9.14).

Figure 9.14 The i-th atom in the block is a distance y_i from the Earth's surface.

We can calculate the total gravitational energy by adding up the gravitational energy associated with the interaction of each of the N atoms with the Earth:

$$U_g = m_1 g y_1 + m_2 g y_2 + m_3 g y_3 + \cdots = g(m_1 y_1 + m_2 y_2 + m_3 y_3 + \cdots)$$

However, the last expression in parentheses appears in the calculation of the y component of the location of the center of mass:

$$M y_{\text{CM}} = m_1 y_1 + m_2 y_2 + m_3 y_3 + \cdots$$

Therefore for a block made up of many atoms, near the Earth's surface, we have this:

GRAVITATIONAL ENERGY U_g

$$U_g = M g y_{\text{CM}} \quad \text{near the Earth's surface}$$

In other words, the gravitational energy of a block of material and the Earth, near the Earth's surface, can be evaluated as though the block were a tiny particle of mass M located at the center of mass of the block.

This simple equation is not valid if the object is so large that the strength of the gravitational field is significantly different at different locations in the object, but in that case mg would not be a good approximation for the gravitational force, nor mgy a good approximation for the gravitational energy associated with each atom.

QUESTION   In a calculation of gravitational energy in the voyage of a spacecraft to the Moon, is it a valid approximation to treat the spacecraft as if all its mass were concentrated at one point?

This is a good approximation, because the spacecraft is very small compared to the distances to the Earth and Moon, so the gravitational force on each atom is nearly the same.

9.3   ROTATIONAL KINETIC ENERGY

A common instance of rotational motion is that in which a rigid system is rotating on an axis, as shown in Figure 9.15. In this situation all the atoms in the system share the same "angular speed" in radians per second. However, they have different linear speeds in meters per second, depending on their distances from the axis, because an atom near the edge must travel farther in one revolution (Figure 9.16).

Angular speed, normally denoted by ω (lowercase Greek omega), is a measure of how fast something is rotating. If an object makes one complete turn of 360 degrees (2π radians) in a time T, we say that its angular speed is $\omega = 2\pi/T$ radians per second. The time T is called the period.

Figure 9.15 A rigid disk rotating about the y axis with angular speed ω.

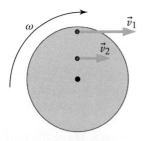

Figure 9.16 A top view of a disk rotating about an axis. A particle near the edge has a higher speed than a particle nearer to the axis. In each case, $v_i = \omega r_i$.

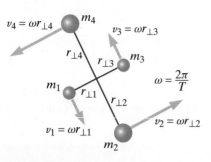

Figure 9.17 A case of rigid rotation about an axis with angular speed ω. Note the different speeds, with $v_i = \omega r_i$.

PERIOD AND ANGULAR SPEED

$$\omega = \frac{2\pi}{T}$$

The period T is the time it takes for a rotating object to make one complete revolution. The angular speed ω has units of radians per second.

An atom rotating at a constant rate around an axis at a distance r from the axis goes a distance of the circumference $2\pi r$ in the period T, so the speed v of the atom can be expressed in terms of the angular speed:

$$v = \frac{2\pi r}{T} = \left(\frac{2\pi}{T}\right) r = \omega r$$

We measure angular speed ω in the "natural" units of radians per second rather than degrees per second, because the fundamental geometrical relationship between angle and arc length (arc length = $r\theta$) is valid only if the angle θ is measured in radians.

The kinetic energy of the system shown in Figure 9.17, which is a rotating rigid object, is $\frac{1}{2}m_1 v_1^2 + \frac{1}{2}m_2 v_2^2 + \frac{1}{2}m_3 v_3^2 + \cdots$. Since the speed of each mass is $v = \omega r_\perp$ (where $r_\perp$ is the perpendicular distance of the mass from the axis of rotation), we can write the rotational kinetic energy like this:

$$K_{rot} = \frac{1}{2}\left[m_1(\omega r_{\perp 1})^2 + m_2(\omega r_{\perp 2})^2 + m_3(\omega r_{\perp 3})^2 + \cdots\right]$$
$$= \frac{1}{2}\left[m_1 r_{\perp 1}^2 + m_2 r_{\perp 2}^2 + m_3 r_{\perp 3}^2 + \cdots\right]\omega^2$$

The quantity in brackets is called the "moment of inertia" and is usually denoted by the letter I:

MOMENT OF INERTIA
$$I = m_1 r_{\perp 1}^2 + m_2 r_{\perp 2}^2 + m_3 r_{\perp 3}^2 + m_4 r_{\perp 4}^2 + \cdots$$

The units of moment of inertia are $\text{kg} \cdot \text{m}^2$.

Using this definition, we have a compact expression for rotational kinetic energy:

ROTATIONAL KINETIC ENERGY
$$K_{\text{rot}} = \frac{1}{2} I \omega^2$$

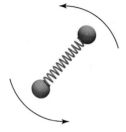

Figure 9.18 A rotating diatomic molecule.

EXAMPLE   **The Moment of Inertia of a Diatomic Molecule**

What is the moment of inertia of a diatomic nitrogen molecule N_2 around its center of mass (Figure 9.18)? The mass of a nitrogen atom is 2.3×10^{-26} kg and the average distance between nuclei is 1.5×10^{-10} m. Use the definition of moment of inertia carefully.

Solution   For two masses, $I = m_1 r_{\perp 1}^2 + m_2 r_{\perp 2}^2$. The distance between masses is d, so the distance of each object from the center of mass is $r_{\perp 1} = r_{\perp 2} = d/2$. Therefore

$$I = M(d/2)^2 + M(d/2)^2 = 2M(d/2)^2$$

$$I = 2 \cdot (2.3 \times 10^{-26}\,\text{kg})(0.75 \times 10^{-10}\,\text{m})^2$$

$$I = 2.6 \times 10^{-46}\,\text{kg} \cdot \text{m}^2$$

EXAMPLE   **The Moment of Inertia of a Bicycle Wheel**

A bicycle wheel has almost all its mass M located in the outer rim at radius R (Figure 9.19). What is the moment of inertia of the bicycle wheel about its center of mass? (*Hint*: It's helpful to think of dividing the wheel into the atoms it is made of and think about how much each atom contributes to the moment of inertia.)

Solution   Let m represent the mass of one atom in the rim. The moment of inertia is

$$I = m_1 r_{\perp 1}^2 + m_2 r_{\perp 2}^2 + m_3 r_{\perp 3}^2 + m_4 r_{\perp 4}^2 + \cdots$$

$$I = m_1 R^2 + m_2 R^2 + m_3 R^2 + m_4 R^2 + \cdots$$

$$I = [m_1 + m_2 + m_3 + m_4 + \cdots] R^2$$

$$I = MR^2$$

Figure 9.19 Most of the mass of a bicycle wheel is in the rim.

We've assumed that the mass of the spokes is negligible compared to the mass of the rim, so that the total mass is just the mass of the atoms in the rim.

EXAMPLE   **Rotational Kinetic Energy and Work**

In Figure 9.20 a wheel is mounted on a stationary axle, which is nearly frictionless so that the wheel turns freely. The wheel has an inner ring with mass 5 kg and radius 10 cm and an outer ring with mass 2 kg and radius 25 cm; the spokes have negligible mass. A string with negligible mass is wrapped around the outer ring and you pull on it, increasing the rotational speed of the wheel. During the time that the wheel's rotation changes from 4 revolutions per second to 7 revolutions per second, how much work do you do?

Solution   System: Wheel and string
Surroundings: Your hand, axle, Earth
Free-body diagram: Figure 9.20
Principle: Energy Principle

$$E_f = E_i + W$$

$$\frac{1}{2}I\omega_f^2 = \frac{1}{2}I\omega_i^2 + W$$

$$W = \frac{1}{2}I\left(\omega_f^2 - \omega_i^2\right)$$

What is the moment of inertia $I = m_1 r_{\perp 1}^2 + m_2 r_{\perp 2}^2 + m_3 r_{\perp 3}^2 + m_4 r_{\perp 4}^2 + \cdots$? Group this sum into a part that includes just the atoms of the inner ring and another part that includes just the atoms of the outer ring:

$$I = (m_1 r_{\perp 1}^2 + m_2 r_{\perp 2}^2 + \cdots)_{inner} + (m_1 r_{\perp 1}^2 + m_2 r_{\perp 2}^2 + \cdots)_{outer}$$
$$I = I_{inner} + I_{outer}$$

This is a general result: The moment of inertia of a composite object is the sum of the moments of inertia of the individual pieces, because we have to add up all the contributions of all the atoms. We already determined that the moment of inertia of a ring is MR^2 (all the atoms are at the same perpendicular distance R from the center), so the moment of inertia of this wheel is

$$I = M_{inner} R_{inner}^2 + M_{outer} R_{outer}^2$$
$$I = (5\,\text{kg})(.1\,\text{m})^2 + (2\,\text{kg})(.25\,\text{m})^2 = (0.050 + 0.125)\,\text{kg} \cdot \text{m}^2 = 0.175\,\text{kg} \cdot \text{m}^2$$

We need to convert revolutions per second into radians per second:

$$\omega_i = \left(4\frac{\text{rev}}{\text{s}}\right)\left(\frac{2\pi\ \text{radians}}{\text{rev}}\right) = 25.1\ \text{radians/s}$$

$$\omega_f = \left(7\frac{\text{rev}}{\text{s}}\right)\left(\frac{2\pi\ \text{radians}}{\text{rev}}\right) = 44.0\ \text{radians/s}$$

You, the Earth, and the axle all exert forces on the system. How much work does the Earth do? Zero, because the center of mass of the wheel doesn't move. How much work does the axle do? If there is negligible friction between the axle and the wheel, the axle does no work, because there is no displacement of the axle's force. Therefore only you do work, and the work that you do is

$$W = \tfrac{1}{2}(0.175\,\text{kg} \cdot \text{m}^2)(44.0^2 - 25.1^2)\frac{\text{radians}^2}{\text{s}^2} = 114\ \text{J}$$

Figure 9.20 A rotating wheel with two rings.

FURTHER DISCUSSION   Although the outer ring has a smaller mass than the inner ring, the outer ring has a larger moment of inertia because its radius is 2.5 times as big, so the factor R^2 is 6.25 times as big. Applying the

Momentum Principle to the wheel, we see that the three forces must add up to zero, because the momentum $M\vec{v}_{\text{CM}}$ of the wheel isn't changing.

9.X.5 If an object has a moment of inertia $19 \text{ kg} \cdot \text{m}^2$ and rotates with an angular speed of 70 radians/s, what is its rotational kinetic energy?

9.X.6 A barbell spins around a pivot at its center (Figure 9.21). The barbell consists of two small balls, each with mass 800 grams, at the ends of a very low mass rod whose length is 35 cm. The barbell spins with angular speed 40 radians/s. Calculate K_{rot}.

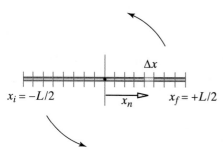

Figure 9.21 A rotating barbell.

Moment of Inertia of a Thin Rod

We've seen how to calculate the moment of inertia in a few simple special cases. We will calculate the moment of inertia of a long thin rod about its center to illustrate a more general technique. Consider a rod of mass M and length L whose thickness is small compared to its length, and that rotates around an axis perpendicular to the page (Figure 9.22). The density of the rod is the same everywhere ("uniform density"); every centimeter along the rod has the same mass. We need to evaluate the sum

$$I = m_1 r_{\perp 1}^2 + m_2 r_{\perp 2}^2 + m_3 r_{\perp 3}^2 + m_4 r_{\perp 4}^2 + \cdots$$

for all the atoms in the rod.

Figure 9.22 A uniform-density rod of length L rotates around an axis perpendicular to the page. We divide it into short slices and choose the origin to be at the center of the rod, with the x axis lying along the rod.

DIVIDE INTO SMALL SLICES   As shown in Figure 9.22, we've chosen the origin of our axes to be at the center of the rod, and we've oriented the x axis to lie along the rod. We mentally divide the rod into N small slices of equal length $\Delta x = L/N$. Each slice has a mass $\Delta M = M/N$, because the rod's density is the same everywhere. We need to add up all the contributions to I made by all the slices.

THE MASS OF ONE SLICE   Concentrate on one representative slice, whose center in Figure 9.22 is located at $\langle x_n, 0, 0 \rangle$ and whose mass is $\Delta M = M/N$. Since $\Delta x = L/N$,

$$N = \frac{L}{\Delta x}$$

$$\Delta M = \frac{M}{N} = \frac{M}{L/\Delta x} = M\frac{\Delta x}{L}$$

Another way to see this is that $\Delta x/L$ is the fraction of the rod contained in this slice of length Δx, so the mass of this slice should be $M(\Delta x/L)$.

THE CONTRIBUTION OF ONE SLICE   Now we're able to write down an expression for the contribution ΔI to the moment of inertia I of the rod that is made by this one representative slice of the rod. We make the approximation that all the atoms of this slice are approximately the same perpendicular distance $r_\perp \approx x_n$ from the center of the rod:

$$\Delta I = (\Delta M)x_n^2 = \left(M\frac{\Delta x}{L}\right)x_n^2 = \frac{M}{L}x_n^2 \Delta x$$

ADDING UP THE CONTRIBUTIONS   We need to add up all such contributions by all such slices. One way to do this is numerically, by dividing the rod into, say, $N = 50$ slices, evaluating $(M/L)x_n^2 \Delta x$ for each slice, and adding up all the contributions ($\sum$ means "summation"):

$$I = \sum_{n=1}^{N} \Delta I = \frac{M}{L} \sum_{n=1}^{N} x_n^2 \Delta x$$

THE FINITE SUM BECOMES A DEFINITE INTEGRAL In some cases, including this case of a long thin rod, it is possible to use integral calculus to evaluate the summation. In the limit of large N (which means small Δx), the sum becomes an integral (the $\int$ symbol is a distorted S for Summation). The finite slice length Δx turns into the infinitesimal quantity dx, and the sum from 1 to N turns into a definite integral with lower and upper limits on x (x_i and x_f):

$$I = \frac{M}{L} \lim_{N \to \infty} \sum_{n=1}^{N} x_n^2 \Delta x = \frac{M}{L} \int_{x_i}^{x_f} x^2 dx$$

THE LIMITS OF INTEGRATION Since we put the origin at the center of the rod in Figure 9.22, the lower limit x_i is $-L/2$ and the upper limit x_f is $+L/2$:

$$I = \frac{M}{L} \int_{-L/2}^{+L/2} x^2 dx$$

QUESTION Try to evaluate this definite integral.

You should find that $I = (1/12)ML^2$. If you are unable to get this result, review integral calculus and make sure that you can do this kind of integral.

Because it is easy to make a mistake in this kind of calculation, it is important to check whether the results are reasonable. First, are the units correct? Yes, $(1/12)ML^2$ has units of $\text{kg} \cdot \text{m}^2$, which are units of moment of inertia. Also, if we double the mass M, I doubles, which makes sense, and if we double the length L, I gets four times bigger, which also makes sense given the factors of $r_\perp^2$ that appear in the definition of moment of inertia.

QUESTION Why does it have to be a *thin* rod for this calculation to be valid?

We made the approximation that all the atoms in one slice had the same perpendicular distance x_n from the center of the rod. If the rod is thick, this isn't a valid approximation, even in the limit of infinitesimally short slices, because now those short slices are thick.

9.X.7 A thin uniform-density rod whose mass is 1.2 kg and whose length is 0.7 m rotates around an axis perpendicular to the rod, with angular speed 50 radians/s. Its center moves with a speed of 8 m/s.

  (a) What is its rotational kinetic energy?

  (b) What is its total kinetic energy?

Moments of Inertia of Simple Shapes

By applying integral calculus as we did above in the case of a long thin uniform-density rod, it is possible to calculate the moments of inertia of some other simple shapes. Here is a summary:

(1)

(2)

(3)

Figure 9.23 Some common uniform-density solids whose moments of inertia are known.

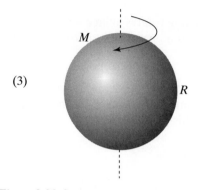

Figure 9.24 These disks and cylinders all have moment of inertia $\frac{1}{2}MR^2$ about their axes.

Figure 9.25 An object with known I_{CM} is connected to a low-mass rod and rotates around an axle. The red stripe on the object rotates at the same rate ω as the rod.

SOME MOMENTS OF INERTIA

For objects of uniform density (see Figure 9.23):

$$(1)\ I_{\text{cylinder}} = \frac{1}{12}ML^2 + \frac{1}{4}MR^2$$

cylinder of length L, radius R, rotating about an axis perpendicular to the cylinder, through center of cylinder.

$$(2)\ I_{\text{disk}} = I_{\text{cylinder}} = \frac{1}{2}MR^2$$

rotating around the axis of the disk or cylinder.

$$(3)\ I_{\text{sphere}} = \frac{2}{5}MR^2$$

rotating around an axis passing through the center of the sphere.

A thin rod is like a cylinder with radius $R \approx 0$, in which case the moment of inertia for rotation about an axis perpendicular to the rod reduces to $\frac{1}{12}ML^2$.

Because only the perpendicular distances of atoms from the axis matter ($r_\perp$), the moment of inertia for rotation about the axis of a long cylinder has exactly the same form as that of a disk, $I_{\text{disk}} = \frac{1}{2}MR^2$. It doesn't matter how long the cylinder is (or to put it another way, how thick the disk is). See Figure 9.24.

> **9.X.8** A uniform-density disk whose mass is 10 kg and radius is 0.4 m makes one complete rotation every 0.2 s. What is the rotational kinetic energy of the disk?
>
> **9.X.9** A uniform-density sphere whose mass is 10 kg and radius is 0.4 m makes one complete rotation every 0.2 s. What is the rotational kinetic energy of the sphere?

Rigid Rotation about a Point Not the Center of Mass

In Figure 9.25 a rigid object with known moment of inertia I_{CM} about its center of mass is connected to a low-mass rod and rotates about an axle that is a distance r_{CM} from the center of mass. What is the kinetic energy of this rotating object?

We know that in general $K_{\text{tot}} = K_{\text{trans}} + K_{\text{rel}}$, and in this case K_{rel} is just K_{rot}, since there is no vibration or explosion. The red line painted through the center of mass on the object in Figure 9.25 rotates with the same angular speed ω as the rod, which rotates around the axle; both the line and the object make one revolution in the same amount of time, so the kinetic energy that is relative to the (moving) center of mass is, as usual, $K_{\text{rot}} = \frac{1}{2}I_{CM}\omega^2$.

> **QUESTION** What is K_{trans}, the kinetic energy associated with the motion of the center of mass?

The center of mass of the object rotates about the axle with angular speed ω at a distance r_{CM} from the axle, so the speed of the center of mass (in meters per second) is $v_{CM} = \omega r_{CM}$. Therefore we have

$$K_{\text{trans}} = \frac{1}{2}Mv_{CM}^2 = \frac{1}{2}M(\omega r_{CM})^2$$

$$K_{\text{trans}} = \frac{1}{2}(Mr_{CM}^2)\omega^2$$

$$K_{\text{tot}} = \frac{1}{2}(Mr_{CM}^2 + I_{CM})\omega^2$$

We see that when the center of mass rotates around a fixed point, K_{trans} brings into the total kinetic energy an additional term $\frac{1}{2}(Mr_{CM}^2)\omega^2$. We can summarize this result by saying that the total kinetic energy is $\frac{1}{2}I\omega^2$, where $I = Mr_{CM}^2 + I_{CM}$. This result is called the parallel axis theorem.

RIGID ROTATION ABOUT A POINT NOT THE CENTER OF MASS

$$K_{tot} = K_{trans} + K_{rot} = \frac{1}{2}(Mr_{CM}^2 + I_{CM})\omega^2$$

This is valid only for a rigid body, one whose rotation rate is the same as the rotation rate about the center of mass.

EXAMPLE

A Rod Rotating Not Around Its Center

A thin rod with mass 140 grams and 60 cm long rotates at an angular speed of 25 radians/s about an axle that is 20 cm from one end of the rod (Figure 9.26). What is its kinetic energy?

Solution   There is translational kinetic energy because the center of mass is moving, and there is rotational kinetic energy because there is rotation about the center of mass. From what we just showed, we have this ($r_{CM} = 0.1$ m):

$$K_{tot} = K_{trans} + K_{rot} = \frac{1}{2}(Mr_{CM}^2 + I_{CM})\omega^2$$

$$K_{tot} = \frac{1}{2}\left(Mr_{CM}^2 + \frac{1}{12}ML^2\right)\omega^2 = \frac{1}{2}M\left(r_{CM}^2 + \frac{1}{12}L^2\right)\omega^2$$

$$K_{tot} = \frac{1}{2}(.140\,\text{kg})\left((.1\,\text{m})^2 + \frac{1}{12}(.6\,\text{m})^2\right)(25\,\text{radians/s})^2$$

$$K_{tot} = 1.75\,\text{J}$$

Figure 9.26 A rod rotates around an axle that is not located at the center of the rod.

9.X.10 A solid uniform-density sphere is tied to a rope and moves in a circle with speed v. The distance from the center of the circle to the center of the sphere is d, the mass of the sphere is M, and the radius of the sphere is R.

(a) What is the angular speed ω?

(b) What is the rotational kinetic energy of the sphere?

(c) What is the total kinetic energy of the sphere?

9.4   THE "POINT PARTICLE SYSTEM"

In analyzing the motion and energy of deformable multiparticle systems (systems whose shape can change), we encounter situations in which it would be extremely convenient to be able to calculate only the change in the translational kinetic energy K_{trans} of the system. Perhaps this might be all we are interested in, or perhaps we might wish to subtract this energy change from the change of the total energy of the system to find out how much energy has gone into vibration or rotation. Fortunately, it is quite simple to calculate ΔK_{trans} by itself.

Because $d\vec{p}_{sys}/dt = \vec{F}_{net}$, the center of mass of a multiparticle system moves exactly like a simple "point" particle whose mass is the mass of the

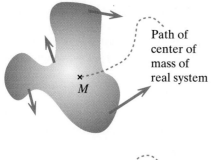

Real system:
Forces act at different locations

Path of
center of
mass of
real system

M

Path of
point-particle
system

M

Point-particle system:
All forces act at the same location

Figure 9.27 The path of the point particle system is exactly the same as the path of the center of mass of the real system.

Figure 9.28 As the jumper leaves the floor, the center of mass has risen h. The system changes shape.

entire system, under the influence of the net external force applied to the entire system. Pretend you could crush the real system down to a very tiny ball and apply to this "point particle" forces with the same magnitudes and directions as those that acted on the real system, but applied directly to the point particle rather than at their original locations. What would the motion of this point particle system look like? It would look exactly like the motion of the center of mass of the real system (Figure 9.27).

These two different systems have the same total mass M, and both are acted on by the same net force, so the two paths are exactly the same. The difference is that the real system may rotate and stretch and vibrate due to the effects of the forces acting at different locations on the extended object. In contrast, the point particle system has no rotational motion, no vibrational motion, no internal energy of any kind. All of the forces act at the location of the point particle, and these forces don't stretch or rotate it. The only energy the point particle system can have is translational kinetic energy, and this is exactly the same as the translational kinetic energy of the real multiparticle system:

$$K_{trans} = \frac{p_{sys}^2}{2M} = \frac{1}{2}Mv_{CM}^2 \quad \text{for the real system or the point particle system}$$

An analysis of the point particle system gives the translational kinetic energy of the real system. This is why the point particle system is a useful concept. The force and energy equations for the fictitious point particle are these:

POINT PARTICLE SYSTEM

A *point particle system* has the same mass as a real multiparticle system, but all its mass is concentrated into a point particle located at the center of mass of the real system. For a point particle system:

$$\frac{d\vec{p}_{sys}}{dt} = \vec{F}_{net}$$

$$\Delta K_{trans} = \Delta\left(\frac{1}{2}Mv_{CM}^2\right) = \int_i^f \vec{F}_{net} \bullet d\vec{r}_{CM}$$

The integral is the work done on the point particle system by the net force, which acts at the location of the point itself. In Chapter 6 we showed that starting from the Momentum Principle for a single point particle we could derive the result that the change in the kinetic energy of the particle is equal to the work done on the particle by the net force. Here, we consider the net force to be applied to the fictitious point particle located at the center-of-mass point, doing work to increase the kinetic energy of the point particle. A more formal derivation of the energy equation for the point particle system may be found at the end of this chapter.

These equations tell us about the motion of the center of mass of a multiparticle system, but they tell us nothing about the vibration or rotation or thermal changes of the system. For a full treatment we need to combine analyses of the point particle system with analyses of the real system.

Application: Jumping Up

As an example of how to exploit the point particle system, we'll analyze jumping (Figure 9.28). You crouch down and then jump straight upward. We want to calculate how fast your center of mass v_{CM} is moving at the instant that your feet leave the floor. Let h be the distance through which your center of mass rises in this process. Note that there is a change of shape of the system.

Let F_N be the electric contact force that the atoms in the floor exert on the bottom of your foot, and let M be your total mass. We don't know exactly how the floor force F_N varies with time, but to get an idea of what happens let's make the crude approximation that it is constant as long as you are in contact with the floor (of course it falls suddenly to zero as your foot leaves the floor). Consider the different point particle system shown in Figure 9.29, which does not change shape.

Jumper: Point Particle System

QUESTION What is the net force on you? What is the energy equation for the point particle system?

As indicated in Figure 9.29 the net force in the upward direction is $(F_N - Mg)$, and the energy equation for the point particle system is

$$\Delta K_{trans} = \frac{1}{2} M v_{CM}^2 - 0 = \int_i^f \vec{F}_{net} \cdot d\vec{r}_{CM} = (F_N - Mg)h$$

As we will see, this is not the same as the energy equation for the real system consisting of you, the jumper! The reason F_N shows up in the point particle energy equation is that it contributes to the net force and impulse, which affects the momentum (which is related to the motion of the center of mass). The work done on the point particle system involves the net force multiplied by the displacement of the center-of-mass point.

Figure 9.29 Force diagram for your body, considered as a point particle. This system does not change shape.

Jumper: Real System

QUESTION What are some of the energy changes that occur during the jump that are not represented in this point particle energy equation?

There is change of kinetic energy of your moving arms and legs relative to the center of mass, decrease of chemical energy inside you, and increase of thermal energy inside you (your temperature rises somewhat).

QUESTION In the real system, how much work is done by the floor force F_N?

The floor force F_N is applied to the bottom of your foot, and the contact point does not move during the entire process leading to lift-off, so there is no work. The definition of the work done by a force involves (the parallel component of) the displacement of the point of contact, at the place where the force is applied to the object. No displacement, no work done on you.

The energy equation for the real system (you, the jumper) looks something like the following, ignoring Q, transfer of energy due to a temperature difference between you and the surrounding air:

$$\Delta E_{sys} = W = \sum \int_i^f \vec{F}_i \cdot d\vec{r}_i \text{ (W is sum of work done by all external forces)}$$

$$\Delta K_{trans} + \Delta K_{other} + \Delta E_{thermal} + \Delta E_{chemical} = -Mgh$$

The floor force F_N does not appear in the energy equation for the real system, because this force does no work (no displacement of the point of contact; $d\vec{r}_{floor} = \vec{0}$). (The floor force does, however, appear in the energy equation for the point particle system.) As before, the gravitational force does negative work on you.

ΔK_{other} includes the rotation of your upper and lower legs and the swinging of your arms. $\Delta E_{thermal}$ corresponds to the higher temperature that your body runs due to the exertion (which ultimately will lead to transfer of energy Q out of your body into the surroundings, due to the higher temperature compared to the cooler air). $\Delta E_{\text{chemical}}$ represents the burning of chemical energy stored in your body, in order to support this activity. None of these terms appear in the energy equation for the point particle system, because the point particle equation deals just with the motion of the mathematical center of mass point, which doesn't rotate or stretch or get hot. A crucial difference is that you change shape (legs unbend, etc.), but the point particle system doesn't.

Finding the Change in Internal Energy

Replace ΔK_{trans} in the energy equation for the real system with the value of ΔK_{trans} we found for the point particle system:

$$\Delta K_{\text{trans}} + \Delta K_{\text{other}} + \Delta E_{\text{thermal}} + \Delta E_{\text{chemical}} = -Mgh$$
$$(F_N - Mg)h + \Delta K_{\text{other}} + \Delta E_{\text{thermal}} + \Delta E_{\text{chemical}} = -Mgh$$
$$\Delta K_{\text{other}} + \Delta E_{\text{thermal}} + \Delta E_{\text{chemical}} = -F_N h$$

Comparison of the Point Particle and Real Energy Equations

The important thing to notice is that the energy equation for the real system and for the point particle system are quite different. Most energy terms do not appear at all in the point particle equation, and the work that appears in the energy equation for the real system is not the same as the work that appears in the energy equation for the point particle system.

> QUESTION Which of the energy changes in the energy equation shown above for the real system are positive? Which are negative?

All of these energy changes of your body are positive except for the change in chemical energy, which decreases to pay for all the other increases.

Application: Stretching a Spring

Let's check to make certain this kind of analysis gives us a sensible answer in a simple situation. You pull on one end of a spring (Figure 9.30) with a force F_L to the left, making a short displacement Δr_L to the left, and you pull on the other end with an equal but opposite force F_R and equal but opposite short displacement Δr_R. The system (the spring) changes shape. Our analysis should yield the expected result: the spring should not gain translational kinetic energy!

Figure 9.30 You pull on both ends of a spring with equal and opposite forces. The system (the spring) changes shape.

Spring: Real System

Consider the spring itself as the system.

> QUESTION Does your right hand do any work on the spring? Of what sign? Does your left hand do any work on the spring? Of what sign? Is there any change in the energy of the spring?

Your left hand exerts a force through a short displacement Δr_L and does some (positive) work $W_L = F_L \Delta r_L \cos(0°) = +F_L \Delta r_L$.

Your right hand also exerts a force through a short displacement Δr_L and does some (positive) work $W_R = F_R \Delta r_R \cos(0°) = +F_R \Delta r_R$.

The total work increases the energy of the spring, in the form of spring (elastic) energy corresponding to increased stretch of the spring. Note carefully that both hands do positive work. The energy equation for the real system undergoing this process (the spring) is this:

$$\Delta \left(\frac{1}{2} k_s s^2 \right) = W_L + W_R$$

$$F_L \longleftarrow \bullet \longrightarrow F_R$$
$$F_{net} = 0$$

Figure 9.31 Force diagram for the "point particle system," which does not change shape.

Spring: Point Particle System

Next consider the "point particle system" consisting of a fictitious particle located at the center of mass of the spring (Figure 9.31). This point particle system does not change shape.

QUESTION What is the net force $\vec{F}_{net}$? How far does the center of mass move? How much work $\sum \vec{F}_{net} \bullet \Delta \vec{r}_{CM}$ is done? What is the change in $K_{trans} = \frac{1}{2}Mv_{CM}^2$?

The net force is zero ($F_R - F_L = 0$). Reassuringly, the center of mass does not move. The work done on the fictitious point particle is zero, so there is no change in the speed of the center of mass, thus no change in the translational kinetic energy $K_{trans} = \frac{1}{2}Mv_{CM}^2$. The energy equation reduces to $0 = 0$, which is certainly correct but provides no detailed information.

Comparison of the Point Particle and Real Energy Equations

The key point here is that in the real system, work is calculated by adding up the amount of work done by *each* force: the work done by your right hand *plus* the work done by your left hand. The work done by each force is the force times the actual displacement of the point of application of that force, $F_L\Delta r_L$ and $F_R\Delta r_R$, both of which are positive.

In the point particle system, however, work is calculated by first adding up all the forces to get the *net* force before calculating the work, and the net force is multiplied by the displacement of the center of mass. In the present example the net force is zero, and the displacement of the center of mass is zero, so the work done on the point particle system is zero.

Note also that the actual work done by an individual force involves the displacement of the point where the force is applied. In contrast, work done on the point particle system involves the displacement of the center of mass point, a point that may be very far away from any of the actual locations where the forces are applied, with a motion that is quite different from the motion of the point of application of an individual force.

Another way of looking at the situation is that the real multiparticle system (the spring) changes shape, and there is potential energy associated with that change of shape. The point particle system does not change shape, and there is no change in potential energy.

The energy equation for the real system and the energy equation for the point particle system are both valid and give correct results for any process, but they provide different kinds of information:

Real system: Information on change of total energy of system

Point particle system: Information on change of translational kinetic energy

The differences are what make it useful to analyze certain kinds of phenomena both ways. Note that the two energy equations are different for this spring example because the system is deformable and changes shape, which leads to very different calculations of the work in the two systems.

9.X.11 If a system such as a large block does not change shape (or rotate faster and faster due to tangential forces, and there are no changes of identity or temperature), the energy equations for the real system and for the point particle system are identical. Explain why.

9.X.12 A runner whose mass is 50 kg accelerates from a stop to a speed of 10 m/s in 3 seconds. (A good sprinter can run 100 meters in about 10 seconds, with an average speed of 10 m/s.)

(a) What is the average horizontal component of the force that the ground exerts on the runner's shoes?

(b) How much displacement is there of the force that acts on the sole of the runner's shoes, assuming that there is no slipping? Therefore, how much work is done on the real system (the runner) by the force you calculated in the previous exercise? How much work is done on the point particle system by this force?

(c) The kinetic energy of the runner increases—what kind of energy decreases? By how much?

9.5   ANALYZING POINT PARTICLE AND REAL SYSTEMS

Here is a procedure to use in analyzing the point particle and real systems.

- Imagine crushing the real system and its forces down to a point at the center of mass of the real system, including moving the tails of the force arrows to that point. Draw diagrams of the initial and final states, representing the point particle system as a dot, and showing all the forces that act on the real system.
- The tails of all the forces must be on the point particle itself.
- Write the Energy Principle for the point particle, paying attention to the fact that each force acts through the distance that the point particle moves, *not* the distance the tail of the force moves in the real system. The energy will include only the term ΔK_{trans}.
- Draw diagrams of the initial and final states of the real system.
- Draw arrows representing the forces acting on the real system. It is very important to place the tail of each arrow at the location where that force acts on an object.
- Look carefully at how far the tails of the force arrows move, from the initial state to the final state. Use this information to calculate the work done by each force. Add up these work terms to find the work done on the real system.
- Write the Energy Principle for the real system. The energy will include the term ΔK_{trans} but may include additional terms.
- Replace ΔK_{trans} in the equation for the real system with the ΔK_{trans} you found from the equation for the point particle.

Figure 9.32 A constant force acts on a thin box containing a spring and a ball of clay. The box moves a distance b, but the ball (center of mass of the system) moves only a distance $b - s$. Tails of arrows representing forces are drawn at the point at which they act.

EXAMPLE

A Box Containing a Spring

A thin box in outer space contains a large ball of clay of mass M, connected to an initially relaxed spring of stiffness k_s (Figure 9.32). The mass of the box and spring are negligible compared to M. The apparatus is initially at rest. Then a force of constant magnitude F is applied to the box. When the box has moved a distance b, the clay makes contact with the left side of the box and sticks there, with the spring stretched an amount s. See the diagram for distances. **(a)** Immediately after the clay sticks to the box, how fast is the box moving? **(b)** What is the increase in internal energy of the clay?

Solution   **(a)** Point particle system
   System: Point particle of mass M
   Surroundings: External object exerting force F
   Initial state: Point particle at rest
   Final state: Point particle moving with (unknown) speed v

Figure 9.33 The point particle system for the apparatus. Tails of arrows representing forces are attached to the point particle, which moves the same distance as the center of mass of the real system.

■ Assume that the process takes place so quickly that there isn't time for any significant energy transfer Q due to a temperature difference between the (real) system and surroundings.

Free-body diagram: Figure 9.33
Energy Principle (point particle only has K_{trans}):

$$\Delta K_{trans} = F\Delta x_{CM}$$
$$\Delta x_{CM} = (b - s) \quad \text{(from the diagram)}$$
$$\frac{1}{2}Mv^2 - 0 = F(b - s)$$
$$v = \sqrt{\frac{2F(b-s)}{M}}$$

(b) Real system
System: Mass, box, and spring
Surroundings: External object exerting force F
Initial state: System at rest
Final state: Box moving with (known) speed v (same as clay)
Free-body diagram: Figure 9.32
Energy Principle:

$$\Delta E_{sys} = W$$
$$\Delta(K_{trans} + U_{spring} + E_{internal}) = Fb$$
$$\Delta K_{trans} = F(b - s) \quad \text{(from part (a))}$$
$$F(b - s) + \frac{1}{2}k_s s^2 + \Delta E_{internal} = Fb$$
$$\Delta E_{internal} = Fs - \frac{1}{2}k_s s^2$$

FURTHER DISCUSSION In the energy equation for the point particle system there is nothing about the spring or thermal energy, because the point particle system doesn't stretch or get hot. In Figure 9.33 the tail of the force vector moves a distance $(b - s)$, moving with the point particle, so the work done on the point particle system is $F(b - s)$.

In the real system, the tail of the force vector moves a longer distance b, so the work done on the real system is Fb. More work is done, and the extra work goes into spring potential energy and internal energy of the clay.

Good labeled diagrams of the two choices of system are extremely important, in order to determine the distances through which forces act in the two cases. Review the procedure at the beginning of the section and see how these steps were implemented in the example given above.

Figure 9.34 You pull up with a force of magnitude F through a distance d on the string of a yo-yo while the yo-yo moves downward a distance h.

EXAMPLE **A Yo-yo**

You're playing with a yo-yo of mass m on a low-mass string (Figure 9.34). You pull up on the string with a force of magnitude F, and your hand moves up a distance d. During this time the mass falls a distance h (and some of the string reels off the yo-yo's axle). **(a)** What is the change in the translational kinetic energy of the yo-yo? **(b)** What is the change in the rotational kinetic energy of the yo-yo, which spins faster?

Solution **(a)** Point particle system
System: Point particle of mass m
Surroundings: Earth and hand
Initial state: Point particle with initial translational kinetic energy
Final state: Point particle with final translational kinetic energy
Free-body diagram: Figure 9.35

Figure 9.35 The point particle system for the yo-yo. Tails of arrows representing forces are attached to the point particle, which moves the same distance as the center of mass of the real system.

Energy Principle (point particle only has K_{trans}):

$$\Delta K_{trans} = (F - mg)\Delta y_{CM}$$

$$\Delta y_{CM} = -h \quad \text{(from diagram)}$$

$$\Delta K_{trans} = (F - mg)(-h) = (mg - F)h$$

(b) Real system
System: Mass and string
Surroundings: Earth and hand
Initial state: Initial rotational and translational kinetic energy
Final state: Final rotational and translational kinetic energy
Free-body diagram: Figure 9.34
Energy Principle:

$$\Delta E_{sys} = W_{hand} + W_{Earth}$$

$$\Delta K_{trans} + \Delta K_{rot} = Fd + (-mg)(-h)$$

$$\Delta K_{trans} = (mg - F)h \quad \text{(from part (a))}$$

$$(mg - F)h + \Delta K_{rot} = Fd + mgh$$

$$\Delta K_{rot} = F(d + h)$$

FURTHER DISCUSSION The key issue is that the distance through which the force F moves is different in the real system and the point particle system. In the real system, your hand applies an upward force of magnitude F through an upward displacement d, doing an amount of work $+Fd$. However, in the other "crushed" system, where all the mass is located at the center of mass of the real system and the tails of all the force vectors are moved to that point, the tail of the upward-pointing force vector for F is applied to the point particle as the point particle drops a distance h. Therefore the work done by the force F on the point particle is $-Fh$. (The work done by the Earth's force is $+mgh$ for both systems, because this force is applied to the center of mass of the real system, which is also the center of mass of the crushed system.) The diagrams labeled with distances are critical to calculating the work correctly for the two different systems.

It is important to understand that the point particle or "crushed" system is a different system, with forces acting on it that are equal in magnitude and direction to the real forces acting on the real system, but applied directly to the point particle, so the displacements through which these forces act are different from the displacements of the real forces acting on the real system.

In this yo-yo example the difference between the two systems is quite striking, because the work done by the force F on the real system is positive, but the work done by the force F on the point particle system is negative. (Recall that the point particle equation was derived from the Momentum Principle; if we think about momentum, we note that the force of the hand acts to oppose the downward motion of the center of mass of the yo-yo.)

Note also that the change of the rotational kinetic energy, $K_{rot} = F(d + h)$, is the same as the work that would be done to increase the purely rotational kinetic energy of a yo-yo whose axle was fixed in space. The distance $(d + h)$ is the length of the extra string that reels off the axle.

9.6 *MODELING FRICTION IN DETAIL

We can use what we've learned about complex systems to model the nature of sliding friction, which is a complicated phenomenon. In Chapter 7 we described a seeming paradox involving friction: that when you drag a block at constant speed across a table it seems that no work is done on the block, yet its temperature increases, indicating an increase in its internal energy. With the new tools developed in this chapter, we are in a position to analyze this phenomenon in more detail and resolve the paradox.

Sliding Block: Point Particle System

A key issue that we will address is the question of how much work is done on the block by the friction force f exerted by the table. One of the new tools is the energy equation for the point particle system (Figure 9.36 and Figure 9.37). You apply a force of magnitude F to the right and the table exerts a force of magnitude f to the left. (Pretend for the moment that you don't know the magnitude of the friction force f.) The change in the kinetic energy of the fictitious center-of-mass point particle is given by the product of the net force $(F - f)$ times the displacement d of the center-of-mass point:

$$\Delta K_{trans} = (F - f)d$$

Since the speed of the block is constant, $\Delta K_{trans} = 0$, so $F - f = 0$, which shows that the friction force is equal and opposite to the force you apply: $f = F$. The thermal energy does not appear in this energy equation for the point particle system, because this equation deals merely with the motion of the (mathematical) center of mass, not with the many kinds of energy in the real system.

We also could have obtained the result $f = F$ directly from the Momentum Principle: $dp_{sys}/dt = F - f = 0$. The Momentum Principle is closely related to the energy equation for the point particle system, because both involve the net force acting on a system and the motion of the center of mass.

Sliding Block: Real System

The energy equation for the real system of the block can be written as follows, where Fd is the work done by you, and W_{fric} is the work done by the table:

$$\Delta K_{trans} + \Delta E_{thermal} = Fd + W_{fric}$$

where $\Delta E_{thermal}$ is the rise in the thermal energy of the block. We're assuming that the process takes a short enough time that there is negligible transfer of energy Q due to a temperature difference between the system of the block and the surroundings (the table), because energy transfer due to a temperature difference between table and block is a relatively slow process. (To remove such energy transfer as a possible complication, we could consider a block sliding not on a table but on an identical block. Then the symmetry of the situation is such that there cannot be any net transfer of energy due to a temperature difference into or out of the upper block. This tactic—reducing the complexity of a model—is a useful approach to complex problems.)

Since the speed of the center of mass v_{CM} does not change, $\Delta K_{trans} = 0$, and the energy equation reduces to

$$\Delta E_{thermal} = Fd + W_{fric}$$

If we conclude that the friction force does an amount of work $-Fd$, then we would have to conclude that the thermal energy of the block doesn't change, which is absurd. The block definitely gets hotter, indicating an increase in its thermal energy. We need to find a way around the entirely plausible but apparently incorrect conclusion that the friction force does an amount of work $-Fd$.

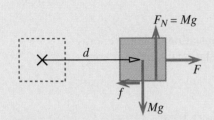

Figure 9.36 The center of mass of the block moves a distance d under the influence of a net force $F - f$.

Figure 9.37 The block considered as a point particle.

QUESTION   Considering the sign of the change of thermal energy, must the magnitude of the work done by the friction force be greater or less than Fd?

Since the thermal energy change $\Delta E_{\text{thermal}}$ is surely positive (the block gets hotter, not colder), the magnitude of the work done by the friction force must be less than Fd. Since the friction force is definitely equal to F, the friction force must act through some effective distance d_{eff} *that must be less than d*!

Evidently the energy equation for the real system has the following form, with d_{eff} less than d:

$$\Delta E_{\text{thermal}} = Fd - Fd_{\text{eff}}$$

9.7   *A PHYSICAL MODEL FOR DRY FRICTION

How can the effective distance through which the friction force does work be less than the distance through which the block moves? We can understand this by looking at a microscopic picture of what happens on the surfaces in contact.

When a metal block slides on a metal surface, the block is supported by as few as three protruding "teeth," called "asperities" in the scientific literature on friction (Figure 9.38). The very high load per unit area on these teeth makes the material partially melt and flow, and high local temperatures produced during sliding lead to adhesion (welding) in the contact regions. The frictional force divided by the tiny contact area corresponds to the large "shear" (sideways) stress required to break these welds. This shearing of contact welds is the dominant friction mechanism for a dry metal sliding on the same metal.

Because the tooth tips can become stronger than the bulk metal due to a process called "work-hardening" (which introduces dislocations in the otherwise regular geometrical arrangements of atoms), shearing often occurs in the weaker regions of the teeth, away from the tip. This is a major effect when the two objects are made of the same material, and chunks of metal can break off and embed in the other surface. Nevertheless, this wear will be ignored in the further discussion. It is in any case a symmetrical effect for identical blocks. If the metal surfaces have oxide coatings, this can reduce the shear stress required to break the temporary weld (which reduces the friction force) and can prevent the breaking off of chunks of metal, if the oxide contact area is the weakest section.

This is the model of dry friction developed in a classic treatise on friction, *The Friction and Lubrication of Solids*, Part 1 and Part II, F. P. Bowden and D. Tabor (Oxford University Press, 1950 and 1964). The physics and chemistry of friction continues to be an active field of research, because the effects of friction can be quite complex, and there is high practical interest in controlling friction. A more recent textbook is *Friction, Wear, Lubrication*, K. C. Ludema (CRC Press, 1996).

Having briefly reviewed a basic model of dry friction, we proceed to use this model to calculate the work done by frictional forces exerted at the contact points. The key issue is that the surface is deformable, which leads to differences in the energy equation for the real system compared with the energy equation for the point particle system.

Figure 9.38 A block slides on a small number of protruding teeth (vertical scale exaggerated).

Figure 9.39 Two projecting teeth of identical blocks, which have temporarily adhered to each other. (Exaggerated vertical scale.)

Actual Work Done by Friction Forces

We'll consider a microscopic picture of the contact region between two identical blocks. Figure 9.39 shows in a schematic way two teeth that have

Figure 9.40 The top block has moved a distance d to the right, but the point of contact has moved only d_{eff}.

Figure 9.41 After the temporary weld breaks, the top tooth catches up to the rest of the block. Now no force acts on the teeth. Both teeth vibrate.

Figure 9.42 A more representative picture of contact between two surfaces, with projecting teeth in contact with flat areas.

Figure 9.43 During this time interval the upper contact moves a distance d, while the lower contact does not move at all.

temporarily adhered to each other. The vertical scale has been greatly exaggerated for clarity—machined surfaces have much gentler slopes.

As the top block is dragged to the right, the teeth continue to stick together for a while. Both teeth must deform as a result; the top tooth is stretched backward, and the bottom tooth is stretched forward. In Figure 9.40 we see that when the top block has moved a distance d to the right, the point where the teeth touch each other has moved a distance of only d_{eff}, because the bottom block hasn't moved. (For two identical blocks made of the same material, on average the two teeth will bend the same amounts, and d_{eff} will be equal to $d/2$.)

This is the key issue: The point of contact where the friction force is applied moves a shorter distance (d_{eff}) than the block itself moves (d). This means that the work done by the friction force is less than one might expect.

Eventually the weld breaks, and the top tooth snaps forward, so all the atoms in the tooth now catch up with the rest of the top block, but during this part of their journey, there is no external force acting on them (Figure 9.41).

The time average of the contact forces is indeed F (as determined by the fact that the net force must be zero, $F - F$), but the effective displacement d_{eff} at the point of contact of the frictional force is less than the displacement of the center of mass of the upper block. For two identical blocks we expect to find that $d_{eff} = d/2$. For other combinations of two materials in contact, all we know is that d_{eff} must be less than d but not necessarily equal to $d/2$.

Internal Energy of the Blocks

Once the weld has broken, the teeth can vibrate, as we saw in Figure 9.41. The vibration of the tooth belonging to the top block and thermal conduction upward from the hot tip into the main body of the block contribute to the increase in the thermal energy of the upper block. Similarly, vibration and thermal conduction in the tooth belonging to the bottom block end up as increase in the thermal energy of the bottom block.

A More Physical Picture

Having two long teeth in contact is rather unphysical. A better picture is the one shown in Figure 9.42 (again with a greatly exaggerated vertical scale), where we show one of the top block's longest teeth in contact with the average surface level of the bottom block, and one of the bottom block's longest teeth similarly in contact with the average surface level of the top block. The two teeth shown here are to be taken as representative of time and space averages of the sliding friction. Take the special case of identical blocks. By symmetry, the frictional force F exerted on the top block is divided on the average into two forces each of magnitude $F/2$ at the ends of the two sets of long teeth.

In Figure 9.43 we see that when the top block moves a distance d to the right the upper contact also moves a distance d, whereas the lower contact does not move at all. The frictional work is therefore $(-F/2)(0) + (-F/2)(d)$, which is $-Fd/2 = -Fd_{eff}$, so d_{eff} is equal to $d/2$ in the special case of two identical blocks.

Summary of Dry Friction

The fundamental reason why the friction force can act through a distance less than d is that the block is deformable. All atoms in the top block eventually

Lower block held at rest

Figure 9.44 The lower block is held at rest, while the upper block slides a distance d. Forces in the vertical direction are not shown.

move the same distance d, but because of the stick/slip contact between the blocks the friction force F acts only for a portion of the displacement ($d/2$ in the special case of identical blocks). The point of contact for the friction force does not move in the same way as the center of mass. Note again that the energy equation for the point particle system is not the same as the energy equation for the real system if the system changes shape.

A Model-Independent Calculation of the Effective Distance

In the special case of a block sliding on an identical block we can calculate d_{eff}, independent of the particular model of the surfaces. Consider a system consisting of both blocks together (Figure 9.44). The bottom block is held stationary by applying a force to the left (to prevent it being dragged to the right). This constraining force acts through no distance (the bottom block stands still), so this force does no work. The only work done on the two-block system is done by you, of magnitude Fd. So the energy equation for the real two-block system is this:

$$\Delta E_{\text{thermal},1} + \Delta E_{\text{thermal},2} = Fd$$

Since the two blocks are identical, half of this increased thermal energy shows up in the upper block ($Fd/2$), and half in the lower block ($Fd/2$). We can plug this into our earlier energy equation for the upper block:

$$\Delta E_{\text{thermal},1} = \frac{Fd}{2} = Fd - Fd_{\text{eff}}$$

QUESTION   Calculate d_{eff}.

We find that the effective distance through which the friction force acts is half the distance d that the block moves: $d_{\text{eff}} = d/2$. This is in agreement with the model of dry friction with bending teeth that we examined earlier, but our result is a general one that applies to any kind of model of friction surfaces for two symmetrical blocks. Bear in mind that if the blocks are not made of the same material, d_{eff} need not be equal to $d/2$.

Lubricated Friction

It is interesting to see that the model-independent result for two identical blocks, $d_{\text{eff}} = d/2$, is also consistent with the case of lubricated friction. We separate the identical two blocks with a film of viscous lubricating oil, so that the two blocks do not make direct contact, as shown in Figure 9.45.

It is a property of simple fluid flow that fluid layers immediately adjacent to the blocks are constrained to share the motion of the blocks. Also, for a common type of flow called "laminar" flow, the displacement profile in the oil is linear. In particular, at the midplane the fluid moves half as far as the top block moves. As a result, as the top block is pulled a distance d to the right, the top layer of the oil is dragged along and moves a distance d to the right, the bottom layer of oil doesn't move, and the layer in the midplane (halfway between the blocks) moves a distance $d/2$.

If we take as the symmetrical systems of interest the top block with the upper half of the oil and the bottom block with the lower half of the oil, we see that the shear force between the two systems (at the midplane in the oil) acts through a distance that is again half the displacement of the top block: $d_{\text{eff}} = d/2$. (Of course the magnitude of the friction force is much reduced by the lubrication, and the applied force F must be much smaller if the velocity is to be constant.)

Figure 9.45 Two objects separated by a film of oil, which is modeled as a stack of fluid layers that can move relative to each other.

This discussion is based on an article "Work and heat transfer in the presence of sliding friction," by B. Sherwood and W. Bernard, *American Journal of Physics* volume 52, number 11, Nov. 1984, pages 1001–1007, which in turn draws on an earlier article "Real work and pseudowork," by B. Sherwood, *American Journal of Physics,* volume 51, number 7, July 1983, pages 597–602.

9.8   *DERIVATION: KINETIC ENERGY OF A MULTIPARTICLE SYSTEM

In this section we derive the important result that $K_{tot} = K_{trans} + K_{rel}$, where $K_{trans} = \frac{1}{2} M v_{CM}^2$ and K_{rel} is the kinetic energy relative to the center of mass. This result sounds entirely plausible, but the formal proof is rather difficult.

As in the case of calculating the gravitational energy of a multiparticle object, the derivation of the kinetic energy of a multiparticle system hinges on the definition of the center of mass point of a collection of atoms. The (vector) location of the i-th atom of the object can be expressed as the sum of two vectors, one from the origin to the center of mass ($\vec{r}_{CM}$) plus another from the center of mass to the i-th atom ($\vec{r}_{i,CM}$): $\vec{r}_i = \vec{r}_{CM} + \vec{r}_{i,CM}$ (Figure 9.46).

The kinetic energy $K_i = \frac{1}{2} m v_i^2$ of the i-th atom is this, where we write $\vec{v}_{i,CM} = d\vec{r}_{i,CM}/dt$, the velocity of the i-th particle relative to the center of mass:

$$K_i = \frac{1}{2} m_i \left| \frac{d}{dt}(\vec{r}_{CM} + \vec{r}_{i,CM}) \right|^2$$

$$= \frac{1}{2} m_i \left| \vec{v}_{CM} + \vec{v}_{i,CM} \right|^2$$

This can be expanded, using the fact that the magnitude squared of a vector can be written as a vector dot product, $|\vec{C}|^2 = \vec{C} \bullet \vec{C} = C^2 \cos(0) = C^2$:

$$K_i = \frac{1}{2} m_i [\vec{v}_{CM} + \vec{v}_{i,CM}] \bullet [\vec{v}_{CM} + \vec{v}_{i,CM}]$$

$$= \frac{1}{2} m_i v_{CM}^2 + 2 \frac{1}{2} m_i \vec{v}_{CM} \bullet \vec{v}_{i,CM} + \frac{1}{2} m_i v_{i,CM}^2$$

Now that we have the kinetic energy of the i-th atom in terms of its velocity $\vec{v}_{i,CM}$ relative to the center of mass, we need to add up the total kinetic energy of all the atoms. The total kinetic energy K_{tot} can be written in the following compact way, where Σ (Greek capital sigma) means "sum," and the sum goes from $i = 1$ through $i = N$:

$$K_{tot} = \sum_{i=1}^{N} \frac{1}{2} m_i v_{CM}^2 + \sum_{i=1}^{N} 2 \left(\frac{1}{2} m_i \right) \vec{v}_{CM} \bullet \vec{v}_{i,CM} + \sum_{i=1}^{N} \frac{1}{2} m_i v_{i,CM}^2$$

First term   **Second term**   **Third term**

First Term

The first term in this summation turns out to be the kinetic energy of a single particle of mass M, moving at the speed of the center of mass:

$$\sum_{i=1}^{N} \frac{1}{2} m_i v_{CM}^2 = \frac{1}{2} v_{CM}^2 \left(\sum_{i=1}^{N} m_i \right) = \frac{1}{2} v_{CM}^2 (m_1 + m_2 + m_3 + \cdots)$$

$$= \frac{1}{2} M v_{CM}^2$$

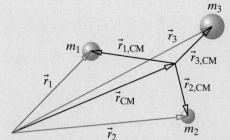

Figure 9.46 Center of mass of a multiparticle system.

Second Term

The second term can be shown to be zero:

$$\sum_{i=1}^{N} 2\left(\frac{1}{2}m_i\right) \vec{v}_{CM} \bullet \vec{v}_{i,CM} = \vec{v}_{CM} \bullet \sum_{i=1}^{N} m_i \vec{v}_{i,CM} = \vec{v}_{CM} \bullet \frac{d}{dt}\left(\sum_{i=1}^{N} m_i \vec{r}_{i,CM}\right) = 0$$

This result follows from the way that we calculate the location of the center of mass of a system:

$$\vec{r}_{CM} = \frac{m_1 \vec{r}_1 + m_2 \vec{r}_2 + m_3 \vec{r}_3 + \cdots}{m_1 + m_2 + m_3 + \cdots} = \frac{\displaystyle\sum_{i=1}^{N} m_i \vec{r}_i}{M}$$

We are measuring the location $\vec{r}_{i,CM}$ of the i-th atom relative to the center of mass, and since the distance from the center of mass to the center of mass is of course zero, we have

$$\vec{0} = \frac{\displaystyle\sum_{i=1}^{N} m_i \vec{r}_{i,CM}}{M} \quad \text{and therefore} \quad \sum_{i=1}^{N} m_i \vec{r}_{i,CM} = \vec{0}$$

Third Term

The third term by definition is the kinetic energy of the atoms relative to the (possibly moving) center of mass:

$$\sum_{i=1}^{N} \frac{1}{2} m_i v_{i,CM}^2 = \frac{1}{2} m_1 v_{1,CM}^2 + \frac{1}{2} m_2 v_{2,CM}^2 + \cdots = K_{rel}$$

Putting the three pieces together, we find that the total kinetic energy splits into two parts: a term associated with the overall motion of the center of mass plus the kinetic energy relative to the center of mass:

$$K_{tot} = K_{trans} + K_{rel}$$

where $K_{trans} = \frac{1}{2} M v_{CM}^2 = \dfrac{p_{sys}^2}{2M}$.

9.9   *DERIVATION: THE POINT PARTICLE ENERGY EQUATION

In this chapter we showed that the energy equation for the point particle version of the system follows from the fact that the motion of the center of mass is just like that of a point particle with the total mass of the real system and subjected to the *net* force acting on the real system. Here we give a more formal derivation of this important result.

Start from the x component of the Momentum Principle for a multiparticle system whose center of mass is moving at nonrelativistic speed:

$$\frac{dp_{sys,x}}{dt} = M_{tot} \frac{dv_{CM,x}}{dt} = F_{net,x}$$

Integrate through the x displacement of the center of mass:

$$M_{tot} \int_{i}^{f} \frac{dv_{CM,x}}{dt} \, dx_{CM} = \int_{i}^{f} F_{net,x} \, dx_{CM}$$

Switch dv and dx:

$$M_{tot} \int_i^f \frac{dx_{CM}}{dt} dv_{CM,x} = \int_i^f F_{net,x} \, dx_{CM}$$

However, dx_{CM}/dt is the x component of the center of mass velocity:

$$M_{tot} \int_i^f v_{CM,x} \, dv_{CM,x} = \int_i^f F_{net,x} \, dx_{CM}$$

The integral on the left can be carried out:

$$M_{tot} \left[\frac{1}{2} v_{CM,x}^2 \right]_i^f = \int_i^f F_{net,x} \, dx_{CM}$$

$$\Delta \left[\frac{1}{2} M_{tot} v_{CM,x}^2 \right] = \int_i^f F_{net,x} \, dx_{CM}$$

We could repeat exactly the same argument for the y and z motions:

$$\Delta \left[\frac{1}{2} M_{tot} v_{CM,y}^2 \right] = \int_i^f F_{net,y} \, dy_{CM}$$

$$\Delta \left[\frac{1}{2} M_{tot} v_{CM,z}^2 \right] = \int_i^f F_{net,z} \, dz_{CM}$$

Note that

$$\frac{1}{2} M_{tot} v_{CM,x}^2 + \frac{1}{2} M_{tot} v_{CM,y}^2 + \frac{1}{2} M_{tot} v_{CM,z}^2 = \frac{1}{2} M_{tot} v_{CM}^2$$

Moreover,

$$F_{net,x} \, dx_{CM} + F_{net,y} \, dy_{CM} + F_{net,z} \, dz_{CM} = \vec{F}_{net} \bullet d\vec{r}_{CM}$$

Adding the three equations together (the sum of the left sides is equal to the sum of the right sides), we have the following equation for the translational kinetic energy of the system:

$$\Delta \left[\frac{1}{2} M_{tot} v_{CM}^2 \right] = \int_i^f \vec{F}_{net} \bullet d\vec{r}_{CM}$$

In words: the change in the translational kinetic energy of a system is equal to the integral of the *net* force acting through the displacement of the center of mass point.

The derivation that we have just carried out shows that although this equation looks like an energy equation, it is actually closely related to the Momentum Principle from which it was derived. The common element is the *net* force.

In contrast, the actual energy equation for the real system involves the work done by each individual force through the displacement of the point of application of that force. If the system deforms or rotates, these displacements of the individual forces need not be the same as the displacement of the center of mass.

SUMMARY

Center of Mass

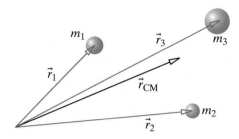

Figure 9.47

$$\vec{r}_{CM} \equiv \frac{m_1\vec{r}_1 + m_2\vec{r}_2 + m_3\vec{r}_3 + \cdots}{m_1 + m_2 + m_3 + \cdots}$$

Problem-Solving Techniques

Combined use of the energy equations for the real system and for the fictitious "point particle system" in analyzing complex phenomena involving multiparticle systems.

> Point particle system: Information on change of translational kinetic energy
> Real system: Information on change of total energy

The Momentum Principle for multiparticle systems, derived from Newton's second and third laws of motion, was extended by defining the center of mass:

$$\frac{d\vec{p}_{sys}}{dt} = \vec{F}_{net}$$

$\vec{p}_{sys} \approx M\vec{v}_{CM}$ (if $v \ll c$, and constant mass).

Gravitational energy of multiparticle systems plus Earth, near the Earth's surface:

$$U_g = Mgy_{CM}$$

Kinetic energy of multiparticle systems:

$$K_{tot} = K_{trans} + K_{rel}$$

where $K_{trans} = \dfrac{p_{sys}^2}{2M} = \dfrac{1}{2}Mv_{CM}^2$.

Kinetic energy relative to center of mass can be split into two terms:

$$K_{rel} = K_{rot} + K_{vib}$$

Moment of inertia about an axis of rotation:

$$I = m_1r_{\perp1}^2 + m_2r_{\perp2}^2 + m_3r_{\perp3}^2 + \cdots$$

Kinetic energy relative to center of mass:

$$K_{rot} = \frac{1}{2}I\omega^2$$

Moments of inertia:

$$I_{disk} = I_{cylinder} = \frac{1}{2}MR^2$$

about center of disk, rotating around axis of cylinder

$$I_{sphere} = \frac{2}{5}MR^2$$

for axis passing through center of sphere.

Uniform solid cylinder of length L, radius R, about axis perpendicular to cylinder, through center of cylinder:

$$I = \frac{1}{12}ML^2 + \frac{1}{4}MR^2$$

For a rigid object, $K_{tot} = K_{trans} + K_{rot} = \frac{1}{2}(Mr_{CM}^2 + I_{CM})\omega^2$
Point particle system

$$K_{trans} = \Delta\left(\frac{1}{2}Mv_{CM}^2\right) = \int_i^f \vec{F}_{net,ext} \bullet d\vec{r}_{CM}$$

Sliding friction can deform the contact surfaces, with the result that the frictional force may act through a distance that is different from the distance through which the center-of-mass point moves. As with any deformable system, the energy equations for the real system and for the point particle system differ (though both are correct).

EXERCISES AND PROBLEMS

Sections 9.1–9.3

9.X.13 Discuss qualitatively the motion of the atoms in a block of steel that falls onto another steel block. Why and how do large-scale vibrations damp out?

9.X.14 Can you give an example of a system that has no atoms located at its center of mass?

9.X.15 Determine the location of the center of mass of an L-shaped object whose thin vertical and horizontal members have the same length L and the same mass M. Use the formal definition to find the x and y coordinates, and check your result by doing the calculation with respect to two different origins,

one in the lower left corner at the intersection of the horizontal and vertical members and one at the right end of the horizontal member.

9.X.16 Two people with different masses but equal speeds slide toward each other with little friction on ice with their arms extended straight out to the side (so each has the shape of a "T"). Her right hand meets his right hand, they hold hands and spin 90°, then release their holds and slide away. Make a rough sketch of the path of the center of mass of the system consisting of the two people, and explain briefly. (It helps to mark equal time intervals along the paths of the two people and of their center of mass.)

9.X.17 The Momentum Principle for multiparticle systems would seem to say that the center of mass of a system moves just as though the system were a point particle. This led Chris to ask, "Wouldn't that mean that a piece of paper ought to fall in the same way as a small metal ball, if they have the same mass?" Explain carefully to Chris the resolution of this puzzle.

9.X.18 Consider the voyage to the Moon that you studied in Chapter 3. Would it make any difference, even a very tiny difference, whether the spacecraft is long or short, if the mass is the same? Explain briefly.

9.X.19 Consider a system consisting of three particles:

$m_1 = 2$ kg, $\vec{v}_1 = \langle 8, -6, 15 \rangle$ m/s

$m_2 = 6$ kg, $\vec{v}_2 = \langle -12, 9, -6 \rangle$ m/s

$m_3 = 4$ kg, $\vec{v}_3 = \langle -24, 34, 23 \rangle$ m/s

(a) What is the total momentum of this system?

(b) What is the velocity of the center of mass of this system?

(c) What is the total kinetic energy of this system?

(d) What is the translational kinetic energy of this system?

(e) What is the kinetic energy of this system relative to the center of mass?

One way to calculate K_{rel} is to calculate the velocity of each particle relative to the center of mass, by subtracting the center-of-mass velocity from the particle's actual velocity to get the particle's velocity relative to the center of mass, then calculating the corresponding kinetic energy, then adding up the three relative kinetic energies. However, there is a much simpler way to determine the specified quantity, without having to do all those calculations; think about what you know about the relationships among the various kinds of kinetic energy in a multiparticle system. (If you wish, you can check your result by doing the complicated calculation just described.)

9.P.20 A man whose mass is 80 kg and a woman whose mass is 50 kg sit at opposite ends of a canoe 5 m long, whose mass is 30 kg.

(a) Relative to the man, where is the center of mass of the system consisting of man, woman, and canoe? (*Hint:* Choose a specific coordinate system with a specific origin.)

(b) Suppose that the man moves quickly to the center of the canoe and sits down there. How far does the canoe move in the water? Explain your work and your assumptions.

9.P.21 Binary stars:

(a) About half of the visible "stars" are actually binary star systems, two stars that orbit each other with no other objects nearby. Describe the motion of the center of mass of a binary star system. Briefly explain your reasoning.

(b) For a particular binary star system, telescopic observations repeated over many years show that one of the stars (whose unknown mass we'll call M_1) has a circular orbit with radius $R_1 = 6 \times 10^{11}$ m, while the other star (whose unknown mass we'll call M_2) has a circular orbit of radius $R_2 = 9 \times 10^{11}$ m about the same point. Make a sketch of the orbits, and show the positions of the two stars on these orbits at some instant. Label the two stars as to which is which, and label their orbital radii. Indicate on your sketch the location of the center of mass of the system, and explain how you know its location, using the concepts and results of this chapter.

(c) This double star system is observed to complete one revolution in 40 years. What are the masses of the two stars? (For comparison, the distance from Sun to Earth is about 1.5×10^{11} m, and the mass of the Sun is about 2×10^{30} kg.) This method is often used to determine the masses of stars. The mass of a star largely determines many of the other properties of a star, which is why astrophysicists need a method for measuring the mass.

9.P.22 By calculating numerical quantities for a multiparticle system, one can get a concrete sense of the meaning of the relationships $\vec{p}_{sys} = M_{tot}\vec{v}_{CM}$ and $K_{tot} = K_{trans} + K_{rel}$. Consider an object consisting of two balls connected by a spring, whose stiffness is 400 N/m. The object has been thrown through the air and is rotating and vibrating as it moves. At a particular instant the spring is stretched 0.3 m, and the two balls at the ends of the spring have the following masses and velocities:

1: 5 kg, $\langle 8, 14, 0 \rangle$ m/s

2: 3 kg, $\langle -5, 9, 0 \rangle$ m/s

(a) For this system, calculate $\vec{p}_{sys}$.

(b) Calculate $\vec{v}_{CM}$.

(c) Calculate K_{tot}.

(d) Calculate K_{trans}.

(e) Calculate K_{rel}.

(f) Here is a way to check your result for K_{rel}. The velocity of a particle relative to the center of mass is calculated by subtracting $\vec{v}_{CM}$ from the particle's velocity. To take a simple example, if you're riding in a car that's moving with $v_{CM,x} = 20$ m/s, and you throw a ball with $v_{rel,x} = 35$ m/s, relative to the car, a bystander on the ground sees the ball moving with $v_x = 55$ m/s. So $\vec{v} = \vec{v}_{CM} + \vec{v}_{rel}$, and therefore we have $\vec{v}_{rel} = \vec{v} - \vec{v}_{CM}$. Calculate $\vec{v}_{rel} = \vec{v} - \vec{v}_{CM}$ for each mass and calculate the corresponding K_{rel}. Compare with the result you obtained in part (e).

Section 9.4

9.X.23 If an object's rotational kinetic energy is 50 J and it rotates with an angular speed of 12 radians/s, what is its moment of inertia?

9.X.24 A uniform-density disk of mass 13 kg, thickness 0.5 m, and radius 0.2 m makes one complete rotation every 0.6 s. What is the rotational kinetic energy of the disk?

9.X.25 A sphere of uniform density with mass 22 kg and radius 0.7 m is spinning, making one complete revolution every 0.5 seconds. The center of mass of the sphere has a speed of 4 m/s.

(a) What is the rotational kinetic energy of the sphere?

(b) What is the total kinetic energy of the sphere?

9.X.26 A cylindrical rod of uniform density is located with its center at the origin, and its axis along the x axis. It rotates about its center in the xy plane, making one revolution every 0.03 seconds. The rod has a radius of 0.08 m, length of 0.7 m, and mass of 5 kg. It makes one revolution every 0.03 seconds. What is the rotational kinetic energy of the rod?

9.X.27 A uniform-density 6 kg disk of radius 0.3 m is mounted on a nearly frictionless axle. Initially it is not spinning. A string is wrapped tightly around the disk, and you pull on the string with a constant force of 25 N through a distance of 0.6 m. Now what is the angular speed?

9.P.28 The Earth is 1.5×10^{11} m from the Sun and takes a year to make one complete orbit. It rotates on its own axis once per day. It can be treated approximately as a uniform-density sphere of mass 6×10^{24} kg and radius 6.4×10^6 m (actually, its center has higher density than the rest of the planet, and the Earth bulges out a bit at the equator). Using this crude approximation, calculate the following:

(a) What is v_{CM}?

(b) What is K_{trans}?

(c) What is ω, the angular speed of rotation around its own axis?

(d) What is K_{rot}?

(e) What is K_{tot}?

9.P.29 Show that the moment of inertia of a disk of mass M and radius R is $\frac{1}{2}MR^2$. Divide the disk into narrow rings, each of radius r and width dr. The contribution to I by one of these rings is simply $r^2 dm$, where dm is the amount of mass contained in that particular ring. The mass of any ring is the total mass times the fraction of the total area occupied by the area of the ring. The area of this ring is approximately $2\pi r\, dr$. Use integral calculus to add up all the contributions.

Sections 9.5–9.6

9.X.30 Under what conditions does the energy equation for the point particle system differ from the energy equation for the real system? Give two examples of such a situation. Give one example of a situation where the two equations look exactly alike.

9.X.31 Consider the acceleration of a car on dry pavement, if there is no slipping. The axle moves at speed v, and the outside of the tire moves at speed v relative to the axle. The instantaneous velocity of the bottom of the tire is zero. How much work is done by the force exerted on the tire by the road? What is the source of the energy that increases the car's translational kinetic energy?

9.X.32 You pull straight up on the string of a yo-yo with a force 0.235 N, and while your hand is moving up a distance 0.18 m, the yo-yo moves down a distance 0.70 m. The mass of the yo-yo is 0.025 kg, and it was initially moving downward with speed 0.5 m/s and angular speed 124 radians/s.

(a) What is the increase in the translational kinetic energy of the yo-yo?

(b) What is the new speed of the yo-yo?

(c) What is the increase in the rotational kinetic energy of the yo-yo?

(d) The yo-yo is approximately a uniform-density disk of radius 0.02 m. What is the new angular speed of the yo-yo?

9.P.33 A string is wrapped around a disk of mass 2.1 kg (its density is not necessarily uniform). Starting from rest, you pull the string with a constant force of 9 N along a nearly frictionless surface. At the instant when the center of the disk has moved a distance 0.11 m, your hand has moved a distance of 0.28 m (Figure 9.48).

Figure 9.48

(a) At this instant, what is the speed of the center of mass of the disk?

(b) At this instant, how much rotational kinetic energy does the disk have relative to its center of mass?

(c) At this instant, the angular speed of the disk is 7.5 radians/s. What is the moment of inertia of the disk?

9.P.34 A chain of metal links with total mass $M = 7$ kg is coiled up in a tight ball on a low-friction table (Figure 9.49). You pull on a link at one end of the chain with a constant force $F = 50$ N. Eventually the chain straightens out to its full length $L = 2.6$ m, and you keep pulling until you have pulled your end of the chain a total distance $d = 4.5$ m.

Figure 9.49

(a) Consider the point particle system. What is the speed of the chain at this instant?

(b) Consider the real system. What is the change in energy of the chain?

(c) In straightening out, the links of the chain bang against each other, and their temperature rises. Assume that the process is so fast that there is insufficient time for significant transfer of energy from the chain to the table due to the temperature difference, and ignore the small amount of energy radiated away as sound produced in the collisions among the links. Calculate the increase in internal energy of the chain.

9.P.35 Tarzan, whose mass is 100 kg, is hanging at rest from a tree limb. Then he lets go and falls to the ground. Just before he lets go, his center of mass is at a height 2.9 m above the ground and the bottom of his dangling feet are at a height 2.1 above the ground. When he first hits the ground he has dropped a distance 2.1, so his center of mass is $(2.9 - 2.1)$ above the ground. Then his knees bend and he ends up at rest in a crouched position with his center of mass a height 0.5 above the ground.

(a) Consider the point particle system. What is the speed v at the instant just before Tarzan's feet touch the ground?

(b) Consider the real system. What is the net change in internal energy for Tarzan from just before his feet touch the ground to when he is in the crouched position?

9.P.36 Here is an experiment on jumping up you can do.

(a) Crouch down and jump straight up, as high as you can. Estimate the location of your center of mass, and measure its height at three stages in this process: in the crouch, at lift-off, and at the top of the jump. Report your measurements. You may need to have a friend help you make the measurements.

(b) Analyze this process as fully as possible, using all the theoretical tools now available to you, especially the concepts in this chapter. Include a calculation of the average force of the floor on your feet, the change in your internal energy, and the approximate time of contact from the beginning of the jump to lift-off. Be sure to explain clearly what approximations and simplifying assumptions you made in modeling the process.

9.P.37 In Section 9.1 we discussed a woman ice skater who pushes away from a wall.

  (a) Estimate the speed she can achieve just after pushing away from the wall. Then estimate the average acceleration during this process. How many g's is this? (That is, what fraction or multiple of 9.8 m/s² is your estimate?) Be sure to explain clearly what approximations and simplifying assumptions you made in modeling the process.

  (b) For this process, choose the woman as the system of interest and discuss the energy transfers, and the changes in the various forms of energy. Estimate the amount of each of these, including the correct signs.

9.P.38 A hoop of mass M and radius R rolls without slipping down a hill, as shown in Figure 9.50. The lack of slipping means that when the center of mass of the hoop has speed v, the tangential speed of the hoop relative to the center of mass is also equal to v_{CM}, since in that case the instantaneous speed is zero for the part of the hoop that is in contact with the ground ($v - v = 0$). Therefore, the angular speed of the rotating hoop is $\omega = v_{CM}/R$.

v of rim
relative to center of mass

v of
center of mass

Figure 9.50

  (a) The initial speed of the hoop is v_i, and the hill has a height h. What is the speed v_f at the bottom of the hill?

  (b) Replace the hoop with a bicycle wheel whose rim has mass M and whose hub has mass m, as shown in Figure 9.51. The spokes have negligible mass. What would the bicycle wheel's speed be at the bottom of the hill?

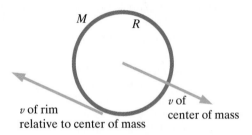

Figure 9.51

9.P.39 A sphere or cylinder of mass M, radius R, and moment of inertia I rolls without slipping down a hill of height h, starting from rest. As explained in Problem 9.P.38, if there is no slipping $\omega = v_{CM}/R$.

  (a) In terms of the given variables (M, R, I, and h), what is v_{CM} at the bottom of the hill?

  (b) If the object is a thin hollow cylinder, what is v_{CM} at the bottom of the hill?

  (c) If the object is a uniform-density solid cylinder, what is v_{CM} at the bottom of the hill?

  (d) If the object is a uniform-density sphere, what is v_{CM} at the bottom of the hill?

An interesting experiment that you can perform is to roll various objects down an inclined board and see how much time each one takes to reach the bottom.

9.P.40 Two disks are initially at rest, each of mass M, connected by a string between their centers, as shown in Figure 9.52. The disks slide on low-friction ice as the center of the string is pulled by a string with a constant force F through a distance d. The disks collide and stick together, having moved a distance b horizontally.

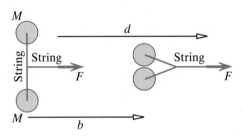

Figure 9.52

  (a) What is the final speed of the stuck-together disks?

  (b) When the disks collide and stick together, their temperature rises. Calculate the increase in internal energy of the disks, assuming that the process is so fast that there is insufficient time for there to be much transfer of energy to the ice due to a temperature difference. (Also ignore the small amount of energy radiated away as sound produced in the collisions between the disks.)

9.P.41 You hang by your hands from a tree limb that is a height L above the ground, with your center of mass a height h above the ground and your feet a height d above the ground, as shown in Figure 9.53. You then let yourself fall. You absorb the shock by bending your knees, ending up momentarily at rest in a crouched position with your center of mass a height b above the ground. Your mass is M. You will need to draw labeled physics diagrams for the various stages in the process.

Figure 9.53

(a) What is the net internal energy change ΔE_{int} in your body (chemical plus thermal)?

(b) What is your speed v at the instant when your feet first touch the ground?

(c) What is the approximate average force F exerted by the ground on your feet during the time when your knees are bending?

(d) How much work is done by this force F?

9.P.42 A box and its contents have a total mass M. A string passes through a hole in the box (Figure 9.54), and you pull on the string with a constant force F (this is in outer space—there are no other forces acting).

Figure 9.54

(a) Initially the speed of the box was v_i. After the box had moved a long distance w, your hand had moved an additional distance d (a total distance of $w + d$), because additional string of length d came out of the box. What is now the speed v_f of the box?

(b) If we could have looked inside the box, we would have seen that the string was wound around a hub that turns on an axle with negligible friction, as shown in Figure 9.55. Three masses, each of mass m, are attached to the hub at a distance r from the axle. Initially the angular speed relative to the axle was ω_i. In terms of the given quantities, what is the final angular speed relative to the axis, ω_f?

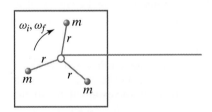

Figure 9.55

9.P.43 Two identical 0.4 kg blocks (labeled 1 and 2) are initially at rest on a nearly frictionless surface, connected by an unstretched spring, as shown in the upper portion of Figure 9.56.

Figure 9.56

Then a constant force of 100 N to the right is applied to block 2, and at a later time the blocks are in the new positions shown in the lower portion of Figure 9.56. At this final time, the system is moving to the right and also vibrating, and the spring is stretched.

MOTION OF THE CENTER OF MASS OF THE REAL SYSTEM

(a) What is the initial location of the center of mass of the real system?

(b) What is the final location of the center of mass of the real system?

(c) How far did the center of mass of the real system move?

POINT PARTICLE SYSTEM

(d) What is the initial location of the point particle system?

(e) What is the final location of the point particle system?

(f) How far did the point particle system move?

(g) In the point particle system, what was the distance through which the point of application of the force moved?

(h) How much work was done on the point particle system?

(i) List all the forms of energy that change in the point particle system during this process.

(j) What was the change in the translational kinetic energy of the point particle system?

(k) What is the final translational kinetic energy of the point particle system?

(l) What is the mass of the point particle system?

(m) What is the final speed of the point particle system?

REAL SYSTEM

(n) The force was applied to a location on the right-hand edge of block 2. Look carefully at the diagrams. In the real system, what was the distance through which the point of application of the force moved?

(o) How much work was done on the real system?

(p) List all the forms of energy that change in the real system during this process (assume that the vibrating spring doesn't get hot).

(q) What is the change in energy of the real system?

(r) What is the final energy for the real system?

(s) What is the final translational kinetic energy of the real system?

(t) What is the final speed of the center of mass of the real system?

(u) What is the final vibrational energy of the real system (spring potential energy plus kinetic energy relative to the center of mass)?

(v) The spring stiffness is 50 N/m. What is the final spring potential energy of the real system?

(w) What is the final vibrational kinetic energy of the real system?

9.P.44 You hold up an object that consists of two blocks at rest, each of mass $M = 5$ kg, connected by a low-mass spring. Then you suddenly start applying a larger upward force of constant magnitude $F = 167$ N (which is greater than $2Mg$). Figure 9.57 shows the situation some time later, when the blocks have moved upward, and the spring stretch has increased.

$F > 2Mg$

$F = 2Mg$
initially

M

M

y_{2i}

y_{2f}

y_{1i}

y_{1f}

Floor

Figure 9.57

The heights of the centers of the two blocks are as follows:

Initial and final positions of block 1: $y_{1i} = 0.3$ m, $y_{1f} = 0.5$ m

Initial and final positions of block 2: $y_{2i} = 0.7$ m, $y_{2f} = 1.2$ m

It helps to show these heights on a diagram. Note that the initial center of mass of the two blocks is $(y_{1i} + y_{2i})/2$, and the final center of mass of the two blocks is $(y_{1f} + y_{2f})/2$.

(a) Consider the point particle system corresponding to the two blocks and the spring. Calculate the increase in the total translational kinetic energy of the two blocks. It is important to draw a diagram showing all of the forces that are acting, and through what distance each force acts.

(b) Consider the real system corresponding to the two blocks and the spring. Calculate the increase of $(K_{vib} + U_s)$, the vibrational kinetic energy of the two blocks (their kinetic energy relative to the center of mass) plus the potential energy of the spring. It is important to draw a diagram showing all of the forces that are acting, and through what distance each force acts.

9.P.45 A box contains machinery that can rotate. The total mass of the box plus the machinery is 7 kg. A string wound around the machinery comes out through a small hole in the top of the box. Initially the box sits on the ground, and the machinery inside is not rotating (left side of Figure 9.58). Then you pull upward on the string with a force of constant magnitude 130 N. At an instant when you have pulled 0.6 m of string out of the box (indicated on the right side of Figure 9.58), the box has risen a distance of 0.2 m and the machinery inside is rotating.

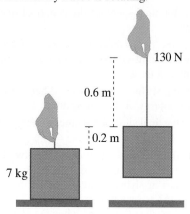

130 N

0.6 m

0.2 m

7 kg

Figure 9.58

POINT PARTICLE SYSTEM

(a) List all the forms of energy that change for the point particle system during this process.

(b) What is the y component of the displacement of the point particle system during this process?

(c) What is the y component of the net force acting on the point particle system during this process?

(d) What is the distance through which the net force acts on the point particle system?

(e) How much work is done on the point particle system during this process?

(f) What is the speed of the box at the instant shown in the right side of Figure 9.58?

(g) Why is it not possible to find the rotational kinetic energy of the machinery inside the box by considering only the point particle system?

REAL SYSTEM

(h) The real system consists of the box, the machinery inside the box, and the string. List all the forms of energy that change for the real system during this process.

(i) What is the translational kinetic energy of the real system, at the instant shown in the right side of Figure 9.58?

(j) What is the distance through which the gravitational force acts on the real system?

(k) How much work is done on the system by the gravitational force?

(l) What is the distance through which your hand moves?

(m) How much work do you do on the real system?

(n) At the instant shown in the right side of Figure 9.58, what is the total kinetic energy of the real system?

(o) What is the rotational kinetic energy of the machinery inside the box?

Sections 9.6–9.7

9.P.46 It is sometimes claimed that friction forces always slow an object down, but this is not true. If you place a box of mass M on a moving horizontal conveyor belt, the friction force of the belt acting on the bottom of the box speeds up the box. At first there is some slipping, until the speed of the box catches up to the speed v of the belt. The coefficient of friction between box and belt is μ.

(a) What is the distance d (relative to the floor) that the box moves before reaching the final speed v? Use energy arguments, and explain your reasoning carefully.

(b) How much time does it take for the box to reach its final speed?

(c) The belt and box of course get hot. Is the effective distance through which the friction force acts on the box greater than or less than d? Give as quantitative an argument as possible. You can assume that the process is quick enough that you can neglect transfer of energy Q due to a temperature difference between the belt and the box. Do not attempt to use the *results* of the friction analysis in this chapter; rather, apply the *methods* of that analysis to this different situation.

(d) Explain the result of part (c) qualitatively from a microscopic point of view, including physics diagrams.

ANSWERS TO EXERCISES

9.X.1 $\langle 0.57, 3.21, 5.64 \rangle$ m

9.X.2 4.6×10^6 m = 4600 km (so it is inside the Earth)

9.X.3 The puck with the string wound around it is rotating and so has additional kinetic energy. You have to pull farther, by the amount of string that unwinds, so you do more work, which corresponds to this puck having more kinetic energy. The Momentum Principle for the motion of the center of mass tells us nothing about the rotational aspects of the motion.

9.X.4 $6 \, \text{kg} \cdot \text{m/s/s}$ to the right; $20 \, \text{m/s}^2$

9.X.5 4.7×10^4 J

9.X.6 39.2 J

9.X.7 **(a)** 61.3 J; **(b)** 99.7 J

9.X.8 395 J

9.X.9 316 J

9.X.10 **(a)** v/d; **(b)** $\frac{1}{2}(\frac{2}{5}MR^2)(v/d)^2$;
(c) $\frac{1}{2}(Md^2 + \frac{2}{5}MR^2)(v/d)^2$

9.X.11 All of the applied forces act through the same displacement, which is also the displacement of the center of mass. So it makes no difference whether you calculate the work done by each individual force and then add them up, or first add up all the forces and then calculate the work. Also, the only energy change in such a system is the overall kinetic energy.

9.X.12 **(a)** 167 N; **(b)** zero displacement, so no work done on the real system, but on the point particle system there is 2500 J of work done; **(c)** internal energy (especially chemical energy) decreases by 2500 J

KEY IDEAS

- Collisions are brief interactions involving large forces.
- To analyze collisions in detail, we must apply both the Energy Principle and the Momentum Principle.
- In analyzing collisions it is useful to choose all colliding objects as the system. With this choice of system:
  - External forces are negligible during the collision.
  - The total momentum of the system is constant.
  - The total energy of the system is constant.
- Collisions in which there is no change in internal energy are called "elastic." Other collisions are "inelastic."
- Changing to a reference frame moving with the center of mass of the system simplifies the analysis of a collision.

10.1 INTERNAL INTERACTIONS IN COLLISIONS

DEFINITION OF COLLISION

A collision is an interaction involving large forces acting for a short time.

Collisions are common events. Neutral molecules in a gas frequently collide with each other—the interatomic electric force is effectively short range, so there is little interaction until the molecules come nearly into contact. Similarly, collisions between macroscopic objects are quite common: before and after a ball hits a wall it has little interaction with the wall, and the contact between ball and wall lasts a very short time.

- The objects need not come into actual contact during the collision, if they interact via long-range forces such as the gravitational force or the electric force.

Choice of System

In analyzing a collision, we typically choose both colliding objects as the system. With this choice of system, the large forces between the colliding objects are internal forces. Because these internal forces are very large—much larger than forces due to objects in the surroundings—the effect of external forces is negligible during the collision. Since a collision occurs during a brief time, energy flow due to any temperature differences between system and surroundings is negligible also. Neglecting external interactions allows us to apply both the Momentum Principle and the Energy Principle in their strongest forms: Conservation of Momentum and Conservation of Energy. Both the total momentum and the total energy of the system must remain constant during a collision, although both momentum and energy can flow between the objects in the system.

$$\Delta \vec{p}_{\text{sys}} + \cancel{\Delta \vec{p}_{\text{surr}}} = 0$$
$$\Delta \vec{p}_{\text{sys}} = 0$$
$$\vec{p}_f = \vec{p}_i$$

$$\Delta E_{\text{sys}} + \cancel{\Delta E_{\text{surr}}} = 0$$
$$\Delta E_{\text{sys}} = 0$$
$$E_f = E_i$$

In Section 2.7 we examined the forces and interaction times involved in two collisions. When a hockey stick struck a hockey puck hard enough that the stick broke, the force exerted by the stick on the puck was about 1000 N, and the interaction lasted only about 0.01 second. (For comparison, the gravitational force by the Earth on a 160 g hockey puck is only about 1.6 N.) We also analyzed a head-on collision between two running students; we found the interaction time to be about 0.005 s, and the force exerted by one student on the other to be about 21000 N—35 times the weight of one student. Such brief, intense interactions are characteristic of collisions.

These characteristics justify an approach to the analysis of collisions that proves fruitful. Up until the collision, and after the collision, we do need to take into account external forces, which affect the total momentum of the system. For example, consider the collision of two baseballs in midair, shown in Figure 10.1. We can tell from the curving trajectories of the balls that the gravitational force by the Earth had a significant effect on the balls before and after the collision. However, during the very short time of the collision, the Earth's gravitational force had a negligible effect, and it is reasonable to say that the total momentum of the system was constant during the impact.

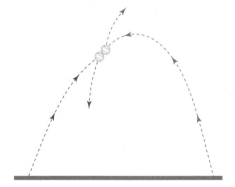

Figure 10.1 Midair collision between two baseballs.

Figure 10.2 Electric or gravitational interactions without contact; these too are considered to be "collisions."

"Collisions" without Contact

Consider the electric interaction of a proton and electron, or the gravitational interaction of two asteroids, as shown in Figure 10.2. They started out far from each other. There is no contact—no touching—yet we call such an interaction a "collision." A collision is any process in which there is little interaction before and after a short time interval, and large interactions during that short time interval. Since electric and gravitational forces depend on $1/r^2$, the forces are small when the objects are far apart.

10.2 ELASTIC AND INELASTIC COLLISIONS

Internal Energy

Often we want to focus on the changes in translational kinetic energy that occur during a collision between two objects. We will use the term "internal energy" to refer to the other kinds of energy objects may have—for example, electronic energy, vibrational energy, rotational energy, or the randomized thermal energy of a solid object whose atoms and chemical bonds are vibrating.

Elastic Collisions

We call a collision "elastic" if there is no change in internal energy of the interacting objects: no thermal energy rise (associated with a temperature increase), no springs newly compressed, no lasting deformations, no new rotations or vibrations, and so on.

ELASTIC COLLISION

In an "elastic" collision, the internal energy of the objects in the system does not change: $\Delta E_{int} = 0$. Since no kinetic energy is converted to E_{int}, in an elastic collision $K_f = K_i$.

When atomic systems with quantized energies collide, the collision will be perfectly elastic if there is insufficient energy available to raise the systems to excited quantum states (Figure 10.3). For example, as discussed in Chapter 8, the energy of the first excited electronic state of a mercury atom is 4.9 eV above the ground state. If an electron whose kinetic energy is only 3 eV collides with a ground-state mercury atom, the collision will be elastic, because there is not enough energy available to change the internal energy of the mercury atom.

With macroscopic systems there are no perfectly elastic collisions, because there is always some dissipation, but many macroscopic collisions are very nearly elastic. For example, collisions between billiard balls or steel balls can be nearly elastic.

Carts used to demonstrate collisions in physics lectures often contain magnets to repel each other without actually making contact, and such collisions are nearly elastic. Collisions between carts that interact through soft springs may also be nearly elastic.

Inelastic Collisions

We call a collision "inelastic" if it isn't elastic; that is, there is some change in the internal energy of the colliding objects. They get hot, or deform, or rotate, or vibrate, and so on.

INELASTIC COLLISION

In an "inelastic" collision, the internal energy of the objects in the system changes: $\Delta E_{int} \neq 0$. Since some kinetic energy is converted to internal energy, in an inelastic collision $K_f \neq K_i$.

As stated above, collisions of macroscopic objects are always at least a bit inelastic, but in many common situations the collision may be nearly elastic. When atomic systems collide, one or more of the colliding objects may undergo a change of quantized energy level, and we call that an inelastic collision. For example, if an electron with kinetic energy 5 eV collides with a mercury atom in its ground state, the mercury atom may be excited to a higher electronic state 4.9 eV above the ground state, leaving only 0.1 eV of kinetic energy in the system.

Maximally Inelastic Collisions ("Sticking Collisions")

The most extreme inelastic collision is the "maximally inelastic" collision, in which there is maximum dissipation. That doesn't necessarily mean that the objects stop dead, because momentum must be conserved. It can be shown that a collision in which the objects stick together is a maximally inelastic collision: the only remaining kinetic energy is present only because the total momentum can't change.

Figure 10.3 Energy levels in an atomic system. If the energy input is too small to reach an excited state, the energy cannot be absorbed.

The simplest example of a maximally inelastic collision is that of two objects that have equal and opposite momenta, so that before the collision the total momentum of the combined system is zero. The momentum afterward must also be zero, and if they are stuck together that means that they must be at rest.

MAXIMALLY INELASTIC COLLISION

In a "maximally inelastic" collision, there is maximum dissipation, and the objects stick together; $K_f \neq K_i$.

MOMENTUM CHANGE IN INELASTIC COLLISIONS

Although the kinetic energy of the system changes in an inelastic collision, the momentum of the system does not, because $\Delta \vec{p}_{\text{sys}} = \vec{F}_{\text{net}} \approx \vec{0}$.

Identifying Inelastic Collisions

When deciding whether or not it is reasonable to treat a collision as approximately elastic, one can look for indications of inelasticity, such as:

- Objects are stuck together after the collision.
- An object is deformed after the collision.
- Objects are hotter after the collision.
- There is more vibration or rotation after the collision.
- An object is in an excited state after the collision.

10.3   A HEAD-ON COLLISION OF EQUAL MASSES

We will start by considering simple 1-D (head-on) collisions in order to see clearly how to apply both the Energy Principle and the Momentum Principle. Consider a head-on collision between two carts rolling or sliding on a track with low friction (Figure 10.4). Similar situations are billiard balls or hockey pucks or vehicles hitting each other head-on. Cart 1 with mass m moves to the right with x momentum p_{1xi}, and it runs into cart 2, which has the same mass m and is initially sitting still.

Taking the two carts as the system, and neglecting the small frictional force exerted by the track, all that the Momentum Principle can tell us is that after the collision the total final x momentum $p_{1xf} + p_{2xf}$ must equal the initial total x momentum p_{1xi}. By itself, the Momentum Principle doesn't tell us how this momentum will be divided between the two carts. We need to consider conservation of energy as well as conservation of momentum in order to predict what will happen.

Before the collision, nonzero energy terms included the kinetic energy of cart 1, K_{1i}, and the internal energies of both carts. After the collision, there is internal energy of both carts and kinetic energy of both carts, $K_{1f} + K_{2f}$.

QUESTION   What types of energy may have changed as a result of the collision?

$K_{1f} + K_{2f}$ might change, and internal energy may have changed. The temperature of the carts might be a bit higher now, with some of the initial kinetic energy having been dissipated into thermal energy. How much energy dissipation occurs depends on the details of the contact between the two interacting carts.

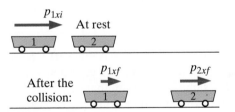

Figure 10.4 Cart 1 runs into stationary cart 2. The carts have equal masses. Friction and air resistance are negligible.

EXAMPLE

Elastic Collision of Two Identical Carts

Cart 1 collides with stationary cart 2, which is identical (Figure 10.4). Suppose that the collision is (nearly) elastic, as it will be if the carts repel each other magnetically or interact through soft springs. In this case there is no change of internal energy. What are the final momenta of the two carts?

Solution

Since the y and z components of momentum don't change, we can work with only x components.

System: Both carts
Surroundings: Earth, track, air (neglect friction and air resistance)

Initial situation: Just before collision
Final situation: Just after collision

Momentum Principle:

$$\vec{p}_f = \vec{p}_i + \vec{F}_{net}\Delta t$$
$$p_{1xf} + p_{2xf} = p_{1xi} + 0$$

Energy Principle:

$$E_f = E_i + W + Q$$
$$K_{1f} + K_{2f} + E_{int1f} + E_{int2f} = K_{1i} + K_{2i} + E_{int1i} + E_{int2i}$$
$$K_{1f} + K_{2f} = K_{1i}$$

- Negligible effect of surroundings. Only x components change.

- Negligible effect of surroundings. Negligible change in internal energy of the carts.

- Note that $p_{1xi}^2 = (p_{1xf} + p_{2xf})^2$, and that

$$K = \frac{1}{2}mv^2 = \frac{1}{2}m\left(\frac{p}{m}\right)^2 = \frac{1}{2}\frac{p^2}{m}$$

Expand the quadratic.
Cancel similar terms.

Combine momentum and energy equations:

$$\frac{p_{1xf}^2}{2m} + \frac{p_{2xf}^2}{2m} = \frac{(p_{1xf} + p_{2xf})^2}{2m}$$
$$p_{1xf}^2 + p_{2xf}^2 = p_{1xf}^2 + 2p_{1xf}p_{2xf} + p_{2xf}^2$$
$$2p_{1xf}p_{2xf} = 0$$

There are two possible solutions to this equation. The term $p_{1xf}p_{2xf}$ can be zero if $p_{1xf} = 0$ or if $p_{2xf} = 0$.

Figure 10.5 If the collision is elastic, cart 1 stops, and cart 2 has all the momentum cart 1 used to have.

FURTHER DISCUSSION If $p_{1xf} = 0$, the physical situation is that cart 1 came to a complete stop. In that case we see from the momentum equation $p_{1xf} + p_{2xf} = p_{2xf} = p_{1xi}$ that cart 2 now has the same momentum that cart 1 used to have. There has been a complete transfer of momentum from cart 1 to cart 2. There has also been a complete transfer of kinetic energy from cart 1 to cart 2. See Figure 10.5.

If $p_{2xf} = 0$, what is the physical situation? In that case $p_{1xf} + p_{2xf} = p_{1xf} = p_{1xi}$, and cart 1 just keeps going, missing cart 2! Of course that won't happen if the carts are on the same narrow track, but the algebra doesn't know that.

It is not possible for both final momenta to be zero, since the total final momentum of the system must equal the nonzero total initial momentum of the system.

EXAMPLE

Maximally Inelastic Collision of Two Identical Carts

Consider the opposite extreme—a maximally inelastic collision of the two identical carts, one initially at rest. That means the carts stick together (perhaps they have sticky material on their ends), and each has the same final momentum $p_{1xf} = p_{2xf}$ (Figure 10.6). **(a)** Find the final momentum, final speed, and final kinetic energy of the carts. **(b)** What is the change in internal energy of the two carts?

Solution

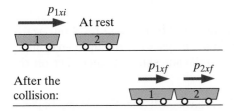

Figure 10.6 In a maximally inelastic collision, cart 1 sticks to cart 2, so the carts have the same final momentum.

■ External forces are negligible during the collision, so the total momentum of the system is constant.

■ Since we know the speed of the carts we can calculate their translational kinetic energy.

■ We know the initial and final translational kinetic energies of the system, so we can use the Energy Principle to find the change in internal energy.

■ The Momentum Principle is still valid even though the collision is inelastic. Fundamental principles apply in all situations.

Since the y and z components of momentum don't change, we can work with only x components.

System: Both carts
Surroundings: Earth, track, air (neglect friction and air resistance)

(a) Momentum Principle (x components):

$$p_{1xf} + p_{2xf} = p_{1xi}$$
$$2p_{1xf} = p_{1xi}$$
$$p_{1xf} = \frac{1}{2}p_{1xi}$$

The final speed of the stuck-together carts is half the initial speed:

$$v_f = \frac{1}{2}v_i.$$

Final translational kinetic energy:

$$(K_{1f} + K_{2f}) = 2\left(\frac{1}{2}mv_f^2\right)$$

$$(K_{1f} + K_{2f}) = 2\left(\frac{1}{2}m\left(\frac{1}{2}v_i\right)^2\right) = \frac{1}{4}mv_i^2$$

$$(K_{1f} + K_{2f}) = \frac{K_{1i}}{2}$$

(b) Energy Principle:

$$K_{1f} + K_{2f} + E_{\text{int},f} = K_{1i} + E_{\text{int},i}$$
$$E_{\text{int},f} - E_{\text{int},i} = K_{1i} - (K_{1f} + K_{2f})$$
$$\Delta E_{\text{int}} = \frac{1}{2}mv_i^2 - \frac{1}{4}mv_i^2$$
$$\Delta E_{\text{int}} = \frac{1}{4}mv_i^2 = \frac{K_{1i}}{2}$$

The final kinetic energy of the system is only half of the original kinetic energy, which means that the other half of the original kinetic energy has been dissipated into increased internal energy ΔE_{int} of the two carts.

FURTHER DISCUSSION In both cases (elastic collision and inelastic collision) we needed to apply both the Momentum Principle and the Energy Principle to find all the unknown quantities. One equation alone was not enough.

We have shown the two extremes of energy conservation in the head-on collision of the moving cart with the stationary cart:

■ If the collision is elastic, cart 1 stops and cart 2 moves with the speed cart 1 used to have.

■ If the collision is maximally inelastic, the carts stick together and move with half the original speed. Half of the original kinetic energy is dissipated into increased internal energy.

Between these two extremes there can be an inelastic but not maximally inelastic collision, in which some amount of the original kinetic energy (less than half) is dissipated into increased internal energy.

10.X.1 A 6 kg mass traveling at speed 10 m/s strikes a stationary 6 kg mass head-on, and the two masses stick together. **(a)** What was the initial total kinetic energy? **(b)** What is the final speed? **(c)** What is the final total kinetic energy? **(d)** What was the increase in internal energy of the two masses?

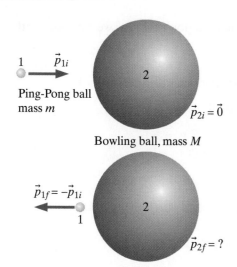

Figure 10.7 A Ping-Pong ball hits a bowling ball head-on and bounces off elastically.

10.4 | HEAD-ON COLLISIONS BETWEEN UNEQUAL MASSES

Suppose that a Ping-Pong ball hits a stationary bowling ball head-on, nearly elastically. Imagine that this takes place in outer space or floating in an orbiting spacecraft, so there is no friction—no significant external forces act on the combined system during the collision.

> **QUESTION** What do you expect? How will the Ping-Pong ball move after hitting the bowling ball? How will the bowling ball move after the collision?

You'll see the Ping-Pong ball bounce straight back with very little change of speed (Figure 10.7), and we'll show this is consistent with conservation of momentum and conservation of energy. (It is possible to show algebraically that this must be true, but we won't go through the messy algebra here.) Less obvious is that the bowling ball does move, although very slowly.

EXAMPLE **A Ping-Pong Ball Hits a Stationary Bowling Ball Head-on**

In an orbiting spacecraft a Ping-Pong ball of mass m (object 1) traveling in the $+x$ direction with initial momentum $\vec{p}_{1i}$ hits a stationary bowling ball of mass M (object 2) head on, as shown in Figure 10.7. What are the **(a)** momentum, **(b)** speed, and **(c)** kinetic energy of each object after the collision? Assume little change in the speed of the Ping-Pong ball, and assume that the collision is elastic.

Solution System: Ping-Pong ball and bowling ball
Surroundings: Nothing that exerts significant forces

Momentum Principle:

$$\vec{p}_{1f} + \vec{p}_{2f} = \vec{p}_{1i} + \vec{p}_{2i}$$

Assume that the speed of the Ping-Pong ball does not change significantly in the collision, so $\vec{p}_{1f} \approx -\vec{p}_{1i}$.

$$-\vec{p}_{1i} + \vec{p}_{2f} = \vec{p}_{1i}$$
$$\vec{p}_{2f} = 2\vec{p}_{1i}$$

■ It may be surprising that the bowling ball ends up with about twice the momentum of the Ping-Pong ball. One way to understand this is that the final momentum of the Ping-Pong ball is approximately $-\vec{p}_{1i}$, so the change in the Ping-Pong ball's momentum is approximately

$$-\vec{p}_{1i} - \vec{p}_{1i} = -2\vec{p}_{1i}$$

The Ping-Pong ball's speed hardly changed, but its momentum changed a great deal. Because momentum is a vector, a change of direction is just as much a change as a change of magnitude. This big change is of course due to the interatomic electric contact forces exerted on the Ping-Pong ball by the bowling ball. By reciprocity, the same magnitude of interatomic contact forces are exerted by the Ping-Pong ball on the bowling ball, which undergoes a momentum change of $+2\vec{p}_{1i}$.

(a) The final momentum of the bowling ball is twice the initial momentum of the Ping-Pong ball (see the marginal note).

(b) Final speed of bowling ball:

$$v_{2f} \approx \frac{p_{2f}}{M} = \frac{2p_{1i}}{M} = \frac{2mv_{1i}}{M} = 2\left(\frac{m}{M}\right)v_{1i}$$

This is a very small speed since $m \ll M$. For example, if the mass of the bowling ball is about 5 kg, and the 2 gram Ping-Pong ball is initially traveling at 10 m/s, the final speed of the bowling ball will be 0.008 m/s.

(c) Kinetic energies:

$$K_2 = \frac{(2p_{1i})^2}{2M} \quad \text{and} \quad K_1 = \frac{p_{1i}^2}{2m}$$

Because the mass of the bowling ball is much larger than the mass of the Ping-Pong ball, the kinetic energy of the bowling ball is much smaller than the kinetic energy of the Ping-Pong ball. The kinetic energy of the 2 g Ping-Pong ball, traveling at 10 m/s, is about 0.1 J, while the 5 kg bowling ball has acquired a kinetic energy of 1.6×10^{-4} J—nearly 1000 times less.

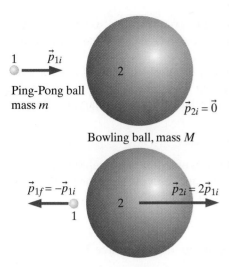

1 $\vec{p}_{1i}$

Ping-Pong ball
mass m

2

$\vec{p}_{2i} = \vec{0}$

Bowling ball, mass M

$\vec{p}_{1f} = -\vec{p}_{1i}$

1

2

$\vec{p}_{2i} = 2\vec{p}_{1i}$

Figure 10.8 The bowling ball gets twice the momentum of the Ping-Pong ball (but its speed and kinetic energy are very small).

FURTHER DISCUSSION Although the kinetic energy of the slowly moving bowling ball is very small, it is not zero, which means that the Ping-Pong ball does lose a little speed in the collision. That's why our assumption that the Ping-Pong ball's speed hardly changes is an approximation.

We made the plausible assumption that the Ping-Pong ball bounced back with nearly its original speed, and showed that this is consistent with momentum conservation if the bowling ball gets twice the momentum of the Ping-Pong ball, and also consistent with energy conservation, because the bowling ball acquires very little kinetic energy. With a lot of algebra, it is also possible to solve the momentum and energy equations exactly for the unknown final momentum of the Ping-Pong ball, and when one makes the approximation that the mass of the bowling ball is very much larger than the mass of the Ping-Pong ball, the results are those given above (see Problem 10.P.30).

When a Ping-Pong ball hits a stationary bowling ball head-on, elastically:

- The Ping-Pong ball bounces back with almost the same speed.
- The Ping-Pong ball has a large change of momentum $-2\vec{p}_{1i}$.
- The bowling ball has a large change of momentum $+2\vec{p}_{1i}$ (Figure 10.8).
- The bowling ball's final speed is very small, because it has large mass but comparable momentum.
- The bowling ball's final kinetic energy is very small, because it has large mass but comparable momentum.

Similar situations include a ball bouncing off a wall, or falling and bouncing off the ground, in which case the entire Earth plays the role of the bowling ball. The Earth gets twice the momentum of the ball but an extremely small speed and extremely small kinetic energy. If a bowling ball sits on a table and is hit by a Ping-Pong ball, and the table exerts a large enough frictional force to prevent the bowling bowl from moving, the bowling ball is essentially part of the Earth.

Although we've looked at the very special case of a head-on collision, it is generally true that when low-mass projectiles collide elastically with stationary large masses, the low-mass projectile bounces off with little change of speed and kinetic energy but may undergo a large change of momentum. The large-mass target can pick up lots of momentum, but little speed or kinetic energy.

> **10.X.2** We can use our results for head-on elastic collisions to analyze the recoil of the Earth when a ball bounces off a wall embedded in the Earth. Suppose that a professional baseball pitcher hurls a baseball ($m = 155$ grams) with a speed of 100 miles per hour ($v_1 = 44$ m/s) at a wall that is securely anchored to the Earth, and the ball bounces back with little loss of kinetic energy. What is the approximate recoil speed of the Earth ($M = 6 \times 10^{24}$ kg)?
>
> **10.X.3** In Exercise 10.X.2, calculate the approximate recoil kinetic energy of the Earth and compare it to the kinetic energy of the baseball. The Earth gets lots of momentum (twice the momentum of the baseball) but very little kinetic energy.

10.5 FRAME OF REFERENCE

A frame of reference is a perspective from which a system is observed. A frame of reference may be stationary or it may move at constant velocity. It is sometimes simpler to analyze phenomena from a moving reference frame.

As a simple example, imagine that you are standing on a street corner and a bus passes you. In your stationary frame of reference, a passenger sitting on the bus is moving with the same velocity as the bus. However, the bus driver

■ If the velocity of a reference frame is not constant, the frame is "noninertial," and the laws of physics as we have stated them do not apply. We'll stick to inertial frames in this textbook.

Figure 10.9 In a stationary reference frame, Jack sees a bowling ball with speed V run into a stationary Ping-Pong ball.

Bowling ball moving with speed V

Ping-Pong ball at rest

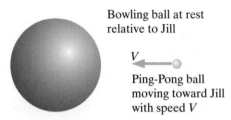

Bowling ball at rest relative to Jill

V

Ping-Pong ball moving toward Jill with speed V

Figure 10.10 Moving with the bowling ball, at speed V, Jill sees the bowling ball at rest, and the Ping-Pong ball moving toward her with speed V.

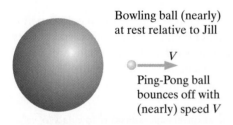

Bowling ball (nearly) at rest relative to Jill

V

Ping-Pong ball bounces off with (nearly) speed V

Figure 10.11 In Jill's moving frame of reference, she sees the Ping-Pong ball bounce off the bowling ball with nearly its original speed V.

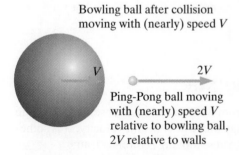

Bowling ball after collision moving with (nearly) speed V

V $2V$

Ping-Pong ball moving with (nearly) speed V relative to bowling ball, $2V$ relative to walls

Figure 10.12 In Jack's stationary reference frame, the Ping-Pong ball has a speed $2V$ after the collision.

may observe the passenger from a moving reference frame "anchored" to the bus. In the moving bus reference frame, the driver sees the passenger sitting still.

Evidently, the velocity $\vec{v}\,'$ ("v-prime") of an object viewed from a moving reference frame is found by subtracting $\vec{v}_{\text{frame}}$, the velocity of the moving frame, from $\vec{v}$, the velocity of the object relative to a stationary, or "laboratory" reference frame:

$$\vec{v}\,' = \vec{v} - \vec{v}_{\text{frame}}$$

This is a very powerful technique that is discussed in more depth later in this chapter. A change of reference frame often makes a problem much simpler to analyze, or reduces the problem to one you've already analyzed.

Bowling Ball Hits Ping-Pong Ball: Change of Reference Frame

In the previous example we analyzed a collision in which a moving Ping-Pong ball hit a stationary bowling ball head-on. What happens when a bowling ball of large mass M with initial x velocity V hits elastically head-on a stationary Ping-Pong ball of small mass m (Figure 10.9)? What is the final speed of the Ping-Pong ball?

System: Ping-Pong ball and bowling ball
Surroundings: Nothing that exerts significant forces

We can perform a change of reference frame that reduces this problem to the previous one. Suppose that this experiment is being done in a spacecraft, away from the walls. The room is dark, and only the glowing bowling ball and the Ping-Pong ball are visible; the walls are not visible.

Jack and Jill are two crew members in the spacecraft. Jack hangs onto a wall, remaining stationary, and watches the bowling ball go by, while Jill launches herself away from the wall with x velocity V and glides along beside the bowling ball. In Jack's stationary reference frame, the Ping-Pong ball is stationary, and the bowling ball is moving, as described above.

What is the speed of the bowling ball as measured by Jill? For Jill, the bowling ball isn't moving. It is always the same distance from her. In Jill's reference frame, the bowling ball is at rest.

What does Jill see the Ping-Pong ball doing? In Jill's reference frame, where the bowling ball is at rest, she sees the Ping-Pong ball moving with x velocity $-V$ (Figure 10.10).

According to the principle of relativity, in any reference frame in uniform motion one can apply physics principles and make correct predictions, even though the details of the calculations look different from the details in some other reference frame in uniform motion. If the walls are dark, and the Ping-Pong ball and bowling ball glow in ultraviolet light, Jill truly cannot tell whether she is moving and the Ping-Pong ball is stationary relative to the walls, or whether she is stationary and the Ping-Pong ball is moving relative to the walls. This is at the heart of the principle of relativity.

By changing reference frames, we've reduced this situation to one very similar to the previous example. In Jill's reference frame, the Ping-Pong ball moves in the $-x$ direction with speed V and hits a stationary bowling ball elastically, head-on. It bounces off in the $+x$ direction with nearly the same speed V, in Jill's reference frame (Figure 10.11).

After the collision, how fast is the Ping-Pong ball moving in Jack's stationary reference frame? Jill and the bowling ball are moving with speed V in the $+x$ direction relative to Jack, and the Ping-Pong ball is moving with speed V in the $+x$ direction relative to Jill and the bowling ball, so it is moving with speed $2V$ relative to Jack and the walls (Figure 10.12). By changing reference frame we've analyzed the problem without doing any significant amount of calculation.

When a bowling ball hits a stationary Ping-Pong ball head-on, elastically:

- The Ping-Pong ball moves with twice the speed of the bowling ball.
- We won't go through the details, but you can appreciate that the change in speed of the bowling ball is insignificant. It just plows straight ahead, having lost very little kinetic energy to the Ping-Pong ball. The small amount of momentum lost by the bowling ball was gained by the Ping-Pong ball ($2mV$).

10.X.4 A uranium atom traveling at speed 4×10^4 m/s collides elastically with a stationary hydrogen molecule, head-on. What is the approximate final speed of the hydrogen molecule?

10.6 SCATTERING: COLLISIONS IN 2-D AND 3-D

Scattering experiments are used to study the structure and behavior of atoms, nuclei, and other small particles. In a scattering experiment, a beam of particles collides with other particles. In atomic or nuclear collisions, we can't observe in detail the curving trajectories inside the tiny interaction region. We only observe the trajectories before and after the collision, when the particles are far apart and their mutual interaction is very weak, so they are traveling in nearly straight lines. An example of scattering is the collision of an alpha particle (helium nucleus) with the nucleus of a gold atom.

Momentum and Energy Equations

An alpha particle collides with a gold nucleus, and is deflected through some final angle θ. The gold nucleus recoils at some other final angle ϕ. The final angles are measured relative to the initial direction of the alpha particle (Figure 10.13 and Figure 10.14). We say that the alpha particle "scattered" through an angle θ.

We take the "before" time early enough and the "after" time late enough that the two particles are sufficiently far away from each other that we can neglect their electric potential energy. As for conservation of momentum, we must write two equations, for the x and y components of momentum conservation.

QUESTION Try to write the two momentum-conservation equations, and the energy-conservation equation. All the speeds are small compared to the speed of light. Choose x in the direction of the incoming alpha particle, and y at right angles to it. Consider ϕ to be a positive angle. Write equations in terms of the magnitudes of the momenta, p_1, p_3, and p_4; the gold nucleus is initially at rest.

We use direction cosines to get the x and y components of momenta:

$$p_1 = p_3 \cos\theta + p_4 \cos\phi, \quad x \text{ momentum conservation}$$
$$0 = p_3 \cos(90° - \theta) + p_4 \cos(90° + \phi), \quad y \text{ momentum conservation}$$
$$\frac{p_1^2}{2m} = \frac{p_3^2}{2m} + \frac{p_4^2}{2M}, \quad \text{energy conservation (scalar)}$$

Make absolutely sure that you understand every detail of these three equations! In particular, pay attention to the fact that since momentum is a vector, the momentum equations are written in terms of x and y components of the momentum.

Figure 10.13 Before the collision, the alpha particle (mass m) approaches the gold nucleus (mass M).

Figure 10.14 After the collision, when the particles are far apart and moving with nearly uniform velocities. Final angles are measured relative to the initial direction of the alpha particle.

QUESTION It is fruitful to reflect on which of the quantities in these three equations are known and which are unknown. We know the masses of the alpha particle and the gold nucleus, and we know the initial kinetic energy of the alpha particle (so we know its speed and momentum). Which quantities in these three equations are unknown?

You should have identified the two final momenta and the two final angles as unknown quantities (p_3, p_4, θ, and ϕ, all of which represent positive numbers). But if there are four unknowns and only three equations, how can we solve for the unknowns? The simple answer is, we can't!

Nevertheless, there are many situations in which these three equations can be exploited. For example, if we measure the final direction of motion of the alpha particle, the angle θ, we can solve for the other three unknown quantities and thereby predict the final momenta p_3 and p_4 of the two particles and the direction of motion ϕ of the gold nucleus, even without measuring all of these quantities.

It is worth mentioning that in experimental particle physics the momentum is often measured in a very direct way. Charged particles curve in a magnetic field due to magnetic forces that act on moving charges, and the radius of curvature is a direct measure of the momentum of the particle, if you know how much charge it has, which you do know if you know what kind of particle it is. Usually the momentum of a particle can be measured more accurately than the velocity, though once you have measured the momentum you can deduce the velocity if you know what kind of particle it is, since in that case you know its mass.

EXAMPLE

p_1

m m

Figure 10.15 Before a collision of two identical particles, one initially at rest.

Elastic Collision: Identical Particles, One Initially at Rest

An interesting special case of an elastic collision is one involving particles of equal mass, such as an elastic collision between two alpha particles, or two gold nuclei, or two billiard balls (whose collision is nearly elastic), in which one of the two objects (the "target") is initially at rest (Figure 10.15 and Figure 10.16). In this special case, show that the angle A must be 90° if the speeds are small compared to the speed of light.

Solution

p_3

m

m

p_4

Angle A

Figure 10.16 After the collision of two identical particles, one initially at rest.

Because the vector momentum of the system is constant, we can write $\vec{p}_1 = \vec{p}_3 + \vec{p}_4$, then calculate p_1^2 by expanding the dot product:

$$p_1^2 = \vec{p}_1 \bullet \vec{p}_1 = (\vec{p}_3 + \vec{p}_4) \bullet (\vec{p}_3 + \vec{p}_4) = p_3^2 + p_4^2 + 2p_3p_4 \cos A$$

Divide by $2m$:

$$\frac{p_1^2}{2m} = \frac{p_3^2}{2m} + \frac{p_4^2}{2m} + \frac{2p_3p_4 \cos A}{2m}$$

Since the kinetic energy K at low speed can be written as $p^2/2m$,

$$K_1 = K_3 + K_4 + \frac{2p_3p_4 \cos A}{2m}$$

However, energy conservation for this elastic collision says that $K_1 = K_3 + K_4$, so we have

$$\cos A = 0 \text{ or } A = 90°$$

FURTHER DISCUSSION In this special situation—an elastic collision at low speeds between equal particles, one of which is initially at rest—the angle between the final velocities must be 90° as long as neither p_3 nor p_4 is zero, in which case the equation reduces to $0 \cos A = 0$, and $\cos A$ is indeterminate. If one of the final momenta is indeed zero, the other particle must have all the momentum of the initial particle. This can happen either because the incoming particle missed the target entirely and kept going in the original direction, or because the incoming particle hit the target dead center, in which case the only consistent solution of the equations is for the incoming particle to come to rest and the target particle to move forward with the initial momentum, as we saw in the head-on collision of two carts earlier in the chapter.

Impact Parameter

The distance between centers perpendicular to the incoming velocity is called the "impact parameter" and is often denoted by b. A head-on collision has an impact parameter of zero. Here in Figures 10.17–10.20 are possible elastic collisions between two billiard balls, for various impact parameters:

Figure 10.17 If the impact parameter, b, is too large, there is no interaction at all between two billiard balls.

Figure 10.18 A grazing impact results in small-angle scattering.

Smaller impact parameters lead to larger scattering unless $b = 0$:

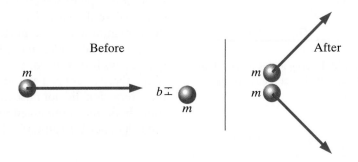

Figure 10.19 A medium impact parameter results in symmetric scattering.

Figure 10.20 In a head-on collision, the impact parameter $b = 0$. In this case all the momentum is transferred to the target.

The smaller the impact parameter, the more severe is the collision, and the larger the deflection angle of the incoming particle (larger "scattering"), except for a head-on collision, where if the masses are equal the incoming ball stops dead and the target ball gets the entire momentum.

Figure 10.21 A collision that is not head-on.

10.7   DISCOVERING THE NUCLEUS INSIDE ATOMS

A famous experiment involving collisions led to the discovery of the nucleus inside the atom. It was carried out in 1911 by a group in England led by the New Zealander Ernest Rutherford. In the experiment, high-speed alpha particles (now known to be helium nuclei, consisting of two protons and two neutrons) were shot at a thin gold foil (whose nuclei consist of 79 protons and 118 neutrons).

When Rutherford and his coworkers decided to perform a scattering experiment in order to probe the internal structure of matter, they were familiar with the relationships of momenta in collisions that we have just studied, and they reasoned that by shooting microscopic particles at a thin layer of metal, they might be able to learn something about the microscopic structure of the metal by looking at patterns of scattering.

In the Rutherford experiment, the velocity of the alpha particles was not high enough to allow them to make actual contact with the nucleus. It is nonetheless appropriate to think of the process as a collision, because there are sizable electric forces acting between the alpha particle and the gold nucleus for a very short time. We call any process of this kind a collision, even if there is no direct contact but only gravitational, electric, or magnetic interactions at a distance. Figure 10.21 shows what the trajectories look like. We say that the alpha particle projectile is "scattered" (deflected) by the interaction with the initially stationary gold nucleus target, and such experiments are called "scattering" experiments.

Energy

2nd excited state

1st excited state

Ground state

Figure 10.22 Energy levels in a nucleus. The potential energy curve is not shown.

Elastic Collisions in the Rutherford Experiment

The conditions of this experiment were such that there was no change in the internal energy of either the alpha particle or the gold nucleus (elastic collision). There do exist "excited" states of nuclei, in which there is a change in the configuration of the nucleons corresponding to a higher energy than the ground state, the normal configuration of the nucleus. The internal energy of the nucleus is quantized, and there are no states with energy intermediate between these quantized energies (Figure 10.22). In the Rutherford experiment, the interaction did not raise the gold nucleus or the alpha particle to an excited state; both nuclei remained in their ground states. This happy accident made it much easier for Rutherford to analyze the experiment.

Alpha particle

Electron
(initially at rest)

Figure 10.23 The alpha particle sweeps away the low-mass electrons and plows straight on ("bowling ball hits Ping-Pong ball").

Electric Potential Energy in the Rutherford Experiment

What about the electric potential energy associated with the electric forces? Long before the collision the two nuclei were far from each other, and long after the collision they are again far from each other. Like gravitational energy, electric energy is proportional to $1/r$, where r is the distance between the two charged objects. Therefore, from long before the collision to long after the collision the change in electric energy would be very nearly zero ($\Delta U_{el} \approx 0$) if the alpha particle and gold nucleus are completely isolated from other particles. However, what about all the electrons surrounding a gold nucleus?

As the alpha particle plows through the gold foil it knocks low-mass electrons away (Figure 10.23), much as a moving bowling ball knocks Ping-Pong balls away (the alpha particle does of course lose a little energy in each such interaction). Therefore we can start our analysis when the alpha particle is deep inside the atom, with the electrons of the gold atom mostly far outside the region of interest. An atomic diameter is about 1×10^{-10} m, whereas the diameter of a gold nucleus is much smaller, less than 1×10^{-14} m, so we might start our analysis of the collision when the particles are about 1×10^{-12} m apart, a distance that is very small compared to an atom and very large compared to a nucleus. At this distance the electric potential energy is

$$U_{\text{elec}} = \frac{1}{4\pi\varepsilon_0} \frac{(2e)(79e)}{r} = \left(9 \times 10^9 \frac{\text{N·m}^2}{\text{C}^2}\right) \frac{(2)(79)(1.6 \times 10^{-19}\,\text{C})^2}{1 \times 10^{-12}\,\text{m}}$$

$$U_{\text{elec}} = (3.6 \times 10^{-14}\,\text{J}) \frac{1\,\text{eV}}{1.6 \times 10^{-19}\,\text{J}} = 0.2 \times 10^6\,\text{eV} = 0.2\,\text{MeV}$$

The radioactive source used by the Rutherford group provided alpha particles whose kinetic energy was 10 MeV, so if we take as the initial state a separation of 1×10^{-12} m the electric energy is only 2% of the kinetic energy. If we also take as the final state a separation of 1×10^{-12} m we have $\Delta U_{\text{elec}} = 0$. Evidently the only significant energy changes in this experiment are changes in the kinetic energies of the alpha particle and the gold nucleus.

The gold nuclei in the gold foil are about 2×10^{-10} apart, the approximate diameter of a typical atom in a metal, so when an alpha particle is 1×10^{-12} m from a gold nucleus, it is comparatively very far from any other gold nucleus. This explains why Rutherford could analyze the experiment as though isolated alpha particles scattered off isolated gold nuclei; electrons and other gold nuclei were far away.

The Plum Pudding Model of an Atom

For about a decade, the prevailing model of an atom had been what was called a "plum pudding" model proposed by J. J. Thomson. Electrons were discovered in 1897 by Thomson, using a cathode ray tube, a device similar to a television picture tube. Thomson was able to show that the "cathode rays" that he and others had observed in electric gas discharges were in fact particles that were negatively charged, by the already existing convention for labeling electrical charges. This was the first conclusive demonstration of the existence of what we now call "electrons," one of the major constituents of all atoms.

Since these "electrons" were extracted from ordinary matter, which is normally not electrically charged, Thomson's discovery implied that matter also contained something having a positive charge. Because the electrons apparently had little mass, it seemed likely that most of the mass in ordinary matter was charged positively. Thomson proposed what was called a "plum pudding" model of atoms: a positively charged substance of uniform, rather low density, with tiny electrons distributed like raisins throughout it. Rutherford's group expected to gather evidence supporting this model from their collision experiments. Rutherford was extremely surprised by the results of these experiments, which led to a radical revision of the model.

The Rutherford Experiment

A new source of fast-moving particles was provided through the discoveries of natural radioactivity by the French scientists Henri Becquerel and Marie and Pierre Curie. These newly discovered particles included high-speed, massive, positively charged "alpha particles," which we now know to be the nuclei of helium atoms, consisting of two protons and two neutrons bound together. Alpha particles are emitted in the radioactive decay of heavy elements. Alpha

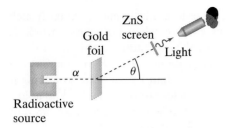

Figure 10.24 Schematic diagram of the Rutherford experiment. A radioactive source emits alpha particles, which are directed at a thin piece of gold foil. Particles scattered by gold nuclei in the gold foil hit a screen coated with ZnS, which fluoresces, giving off light when hit by an energetic particle. The light was observed by the experimenter.

Figure 10.25 This diagram of an atom is not to scale; the radius of the nucleus is around 1×10^{-15} m, which is only 1/100,000 the radius of the electron cloud (about 1×10^{-10} m), so the nucleus would not be visible at all on this scale.

particles were convenient microscopic projectiles, and Rutherford's group decided to shoot alpha particles at a very thin piece of gold foil and observe where they reappeared.

QUESTION Based on the plum pudding model, what did Rutherford and his colleagues expect to see in their collision experiments?

The researchers expected the alpha particles to be deflected only very slightly by interactions with the low-mass electrons and with the low-density "pudding" of positively charged matter in the gold atoms. They could detect a deflected ("scattered") alpha particle by a pulse of light that was emitted when the particle struck a fluorescent material (zinc sulfide). They moved their detector around at various angles to the incoming beam of alpha particles and measured what fraction of the alpha particles were scattered through various angles (Figure 10.24).

Rutherford and his coworkers were astounded to observe that an alpha particle, which is much more massive than an electron, sometimes bounced straight backward, as though it had encountered an even more massive yet very tiny positive particle inside the gold foil, rather than being slightly deflected in its passage through the "pudding" of positive charge. In other words, the experimenters thought they were throwing bowling balls at Ping-Pong balls, and they were amazed to see what they thought were bowling balls come bouncing back.

"It was quite the most incredible event that has ever happened to me in my life," Rutherford said. "It was almost as incredible as if you fired a 15-inch [artillery] shell at a piece of tissue paper and it came back and hit you."

He continued, "On consideration, I realized that this scattering backward must be the result of a single collision, and when I made the calculations I saw that it was impossible to get anything of that order of magnitude unless you took a system in which the greater part of the mass of the atom was concentrated in a minute nucleus. It was then that I had the idea of an atom with a minute massive center, carrying a charge." (See Figure 10.25.)

Computer Modeling of the Rutherford Experiment

We can make a computational model of the Rutherford experiment; see Problem 10.P.38. Remember that Coulomb's law for electric forces is very similar to Newton's law of gravitation. The electric force $\vec{F}_{elec}$ between two small charged particles carrying amounts of charge q_1 and q_2, and located a distance r apart, is

$$\vec{F}_{elec} = \frac{1}{4\pi\varepsilon_0} \frac{q_1 q_2}{r^2} \hat{r}$$

The charges q_1 and q_2 are measured in units of coulombs, and the proportionality factor is measured to be $1/(4\pi\varepsilon_0) = 9 \times 10^9$ N·m²/coulomb². The electric charge of the proton is $e = +1.6 \times 10^{-19}$ coulomb, and the electric charge of the electron is $-e = -1.6 \times 10^{-19}$ coulomb. In the Rutherford experiment the charge on the alpha particle is $2e$ (2 protons and 2 neutrons), and the charge on the gold nucleus is $79e$ (79 protons and 118 neutrons).

10.8	***DISTRIBUTION OF SCATTERING ANGLES**

In general, we have three equations describing an elastic collision—two momentum equations (in the x and y directions) and one energy equation, but there are four unknown quantities (the outgoing speeds and directions for each particle). The concept of impact parameter provides an additional quantity, and you saw earlier how the impact parameter determines the scattering angle θ of the incoming particle, from which you can predict the other final values (p_3, p_4, and ϕ).

In a macroscopic situation, such as shooting billiard balls, you can choose an impact parameter simply by carefully aiming at a point a perpendicular distance b from the center of the target. Unfortunately, in experiments where you fire alpha particles or other subatomic projectiles at nuclei, you can't aim at a specific impact parameter because the target is so incredibly small. All you can do is fire lots of projectiles at the subatomic target and expect a probabilistic distribution of impact parameters.

When we fire a projectile at this microscopic target, the probability of hitting one particular ring (whose radius corresponds to the impact parameter) is proportional to the area of that ring (Figure 10.26). If you calculate the range of scattering angles that corresponds to hitting somewhere within a particular ring, you can predict the angular distribution of scattering angles that will be observed.

Figure 10.26 The probability of hitting a particular ring is proportional to the area of the ring.

*Cross Section

If a target particle such as a gold nucleus has a radius R, we say that its "geometrical cross section" is πR^2. A high-speed alpha particle that hits within this cross-sectional area will interact strongly with the gold nucleus. Quantum-mechanical effects can make the effective target size bigger or smaller than the geometrical cross section, and the effective target size is called the "cross section" for interactions of this kind. Cross sections are measured experimentally by observing how often projectiles are deflected by an interaction, instead of passing straight through the material without interacting. Cross sections have units of square meters.

A "differential" cross section for scattering within a particular range of angles is the area in square meters of a ring bounded by impact parameters corresponding to the given angular range.

*Conservation Laws vs. Details of the Interaction

We used general principles to determine the final momenta of two colliding particles (conservation of momentum, conservation of energy). We didn't have to say what kind of force was involved. However, doesn't it make a difference whether the interaction is electric or nuclear? Yes, the type of interaction determines the *distribution* of scattering angles. For example, in what fraction of the collisions is the scattering angle between 0 and 10 degrees? Between 10 and 20 degrees? Between 20 and 30 degrees? These fractions are different for different forces.

To put it another way, if the scattering angle is 20 degrees, the momenta are determined. The question is, how often is the scattering angle in fact near 20 degrees? Measurement of the distribution of scattering angles in collisions at modern particle accelerators is one of the ways used to study the nature of the strong and weak interactions.

Rutherford used Coulomb's law of electric force to make a detailed prediction of the distribution of scattering angles, assuming that the positive nucleus was pointlike and very massive. That is, he predicted what fraction of the alpha particles would be scattered by less than 10 degrees, by an angle between 10 and 20 degrees, by an angle between 20 and 30 degrees, and so on.

The observed distribution of scattering angles agreed with the predicted distribution (Figure 10.27 shows a portion of this distribution).

The calculation that Rutherford did involves a lot of complicated geometry, algebra, and probability (with a random distribution of impact parameters), so we won't go through it. The main features of Figure 10.27 are easily understood, however. Because the gold nucleus is very small, most of the time the alpha particle will miss the nucleus by a large distance. With a large impact parameter the electric force is very small, and the deflection (scattering angle) is very small. For that reason small scattering angles are common. Large scattering angles occur only for nearly head-on collisions (small impact parameter), and these events are rare.

Since Rutherford's prediction assumed a pointlike nucleus, the observations were consistent with the nucleus being very small. Also, the good fit between prediction and observation strengthened the belief that the interaction was purely electric in nature (the strong interaction played no role because it is a very short-range interaction, and the alpha particle was not going fast enough to overcome the electric forces and contact the gold nucleus).

If the alpha particle has high enough energy, it can overcome the electric repulsion and make contact with the gold nucleus. In that case the strong interaction comes into play, and the angular distribution changes. The radius of the gold nucleus can be determined by finding out how much energy an alpha particle has to have in order to obtain an angular distribution that deviates from what is expected due just to the electric interaction.

Figure 10.27 For an electric interaction, the fraction of scatters at different scattering angles falls rapidly with angle.

QUESTION For a particular scattering angle, can we use momentum and energy conservation to determine the unknown angle and momenta, even if the alpha particle makes contact with the gold nucleus?

Yes. These conservation principles are completely general and apply to all types of interactions. However, they cannot by themselves predict the angular distribution—that is, how often a particular range of scattering angles will occur. To do that you have to have a mathematical description for the specific interaction, such as the $1/r^2$ force law for electric interactions.

10.9 RELATIVISTIC MOMENTUM AND ENERGY

Rutherford's original experiments were carried out with alpha particles whose kinetic energy was about 10 MeV (10 million electron-volts), with a corresponding speed that was small compared to the speed of light. In order to probe the nucleus more deeply, physicists later built special machines, called "particle accelerators," which use electric forces to accelerate charged particles to very much higher energies, with speeds that may be as much as 99.9999% of the speed of light. For such high speeds we cannot use the low-speed approximate formulas for momentum and energy.

Physicists analyzing collisions produced by high-energy accelerators routinely do calculations based on the relativistic energy and momentum formulas. Consider a collision in which a particle of mass m_1 and magnitude of momentum p_1 strikes a stationary particle of mass m_2 (Figure 10.28). There is a change of identity, with the production of two new particles of mass m_3 and m_4, with momenta p_3 and p_4, at angles θ and ϕ to the horizontal (Figure 10.29).

Figure 10.28 Before a high-energy collision.

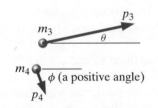

Figure 10.29 After the collision.

We will write down relativistically correct equations for momentum conservation and energy conservation, in terms of momentum p and energy E, knowing that we can determine E from p, since $E^2 - (pc)^2 = (mc^2)^2$:

$$p_1 = p_3 \cos\theta + p_4 \cos\phi, \; x \text{ momentum conservation}$$

$$0 = p_3 \cos(90° - \theta) - p_4 \cos(90° - \phi), \; y \text{ momentum conservation}$$

$$E_1 + m_2 c^2 = E_3 + E_4, \; \text{energy conservation (scalar)}$$

The energy equation is equivalent to

$$\sqrt{(p_1 c)^2 + (m_1 c^2)^2} + m_2 c^2 = \sqrt{(p_3 c)^2 + (m_3 c^2)^2} + \sqrt{(p_4 c)^2 + (m_4 c^2)^2}$$

QUESTION | Suppose that the initial momentum p_1 of the projectile is known, and all four masses are known (because we were able to identify the particles). How many unknown quantities are there, considering that we can write the energy equation in terms of p?

There are four unknown quantities, the magnitudes of the momenta and the directions of the two outgoing particles (p_3, p_4, θ, and ϕ), but only three equations. A typical situation in a high-energy particle experiment is that one of these unknown quantities is measured, and momentum and energy conservation are used to calculate the other three quantities.

As in the Rutherford experiment, measurements of the outgoing particle momenta alone do not tell us anything about the detailed nature of the interaction involved in the collision, because they are determined by general momentum and energy conservation laws. The identity changes of the particles tell us a lot, and the distribution of scattering angles also tells us a lot.

EXAMPLE | **Photodissociation of the Deuteron**

A photon of energy E is absorbed by a stationary deuteron. As a result, the deuteron breaks up into a proton and a neutron. The photon energy is sufficiently low that the proton and neutron have speeds small compared to the speed of light, but there is a change of rest mass. The proton is observed to move in a direction at an angle θ to the direction of the incoming photon. Write equations that could be solved to determine the unknown magnitudes of the proton momentum p_p and neutron momentum p_n and the positive angle α of the neutron to the direction of the incoming photon. Let the precise masses of the particles be M_d, M_p, and M_n for the deuteron, proton, and neutron. Explain clearly on what principles your equations are based. Provide a physics diagram of the reaction and label the angles. Do not attempt to solve the equations.

Solution | System: All of the particles (Figure 10.30)
Surroundings: Nothing significant

Momentum conservation (Figure 10.30): $\vec{p}_{\text{photon}} = \vec{p}_p + \vec{p}_n$

A photon with energy E has momentum $p = E/c$:

$$\langle E/c, 0, 0 \rangle = \langle p_p \cos\theta + p_n \cos\alpha, \; p_p \cos(90° - \theta) - p_n \cos(90° - \alpha), 0 \rangle$$

This is equivalent to two component equations:

$$E/c = p_p \cos\theta + p_n \cos\alpha$$

$$0 = p_p \cos(90° - \theta) - p_n \cos(90° - \alpha)$$

Figure 10.30 Physics diagram for photodissociation of deuteron. The angle θ was measured.

Energy conservation (small proton and neutron speeds):

$$E + M_d c^2 = \left(M_p c^2 + \frac{p_p^2}{2M_p} \right) + \left(M_n c^2 + \frac{p_n^2}{2M_n} \right)$$

We have three equations (two momentum component equations and one energy equation) in three unknowns: p_p, p_n, and α (θ was measured). In principle we can solve for the unknown quantities, though the algebra is messy.

We could also have written the energy equation relativistically, using the relationship $E^2 - (pc)^2 = (mc^2)^2$, which gives $E = \sqrt{(pc)^2 + (mc^2)^2}$:

$$E + M_d c^2 = \sqrt{(p_p c)^2 + (M_p c^2)^2} + \sqrt{(p_n c)^2 + (M_n c^2)^2}$$

Identifying a Particle

A variation on the relativistic collision analysis can be used to identify one of the outgoing particles. Suppose that we measure p_3 and θ, and we know m_3 but not m_4. We can use our three equations in the three unknowns (m_4, p_4, and ϕ) to solve for the unknown mass m_4. This is a very powerful technique when we are unable to observe the particle directly. For example, neutrinos don't have electromagnetic or strong interactions and therefore typically fail to register in ordinary detectors. However, if we measure the other outgoing particle we can determine the properties of the unseen particle.

The Discovery of the Neutrino

A specific example of such analyses was the original motivation for proposing the existence of the neutrino. The neutron (either free or when bound into some nuclei) was observed to decay with the creation of a proton and an electron, both charged particles that are easy to detect. However, the amount of energy and momentum carried by the proton and electron varied from one decay to the next, which made no sense:

$$n \rightarrow p^+ + e^-, \text{ violating momentum and energy conservation?}$$

It was proposed that there was an unseen third particle, called the neutrino (or antineutrino in neutron decay), which carried energy and momentum that would salvage the conservation laws:

$$n \rightarrow p^+ + e^- + \bar{v}, \text{ conserving momentum and energy}$$

It was possible to solve for the properties of the unseen particle by the techniques outlined above, and it was found that the neutrino must have nearly zero mass and must travel at nearly the speed of light. (Recent experiments show that the neutrino has a very small nonzero mass and must travel at slightly less than the speed of light.)

For many years the neutrino was just a theoretical construct that balanced the accounts, but eventually the neutrino was detected directly. The neutrino is very hard to detect because it does not have electromagnetic or strong interactions, only weak interactions, so the most probable thing for it to do is to pass right through any detector you place in its way. In fact, neutrinos emitted in nuclear reactions in the Sun mostly pass right through the entire Earth! However, with extremely intense neutrino beams produced in nuclear reactors or accelerators, or with gigantic underground detectors used to identify neutrinos coming from the Sun, it is possible to have a measurable rate of neutrino reactions.

Figure 10.31 Energy levels for a gold nucleus. A sufficiently energetic alpha particle may have enough energy to raise the nucleus above its ground state.

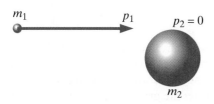

Figure 10.32 You measure the incoming and outgoing momenta of the alpha particle and deduce something about the energy levels in the gold nucleus.

INELASTIC COLLISIONS AND QUANTIZED ENERGY

The Rutherford experiment involved elastic collisions, because the internal energies of the gold nucleus and the alpha particle didn't change during the interaction. When there is an internal energy change, or the production of additional particles or the emission of light, we call the collision "inelastic."

If we shoot alpha particles at a gold nucleus with higher energy than the energy of alpha particles in the Rutherford experiment, we may be able to excite the gold nucleus to a higher quantum energy level, an amount ΔE above the ground state of the gold nucleus (Figure 10.31). Such a state corresponds to a different configuration of the nucleons inside the gold nucleus.

Suppose that you know the incoming momentum p_1 of the alpha particle, and you are able to measure its outgoing momentum p_3 and scattering angle θ (Figure 10.32).

QUESTION   What is the form of the energy conservation and momentum conservation equations in this case? Don't assume that the speeds are small compared to the speed of light. Note that the rest energy of the recoiling gold nucleus is $M_4c^2 + \Delta E$ because it is in an excited state.

Here are the relevant equations:

$$p_1 = p_3 \cos\theta + p_4 \cos\phi,$$
where p_4 and ϕ refer to the gold nucleus, which is not observed in the experiment

$$0 = p_3 \cos(90° - \theta) - p_4 \cos(90° - \phi)$$

$$E_1 + Mc^2 = E_3 + E_4, \text{ where } E_4 = \sqrt{(p_4c)^2 + (M_4c^2 + \Delta E)^2}$$

QUESTION   How many unknown quantities are there in the energy and momentum equations? Is there enough information to be able to determine the internal energy difference ΔE? (This is one of the methods used to study the energy levels of nuclei. This energy level determination can be verified by measuring the energy of a gamma-ray photon emitted when the nucleus drops back to the ground state.)

There are three unknowns (p_4, ϕ, and ΔE), and there are three equations (energy, x and y momentum), so we could determine ΔE.

Change of Identity

Instead of merely exciting the target particle to a higher internal energy state, suppose that there is a change of identity in which the target particle turns into some other particle. An example is the following particle reaction, which was used in the 1950s to discover the Δ^+ particle of mass m_Δ, which is a particle more massive than a proton:

$$\pi^- + p^+ \rightarrow \pi^- + \Delta^+$$

A beam of negative pions (π^-) of known mass m_π and known kinetic energy, produced by a particle accelerator, is directed at a container of hydrogen gas, and the targets for the collisions are the protons of mass m_p that are the nuclei of the hydrogen atoms. Sometimes an incoming pion interacts through the strong interaction with a proton to produce a Δ^+ particle. The momentum of the scattered pion can be measured by measuring the radius of curvature of its curving trajectory in a magnetic field.

QUESTION What is the form of the energy conservation and momentum conservation equations in this case? (Don't assume that the speeds are small compared to the speed of light.)

Here are the relativistically correct equations:

$$p_1 = p_3 \cos\theta + p_4 \cos\phi$$
$$0 = p_3 \cos(90° - \theta) - p_4 \cos(90° - \phi)$$
$$E_1 + m_p c^2 = E_\pi + E_\Delta, \text{ where } E^2 - (pc)^2 = (mc^2)^2$$

QUESTION How many unknown quantities are there in the energy and momentum equations? Is there enough information to be able to determine the mass of the Δ^+ particle?

There are three unknowns (p_4, ϕ, and E_Δ) and three equations, so it is possible to determine the mass of the Δ^+ particle.

What We Learn from Inelastic Collisions

The analysis for finding the mass change in the preceding exercises is essentially the same as the analysis for determining the internal energy change of an excited gold nucleus. In fact, we can say that the Δ^+ particle is an excited state of three quarks, with the proton representing the ground state of three quarks (Figure 10.33). The excited state (the Δ^+ particle) can decay to the ground state (the proton) with the emission of a neutral pion (π^0), which is a quark–antiquark pair. The extra kinetic energy is equal to $(m_\Delta c^2 - m_p c^2)$. Scattering experiments have been an important tool for discovering new particles (or new excited states of quark systems).

Δ^+, an excited state of three quarks

$$\text{———————} \quad E_1 = m_\Delta c^2$$

$\Delta^+ \rightarrow p^+ + \pi^0$

p^+, the ground state of three quarks $\quad E_0 = m_p c^2$

Figure 10.33 The Δ^+ particle can be considered to be an excited state of three quarks, with the proton representing the ground state. A decay to the ground state can occur with the emission of a neutral pion (π^0).

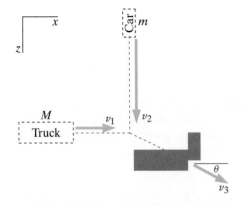

x

z

Car m

M v_1 v_2

Truck

θ

v_3

Figure 10.34 A sticking collision between a car and a truck on an icy road. This is a top view, looking down on the xz plane from a location on the $+y$ axis.

| 10.11 | **COLLISIONS IN OTHER REFERENCE FRAMES** |

In the case of a bowling ball colliding head-on with a Ping-Pong ball, we saw that it can be useful to carry out the analysis in a moving reference frame. In this section we will explore this further by analyzing a two-dimensional sticking collision in two different reference frames.

A Sticking Collision

We'll consider an inelastic collision in which a truck and a car collide in an icy intersection and stick together (Figure 10.34). In the figure we have shown velocity vectors rather than momentum vectors; a typical truck would have much more momentum than the car, given the truck's much greater mass.

Let M be the mass of the truck, m the mass of the car, and θ the angle of their mutual final velocity, as shown in the diagram. Also let v_1 be the initial speed of the truck, v_2 the initial speed of the car, and v_3 the final speed when they're stuck together and sliding on the ice.

QUESTION How does the fact that the intersection is icy affect our ability to apply conservation of momentum to the truck–car collision?

Friction with the road is small if the intersection is icy, so we can presumably ignore road friction and air resistance during the short time of the actual collision. Therefore the total momentum of the truck + car system should be (approximately) unchanged. Momentum conservation is really three equations, one each for the x, y, and z directions. In the y direction (vertically upward) the ice simply supports the weights of the vehicles, so the net force is zero, and the zero y component of momentum remains zero throughout the collision.

QUESTION   In terms of the named quantities, write two equations for the x and z components of the momentum conservation equation, equating momentum components before the collision to momentum components after the vehicles stick together.

The unit vector $\hat{v}_3$ giving the final direction of motion is:

$$\hat{v}_3 = \langle \cos\theta, 0, \cos(90° - \theta) \rangle = \langle \cos\theta, 0, \sin\theta \rangle$$

so

$$\vec{v}_3 = v_3 \langle \cos\theta, 0, \sin\theta \rangle = \langle v_3 \cos\theta, 0, v_3 \sin\theta \rangle$$

The momentum conservation equation is:

$$\langle Mv_1, 0, 0 \rangle + \langle 0, 0, mv_2 \rangle = \langle (M+m)v_3 \cos\theta, 0, (M+m)v_3 \sin\theta \rangle$$

Separating this equation into x and z components:

$$x: \quad Mv_1 = (M+m)v_3 \cos\theta$$

$$z: \quad mv_2 = (M+m)v_3 \sin\theta$$

Assume that we know the initial speeds of the truck and car (v_1 and v_2). Then we have two equations in two unknowns (v_3 and θ), and we can solve for the final speed and angle. To find θ, divide the z equation by the x equation:

$$\frac{(M+m)v_3 \sin\theta}{(M+m)v_3 \cos\theta} = \frac{mv_2}{Mv_1} = \tan\theta$$

so

$$\theta = \arctan\left(\frac{mv_2}{Mv_1}\right)$$

To find v_3, square the equations and add them:

$$(Mv_1)^2 + (mv_2)^2 = [(M+m)v_3]^2 [(\sin\theta)^2 + (\cos\theta)^2]$$

$$\sqrt{(Mv_1)^2 + (mv_2)^2} = (M+m)v_3$$

$$v_3 = \frac{\sqrt{M^2 v_1^2 + m^2 v_2^2}}{M+m}$$

We used the trig identity $\sin^2\theta + \cos^2\theta = 1$, which corresponds to the Pythagorean theorem applied to a triangle whose hypotenuse has length 1.

Energy in a Sticking Collision

A sticking collision is unusual in that momentum conservation alone gives us enough equations (two) to be able to solve for the unknown final speed and unknown direction. If the car and truck bounce off each other instead of sticking, there are four unknowns instead of two, because there are two final velocities, each with a speed and a direction to be calculated. In that case momentum conservation alone doesn't provide enough equations to be able to analyze the situation fully.

There is an important issue associated with energy in a sticking collision such as this one. Using our result for the final speed, we can write the kinetic energy after the collision:

$$\frac{1}{2}(M+m)v_3^2 = \frac{1}{2}(M+m)\left[\frac{(Mv_1)^2 + (mv_2)^2}{(M+m)^2}\right]$$

After minor rearrangements we get the following:

$$\frac{1}{2}(M+m)v_3^2 = \left(\frac{M}{M+m}\right)\left(\frac{1}{2}Mv_1^2\right) + \left(\frac{m}{M+m}\right)\left(\frac{1}{2}mv_2^2\right)$$

Since $M/(M+m)$ is less than 1, and $m/(M+m)$ is also less than 1, the final kinetic energy is less than the initial kinetic energy.

QUESTION Where is the "missing" kinetic energy?

Part of the energy was radiated away as sound waves in the air, produced when the vehicles collided with a bang. Some of the energy went into raising the temperature of the vehicles, so that their thermal energy is now higher than before (larger random motions of the atoms).

Some of the energy went into deforming the metal bodies of the truck and car. A bent fender has a higher internal energy, associated with a change in the configuration of the atoms. Sometimes you can get such energy back (as when you compress a spring, changing the configuration of the atoms in the spring, and later let it push back, returning to its original configuration), and we speak of "elastic" energy. Sometimes, however, you can't get the energy back, because the configuration change gets "locked in."

A sticking collision is an example of an "inelastic" collision, referring to the effective "loss" of kinetic energy into forms that are inaccessible.

10.X.5 Car 1 headed north and car 2 headed west collide. They stick together and leave skid marks on the pavement, which show that car 1 was deflected 30° (so car 2 was deflected 60°). What can you conclude about the cars before the collision?

The Collision in a Different Reference Frame

We can show that a truck–car collision process looks a lot simpler from the vantage point of a helicopter flying over the scene with the final velocity of the truck + car combined system. This will lead to new insights. Suppose the accident takes place at night, and from the helicopter all you can see are lights mounted on top of the two vehicles. Figure 10.35 shows a multiple exposure of the scene that you would see from a helicopter that is hovering over the intersection.

Suppose instead that your helicopter flies over the intersection with the same velocity that the stuck-together truck and car share after their collision, as shown in Figure 10.36. The view thus obtained is in an important sense "simpler," as shown in Figure 10.37. From the moving observation platform you see the truck and car run into each other and then stand still (since you and they now have the same velocity). In particular, in this (moving) reference frame it is obvious that there is a loss of kinetic energy into other forms of energy, because after the collision the kinetic energy is clearly zero. In contrast, we had to go through a fair amount of algebra in the nonmoving reference frame in order to establish the loss of kinetic energy.

The reason that we imagine making these observations at night is so that you won't be distracted by the moving background. If you actually view this collision from a moving helicopter in daylight, there is a strong tendency to compensate mentally for your motion and to put yourself (mentally) into the reference frame of the road, in which case you could miss the important fact that *relative to you*, the car and truck are stationary after the collision.

You may have had an experience related to this situation. You're sitting in a bus or train, waiting to leave. You think you've started forward, but then you realize that you were fooled by a neighboring bus or train moving backward. For a moment you were mentally in the reference frame of the moving vehicle.

Total Momentum Is Zero in the Center-of-Mass Frame

Viewing a collision from a moving reference frame can provide insights into all kinds of collisions, not just sticking collisions. The key point that leads to a useful generalization is that the total momentum of the truck + car system after the collision is zero when the vehicles are motionless relative to your helicopter.

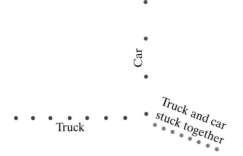

Figure 10.35 View of collision from hovering (stationary) helicopter.

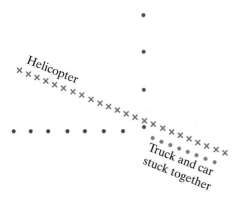

Figure 10.36 Path of moving helicopter that produces the simple view seen in Figure 10.37.

Figure 10.37 View of collision from helicopter moving with velocity of stuck-together wreck.

QUESTION   In this moving reference frame, if the total momentum is zero after the collision, what was the total momentum before the collision? Why?

The helicopter is moving at constant velocity (constant speed and direction), and we know that the Momentum Principle is valid in this uniformly moving reference frame. Thus, the principle of momentum conservation still holds, and the initial total momentum of the car + truck system must be zero if the final total momentum is zero. That's what makes the process look so simple: the total momentum is zero before and after the collision in the moving reference frame.

In viewing the scene from the moving helicopter, we have subtracted from each of the velocities an unchanging, constant vector velocity, the velocity $\vec{v}_h$ of the helicopter, and this subtraction makes no change in the Momentum Principle for the car (or truck):

$$\frac{d[m(\vec{v} - \vec{v}_h)]}{dt} = \frac{d[m\vec{v}]}{dt} = \vec{F}_{\text{net}}$$

The net force is the same in both reference frames because the contact electric interaction forces depend on interatomic distances, which are the same in both reference frames.

Evidently we gain insight into a collision by viewing the collision in a moving reference frame where the total momentum is zero. How do we determine the appropriate velocity for such a reference frame? Let this velocity be $\vec{v}_{\text{fr}}$, which is to be subtracted from all the original velocities to obtain the velocities as viewed in the moving reference frame. We want the total momentum to add up to zero in the moving frame:

$$\vec{p}_{\text{sys,new frame}} = m_1(\vec{v}_1 - \vec{v}_{\text{fr}}) + m_2(\vec{v}_2 - \vec{v}_{\text{fr}}) = \vec{0}$$
$$(m_1 + m_2)\vec{v}_{\text{fr}} = m_1\vec{v}_1 + m_2\vec{v}_2$$
$$\vec{v}_{\text{fr}} = \frac{m_1\vec{v}_1 + m_2\vec{v}_2}{m_1 + m_2}$$

We have solved for a particular velocity for the moving reference frame that makes the total momentum in that frame be zero—but look at the form of this equation.

QUESTION   Does this equation look familiar?

Evidently $\vec{v}_{\text{fr}} = \vec{v}_{\text{CM}}$, the velocity of the center of mass of the system, since

$$\vec{r}_{\text{CM}} = \frac{m_1\vec{r}_1 + m_2\vec{r}_2}{m_1 + m_2}$$

We have obtained an important result. In a reference frame moving with the center of mass of a system, the total momentum of the system is zero. If no external forces act, the total momentum is unchanged, and the total momentum is zero before and after a collision.

We can also express the frame velocity directly in terms of the total momentum:

$$\vec{v}_{\text{CM}} = \frac{m_1\vec{v}_1 + m_2\vec{v}_2}{m_1 + m_2} = \frac{\vec{p}_{\text{sys}}}{M_{\text{tot}}}$$

This provides another way to calculate the frame velocity and shows why this is often called the "center-of-momentum" velocity.

Let's reanalyze the car–truck collision from this new point of view. Again let M be the mass of the truck, m the mass of the car, and θ the angle of their mutual final velocity, as shown in Figure 10.38. Also let v_1 be the initial speed of the truck, v_2 the initial speed of the car, and v_3 the final speed when they're stuck together and sliding on the ice.

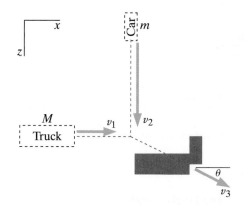

Figure 10.38 Collision of car and truck.

QUESTION What are the x and z components of $\vec{v}_{CM}$ in terms of v_1 and v_2?

$$\vec{v}_{CM} = \left\langle \frac{Mv_1}{M+m}, 0, \frac{mv_2}{M+m} \right\rangle$$

QUESTION In the moving center-of-momentum reference frame, what are the x and z components of the car's velocity before the collision? Remember that you find the car's velocity (or its components) in the center-of-momentum frame by subtracting the center-of-momentum velocity (or its components) from that of the car. (That is, the view from the helicopter is obtained by subtracting the helicopter's velocity from the velocities of the car and truck.)

$$\left\langle -\frac{Mv_1}{M+m}, 0, v_2 - \frac{mv_2}{M+m} \right\rangle = \left\langle -\frac{Mv_1}{M+m}, 0, \frac{Mv_2}{M+m} \right\rangle$$

QUESTION In the moving center-of-momentum reference frame, what are the x and z components of the truck's velocity before the collision? Show that this velocity is in a direction opposite to the velocity of the car, and that the total momentum is zero.

$$\left\langle v_1 - \frac{Mv_1}{M+m}, 0, -\frac{mv_2}{M+m} \right\rangle = \left\langle \frac{mv_1}{M+m}, 0, -\frac{mv_2}{M+m} \right\rangle$$

The x and z components of velocity are in the opposite direction to those of the car (multiply the car's velocity by $-m/M$ and you have the truck's velocity). Also, the total momentum is zero in the center-of-momentum frame:

$$\left\langle \left[m\left(-\frac{Mv_1}{M+m}\right) + M\left(\frac{mv_1}{M+m}\right) \right], 0, \left[m\left(\frac{Mv_2}{M+m}\right) + M\left(-\frac{mv_2}{M+m}\right) \right] \right\rangle = \vec{0}$$

QUESTION In the moving center-of-momentum reference frame, what is the speed of the car after the collision? Of the truck? Why?

Both are zero; in the center-of-mass frame the total momentum is zero, and with the car and truck stuck together the only way for this to be true is for them to be at rest. Because of this, the kinetic energy loss to other kinds of energy is equal to the total kinetic energy in the center-of-momentum frame before the collision.

QUESTION Now, after the collision, transform back to the (non-moving) reference frame of the intersection, by adding back in the $\vec{v}_{CM}$ that had been subtracted. What are the speed and direction of the car after the collision? Of the truck? Compare with the results obtained in our earlier analysis of this collision.

$$\vec{v}_3 = \left\langle \frac{Mv_1}{M+m}, 0, \frac{mv_2}{M+m} \right\rangle$$

which yields $\qquad v_3 = \dfrac{\sqrt{M^2 v_1^2 + m^2 v_2^2}}{M+m}; \qquad \theta = \arctan\left(\dfrac{mv_2}{Mv_1}\right)$

These are the same results we obtained in our earlier analysis.

Energy in the Two Reference Frames

As far as energy is concerned, note that the kinetic energy of the car + truck system is as usual

$$K_{tot} = K_{trans} + K_{rel}$$

where $K_{trans} = \dfrac{p_{sys}^2}{2M_{tot}}$. In the original reference frame, $K_{trans} > 0$ but is the same before and after the collision, since p_{sys} doesn't change. In the center-of-momentum reference frame, $K_{trans} = 0$ both before and after the collision, because $p_{sys} = 0$ in the center-of-momentum reference frame.

In either reference frame, the kinetic energy relative to the center of mass, K_{rel}, decreases to zero in the collision, because the car and truck after the collision are traveling with the center of mass. It is ΔK_{rel} that represents the loss of kinetic energy in this inelastic collision, and this quantity is the same in both reference frames. The increase in internal energy is relatively easy to calculate in the center-of-momentum reference frame, as it is simply

$$\Delta E_{int} = -\Delta K_{rel} = 0 - (K_{truck,i} + K_{car,i})$$

Also note that distances between atoms are the same in both reference frames, so potential energy associated with bending of metal parts is the same in both frames.

10.X.6 A 1000 kg car moving east at 30 m/s runs head-on into a 3000 kg truck moving west at 20 m/s. The vehicles stick together. Use the concept of the center-of-momentum frame to determine how much kinetic energy is lost.

SUMMARY

New concepts
Elastic and inelastic collisions
Scattering, and scattering distributions
Impact parameter (0 if head-on)
The center-of-mass reference frame

$$\vec{v}_{cm} = \frac{m_1\vec{v}_1 + m_2\vec{v}_2}{m_1 + m_2} = \frac{\vec{p}_{sys}}{M_{tot}}$$

Simple results for head-on elastic collisions between Ping-Pong balls and bowling balls
Evidence for the nuclear model of an atom

Problem-solving techniques
Conservation of momentum and conservation of energy applied to collisions
Simplification of calculations in the center-of-mass reference frame

EXERCISES AND PROBLEMS

Sections 10.1–10.2

10.X.7 A force $\langle 30, 0, 0 \rangle$ N acts for 0.2 seconds on an object of mass 1.2 kg whose initial velocity was $\langle 0, 20, 0 \rangle$ m/s. What is the new velocity?

10.X.8 Under what conditions is the momentum of a system constant? Can the x component of momentum be constant even if the y component is changing? In what circumstances? Give an example of such behavior.

10.X.9 A ball whose mass is 0.2 kg hits the floor with a speed of 8 m/s and rebounds upward with a speed of 7 m/s. The ball was in contact with the floor for 0.5 ms $(0.5 \times 10^{-3}$ s).

(a) What was the average magnitude of the force exerted on the ball by the floor?

(b) Calculate the magnitude of the gravitational force that the Earth exerts on the ball.

(c) In a collision, for a brief time there are forces between the colliding objects that are much greater than external forces. Compare the magnitudes of the forces found in parts (a) and (b).

10.X.10 In a collision between an electron and a hydrogen atom, why is it useful to select both objects as the system? Pick all that apply:

(A) The total momentum of the system does not change during the collision.

(B) The sum of the final kinetic energies must equal the sum of the initial kinetic energies for a two-object system.

(C) The kinetic energy of a two-object system is nearly zero.

(D) The forces the objects exert on each other are internal to the system and don't change the total momentum of the system.

(E) During the time interval from just before to just after the collision, external forces are negligible.

10.X.11 Two asteroids in outer space collide and stick together. The mass of each asteroid, and the velocity of each asteroid before the impact, are known. To find the momentum of the stuck-together asteroids after the impact, what approach would be useful?

(A) Use the Energy Principle.

(B) Use the Momentum Principle.

(C) It depends on whether or not the speed of the asteroids was near the speed of light.

(D) Use the relationship among velocity, displacement, and time.

(E) It depends on whether the collision was elastic or inelastic.

10.X.12 You know that a collision must be "elastic" if:

(A) The colliding objects stick together.

(B) The colliding objects are stretchy or squishy.

(C) The sum of the final kinetic energies equals the sum of the initial kinetic energies.

(D) There is no change in the internal energies of the objects (thermal energy, vibrational energy, etc.).

(E) The momentum of the two-object system doesn't change.

10.P.13 A spring has an unstretched length of 0.32 m. A block with mass 0.2 kg is hung at rest from the spring, and the spring becomes 0.4 m long (Figure 10.39). Next the spring is stretched to a length of 0.43 m and the block is released from rest. Air resistance is negligible.

Figure 10.39

(a) How long does it take for the block to return to where it was released?

(b) Next the block is again positioned at rest, hanging from the spring (0.4 m long) as shown in Figure 10.40. A bullet of mass 0.003 kg traveling at a speed of 200 m/s straight upward buries itself in the block, which then reaches a maximum height h above its original position. What is the speed of the block immediately after the bullet hits?

(c) Now write an equation that could be used to determine how high the block goes after being hit by the bullet (a height h), but you need not actually solve for h.

Figure 10.40

Sections 10.3–10.5

10.X.14 What happens to the velocities of the two objects when a high-mass object hits a low-mass object head-on? When a low-mass object hits a high-mass object head-on?

10.X.15 Object A has mass $m_A = 7$ kg and initial momentum $\vec{p}_{A,i} = \langle 17, -5, 0 \rangle$ kg $\cdot$ m/s, just before it strikes object B, which has mass $m_B = 11$ kg. Just before the collision object B has initial momentum $\vec{p}_{B,i} = \langle 4, 6, 0 \rangle$ kg $\cdot$ m/s.

(a) Consider a system consisting of both objects A and B. What is the total initial momentum of this system, just before the collision?

(b) The forces that A and B exert on each other are very large but last for a very short time. If we choose a time interval from just before to just after the collision, what is the approximate value of the impulse applied to the two-object system due to forces exerted on the system by objects outside the system?

(c) Therefore, what does the Momentum Principle predict that the total final momentum of the system will be, just after the collision?

(d) Just after the collision, object A is observed to have momentum $\vec{p}_{A,f} = \langle 13, 3, 0 \rangle$ kg $\cdot$ m/s. What is the momentum of object B just after the collision?

At this point we've learned all that we can from applying the Momentum Principle. Next we'll see what additional information we can obtain by using the Energy Principle.

(e) Before the collision, what was the magnitude of the momentum of object A?

(f) Before the collision, what was the kinetic energy of object A? Remember that you can calculate kinetic energy not only from $K = \frac{1}{2}m|\vec{v}|^2$ but more directly from $K = \frac{1}{2}|\vec{p}|^2/m$.

(g) Before the collision, what was the magnitude of the momentum of object B?

(h) Before the collision, what was the kinetic energy of object B?

(i) After the collision, what was the magnitude of the momentum of object A?

(j) After the collision, what was the kinetic energy of object A?

(k) After the collision, what was the magnitude of the momentum of object B?

(l) After the collision, what was the kinetic energy of object B?

(m) Before the collision, what was the total kinetic energy of the system?

(n) After the collision, what was the total kinetic energy of the system?

(o) Was this collision elastic, inelastic, or is there not enough information to decide? (Remember that an "elastic" collision is one in which the final value of the total kinetic energy is equal to the initial value of the total kinetic energy.)

(p) Assume that all of the energy is either kinetic energy or thermal energy. Calculate the increase of thermal energy of the two objects.

(q) In the preceding analysis, what assumption did you make about Q (energy flow from surroundings into the system due to a temperature difference)?

  (A) $Q = \Delta E_{thermal}$

  (B) Q was negligible because the process lasted only a short time.

  (C) Q can be ignored because colliding objects never get hot.

10.X.16 In outer space a rock whose mass is 3 kg and whose velocity was $\langle 3900, -2900, 3000 \rangle$ m/s struck a rock with mass 13 kg and velocity $\langle 220, -260, 300 \rangle$ m/s. After the collision, the 3 kg rock's velocity is $\langle 3500, -2300, 3500 \rangle$ m/s.

(a) What is the final velocity of the 13 kg rock?

(b) What is the change in the internal energy of the rocks?

(c) Which of the following statements about Q (transfer of energy into the system because of a temperature difference between system and surroundings) are correct?

  (A) $Q \approx 0$ because the duration of the collision was very short.

  (B) $Q = \Delta E_{thermal}$ of the rocks.

  (C) $Q \approx 0$ because there are no significant objects in the surroundings.

  (D) $Q = \Delta K$ of the rocks.

10.X.17 In outer space, rock 1 whose mass is 5 kg and whose velocity was $\langle 3300, -3100, 3400 \rangle$ m/s struck rock 2, which was at rest. After the collision, rock 1's velocity is $\langle 2800, -2400, 3700 \rangle$ m/s.

(a) What is the final momentum of rock 2?

(b) Before the collision, what was the kinetic energy of rock 1?

(c) Before the collision, what was the kinetic energy of rock 2?

(d) After the collision, what is the kinetic energy of rock 1?

(e) Suppose that the collision was elastic (that is, no change in kinetic energy and therefore no change in thermal or other internal energy of the rocks). In that case, after the collision, what is the kinetic energy of rock 2?

(f) On the other hand, suppose that in the collision some of the kinetic energy is converted into thermal energy of the two rocks, where $\Delta E_{thermal,1} + \Delta E_{thermal,2} = 7.16 \times 10^6$ J. What is the final kinetic energy of rock 2?

(g) In this case (some of the kinetic energy being converted to thermal energy), what was the transfer of energy Q (microscopic work) from the surroundings into the two-

rock system during the collision? (Remember that Q represents energy transfer due to a temperature difference between a system and its surroundings.)

10.X.18 Just before a baseball is struck by a bat, the baseball has momentum $\vec{p}_{ball}$, and the bat has momentum $\vec{p}_{bat}$. What is the total momentum of the system consisting of the baseball and the bat?

10.X.19 Consider a head-on collision between two objects. Object 1, which has mass m_1, is initially in motion, and collides head-on with object 2, which has mass m_2 and is initially at rest. Which of the following statements about the collision are true?

  (A) $\vec{p}_{1,initial} = \vec{p}_{1,final} + \vec{p}_{2,final}$

  (B) $|\vec{p}_{1,final}| < |\vec{p}_{1,initial}|$.

  (C) If $m_2 >> m_1$, then $|\Delta \vec{p}_1| > |\Delta \vec{p}_2|$.

  (D) If $m_1 >> m_2$, then the final speed of object 2 is less than the initial speed of object 1.

  (E) If $m_2 >> m_1$, then the final speed of object 1 is greater than the final speed of object 2.

10.X.20 A bullet of mass 0.102 kg traveling horizontally at a speed of 300 m/s embeds itself in a block of mass 3.5 kg that is sitting at rest on a nearly frictionless surface.

(a) What is the speed of the block after the bullet embeds itself in the block?

(b) Calculate the kinetic energy of the bullet plus the block before the collision.

(c) Calculate the kinetic energy of the bullet plus the block after the collision.

(d) Was this collision elastic or inelastic?

(e) Calculate the rise in internal energy of the bullet plus block as a result of the collision.

10.P.21 A car of mass 2300 kg collides with a truck of mass 4300 kg, and just after the collision the car and truck slide along, stuck together, with no rotation. The car's velocity just before the collision was $\langle 38, 0, 0 \rangle$ m/s, and the truck's velocity just before the collision was $\langle -16, 0, 27 \rangle$ m/s.

(a) Your first task is to determine the velocity of the stuck-together car and truck just after the collision. What system and principle should you use?

  (A) Energy Principle

  (B) Car plus truck

  (C) Momentum Principle

  (D) Car alone

  (E) Truck alone

(b) What is the velocity of the stuck-together car and truck just after the collision?

(c) In your analysis in part (b), why can you neglect the effect of the force of the road on the car and truck?

(d) What is the increase in internal energy of the car and truck (thermal energy and deformation)?

(e) Is this collision elastic or inelastic?

10.P.22 It has been proposed to propel spacecraft through the Solar System with a large sail that is struck by photons from the Sun.

(a) Which would be more effective, a black sail that absorbs photons or a shiny sail that reflects photons back toward the Sun? Explain briefly.

(b) Suppose that N photons hit a shiny sail per second, perpendicular to the sail. Each photon has energy E. What is the force on the sail? Explain briefly.

10.P.23 In a nuclear fission reactor, each fission of a uranium nucleus is accompanied by the emission of one or more high-speed neutrons, which travel through the surrounding material. If one of these neutrons is captured in another uranium nucleus, it can trigger fission, which produces more fast neutrons, which could make possible a chain reaction (Figure 10.41). However, fast neutrons have low probability of capture and usually scatter off uranium nuclei without triggering fission. In order to sustain a chain reaction, the fast neutrons must be slowed down in some material, called a "moderator." For reasons having to do with the details of nuclear physics, slow neutrons have a high probability of being captured by uranium nuclei.

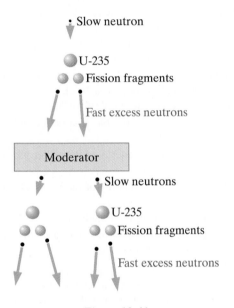

Figure 10.41

A slow neutron induces fission of U-235, with the emission of additional (fast) neutrons. The moderator is some material that slows down the fast neutrons, enabling a chain reaction.

In the following analyses, remember that neutrons have almost no interaction with electrons. Neutrons do, however, interact strongly with nuclei, either by scattering or by being captured and made part of the nucleus. Therefore you should think about neutrons interacting with nuclei (through the strong force), not with entire atoms.

(a) Based on what you now know about collisions, explain why fast neutrons moving through a block of uranium experience little change in speed.

(b) Explain why carbon should be a much better moderator of fast neutrons than uranium.

(c) Should water be a better or worse moderator of fast neutrons than carbon? Explain briefly.

Background: The first fission reactor was constructed in 1941 in a squash court under the stands of Stagg Field at the University of Chicago by a team led by the physicist Enrico Fermi. The moderator consisted of blocks of graphite, a form of carbon. The graphite had to be exceptionally pure because certain kinds of impurities have nuclei that capture neutrons with

high probability, removing them from contributing to the chain reaction. Many reactors use ordinary "light" water as a moderator, though sometimes a proton captures a neutron and forms a stable deuterium nucleus, in which case the neutron is lost to contributing to the chain reaction. Heavy water, D_2O, in which the hydrogen atoms are replaced by deuterium atoms, actually works better as a moderator than light water, because the probability of a deuterium nucleus capturing a neutron to form tritium is quite small.

Sections 10.6–10.10

10.X.24 In an elastic collision involving known masses and initial momenta, how many unknown quantities are there after the collision? How many equations are there? In a sticking collision involving known masses and initial momenta, how many unknown quantities are there after the collision? Explain how you can determine the amount of kinetic energy change.

10.X.25 In order to close a door, you throw an object at the door. Which would be more effective in closing the door, a 50 gram tennis ball or a 50 gram lump of sticky clay? Explain clearly what physics principles you used to draw your conclusion.

10.X.26 A bullet of mass m traveling horizontally at a very high speed v embeds itself in a block of mass M that is sitting at rest on a nearly frictionless surface.

(a) What is the speed of the block after the bullet embeds itself in the block?

(b) Calculate the total translational kinetic energy before and after the collision. Compare the two results and explain why there is a difference.

10.X.27 What properties of the alpha particle and the gold nucleus in the original Rutherford experiment were responsible for the collisions being elastic collisions?

10.X.28 Give an example of what we can learn about matter through the use of momentum and energy conservation applied to scattering experiments. Explain what it is that we cannot learn this way, for which we need to measure the distribution of scattering angles.

10.X.29 What is it about analyzing collisions in the center-of-mass frame that simplifies the calculations?

10.P.30 A projectile of mass m_1 moving with speed v_1 in the $+x$ direction strikes a stationary target of mass m_2 head-on. The collision is elastic. Use the Momentum Principle and the Energy Principle to determine the final velocities of the projectile and target, making no approximations concerning the masses. After obtaining your results, see what your equations would predict if $m_1 \gg m_2$, or if $m_2 \gg m_1$. Verify that these predictions are in agreement with the analysis in this chapter of the Ping-Pong ball hitting the bowling ball, and of the bowling ball hitting the Ping-Pong ball.

10.P.31 A gold nucleus contains 197 nucleons packed tightly against each other. A single nucleon (proton or neutron) has a radius of about 1×10^{-15} m.

(a) Calculate the approximate radius of the gold nucleus.

(b) Rutherford correctly predicted the angular distribution for 10-MeV (kinetic energy) alpha particles colliding with gold nuclei. What kinetic energy must alpha particles have in order to make contact with a gold nucleus? (In this case the angular distribution will deviate from that predicted by Rutherford, which was based solely on electric interactions.)

10.P.32 An alpha particle (a helium nucleus, containing 2 protons and 2 neutrons) starts out with kinetic energy of 10 MeV (10×10^6 eV), and heads in the $+x$ direction straight toward a gold nucleus (containing 79 protons and 118 neutrons). The particles are initially far apart, and the gold nucleus is initially at rest. Answer the following questions about the collision. If you do Problem 10.P.38, this problem will help you get started.

(a) What is the initial momentum of the alpha particle? (You may assume that its speed is small compared to the speed of light).

(b) What is the final momentum of the alpha particle, long after it interacts with the gold nucleus?

(c) What is the final momentum of the gold nucleus, long after it interacts with the alpha particle?

(d) What is the final kinetic energy of the alpha particle?

(e) What is the final kinetic energy of the gold nucleus?

(f) Assuming that the movement of the gold nucleus is negligible, calculate how close the alpha particle will get to the gold nucleus in this head-on collision.

(g) What principle did you use to calculate the distance of closest approach?

(h) The radius of the alpha particle is about 2×10^{-15} m, and the radius of the gold nucleus is about 8×10^{-15} m. At the point of closest approach, do the two particles touch each other?

(i) A gold nucleus contains 197 nucleons packed tightly against each other. A single nucleon (proton or neutron) has a radius of about 1.3×10^{-15} m. Calculate the approximate radius of the gold nucleus. (*Hint*: Remember that the volume of a sphere is $\frac{4}{3}\pi r^3$.)

(j) Also calculate the approximate radius of the alpha particle, which consists of 4 nucleons.

(k) To choose initial conditions for a program you may write to model collisions between an alpha particle and a gold nucleus, you will need to choose an initial location, and Δt.

(l) Estimating Δt: What is the initial speed of the alpha particle?

(m) We suggest that in a computational model of this process that you start the alpha particle at a distance of 1×10^{-13} m from the gold nucleus, because when it is much farther away than this, its trajectory is nearly a straight line at constant speed. If the alpha particle did not slow down, how long would it take to travel all the way to the location of the nucleus (a distance of 1×10^{-13} m)?

For an accurate computer calculation you would need to take many steps during the time the alpha particle is near the gold nucleus. Considering your estimate above, we suggest that you use $\Delta t = 1 \times 10^{-23}$ s in a computational model.

10.P.33 Redo the analysis of the Rutherford experiment, this time using the concept of the center-of-momentum reference frame. Let m = the mass of the alpha particle and M = the mass of the gold nucleus. Consider the specific case of the alpha particle rebounding straight back. The incoming alpha particle has a momentum p_1, the outgoing alpha particle has a momentum p_3, and the gold nucleus picks up a momentum p_4.

(a) Determine the velocity of the center of momentum of the system.

(b) Transform the initial momenta to that frame (by subtracting the center-of-momentum velocity from the original velocities).

(c) Show that if the momenta in the center-of-momentum frame simply turn around ($180°$), with no change in their magnitudes, both momentum and energy conservation are satisfied, whereas no other possibility satisfies both conservation principles. (Try drawing some other momentum diagrams.)

(d) After the collision, transform back to the original reference frame (by adding the center-of-momentum velocity to the velocities of the particles in the center-of-mass frame). Although using the center-of-momentum frame may be conceptually more difficult, the algebra for solving for the final speeds is much simpler.

10.P.34 A Fe-57 nucleus is at rest and in its first excited state, 14.4 keV above the ground state (14.4×10^3 eV, where 1 eV = 1.6×10^{-19} J). The nucleus then decays to the ground state with the emission of a gamma ray (a high-energy photon).

(a) What is the recoil speed of the nucleus?

(b) Calculate the slight difference in eV between the gamma-ray energy and the 14.4 keV difference between the initial and final nuclear states.

(c) The "Mössbauer effect" is the name given to a related phenomenon discovered by Rudolf Mössbauer in 1957, for which he received the 1961 Nobel Prize for physics. If the Fe-57 nucleus is in a solid block of iron, occasionally when the nucleus emits a gamma ray the entire solid recoils as one object. This can happen due to the fact that neighboring atoms and nuclei are connected by the electric interatomic force. In this case, repeat the calculation of part (b) and compare with your previous result. Explain briefly.

10.P.35 There is an unstable particle called the "sigma-minus" (Σ^-), which can decay into a neutron and a negative pion (π^-): $\Sigma^- \rightarrow n + \pi^-$. The mass of the Σ^- is 1196 MeV/c^2, the mass of the neutron is 939 MeV/c^2, and the mass of the π^- is 140 MeV/c^2. Write equations that could be used to calculate the momentum and energy of the neutron and the pion. You do not need to solve the equations, which would involve some messy algebra. However, be clear in showing that you have enough equations that you could in principle solve for the unknown quantities in your equations.

It is advantageous to write the equations not in terms of v but rather in terms of E and p; remember that $E^2 - (pc)^2 = (mc^2)^2$.

10.P.36 A beam of high-energy π^- (negative pions) is shot at a flask of liquid hydrogen, and sometimes a pion interacts through the strong interaction with a proton in the hydrogen, in the reaction $\pi^- + p^+ \rightarrow \pi^- + X^+$, where X^+ is a positively charged particle of unknown mass.

The incoming pion momentum is 3 GeV/c (1 GeV = 1000 MeV = 1×10^9 electron-volts). The pion is scattered through $40°$, and its momentum is measured to be 1510 MeV/c (this is done by observing the radius of curvature of its circular trajectory in a magnetic field). A pion has a rest energy of 140 MeV, and a proton has a rest energy of 938 MeV.

What is the rest mass of the unknown X^+ particle, in MeV/c^2? Explain your work carefully.

It is advantageous to write the equations not in terms of v but rather in terms of E and p; remember that $E^2 - (pc)^2 = (mc^2)^2$.

10.P.37 At the PEP II facility at the Stanford Linear Accelerator Center (SLAC) in California and at the KEKB facility in Japan, electrons with momentum 9.03 GeV/c were made to collide head-on with positrons whose momentum is 3.10 GeV/c (1 GeV = 10^9 eV); see Figure 10.42. That is, pc for the electron is 9.03 GeV and pc for the positron is 3.10 GeV. The values of pc and the corresponding energies are so large with respect to the electron or positron rest energy (0.5 MeV = 0.0005 GeV) that for the purposes of this analysis you may, if you wish, safely consider the electron and positron to be massless.

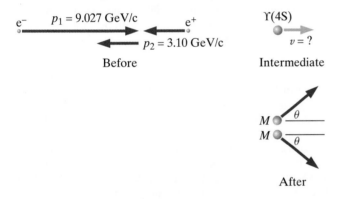

Figure 10.42

(a) The electron-positron collision produces in an intermediate state a particle called the $\Upsilon(4S)$ ("Upsilon 4S"), in the reaction $e^- + e^+ \rightarrow \Upsilon(4S)$. Show that the rest energy of the $\Upsilon(4S)$ is 10.58 GeV.

(b) What is the speed of the $\Upsilon(4S)$ produced in the collision?

(c) The $\Upsilon(4S)$ decays almost immediately into two "B" mesons: $\Upsilon(4S) \rightarrow B^0 + \overline{B^0}$. The $\overline{B^0}$ is the antiparticle of the B^0 and has the same rest energy $Mc^2 = 5.28$ GeV as the B^0. Consider the case in which both "B" mesons are emitted at the same angle θ to the direction of the moving $\Upsilon(4S)$, as shown in Figure 10.43. Calculate this angle θ.

Figure 10.43

Computational Problems

10.P.38 Use numerical integration techniques to explore possible trajectories of an alpha particle when it collides with a gold nucleus at different impact parameters. Your program should display the paths of the alpha particle and of the gold nucleus, and also plot the x and y components of momentum of both particles as a function of time. We will consider motion only in two dimensions (Figure 10.44; a slice through a 3-D beam of alpha particles). You may find it useful to work through Problem 10.P.32 as background for this computational model.

An alpha particle is a helium nucleus (two protons and two neutrons, charge $+2e$); in the Rutherford experiment its initial kinetic energy was about 10 MeV (remember that 1 eV = 1.6 × 10^{-19} J). The gold nucleus is inside a piece of gold foil and is initially stationary; its charge is $+79e$ and it contains 197 nucleons. Initially we will ignore the interactions of the alpha particle with the cloud of electrons surrounding the nucleus, which are in fact small.

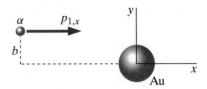

Figure 10.44

(a) In this model it is important to pick reasonable initial conditions and display scales—otherwise you may not see anything meaningful. Before beginning your program, calculate or estimate reasonable values for the following and report your results:

- *Closest approach.* For a head-on collision, calculate approximately the distance of closest approach, given that the alpha particle and the gold nucleus repel each other. Assume that the gold atom's electrons are scattered away by the incoming alpha particle.

- *Initial position of the alpha particle.* Given the distance of closest approach, what would be a reasonable initial x for the alpha particle? What would be a good range to explore for the impact parameter b (initial y coordinate of the alpha particle)? Use these estimates to decide on a scale for the coordinates used to display the trajectories.

- *Value of Δt.* Trial and error alone will not be very helpful in estimating Δt. You can calculate the initial speed of the alpha particle, and you know the approximate size scale for the collision. Choose a value of Δt such that the alpha particle doesn't move very far in one step.

- *Time scale.* To plot p_x and p_y vs. t during the interaction, you must select a scale for the time axis that allows you to observe the entire interaction. About how long should the time axis be?

- *Momentum scale.* What should the maximum value of the p_x axis be? You should scale the p_y axis to the same value, to make comparisons easy.

- *Exit criterion.* You will need to exit from your computational loop when the particles have moved far apart. How will you test for this?

(b) Write a program that displays:

- The paths of the alpha particle and the gold nucleus.
- The value of the impact parameter b, and the value of the corresponding scattering angle (angle between the initial path of the alpha particle and its final path). The angle can be calculated by taking the arctangent of p_y/p_x, using the final values.
- A plot of p_x for each particle and total p_x, and a similar plot for p_y. (Consider what these plots should look like if the program is working properly.) Report what value of Δt gives adequate accuracy. What is your criterion? Report values of the impact parameter b that produce the following:
- A very small scattering angle
- An angle around 90 degrees
- A large deflection angle (back-scattering)

(c) What other physical quantity could you plot that would provide a check on the accuracy of your calculation? Carry out this check and report what value of Δt provides good accuracy according to this check.

(d) To see what a different model of matter would have predicted, replace the gold nucleus in your model by a pro-

ton. Describe briefly what you observe. (Note that if the protons were uniformly spread out throughout the volume of the atom, they would be far enough apart that on average an alpha particle would interact with only one of them.)

ANSWERS TO EXERCISES

10.X.1 (a) 300 J; **(b)** 5 m/s; **(c)** 150 J; **(d)** 150 J
10.X.2 2.3×10^{-24} m/s (!) So it is an excellent approximation to say that the Earth doesn't recoil.
10.X.3 1.6×10^{-23} J; $K_{\text{baseball}} = 150$ J
10.X.4 8×10^4 m/s
10.X.5 Car 1's momentum was $\sqrt{3}$ greater than the momentum of car 2.
10.X.6 We outline the entire solution.

First find $v_{\text{cm},x}$:

$$v_{\text{cm},x} = \frac{(1000\,\text{kg})(30\,\text{m/s}) + (3000\,\text{kg})(-20\,\text{m/s})}{(4000\,\text{kg})} = -7.5\,\text{m/s}$$

Transform to center-of-momentum frame:

$$v_{\text{car},x} = (30\,\text{m/s}) - (-7.5\,\text{m/s}) = 37.5\,\text{m/s}$$
$$v_{\text{truck},x} = (-20\,\text{m/s}) - (-7.5\,\text{m/s}) = -12.5\,\text{m/s}$$

Check to make sure that the momenta are equal and opposite in this frame:

$$p_{\text{car},x} = (1000\,\text{kg})(37.5\,\text{m/s}) = 37500\,\text{kg} \cdot \text{m/s}$$
$$p_{\text{truck},x} = (3000\,\text{kg})(-12.5\,\text{m/s}) = -37500\,\text{kg} \cdot \text{m/s}$$

In this frame the kinetic energy before the collision is this:

$$K = \frac{(37500\,\text{kg} \cdot \text{m/s})^2}{2(1000\,\text{kg})} + \frac{(37500\,\text{kg} \cdot \text{m/s})^2}{2(3000\,\text{kg})} = 9.4 \times 10^5\,\text{J}$$

After the collision, the velocities are zero in this frame, so kinetic energy in this frame goes to zero, and there is an increase in the internal energy of the mangled car and truck of amount 9.4×10^5 J. There must be the same internal energy increase in the original reference frame, so the kinetic energy lost in the original reference frame was 9.4×10^5 J.

Angular Momentum

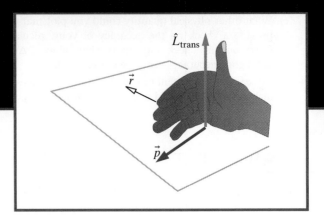

- **Angular momentum is a measure of rotational motion.**
- **Angular momentum has two parts:**
  - *Rotational* or *spin* angular momentum (around the center of mass)
  - *Translational* or *orbital* angular momentum (relative to a specific location)
    * Even an object moving in a straight line can have translational angular momentum
- **The change in angular momentum of a system is equal to the net angular impulse applied, which involves:**
  - The net torque (twist) exerted by all objects not in the system
  - The time interval over which the interaction occurs
- **Angular momentum and torque are vectors involving cross products.**
- **Angular momentum is conserved: The change in angular momentum of a system plus the change in angular momentum of the surroundings is zero.**
- **Angular momentum in microscopic systems is quantized.**

Rotational Motion

Angular momentum is a measure of rotational motion. It is useful to divide angular momentum into two parts: translational (or "orbital") angular momentum, and rotational (or "spin") angular momentum. A particular object can have either or both kinds of angular momentum. For example, the orbit of the Earth around the Sun can be described in terms of translational (orbital) angular momentum, while the revolution of the Earth around its own axis can be described in terms of rotational (spin) angular momentum (Figure 11.1). We will see that we can relate translational angular momentum to translational kinetic energy, and rotational angular momentum to rotational kinetic energy.

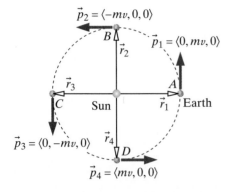

Figure 11.1 The Earth orbits the Sun, and it also spins about its own axis (which is tilted with respect to the orbital plane). The sizes of the Earth and Sun are exaggerated in this diagram.

Figure 11.2 The Earth goes around the Sun in a nearly circular orbit. The position vector and the momentum vector change continuously.

11.1 TRANSLATIONAL ANGULAR MOMENTUM

As the Earth goes around the Sun in a nearly circular orbit, the position vector (from Sun to Earth) is constantly changing in direction, and the momentum of the Earth is also constantly changing in direction, as shown in Figure 11.2. Both the position vector and the momentum vector continually change, though their magnitudes are (nearly) constant. In Figure 11.2, as the Earth goes from A to B in three months (one quarter of a year), the Earth's momentum changes from $\langle 0, mv, 0 \rangle$ to $\langle -mv, 0, 0 \rangle$, and the change in the momentum is $\Delta \vec{p} = \langle -mv, 0, 0 \rangle - \langle 0, mv, 0 \rangle = \langle -mv, -mv, 0 \rangle$. This change in the Earth's momentum is of course due to the gravitational force exerted on the Earth by the Sun.

Nevertheless, it feels like something is constant here. For example, the angle between the position vector and the momentum vector is always 90° for this circular orbit. Also, the direction of rotation of the circular motion stays constant (counterclockwise in Figure 11.2). The quantity that is constant is $\vec{L}_{\text{trans,Sun}}$, the translational angular momentum of the Earth relative to the Sun.

Magnitude of Translational Angular Momentum

The magnitude of the Earth's orbital or translational angular momentum, relative to the center of the Sun, involves $|\vec{r}|$, $|\vec{p}|$, and the angle θ between the two. It is defined like this:

$$\left| \vec{L}_{\text{trans,Sun}} \right| = |\vec{r}|\,|\vec{p}| \sin \theta$$

EXAMPLE

Earth's Translational Angular Momentum

Calculate the magnitude of the Earth's translational (orbital) angular momentum relative to the Sun when the Earth is at location A and when the Earth is at location B as shown in Figure 11.2. The mass of the Earth is 6×10^{24} kg and its distance from the Sun is 1.5×10^{11} m (see the table of values at the end of this book.)

Solution

The Earth makes one complete orbit of the Sun in 1 year, so its average speed is:

$$v = \frac{2\pi (1.5 \times 10^{11}\,\text{m})}{(365)(24)(60)(60)\,\text{s}} = 3.0 \times 10^4\,\text{m/s}$$

At location A

$$\vec{p} = \langle 0, 6 \times 10^{24}\,\text{kg} \cdot 3.0 \times 10^4\,\text{m/s}, 0 \rangle = \langle 0, 1.8 \times 10^{29}, 0 \rangle\,\text{kg} \cdot \text{m/s}$$

$$|\vec{p}| = 1.8 \times 10^{29}\,\text{kg} \cdot \text{m/s}$$

$$\left| \vec{L}_{\text{trans,Sun}} \right| = (1.5 \times 10^{11}\,\text{m})(1.8 \times 10^{29}\,\text{kg} \cdot \text{m/s}) \sin 90°$$

$$= 2.7 \times 10^{40}\,\text{kg} \cdot \text{m}^2/\text{s}$$

At location B, $|\vec{r}|$, $|\vec{p}|$, and θ are the same as they were at location A, so $\left| \vec{L}_{\text{trans,Sun}} \right|$ also has the same value it had at location A.

Direction of Translational Angular Momentum

The magnitude of the translational angular momentum doesn't tell us everything we need to know about the orbital motion of the Earth. To describe the motion indicated in Figure 11.2 we could say something like this:

> If you stand at a location on the $+z$ axis and look toward the origin, the Earth appears to be moving counterclockwise in the xy plane.

However, this is annoyingly verbose and not useful mathematically; for example, one can't add "counterclockwise in the xy plane" to "clockwise in the yz plane." Instead, we specify the direction this way:

- The plane in which both $\vec{r}$ and $\vec{p}$ lie can be indicated by a unit vector perpendicular to that plane.
- The direction of motion within the plane (clockwise or counterclockwise) can be indicated by establishing a "right-hand" rule for this unit vector.
- Right-hand rule: Curl the fingers of your right hand in the direction of the rotational motion in the plane, and extend your thumb, as shown in Figure 11.3. The unit vector representing the direction of the angular momentum is defined to point in the direction of your thumb.
- If the rotational motion is counterclockwise, the unit vector (and your right thumb) point out of the plane, as in Figure 11.3. If the rotational motion is clockwise, the unit vector (and your right thumb) point into the plane, as in Figure 11.4.

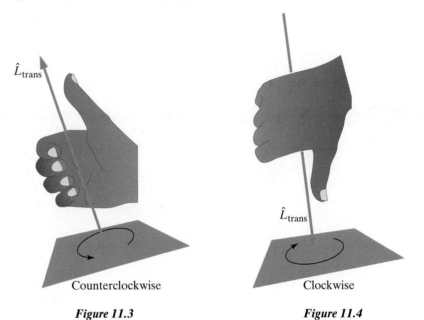

Counterclockwise   Clockwise

Figure 11.3   *Figure 11.4*

EXAMPLE   **Direction of Translational Angular Momentum**

Assuming the usual coordinate system, what is the direction of the Earth's translational angular momentum, as indicated in Figure 11.2?

Solution   Applying the right-hand rule we find that the Earth's translational angular momentum is in the $+z$ direction: out of the page toward you.

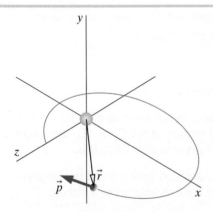

Figure 11.5 A comet orbits the Sun. The momentum vector and relative position vector lie in the xz plane.

Angular Momentum as a Vector

Putting together the results from the previous two examples, we can express the translational angular momentum of the Earth with respect to the Sun as a vector:

$$\vec{L}_{\text{trans,Sun}} = (2.7 \times 10^{40}\,\text{kg} \cdot \text{m}^2/\text{s}) \cdot \langle 0, 0, 1 \rangle = \langle 0, 0, 2.7 \times 10^{40} \rangle\,\text{kg} \cdot \text{m}^2/\text{s}$$

11.X.1 What is the direction of the orbital (translational) angular momentum of the comet shown in Figure 11.5, relative to the Sun?

An Elliptical Orbit: Halley's Comet

The previous exercise raises an interesting question: What if $|\vec{r}|$, $|\vec{p}|$, and θ are not constant (as they were in a circular orbit)? Is the angular momentum, relative to the Sun, of a comet in an elliptical orbit around the Sun constant, or does it change? During this orbit, all three parameters ($|\vec{r}|$, $|\vec{p}|$, and θ) change continuously.

Halley's comet orbits the Sun with a period of 75.3 years, and will next be visible on Earth in the year 2061. Figure 11.6 displays two views of the orbit, here shown in the xy plane. From the upper diagram it is clear that the orbit of the comet is highly elliptical. The lower diagram shows a magnified view of the portion of the orbit nearest to the Sun, and displays the momentum vectors and position vectors for the comet at two different points in its orbit. From available data we can calculate the angular momentum of Halley's comet relative to the Sun at two different locations.

EXAMPLE	Angular Momentum of Halley's Comet

The highly elliptical orbit of Halley's comet is shown in Figure 11.6. When the comet is closest to the Sun, at the location specified by the position vector $\vec{r}_1$ ("perihelion"), it is 8.77×10^{10} m from the Sun, and its speed is 5.46×10^4 m/s. When the comet is at the location specified by the position vector $\vec{r}_2$, its speed is 1.32×10^4 m/s. At that location the distance between the comet and the Sun is 1.19×10^{12} m, and the angle θ is $17.81°$. The mass of the comet is estimated to be 2.2×10^{14} kg. Calculate the translational (orbital) angular momentum of the comet, relative to the Sun, at both locations.

Solution Direction: At both locations, the direction of the translational angular momentum of the comet is in the $-z$ direction (into the page); determined by using the right-hand rule.

At location 1:

$$\left| \vec{L}_{\text{trans,Sun}} \right| = (8.77 \times 10^{10}\,\text{m})(2.2 \times 10^{14}\,\text{kg})(5.46 \times 10^4\,\text{m/s}) \sin 90°$$

$$= 1.1 \times 10^{30}\,\text{kg} \cdot \text{m}^2/\text{s}$$

$$\vec{L}_{\text{trans,Sun}} = \langle 0, 0, -1.1 \times 10^{30} \rangle\,\text{kg} \cdot \text{m}^2/\text{s}$$

At location 2:

$$\left| \vec{L}_{\text{trans,Sun}} \right| = (1.19 \times 10^{12}\,\text{m})(2.2 \times 10^{14}\,\text{kg})(1.32 \times 10^4\,\text{m/s}) \sin 17.81°$$

$$= 1.1 \times 10^{30}\,\text{kg} \cdot \text{m}^2/\text{s}$$

$$\vec{L}_{\text{trans,Sun}} = \langle 0, 0, -1.1 \times 10^{30} \rangle\,\text{kg} \cdot \text{m}^2/\text{s}$$

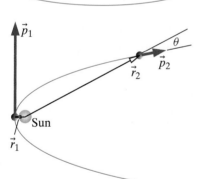

Figure 11.6 The orbit of Halley's comet, shown here in the xy plane. The top figure shows the full orbit. The bottom figure shows a magnified view of the portion of the orbit nearest the Sun.

FURTHER DISCUSSION Even in this highly elliptical orbit, the comet's translational angular momentum is constant throughout the orbit, despite the fact that its position, its momentum, and the angle between them change continuously. It was observations like this that first suggested that angular momentum might be a conserved quantity, and should not change if there is no twist exerted on the system by the surroundings.

A Straight Path

In calculating translational (orbital) angular momentum, we use the position and momentum of an object at a particular instant. We do not necessarily have any information about how the object's path will curve in the future or whether it has curved in the past. Figure 11.7 illustrates this point by showing many possible paths of a spacecraft passing an asteroid. It is possible that the spacecraft will be nearly unaffected by the asteroid, and will travel nearly in a straight line, but we can still calculate its orbital angular momentum at the instant shown.

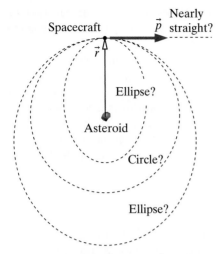

Figure 11.7 A snapshot of a spacecraft passing an asteroid. Its future motion might be a circle or an ellipse, or even a nearly straight line if the asteroid's mass is small.

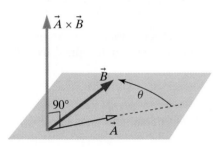

Figure 11.8 The cross product produces a vector that is perpendicular to the two original vectors.

Later in this chapter we will see why it may be physically meaningful to attribute angular momentum to an object that moves in a straight line.

Choosing a Location

Translational or orbital angular momentum is defined relative to a particular location. If we choose a location A, the magnitude of the angular momentum includes the distance r_A to the moving object. The translational angular momentum relative to some other location B would have a different magnitude because r_B would be a different distance. We will see later that it is usually advantageous to choose a location around which there is little or no twist applied by objects in the surroundings.

The Vector Cross Product

The formal definition of translational angular momentum involves a vector cross product. In the mathematics of vectors, the cross product is an operation that combines two vectors to produce a third vector that is perpendicular to the plane defined by the original vectors (Figure 11.8). The cross product of two vectors $\vec{A}$ and $\vec{B}$ is written as $\vec{A} \times \vec{B}$.

We already know one way of evaluating a vector cross product, by finding the magnitude and direction separately. The quantity

$$\left| \vec{L}_{trans,Sun} \right| = |\vec{r}| \, |\vec{p}| \sin\theta$$

is the magnitude of the cross product $\vec{r} \times \vec{p}$:

$$|\vec{r} \times \vec{p}| = \left| \vec{L}_{trans,Sun} \right| = |\vec{r}| \, |\vec{p}| \sin\theta$$

where θ is the angle between the two vectors when they are placed tail to tail. θ is always less than or equal to 180°. The direction of $\vec{r} \times \vec{p}$ is given by a right-hand rule:

- Draw the vectors $\vec{r}$ and $\vec{p}$ so their tails are in the same location.
- Point the fingers of your right hand in the direction of $\vec{r}$ (Figure 11.9), then fold them through the angle θ toward $\vec{p}$ (Figure 11.10). You may need to turn your hand over to be able to do this.
- Stick your thumb out. Your thumb points in the direction of the angular momentum vector.

Figure 11.9 With the tails of $\vec{r}$ and $\vec{p}$ together, point your fingers along $\vec{r}$.

Figure 11.10 Fold your fingers toward $\vec{p}$. Your extended thumb points in the direction of $\vec{r} \times \vec{p}$.

Figure 11.11 $r_\perp = r \sin\theta$ is the perpendicular distance from the tail of $\vec{r}$ to a line collinear with $\vec{p}$.

The Perpendicular Distance $r_\perp$

The quantity $r \sin\theta$ has a geometrical meaning. As shown in Figure 11.11, it is the component of $\vec{r}$ perpendicular to $\vec{p}$. If you draw a line perpendicular to $\vec{p}$, through the point about which $\vec{L}$ is to be calculated, $r_\perp = r \sin\theta$ is the distance along this line from the tail of $\vec{r}$ to the line along which the momentum or velocity points. $r_\perp$ is closely related to the impact parameter introduced in the previous chapter on collisions.

PERPENDICULAR DISTANCE $r_\perp$

$$r_\perp = r \sin\theta$$

Vector Components in the Cross Product

Given the components of two vectors, it is possible to evaluate the cross product in terms of unit vectors along the three axes (Figure 11.12). Recall that $\hat{\imath} = \langle 1, 0, 0 \rangle$, and so on.

Figure 11.12 Cross products of unit vectors.

First, note that $\hat{\imath} \times \hat{\imath} = 0$, $\hat{\jmath} \times \hat{\jmath} = 0$, and $\hat{k} \times \hat{k} = 0$, since when we cross a vector with itself the angle between the two vectors is zero, and $\sin 0° = 0$.

Second, $\hat{\imath} \times \hat{\jmath} = \hat{k}$, since the angle is 90° and the right-hand rule gives a result in the $+z$ direction (out of the page; Figure 11.12). On the other hand, $\hat{\jmath} \times \hat{\imath} = -\hat{k}$, because the right-hand rule gives a result in the $-z$ direction (into the page). Similarly, $\hat{\jmath} \times \hat{k} = \hat{\imath}$, $\hat{k} \times \hat{\jmath} = -\hat{\imath}$, $\hat{k} \times \hat{\imath} = \hat{\jmath}$, and $\hat{\imath} \times \hat{k} = -\hat{\jmath}$.

If $\vec{A}$ and $\vec{B}$ lie in the xy plane, we can use these results for the unit vectors to calculate the cross product, which will be in the $+z$ or $-z$ direction:

$$
\begin{aligned}
\vec{A} \times \vec{B} &= (A_x \hat{\imath} + A_y \hat{\jmath}) \times (B_x \hat{\imath} + B_y \hat{\jmath}) \\
&= (A_x B_x)\hat{\imath} \times \hat{\imath} + (A_y B_y)\hat{\jmath} \times \hat{\jmath} + (A_x B_y)\hat{\imath} \times \hat{\jmath} + (A_y B_x)\hat{\jmath} \times \hat{\imath} \\
&= 0 + 0 + A_x B_y \hat{k} - A_y B_x \hat{k} \\
&= (A_x B_y - A_y B_x)\hat{k}
\end{aligned}
$$

Similar results hold for components of the cross product in the x and y directions, so that in general we have the following result:

CROSS PRODUCT $\vec{A} \times \vec{B}$

$$\vec{A} \times \vec{B} = \langle A_y B_z - A_z B_y, \ A_z B_x - A_x B_z, \ A_x B_y - A_y B_x \rangle$$
$$|\vec{A} \times \vec{B}| = |\vec{A}||\vec{B}| \sin\theta$$

The direction of the vector is given by the right-hand rule.

For example,

$$
\begin{aligned}
\langle 2, 3, 4 \rangle \times \langle 5, 6, 7 \rangle &= \langle 3 \cdot 7 - 4 \cdot 6, \ 4 \cdot 5 - 2 \cdot 7, \ 2 \cdot 6 - 3 \cdot 5 \rangle \\
&= \langle -3, 6, -3 \rangle
\end{aligned}
$$

This approach to calculating a cross product is particularly useful in computer calculations. Note the cyclic nature of the subscripts: xyz, yzx, zxy.

If you have studied determinants, you may recognize that the components of $\vec{A} \times \vec{B}$ can be thought of as the "minors" of a 3 by 3 determinant in which $\vec{A}$ and $\vec{B}$ are the second and third rows, and the first row is $\hat{\imath}, \hat{\jmath}, \hat{k}$.

Applying this general formalism to translational angular momentum, we can define the angular momentum of a particle, relative to location A, as:

TRANSLATIONAL ANGULAR MOMENTUM OF AN OBJECT RELATIVE TO LOCATION A

$$\vec{L}_{\text{trans,A}} = \vec{r}_A \times \vec{p}$$
$$\vec{L}_{\text{trans,A}} = \langle r_y p_z - r_z p_y, \ r_z p_x - r_x p_z, \ r_x p_y - r_y p_x \rangle$$
$$\left| \vec{L}_{\text{trans,A}} \right| = |\vec{r}| \, |\vec{p}| \sin\theta$$

The direction of $\vec{L}_{\text{trans,A}}$ is given by the right-hand rule.

Note that this formal definition allows us to calculate the translational (orbital) angular momentum of an object at any instant, even if we do not know the direction of the path it will follow in the future.

Two-Dimensional Projections

Because it is more difficult to sketch a situation in three dimensions, whenever possible we will work with two-dimensional projections onto a plane. If $\vec{A}$ and $\vec{B}$ lie in the xy plane, the cross product vector $\vec{A} \times \vec{B}$ points in the $+z$ direction (out of the page, $\odot$) or in the $-z$ direction (into the page, $\otimes$). The symbol $\odot$ is supposed to suggest the tip of an arrow pointing toward you (angular momentum pointing toward you). The symbol $\otimes$ is supposed to suggest the feathers on the end of an arrow that is heading away from you (angular momentum pointing away from you).

EXAMPLE

Calculating Angular Momentum

Use three different methods to calculate the angular momentum of the particle in Figure 11.13 relative to location B. How does this value compare to the angular momentum of the same particle relative to location A and to location C?

Solution

Method 1: Vector cross product

$$\vec{r}_B = \langle -4, 3, 0 \rangle \text{ m}$$
$$\vec{p} = \langle 10, 0, 0 \rangle \text{ kg} \cdot \text{m/s}$$
$$\vec{L}_B = \vec{r}_B \times \vec{p}$$
$$= \langle r_y p_z - r_z p_y, \ r_z p_x - r_x p_z, \ r_x p_y - r_y p_x \rangle$$
$$= \langle (3)(0) - (0)(0), \ (0)(10) - (-4)(0), \ (-4)(0) - (3)(10) \rangle$$
$$= \langle 0, 0, -30 \rangle \text{ kg} \cdot \text{m}^2/\text{s}$$

Method 2: Direction and magnitude using $|\vec{r}_B| |\vec{p}| \sin\theta$

In Figure 11.14 we have placed $\vec{r}_B$ and $\vec{p}$ tail to tail. The vectors $\vec{r}_B$ and $\vec{p}$ define a plane, which is the xy plane (the plane of the paper). Now fold the fingers of your right hand from $\vec{r}_B$ toward $\vec{p}$ (Figure 11.14). To be able to do this, your thumb must point into the page ($\otimes$), so the angular momentum is in the $-z$ direction (into the plane of the paper). Note too that you fold your fingers clockwise, which is associated with angular momentum into the page.

$$|\vec{L}_B| = |\vec{r}_B| |\vec{p}| \sin\theta$$
$$|\vec{r}_B| = \sqrt{3^2 + 4^2} = 5 \text{ m}$$
$$|\vec{p}| = 10 \text{ kg} \cdot \text{m/s}$$
$$\sin\theta = 3/5 \quad \text{See Figure 11.15}$$
$$|\vec{L}_B| = (5)(10)(3/5) = 30 \text{ kg} \cdot \text{m}^2/\text{s}$$
$$\vec{L}_B = \langle 0, 0, -30 \rangle \text{ kg} \cdot \text{m}^2/\text{s}$$

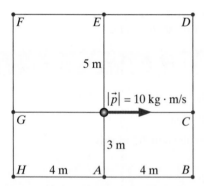

Figure 11.13 Determine the magnitude and direction of the angular momentum of the particle relative to various locations. All points lie in the xy plane.

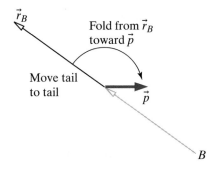

Figure 11.14 Place $\vec{r}_B$ and $\vec{p}$ tail to tail, then fold your right fingers from $\vec{r}_B$ toward $\vec{p}$.

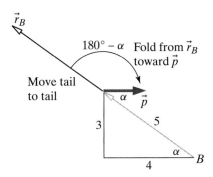

Figure 11.15 Place $\vec{r}_B$ and $\vec{p}$ tail to tail, then rotate $\vec{r}_B$ toward $\vec{p}$. The sine of this angle is $\sin(180° - \alpha) = \sin\alpha = 3/5$.

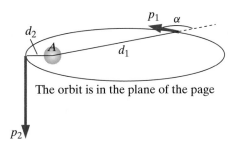

The orbit is in the plane of the page

Figure 11.16 A comet orbits the Sun. Location A is at the center of the Sun.

Figure 11.17 A bicycle wheel spinning with angular speed ω. Almost all the mass of the wheel is in the rim.

Method 3: Direction and magnitude using $r_\perp$
Determine the direction using the right-hand rule, as in Method 2 above. Figure 11.15 shows that $r_\perp = 3$ m. Therefore

$$|\vec{L}_B| = (3)(10) = 30\,\text{kg}\cdot\text{m}^2/\text{s}$$

FURTHER DISCUSSION   There are several different ways to calculate angular momentum, all giving the same answers, but some are easier to use than others in particular geometrical situations. As a rule, look first to see whether you know $r_\perp$, in which case you can calculate the magnitude from the simple formula $|\vec{L}| = r_\perp |\vec{p}|$.

We see immediately that the translational angular momentum relative to location A is exactly the same as the angular momentum relative to location B, because the perpendicular distance $r_\perp$ is the same (3 m).

Watch especially for the important special case of $r_\perp$ being zero. About location C, $r_\perp$ is zero because the momentum points straight toward C; the impact parameter is zero.

Be sure to do the following exercises before going on, because you need to be able to use the concept of cross product quickly and easily in later discussions. One way to check your work is to try more than one method to find the angular momentum.

11.X.2 Determine both the direction and magnitude of the angular momentum of the particle in Figure 11.13, relative to the locations D, E, F, G, and H. We've already analyzed the angular momentum relative to A, B, and C in the example given above. Notice how the magnitude and direction of the angular momentum relative to the different locations differ in magnitude and direction.

11.X.3 A comet orbits the Sun (Figure 11.16). When it is at location 1 it is a distance d_1 from the Sun, and has magnitude of momentum p_1. Location A is at the center of the Sun. When the comet is at location 2, it is a distance d_2 from the Sun, and has magnitude of momentum p_2.

(a) When the comet is at location 1, what is the direction of $\vec{L}_A$?

(b) When the comet is at location 1, what is the magnitude of $\vec{L}_A$?

(c) When the comet is at location 2, what is the direction of $\vec{L}_A$?

(d) When the comet is at location 2, what is the magnitude of $\vec{L}_A$?

Later we'll see that the Angular Momentum Principle tells us that the angular momentum at location 1 must be equal to the angular momentum at location 2.

11.2   ROTATIONAL ANGULAR MOMENTUM

Although its center of mass may be stationary, an object that spins around an axis through its center also has angular momentum. The Earth rotates around its axis once each day, a bicycle wheel spins around its axle, a gyroscope spins on its axis—each of these objects has rotational, or spin, angular momentum. We can calculate the angular momentum of a spinning object by dividing the object into small pieces, and adding the translational angular momentum of each piece.

As a simple example, consider a bicycle wheel spinning about its center of mass with angular speed ω (Figure 11.17). Most of the mass is in the rim of the wheel, so we will neglect the low-mass spokes. Imagine dividing the rim into 20 equal-sized pieces, each with mass $M/20$, and consider each piece to be an object orbiting the axis. The speed of one piece is

$$v = \omega R$$

so the magnitude of the translational angular momentum of one piece, relative to the center of mass, is:

$$\left|\vec{L}_{\text{CM, one piece}}\right| = R(M/20)v\sin 90°$$
$$= R(M/20)\omega R = (M/20)R^2\omega$$

By the right-hand rule, the direction of the translational angular momentum of each piece is into the page (the $-z$ direction), so to get the magnitude of the angular momentum of the entire wheel, we can add the contribution of each piece:

$$\left|\vec{L}_{\text{CM}}\right| = 20 \cdot (M/20)R^2\omega$$
$$= MR^2\omega$$

Recall that in Chapter 9 we found that $I = MR^2$ is the moment of inertia of a bicycle wheel. We can therefore rewrite the angular momentum of the wheel as

$$\left|\vec{L}_{\text{CM}}\right| = I\omega$$

The direction of the angular momentum of the entire wheel is $-z$ (into the page). Note that the result would have been the same if we had divided the wheel into 1000 pieces, or into its component atoms.

To convert the result to a vector, we define the angular velocity $\vec{\omega}$ to be a vector with magnitude ω and direction given by a variant of the right-hand rule (Figure 11.18):

Figure 11.18 Finding the direction of the angular velocity vector. Curl your fingers in the direction of rotation; your thumb sticks out in the direction of $\vec{\omega}$.

ANGULAR VELOCITY VECTOR

$$\vec{\omega}$$

Magnitude: radians/second
Direction: see Figure 11.18

We can now express the rotational angular momentum of an object spinning around an axis through its center of mass:

$$\vec{L}_{\text{rot}} = I\vec{\omega}$$

A formal derivation of this result is given in the following section.

EXAMPLE **Rotational Angular Momentum of a Bicycle Wheel**

A bicycle wheel has a mass of 0.8 kg and a radius of 32 cm. If the wheel rotates in the xz plane, spinning clockwise when viewed from the $+y$ axis, and making one full revolution in 0.75 seconds, what is the rotational angular momentum of the wheel?

Solution The direction of $\vec{\omega}$ is $-y$.

$$I = MR^2 = (0.8\,\text{kg})(0.32\,\text{m})^2 = 0.082\,\text{kg}\cdot\text{m}^2$$
$$\omega = \frac{2\pi}{0.75\,\text{s}} = 8.38\,\text{s}^{-1}$$
$$\left|\vec{L}_{\text{rot}}\right| = (0.082\,\text{kg}\cdot\text{m}^2)(8.38\,\text{s}^{-1}) = 0.69\,\text{kg}\cdot\text{m}^2/\text{s}$$
$$\vec{L}_{\text{rot}} = \langle 0, -0.69, 0 \rangle\,\text{kg}\cdot\text{m}^2/\text{s}$$

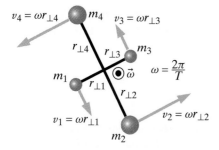

Figure 11.19 A case of rigid rotation about an axis with angular speed ω. Note the different speeds, with $v_i = \omega r_{\perp i}$.

Derivation: Rotational Angular Momentum

Consider a system consisting of a rigid object that is rotating on an axis, with all the atoms in the system sharing the same angular speed ω in radians per second but with different linear speeds in meters per second, depending on their distances from the axis (Figure 11.19). The angular velocity vector $\vec{\omega}$ points out of the page.

The magnitude of the angular momentum of each piece is this, where $r_\perp$ is the perpendicular distance of a piece from the axis:

$$|\vec{r} \times \vec{p}| = r_\perp mv = r_\perp m(\omega r_\perp) = mr_\perp^2 \omega$$

So the magnitude of the rotational angular momentum is this:

$$|\vec{L}_{\text{rot}}| = [m_1 r_{\perp 1}^2 + m_2 r_{\perp 2}^2 + m_3 r_{\perp 3}^2 + m_4 r_{\perp 4}^2]\omega$$

As we saw in Chapter 9, the quantity in brackets is called the "moment of inertia" and is usually denoted by the letter I:

$$I = m_1 r_{\perp 1}^2 + m_2 r_{\perp 2}^2 + m_3 r_{\perp 3}^2 + m_4 r_{\perp 4}^2 + \cdots$$

Since the rotational angular momentum points in the direction of $\vec{\omega}$, we have the following for the angular momentum relative to the center of mass, even if the center of mass is moving in some complicated way:

ROTATIONAL ANGULAR MOMENTUM

$$\vec{L}_{\text{rot}} = I\vec{\omega}$$

This way of evaluating rotational angular momentum is often easier to use than the more basic expression in terms of cross products $\vec{r} \times \vec{p}$, especially because the moments of inertia for common objects can be looked up in reference books. For example, the moment of inertia about an axis passing through the center of a uniform-density solid disk of radius R is $\frac{1}{2}MR^2$, and the moment of inertia about an axis passing through the center of a uniform-density solid sphere of radius R is $\frac{2}{5}MR^2$.

11.X.4 The moment of inertia of a sphere of uniform density rotating on its axis is $\frac{2}{5}MR^2$. Use data given at the end of this book to calculate the magnitude of the rotational angular momentum of the Earth.

If the Masses Don't Lie in a Plane

In Figure 11.19 all the masses lie in perpendicular to the Axis, a plane that is perpendicular to the axis of rotation. What if they don't all lie in this plane? We'll show that for simple rotating systems the moment of inertia is still calculated by adding up terms like $mr_\perp^2 \omega$, where the perpendicular distances are measured from the axis of rotation.

In Figure 11.20 two equal masses are mounted to a rotating shaft, and we want to determine the angular momentum relative to the center of the device, marked C. We show the directions of the angular momentum $\vec{L}_C = \vec{r}_C \times \vec{p}$ of each mass. You can see that the individual contributions have components that are not along the shaft. However, these components cancel for pairs of masses above and below each other. The effect is that the total angular momentum vector lies along the shaft, and can be calculated by adding up $mr_\perp^2 \omega$ for each mass, where $r_\perp$ is the perpendicular distance from the axis.

Here is the proof: the magnitude of the angular momentum of mass 1 about location C is $|\vec{L}_1| = r_1 mv_1 = r_1 m\omega r_\perp$. The y component of this angular momentum is $|\vec{L}_1| \cos(90° - \theta) = r_1 m\omega r_\perp \sin\theta = m\omega r_\perp^2$ since $r_1 \sin\theta = r_\perp$.

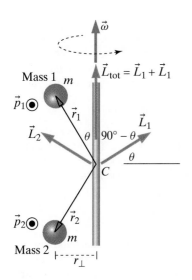

Figure 11.20 The off-axis components of the angular momenta of the two masses cancel. Their total angular momentum is along the axis of rotation.(The two masses at this instant are in the xy plane.)

Similarly, the y component of the angular momentum of mass 2 about location C is also $m\omega r_\perp^2$. Therefore the total angular momentum of the two masses is in the y direction and is given by

$$L_{C,y} = mr_\perp^2\omega + mr_\perp^2\omega = (mr_\perp^2 + mr_\perp^2)\omega = I\omega$$

since
$$I = mr_\perp^2 + mr_\perp^2$$

For example, the moment of inertia of a uniform-density thin disk rotating on its axle turns out to be $I_{\text{disk}} = \frac{1}{2}MR^2$, where M is the mass of the disk and R is its radius. Because only the distances of atoms from the axle matter, the moment of inertia of a long cylinder has exactly the same form, $I_{\text{disk}} = \frac{1}{2}MR^2$, and it doesn't matter how long the cylinder is (or to put it another way, how thick the disk is).

In this introductory treatment of angular momentum we will deal with simple, symmetrical rotating systems for which the moment of inertia can be calculated simply as $I = m_1 r_{\perp 1}^2 + m_2 r_{\perp 2}^2 + m_3 r_{\perp 3}^2 + m_4 r_{\perp 4}^2 + \cdots$. For a more complicated situation, see the optional Section 11.12.

Since $L_{\text{rot}} = I\omega$, we can express K_{rot} in terms of the rotational angular momentum:

K_{rot} and $\vec{L}_{\text{rot}}$

$$K_{\text{rot}} = \frac{1}{2}I\omega^2 = \frac{1}{2}\frac{(I\omega)^2}{I} = \frac{L_{\text{rot}}^2}{2I}$$

Figure 11.21 A rotating barbell.

11.X.5 A barbell spins around a pivot at its center at A (Figure 11.21). The barbell consists of two small balls, each with mass $m = .4\,\text{kg}$, at the ends of a very low mass rod of length $d = 0.6\,\text{m}$. The barbell spins clockwise with angular speed $\omega_0 = 20$ radians/s.

(a) Consider the two balls separately, and calculate $\vec{L}_{\text{trans},1,A}$ and $\vec{L}_{\text{trans},2,A}$ (both direction and magnitude in each case).

(b) Calculate $\vec{L}_{\text{tot},A} = \vec{L}_{\text{trans},1,A} + \vec{L}_{\text{trans},2,A}$ (both direction and magnitude).

(c) Next, consider the two balls together and calculate I for the barbell.

(d) What is the direction of the angular velocity $\vec{\omega}_0$?

(e) Calculate $\vec{L}_{\text{rot}} = I\vec{\omega}_0$ (both direction and magnitude).

(f) How does $\vec{L}_{\text{rot}}$ compare to $\vec{L}_{\text{tot},A}$? The point is that the form $I\omega$ is just a convenient way of calculating the (rotational) angular momentum of a multiparticle system. In principle one can always calculate the angular momentum simply by adding up the individual angular momenta of all the particles.

(g) Calculate K_{rot}.

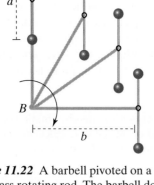

Figure 11.22 A barbell pivoted on a low-mass rotating rod. The barbell does not rotate.

11.X.6 The barbell in the previous exercise is mounted on the end of a low-mass rigid rod of length $b = 0.9\,\text{m}$ (Figure 11.22). The apparatus is started in such a way that although the rod rotates clockwise with angular speed $\omega_1 = 15\,\text{rad/s}$, the barbell maintains its vertical orientation.

(a) Calculate $\vec{L}_{\text{rot}}$ (both direction and magnitude).

(b) Calculate $\vec{L}_{\text{trans},B}$ (both direction and magnitude).

(c) Calculate $\vec{L}_{\text{tot},B}$ (both direction and magnitude).

11.X.7 The apparatus in the previous exercise is restarted in such a way that it again rotates clockwise with angular speed $\omega_1 = 15\,\text{rad/s}$, but in addition, the barbell rotates clockwise about its center, with an angular speed $\omega_2 = 20\,\text{rad/s}$ (Figure 11.23).

(a) Calculate $\vec{L}_{\text{rot}}$ (both direction and magnitude).

(b) Calculate $\vec{L}_{\text{trans},B}$ (both direction and magnitude).

(c) Calculate $\vec{L}_{\text{tot},B}$ (both direction and magnitude).

Figure 11.23 A barbell pivoted on a low-mass rotating rod. The barbell rotates.

11.3 TRANSLATIONAL PLUS ROTATIONAL ANGULAR MOMENTUM

Recall from Chapter 9 that the kinetic energy of a multiparticle system can be split into overall translational kinetic energy plus rotational and vibrational kinetic energy relative to the center of mass:

$$K_{\text{tot}} = K_{\text{trans}} + K_{\text{rel}}$$
$$= K_{\text{trans}} + K_{\text{rot}} + K_{\text{vib}}$$

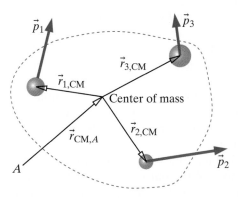

Figure 11.24 Position vectors relative to the center of mass.

In an analogous fashion, angular momentum can be split into a term associated with the motion of the center of mass (which we will call translational angular momentum, also called "orbital" angular momentum) plus a term associated with rotation around the center of mass (which we will call rotational angular momentum, also called "spin" angular momentum):

$$\vec{L}_A = \vec{L}_{\text{trans}, A} + \vec{L}_{\text{rot}}$$

We will calculate the angular momentum relative to location A of a three-mass system (Figure 11.24). We specify the positions of each mass relative to location A by going from A to the center of mass (vector $\vec{r}_{\text{CM}}$) and then to the individual masses. (The location A need not be in the same plane as the masses, nor are the momentum vectors necessarily in the plane of the masses.)

The position of mass m_1 relative to location A is $\vec{r}_1 = \vec{r}_{\text{CM}} + \vec{r}_{1,\text{CM}}$, the position of mass m_2 is $\vec{r}_2 = \vec{r}_{\text{CM}} + \vec{r}_{2,\text{CM}}$, and the position of mass m_3 is $\vec{r}_3 = \vec{r}_{\text{CM}} + \vec{r}_{3,\text{CM}}$ (Figure 11.25). Using these vectors we calculate the total angular momentum relative to location A:

$$\vec{L}_A = (\vec{r}_{\text{CM}} + \vec{r}_{1,\text{CM}}) \times \vec{p}_1 + (\vec{r}_{\text{CM}} + \vec{r}_{2,\text{CM}}) \times \vec{p}_2 + (\vec{r}_{\text{CM}} + \vec{r}_{3,\text{CM}}) \times \vec{p}_3$$
$$\vec{L}_A = [\vec{r}_{\text{CM}} \times (\vec{p}_1 + \vec{p}_2 + \vec{p}_3)] + [\vec{r}_{1,\text{CM}} \times \vec{p}_1 + \vec{r}_{2,\text{CM}} \times \vec{p}_2 + \vec{r}_{3,\text{CM}} \times \vec{p}_3]$$

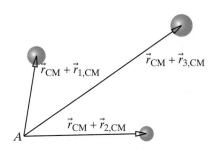

Figure 11.25 Position vectors relative to location A.

Since $(\vec{p}_1 + \vec{p}_2 + \vec{p}_3)$ is the total momentum $\vec{P}_{\text{tot}}$, we have the following:

$$\vec{L}_A = \underbrace{[\vec{r}_{\text{CM}} \times \vec{P}_{\text{tot}}]}_{\text{"translational"}} + \underbrace{[\vec{r}_{1,\text{CM}} \times \vec{p}_1 + \vec{r}_{2,\text{CM}} \times \vec{p}_2 + \vec{r}_{3,\text{CM}} \times \vec{p}_3]}_{\text{"rotational"}}$$

The first term in the total angular momentum

$$\vec{r}_{\text{CM}} \times \vec{P}_{\text{tot}}$$

is the angular momentum the system would have if all the mass were concentrated at the center of mass. It is the angular momentum that a "collapsed" point-particle version of the system would have. This is the translational (or orbital) angular momentum $\vec{L}_{\text{trans}, A}$, which is already familiar.

The second term in the total angular momentum,

$$\vec{r}_{1,\text{CM}} \times \vec{p}_1 + \vec{r}_{2,\text{CM}} \times \vec{p}_2 + \vec{r}_{3,\text{CM}} \times \vec{p}_3$$

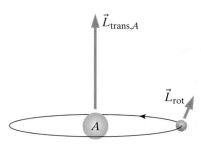

Figure 11.26 Translational angular momentum of the Earth (relative to the location of the Sun), and rotational angular momentum of the Earth (relative to the Earth's center of mass). At this instant the *linear* momentum $\vec{p}$ of the Earth is into the page. The Earth's axis is tilted 23° away from a perpendicular to the plane of the orbit, so the two angular momenta are not parallel.

is the angular momentum of the system relative to the center of mass of the system. We call this term the "rotational" angular momentum $\vec{L}_{\text{rot}}$. An example of rotational angular momentum is the angular momentum due to the rotation of the Earth on its axis. This term is also called the "spin" angular momentum. In Figure 11.26 we show both the translational (orbital) angular momentum of the Earth relative to the location of the Sun, and the rotational (spin) angular momentum of the Earth relative to the Earth's center of mass. The total angular momentum of the Earth relative to the Sun is the vector sum of the Earth's translational and rotational angular momenta.

TRANSLATIONAL PLUS ROTATIONAL ANGULAR MOMENTUM

$$\vec{L}_A = \vec{L}_{\text{trans},A} + \vec{L}_{\text{rot}}$$

Translational: $\vec{L}_{\text{trans},A} = \vec{r}_{\text{CM},A} \times \vec{P}_{\text{tot}}$

Rotational: $\vec{L}_{\text{rot}} = \vec{r}_{1,\text{CM}} \times \vec{p}_1 + \vec{r}_{2,\text{CM}} \times \vec{p}_2 + \cdots$

The translational angular momentum differs for different choices of the location A, so we need the "A" subscript on the translational angular momentum and the total angular momentum to remind us of this dependence. The rotational angular momentum doesn't need a subscript because it is calculated relative to the center of mass, and this calculation is unaffected by our choice of the point A, and by the motion of the center of mass, which may even be accelerating.

For an ideal point particle, or the "collapsed" point-particle version of a multiparticle system, the translational angular momentum is the only kind of angular momentum there is, just as translational kinetic energy is the only kind of kinetic energy a point particle can have.

In Chapter 9 we saw that a rigid object that rotates about an axle that is a distance r_{CM} from the center of mass of the object has translational plus rotational kinetic energy equal to $\frac{1}{2}I\omega^2$, where $I = Mr_{\text{CM}}^2 + I_{\text{CM}}$ (this is also referred to as the "parallel axis theorem" for moment of inertia). The same is true for the angular momentum of a rotating rigid object: the translational plus rotational angular momentum is $(Mr_{\text{CM}}^2 + I_{\text{CM}})\omega$.

Figure 11.27 A rod rotating in the vertical plane with a wheel attached.

11.X.8 A rod rotates in the vertical plane around a horizontal axle. A wheel is free to rotate on the rod, as shown in Figure 11.27. A vertical stripe is painted on the wheel. As the rod rotates clockwise, the vertical stripe on the wheel remains vertical. Is the translational angular momentum of the wheel relative to location A zero or nonzero? If nonzero, what is its direction? Is the rotational angular momentum of the wheel zero or nonzero? If nonzero, what is its direction?

11.X.9 Consider a system similar to that in the previous exercise, but with the wheel welded to the rod (not free to turn). As the rod rotates clockwise, does the stripe on the wheel remain vertical? Is the translational angular momentum of the wheel relative to location A zero or nonzero? If nonzero, what is its direction? Is the rotational angular momentum of the wheel zero or nonzero? If nonzero, what is its direction?

11.X.10 Pinocchio rides a horse on a merry-go-round turning counter-clockwise as viewed from above, with his long nose always pointing forward, in the direction of his velocity. Is Pinocchio's translational angular momentum relative to the center of the merry-go-round zero or nonzero? If nonzero, what is its direction? Is his rotational angular momentum zero or nonzero? If nonzero, what is its direction?

11.4 | THE ANGULAR MOMENTUM PRINCIPLE

So far we have concentrated on defining and calculating angular momentum. Next we will show that an off-center force that applies a twist to a system can change the angular momentum of the system.

For a single particle, the Momentum Principle says that the time derivative of the momentum is the net force: $d\vec{p}/dt = \vec{F}_{\text{net}}$. Let's see what the time derivative of angular momentum turns out to be. By applying the product rule for derivatives, we have

$$\frac{d(\vec{r}_A \times \vec{p})}{dt} = \frac{d\vec{r}_A}{dt} \times \vec{p} + \vec{r}_A \times \frac{d\vec{p}}{dt}$$

Since $\dfrac{d\vec{r}_A}{dt} = \vec{v}$ and $\vec{p} = \dfrac{m\vec{v}}{\sqrt{1-(v^2/c^2)}}$, we have

$$\frac{d\vec{r}_A}{dt} \times \vec{p} = \frac{\vec{v} \times m\vec{v}}{\sqrt{1-(v^2/c^2)}} = \vec{0}$$

since $|\vec{v} \times \vec{v}| = v^2 \sin(0°) = 0$. Therefore we have proved that the following is true for a single particle:

$$\frac{d(\vec{r}_A \times \vec{p})}{dt} = \vec{r}_A \times \vec{F}_{net}$$

We have previously defined the quantity $\vec{L}_A = \vec{r}_A \times \vec{p}$ to be the angular momentum of a particle relative to location A. We call the quantity $\vec{\tau}_A = \vec{r}_A \times \vec{F}$ the "torque" exerted about location A (Figure 11.28). Torque is usually symbolized by τ, the Greek lowercase letter tau. Torque is a word of Latin origin and simply means "twist."

Figure 11.28 Torque relative to location A is defined as $\vec{\tau}_A = \vec{r}_A \times \vec{F}$.

DEFINITION OF TORQUE

$$\vec{\tau}_A \equiv \vec{r}_A \times \vec{F}$$

Torque due to a force $\vec{F}$ is defined about a particular location, A. $\vec{r}_A$ is a vector from location A to the point of application of the force (Figure 11.28).

Applying a torque, or twist, to a system changes the angular momentum of the system. Just as we call the vector sum of all the forces acting on a system $\vec{F}_{net}$, we call the vector sum of all the torques acting on a system $\vec{\tau}_{net,A}$.

THE ANGULAR MOMENTUM PRINCIPLE FOR A PARTICLE

$$\frac{d\vec{L}_A}{dt} = \vec{\tau}_A \quad \text{(derivative form)}$$

The rate of change of the angular momentum of a particle relative to location A is equal to the net torque applied to the particle about location A.

$$\Delta \vec{L}_A = \vec{\tau}_{net,A}\Delta t \quad \text{(finite time form)}$$

The change in angular momentum of a particle relative to location A is equal to the net angular impulse $\vec{\tau}_{net,A}\Delta t$ applied.

The Angular Momentum Principle is similar to the Momentum Principle $d\vec{p}/dt = \vec{F}_{net}$ or $\Delta\vec{p} = \vec{F}_{net}\Delta t$. The term $\vec{\tau}_{net,A}\Delta t$ is the net "angular impulse." We will apply this new principle after practicing how to calculate torques.

Calculating Torque

Torque and angular momentum are both calculated by using the cross product. In this section we offer some additional ways to think about cross products, in the context of practice in calculating torques.

Figure 11.29 Twist a nut, or push straight at it?

> QUESTION In Figure 11.29, which way of pushing on a wrench is more effective in twisting the nut?

You can see that a force parallel to the handle is completely ineffective in twisting the nut. Evidently it is only the perpendicular component of a force that is effective in twisting an object.

Figure 11.30 Push at the end, or near the nut?

> QUESTION In Figure 11.30, where should you push on the wrench—at the end of the wrench or near the nut?

Figure 11.31 Push on the wrench at an angle.

The farther away from the nut we push, the more effective we are in twisting the nut. The advantage of a long wrench is to provide more leverage in twisting.

Suppose that we apply a force at an angle θ to the radius (Figure 11.31). Evidently the component of the force that is effective in twisting is $F\sin\theta$, the perpendicular component. Moreover, the bigger the lever arm (r_A, the distance from the nut at location A), the more effective we'll be. To capture both effects, we define the magnitude of "torque" (which means "twist") exerted by a force $\vec{F}$ relative to a location A as follows:

MAGNITUDE OF TORQUE RELATIVE TO LOCATION A

$$\tau = r_A F \sin\theta$$

For a purely perpendicular force, $\theta = 90°$, $\sin\theta = 1$, and the torque is $r_A F$. For a force that is parallel to the lever arm, $\theta = 0°$, $\sin\theta = 0$, and the torque is zero. See Figure 11.29 for these extreme examples.

Direction of Torque

The direction of the torque is given by the cross product and associated right-hand rule.

QUESTION In Figure 11.31, is the torque vector into the page or out of the page? Use the right-hand rule.

Move the $\vec{r}$ and $\vec{F}$ vectors tail to tail, and fold the fingers of your right hand from $\vec{r}$ toward $\vec{F}$. The only way you can do this results in your extended right thumb pointing into the page. If you try to keep your extended right thumb pointing out of the page, you will find that you can't fold your fingers toward the direction of the force. To put it another way, folding your fingers from the direction of $\vec{r}$ toward the direction of $\vec{F}$ is a clockwise rotation, which means the angular momentum points into the page.

Most nuts and bolts have what are called "right-handed threads." In Figure 11.31 a right-handed nut will advance in the direction of the torque vector, into the page. If you reverse the direction of the force, so the torque vector points out of the page, a right-handed nut will advance out of the page. (Exceptions include left-handed threads on gas cylinders containing explosive gases; the unexpected behavior alerts users to danger.)

The fact that a physical right-handed nut advances in the direction of the torque vector is another justification for drawing torque cross-product vectors along an axis of rotation, perpendicular to the plane of the motion.

11.X.11 In Figure 11.31, if $r_A = 3\,\text{m}$, $F = 4\,\text{N}$, and $\theta = 30°$, what is the magnitude of the torque about location A, including units? If the force in Figure 11.31 were perpendicular to $\vec{r}_A$ but gave the same torque as before, what would its magnitude be?

11.X.12 At $t = 15\,\text{s}$, a particle has angular momentum $\langle 3, 5, -2 \rangle$ kg·m²/s relative to location A. A constant torque $\langle 10, -12, 20 \rangle$ N·m relative to location A acts on the particle. At $t = 15.1\,\text{s}$, what is the angular momentum of the particle?

Comets and the Angular Momentum Principle

If the net torque around a location A is zero, the angular momentum about that location doesn't change. There may be forces acting, causing changes in the linear momentum, but if these forces don't exert any torques, the rate of change of angular momentum is zero, in which case angular momentum is constant.

Comet orbits provide a nice example. Most comets have very long elliptical orbits around the Sun (Figure 11.32). In our earlier discussion of Halley's

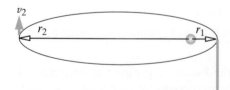

The orbit is in the plane of the page

Figure 11.32 A long elliptical orbit of a comet around the Sun.

comet we saw that the translational angular momentum of the comet relative to the Sun had the same value at different times along the orbit. We can now understand this observation in terms of the Angular Momentum Principle.

| EXAMPLE | **Torque on a Comet** |

Relative to the center of the Sun, explain why the torque exerted by the Sun's gravitational force on the comet is zero at every point along the orbit (Figure 11.32). What does that say about the angular momentum of the comet relative to the center of the Sun?

Solution The gravitational force on the comet always points directly at the Sun, at location A (Figure 11.33). In $|\vec{\tau}_A| = |r_\perp \times \vec{F}_{grav}|$, the lever arm $r_\perp$ is always zero, so there is no torque relative to location A that can change the angular momentum of the comet. Therefore the angular momentum of the comet, relative to the location of the Sun, cannot change.

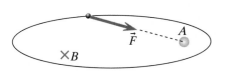

Figure 11.33 The gravitational force acting on the comet points directly at the center of the Sun (location A).

FURTHER DISCUSSION About location B in Figure 11.33, the torque acting on the comet is *not* zero, and the angular momentum of the comet does *not* stay constant. If on the other hand you consider the system consisting of both Sun and comet, the gravitational forces that the two objects exert on each other make torques about location B, or any location, that vectorially add up to zero, and the total angular momentum of the combined system will stay constant. The Sun actually moves a little due to the force that the comet exerts on it, so the Sun's angular momentum relative to location B changes, but its change is the opposite of the change of the angular momentum relative to B of the comet.

Angular momentum is a conserved quantity:

CONSERVATION OF ANGULAR MOMENTUM

$$\Delta \vec{L}_{A,\text{system}} + \Delta \vec{L}_{A,\text{surroundings}} = 0$$

If a system gains angular momentum, the surroundings lose that amount. For the system consisting of the comet, the Sun is the surroundings. About location B in Figure 11.33, the angular momentum of the system (the comet) changes, but the angular momentum of the surroundings (the Sun) has an opposite change, because the torque that the comet exerts on the Sun has the same magnitude but opposite direction.

> **11.X.13** In Figure 11.32, in terms of the comet's mass m and speed v_1 when nearest the Sun (distance r_1 from the Sun), what is the comet's angular momentum relative to the center of the Sun (direction and magnitude)? What is the comet's angular momentum relative to the center of the Sun (direction and magnitude) when it is farthest from the Sun (distance r_2 from the Sun) and traveling at speed v_2? What can you conclude about the relationship between v_1 and v_2?

Comments about Comets

Some comets go far beyond Pluto and spend most of their time there, because they're traveling very slowly, as you have just shown. We see them when they come near the Sun, but only for a few months, because they're now traveling fast.

The result $r_1 v_2 = r_2 v_2$ at the closest and farthest points in the orbit is correct no matter what kind of "central" force is involved. For any force that acts along a line connecting two objects, there is no torque about a point at the center of one of the objects, so angular momentum is constant about that location. In the particular case of the gravitational force, we also have an energy equation relating v_1 and v_2:

$$\frac{1}{2}mv_1^2 - \frac{GMm}{r_1} = \frac{1}{2}mv_2^2 - \frac{GMm}{r_2}$$

Historical Note: Kepler and Elliptical Orbits

Based on careful, accurate naked-eye measurements made by Tycho Brahe before the invention of the telescope, Johannes Kepler in 1609 announced his discovery that the planets follow elliptical orbits around the Sun.

Kepler also stated that he had found that "a radius vector joining any planet to the Sun sweeps out equal areas in equal lengths of time." This is equivalent to conservation of angular momentum, as can be seen with the aid of Figure 11.34. The area swept out in a time Δt is the area of a triangle whose base is $v\Delta t$ and whose altitude is $r\sin\theta$. This area is $\frac{1}{2}(rv\sin\theta)\Delta t$, which is proportional to what we now call the angular momentum $rmv\sin\theta$.

In 1618 Kepler announced his discovery that the square of the time it takes a planet to go around the Sun is proportional to the cube of the mean distance from the Sun (a result you could derive in Chapter 5 for the simpler case of circular orbits).

Later in the 1600s Newton explained all three of these discoveries as derivable from his second law of motion (the Momentum Principle) plus his universal law of gravitation. Kepler's insights provided important tests for Newton's theories.

Figure 11.34 "Equal area in equal time."

11.5 | MULTIPARTICLE SYSTEMS

The Angular Momentum Principle has already given us new insight into the motion of a single particle acted on by a torque due to a single force. The equation becomes a really powerful tool when we extend it to apply to multiparticle systems acted on by multiple torques. For example, we will be able to understand the counterintuitive behavior of spinning tops and gyroscopes.

We will now derive a multiparticle version of the Angular Momentum Principle, following closely the derivation in Chapter 3 of a multiparticle version of the Momentum Principle.

To be as concrete as possible, we'll consider a system consisting of just three particles. It is easy to see how this generalizes to larger systems, including a block consisting of an astronomically huge number of atoms.

In Figure 11.35 we show all of the forces acting on each particle, where the lowercase $\vec{f}$'s are internal forces, and the upper case $\vec{F}$'s are external forces applied by objects in the surroundings, such as the Earth that applies a gravitational attraction. We write the Angular Momentum Principle applied to each of the three particles. We measure angular momenta and torques relative to location A, but to avoid clutter we don't write the subscript A in any of these equations until the end. ($\vec{L}_1$ is the angular momentum of m_1 relative to A.)

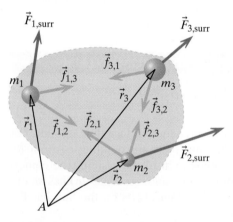

Figure 11.35 The Angular Momentum Principle in a multiparticle system.

$$\frac{d\vec{L}_1}{dt} = \vec{r}_1 \times (\vec{F}_{1,\text{surr}} + \vec{f}_{1,2} + \vec{f}_{1,3})$$

$$\frac{d\vec{L}_2}{dt} = \vec{r}_2 \times (\vec{F}_{2,\text{surr}} + \vec{f}_{2,1} + \vec{f}_{2,3})$$

$$\frac{d\vec{L}_3}{dt} = \vec{r}_3 \times (\vec{F}_{3,\text{surr}} + \vec{f}_{3,1} + \vec{f}_{3,2})$$

Nothing new so far–but now we add up these three equations. That is, we create a new equation by adding up all the terms on the left sides of the three equations, adding up all the terms on the right sides, and setting them equal to each other. In doing so, we take into account the fact that many of these terms cancel. By the reciprocity of electric and gravitational forces (Newton's third law), we have

$$\vec{f}_{1,2} = -\vec{f}_{2,1}$$
$$\vec{f}_{1,3} = -\vec{f}_{3,1}$$
$$\vec{f}_{2,3} = -\vec{f}_{3,2}$$

Therefore the torques of the internal forces cancel. For example, consider this piece of the sum:

$$\vec{r}_1 \times \vec{f}_{1,2} + \vec{r}_2 \times \vec{f}_{2,1} = \vec{r}_1 \times \vec{f}_{1,2} - \vec{r}_2 \times \vec{f}_{1,2} = (\vec{r}_1 - \vec{r}_2) \times \vec{f}_{1,2}$$

The vector $\vec{r}_1 - \vec{r}_2$ points from m_2 toward m_1, so the angle between it and $\vec{f}_{1,2}$ is either $0°$ or $180°$, and $\sin\theta$ is zero. Consequently all of the torques associated with the internal forces cancel, and all that remains is this:

$$\frac{d(\vec{L}_1 + \vec{L}_2 + \vec{L}_3)}{dt} = \vec{r}_1 \times \vec{F}_{1,\text{surr}} + \vec{r}_2 \times \vec{F}_{2,\text{surr}} + \vec{r}_3 \times \vec{F}_{3,\text{surr}}$$

The right side of this equation represents the net torque due to external forces, $\vec{\tau}_{\text{net},A}$. The great importance of this equation is that the reciprocity of gravitational and electric forces has allowed us to get rid of all the torques due to internal forces, the torques that the particles in the system exert on each other. We rewrite our result like this:

THE ANGULAR MOMENTUM PRINCIPLE FOR A MULTIPARTICLE SYSTEM

$$\frac{d\vec{L}_{\text{tot},A}}{dt} = \vec{\tau}_{\text{net},A}$$

or

$$\Delta\vec{L}_{\text{tot},A} = \vec{\tau}_{\text{net},A}\Delta t$$

In words, the rate of change of the "total angular momentum" of a system relative to a location A, $\vec{L}_{\text{tot},A} = \vec{L}_{1,A} + \vec{L}_{2,A} + \vec{L}_{3,A} + \cdots$, is equal to the net torque due to forces exerted on that system by objects in the surroundings, relative to location A. (Or, the change in the angular momentum is equal to the angular impulse.) The internal forces and torques do not appear in this multiparticle version of the Angular Momentum Principle.

Often it pays to choose location A to be at the place where the center of mass happens to be at that instant. In that case, the Angular Momentum Principle simplifies:

$$\frac{d\vec{L}_{\text{CM}}}{dt} = \frac{d}{dt}[(\vec{r}_{\text{CM,CM}} \times \vec{P}_{\text{tot}}) + \vec{L}_{\text{rot}}] = \frac{d\vec{L}_{\text{rot}}}{dt}$$

since $\vec{r}_{\text{CM,CM}} = \vec{0}$. That is, the position of the center of mass, relative to the center of mass itself, is of course a zero vector. Therefore we have the following important and useful special case for the angular momentum relative to the center of mass:

THE ANGULAR MOMENTUM PRINCIPLE RELATIVE TO THE CENTER OF MASS

$$\frac{d\vec{L}_{\text{rot}}}{dt} = \vec{\tau}_{\text{net,CM}}$$

or

$$\Delta\vec{L}_{\text{rot}} = \vec{\tau}_{\text{net,CM}}\Delta t$$

Figure 11.36 A child jumps onto a playground ride, and the disk spins. What is the rate of spin?

A Child Jumps on to a Playground Ride

A playground ride consists of a disk mounted on a low-friction axle (Figure 11.36). A child runs on a line tangential to the disk and jumps onto the outer edge of the disk. The disk spins, but at what rate? How many revolutions per minute, or radians per second?

Neither the Momentum Principle nor the Energy Principle are sufficient to analyze this collision. If we know the speed of the child just after jumping onto the disk, we can use the Momentum Principle to determine the impulse exerted by the axle on the disk, which prevents the disk from moving sideways, but we need to know the spin rate in order to be able to calculate the new speed of the child. The Energy Principle can be used to determine how much internal energy is produced in the inelastic sticking collision of the child and the disk, but only if we know the spin rate. We need another principle that will yield the spin rate: the Angular Momentum Principle.

It is important to see that even though the child is running in a straight line in Figure 11.36, the child does have angular momentum, because the disk is caused to rotate when hit off-center by the child.

EXAMPLE   **A Child Jumps onto a Playground Ride**

A playground ride consists of a uniform-density disk of mass 300 kg and radius 2 m mounted on a low-friction axle (Figure 11.36). Starting from a distance of 5 m from the disk, a child of mass 40 kg runs at 3 m/s on a line tangential to the disk and jumps onto the outer edge of the disk. If the disk was initially at rest, how fast does it rotate just after the collision?

Solution   System: Disk plus child
Surroundings: Earth, ground, axle
Initial state: Child 5 m away, running at 3 m/s; disk not rotating
Final state: Just after collision; child now stuck to the disk
Free-body diagram: Figure 11.37. The axle exerts a large impulsive force at the time of impact in response to the child ramming the disk against the axle; this force prevents the disk from moving to the right, and it changes the total momentum of the combined system.

Principle: The Angular Momentum Principle $\vec{L}_{A,f} = \vec{L}_{A,i} + \vec{\tau}_{net}\Delta t$

We can choose any point A about which to calculate torque and angular momentum. To simplify the analysis, we choose A to be at the location of the axle, so that the large impulsive force of the axle points toward A and therefore exerts no torque. The force of the Earth is downward and the force of the ground is upward, with equal magnitude, so the net torque on our chosen system is zero, and $\vec{L}_{A,f} = \vec{L}_{A,i}$. We consider just the component of angular momentum that is perpendicular to the xz plane of the motion, $L_{A,y}$.

The initial angular momentum of the system consists solely of the child's angular momentum, since the disk isn't rotating. Here is the magnitude:

$$|L_{A,y,i}| = |\vec{r}_A||\vec{p}|\sin\theta = (|\vec{r}_A|\sin\theta)|\vec{p}|$$
$$= (2\,\text{m})(40\,\text{kg})(3\,\text{m/s})$$
$$= 240\,\text{kg}\cdot\text{m}^2/\text{s}$$

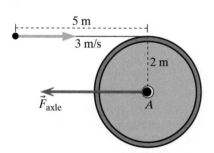

Figure 11.37 The free-body diagram for the system of disk plus child. The axle is in the surroundings and exerts a large impulsive force to the left when the child rams the disk against the axle.

Here we used the fact that in Figure 11.38 ($|\vec{r}_A|\sin\phi$) is the side opposite the angle ϕ, and that $\sin\phi = \sin(180° - \theta) = \sin\theta$.

Another way to calculate the magnitude of the initial angular momentum for this motion in the xz plane is to use the general formula for components of angular momentum, which gives the same result and is a check on our work:

$$|L_{A,y,i}| = |zp_x - xp_z| = |zp_x| = (2\,\text{m})(40\,\text{kg})(3\,\text{m/s}) = 240\,\text{kg}\cdot\text{m}^2/\text{s}$$

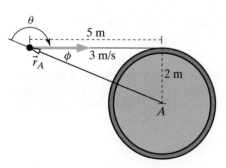

Figure 11.38 Applying the definition $L_A = r_{AP}\sin\theta$.

The final angular momentum of the system consists of spin (rotational) angular momentum of the disk plus orbital (translational) angular momentum of the child on the disk. If the disk rotates with angular speed ω, the child's speed is $\omega r = \omega(2\,\text{m})$.

$$|L_{A,y,f}| = |L_{A,y,i}| = 240\,\text{kg} \cdot \text{m}^2/\text{s}$$

$$= I\omega + Rmv = \frac{1}{2}MR^2\omega + Rm(\omega R)$$

$$= \frac{1}{2}(300\,\text{kg})(2\,\text{m})^2\omega + (40\,\text{kg})(2\,\text{m})^2\omega$$

$$\omega = \frac{240\,\text{kg} \cdot \text{m}^2/\text{s}}{\frac{1}{2}(300\,\text{kg})(2\,\text{m})^2 + (40\,\text{kg})(2\,\text{m})^2}$$

$$\omega = 0.316\,\text{radians/s}$$

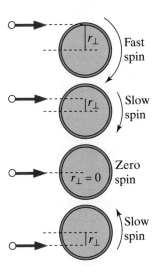

Figure 11.39 How much the disk spins, and in what direction, depends on the impact parameter $r_\perp$.

FURTHER DISCUSSION Before the collision there is only straight-line motion of the child, who nevertheless has (translational) angular momentum relative to the axle. There is obviously angular momentum after the collision, so it makes sense that there had to have been angular momentum before the collision, even though the child was running in a straight line.

Because the axle exerts a large impulsive force to the left, the Momentum Principle predicts that the child's speed should be smaller just after the collision (the rotating disk has zero linear momentum, because its center of mass doesn't move). In fact, the child's speed just after the collision is ωR, which is only 0.35 m/s.

Concerning the Energy Principle, this is an inelastic collision. The kinetic energy decreases and the internal energy increases; the child's feet and the disk get warmer.

Note that the perpendicular distance from A to the line of the child's momentum is $r_\perp = 2\,\text{m}$, and $|L_{A,y}| = r_\perp|\vec{p}| = 240\,\text{kg} \cdot \text{m}^2/\text{s}$.

QUESTION What was the magnitude of the (translational) angular momentum of the child just before hitting the disk?

It was $240\,\text{kg} \cdot \text{m}^2/\text{s}$, because $r_\perp = 2\,\text{m}$ all along the path, and the child's momentum $\vec{p}$ didn't change until hitting the disk.

In Figure 11.39 we see that how much the disk spins, and in what direction, depends on the impact parameter—how far off center the child hits the disk.

The Child Has Rotational Angular Momentum

We've skipped over a subtle point in calculating the final angular momentum of the system consisting of the disk plus the child. After the collision the child is stuck to the disk and has not only orbital (translational) angular momentum $Rmv = Rm(R\omega) = mR^2\omega$ but also spin (rotational) angular momentum. As the disk rotates about the axle at a rate ω, the child also rotates about the child's own center of mass at a rate ω. If the child's moment of inertia about the child's center of mass is I_{CM}, there is an additional contribution to the final angular momentum of amount $I_{CM}\omega$.

We can make an approximate model of the child as a cylinder whose radius is about $r = 15$ cm, in which case the magnitude of the child's translational plus rotational angular momentum is

$$mR^2\omega + \frac{1}{2}mr^2\omega = m(2^2 + 0.5(0.15)^2)\omega$$

The quantity $(2^2 + 0.5(0.15)^2) = 4 + .011 \approx 4$, which is why in this case it makes only a very slight difference whether we include the child's rotational angular momentum around the child's own center of mass.

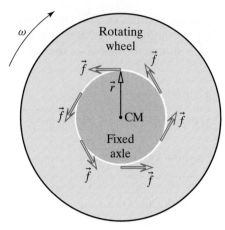

Figure 11.40 A wheel rotates on a fixed axle. There are frictional forces exerted by the axle on the inner surface of the wheel (other forces are omitted).

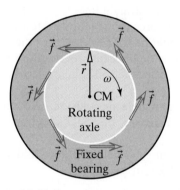

Figure 11.41 An axle rotates inside a fixed bearing. There are frictional forces exerted by the bearing on the outer surface of the axle (other forces are omitted).

Frictional Torque

After the child has jumped onto the disk and made it rotate, the rotation rate will slowly decrease due to friction between the disk and the axle. Figure 11.40 shows a wheel that rotates on a fixed axle. There are frictional forces exerted by the axle on the inner surface of the wheel (other forces are omitted from the diagram). The frictional forces exert torques ($\vec{r} \times \vec{f}$) about the center of mass of the wheel, which make the angular speed ω of the wheel steadily decrease.

Figure 11.41 shows the opposite case: an axle rotates inside a bearing, which is fixed in place. For example, the axle of a wagon may pass through bearings attached to the bottom of the wagon, and there are wheels rigidly attached to each end of the axle. There are frictional forces exerted by the bearing on the outer surface of the axle (other forces are omitted from the diagram). We again see that frictional forces exert torques ($\vec{r} \times \vec{f}$) about the center of the axle, which make the angular speed ω of the axle steadily decrease.

There are three situations in which it may be a good approximation to neglect frictional torques as we will often do in this chapter:

- If the coefficient of friction μ between the axle and the wheel is small, the frictional force may be negligible, so the frictional torque may be negligible. In rotating machinery μ is often reduced with lubricating oil.
- In Figure 11.40, if the axle and the inside surface of the wheel have small radii, the distance $|\vec{r}|$ from the center of mass to the point of contact between axle and wheel is small, and the frictional torque $\vec{r} \times \vec{f}$ is small. For example, in a magnetic compass the needle sits on a sharp point, which has a very small radius, so any frictional force has a very small lever arm and torque. When an ice skater spins on the tip of the skate, the frictional torque may be very small because the lever arm $|\vec{r}|$ is very small. Although a thin axle has the advantage of reducing frictional torque on a wheel, there may be practical limits imposed by the requirement that the axle be able to support the wheel or the weight of a vehicle.
- In a rotational collision like that of the child jumping on the playground ride, during the brief time interval of the collision any frictional angular impulse $\vec{\tau}\Delta t$ of the axle is small because Δt is small.

> **11.X.14** Consider a rotating star far from other objects. Its rate of spin stays constant, and its axis of rotation keeps pointing in the same direction. Why?
>
> **11.X.15** Because the Earth is nearly perfectly spherical, gravitational forces act on it effectively through its center. Explain why the Earth's axis points at the North Star all year long. Also explain why the Earth's rotation speed stays the same throughout the year (one rotation per 24 hours). In your analysis, does it matter that the Earth is going around the Sun?
>
> In actual fact, the Earth is not perfectly spherical. It bulges out a bit at the equator, and tides tend to pile up water at one side of the ocean. As a result, there are small torques exerted on the Earth by other bodies, mainly the Sun and the Moon. Over many thousands of years there are changes in what portion of sky the Earth's axis points toward (change of direction of rotational angular momentum), and changes in the length of a day (change of magnitude of rotational angular momentum). See Section 11.11.

Summary: Angular Momentum and Torque

At last we have all the tools we need to investigate a variety of situations involving angular momentum. Let's summarize these new tools and concepts:

- A force exerts a torque relative to a chosen location A: $\vec{\tau}_A = \vec{r}_A \times \vec{F}$.

- Torque causes changes in angular momentum:

$$\frac{d\vec{L}_{\text{tot},A}}{dt} = \vec{\tau}_{\text{net},A} \text{ or } \Delta\vec{L}_{\text{tot},A} = \vec{\tau}_{\text{net},A}\Delta t$$

- If the net torque relative to A is zero, the angular momentum relative to A does not change. (And if we know the angular momentum isn't changing about some location, as in equilibrium, we know that the net torque must be zero about that location.)
- The angular momentum can be divided into two pieces: translational angular momentum of the system as a whole (as though all the mass were concentrated at the center-of-mass point), plus rotational angular momentum relative to the center of mass: $\vec{L}_A = \vec{L}_{\text{trans},A} + \vec{L}_{\text{rot}}$.
- Torque about the center of mass changes the rotational angular momentum:

$$\frac{d\vec{L}_{\text{rot}}}{dt} = \vec{\tau}_{\text{net},\text{CM}} \text{ or } \Delta\vec{L}_{\text{rot}} = \vec{\tau}_{\text{net},\text{CM}}\Delta t$$

- Rotational angular momentum is given by moment of inertia times angular velocity: $\vec{L}_{\text{rot}} = I\vec{\omega}$, where $I = m_1 r_{\perp 1}^2 + m_2 r_{\perp 2}^2 + m_3 r_{\perp 3}^2 + \cdots$.
- Rotational kinetic energy about the center of mass can be calculated as

$$K_{\text{rot}} = \frac{1}{2}I\omega^2 = \frac{1}{2}\frac{(I\omega)^2}{I} = \frac{L_{\text{rot}}^2}{2I}$$

11.6 THREE FUNDAMENTAL PRINCIPLES OF MECHANICS

We now have three fundamental principles that relate a change in some property of a system to the interactions of the system with external objects. In the special case in which there are no external interactions, these principles indicate that a property of the system doesn't change (momentum or energy or angular momentum).

Momentum	Angular momentum	Energy
$\dfrac{d\vec{p}}{dt} = \vec{F}_{\text{net}}$	$\dfrac{d\vec{L}_A}{dt} = \vec{\tau}_{\text{net},A}$	$\Delta E = W + Q$
If there are external forces, momentum changes.	If there are external torques, angular momentum changes.	If there are energy inputs, energy changes.
Location of object does not matter.	Location of object relative to point A is important.	Location of object does not matter.

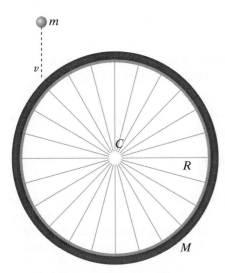

Figure 11.42 A lump of clay falls onto a wheel mounted on a frictionless axle.

11.X.16 A stationary bicycle wheel of radius R is mounted in the vertical plane on a horizontal low-friction axle (Figure 11.42). The wheel has mass M, all concentrated in the rim (the spokes have negligible mass). A lump of clay with mass m falls and sticks to the outer edge of the wheel at the location shown. Just before the impact the clay has a speed v.

  (a) Just before the impact, what is the angular momentum of the combined system of wheel plus clay about the center C?

  (b) Just after the impact, what is the angular momentum of the combined system of wheel plus clay about the center C, in terms of the angular speed of the wheel?

  (c) Just after the impact, what are the magnitude and direction of the angular velocity of the wheel?

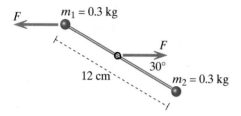

Figure 11.43 Pull a disk on a frictionless surface, unwinding the string.

Figure 11.44 Forces applied to a barbell.

(d) Qualitatively, what happens to the linear momentum of the combined system? Why?

11.X.17 A disk of radius 8 cm is pulled along a frictionless surface with a force of 10 N by a string wrapped around the edge (Figure 11.43). 24 cm of string has unwound off the disk. What are the magnitude and direction of the torque exerted about the center of the disk at this instant?

11.X.18 In Figure 11.43, the uniform solid disk has mass 0.4 kg (moment of inertia $I = \frac{1}{2}MR^2$). At the instant shown, the angular velocity is 20 radians/s into the page.

(a) At this instant, what are the magnitude and direction of the angular momentum about the center of the disk?

(b) At a time 0.2 s later, what are the magnitude and direction of the angular momentum about the center of the disk?

(c) At this later time, what are the magnitude and direction of the angular velocity?

11.X.19 A barbell is mounted on a nearly frictionless axle through its center (Figure 11.44). At this instant, there are two forces of equal magnitude applied to the system as shown, with the directions indicated, and at this instant the angular velocity is 60 radians/s, counterclockwise. In the next 0.001 s, the angular momentum relative to the center increases by an amount 2.5×10^{-4} kg·m²/s. What is the magnitude of each force? What is the net force?

11.7 SYSTEMS WITH ZERO TORQUE

In a situation in which there is zero external torque, the angular momentum of a system does not change, even if the moment of inertia of the system does change, due to a change in the shape or configuration of the system. Athletes like figure skaters, divers, and dancers can use this fact to change the rotation of their bodies.

A Spinning Skater

Figure 11.45 The skater spins slowly with leg stretched out.

You may have seen an ice skater spin vertically on the tip of one skate, with her arms and one leg outstretched, then pull her leg in and bring her arms to a vertical position above her head. She then spins much faster (Figures 11.45 and 11.46).

What's going on? There is some frictional force of the ice on the tip of the skate, but this force is applied so close to the axis of rotation that the torque is small.

> **QUESTION** During the short time when the skater quickly changes her configuration, what can you say about the rotational angular momentum? Why?

Small torque implies small change in angular momentum per unit time, so in a short time the angular momentum of the skater will hardly change.

> **QUESTION** If the rotational angular momentum hardly changes, how can the skater spin faster?

When she changes her configuration, the moment of inertia decreases, because some parts of her body are now closer to the axis of rotation. For the rotational angular momentum not to change, the angular velocity must increase to compensate for the decreased moment of inertia: $I_1\omega_1 = I_2\omega_2$, and therefore $\omega_2 = I_1\omega_1/I_2$.

Now you may quite legitimately be puzzled about how this actually works! *Why* does moving her arms and leg closer to the spin axis increase her angular

Figure 11.46 The skater pulls in her leg and spins much faster.

speed? At one level of discussion, you can close your eyes (and maybe hold your nose) and say, "Well, we did all that general analysis of the effect of torques on the angular momentum of multiparticle systems, so if that's what we get when we apply these general principles, I guess that's that." At another level, though it would be very nice to get a better sense of the detailed mechanisms involved in this odd phenomenon.

One approach is to analyze the changes in energy involved in this skating maneuver. The angular momentum L_{rot} of the skater doesn't change, so as her moment of inertia decreases, the kinetic energy of the skater $\frac{1}{2}I\omega^2 = L_{rot}^2/(2I)$ actually increases.

QUESTION | Where does this increased energy come from?

Evidently the skater has to expend chemical energy in order to increase her kinetic energy. In fact, at high spin rates it takes a noticeable effort to pull her arms and leg toward the spin axis. This is even more dramatic when holding heavy weights.

A popular physics demonstration is to sit on a rotating stool holding a dumbbell in each hand. Start spinning slowly with the arms held out, then pull your hands in toward your chest. It requires considerable effort to pull the dumbbells in. After all, the dumbbells would tend to move in straight lines, and you must exert a radial force just to keep turning them in a circle (though in that case you do no mechanical work, because the motion of the dumbbells has no radial component in the direction of the force you apply). To move the dumbbells into a smaller radius requires applying an even larger force that does work on the dumbbells, because there is now a radial component of the motion, in the direction of the radial force you exert.

A High Dive

You may have seen a skilled diver leap off a high board, tuck himself into a tight ball (holding onto his ankles), and rotate quite fast for a few turns. Then he straightens out and enters the water like a knife, hardly ruffling the surface (Figure 11.47).

QUESTION | What can you say about the diver's rotational angular momentum while he is in the air? Why?

We can probably neglect air resistance, because the diver is very far from reaching terminal speed. Although air resistance is distributed over the entire body, and there is a gravitational force acting on every atom in the body, torques to the left and right of the center of mass tend to cancel each other out (this cancellation is exact for the gravitational forces). Therefore there is negligible torque about the center of mass, and the rotational angular momentum $\vec{L}_{rot}$ does not change.

QUESTION | Early in the dive the diver is spinning rapidly (Figure 11.48). If rotational angular momentum is constant during the dive, where did that rapid spin come from?

When the diver jumps off the board, he must thrust with his feet in such a way that the force of the board on his feet has a sizable lever arm about his center of mass, so that by the time his feet lose contact with the board, the diver already has acquired a sizable rotational angular momentum.

QUESTION | When the diver pulls out of the tucked position, why does he stop spinning rapidly?

There is a large increase in the moment of inertia, because many atoms are now much farther from the center of mass than they were when the diver was in the tucked position. Larger moment of inertia implies smaller angular speed, since the magnitude of the rotational angular momentum $I\omega$ does not change.

Figure 11.47 A diver's moment of inertia is large when his body is extended.

Figure 11.48 A diver's moment of inertia is smaller when he is curled into a ball.

QUESTION   Can his body go straight from then on?

His body can't really go completely straight, because he still has rotational angular momentum. However, his angular speed may be so small that you hardly notice it, especially in comparison with the very rapid spin in the preceding tucked position. Moreover, his body could approximately follow the curving path of his center of mass, with the body rotating just enough to stay tangent to the trajectory. This enhances the illusion of straight motion. The most important aspect for good form is to arrange for the body to rotate into the vertical position at the time of entering the water, so as to make little splash.

See Problem 11.P.69, which asks you to analyze a diver's motion.

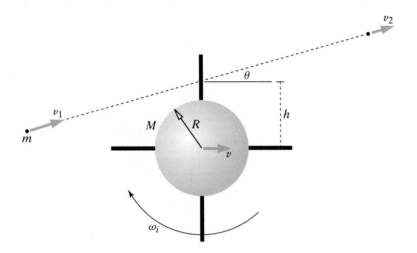

Figure 11.49 A satellite is struck by a tiny meteor.

EXAMPLE

A Satellite with Solar Panels

A satellite has four low-mass solar panels sticking out (Figure 11.49). The satellite can be considered to be approximately a uniform solid sphere. Originally it is traveling to the right with speed v and rotating clockwise with angular speed ω_i.

A tiny meteor traveling at high speed v_1 rips through one of the solar panels and continues in the same direction but at reduced speed v_2. Afterward calculate the v_x and v_y components of the center-of-mass velocity of the satellite and its angular velocity ω_f (magnitude and direction). Additional data are provided on the diagram.

Solution   What fundamental principles are useful starting points? The Momentum Principle and the Angular Momentum Principle. The Energy Principle would not be a useful starting point, because we don't know how much internal energy increase is associated with the meteor ripping through the panel. Take the satellite and meteor together as the system of interest, in which case the momentum and the angular momentum will remain constant, there being no external forces or torques.

System: Both the satellite and the meteor
Surroundings: Nothing during the collision

Initial State: Just before collision
Final state: Just after collision

Momentum Principle:

x components:

$$p_{xf} = p_{xi}$$

$$Mv_x + mv_2 \cos\theta = Mv + mv_1 \cos\theta$$

$$v_x = v + \frac{m}{M}(v_1 - v_2) \cos\theta$$

■ Choosing times just before and just after the collision makes it particularly easy to see how to calculate the translational angular momentum of the meteor, since it is directly above the center of the satellite at those instants.

y components:

$$p_{yf} = p_{yi}$$

$$Mv_y + mv_2 \cos(90° - \theta) = m_1 v_1 \cos(90° - \theta)$$

$$v_y = \frac{m}{M}(v_1 - v_2) \sin\theta$$

Angular Momentum Principle, about the location in space where the center of the satellite was initially

Component into the page ($-z$):

$$I\omega_i + hmv_1 \sin(90° - \theta) = I\omega_f + hmv_2 \sin(90° - \theta)$$

$$I = \frac{2}{5}MR^2$$

$$\omega_f = \omega_i + \frac{hm}{\left(\frac{2}{5}MR^2\right)}(v_1 - v_2) \cos\theta, \text{ clockwise, into the page}$$

11.8   SYSTEMS WITH NONZERO TORQUES

In the remainder of this chapter we will analyze systems in which there are external torques applied to the system, which can change the angular momentum of the system.

EXAMPLE

A Meter Stick on the Ice

Consider a meter stick whose mass is 300 grams and that lies on ice (in Figure 11.50 we're looking down on the meter stick). You pull at one end of the meter stick, at right angles to the stick, with a force of 6 newtons. Assume that friction with the ice is negligible. What is the rate of change of the center-of-mass speed v_{CM}? What is the rate of change of the angular speed ω?

Figure 11.50 Pull one end of a meter stick that is lying on ice (negligible friction).

■ In Chapter 9 we showed that the moment of inertia I around the center of mass of a uniform rod of mass M and length L is $ML^2/12$; $L = 1$ m here.

Solution

System: Stick
Surroundings: Your hand (pulling); ice (negligible effect)

Momentum Principle:
$$d\vec{P}/dt = d(m\vec{v}_{CM})/dt = \vec{F}_{net}$$
$$dv_{CM}/dt = (6\,N)/(0.3\,kg) = 20\,m/s^2$$

Angular Momentum Principle about center of mass:
$$d\vec{L}_{rot}/dt = \vec{\tau}_{net,CM}$$

Component into page ($-z$ direction):
$$I\,d\omega/dt = (0.5\,m)(6\,N) \sin 90° = 3\,N \cdot m$$
$$d\omega/dt = (3\,N \cdot m)/[(0.3\,kg \cdot m^2)/12] = 120\,radians/s^2$$

In vector terms, $d\vec{\omega}/dt$ points into the page, corresponding to the fact that the angular velocity points into the page and is increasing.

Figure 11.51 Pull one end of a meter stick that is lying on ice (negligible friction).

■ There is zero torque about location A, because there is zero distance from A to where the force is applied.

Figure 11.52 Wrap a string around a hockey puck, then pull it along the ice with negligible friction, with a constant tension F_T.

Alternative Analysis—Taking Torques Around the End of the Stick

It is interesting to re-analyze the motion of the meter stick by calculating torque and angular momentum about the end of the stick where the force is applied (location A in Figure 11.51), rather than about the center of mass. To be cautious and correct, we should say that we are taking torques about a location fixed in the ice next to the place where the end of the stick is momentarily located. This is a fixed location, not tied to the moving stick.

System: Stick
Surroundings: Your hand (pulling); ice (negligible effect)

Momentum Principle unchanged, which gives this:
$dv_{CM}/dt = (6\,\text{N})/(0.3\,\text{kg}) = 20\,\text{m/s}^2$

Angular Momentum Principle about location A:
$d\vec{L}_A/dt = \vec{\tau}_{\text{net},A}$

Component into page ($-z$ direction):
$$\frac{d}{dt}\left[\frac{ML^2}{12}\omega - \left(\frac{L}{2}\right)Mv_{CM}\right] = 0$$
$$\frac{d\omega}{dt} = \frac{6}{L}\frac{dv_{CM}}{dt} = \frac{6}{(1\,\text{m})}(20\,\text{m/s}^2) = 120\,\text{radians/s}^2$$

This agrees with our analysis where we took torques about the center of mass of the stick, but the details of the calculation are rather different. It is a good check on an analysis involving the Angular Momentum Principle to do the problem for two different choices of the location about which to calculate the net torque.

A Puck with String Wound Around It

Wrap a string around the outside of a hockey puck. Then pull on the string with a constant tension F_T (Figure 11.52). The puck has mass M and radius R. Assume that friction with the ice rink is negligible. Evidently $dv_{CM}/dt = F_T/M$.

> QUESTION The moment of inertia of a uniform solid puck of mass M and radius R is $MR^2/2$. What is the initial rate of change $d\omega/dt$ of the angular speed ω?

The torque about the center of mass is RF_T, into the page (down into the ice). The rotational angular momentum is $(MR^2/2)\omega$, into the page (down into the ice). Therefore $(MR^2/2)d\omega/dt = RF_T$, and we have $d\omega/dt = (2F_T)/(MR)$.

> **11.X.20** Redo the analysis, calculating torque and angular momentum relative to a fixed location in the ice anywhere underneath the string (similar to the analysis of the meter stick around one end). Show that the two analyses of the puck are consistent with each other.

Equilibrium

If the net torque on a system is zero, its angular momentum is constant. Conversely, if we know that the angular momentum is not changing, we can conclude that the net torque must be zero. For example, if a system is in equilibrium, not only must the net force on the system be zero; the net torque must also be zero. This allows us to make conclusions about the individual torques whose vector sum is zero.

EXAMPLE

Two People on a Seesaw

Figure 11.53 shows two persons on a seesaw. The person on the left has mass $M_1 = 90$ kg and sits at a distance $d_1 = 1.2$ m from the nearly frictionless axle. The person on the right has mass $M_2 = 40$ kg. What is the upward force that the axle must exert? Where must the person on the right sit in order that the seesaw not rotate? The seesaw itself has negligible mass.

Solution

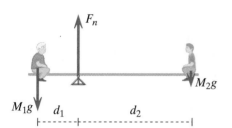

Figure 11.53 A seesaw in equilibrium.

System: The two persons and the seesaw
Surroundings: Earth and axle

Principle: The Momentum Principle $\dfrac{d\vec{p}_{sys}}{dt} = \vec{F}_{net}$

Since the system is not moving, the momentum of the system isn't changing, so $d\vec{p}_{sys}/dt = \vec{0}$, and therefore the net force $\vec{F}_{net} = \vec{0}$. Consider the y forces:

$$F_N - M_1 g - M_2 g = 0$$
$$F_N = (M_1 + M_2)g = ((90 + 40)\,\text{kg})(9.8\,\text{N/kg}) = 1274\,\text{N}$$

It isn't surprising to find that the axle must exert an upward force equal to the combined weights of the two persons. Next we use a different principle to determine where the person on the right should sit.

Principle: The Angular Momentum Principle $\dfrac{d\vec{L}_A}{dt} = \vec{\tau}_{net}$

Since the angular momentum isn't changing, $d\vec{L}_A/dt = \vec{0}$ about any point whatsoever, and therefore $\vec{\tau}_{net} = \vec{0}$ about any point whatsoever. It is convenient to choose point A to be at the location of the axle, because then the upward force of the axle exerts no torque about point A. Consider the z component of the net torque:

$$M_1 g d_1 - M_2 g d_2 = 0$$
$$d_2 = \frac{M_1}{M_2}d_1 = \frac{90\,\text{kg}}{40\,\text{kg}}(1.2\,\text{m}) = 2.7\,\text{m}$$

FURTHER DISCUSSION This analysis would also be valid if the seesaw were rotating at a constant angular velocity, because it would still be the case that the angular momentum would not be changing, and the net torque would be zero.

This is an example of a category of "statics" problems in which it is necessary to consider both the Momentum Principle and the Angular Momentum Principle in order to carry out a complete analysis.

11.X.21 Write a different torque equation around a location fixed in space where the person on the left is sitting, and show that it is in fact equivalent to the torque equation around the axle, when you take the force equation into consideration. It is often convenient to choose your fixed location so that some of the forces create no torque around that location and therefore don't appear in the Angular Momentum Principle.

11.9 **PREDICTING POSITIONS WHEN THERE IS ROTATION**

Here we show how to predict the position of something when rotation is involved.

| EXAMPLE | **Predicting Position with Rotation** |

A wheel of radius R and moment of inertia I is mounted on a low-friction axle (Figure 11.54). A string is wrapped around the edge, and you pull on it with a force F. At a certain time the angular speed is ω_i. **(a)** After a time interval Δt, what is the angular speed ω_f? **(b)** How far did your hand move during this time interval?

Solution

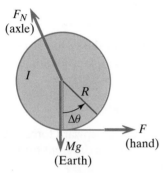

Figure 11.54 A wheel rotates on a low-friction axle, pulled by a string wrapped around the edge.

System: Wheel plus string
Surroundings: Earth, axle, hand
Free-body diagram: Figure 11.54
   Principle: The Momentum Principle tells us that the net force must be zero, because the center of mass does not move. To obtain information about rotation, we need to use the Angular Momentum Principle. We'll use the up-date form, about the center of mass:

$$\vec{L}_{\text{rot}, f} = \vec{L}_{\text{rot}, i} + \vec{\tau}_{\text{net}} \Delta t$$

(a) Calculate the final angular speed (only the force of the hand exerts a nonzero torque about the axle):

$$I\omega_f = I\omega_i + RF\Delta t$$

$$\omega_f = \omega_i + \frac{RF\Delta t}{I}$$

(b) Calculate how far your hand moved:
   To answer this question, first we will determine through how large an angle the wheel turned during the time interval Δt. Suppose that there is a red line drawn radially from the center of the wheel to the edge. Let θ be the angle that this red line makes with the y axis. This angle changes at an instantaneous rate of $\omega = d\theta/dt$, in radians per second. The change in the angle θ is

$$\Delta\theta = \omega_{\text{avg}} \Delta t$$

Because the angular speed ω changes at a constant rate RF/I, the average angular speed is

$$\omega_{\text{avg}} = \frac{\omega_i + \omega_f}{2}$$

Therefore the change in the angle of the red line is

$$\Delta\theta = \frac{\omega_i + \omega_f}{2} \Delta t$$

Since we calculated ω_f from the Angular Momentum Principle, now we know through how big an angle the wheel turned. What remains to be calculated is how far your hand moved. If the angle through which the wheel rotates is 2π radians (360°), an amount of string $2\pi R$ will come off the edge of the wheel, since the circumference of the wheel is $2\pi R$. More generally, if the angle through which the wheel rotates is $\Delta\theta$ (in radians), the amount of string that comes off is $R\Delta\theta$.
   The length of string that comes off the wheel is equal to the distance your hand moves:

$$\Delta x_{\text{hand}} = R\frac{\omega_i + \omega_f}{2} \Delta t$$

FURTHER DISCUSSION   The rate of change of the angular velocity $d\vec{\omega}/dt$ is often called the "angular acceleration" $\vec{\alpha}$, and the Angular Momentum Principle applied to pure rotation of an object whose moment of inertia is not changing is written as $I\vec{\alpha} = \vec{\tau}_{\text{net}}$. In this form, the equation looks very similar to

$M\vec{a} = \vec{F}_{net}$. Just as a constant acceleration $\vec{a}$ means that the average velocity is $(\vec{v}_i + \vec{v}_f)/2$, so also if the angular acceleration $\vec{\alpha}$ is constant the average angular velocity is $(\vec{\omega}_i + \vec{\omega}_f)/2$.

We could also determine the distance the hand moves by using the Energy Principle in combination with our result for ω_f which we obtained using the Angular Momentum Principle:

$$\frac{1}{2}I\omega_f^2 = \frac{1}{2}I\omega_i^2 + F\Delta x_{hand}$$

11.X.22 A uniform-density wheel of mass 6 kg and radius 0.3 m rotates on a low-friction axle. Starting from rest, a string wrapped around the edge exerts a constant force of 15 N for 0.6 s. **(a)** What is the final angular speed? **(b)** What is the average angular speed? **(c)** Through how big an angle did the wheel turn? **(d)** How much string came off the wheel?

11.10 ANGULAR MOMENTUM QUANTIZATION

Many elementary particles have rotational angular momentum (or, if you wish to be very cautious, these particles behave as though they have rotational angular momentum, which comes down to essentially the same thing). Electrons bound in an atom may have translational angular momentum relative to the nucleus as well as their own rotational angular momentum. Atoms as a whole may have angular momentum, as do many nuclei. The total angular momentum is as expected the sum of the translational angular momenta plus the rotational angular momenta.

The surprising thing about angular momentum at the atomic and subatomic level is that it is quantized. The basic quantum of angular momentum is this:

THE ANGULAR MOMENTUM QUANTUM

$$\hbar = \frac{h}{2\pi} = 1.05 \times 10^{-34} \text{ joule} \cdot \text{second}$$

Here h is "Planck's constant," 6.63×10^{-34} joule-second. Whenever you measure a vector component of the angular momentum (that is, along the x or y or z axis), you get either a half-integer or integer multiple of this quantum of angular momentum.

11.X.23 Show that $\hbar$ and angular momentum have the same units.

The Bohr Model of the Hydrogen Atom

As discussed in the previous chapter, in 1911 Rutherford and his group discovered the nucleus in atoms. Stimulated by this discovery, in 1913 the Danish physicist Niels Bohr made a bold conjecture that a hydrogen atom could be modeled as an electron going around a proton in circular orbits, but only in those orbits whose translational angular momentum is an integer multiple of $\hbar$. Consequently, these orbits would have only certain radii, and only certain values of energy. The differences in energies between these quantized orbits match the observed energies of photons emitted by atomic hydrogen (Figure 11.55).

Bohr's basic hypothesis—that the angular momentum of electrons in an atom is quantized—has proven to be an important insight, and has been retained as a fundamental tenet of the quantum mechanical model of an atom. As this model has been refined, a probabilistic view of the motion of electrons

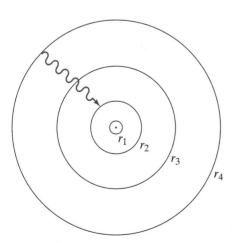

Figure 11.55 In the Bohr model of the hydrogen atom, electrons can be in only those circular orbits for which angular momentum is a multiple of $\hbar$. In a transition between allowed orbits, a photon is emitted.

has been adopted. In more sophisticated models electrons do not have precise trajectories, but only a "probability density"—a probability of being found at a particular location. These more complex quantum mechanical models explain a much wider range of atomic and molecular phenomena than does Bohr's original model.

The simple Bohr model does, however, predict correctly the allowed electronic energy levels for atomic hydrogen (as determined from the emission and absorption spectra of hydrogen atoms). We will work with this model to see how quantization of angular momentum leads to the prediction of electronic energy levels in a hydrogen atom.

Allowed Radii of Electron Orbits

Because the proton has much more mass than an electron (about 2000 times as much), we'll make the approximation that the proton is at rest, with the electron in a circular orbit around the proton.

> QUESTION   If the electron momentum is p, and the radius of the circular orbit is r, what is the translational angular momentum?

The translational angular momentum of the electron relative to the location of the proton is $L_{\text{trans},C} = rp\sin(90°) = rp$ (Figure 11.56). Bohr proposed that the only possible states of the hydrogen atom are those where the electron is in a circular orbit whose translational angular momentum is an integer multiple of $\hbar$:

BOHR: ANGULAR MOMENTUM IS QUANTIZED

$$|\vec{L}_{\text{trans},C}| = rp = N\hbar,$$

where N is an integer $(1, 2, 3, \ldots)$

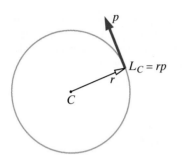

Figure 11.56 The angular momentum of the electron in a circular Bohr orbit relative to the proton. L_C is out of the page.

The Electric Force and a Circular Orbit

> QUESTION   What is the magnitude of the electric force that the proton exerts on the electron?

$$F_{\text{el}} = \frac{1}{4\pi\varepsilon_0}\frac{e^2}{r^2}$$

where $+e$ is the electric charge on the proton ($-e$ for the electron).

> QUESTION   Use the Momentum Principle to relate this force to the circular motion of the electron.

In a circular orbit at constant speed we know that the rate of change of the momentum is $(v/r)p$, so we have in the approximation that $v \ll c$, where m is the mass of the electron,

$$\left(\frac{v}{r}\right)p = \frac{mv^2}{r} = F = \frac{1}{4\pi\varepsilon_0}\frac{e^2}{r^2}$$

Solving for r

We are looking for the allowed values of the orbit radius r, so we look for ways to express v in terms of r. Bohr's angular momentum condition gives us v in terms of r:

$$|\vec{L}_{\text{trans},C}| = rp = N\hbar \quad \text{leads to} \quad v = \frac{p}{m} = \frac{N\hbar}{mr}$$

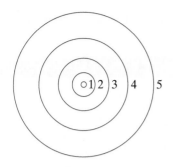

Figure 11.57 Bohr radii for $N = 1$ through 5. The nucleus is not shown.

Putting this result for r into the result obtained from the Momentum Principle, $mv^2/r = F$, we have

$$\frac{m}{r}\left(\frac{N\hbar}{mr}\right)^2 = \frac{1}{4\pi\varepsilon_0}\frac{e^2}{r^2}$$

Solving the latter relation for r, we obtain the following result (Figure 11.57):

ALLOWED BOHR RADII FOR ELECTRON ORBITS

$$r = N^2\frac{\hbar^2}{\left(\dfrac{1}{4\pi\varepsilon_0}\right)e^2 m}$$

Let's evaluate this result numerically, so that later we can compare with measurements of the spectrum of light emitted by excited atomic hydrogen. Using

$\hbar = 1.05 \times 10^{-34}\,\text{J}\cdot\text{s},$
$1/4\pi\varepsilon_0 = 8.99 \times 10^9\,\text{N}\cdot\text{m}^2/\text{coulomb}^2,$
$e = 1.60 \times 10^{-19}\,\text{coulomb},$
and the mass of the electron $m = 9.11 \times 10^{-31}$,

we find this:

$$r = N^2(0.53 \times 10^{-10}\,\text{m})$$

This is a striking result. The simple Bohr model predicts that the smallest permissible electron radius ($N = 1$) is $0.53 \times 10^{-10}\,\text{m}$, and atoms are in fact observed to have radii of approximately this size.

Energy for a Circular Orbit

We cannot observe directly the radius of the orbit of an electron in an atom. What we can observe are photons emitted by excited atoms, which tell us the differences in energies between various energy levels of the atom. Using the Bohr model we can predict the energies of these energy levels, and see whether the differences between these levels match the energies of photons observed in an atomic spectrum.

Given a set of possible values for r, we can calculate the possible values for energy of the electron, starting from this observation:

The result $mv^2/r = F$ leads to $mv^2 = \dfrac{1}{4\pi\varepsilon_0}\dfrac{e^2}{r}$.

QUESTION   Now write a formula for the kinetic energy plus electric potential energy of the hydrogen atom, using the Bohr model.

Kinetic energy:

$$K = \frac{1}{2}mv^2 = \frac{1}{2}\left(\frac{1}{4\pi\varepsilon_0}\frac{e^2}{r}\right)\quad\text{(from the previous equation)}$$

Electric potential energy:

$$U_{\text{el}} = -\frac{1}{4\pi\varepsilon_0}\frac{e^2}{r}$$

Therefore the energy is this (omitting the rest energies):

$$E = K + U_{\text{el}} = -\frac{1}{2}\left(\frac{1}{4\pi\varepsilon_0}\frac{e^2}{r}\right)$$

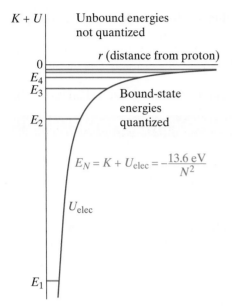

$K + U$

Unbound energies
not quantized

r (distance from proton)

0

E_4
E_3

Bound-state
energies
quantized

E_2

$$E_N = K + U_{\text{elec}} = -\frac{13.6\,\text{eV}}{N^2}$$

U_{elec}

E_1

Figure 11.58 Bohr model prediction of electronic energy levels for a hydrogen atom. The potential energy curve for the electron–proton system is also shown.

According to the Bohr model only certain radii will actually occur. Inserting our previous expression for r, we get:

BOHR MODEL ENERGY LEVELS
$$E = -\frac{\left(\dfrac{1}{4\pi \varepsilon_0}\right)^2 e^4 m}{2N^2 \hbar^2}$$

Evaluating this expression we get

$$E = -\frac{2.17 \times 10^{-18}\,\text{J}}{N^2}$$

Converting to electron-volts ($1\,\text{eV} = 1.6 \times 10^{-19}\,\text{J}$) we have (Figure 11.58)

$$E = -\frac{13.6\,\text{eV}}{N^2}, \qquad N = 1, 2, 3, \dots$$

This prediction for the quantized energies of atomic hydrogen agrees well with the observed electronic spectrum of hydrogen. We quoted this result for the quantized energy levels in Chapter 8. The energies are negative, corresponding to bound states.

Photons Emitted by Atomic Hydrogen

Bohr proposed that electromagnetic radiation would be emitted when there was a sudden change from a higher energy level to a lower one. We saw in Chapter 8 that the Bohr energy formula does correctly predict the energies of photons emitted by excited atomic hydrogen.

QUESTION   In the Bohr model, does photon emission correspond to the radius of the circular orbit getting larger or smaller? Does the quantum number N increase or decrease?

The kinetic plus potential energy is negative, corresponding to a bound state, so a higher energy is one that is less negative and therefore corresponds to a larger radius (since the energy is proportional to $-1/r$). Also, higher energy corresponds to a larger value of N, which is also associated with a larger radius. Therefore photon emission is associated with a decrease in r and a decrease in N (Figure 11.59). Photon absorption involves an increase in r and N.

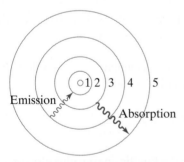

Emission

1 2 3   4   5

Absorption

Figure 11.59 A photon is emitted in a transition from a higher to a lower energy level. A photon is absorbed in a transition from a lower to a higher energy level.

Limitations of the Bohr Model

The Bohr model was an important predecessor to current quantum mechanical models of the atom. It predicts the main aspects of the hydrogen energy levels correctly (neglecting various small effects). The idea that photon emission is associated with a drop from a higher energy level to a lower energy level, with the photon energy equal to the difference in the hydrogen energy levels, is fundamental to the extended quantum-mechanical treatment.

The notion of deterministic circular orbits has been abandoned and replaced with a picture of a probabilistic electron "cloud," but even in more sophisticated models the average radius of this electron cloud is larger for higher energies, just as in the Bohr model. From your study of chemistry you are probably familiar with the shapes of the "probability density" functions or "orbitals" describing the electron's location in the atom in different energy states, some of which are spherical, others dumbbell shaped.

The idea that angular momentum is quantized has far-reaching consequences and is a major element in modern quantum mechanics. However, the actual quantization rules in a hydrogen atom are more complex than those that are assumed in the Bohr model. For example, the translational angular momentum in the ground state ($N = 1$) is actually zero, not $\hbar$, and for the next higher state ($N = 2$), the z component of translational angular momentum can be either 0 or $\hbar$.

The photon itself has angular momentum $\hbar$, so there are "selection rules" for the emission of a photon from an excited hydrogen atom. For example, the change in the angular momentum of the hydrogen atom cannot be zero, because angular momentum must be conserved. If the hydrogen atoms's angular momentum doesn't change, a single photon cannot be emitted. It is possible in this case to emit *two* photons whose angular momentum adds up to zero, but this is a rare event.

Particle Spin

An important example of angular momentum quantization is the observation that many elementary particles have rotational angular momentum, and it is quantized. We know this in part because of observations of the interaction of these particles with other particles having nonzero angular momentum; in these interactions total angular momentum is constant.

For example, the electron, muon, and neutrino are said to have spin 1/2, because measurement of a component of their rotational angular momentum always yields $\pm(1/2)\hbar$. Every quark has spin 1/2, and particles built out of three quarks, such as protons and neutrons, necessarily have half-integral spin. The spin of the proton and neutron is 1/2, as though two of the quarks have their spins opposed ($\Uparrow\Downarrow\Uparrow$), with no translational angular momentum of the quarks. Some short-lived three-quark particles have spin 3/2. One way to get this is for all the quarks to have their spins aligned ($\Uparrow\Uparrow\Uparrow$), or there can be some translational angular momentum of the quarks. While rotational angular momentum can be half-integral, translational angular momentum is always integral (0, 1, 2, etc.).

Particles made out of one quark and one antiquark, called mesons, have integral spin. For example, the pion spin is zero (quark and antiquark spins opposed, $\Uparrow\Downarrow$), and the spin of the "rho" meson is 1 (quark and antiquark spins aligned, $\Uparrow\Uparrow$).

Presumably the angular momentum of a macroscopic object such as a baseball is also quantized, but the angular momentum of a baseball is huge compared to Planck's constant, and you don't notice the quantization, which is on an exceedingly fine scale (1×10^{-34} J·s!). Only at the atomic level are the effects of quantization really evident.

Consequences of Angular Momentum Quantization

The quantization of angular momentum has far-reaching consequences. Angular momentum quantization plays a major role in determining the structure of atoms and the nature of the chemical periodic table. The two lowest-energy-state electrons in an atom always have zero translational angular momentum and zero rotational angular momentum (because the two rotational angular momenta are always oppositely aligned and add up exactly to zero).

There is a very deep connection between angular momentum and the statistical behavior of particles. Particles such as the electron and the proton that have half-integral spin are called "fermions" and exhibit the peculiar behavior that two fermions cannot be in the exact same quantum state. This is why only two electrons can be in the lowest energy state of an atom, with their spins opposed. Their spin directions are forbidden to be the same, because then their energy and angular momentum would be exactly the same, which is forbidden for fermions. This prohibition is called the "Pauli exclusion principle."

On the other hand, particles with integral spin, called "bosons," are not subject to the Pauli exclusion principle, and there is no limit to the number of bosons that can be in the same energy state. In fact, there is a special state of matter called a "Bose–Einstein condensation" in which very large numbers of particles end up in exactly the same quantum state. This state of matter was predicted long before it was actually created and observed in 1995.

In a diatomic gas molecule such as oxygen (O_2), the kinetic energy associated with rotation of the molecule is of course $L_{rot}^2/(2I)$. However, the rotational angular momentum L_{rot} is quantized, so the rotational energies of oxygen are quantized. The phenomenon of angular momentum quantization affects the specific heat of diatomic gases at low temperatures, as will be discussed in a later chapter. (You can of course refer to the angular momentum of an oxygen molecule as translational angular momentum of the two atoms instead of rotational angular momentum of the molecule.)

It is an important prediction of quantum mechanics that the x, y, or z component of the angular momentum (L_x, L_y, or L_z) can only be an integer or half-integer multiple of $\hbar$, whereas the square magnitude of the angular momentum has the quantized values $l(l+1)\hbar^2$, where l has integer or half-integer values, depending on whether a component has integer or half-integer values:

QUANTIZED VALUES OF L^2

$$L^2 = l(l+1)\hbar^2$$

where l is integer or half-integer

All nuclei that have an even number of protons and an even number of neutrons ("even-even" nuclei) have a total angular momentum of zero in their nuclear ground state. Examples include carbon-12 (6 protons and 6 neutrons) and oxygen-16 (8 protons and 8 neutrons). The spins of protons and neutrons in even-even nuclei are paired up in the lowest-energy state of a nucleus with each other in such a way as to produce zero net angular momentum. This would be an exceedingly unlikely outcome if angular momentum weren't quantized.

11.11 *GYROSCOPES

A gyroscope is a fascinating device, and its unusual properties have been exploited to stabilize ships and spacecraft. Its behavior provides a good model and analogy for some important aspects of the quantum mechanical behavior of atoms and nuclei. Magnetic resonance imaging (MRI) is based on the gyroscope-like behavior of spinning nuclei.

A gyroscope has a spinning disk mounted on an axle. If you place one end of the axle on a vertical support, you may observe a very complex motion (Figure 11.60). The gyroscope rises and falls ("nutation") as it revolves around the support ("precession").

One fruitful approach in modeling a complex physical system is to try to analyze the simplest motion that the system is capable of. For example, we

Figure 11.60 A gyroscope can exhibit both "nutation" and "precession."

Figure 11.61 A gyroscope that is precessing horizontally, with no nutation.

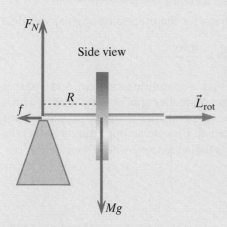

Figure 11.62 Side view showing external forces acting on the gyroscope and the rotational angular momentum vector.

can observe a pure precession of a gyroscope, with no nutation. It is possible with care to start the gyroscope with special initial conditions so that it merely precesses, without bobbing up and down. We'll try to analyze this special kind of gyroscope motion. In fact, we'll start with a particularly simple form of precession, with the rotational axis horizontal (Figure 11.61).

We define some variables to make it easier to describe and discuss the situation. The gyroscope disk rotates around its own axis with some spin angular speed that we'll call ω, and the disk has a moment of inertia I. The rotational angular momentum $\vec{L}_{rot}$ of the disk always points horizontally but is continually changing direction as the gyroscope precesses (if we neglect friction, the angular speed ω is constant). The gyroscope revolves around the support with angular speed Ω (Greek uppercase omega), constant in magnitude and direction (if we neglect friction). Note that $v_{CM} = \Omega r$ and $\Omega = (v_{CM}/r)$, where r is the distance from the support to the center of mass.

In Chapter 5 we derived the formula for the time rate of change of the unit momentum vector $\hat{p}$, which had the magnitude (v/r), which is the same as angular speed ω. In general, the magnitude of the time rate of change of any rotating unit vector $\hat{X}$ is $|d\hat{X}/dt| = \omega$, and the magnitude of the rate of change of a rotating vector $\vec{X} = |\vec{X}|\hat{X}$ whose magnitude isn't changing is $\omega|\vec{X}|$. Since the angular speed of the gyroscope's precession is Ω, the magnitude of the rate of change of the (linear) momentum of the gyroscope is ΩP, and the magnitude of the rate of change of the angular momentum is ΩL.

QUESTION What forces act on the gyroscope? What relationships are there among these forces?

The support pushes up with a force we'll call F_N, and the Earth pulls down with a force Mg through the center of mass (Figure 11.62). Since the center of mass stays at the same height all the time in the case of simple precession, the vertical component of the net force must be zero. Therefore $F_N = Mg$.

This isn't quite the whole story, though. Note that the center of mass of the gyroscope is moving in a circle, which requires that there be a radially inward force to make the momentum vector rotate (with angular speed Ω). The only object that can exert this force is the support. There must be a small horizontal frictional force f such that $|d\vec{P}/dt| = \Omega P = f$, where $P = Mv_{CM} = M(\Omega R)$; R is the radius of the circle traveled by the center of mass. This force, $f = MR\Omega^2$, is small if the precession rate Ω is small.

It is easy to observe with a toy gyroscope that as the spin of the disk slows down due to friction (smaller ω), the precession actually speeds up (larger Ω). If we could predict the precession speed for a given spin, we would largely "understand" this simple example of gyroscope motion. What can we use to attempt a prediction?

We've already used the Momentum Principle to conclude something about the forces F_N and f. What about energy? If the gyroscope precesses in the horizontal plane, there's no change in gravitational energy, because the center of mass remains always at the same height. There's no change in kinetic energy, either. So energy considerations don't seem to tell us anything interesting about precession (though they would be useful in analyzing the more complex rise and fall of nutation). The remaining principle to apply is the Angular Momentum Principle (no wonder we left gyroscopes until this chapter).

QUESTION What is the magnitude of the rotational angular momentum of the gyroscope? What is the magnitude of the translational angular momentum of the gyroscope around the support?

We said that the rotating disk has moment of inertia I and angular speed ω, so the rotational angular momentum is simply $L_{\text{rot}} = I\omega$. The magnitude of the translational angular momentum is this:

$$L_{\text{support}} = |\vec{R} \times \vec{P}| = RP$$

Evidently the magnitudes of both the rotational and translational angular momenta are constant, not changing with time. However, what about the directions of these angular momenta? Do they change with time?

QUESTION How does the direction of the translational angular momentum change with time? How does the direction of the rotational angular momentum change with time?

Neither the magnitude nor the direction of the translational angular momentum changes (neglecting friction): it has constant magnitude and points vertically upward at all times. However, the direction of the rotational angular momentum is constantly changing as the gyroscope precesses. We need to calculate the rate at which the rotational angular momentum vector is changing, $d\vec{L}_{\text{rot}}/dt$. Since the rotational angular momentum vector is a rotating vector, which rotates with the precession angular speed Ω, we have this:

$$\left| \frac{d\vec{L}_{\text{rot}}}{dt} \right| = \Omega L_{\text{rot}}$$

QUESTION Therefore, what is the Angular Momentum Principle for the gyroscope?

The remaining element we need for the Angular Momentum Principle is the torque that acts around the center of mass, at the center of the spinning disk. Look again at the force diagram (Figure 11.63), and you see that the magnitude of the torque about the center of mass is $RF_N = RMg$, and the direction of the torque is into the page. Seen from above, the torque points at a right angle to the rotational angular momentum (Figure 11.64). Therefore the Angular Momentum Principle yields

$$\left| \frac{d\vec{L}_{\text{rot}}}{dt} \right| = \Omega L_{\text{rot}} = \tau_{\text{CM}}$$

where $L_{\text{rot}} = I\omega$, and $\tau_{\text{CM}} = RMg$. Solving for the precession angular speed Ω, we have

$$\Omega = \frac{\tau_{\text{CM}}}{L_{\text{rot}}} = \frac{RMg}{I\omega}$$

This is a surprisingly simple result for such a complicated system, and it agrees at least qualitatively with observations. A smaller spin ω is associated with a larger precession rate Ω. Conversely, a gyroscope that spins very fast precesses very slowly. If possible, your instructor will provide an opportunity to test the theory quantitatively by observing an actual gyroscope whose spin and precession rates can be measured.

*Reflection

Reflect on how we arrived at this result. Look over the steps involved. The key point is that the rotational angular momentum varies in direction without varying in magnitude, and the magnitude of the rate of change of the rotational angular momentum is simply Ω times L_{rot}. This rate of change of the rotational angular momentum is equal to the net torque acting on the system (around the center of mass).

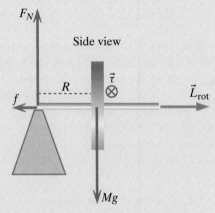

Figure 11.63 Side view showing external forces acting on the gyroscope and the rotational angular momentum vector. There is a torque around the center of mass, into the page, due to the F_N force.

View from above

Figure 11.64 View from above, showing the torque vector and the rotational angular momentum vector.

Figure 11.65 What is the precession rate when the spin axis is at an angle θ to the vertical?

11.X.24 To complete this reflection, determine the relationship between ω and Ω for the case of pure precession, but with the spin axis at an arbitrary angle θ to the vertical (Figure 11.65; $\theta = 90°$ is the case of horizontal precession we treated). If you have the opportunity, see whether this relationship holds for a real gyroscope.

*Uses of Gyroscopes

The precession we calculated is the result of a nonzero torque acting around the center of mass. Through clever mounting of the gyroscope it is possible to make this torque vanishingly small, in which case the axis does not precess but always maintains the same direction, which has been useful in navigation. The satellite-based Global Positioning System (GPS) can tell you where your airplane is; gyroscopically based instruments can tell you the orientation and attitude of your airplane at this known location.

Large gyroscopes have been used to stabilize ships against rolling in the sea. A gyroscope with very high rotational angular momentum (large moment of inertia and large angular speed) is hard to turn quickly, and this can provide mechanical stability. Gyroscopes are also used to stabilize the orientation of spacecraft and satellites.

*Friction

We deliberately neglected friction in the calculation. Friction has two main effects on a gyroscope. Friction in the spin axis makes ω decrease with time, which leads to an increased precession rate Ω. Friction on the top of the support slows down the precession, which means that Ω is no longer equal to RMg/L_{rot} as it should be for pure precession. As a result of these effects, what you observe with a gyroscope that starts out in pure precession is that the spin axis eventually starts to tip lower, accompanied by faster precession. The gyroscope starts out with a stately, dignified, slow precession, but toward the end gives the impression of motion that is more and more frantic.

*Magnetic Resonance Imaging (MRI)

Electrically charged atomic and subatomic objects that have angular momentum act like little bar magnets with north and south poles along the angular momentum axis (we say that they have "magnetic moments"). It is as though the spinning charged particle constituted little loops of current, and current loops produce a magnetic field. Even some electrically neutral particles such as the neutron have magnetic moments, because they are built out of electrically charged quarks that have magnetic moments.

In the presence of an applied magnetic field (typically produced by other current-carrying coils of wire), a bar magnet tends to twist to line up with the applied magnetic field. That is, a magnetic field exerts a torque on a bar magnet. A well-known example is that a compass needle aligns with the Earth's magnetic field, which is useful for navigation. (The compass needle is magnetized and is itself a little bar magnet.)

Since a magnetic field applies a torque to a bar magnet, or to atomic or subatomic particles that have angular momentum, such particles precess in the presence of an applied magnetic field (Figure 11.66). This phenomenon is exploited in a technique important in physics, chemistry, and biology, called NMR—nuclear magnetic resonance. A particularly useful application of NMR is in magnetic resonance imaging (MRI).

Figure 11.66 Precession of proton spin in an external magnetic field.

In MRI, a large, steady, uniform magnetic field created by large direct-current-carrying coils of wire surrounding the patient makes protons in the person's body precess. Many of the most common nuclei in the body are even-even nuclei (in particular, the most common isotopes of carbon and oxygen) and have zero angular momentum and therefore no magnetic moment, and these do not precess. However, hydrogen nuclei (protons) are very common in the body, have rotational angular momentum, and do precess. The stronger the magnetic field, the larger the torque acting on the proton nuclear magnets, and the faster their precession. (Note that for a gyroscope, $\Omega = \tau/L_{\text{rot}}$; the precession rate is proportional to the applied torque τ.)

With the protons precessing in the presence of the large steady magnetic field, a small high-frequency (time-varying) magnetic field tuned to match exactly the precession frequency of the protons will flip the spins upside down. This is called a "resonance" phenomenon. The high-frequency signal is turned off, and the protons revert back to being aligned with the steady magnetic field. The act of flipping back emits radiation (at the precession frequency) that can be detected by coils connected to a receiver, and the strength of the signal indicates how many protons were affected.

That's the basic physical mechanism, but this by itself would not yield spatial detail about the interior of the body. The trick is to superimpose on the large steady magnetic field a small nonuniform magnetic field, which has the effect of establishing slightly different torques on protons at different locations in the body. As a result, when the protons flip back into alignment they radiate signals whose frequencies indicate their locations. A computer algorithm calculates how much of each frequency is present in the signal, and this indicates how many protons are at each location. In this way a very detailed image is built up of the interior of the body.

*Precession of Spin Axes in Astronomy

As we mentioned earlier, the Earth is subject to small torques due to gravitational forces of the Sun and Moon acting on its nonspherical, "oblate" shape. There is a torque on the Earth's equatorial bulge, perpendicular to the spin rotation axis, and this causes the Earth's axis to precess very slowly, once around about every 26,000 years, so that the "North Star" hasn't always been the star we call Polaris. This effect is called the "precession of the equinoxes" because it leads to a change in what month the spring and fall equinoxes occur. To see why there is such an effect, consider Figure 11.67.

Figure 11.67 Forces on the Earth due to the Sun. The diagram is not to scale; the asymmetry of the Earth is greatly exaggerated. The net torque about the center of mass of the Earth is out of the page.

The Sun (or Moon) exerts a slightly larger force on the closer bulge due to the $1/r^2$ dependence of the force, so there is a nonzero net torque around the center of mass of the Earth, pointing toward you, out of the page. This makes the rotational angular momentum $\vec{L}_{rot}$ of the Earth precess in a "retrograde" way—that is, in the opposite direction to most rotations in our Solar System. The direction of the axis of the Earth slowly changes due to the torque. (The size of the equatorial bulge, and the differences in F_{near} and F_{far}, have been greatly exaggerated in Figure 11.67).

Since the Earth's axis is tipped 23° away from perpendicular to the Earth–Sun orbit, the precession of the axis makes a big change in the location of the "North Star." Thirteen thousand years from now the "North Star" will be a star that is 46° away from Polaris in the night sky.

The Moon going around the Earth is a kind of gyroscope. The Earth–Moon orbit is inclined a few degrees to the plane of the Earth–Sun orbit, and the Sun exerts a nonzero torque perpendicular to the angular momentum of the Earth–Moon system. To see why, consider a simpler case in which the Earth and Moon have equal mass, and you see the same effect as with the equatorial bulge of the Earth (Figure 11.68).

Figure 11.68 Forces on the Earth–Moon system due to the Sun. The diagram is not to scale; the Earth and Moon are shown as if their masses were equal. The net torque on the Earth–Moon system about its center of mass is out of the page.

This torque varies in magnitude during a month, but the averaged effect is that the Earth–Moon system precesses once around in about 18 years (this precession is also "retrograde"). This has an effect on the timing of eclipses, which can occur only when the Moon in its orbit is passing through the plane of the Earth–Sun orbit.

*Tidal Torques

If the Earth did not rotate, the Moon would create tides in the oceans that would pile up in line with the Moon. However, the rotating Earth drags these tidal bulges so that they are no longer in line with the Moon, and the Moon exerts a small torque (Figure 11.69). This torque is directed along the axis and acts to slow down the spin rate, so that the day is getting longer than 24 hours. This effect is called "tidal friction." This interaction with the Moon has the effect that as the rotational angular momentum of the Earth decreases, the translational angular momentum of the Earth–Moon system increases (conservation of total angular momentum), with the result that the Earth and Moon are getting farther apart. (These tidal forces have a small net tangential component that acts on both Earth and Moon to increase their translational angular momenta.)

Something like this presumably happened in the past to the Moon, when it was molten. The rotational angular momentum of the Moon decreased

Figure 11.69 Tidal torques due to the Moon, viewed from a point above the North Pole. In the top image the Earth is not rotating; in the bottom image it is rotating. (Not to scale; the tidal effects are greatly exaggerated.)

Figure 11.70 If the object lacks axial symmetry, the angular momentum and angular velocity may have different directions!

Figure 11.71 If the object is symmetric, the angular momentum is in the direction of the angular velocity.

Figure 11.72 In circular motion, both $\vec{r}$ and $\vec{v}$ are rotating vectors. $\vec{\omega}$ is out of the page. At any instant, $\vec{\omega} \times \vec{r}$ is in the direction of $\vec{v}$, while $\vec{\omega} \times \vec{v}$ is toward the center, in the direction of $\vec{a}$, as predicted by the relation $d\vec{X}/dt = \vec{\omega} \times \vec{X}$.

due to tidal torques exerted by the Earth, which are much larger than the tidal torques that the low-mass Moon exerts on the Earth. When the Moon's spin angular speed had decreased to be the same as its translational angular speed (currently about 2π radians or $360°$ per month), the process terminated. As a result, now the Moon always displays nearly the same face to the Earth.

11.12   *MORE COMPLEX ROTATIONAL SITUATIONS

In more advanced courses you may study nonsymmetric rotational situations like the one in Figure 11.70. The angular velocity points along the axis of rotation, but the (translational) angular momentum of each mass is $\vec{r}_A \times \vec{p}$, so the total angular momentum $\vec{L}_A$ does *not* point along the axis! Moreover, the angular momentum vector continually changes direction. This change of angular momentum requires a nonzero torque, which is applied to the axle by the bearings in which the axle rotates. The effect of this "dynamic imbalance" is to cause severe wear on the axle and bearings. Car tires must be carefully balanced to prevent this.

A symmetric object like the one shown in Figure 11.71 has an angular momentum that does point along the axis, because the perpendicular components of the angular momenta cancel each other. This is the simpler kind of situation we have dealt with in this chapter. In the symmetric case shown in Figure 11.71 it is true that $I = m_1 r_{\perp 1}^2 + m_2 r_{\perp 2}^2 + m_3 r_{\perp 3}^2 + m_4 r_{\perp 4}^2 + \cdots$ about the axis of rotation, and the simple formulas $\vec{L}_{\text{rot}} = I\vec{\omega}$ and $K_{\text{rot}} = \frac{1}{2}I\omega^2$ are valid.

We calculated moment of inertia as a sum (integral) over $r_\perp^2$, the perpendicular distance to an axis. To deal with rotational motion in general requires expressing the moment of inertia as a "tensor," a 3 by 3 array of numbers representing integrals over x^2, xy, xz, y^2, yx, yz, z^2, zx, and zy. Matrix multiplication of this tensor times the angular velocity vector $\vec{\omega}$ yields an angular momentum vector that need not point in the same direction as $\vec{\omega}$. This concept is beyond the scope of this introductory textbook.

11.13   *RATE OF CHANGE OF A ROTATING VECTOR

A minor comment: In Chapter 5 we derived the formula for the time rate of change of the unit momentum vector $\hat{p}$, which had the magnitude (v/r), which is the same as ω. In general, the magnitude of the time rate of change of any rotating unit vector $\hat{X}$ is $|d\hat{X}/dt| = \omega$, and the magnitude of the rate of change of a rotating vector $\vec{X} = |\vec{X}|\hat{X}$ whose magnitude isn't changing is $\omega|\vec{X}|$.

If you look at a diagram involving the angular velocity vector $\vec{\omega}$ you can see that the direction of $d\vec{X}/dt$ is given by the cross product $\vec{\omega} \times \vec{X}$. It can be shown rather easily that

$$\frac{d\vec{X}}{dt} = \vec{\omega} \times \vec{X}$$

gives both the magnitude and direction of a rotating vector (Figure 11.72). This is another illustration of the usefulness of cross products.

SUMMARY

The Angular Momentum Principle for a system:

$$\frac{d\vec{L}_{tot,A}}{dt} = \vec{\tau}_{net,A} \quad \text{or} \quad \Delta\vec{L}_{tot,A} = \vec{\tau}_{net,A}\Delta$$

Angular momentum of a particle about location A (see Figure 11.73):

$$\vec{L}_A = \vec{r}_A \times \vec{p}$$
$$L_A = r_\perp p = r_A p \sin\theta$$

The direction is given by the right-hand rule.

$$\vec{L}_A = \langle (yp_z - zp_y), (zp_x - xp_z), (xp_y - yp_x) \rangle$$

Figure 11.73

Torque is a measure of the twist imparted by a force about location A:

$$\vec{\tau}_A = \vec{r}_A \times \vec{F}$$

Angular momentum for a multiparticle system about some location A can be divided into translational and rotational angular momentum

$$\vec{L}_A = \vec{L}_{trans,A} + \vec{L}_{rot}$$
$$\vec{L}_{trans,A} = \vec{r}_{CM,A} \times \vec{P}_{tot}$$
$$\vec{L}_{rot} = \vec{r}_{1,CM} \times \vec{p}_1 + \vec{r}_{2,CM} \times \vec{p}_2 + \cdots$$

(See Figure 11.74.)

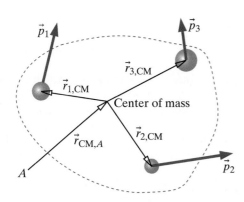

Figure 11.74

Rotational angular momentum in terms of moment of inertia:

$$\vec{L}_{rot} = I\vec{\omega}$$

Rate of change of rotational angular momentum:

$$\frac{d\vec{L}_{rot}}{dt} = \vec{\tau}_{net,CM}$$

Kinetic energy relative to center of mass in terms of moment of inertia and rotational angular momentum:

$$K_{rot} = \frac{1}{2}I\omega^2 = \frac{1}{2}\frac{(I\omega)^2}{I} = \frac{L_{rot}^2}{2I}$$

Moments of inertia:

$$I_{disk} = I_{cylinder} = \frac{1}{2}MR^2$$

about center of disk, rotating around axis of cylinder

$$I_{sphere} = \frac{2}{5}MR^2$$

for axis passing through center of sphere.

Uniform solid cylinder of length L, radius R, about axis perpendicular to cylinder, through center of cylinder:

$$I = \frac{1}{12}ML^2 + \frac{1}{4}MR^2$$

For a rigid object rotating about an axle that is not the center of mass,

$$K_{tot} = K_{trans} + K_{rot} = \frac{1}{2}(Mr_{CM}^2 + I_{CM})\omega^2$$

$$\vec{L}_A = \vec{L}_{trans,A} + \vec{L}_{rot} = (Mr_{CM}^2 + I_{CM})\omega$$

Angular momentum is quantized in microscopic systems:

A component of angular momentum is an integer or half-integer multiple of $\hbar$.

The square magnitude of angular momentum has quantized values

$$L^2 = l(l+1)\hbar^2$$

where l has integer or half-integer values.

Bohr model of hydrogen:

$$E = -\frac{13.6\,eV}{N^2}, \quad N = 1, 2, 3, \ldots$$

Angular speed of precession for a gyroscope:

$$\Omega = \frac{\tau}{L_{rot}} = \frac{RMg}{I\omega}$$

EXERCISES AND PROBLEMS

Section 11.1

11.X.25 Give an example of a situation in which an object is traveling in a straight line, yet has nonzero angular momentum.

11.X.26 Evaluate the cross product $(5\hat{\imath} + 3\hat{\jmath}) \times (-4\hat{\imath} + 2\hat{\jmath})$, which expands to $-20\hat{\imath} \times \hat{\imath} + 10\hat{\imath} \times \hat{\jmath} - 12\hat{\jmath} \times \hat{\imath} + 6\hat{\jmath} \times \hat{\jmath}$.

11.X.27 Figure 11.75 shows seven particles, each with the same magnitude of momentum $|\vec{p}| = 25\,\text{kg} \cdot \text{m/s}$ but with different directions of momentum and different positions relative to location A. The distances shown in the diagram have these values: $w = 18$ m, $h = 28$ m, and $d = 27$ m.

Calculate the z component of angular momentum L_{Az} for each particle (x to the right, y up, z out of the page). Make sure you give the correct sign.

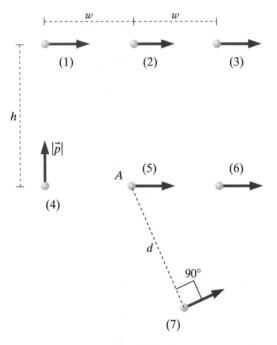

Figure 11.75

11.X.28 What is the angular momentum $\vec{L}_A$ if $\vec{r}_A = \langle 9, -9, 0 \rangle$ m and $\vec{p} = \langle 12, 10, 0 \rangle$ kg $\cdot$ m/s?

11.X.29 At a particular instant the location of an object relative to location A is given by the vector $\vec{r}_A = \langle 6, 6, 0 \rangle$ m. At this instant the momentum of the object is $\vec{p} = \langle -11, 13, 0 \rangle$ kg $\cdot$ m/s. What is the angular momentum of the object about location A?

11.X.30 What are the units of moment of inertia? Of angular speed ω? Of angular momentum? Of linear momentum?

11.X.31 A common amusement park ride is a Ferris wheel (see Figure 11.76, which is not drawn to scale). Riders sit in chairs that are on pivots so they remain level as the wheel turns at a constant rate. A particular Ferris wheel has a radius of 24 meters, and it makes one complete revolution around its axle (at location A) in 20 seconds. In all of the following questions, consider location A (at the center of the axle) as the location around

which we will calculate the angular momentum. At the instant shown in the diagram, a child of mass 40 kg, sitting at location F, is traveling with velocity $\langle 7.5, 0, 0 \rangle$ m/s.

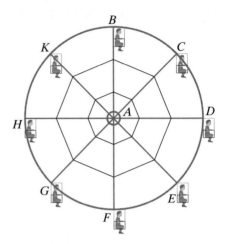

Figure 11.76

(a) What is the linear momentum of the child?

(b) In the definition $\vec{L} = \vec{r} \times \vec{p}$, what is the vector $\vec{r}$?

(c) What is $\vec{r}_\perp$?

(d) What is the magnitude of the angular momentum of the child about location A?

(e) What is the plane defined by $\vec{r}$ and $\vec{p}$ (that is, the plane containing both of these vectors)?

(f) Use the right-hand rule to determine the z component of the angular momentum of the child about location A.

(g) You used the right-hand rule to determine the z component of the angular momentum, but as a check, calculate in terms of position and momentum: What is xp_y? What is yp_x? Therefore, what is the z component of the angular momentum of the child about location A?

(h) The Ferris wheel keeps turning, and at a later time, the same child is at location E, with coordinates $\langle 16.971, -16.971, 0 \rangle$ m relative to location A, moving with velocity $\langle 5.303, 5.303, 0 \rangle$ m/s. Now what is the magnitude of the angular momentum of the child about location A?

Section 11.2

11.X.32 Calculate the angular momentum for a rotating disk, sphere, and rod:

(a) A uniform disk of mass 13 kg, thickness 0.5 m, and radius 0.2 m is located at the origin, oriented with its axis along the y axis. It rotates clockwise around its axis when viewed from above (that is, you stand at a point on the $+y$ axis and look toward the origin at the disk). The disk makes one complete rotation every 0.6 s. What is the rotational angular momentum of the disk? What is the rotational kinetic energy of the disk?

(b) A sphere of uniform density, with mass 22 kg and radius 0.7 m is located at the origin, and rotates around an axis parallel with the x axis. If you stand somewhere on the $+x$ axis and look toward the origin at the sphere, the sphere spins counterclockwise. One complete revolution takes 0.5 seconds. What is the rotational angular momentum of the sphere? What is the rotational kinetic energy of the sphere?

(c) A cylindrical rod of uniform density is located with its center at the origin, and its axis along the z axis. Its radius is 0.06 m, its length is 0.7 m, and its mass is 5 kg. It makes one revolution every 0.03 seconds. If you stand on the $+x$ axis and look toward the origin at the rod, the rod spins clockwise. What is the rotational angular momentum of the rod? What is the rotational kinetic energy of the rod?

11.X.33 If an object has a moment of inertia 19 kg $\cdot$ m^2 and the magnitude of its rotational angular momentum is 36 kg $\cdot$ m^2/s, what is its rotational kinetic energy?

11.X.34 The moment of inertia of a uniform-density disk rotating about an axle through its center can be shown to be $\frac{1}{2}MR^2$. This result is obtained by using integral calculus to add up the contributions of all the atoms in the disk. The factor of 1/2 reflects the fact that some of the atoms are near the center and some are far from the center; the factor of 1/2 is an average of the square distances. A uniform-density disk whose mass is 16 kg and radius is 0.15 m makes one complete rotation every 0.5 s.

(a) What is the moment of inertia of this disk?

(b) What is its rotational kinetic energy?

(c) What is the magnitude of its rotational angular momentum?

11.X.35 Mounted on a low-mass rod of length 0.32 m are four balls (Figure 11.77). Two balls (shown in red on the diagram), each of mass 0.82 kg, are mounted at opposite ends of the rod. Two other balls, each of mass 0.29 kg (shown in blue on the diagram), are each mounted a distance 0.08 m from the center of the rod. The rod rotates on an axle through the center of the rod (indicated by the "x" in the diagram), perpendicular to the rod, and it takes 0.9 seconds to make one full rotation.

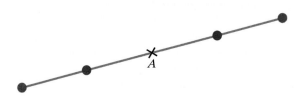

Figure 11.77

(a) What is the moment of inertia of the device about its center?

(b) What is the angular speed of the rotating device?

(c) What is the magnitude of the angular momentum of the rotating device?

11.X.36 A low-mass rod of length 0.30 m has a metal ball of mass 1.7 kg at each end. The center of the rod is located at the origin, and the rod rotates in the yz plane about its center. The rod rotates clockwise around its axis when viewed from a point on the $+x$ axis, looking toward the origin. The rod makes one complete rotation every 0.5 s.

(a) What is the moment of inertia of the object (rod plus two balls)?

(b) What is the rotational angular momentum of the object?

(c) What is the rotational kinetic energy of the object?

11.X.37 In Figure 11.78 a barbell spins around a pivot at its center at A. The barbell consists of two small balls, each with mass 500 grams (0.5 kg), at the ends of a very low mass rod of length $d = 20$ cm (0.2 m; the radius of rotation is 0.1 m). The barbell spins clockwise with angular speed 80 radians/s.

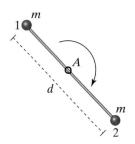

Figure 11.78

We can calculate the angular momentum and kinetic energy of this object in two different ways, by treating the object as two separate balls or as one barbell.

I: Treat the object as two separate balls

(a) What is the speed of ball 1?

(b) What are the magnitude and direction of the *translational* angular momentum $\vec{L}_{\text{trans},1,A}$ of just one of the balls (ball 1)?

(c) What are the magnitude and direction of the *translational* angular momentum $\vec{L}_{\text{trans},2,A}$ of the other ball (ball 2)?

(d) By adding the translational angular momentum of ball 1 and the translational angular momentum of ball 2, calculate the total angular momentum of the barbell, $\vec{L}_{\text{tot},A}$ (magnitude and direction).

(e) Calculate the *translational* kinetic energy of ball 1.

(f) Calculate the *translational* kinetic energy of ball 2.

(g) By adding the translational kinetic energy of ball 1 and the translational kinetic energy of ball 2, calculate the total kinetic energy of the barbell.

II: Treat the object as one barbell

(h) Calculate the moment of inertia I of the barbell.

(i) What is the direction of the angular velocity vector $\vec{\omega}$?

(j) Use the moment of inertia I and the angular speed $|\omega| = 80$ rad/s to calculate the rotational angular momentum of the barbell (magnitude and direction).

(k) How does this value, $|\vec{L}_{\text{rot}}|$, compare to the magnitude of the angular momentum calculated earlier, $|\vec{L}_{\text{tot},A}|$, by adding the translational angular momenta of the two balls?

(l) Use the moment of inertia I and the angular speed $|\omega| = 80$ rad/s to calculate the rotational kinetic energy of the barbell:

(m) How does this value, K_{rot}, compare to the kinetic energy K_{total} calculated earlier by adding the translational kinetic energies of the two balls?

11.P.38 Calculate the angular momentum of the Earth:

(a) Calculate the magnitude of the translational angular momentum of the Earth relative to the center of the Sun. See the data on inside back cover.

(b) Calculate the magnitude of the rotational angular momentum of the Earth. How does this compare to your result in part (a)?

(The angular momentum of the Earth relative to the center of the Sun is the sum of the translational and rotational angular momenta. The rotational axis of the Earth is tipped 23.5° away from a perpendicular to the plane of its orbit.)

11.P.39 In Figure 11.79 two small objects each of mass $m = 0.3$ kg are connected by a lightweight rod of length $d = 1.5$ m. At a particular instant they have velocities whose magnitudes are $v_1 = 38$ m/s and $v_2 = 60$ m/s and are subjected to external forces whose magnitudes are $F_1 = 41$ N and $F_2 = 26$ N. The distance $h = 0.3$ m, and the distance $w = 0.7$ m. The system is moving in outer space.

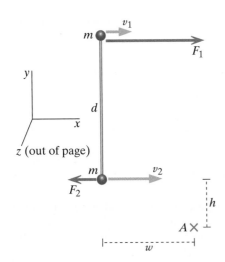

Figure 11.79

(a) What is the total (linear) momentum $\vec{p}_{total}$ of this system?

(b) What is the velocity $\vec{v}_{CM}$ of the center of mass?

(c) What is the total angular momentum $\vec{L}_A$ of the system relative to point A?

(d) What is the rotational angular momentum $\vec{L}_{rot}$ of the system?

(e) What is the translational angular momentum $\vec{L}_{trans}$ of the system relative to point A?

(f) After a short time interval $\Delta t = 0.23$ s, what is the total (linear) momentum $\vec{p}_{total}$ of the system?

11.P.40 In Figure 11.80 two small objects each of mass m are connected by a lightweight rod of length d. At a particular instant they have velocities as shown and are subjected to external forces as shown. The system is moving in outer space.

Figure 11.80

(a) What is the total (linear) momentum $\vec{p}_{total}$ of this system?

(b) What is the velocity $\vec{v}_{CM}$ of the center of mass?

(c) What is the total angular momentum $\vec{L}_A$ of the system relative to point A?

(d) What is the rotational angular momentum $\vec{L}_{rot}$ of the system?

(e) What is the translational angular momentum $\vec{L}_{trans}$ of the system relative to point A?
After a short time interval Δt,

(f) What is the total (linear) momentum $\vec{p}_{total}$ of the system?

(g) What is the rotational angular momentum $\vec{L}_{rot}$ of the system?

Sections 11.3–11.6

11.X.41 Under what circumstances is angular momentum constant? Give an example of a situation in which the x component of angular momentum is constant, but the y component isn't.

11.X.42 Give examples of translational angular momentum and rotational angular momentum in our Solar System.

11.X.43 Give an example of a physical situation in which the angular momentum is zero yet the translational and rotational angular momenta are both nonzero.

11.X.44 Calculating torque in Figure 11.81:

(a) If $r_A = 3$ m, $F = 8$ N, and $\theta = 51°$, what is the magnitude of the torque about location A, including units?

(b) If the force were perpendicular to $\vec{r}_A$ but gave the same torque as in the preceding question, what would its magnitude be?

Figure 11.81

11.X.45 Under what conditions is the torque about some location equal to zero?

11.X.46 Make a sketch showing a situation in which the torque due to a single force about some location is 20 N · m in the positive z direction, whereas about another location the torque is 10 N · m in the negative z direction.

11.X.47 As shown in Figure 11.82, seven forces all with magnitude $|\vec{F}| = 25$ N are applied to an irregularly shaped object. Each force is applied at a different location on the object, indicated by the tail of the arrow; the directions of the forces differ. The distances shown in the diagram have these values: $w = 9$ m, $h = 14$ m, and $d = 13$ m.

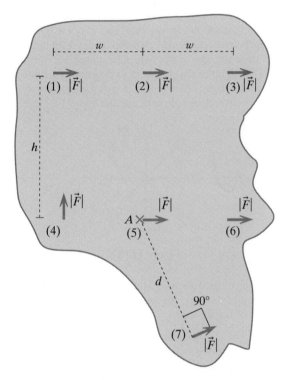

Figure 11.82

For each force, calculate the z component of the torque due to that force, relative to location A (x to the right, y up, z out of the page). Make sure you give the correct sign.

Relative to location A, what is the z component of the net torque acting on this object?

11.X.48 What is required for the angular momentum of a system to be constant?

(A) zero net torque
(B) zero impulse
(C) no energy transfers
(D) zero net force

11.P.49 Let's compare the Momentum Principle and the Angular Momentum Principle in a simple situation. Consider a mass m falling near the Earth (Figure 11.83). We choose a location A off to the side, on the ground. Neglecting air resistance, the Momentum Principle gives $dp_y/dt = -mg$, yielding $dv_y/dt = -g$ (nonrelativistic). What does the Angular Momentum Principle yield?

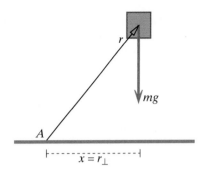

Figure 11.83

11.P.50 A small rock passes a massive star, following the path shown in red on the diagram. When the rock is a distance 4.5×10^{13} m (indicated as d_1 in Figure 11.84) from the center of the star, the magnitude p_1 of its momentum is 1.35×10^{17} kg · m/s, and the angle is 126°. At a later time, when the rock is a distance $d_2 = 1.3 \times 10^{13}$ m from the center of the star, it is heading in the $-y$ direction. There are no other massive objects nearby. What is the magnitude p_2 of the final momentum?

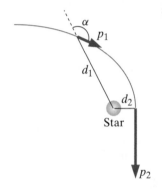

Figure 11.84

Section 11.7

11.X.51 A stationary bicycle wheel of radius 0.9 m is mounted in the vertical plane (Figure 11.85). The axle is held up by supports that are not shown, and the wheel is free to rotate on the nearly frictionless axle. The wheel has mass 4.8 kg, all concentrated in

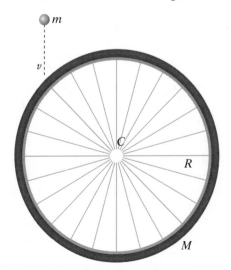

Figure 11.85

the rim (the spokes have negligible mass). A lump of clay with mass 0.5 kg falls and sticks to the outer edge of the wheel at the location shown. Just before the impact the clay has speed 5 m/s, and the wheel is rotating clockwise with angular speed 0.33 radians/s.

(a) Just before the impact, what is the angular momentum (magnitude and direction) of the combined system of wheel plus clay about the center C? (As usual, x is to the right, y is up, and z is out of the screen, toward you.)

(b) Just after the impact, what is the angular momentum (magnitude and direction) of the combined system of wheel plus clay about the center C?

(c) Just after the impact, what is the angular velocity (magnitude and direction) of the wheel?

(d) Qualitatively, what happens to the linear momentum of the combined system? Why?

  (A) The downward linear momentum decreases because the axle exerts an upward force.

  (B) Some of the linear momentum is changed into angular momentum.

  (C) Some of the linear momentum is changed into energy.

  (D) There is no change because linear momentum is always conserved.

11.X.52 A rotating uniform-density disk of radius 0.6 m is mounted in the vertical plane, as shown in Figure 11.86. The axle is held up by supports that are not shown, and the disk is free to rotate on the nearly frictionless axle. The disk has mass 5 kg. A lump of clay with mass 0.4 kg falls and sticks to the outer edge of the wheel at the location $\langle -0.36, 0.480, 0 \rangle$ m, relative to an origin at the center of the axle. Just before the impact the clay has speed 8 m/s, and the disk is rotating clockwise with angular speed 0.51 radians/s.

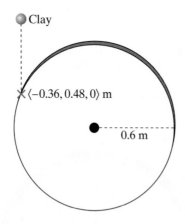

Figure 11.86

(a) Just before the impact, what is the angular momentum (magnitude and direction) of the combined system of wheel plus clay about the center C? (As usual, x is to the right, y is up, and z is out of the screen, toward you.)

(b) Just after the impact, what is the angular momentum (magnitude and direction) of the combined system of wheel plus clay about the center C?

(c) Just after the impact, what is the angular velocity (magnitude and direction) of the wheel?

(d) Qualitatively, what happens to the linear momentum of the combined system? Why?

  (A) There is no change because linear momentum is always conserved.

  (B) Some of the linear momentum is changed into angular momentum.

  (C) Some of the linear momentum is changed into energy.

  (D) The downward linear momentum decreases because the axle exerts an upward force.

11.X.53 A device consists of eight balls, each of mass 0.8 kg attached to the ends of low-mass spokes of length 1.4 m, so the radius of rotation of the balls is 0.7 m. The device is mounted in the vertical plane, as shown in Figure 11.87. The axle is held up by supports that are not shown, and the wheel is free to rotate on the nearly frictionless axle. A lump of clay with mass 0.20 kg falls and sticks to one of the balls at the location shown, when the spoke attached to that ball is at 45 degrees to the horizontal. Just before the impact the clay has a speed 9 m/s, and the wheel is rotating counterclockwise with angular speed 0.23 radians/s.

Figure 11.87

(a) Which of the following statements are true about the device and the clay, for angular momentum relative to the axle of the device?

  (A) The angular momentum of the device + clay just after the collision is equal to the angular momentum of the device + clay just before the collision.

  (B) The angular momentum of the falling clay is zero because the clay is moving in a straight line.

  (C) Just before the collision, the angular momentum of the wheel is 0.

  (D) The angular momentum of the device is the sum of the angular momenta of all eight balls.

  (E) The angular momentum of the device is the same before and after the collision.

(b) Just before the impact, what is the (vector) angular momentum of the combined system of device plus clay about the center C? (As usual, x is to the right, y is up, and z is out of the screen, toward you.)

(c) Just after the impact, what is the angular momentum of the combined system of device plus clay about the center C?

(d) Just after the impact, what is the (vector) angular velocity of the device?

(e) Qualitatively, what happens to the total linear momentum of the combined system? Why?

  (A) Some of the linear momentum is changed into energy.

(B) Some of the linear momentum is changed into angular momentum.

(C) There is no change because linear momentum is always conserved.

(D) The downward linear momentum decreases because the axle exerts an upward force.

(f) Qualitatively, what happens to the total kinetic energy of the combined system? Why?

(A) Some of the kinetic energy is changed into linear momentum.

(B) Some of the kinetic energy is changed into angular momentum.

(C) The total kinetic energy decreases because there is an increase of internal energy in this inelastic collision.

(D) There is no change because kinetic energy is always conserved.

11.X.54 Figure 11.87 depicts a device that can rotate freely with little friction with the axle. The radius is 0.4 m, and each of the eight balls has a mass of 0.3 kg. The device is initially not rotating. A piece of clay falls and sticks to one of the balls as shown in the figure. The mass of the clay is 0.066 kg and its speed just before the collision is 10 m/s.

(a) Which of the following statements are true, for angular momentum relative to the the axle of the wheel?

(A) Just before the collision, $r_\perp = 0.4\sqrt{2} = 0.4\cos(45°)$ (for the clay).

(B) The angular momentum of the wheel is the same before and after the collision.

(C) Just before the collision, the angular momentum of the wheel is 0.

(D) The angular momentum of the wheel is the sum of the angular momenta of all eight balls.

(E) The angular momentum of the wheel + clay after the collision is equal to the initial angular momentum of the clay.

(F) The angular momentum of the falling clay is zero because the clay is moving in a straight line.

(b) Just after the collision, what is the speed of one of the balls?

11.P.55 A stick of length L and mass M hangs from a low-friction axle (Figure 11.88). A bullet of mass m traveling at a high speed v strikes near the bottom of the stick and quickly buries itself in the stick.

(a) During the brief impact, is the linear momentum of the stick + bullet system constant? Explain why or why not. Include in your explanation a sketch of how the stick shifts on the axle during the impact.

(b) During the brief impact, around what point does the angular momentum of the stick + bullet system remain constant?

(c) Just after the impact, what is the angular speed ω of the stick (with the bullet embedded in it)? (Note that the center of mass of the stick has a speed $\omega L/2$. The moment of inertia of a uniform rod about its center of mass is $\frac{1}{12}ML^2$.)

(d) Calculate the change in kinetic energy from just before to just after the impact. Where has this energy gone?

(e) The stick (with the bullet embedded in it) swings through a maximum angle θ_{max} after the impact, then swings back. Calculate θ_{max}.

11.P.56 A space station has the form of a hoop of radius R, with mass M (Figure 11.89). Initially its center of mass is not moving, but it is spinning with angular speed ω_0. Then a small package of mass m is thrown by a spring-loaded gun toward a nearby spacecraft as shown; the package has a speed v after launch. Calculate the center-of-mass velocity of the space station (v_x and v_y) and its rotational speed ω after the launch.

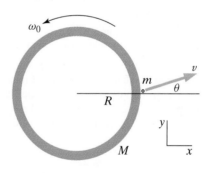

Figure 11.89

11.P.57 A certain comet of mass m at its closest approach to the Sun is observed to be at a distance r_1 from the center of the Sun, moving with speed v_1 (Figure 11.90). At a later time the comet is observed to be at a distance r_2 from the center of the Sun, and the angle between $\vec{r}_2$ and the velocity vector is measured to be θ. What is v_2? Explain briefly.

Figure 11.88

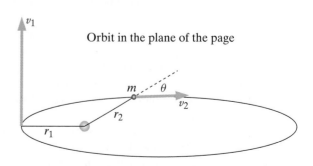

Figure 11.90

11.P.58 A playground ride consists of a disk of mass M and radius R mounted on a low-friction axle (Figure 11.91). A child of mass m runs at speed v on a line tangential to the disk and jumps onto the outer edge of the disk.

Figure 11.91

(a) If the disk was initially at rest, now how fast is it rotating? (The moment of inertia of a uniform disk is $\frac{1}{2}MR^2$.)

(b) If you were to do a lot of algebra to calculate the kinetic energies in part (a), you would find that $K_f < K_i$. Where has this energy gone?

(c) Calculate the change in *linear* momentum of the system consisting of the child plus the disk (but not including the axle), from just before to just after impact. What caused this change in the linear momentum?

(d) The child on the disk walks inward on the disk and ends up standing at a new location a distance $R/2$ from the axle. Now what is the angular speed?

(e) If you were to do a lot of algebra to calculate the kinetic energies in part (d), you would find that $K_f > K_i$. Where has this energy come from?

(f) In part (a), estimate numerical values for all of the quantities and determine a value for the spin rate ω. Is your result reasonable?

11.P.59 A playground ride consists of a disk of mass $M = 43$ kg and radius $R = 1.7$ m mounted on a low-friction axle (Figure 11.92). A child of mass $m = 25$ kg runs at speed $v = 2.3$ m/s on a line tangential to the disk and jumps onto the outer edge of the disk.

Figure 11.92

ANGULAR MOMENTUM

(a) Consider the system consisting of the child and the disk, but not including the axle. Which of the following statements are true, from just before to just after the collision?

  (A) The axle exerts a force on the system but nearly zero torque.

  (B) The momentum of the system changes.

  (C) The momentum of the system doesn't change.

  (D) The angular momentum of the system about the axle hardly changes.

  (E) The torque exerted by the axle is nearly zero even though the force is large, because $|\vec{r}|$ is nearly zero.

  (F) The angular momentum of the system about the axle changes.

  (G) The torque exerted by the axle is zero because the force exerted by the axle is very small.

(b) Relative to the axle, what was the magnitude of the angular momentum of the child before the collision?

(c) Relative to the axle, what was the angular momentum of the system of child plus disk just after the collision?

(d) If the disk was initially at rest, now how fast is it rotating? That is, what is its angular speed? (The moment of inertia of a uniform disk is $\frac{1}{2}MR^2$.)

(e) How long does it take for the disk to go around once?

ENERGY

(f) What is the change in the kinetic energy of the child plus the disk?

(g) Where has most of this kinetic energy gone?

  (A) Increased chemical energy in the child

  (B) Increased internal energy of the disk and child

  (C) Increased translational kinetic energy of the disk

MOMENTUM

(h) What was the speed of the child just after the collision?

(i) What was the speed of the center of mass of the disk just after the collision?

(j) What was the magnitude of the linear momentum of the disk just after the collision?

(k) Calculate the change in the x component of linear momentum Δp_x of the system consisting of the child plus the disk (but not including the axle), from just before to just after impact. Take the x axis to be in the direction of the initial velocity of the child.

(l) Why has the linear momentum changed?

  (A) Some of the original linear momentum changed into angular momentum.

  (B) Some of the original linear momentum changed into energy.

  (C) The axle exerted a force on the system.

ANGULAR MOMENTUM AGAIN

(m) Next, the child on the disk walks inward on the disk and ends up standing at a new location a distance $R = 0.7$ m from the axle. Now what is the angular speed?

(n) What is the change in the kinetic energy of the child plus the disk, from the beginning to the end of the walk on the disk?

(o) What was the source of this increased kinetic energy?

11.P.60 You sit on a rotating stool and hold barbells in both hands with your arms fully extended horizontally. You make one complete turn in 2 seconds. You then pull the barbells in close to your body.

(a) Estimate how long it now takes you to make one complete turn. Be clear and explicit about the principles you apply and about your assumptions and approximations.

(b) About how much energy did you expend?

11.P.61 An ice skater whirls with her arms and one leg stuck out as shown in Figure 11.93, making one complete turn in one second. Then she quickly moves her arms up above her head and pulls her leg in as shown in Figure 11.94.

Figure 11.93

Figure 11.94

(a) Estimate how long it now takes for her to make one complete turn. Explain your calculations, and state clearly what approximations and estimates you make.

(b) Estimate the minimum amount of chemical energy she must expend to change her configuration.

11.P.62 In Figure 11.95 two small objects each of mass m_1 are connected by a lightweight rod of length L. At a particular instant the center of mass speed is v_1 as shown, and the object is rotating counterclockwise with angular speed ω_1. A small object of mass m_2 traveling with speed v_2 collides with the rod at an angle θ_2 as shown, at a distance b from the center of the rod. After being struck, the mass m_2 is observed to move with speed v_4, at angle θ_4. All the quantities are positive magnitudes. This all takes place in outer space.

For the object consisting of the rod with the two masses, write equations that, in principle, could be solved for the center of mass speed v_3, direction θ_3, and angular speed ω_3 in terms of the given quantities. State clearly what physical principles you use to obtain your equations.

Don't attempt to solve the equations; just set them up.

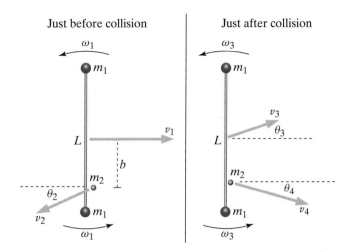

Just before collision   Just after collision

Figure 11.95 A spinning rod is struck by a small object.

11.P.63 In Figure 11.96 a spherical nonspinning asteroid of mass M and radius R moving with speed v_1 to the right collides with a similar nonspinning asteroid moving with speed v_2 to the left, and they stick together. The impact parameter is d. Note that $I_{sphere} = \frac{2}{5} MR^2$.

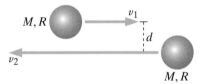

Figure 11.96 Two asteroids collide and stick together.

After the collision, what is the velocity v_{CM} of the center of mass and the angular velocity ω about the center of mass? (Note that each asteroid rotates about its own center with this same ω.)

11.P.64 A tiny piece of space junk of mass m strikes a glancing blow to a spherical satellite (Figure 11.97). After the collision the space junk is traveling in a new direction and moving more slowly. The space junk had negligible rotation both before and after the collision. The velocities of the space junk before and after the collision are shown in the diagram. The center of mass of the satellite is at its geometrical center. The satellite has mass M, radius R, and moment of inertia I about its center. Before the collision the satellite was moving and rotating as shown in the diagram.

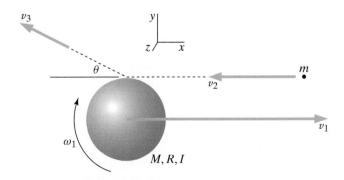

Figure 11.97 Space junk strikes a satellite.

(a) Just after the collision, what are the components of the center-of-mass velocity of the satellite (v_x and v_y) and its rotational speed ω?

(b) Calculate the rise in the internal energy of the satellite and space junk combined. You do *not* need to substitute in the values for quantities you already calculated in part (a).

11.P.65 A spherical satellite of radius 4.8 m and mass 205 kg is originally moving with velocity $\langle 2600, 0, 0 \rangle$ m/s, and is originally rotating with an angular speed 2 radians/second, in the direction shown in the diagram. A small piece of space junk of mass 4.1 kg is initially moving toward the satellite with velocity $\langle -2200, 0, 0 \rangle$ m/s. The space junk hits the edge of the satellite at location C as shown in Figure 11.98, and moves off with a new velocity $\langle -1300, 480, 0 \rangle$ m/s. Both before and after the collision, the rotation of the space junk is negligible.

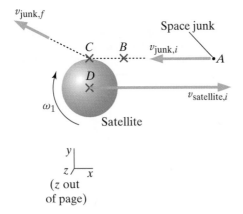

Figure 11.98

MOMENTUM

(a) After the collision, what is the final momentum of the satellite?

(b) What is the final center-of-mass velocity of the satellite?

ANGULAR MOMENTUM

(c) Assuming that density of the satellite is uniform throughout, what is the moment of inertia of the satellite? (The moment of inertia of a uniform-density sphere is $\frac{2}{5}MR^2$.)

(d) What is the initial (vector) rotational angular momentum of the satellite, around location D (its center of mass)? (Be sure your signs are correct).

(e) Which of the following statements about the translational angular momentum of the space junk about location D are true? Check all that apply:

(A) $xp_y - yp_x$ is the same when the space junk is at locations A, B, and just before getting to C.

(B) The translational angular momentum of the space junk is the same when the space junk is at locations A, B, and just before getting to C.

(C) Because the space junk is traveling in a straight line, its angular momentum is zero.

(D) $r_\perp$ is the same when the space junk is at locations A, B, and just before getting to C.

(E) The translational angular momentum of the space junk is in the $-z$ direction.

(f) An instant before the collision, when the space junk is almost at location C, what is the (vector) translational angular momentum of the space junk about location D?

(g) An instant after the collision, when the space junk is just slightly beyond location C, what is the (vector) translational angular momentum of the space junk about location D?

(h) At the same instant after the collision, what is the (vector) rotational angular momentum of the satellite?

(i) What is the new angular speed of the satellite?

(j) In what direction is the satellite now rotating?

(A) counterclockwise (opposite to original direction)

(B) clockwise (same as original direction)

(C) it is not rotating

ENERGY

Consider a system consisting of both objects (space junk and satellite).

(k) Before the collision, what was the total kinetic energy (translational + rotational) of the system?

(l) After the collision, what is the total kinetic energy of the system?

(m) What was the rise in internal energy (thermal and distortion) for the space junk and satellite combined?

11.P.66 A stick of mass M and length L is lying on ice (Figure 11.99). A small mass m traveling at high speed v_i strikes the stick a distance d from the center and bounces off with speed v_f as shown in the diagram, which is a top view of the situation. The magnitudes of the initial and final angles to the x axis of the small mass's velocity are θ_i and θ_f. All of the symbols in the diagram represent positive numbers.

View from above

Figure 11.99 A stick lying on ice is struck by a small mass.

(a) Afterward, what are the velocity components v_x and v_y of the center of the stick? Explain briefly.

(b) Afterward, what are the magnitude and direction of the angular velocity $\vec{\omega}$ of the stick?

(c) What is the increase in internal energy of the objects? You can leave your expression in terms of the initial quantities and v_x, v_y, and ω.

11.P.67 A stick of mass 1.3 kg and length 0.4 m is lying at rest on ice (Figure 11.100). A small object with mass 0.06 kg traveling at a high speed of 200 m/s strikes the stick a distance 0.2 m from the center and bounces off with speed 60 m/s as shown in the diagram, which is a top view of the situation, where the motion is in the xy plane and the z axis points upward from the ice. The magnitudes of the initial and final angles to the x axis of the small mass's velocity are $\theta_i = 26$ degrees and $\theta_f = 82$ degrees. The moment of inertia about the center of a uniform stick of mass M and length L is $I = ML^2/12$.

11.P.69 A diver dives from a high platform (Figure 11.103). When he leaves the platform, he tucks tightly and performs three complete revolutions in the air, then straightens out with his body fully extended before entering the water (Figure 11.102). He is in the air for a total time of 1.4 seconds. What is his angular speed ω just as he enters the water? Give a numerical answer. Be explicit about details of your model, and include (brief) explanations. You will need to estimate some quantities.

Figure 11.102

View from above

Figure 11.100

Figure 11.103

(a) Afterward, what is the velocity of the center of the stick?

(b) Afterward, what is the angular velocity $\vec{\omega}$ of the stick?

(c) What is the increase in internal energy of the objects?

11.P.68 A device consisting of four heavy balls connected by low-mass rods is free to rotate about an axle, as shown in Figure 11.101. It is initially not spinning. A small bullet traveling very fast buries itself in one of the balls. $m = 0.002$ kg, $v = 550$ m/s, $M_1 = 1.2$ kg, $M_2 = 0.4$ kg, $R_1 = 0.6$ m, and $R_2 = 0.2$ m. The axle of the device is at the origin $\langle 0, 0, 0 \rangle$, and the bullet strikes at location $\langle 0.155, 0.580, 0 \rangle$ m. Just after the impact, what is the angular speed of the device? Note that this is an inelastic collision; the system's temperature increases.

11.P.70 Suppose that an asteroid of mass 2×10^{21} kg is nearly at rest outside the solar system, far beyond Pluto. It falls toward the Sun and crashes into the Earth at the equator, coming in at an angle of 30 degrees to the vertical as shown, against the direction of rotation of the Earth (Figure 11.104). It is so large that its motion is barely affected by the atmosphere.

Figure 11.101

Figure 11.104 An asteroid crashes into the Earth (not to scale).

(a) Calculate the impact speed.

(b) Calculate the change in the length of a day due to the impact.

Section 11.8

11.X.71 Two people of different masses sit on a seesaw (Figure 11.105). M_1, the mass of person 1, is 90 kg, M_2 is 42 kg, $d_1 = 0.8$ m, and $d_2 = 1.3$ m. The mass of the board is negligible.

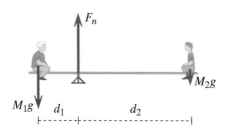

Figure 11.105

(a) What are the magnitude and direction of the torque about the pivot due to the gravitational force on person 1?

(b) What are the magnitude and direction of the torque about the pivot due to the gravitational force on person 2?

(c) Since at this instant the linear momentum of the system may be changing, we don't know the magnitude of the "normal" force exerted by the pivot. Nonetheless, it is possible to calculate the torque due to this force. What are the magnitude and direction of the torque about the pivot due to the force exerted by the pivot on the board?

(d) What are the magnitude and direction of the net torque on the system (board + people)?

(e) Because of this net torque, what will happen?

  (A) The seesaw will begin to rotate clockwise.

  (B) The seesaw will begin to rotate counterclockwise.

  (C) The seesaw will not move.

(f) Person 2 moves to a new position, in which the magnitude of the net torque about the pivot is now 0, and the seesaw is balanced. What is the new value of d_2 in this situation?

11.X.72 A board of length $2d = 6$ m rests on a cylinder (the "pivot"). A ball of mass 5 kg is placed on the end of the board. Figure 11.106 shows the objects at a particular instant.

(a) On a free-body diagram, show the forces acting on the ball + board system, in their correct locations.

(b) Take the point at which the board touches the cylinder as location A. What is the magnitude of the torque on the system of (ball + board) about location A?

(c) Which of the following statements are correct?

  (A) Because there is a torque, the angular momentum of the system will change in the next tenth of a second.

(B) The forces balance, so the angular momentum of the system about location A will not change.

(C) The force by the cylinder on the board contributes nothing to the torque about location A.

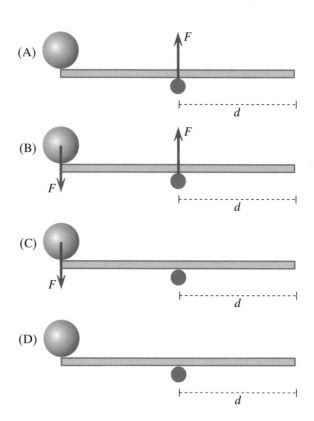

Figure 11.106

11.X.73 A barbell is mounted on a nearly frictionless axle through its center. The low-mass rod has a length $d = 0.22$ m, and each ball has a mass $m = 0.1$ kg. At this instant, there are two forces of equal magnitude F applied to the system as shown in Figure 11.107, with the directions indicated, and at this instant the angular velocity is 80 radians/s, counterclockwise, and the angle $\theta = 35°$. In the next 0.004 s, the angular momentum relative to the center increases by an amount 3.5×10^{-4} kg·m^2/s. What is the magnitude of each force?

Figure 11.107

11.P.74 A disk of radius 0.2 m and moment of inertia 1.5 kg·m^2 is mounted on a nearly frictionless axle (Figure 11.108). A string is wrapped tightly around the disk, and you pull on the string with a constant force of 25 N. After a while the disk has reached an angular speed of 2 radians/s. What is its angular speed 0.1 seconds later? Explain briefly.

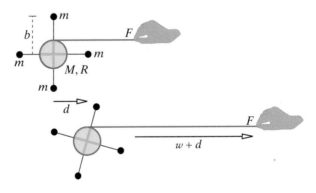

Figure 11.108

11.P.75 String is wrapped around an object of mass M and moment of inertia I. With your hand you pull the string straight up with some constant force F such that the center of the object does not move up or down, but the object spins faster and faster (Figure 11.109). This is like a yo-yo; nothing but the vertical string touches the object.

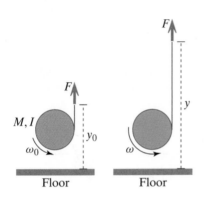

Figure 11.109

When your hand is a height y_0 above the floor, the object has an angular speed ω. When your hand has risen to a height y above the floor, what is the angular speed ω of the object?

Your result should not contain F or the (unknown) radius of the object. Explain the physics principles you are using.

11.P.76 A yo-yo is constructed of three disks: two outer disks of mass M, radius R, and thickness d, and an inner disk (around which the string is wrapped) of mass m, radius r, and thickness d. The yo-yo is suspended from the ceiling and then released with the string vertical (Figure 11.110).

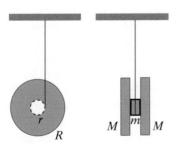

Figure 11.110

Calculate the tension in the string as the yo-yo falls. Note that when the center of the yo-yo moves down a distance y, the yo-yo turns through an angle y/r, which in turn means that the angular

speed ω is equal to v_{CM}/r. The moment of inertia of a uniform disk is $\frac{1}{2}MR^2$.

11.P.77 String is wrapped around an object of mass 1.2 kg and moment of inertia $0.06 \text{ kg} \cdot \text{m}^2$. With your hand you pull the string straight up with some constant force F such that the center of the object does not move up or down, but the object spins faster and faster (Figure 11.111). This is like a yo-yo; nothing but the vertical string touches the object. When your hand is a height $y_0 = 0.25$ m above the floor, the object has an angular speed $\omega_0 = 12$ radians/s. When your hand has risen to a height $y = 0.66$ m above the floor, what is the angular speed ω of the object? Your answer must be numeric and not contain the symbol F.

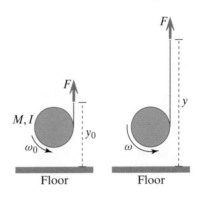

Figure 11.111

11.P.78 A string is wrapped around a uniform disk of mass M and radius R. Attached to the disk are four low-mass rods of radius b, each with a small mass m at the end (Figure 11.112).

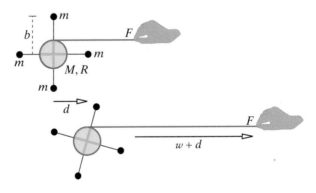

Figure 11.112

The apparatus is initially at rest on a nearly frictionless surface. Then you pull the string with a constant force F. At the instant when the center of the disk has moved a distance d, a length w of string has unwound off the disk.

(a) At this instant, what is the speed of the center of the apparatus? Explain your approach.

(b) At this instant, what is the angular speed of the apparatus? Explain your approach.

(c) You keep pulling with constant force F for an additional time Δt. By how much ($\Delta\omega$) does the angular speed of the apparatus increase in this time interval Δt?

11.P.79 A solid object of uniform density with mass M, radius R, and moment of inertia I rolls without slipping down a ramp at an angle θ to the horizontal. The object could be a hoop, a disk, a sphere, etc.

(a) Carefully follow the complete analysis procedure explained in earlier chapters, but with the addition of the Angular Momentum Principle about the center of mass. Note that in your force diagram you must include a small frictional force f that points *up* the ramp. Without that force the object will slip. Also note that the condition of nonslipping implies that the instantaneous velocity of the atoms of the object that are momentarily in contact with the ramp is zero, so $f < \mu F_N$ (no slipping). This zero-velocity condition also implies that $v_{CM} = \omega R$, where ω is the angular speed of the object, since the instantaneous speed of the contact point is $v_{CM} - \omega R$.

(b) The moment of inertia about the center of mass of a uniform hoop is MR^2, for a uniform disk it is $(1/2)MR^2$, and for a uniform sphere it is $(2/5)MR^2$. Calculate the acceleration dv_{CM}/dt for each of these objects.

(c) If two hoops of different mass are started from rest at the same time and the same height on a ramp, which will reach the bottom first? If a hoop, a disk, and a sphere of the same mass are started from rest at the same time and the same height on a ramp, which will reach the bottom first?

(d) Write the energy equation for the object rolling down the ramp, and for the point-particle system. Show that the time derivatives of these equations are compatible with the force and torque analyses.

11.P.80 A string is wrapped around a uniform disk of mass $M = 1.2$ kg and radius $R = 0.11$ m (Figure 11.113). (Recall that the moment of inertia of a uniform disk is $\frac{1}{2}MR^2$.) Attached to the disk are four low-mass rods of radius $b = 0.14$ m, each with a small mass $m = 0.4$ kg at the end. The device is initially at rest on a nearly frictionless surface. Then you pull the string with a constant force $F = 21$ N. At the instant that the center of the disk has moved a distance $d = 0.037$ m, a length $w = 0.032$ m of string has unwound off the disk.

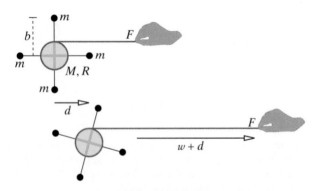

Figure 11.113

(a) At this instant, what is the speed of the center of the apparatus?

(b) At this instant, what is the angular speed of the apparatus?

(c) You keep pulling with constant force 21 N for an additional 0.045 s. Now what is the angular speed of the apparatus?

11.P.81 Design a decorative "mobile" to consist of a low-mass rod of length 0.49 m suspended from a string so that the rod is horizontal, with two balls hanging from the ends of the rod. At the left end of the rod hangs a ball with mass 0.484 kg. At the right end of the rod hangs a ball with mass 0.273 kg. You need to decide how far from the left end of the rod you should attach the string that will hold up the mobile, so that the mobile hangs motionless with the rod horizontal ("equilibrium"). You also need to determine the tension in the string supporting the mobile.

(a) What is the tension in the string that supports the mobile?

(b) How far from the left end of the rod should you attach the support string?

11.P.82 A rod of length L and negligible mass is attached to a uniform disk of mass M and radius R (Figure 11.114). A string is wrapped around the disk, and you pull on the string with a constant force F. Two small balls each of mass m slide along the rod with negligible friction. The apparatus starts from rest, and when the center of the disk has moved a distance d, a length of string s has come off the disk, and the balls have collided with the ends of the rod and stuck there. The apparatus slides on a nearly frictionless table. Here is a view from above:

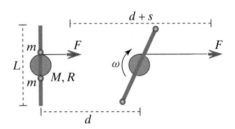

Figure 11.114

(a) At this instant, what is the speed v of the center of the disk?

(b) At this instant the angular speed of the disk is ω. How much internal energy change has there been?

(c) In the next short amount of time Δt, by how much will the angular speed change ($\Delta\omega$)?

Section 11.10

11.X.83 What features of the Bohr model of hydrogen are consistent with the later, full quantum mechanical analysis? What features of the Bohr model had to be abandoned?

11.X.84 According to the Bohr model of the hydrogen atom, what is the magnitude of the translational angular momentum of the electron (relative to the location of the proton) when the atom is in the 2nd excited state above the ground state ($N = 3$)?

11.X.85 Give some examples of atomic or nuclear phenomena associated with particle spin.

11.P.86 Calculate rotational energy levels and photon emission energies for a carbon monoxide molecule (CO).

(a) Calculate the energies of the quantized rotational energy levels for the CO molecule. Estimate any quantities you need. See discussion of diatomic molecules in the section on quantized angular momentum; the parameter l has values $0, 1, 2, 3 \ldots$.

(b) Describe the emission spectrum for electromagnetic radiation emitted in transitions among the rotational CO energy levels. Include a calculation of the lowest-energy emission in electron volts (1.6×10^{-19} J).

(c) It is transitions among "electronic" states of atoms that produce visible light, with photon energies on the order of a couple of electron-volts. Each electronic energy level has quantized rotational and vibrational (harmonic oscillator) energy sublevels. Explain why this leads to a visible spectrum that contains "bands" rather than individual energies.

11.P.87 The Bohr model correctly predicts the main energy levels not only for atomic hydrogen but also for other "one-electron" atoms where all but one of the atomic electrons has been removed, such as in He$^+$ (one electron removed) or Li^{++} (two electrons removed).

(a) Predict the energy levels in eV for a system consisting of a nucleus containing Z protons and just one electron. You need not recapitulate the entire derivation for the Bohr model, but do explain the changes you have to make to take into account the factor Z.

(b) The negative muon (μ^-) behaves like a heavy electron, with the same charge as the electron but with a mass 207 times as large as the electron mass. As a moving μ^- comes to rest in matter, it tends to knock electrons out of atoms and settle down onto a nucleus to form a "one-muon" atom. For a system consisting of a lead nucleus (Pb^{208} has 82 protons and 126 neutrons) and just one negative muon, predict the energy in eV of a photon emitted in a transition from the first excited state to the ground state. The high-energy photons emitted by transitions between energy levels in such "muonic atoms" are easily observed in experiments with muons.

(c) Calculate the radius of the smallest Bohr orbit for a μ^- bound to a lead nucleus (Pb^{208} has 82 protons and 126 neutrons). Compare with the approximate radius of the lead nucleus (remember that the radius of a proton or neutron is about 1×10^{-15} m, and the nucleons are packed closely together in the nucleus).

Comments: This analysis in terms of the simple Bohr model hints at the result of a full quantum-mechanical analysis, which shows that in the ground state of the lead–muon system there is a rather high probability for finding the muon *inside* the lead nucleus. Nothing in quantum mechanics forbids this penetration, especially since the muon does not participate in the strong interaction. Electrons in an atom can also be found inside the nucleus, but the probability is very low, because on average the electrons are very far from the nucleus, unlike the muon.

The eventual fate of the μ^- in a muonic atom is that it either decays into an electron, neutrino, and antineutrino, or it reacts through the weak interaction with a proton in the nucleus to produce a neutron and a neutrino. This "muon capture" reaction is more likely if the probability is high for the muon to be found inside the nucleus, as is the case with heavy nuclei such as lead.

11.P.88 The nucleus dysprosium-160 (containing 160 nucleons) acts like a spinning object with quantized angular momentum, $L^2 = l(l+1)\hbar^2$, and for this nucleus it turns out that l must be an even integer $(0, 2, 4 \ldots)$. When a Dy-160 nucleus drops from the $l = 2$ state to the $l = 0$ state, it emits an 87 keV photon (87×10^3 eV).

(a) What is the moment of inertia of the Dy-160 nucleus?

(b) Given your result from part (a), find the approximate radius of the Dy-160 nucleus, assuming it is spherical. (In fact, these and similar experimental observations have shown that some nuclei are not quite spherical.)

(c) The radius of a (spherical) nucleus is given approximately by $(1.3 \times 10^{-15}$ m$)A^{1/3}$, where A is the total number of protons and neutrons. Compare this prediction with your result in part (b).

Section 11.11

11.X.89 Two gyroscopes are made exactly alike except that the spinning disk in one is made of low-density aluminum, whereas the disk in the other is made of high-density lead. If they have the same spin angular speeds and the same torque is applied to both, which gyroscope precesses faster?

11.P.90 This problem requires that you have a toy gyroscope available. The purpose of this problem is to make as concrete as possible the unusual motions of a gyroscope and their analysis in terms of fundamental principles. In all of the following studies, the effects are most dramatic if you give the gyroscope as large a spin angular speed as possible.

(a) Hold the spinning gyroscope firmly in your hand, and try to rotate the spin axis quickly to point in a new direction. Explain qualitatively why this feels "funny." Also explain why you *don't* feel anything odd when you move the spinning gyroscope in any direction *without* changing the direction of the spin axis.

(b) Support one end of the spinning gyroscope (on a pedestal or in an open loop of the string) so that the gyroscope precesses *counterclockwise* as seen from above. Explain this counterclockwise precession direction; include sketches of top and side views of the gyroscope.

(c) Again support one end of the spinning gyroscope so that the gyroscope precesses *clockwise* as seen from above. Explain this clockwise precession direction; include sketches of top and side views of the gyroscope.

11.P.91 This problem requires that you have a toy gyroscope available. The purpose of this problem is to make as concrete as possible the unusual motions of a gyroscope and their analysis in terms of fundamental principles. In all of the following studies, the effects are most dramatic if you give the gyroscope as large a spin angular speed as possible.

(a) If you knew the spin angular speed of your gyroscope, you could predict the precession rate. Invent an appropriate experimental technique and determine the spin angular speed approximately. Explain your experimental method and your calculations. Then predict the corresponding precession rate, and compare with your measurement of the precession rate. You will have to measure and estimate some properties of the gyroscope and how it is constructed.

(b) Make a *quick* measurement of the precession rate with the spin axis horizontal, then make another quick measurement of the precession rate with the spin axis nearly vertical. (If you make quick measurements, friction on the spin axis doesn't have much time to change the spin angular speed.) Repeat, this time with the spin axis initially nearly vertical, then horizontal. Making all four of these measurements gives you some indication of how much the spin unavoidably changes due to friction while you are quickly changing the angle. What do you conclude about the dependence of the precession rate on the angle, assuming the same spin rate at these different angles? What is the theoretical prediction for the dependence of the precession rate on angle (for the same spin rate)?

11.P.92 A bicycle wheel with a heavy rim is mounted on a lightweight axle, and one end of the axle rests on top of a post. The wheel is observed to precess in the horizontal plane. With the spin direction shown in Figure 11.115, does the wheel precess clockwise or counterclockwise? Explain in detail, including appropriate diagrams.

Figure 11.115 A bicycle wheel on a pivot.

11.P.93 A solid wood top spins at high speed on the floor, with a spin direction shown in Figure 11.116.

Figure 11.116 A wood or steel top.

(a) Using appropriately labeled diagrams, explain the direction of motion of the top (you do not need to explain the magnitude).

(b) How would the motion change if the top had a higher spin rate? Explain briefly.

(c) If the top were made of solid steel instead of wood, explain how this would affect the motion (for the same spin rate).

11.P.94 The axis of a gyroscope is tilted at an angle of 30° to the vertical (Figure 11.117). The rotor has a radius of 15 cm, mass 3 kg, moment of inertia 0.06 kg·m², and spins on its axis

at 30 radians/s. It is supported in a cage (not shown) in such a way that without an added weight it does not precess. Then a mass of 0.2 kg is hung from the axis at a distance of 18 cm from the center of the rotor.

Figure 11.117 A gyroscope.

(a) Viewed from above, does the gyroscope precess in a (1) clockwise or a (2) counterclockwise direction? That is, does the top end of the axis move (1) out of the page or (2) into the page in the next instant? Explain your reasoning.

(b) How long does it take for the gyroscope to make one complete precession?

Computational Problem

11.P.95 In Chapter 3 you may have written a program to model the motion of a planet going around a fixed star. In this problem you will build on that program.

(a) Use initial conditions that produce an elliptical orbit. At each step calculate the translational angular momentum $\vec{L}_{trans,A}$ of the planet with respect to a location A chosen to be in the orbital plane but outside the orbit. Display this in two ways (i and ii below), and briefly describe in words what you observe:

  (i) Display $\vec{L}_{trans,A}$ as an arrow with its tail at location A, throughout the orbit. Since the magnitude of $\vec{L}_{trans,A}$ is quite different from the magnitudes of the distances involved, you will need to scale the arrow by some factor to fit it on the screen.

  (ii) Graph the component of $\vec{L}_{trans,A}$ perpendicular to the orbital plane as a function of time.

(b) Repeat part (a), but this time choose a different location B at the center of the fixed star, and calculate and display $\vec{L}_{trans,B}$ relative to that location B. As in part (a), display $\vec{L}_{trans,B}$ as an arrow (scaled appropriately), and also graph the component of $\vec{L}_{trans,B}$ perpendicular to the orbital plane, as a function of time, and briefly describe in words what you observe.

(c) Choose a location C which is not in the orbital plane, and calculate and display $\vec{L}_{trans,C}$ as an arrow throughout the orbit. (You do not need to make a graph.) Briefly describe in words what you observe.

ANSWERS TO EXERCISES

11.X.1 $-y$

11.X.2 D 50 kg·m^2/s out of page; E 50 kg·m^2/s out of page; F 50 kg·m^2/s out of page; G 0; H 30 kg·m^2/s into page

11.X.3 **(a)** out of page; **(b)** $d_1 p_1 \sin \alpha$; **(c)** out of page; **(d)** $d_2 p_2$

11.X.4 7.1×10^{33} kg·m^2/s

11.X.5 **(a)** $L_1 = 0.72$ kg·m^2/s and $L_2 = 0.72$ kg·m^2/s, both into page; **(b)** $L_{\text{rot}} = 1.44$ kg·m^2/s into page; **(c)** $I = 0.072$ kg·m^2; **(d)** $\vec{\omega}_0$ into page; **(e)** $I\omega_0 = 1.44$ kg·m^2/s into page; **(f)** They're the same; **(g)** $K_{\text{rot}} = 14.4$ J

11.X.6 **(a)** $\vec{L}_{\text{rot}} = \vec{0}$; **(b)** $L_{\text{trans}} = 9.72$ kg·m^2/s into page; **(c)** $L_{\text{tot}} = 9.72$ kg·m^2/s into page

11.X.7 **(a)** $L_{\text{rot}} = 1.44$ kg·m^2/s into page; **(b)** $L_{\text{trans}} = 9.72$ kg·m^2/s into page; **(c)** $L_{\text{tot}} = 11.16$ kg·m^2/s into page

11.X.8 Nonzero; into page; zero; (no direction)

11.X.9 No; nonzero; into page; nonzero; into page

11.X.10 Nonzero; up (toward the sky); nonzero; up

11.X.11 6 N·m; 2 N

11.X.12 $\langle 4, 3.8, 0 \rangle$ kg·m^2/s

11.X.13 $r_1 m v_1$ into page; $r_2 m v_2$ into page; $v_2 = (r_1/r_2)v_1$

11.X.14 Negligible external forces means negligible external torques. Therefore magnitude and direction of angular momentum are constant.

11.X.15 No torques around center of mass means no change in rotational angular momentum, so rotational angular momentum stays constant in magnitude (which determines length of day) and direction (which determines what "North Star" the axis points at). Doesn't matter that Earth is going around Sun; rotational angular momentum affected solely by torque around center of mass.

11.X.16 **(a)** $R\cos(45°)mv$ out of page; **(b)** $(M + m)R^2\omega$ out of page; **(c)** $R\cos(45°)mv/[(M + m)R^2]$ out of page; **(d)** The linear momentum of the clay decreases because the axle exerts an impulsive force upward (the wheel always has zero linear momentum because its center of mass doesn't move).

11.X.17 0.8 N·m into page

11.X.18 **(a)** 0.0256 kg·m^2/s into page; **(b)** 0.1856 kg·m^2/s into page; **(c)** 145 radians/s into page

11.X.19 8.33 N; the net force is zero, so there is no change in the linear momentum of the system, and the center of mass does not move.

11.X.20 $dL_z/dt = \frac{d}{dt}[RMv_{\text{CM}} - \frac{1}{2}MR^2\omega] = 0$, since there is no torque about a point under the string. Differentiating, we get $d\omega/dt = (2/R)(dv_{\text{CM}}/dt) = 2F_T/(MR)$, as before.

11.X.21 $dL_z/dt = d_1 F_N - (d_1 + d_2)M_2g = 0$ (+z out of page), and since $F_N = M_1g + M_2g$, this is equivalent to $d_1 M_1g - d_2 M_2g$, as before.

11.X.22 **(a)** 10 radians/s; **(b)** 5 radians/s; **(c)** 3 radians; **(d)** 0.9 m

11.X.24 Same as in horizontal case: $\Omega = RMg/(I\omega)$

Entropy: Limits on the Possible

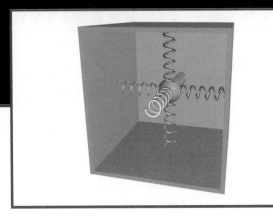

KEY IDEAS

- Thermal energy in a macroscopic object can be randomly distributed among the particles of the system in many different ways, and each specific arrangement is called a "microstate."
- When two objects are placed in thermal contact, energy becomes distributed between the two objects in such a way as to maximize the number of possible ways of arranging the available energy (number of microstates).
- The fundamental assumption of statistical mechanics is that all microstates corresponding to a particular total energy are equally probable.
- Entropy is determined by the number of ways of arranging the energy (number of microstates). The entropy of a system containing two objects is the sum of the two entropies. Like the number of microstates, the entropy tends to be maximized.
- Temperature is related to the rate at which entropy increases as energy increases.
- With a simple ball-and-spring model of a solid, for a particular thermal energy one can calculate the number of microstates, the entropy, temperature, and specific heat capacity.
- The probability of finding a microscopic system in an excited state increases with temperature.

12.1   STATISTICAL ISSUES

A pen lying on a table doesn't suddenly jump upward, despite the fact that there is plenty of energy in the table in the form of microscopic kinetic and potential energy. Why doesn't this happen? It wouldn't be a violation of the Energy Principle for energy to flow from the table into the pen. In fact, if instead of a pen you place a single atom on a table at room temperature, on average it will get so much energy from the table that it could jump thousands of meters upward (if there were no air or other obstacles in the way). This chapter deals with a statistical analysis of microscopic energy that puts limits on what is possible.

Much of this chapter is based on an article by Thomas A. Moore and Daniel V. Schroeder, "A different approach to introducing statistical mechanics," *American Journal of Physics*, vol. 65, pp. 26–36 (January 1997).

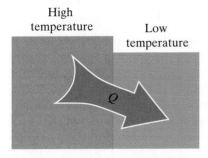

Figure 12.1 We observe that energy flows from a hotter object to a cooler one.

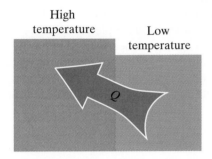

Figure 12.2 Could energy be transferred from the cooler object to the hotter one?

What Is Temperature?

We have repeatedly encountered a connection between the temperature of an object and the average motion of the atoms that make up the object, but our understanding of the meaning and role of temperature is incomplete. What is the quantitative relationship between temperature and microscopic energy?

Why Does Energy Flow from Hot to Cold?

When two blocks of different temperatures are placed in contact with each other, we observe that energy flows from the hotter block to the colder block (Figure 12.1). We can understand this in a rough way. The average energy of atoms in the hotter block is greater than the average energy of atoms in the colder block. In the interface region, where atoms of the two blocks are in contact with each other, it seems more likely that an atom in the hotter block will lose energy rather than gain energy when it collides with an atom in the colder block.

Couldn't energy flow the other way? After all, sometimes an individual atom in the hotter block might happen to have a lot less than its average energy when it collides with an individual atom in the colder block that happens to have a lot more than its average energy. In that case energy would be transferred from the colder block to the hotter block (Figure 12.2).

This "uphill" movement of energy seems unlikely, but how unlikely is it? Should we occasionally observe a hot block get hotter when we put it in contact with a cold block? If you put an ice cube in your drink, could your drink get warmer and the ice cube get colder? This would not violate conservation of energy, because the total energy of the two objects would still be the same. So what physical principle would be violated?

Why Do Things Run Down?

A related question, and a very deep one, has to do with the "reversibility" of processes. The fundamental physical interactions seem to be completely "reversible" in the following technical, physics sense: Make a movie of an alpha particle scattering off a gold nucleus, then run the movie forward or reversed. Both views look entirely reasonable. In contrast, make a movie of a ball bouncing on the floor with decreasing height on each bounce, then run the movie forward or reversed. The reversed movie looks silly: you see a ball bouncing higher and higher with each bounce! Would such a motion violate conservation of energy? Not necessarily. It could be that the atoms in the floor happen on average to give energy to the ball, and the floor is getting colder as a result of this continuing loss of energy. So what physical principle is violated by a process represented by the reversed movie?

Statistical Models

This chapter addresses these questions by applying probabilistic and statistical ideas to the behavior of systems. This subject is called "statistical mechanics." We will first apply statistical mechanics to a simple model of a solid, and we'll find that we can go surprisingly far with just a few new concepts. The main concepts that we will develop for the specific case of a solid apply in general to a wide variety of systems.

How will we know whether our statistical model explains anything? Our criterion for understanding is whether the predictions of our microscopic model agree with measurements of macroscopic systems, such as measurements of heat capacity (the amount of energy required to increase the temperature of an object by one degree).

Figure 12.3 The familiar model of a solid as massive balls connected by springs.

Figure 12.4 A single atom can oscillate in three dimensions. We simplify our model by assuming that each atom moves independently of the surrounding atoms—in effect, that it is connected to rigid walls instead of moving atoms.

A STATISTICAL MODEL OF SOLIDS

We have noted in previous chapters that the interactions between atoms in a solid are electric in nature, but that a detailed model of them involves quantum mechanics. The interatomic potential energy function encountered in Chapter 7 provides a reasonably accurate description of these interactions. Because of the similarity of this function to the potential energy curve for a harmonic oscillator (a mass on a spring), in previous chapters we have modeled a solid as a large number of tiny masses (the atoms) connected to their neighbors by springs (the interatomic bonds), as shown in Figure 12.3. This model has allowed us to understand qualitatively how solids interact with other objects. We would now like to use this model to ask detailed quantitative questions about the distribution of energy in a solid.

In this chapter we focus on calculating how probable a particular distribution of speeds or energies in a solid would be. We do this for a solid because it happens to be easier to calculate probabilities for a simple model of a solid than it is for a gas or a liquid. The atoms in a solid are nearly fixed in position, so we don't have to consider how likely different spatial arrangements might be (unlike the situation for a gas). Reasoning about the energy distribution in a solid will also enable us to draw conclusions about the transfer of thermal energy from one solid object to another.

The Einstein Model of a Solid

Because our goal is to calculate the probability of particular distributions of energy among atoms, we can simplify our model of a solid even further. We can think of each atom as moving independently of its neighbors, as though it were connected to rigid walls rather than to other atoms (Figure 12.4). Of course in a real solid energy is exchanged with neighboring atoms. However, to address our current questions we do not need to worry about the mechanism of energy exchange between atoms, since we will focus on calculating probabilities for various distributions of energy among the atoms of the solid.

This is a different kind of model from those we have so far constructed, because this very simple model would not allow us to predict in detail the dynamic motion of each atom in the solid, nor to ask how long it would take for energy to flow from one end of a solid to the other. In fact, there could not be propagation of sound in such a solid, with its "rigid walls." The model will not shed light on the details of a process, but it does allow us to ask questions about initial and final states. Since at the moment the questions we want to address involve initial and final states, this model is useful, and it is mathematically much simpler than the connected-atoms model.

Einstein proposed this simple model in 1907, and he found that some basic properties of solids such as heat capacity could in fact be understood using this model. This model also allows us to understand in detail the statistical nature of energy transfer between a hot object and a cold object, and why two objects come to "thermal equilibrium" (the same final temperature). This simple model of a solid will help us gain a more sophisticated and powerful understanding of the meaning of temperature.

A Three-Dimensional Oscillator—Three One-Dimensional Oscillators

We will consider each atom in a solid to be connected by springs to immovable walls. Each isolated atom is a three-dimensional spring–mass system, with $\vec{s}$ representing the three-dimensional vector displacement away from a fixed equilibrium position. Since $p^2 = p_x^2 + p_y^2 + p_z^2$ and $s^2 = s_x^2 + s_y^2 + s_z^2$, we can write

Figure 12.5 Since we are considering the atoms to be independent, in our model we can replace a single three-dimensional oscillator (one atom) by three independent one-dimensional oscillators.

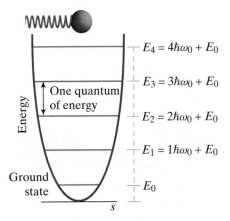

Figure 12.6 The quantized energy levels of a one-dimensional atomic oscillator.

the energy of a three-dimensional classical oscillator as the sum of three parts, corresponding to the x, y, and z oscillations:

$$K_{\text{vib}} + U_s = \left(\frac{p_x^2}{2m} + \frac{1}{2}k_s s_x^2 \right) + \left(\frac{p_y^2}{2m} + \frac{1}{2}k_s s_y^2 \right) + \left(\frac{p_z^2}{2m} + \frac{1}{2}k_s s_z^2 \right)$$

Recall from Chapter 8 that a quantized oscillator (a quantum mechanical "ball and spring") has evenly spaced energy levels. A complete quantum mechanical analysis of an oscillator that is free to oscillate in three dimensions rather than just one dimension leads to the conclusion that the motion of a 3-D oscillator can be separated into x, y, and z components, each of which has the same energy level structure as the familiar one-dimensional oscillator. This is mathematically equivalent to replacing each three-dimensional oscillator (an atom) with three ordinary one-dimensional oscillators, which we will do to simplify our model. We will think of a block as containing N one-dimensional oscillators, corresponding to $N/3$ atoms (Figure 12.5).

In Chapter 8 we discussed energy quantization in atomic spring–mass systems. As predicted by quantum mechanics and abundantly confirmed by experiments, energy can be added to a one-dimensional atomic oscillator only in multiples of one "quantum" of energy $\hbar\omega_0$, where $\omega_0 = \sqrt{k_{s,i}/m_a}$ (Figure 12.6):

Added energy can be 0, $1\hbar\omega_0$, $2\hbar\omega_0$, $3\hbar\omega_0$, etc.

Here $\hbar = h/(2\pi) = 1.05 \times 10^{-34}$ joule · second, $k_{s,i}$ is the interatomic spring stiffness, and m_a is the mass of the atom. The "ground state" (lowest energy level) of the quantum oscillator is not at the bottom of the potential energy well but $\frac{1}{2}\hbar\omega_0$ above the bottom. This offset doesn't matter for what we are going to do; it just sets the baseline energy level, and we will measure energies starting from the ground state.

Distributing Energy Among Objects

When two identical blocks are brought into contact, their total energy is shared among all the oscillators (atoms) in both blocks. It seems plausible that the most likely outcome would be that the total thermal energy be shared equally between identical blocks. However, might there be some probability that the first block would have more energy than the other, or even have all of it? With a large number of atoms the probability of even a small deviation from the most probable distribution turns out to be hugely unlikely. In order to look at this question in detail, we need to find a way to calculate the probability of various possible distributions of the energy. If we can find the most probable energy distribution, we can predict the eventual equilibrium distribution of energy between two objects brought into thermal contact with each other.

Distributing Energy: A Single Atom Inside a Solid

Consider a single atom inside a solid, whose energy we model in terms of the energy of three one-dimensional oscillators (x, y, z), neglecting interactions with the neighbors. Each of these three oscillators can have 0, 1, 2, or more "quanta" of vibrational energy ($\hbar\omega_0$) added to its ground state.

Suppose that the total vibrational energy added to the atom is four quanta, and we ask how we might distribute this energy among the three oscillators. We will soon see that enumerating all the possible ways of distributing the energy among the oscillators leads to a deeper understanding of the statistical behavior of a solid. We could give all the energy to the first one-dimensional oscillator and none to the others, or all to the second, or all to the third, as shown in Figure 12.7.

Figure 12.7 Three ways of distributing four quanta of vibrational energy among three one-dimensional oscillators.

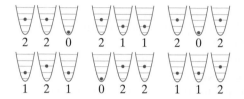

Figure 12.8 Six more ways of distributing four quanta of vibrational energy among three one-dimensional oscillators.

Figure 12.9 Still more ways of distributing four quanta of vibrational energy among three one-dimensional oscillators.

■ In principle you could prepare a system in one particular microstate and hope it would stay that way, but if there is the slightest bit of interaction among the pieces of the system, or with the surroundings, over time the fundamental assumption of statistical mechanics claims that all microstates will occur with equal probability. The Einstein model is simplified in that it doesn't provide a mechanism for energy to transfer from one atom to another (a change of microstate) or even from one oscillator to another within the same atom. The Einstein model is useful for counting microstates, but for other purposes it needs to be supplemented by the notion that some slight interaction does occur.

Figure 12.10 Two small blocks in thermal contact.

Atom 1	Atom 2	# ways
4	0	15 · 1
0	4	1 · 15

Figure 12.11 If we give all four quanta to one atom or the other, there are 30 ways of distributing four quanta of energy between the two atoms (that is, among six independent oscillators).

We could also give three quanta to one oscillator, and give the remaining quantum to one of the others, as shown in Figure 12.8.

Or we could give two quanta to one oscillator, and distribute the other two to the others, as shown in Figure 12.9.

That's it—there aren't any more ways to distribute the energy. (Check to make sure that you can't think of any other arrangements.) By explicitly listing all the possible arrangements, we see that there are 15 different ways that the four energy quanta could be distributed among the three one-dimensional oscillators.

In each of these 15 cases the total energy of the three-oscillator atom is exactly the same. When we make macroscopic measurements of the energy of a block, we don't know and we usually don't care exactly how the energy is distributed among the atoms that make up the block, because the internal energy of the block is the same for all of these distributions. However, the number of different arrangements influences interactions with other objects, as we will see.

"Microstate" and "Macrostate"

Some terminology: We say that there are 15 "microstates," such as the microstate with three quanta in the first oscillator, zero in the second, and one in the third (the second microstate in Figure 12.8). The 15 microstates correspond to one "macrostate," which is characterized by the total energy being equal to four quanta of energy, no matter how distributed.

> ### THE FUNDAMENTAL ASSUMPTION OF STATISTICAL MECHANICS
>
> The fundamental assumption of statistical mechanics is that, over time, an isolated system in a given macrostate is equally likely to be found in any of its possible microstates.

For example, we isolate three oscillators (corresponding to one atom), containing a total energy of four quanta, which is a macrostate (we haven't said which oscillators have how many quanta). The fundamental assumption of statistical mechanics implies that if we repeatedly observe the detailed arrangement of the four quanta among the three oscillators (the microstate), over time we would find on the average that each of these microstates would occur 1/15th of the times that we looked. This fundamental assumption is plausible, but ultimately it is justified by the fact that deductions based on this assumption do agree with experimental observations.

Distributing Energy: Two Interacting Atoms

We can now consider the case of two blocks in thermal contact (Figure 12.10). The importance of counting arrangements ("microstates") can be seen when we have systems of oscillators interacting with each other. Consider the smallest possible "blocks," two neighboring atoms (a total of six one-dimensional oscillators), and suppose four quanta of vibrational energy, $4\hbar\omega_0$, are distributed among these six oscillators. We already know that there are 15 ways to distribute the four quanta among the three one-dimensional oscillators of the first atom, and in that case there is only one way to distribute zero quanta to the second atom. Similarly, in Figure 12.11 we see that there are 15 ways to distribute the four quanta among the three one-dimensional oscillators of the second atom, and in that case there is only one way to distribute zero quanta to the first atom.

Atom 1	Atom 2	# ways
3	1	$10 \cdot 3$
1	3	$3 \cdot 10$

Figure 12.12 We find 60 more ways to distribute four quanta among two atoms (six oscillators) if we give three quanta to one atom and one quantum to the other.

Atom 1	Atom 2	# ways
2	2	$6 \cdot 6$

Figure 12.13 We find 36 ways of distributing four quanta of energy by giving two quanta to each atom.

q_1	q_2	# ways 1	# ways 2	# ways 1 · # ways 2
0	4	1	15	15
1	3	3	10	30
2	2	6	6	36
3	1	10	3	30
4	0	15	1	15

Figure 12.14 Summary of the 126 different ways to distribute four quanta of vibrational energy between two atoms (each consisting of three independent oscillators).

Figure 12.15 Bar graph showing the total number of ways of distributing four quanta of vibrational energy between two atoms (six one-dimensional oscillators).

However, since the two atoms are in contact with each other, the four quanta could also be shared between the atoms. For example, the first atom might have only three quanta, with the second atom having the other quantum.

> **QUESTION** How many ways are there to distribute three quanta among three one-dimensional oscillators? Try to list all the possibilities, as we did for the case of four quanta.

You should have found ten ways, which we could list as 300 (that is, three quanta on the first oscillator, zero on the second, and zero on the third), 030, 003, 201, 210, 021, 120, 012, 102, and 111.

> **QUESTION** For each of these arrangements, how many ways are there to distribute the one remaining quantum among the other three one-dimensional oscillators?

There are just three ways to arrange one quantum among the other three oscillators: 100, 010, and 001.

The product of these two numbers is the total number of ways of distributing the four quanta in such a way that there are three quanta on the first atom and one on the other. This product is $10 \cdot 3 = 30$ ways. These results are summarized in Figure 12.12.

Now let's give just two quanta to the first atom (leaving two quanta for the other atom).

> **QUESTION** How many ways are there to distribute two quanta among three one-dimensional oscillators?

You should find that there are six ways: 200, 020, 002, 101, 110, and 011.

> **QUESTION** For each of these arrangements, how many ways are there to distribute the two remaining quanta among the other three one-dimensional oscillators?

Clearly there are also six ways to arrange the remaining two quanta among the other three oscillators. The total number of ways of distributing the four quanta in such a way that there are two quanta on the first atom and two on the other is $6 \cdot 6 = 36$ (Figure 12.13).

Now we have carried out enough calculations to be able to make a table of all the ways of distributing four quanta between two atoms (that is, among six one-dimensional oscillators). In this table, q_1 is the number of quanta (0 to 4) assigned to the first atom, and q_2 is the number of quanta assigned to the second (with $q_1 + q_2$ necessarily equal to 4). In Figure 12.14 we show the number of ways to arrange the quanta among the first three one-dimensional oscillators, among the second three oscillators, and the product, which is the total number of ways of distributing q_1 and q_2 quanta.

Figure 12.15 is a graph of the total number of ways of distributing the quanta vs. q_1. Remember that the basic assumption of statistical mechanics is that all of these 126 different arrangements ($15 + 30 + 36 + 30 + 15 = 126$ microstates) with a total energy of four quanta are equally probable.

> **QUESTION** Therefore, if two atoms share four quanta, what is the most likely division of the energy between the two? What is the probability that at some instant the energy is equally divided? What is the probability that at some instant the first atom has all the energy?

Evidently the most probable division is 2 and 2, since there are 36 ways for this to be done, whereas all other divisions have fewer ways for this to happen. The probability for a 2-2 division is 36/126, or 29%, so even though this is the most probable division, it happens less than one-third of the time. The probability that the first atom has all the energy is the number of ways for this to happen (15) divided by 126: 15/126 = 12%.

This is a microcosm of what happens when two identical blocks are brought into contact: the most probable division of the thermal energy is that it is shared equally. In order to understand in detail how this works in a real macroscopic system, we need to consider large numbers of atoms and large numbers of quanta. For numbers of atoms or quanta even slightly larger than the few we've been considering, it becomes practically impossible to figure out the number of arrangements of quanta by explicitly listing all the possibilities as we did up to now. We need a formula for calculating the number of arrangements.

> **12.X.1** For practice in counting microstates, determine how many ways there are to arrange three quanta among four one-dimensional oscillators. (This would represent one-and-a-third atoms, so this doesn't make physical sense.)

One System or Many?

The results for distributing four quanta among two atoms (six oscillators) can be thought of in two complementary ways. You can say, "I will make frequent observations of my isolated two-atom system, and I expect that in 29% of these observations I'll find the energy split evenly (2-2)." Alternatively, you can set up 100 of these isolated two-atom systems and say, "Whenever I look at my 100 systems, I expect that 29 of them will have the energy split evenly (2-2)." Sometimes the "one system, many observations" view is particularly helpful, and sometimes the "many systems, one observation" view is the more useful way to think about the statistical nature of a phenomenon.

A Formula for the Number of Arrangements of Quanta

Clearly it would be not only tedious but also impractical to keep on counting states as we did above for systems involving very large numbers of atoms. We need a formula for calculating how many different ways a set of objects can be arranged. We can develop this formula in a general way, using concepts of probability, and then apply it to solids made up of atomic oscillators.

Suppose you have five billiard balls in a bag, numbered 1 through 5. You draw them out of the bag, one at a time, and record the sequence of numbers, such as 34152 (Figure 12.16). Then you put them back in the bag and repeat. How many different number sequences are possible?

Figure 12.16 Five numbered billiard balls are taken one at a time from a bag. How many different number sequences are possible?

- The first ball might be any one of the five balls.
- For each of these five possibilities, there are four possibilities for the second ball, or $5 \cdot 4 = 20$ choices so far.
- For each of the $5 \cdot 4 = 20$ choices of the first two balls, there are three different possible choices for the third ball, or $5 \cdot 4 \cdot 3$ choices so far.
- For each of the $5 \cdot 4 \cdot 3 = 60$ choices of the first three balls, there are two possible choices for the fourth ball, for a total of $5 \cdot 4 \cdot 3 \cdot 2 \cdot 1 = 120$ possibilities, since there is only one remaining ball to choose.

Evidently there are 120 different "permutations" of the five integers. It would be exceedingly tedious to list all these different arrangements, but we have a simple formula to calculate how many there are: 5!, which is the standard notation for "5 factorial," meaning $5 \cdot 4 \cdot 3 \cdot 2 \cdot 1$. Fortunately, for analyzing a solid all we care about is how many arrangements (microstates) there are for a particular total energy, not the details about how much of the energy is assigned to which oscillators. So all we need from a formula is the number of arrangements.

12.X.2 We can easily check this factorial formula for the case of three balls. Explicitly list all possible permutations of the numbers 1, 2, and 3, and verify that there are indeed $3! = 3 \cdot 2 \cdot 1 = 6$ possible arrangements.

Figure 12.17 Five billiard balls are taken one at a time from a bag. Three are red and two are green. How many different color sequences are possible?

Making Fewer Distinctions

Now suppose that of the five balls, three are red (R) and two are green (G), and we ask how many different arrangements of the colors are possible, such as RRGGR shown in Figure 12.17. We know that there are $5! = 120$ different arrangements of the numbered balls, but if we're only interested in the color sequence, the numerical order of the red balls is irrelevant, as is the numerical order of the green balls.

There are $3! = 6$ permutations of the red balls among each other, and $2! = 2$ permutations of the green balls among each other. Therefore there are many fewer than 120 distinctively different color sequences, and we need to correct for this by dividing by the extra permutations:

$$\text{number of color sequences} = \frac{5!}{3!2!} = \frac{120}{(6)(2)} = 10$$

12.X.3 Check this result by listing all the different ways of ordering three R's and two G's.

· · **|** · **|**·

Figure 12.18 A representation of a situation where the first oscillator has 2 quanta, the second oscillator has 1 quantum, and the third oscillator has 1 quantum.

Arranging Quanta Among Oscillators

By an appropriate choice of visual representation, we can convert our problem of calculating the number of arrangements of q quanta among N one-dimensional oscillators ($N/3$ atoms) into the problem we just solved, for which we have a formula. Consider again the specific case of $q = 4$ quanta distributed among $N = 3$ one-dimensional oscillators. We'll represent a quantum of energy by the symbol •, and a (fictitious) boundary between oscillators by a vertical bar **|**. Figure 12.18 is a picture of a particular situation where the first oscillator has 2 quanta, the second oscillator has 1 quantum, and the third oscillator has 1 quantum; together with the boundaries between oscillators we have 6 objects arranged in a particular sequence.

This sequence represents a total energy of 4 quanta and 2 boundaries (a total of 6 things, if we consider a boundary to be a "thing"). Figure 12.19 illustrates two such sequences. Notice that we need $N - 1 = 2$ vertical bars to be able to indicate which oscillators have how much energy. How many such sequences like this are there? Rearranging quanta and boundaries is equivalent to moving quanta between oscillators. In this pictorial form, the problem is like having 4 red balls and 2 green ones. Therefore we can write a formula for the number of different arrangements (number of different microstates):

$$\frac{6!}{4!2!} = 15 \text{ ways of arranging 4 quanta among 3 oscillators}$$

Figure 12.19 Representations of two situations for the number of quanta in 3 oscillators.

This agrees with our earlier calculations. Generalizing to arbitrary numbers of quanta distributed among arbitrary numbers of oscillators, we have this important result for the number of microstates, written as Ω (Greek capital omega):

NUMBER OF WAYS Ω TO ARRANGE q QUANTA AMONG N ONE-DIMENSIONAL OSCILLATORS

$$\Omega = \frac{(q + N - 1)!}{q!(N - 1)!}$$

12.X.4 Verify that this formula gives the correct number of ways to arrange 0, 1, 2, 3, or 4 quanta among 3 one-dimensional oscillators, given in earlier tables (1, 3, 6, 10, 15).

Very Big Numbers

As you increase q and/or N, this formula gets very big very fast. For example, the number of ways to distribute 100 quanta among 300 oscillators (100 atoms) is about 1.7×10^{96}, which is 17 followed by 95 zeros!

In striking contrast, there is only one way to arrange to have all 100 energy quanta be placed on just one particular oscillator out of all the 300 oscillators. Although this arrangement would satisfy the requirement that the total energy of the system be 100 quanta, this is extremely improbable. The fundamental assumption of statistical mechanics is that in an isolated system, over time, all microstates are equally probable, so the odds that the actual microstate is the one with all the energy given to just one particular oscillator, of your choice, is 1 in 1.7×10^{96}, which makes this unlikely event essentially impossible.

A typical macroscopic object such as a block of ordinary size contains 1×10^{23} or more atoms, not a mere hundred, and the number of quanta is even larger. How likely is it that all the energy will be found concentrated on one particular atom? The mind boggles at the astronomically huge odds against this ever happening. Could it happen? Yes. Is any human ever likely to observe it? No.

12.X.5 Suppose that you look once every second at a system with 300 oscillators and 100 energy quanta, to see whether your favorite oscillator happens to have all the energy (all 100 quanta) at the instant you look. You expect that just once out of 1.7×10^{96} times you will find all of the energy concentrated on your favorite oscillator. On the average, about how many years will you have to wait? Compare this to the age of the Universe, which is thought to be about 1×10^{10} years. (1 year $\approx \pi \times 10^7$ seconds.)

12.3 THERMAL EQUILIBRIUM OF BLOCKS IN CONTACT

We now have the tools necessary for analyzing in some detail what will happen when two blocks are brought into contact and approach thermal equilibrium. We'll choose two blocks made of the same material, so a quantum of energy is the same for the atomic oscillators in both blocks (same atomic mass m; same interatomic forces, so same "spring stiffness" k_s). We'll make the analysis somewhat general by choosing blocks of different sizes. One block contains N_1 one-dimensional oscillators ($N_1/3$ atoms) and initially contains q_1 quanta of energy, and the other block contains N_2 one-dimensional oscillators ($N_2/3$ atoms) and initially contains q_2 quanta of energy.

QUESTION We want to treat the simple case where the total energy $q_1 + q_2$ of the two-block system remains fixed at all times. What simplifying assumption should we make about the situation?

During the entire process we assume that there is little energy transferred into or out of the surrounding air or the supports for the blocks, so that the total energy of the two blocks doesn't change. However, the number of quanta in each block, q_1 or q_2, need not stay fixed, since energy can flow back and forth between the two blocks.

Consider a very concrete example. Suppose that $N_1 = 300$ (100 atoms), $N_2 = 200$ (about 67 atoms; or choose 201 oscillators if you wish to be exact),

Figure 12.20 A total of 100 quanta of vibrational energy are available to be distributed between two systems, one consisting of 300 oscillators (100 atoms), the other consisting of 200 oscillators (about 67 atoms).

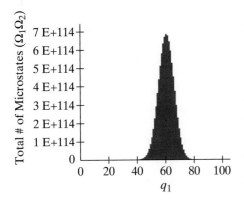

Figure 12.21 The number of ways of distributing 100 quanta of vibrational energy between two blocks having 300 and 200 oscillators, respectively. q_1 is the number of quanta in the first (larger) block.

■ At this point it is very useful to do Problem 12.P.69, in which these ideas are made very concrete.

and the total energy distributed throughout the two blocks corresponds to $q_1 + q_2 = 100$ quanta (Figure 12.20).

We use a computer to calculate the number of ways that q_1 quanta can be distributed among the $N_1 = 300$ oscillators of the first block (using the formula we developed earlier), and we multiply this number times the number of ways that $q_2 = (100 - q_1)$ quanta can be distributed among the $N_2 = 200$ oscillators of the second block.

The product of these two calculations is the number of ways that we can arrange the 100 quanta so that the first block has q_1 quanta and the second block has $q_2 = (100 - q_1)$ quanta, for a total energy shared between the two blocks of 100 quanta. We have the computer do this calculation for $q_1 = 0, 1, 2, 3, \ldots 99, 100$ quanta, which corresponds to $q_2 = 100, 99, 98, \ldots 1, 0$ quanta. The following table shows the first few results, where the number of microstates is denoted by Ω (Greek uppercase omega):

q_1	$q_2 = (100 - q_1)$	$\Omega_1 = \dfrac{(q_1 + 300 - 1)!}{q_1!(300 - 1)!}$	$\Omega_2 = \dfrac{(q_2 + 200 - 1)!}{q_2!(200 - 1)!}$	Total # of Ways $\Omega_1 \Omega_2$
0	100	1	2.772 E+81	2.772 E+81
1	99	300	9.271 E+80	2.781 E+83
2	98	4.515 E+04	3.080 E+80	1.391 E+85
3	97	4.545 E+06	1.016 E+80	4.619 E+86
4	96	3.443 E+08	3.331 E+79	1.147 E+88
…	…	…	…	…

We see in the last column that there are an enormous number of ways of arranging 100 quanta among 500 oscillators, and that this number is growing rapidly in the table. In Figure 12.21 the possible number of arrangements of quanta is plotted on the y axis, and q_1, the number of quanta assigned to the first block, is plotted on the x axis.

The most probable arrangement (indicated by the highest point of the peak on the graph) is that 60 of the 100 quanta will be found in the first block, which contains 300 oscillators out of the total of 500 oscillators. This seems reasonable, since it does seem most probable that the energy would be distributed uniformly throughout the two blocks (since they're made of the same material), and that would give 3/5 (60%) of the energy to the first block. It is gratifying that our statistical analysis leads to this plausible result.

It appears from the width of the peak shown in the graph that we shouldn't be too surprised if occasionally we would find that the first block contains anywhere between 40 to 80 of the 100 quanta, but it appears that it is very unlikely to find fewer than 40 or more than 80 of the quanta in the first block.

Relatively speaking, how likely is it for *none* of the energy to be in the first block? In that case, $q_1 = 0$, and from the table we see that the number of ways to arrange the 100 quanta this way is 2.772 E+81 (2.772×10^{81}). That's an awfully big number, but how big is it compared to the most probable arrangement, where $q_1 = 60$, for which there are about 7×10^{114} ways according to the graph? Evidently it is less likely by about a factor of 10^{33} (!) to find the energy split 0-100 rather than 60-40. Is it possible according to the laws of physics for none of the energy to be in the first block? Yes. Is it likely that we would ever observe such an unusual distribution? Most emphatically not!

Note that the number of ways to arrange the quanta 0-100 (2.772×10^{81}) isn't actually visible on the graph, because it is 1×10^{33} times smaller than the peak of the graph. Values of q_1 outside the 40 to 80 range are invisible on the graph, because they are relatively so very small compared to the peak.

Figure 12.22 The number of ways of distributing 1000 quanta between two blocks containing 3000 and 2000 oscillators, respectively.

Width of the Distribution

Compared to the graph in Figure 12.15 of our earliest calculation (6 oscillators and 4 quanta), the peak shown in Figure 12.21 occupies a much narrower range of the graph, reflecting the sharply decreased relative probability of observing extreme distributions of the energy.

For further comparison, Figure 12.22 is a graph of the number of ways of distributing 1000 quanta among two blocks containing 3000 and 2000 oscillators. This peak is much narrower than the peak in the previous graph (where there were only 100 quanta distributed among only 300 and 200 oscillators). This is a general trend. The larger the number of quanta and oscillators, the narrower is the peak around the most probable distribution.

The fractional width of the peak is the width of the peak at half height divided by the value of q_1 that gives the maximum probability (in the present case, the width divided by 600). It can be shown that the fractional width of the peak is proportional to $1/\sqrt{q}$ or to $1/\sqrt{N}$, whichever is larger. That is, if you quadruple the number of quanta or the number of oscillators, the fractional width of the peak decreases by a factor of 2.

QUESTION | Consider two blocks that are of ordinary macroscopic size, containing 3×10^{23} atoms and 2×10^{23} atoms, and many quanta per atom. Qualitatively, what would you expect about the width of the peak, if you could calculate it? How likely is it that you would ever observe a significant fluctuation away from the most probable 60-40 split?

You would expect the peak to be extremely narrow, and hence the probability of significant fluctuations would be very low.

The Most Probable Is the Only Real Possibility

These considerations show that in the world of macroscopic objects such as ordinary-sized blocks, the most probable arrangement is essentially the *only* arrangement that is ever observed. That is why you do not see a block suddenly leap up from the table when all of the thermal energy of the table floods into the block. On the other hand, at the microscopic level we should not be surprised if the energy of one of the atoms in a block varies a lot, since the most probable distribution includes many different arrangements of the quanta, with varying numbers of quanta on one particular atom.

Entropy and Equilibrium

We now have a good statistical description of the thermal equilibrium of two blocks in contact. At equilibrium, energy is distributed between the two blocks in the most probable manner, based on having the largest number of ways of achieving this distribution (largest number of microstates for the macrostate of given total energy). We have been studying very small systems consisting of a hundred or more atoms, and we found that in small systems there is some significant probability of finding the energy distributed somewhat differently than the most probable 60-40 division. (Objects consisting of only a few hundred atoms are called "nanoparticles" and are currently the subject of intense research, because many of their properties and behaviors are intermediate between those of atoms and those of large-scale objects.) For large macroscopic objects containing 10^{20} atoms or more, the most probable distribution of the energy is essentially the only energy distribution we will ever observe, because the probability of distributions that are only slightly different is very small.

Next we will study more deeply the details of why a particular thermal equilibrium becomes established. Suppose for example that the first block

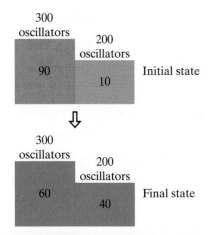

Figure 12.23 If the initial energy distribution between two systems in thermal contact is not the most probable energy distribution, energy will be exchanged until the most probable distribution is reached.

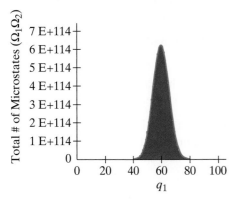

Figure 12.24 Ways of distributing 100 quanta of vibrational energy between a system of 300 oscillators and a system of 200 oscillators.

(a nanoparticle with 300 oscillators, or 100 atoms) starts out with 90 quanta and the second (200 oscillators, or about 67 atoms) with only 10 quanta (Figure 12.23). When we put them together, we expect this to shift toward a 60-40 distribution (Figure 12.24). Studying the details of why this shift occurs will lead us to a deep understanding of the concepts of temperature and something called "entropy."

Since we start from a nonequilibrium energy distribution, it would be nice if we could make some kind of graph where we could see the total number of ways to arrange the quanta even when we are far from equilibrium. As we have seen, the graph of the total number of microstates is so strongly peaked that we don't see anything outside the peak. For example, on the graph in Figure 12.24 where we can see 1×10^{114} we can't see the relatively much smaller value of 1×10^{81}. A way to get around this problem is to plot the logarithm of the data, which makes the data visible across the whole range of energy distributions.

In statistical mechanics it is standard practice to use the base-e or "natural" logarithm ("ln"). For the case of distributing 100 quanta among 300 and 200 one-dimensional oscillators, which we studied before, Figure 12.25 shows a plot of the natural logarithms of the number of ways Ω_1 to arrange q_1 quanta in block 1, the number of ways Ω_2 to arrange $q_2 = (100 - q_1)$ quanta in block 2, and the total number of ways $\Omega_1 \Omega_2$ to arrange the 100 quanta among the 500 oscillators. These quantities are plotted against q_1, of course running left to right, 0 to 100. You can think of q_2 as running right to left.

Because we are considering discrete microstates, each of these curves is really a set of 100 dots, but we connect the dots and make continuous curves. This is particularly appropriate when we deal with macroscopic objects, where an increment of one quantum along the x axis is practically an infinitesimal fraction of the axis.

By taking (natural) logarithms, we have converted a highly-peaked graph into a slowly varying one. To make sure you understand how this graph is related to the peaked graph, do the following:

> **12.X.6** For this situation, we calculated earlier that there is just one way to arrange 0 quanta in block 1, and 2.772×10^{81} ways to arrange 100 quanta in block 2. Take the natural logarithm of this number and verify that all three curves on the graph make sense to you at $q_1 = 0$ ($q_2 = 100$). Note that at the right end of the graph, at $q_1 = 100$, the curve for $\ln(\Omega_1)$ goes higher than $\ln(2.772 \times 10^{81})$, because block 1 has a larger number of oscillators among which to distribute the 100 quanta (300 oscillators compared with 200).

Definition of Entropy

We will deal repeatedly with the natural logarithm of the number of ways to arrange energy among a group of atoms (the number of microstates corresponding to a particular macrostate of specified energy). This quantity, $\ln(\Omega)$, when multiplied by the Boltzmann constant k_B, is called the "entropy" of the object and is denoted by the letter "S" (the triple equal sign means "is defined as"):

DEFINITION OF ENTROPY S
$S \equiv k_B \ln \Omega$
The Boltzmann constant is $k_B = 1.38 \times 10^{-23}$ J/K.

Because the Boltzmann constant is $k_B = 1.38 \times 10^{-23}$ J/K, entropy has units of joules per kelvin. The Boltzmann constant is included in the

Figure 12.25 A logarithmic plot of the ways of distributing 100 quanta of vibrational energy between two systems, plotted against q_1, the number of quanta in object 1.

■ At this point it is very useful to do Problem 12.P.70, to see in detail how to produce Figure 12.25 or Figure 12.26.

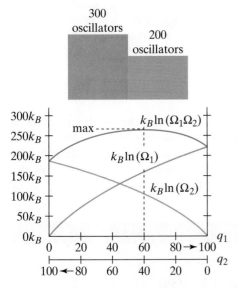

Figure 12.26 A plot of the entropies S_1, S_2, and $S_1 + S_2$ plotted against q_1, the number of quanta in object 1. Entropy is $k_B \ln \Omega$.

definition for consistency with an older, macroscopically based definition of entropy.

Since $k_B \ln(\Omega_1 \Omega_2) = k_B \ln(\Omega_1) + k_B \ln(\Omega_2)$ (this is a property of logarithms), the entropy S of the two-block system is equal to the entropy S_1 of the first block plus the entropy S_2 of the second block. A consequence of defining entropy in terms of a logarithm is that we can consider entropy as describing a property of a system, and when there is more than one object in a system we get the total entropy simply by adding up the individual entropies, $S_{tot} = S_1 + S_2$, as is the case with energy ($E_{tot} = E_1 + E_2$).

Figure 12.26 is a plot of the entropies of each block, and their sum. Since both $k_B \ln(\Omega_1 \Omega_2)$ and $\Omega_1 \Omega_2$ go through a maximum at the same value of q_1, we can state the condition for thermal equilibrium of the two blocks, in terms of entropy:

CONDITION FOR THERMAL EQUILIBRIUM

In equilibrium the most probable energy distribution is the one that maximizes the total entropy $S_{tot} = S_1 + S_2$ of two objects in thermal contact.

QUESTION   In Figure 12.26, what is the physical significance of the fact that the top curve goes through a maximum for $q_1 = 60$?

The most probable energy distribution is a 60/40 sharing, corresponding to the 300/200 sizes of the two objects.

QUESTION   In Figure 12.26, what is the physical significance of the point where the two lower curves cross each other?

Nothing! This point isn't anything special. It is the point where the entropy in one object is equal to the entropy in the other object (and equal to half the entropy of the combined system). It turns out that this doesn't have any real physical significance.

12.4   THE SECOND LAW OF THERMODYNAMICS

The "first law of thermodynamics" is another name for the Energy Principle $\Delta E = W + Q$, which by now should be very familiar to you. The "second law of thermodynamics" is, however, something new. It can be stated in a number of equivalent forms, but there is a particularly useful formulation in terms of entropy:

THE SECOND LAW OF THERMODYNAMICS

If a closed system is not in equilibrium, the most probable consequence is that the entropy of the system will increase.

In other words, a closed system will tend toward maximum entropy.

As a specific example of the second law of thermodynamics, consider our two blocks. The most probable energy distribution is the one for which the total entropy is a maximum, and if initially the energy distribution is something else, it is highly likely that the entropy will increase. For nanoparticles there can be significant fluctuations away from this state of maximum entropy (with accompanying decrease in total entropy), but for ordinary-sized systems these fluctuations are extremely small, and for practical purposes the entropy of a closed macroscopic system never decreases.

Loosely, one can restate the second law of thermodynamics like this: "A closed system tends toward increasing disorder." This is overly vague without precise definitions of "order" and "disorder," and it is the definition of entropy as $k_B \ln \Omega$ that provides the needed precision. A closed system tends toward increasing entropy, which is a measure of the number of microstates corresponding to a particular macrostate of given energy.

Irreversibility

Any process in which the entropy of the Universe increases is "irreversible" in the technical, physics sense: a reversed movie of the process looks odd. However, any process in which the entropy of the Universe doesn't change is in principle "reversible," and there do exist processes that are approximately or nearly reversible. For example, a steel ball bearing that bounces vertically on a steel plate rebounds almost to the same height from which it was dropped, and this represents a (nearly) reversible process, because the system returns (nearly) to its original state.

However, if we watch for a while, the ball bearing returns to lower and lower heights and eventually settles down to rest on the plate. If the ball bearing kept returning to its original height, this would not be a violation of energy conservation. It is the second law of thermodynamics that says that we cannot expect the process to be completely reversible. There are a larger number of ways to share the energy between the ball and the plate than the number of ways for the ball to keep all the energy.

Suppose we make a movie of the ball bearing bouncing lower and lower and coming to rest on the plate. Then we run it backwards, and our friends see the ball starting to bounce, and bouncing higher and higher (Figure 12.27). This needn't violate energy conservation: energy in the plate could be flowing into the ball. This reversed movie certainly looks odd, though, presumably because we have an instinctive sense, based on lots of experience, that the entropy of the Universe increases rather than decreases.

Time running forward

Time running in reverse

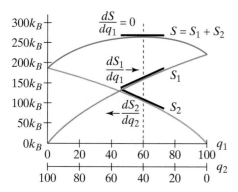

Figure 12.27 If the height of a bouncing ball increased with time instead of decreasing with time, this would not necessarily violate conservation of energy.

REVERSIBLE AND IRREVERSIBLE PROCESSES

Reversible process: $\Delta S_{sys} + \Delta S_{surroundings} = 0$

Irreversible process: $\Delta S_{sys} + \Delta S_{surroundings} > 0$

12.X.7 You see a movie in which a shallow puddle of water coalesces into a perfectly cubical ice cube. How do you know the movie is being played backwards? Otherwise, what physical principle would be violated?

Figure 12.28 Entropy vs. number of quanta of energy in system 1.

12.5 WHAT IS TEMPERATURE?

Until now, we have loosely associated temperature with the average energy of a molecule. The concept of entropy makes possible a deeper connection between our macroscopic measurements of temperature and a fundamental, atomic, statistical view of matter and energy. We will develop a statistically based definition of temperature.

Consider again our two blocks. On the graph in Figure 12.28 we plot entropy $S = k_B \ln \Omega$. Remember that we are approximating a set of dots by a smooth curve, since the dots are very close together (tiny energy spacing). At the maximum of the total entropy $S = S_1 + S_2$, the total entropy curve is horizontal (slope = 0), and the following relationship is true at equilibrium:

$$\frac{dS}{dq_1} = \frac{dS_1}{dq_1} + \frac{dS_2}{dq_1} = 0$$

Since $q_2 = (100 - q_1)$, this leads to

$$\frac{dS_1}{dq_1} - \frac{dS_2}{dq_2} = 0 \text{ (at equilibrium)}$$

dS_2/dq_2 is the slope moving from right to left and is positive, since q_2 and S_2 increase from right to left. Therefore we see that these slopes are the same at equilibrium:

$$\frac{dS_1}{dq_1} = \frac{dS_2}{dq_2} \quad \text{(at equilibrium; maximum total entropy)}$$

dS_1/dq_1 is a measure of the state of the first block, and dS_2/dq_2 is a measure of the state of the second block.

QUESTION What physical property can these derivatives represent? What physical property of the blocks is the same when the two blocks reach thermal equilibrium?

The blocks reach the same temperature. It must be that the derivative of the entropy with respect to the energy is somehow related to temperature.

To see what this relationship is, suppose that when the two blocks were initially brought into contact, the first block contained $q_1 = 90$ quanta and the second block only $q_2 = 100 - 90 = 10$ quanta, as illustrated in Figure 12.29. Note that in Figure 12.29 the initial slope of the entropy curve for block 2 is steeper than the initial slope of the entropy curve for block 1. (Remember: dS_2/dq_2 is the slope moving from right to left and is positive, since q_2 and S_2 increase from right to left.) Therefore, if we remove one quantum of energy from block 1 and give it to block 2, we'll increase the entropy in block 2 more than we'll decrease the entropy in block 1.

QUESTION Will this result in a net increase or a net decrease in the total entropy S? (See Figure 12.29.)

Since the entropy of block 2 increases more than the entropy of block 1 decreases, there is a net increase of the entropy of the two blocks together. We already established that the state of the two blocks will evolve toward greater total entropy, not less total entropy.

QUESTION Consider which block gives up energy, and which block takes it up. Which block is initially at a higher temperature?

Evidently block 1 must be at the higher temperature, because on average it will give energy to block 2 in order that the total entropy increase.

QUESTION Does a steeper or a less steep slope of entropy vs. energy correspond to higher temperature?

Initially, block 1 has the higher temperature, and S_1 vs. q_1 has the smaller slope. Initially, block 2 has the lower temperature, and S_2 vs. q_2 has the larger slope.

QUESTION Therefore, which of the following looks like a better guess for relating temperature to entropy: $T_1 \propto dS_1/dq_1$ or $1/T_1 \propto dS_1/dq_1$?

Since block 1 has the higher temperature and the smaller slope (smaller dS/dq), it seems possible that $1/T_1 \propto dS_1/dq_1$.

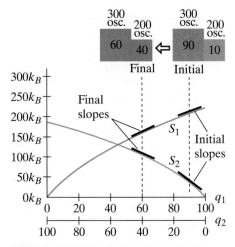

Figure 12.29 Location of the initial and final states of the two-block system on a plot of entropy vs. number of energy quanta in system 1.

Definition of Temperature

For these reasons, we *define* temperature in terms of dS/dq. However, the magnitude of one quantum of energy varies for different systems, so we can't compare different systems if we use q. So instead of q we use E_{int}, the internal energy above the ground state of a system measured in joules. For a group of oscillators, E_{int} is the number of quanta q times the energy per quantum: $E_{int} = q\hbar\sqrt{k_s/m}$ for a system of harmonic oscillators, where k_s is the spring stiffness.

DEFINITION OF TEMPERATURE T
$$\frac{1}{T} \equiv \frac{dS}{dE_{int}}$$

The Boltzmann constant k_B in the definition $S \equiv k_B \ln \Omega$ makes the units come out right. With the internal energy E_{int} of a system measured in joules (J), and entropy S measured in joules per kelvin (J/K), temperature T is measured in kelvins (K). (Remember that we're approximating a set of closely spaced dots by a curve, so taking a derivative makes sense.)

This is a highly sophisticated and abstract way of defining temperature. How does this relate to the temperature that is measured by an ordinary thermometer? We will see in the next chapter that the temperature defined by $1/T = dS/dE_{int}$ is the same as the "absolute" temperature, which appears for example in the ideal gas law $PV = N_{moles}RT$, which you have probably encountered in previous chemistry or physics studies. On this temperature scale, ice melts at a temperature of +273.15 K (0 °C), and water boils at +373.15 K (100 °C). To put it another way, absolute zero Kelvin is at −273.15° Celsius. For our purposes, it will be adequate to say ice melts at 273 K. Room temperature is about 20 °C or about 293 K.

To recapitulate, the greater the dependence of the entropy on the energy for an object (the steeper the slope of S vs. E_{int}), the more "eager" the object is to take in energy to contribute to increasing the total entropy of the total system. It will do this if it can take energy from another object that has a smaller dependence of entropy on energy (smaller slope), since the second object's entropy will decrease less than the first object's entropy will increase. An object with a large dS/dE_{int} has a low temperature, since it is low-temperature objects that take in energy from high-temperature objects.

Conversely, the smaller the dependence of the entropy on the energy, the less reluctant the object is to give up energy and decrease its own entropy, if it can give the energy to an object with a greater dependence of entropy on energy. An object with a small dS/dE_{int} has a high temperature, since it is high-temperature objects that donate energy to low-temperature objects.

■ At this point it is useful to do Problem 12.P.71, in which you calculate the temperature of an object as a function of the amount of energy in the object.

W, Q, and Entropy

More formally, we should write $1/T = \partial S/\partial E_{int}$, involving a "partial derivative," which means that we hold everything but S and E constant when we take the derivative. In particular, we hold the volume constant, which means that we do no work on the system to compress it or stretch it, so we should write $1/T = (\partial S/\partial E_{int})_V$, where the subscript V means "take the (partial) derivative holding the volume constant."

One might ask whether there is any entropy change of the surroundings associated with energy exchange in the form of work. The answer turns out to be no for processes that are reversible (there do exist irreversible processes in which work can be associated with an increase in entropy). Energy transfer caused by a temperature difference between objects in contact, Q, is

Figure 12.30 Mechanical work (compression or stretching of an oscillator system) compared to energy transfer due to a temperature difference to the same system.

"disorganized" energy transfer and is associated with entropy change. Work is "organized" energy transfer and in many situations doesn't affect the entropy. It is beyond the scope of this textbook to prove this rigorously, but we can illustrate the basic issues in terms of an Einstein solid.

If you mechanically squeeze or compress a block, doing work on it to raise its energy, it can be shown that you shift the energy levels upward without changing which state an oscillator is in, as shown in Figure 12.30. The "spring" stiffnesses don't change, but there is more energy stored in the "springs." On the other hand, if there is transfer of energy into the system due to a temperature difference, Q, there is no change in the energy level, but an oscillator will jump to a higher level. This jump from one level to another affects the probability calculations and the entropy, which is the natural log of the number of ways to arrange the energy quanta.

Roughly speaking, transfer of energy Q due to a temperature difference alters which state the system is in and affects the entropy. Work W alters the energy levels without changing which level the system is in, and doesn't affect the entropy. A small amount of transfer of energy into a system due to a temperature difference of amount Q, with $\Delta E_{int} = Q$, leads to an entropy change of that system.

Since $\dfrac{1}{T} = \dfrac{\Delta S}{\Delta E_{int}} = \dfrac{\Delta S}{Q}$, we conclude the following:

ENTROPY CHANGE ASSOCIATED WITH SMALL Q

$$\Delta S = \frac{Q}{T} \quad \text{(small } Q, \text{ associated with nearly constant } T)$$

If there is a great deal of energy transfer due to a temperature difference, causing a large temperature change in a system, we must add up (integrate) the contributions to the entropy change.

> **12.X.8** There was transfer of energy of 5000 J into a system due to a temperature difference, and the entropy increased by 10 J/K. What was the approximate temperature of the system?
>
> **12.X.9** It takes about 335 joules to melt one gram of ice. During the melting, the temperature stays constant. Which has higher entropy, a gram of liquid water at $0°$ C or a gram of ice at $0°$ C? Does this make sense? How large is the entropy difference?

Can the Entropy of an Object Decrease?

QUESTION We analyzed two blocks that initially had a 90-10 energy distribution. In the process of changing to a 60-40 distribution, was it possible for the entropy of *one* of the blocks to decrease significantly? Is this a violation of the second law of thermodynamics?

The entropy of block 1 decreased, but the entropy of block 2 increased even more. We believe that the Universe is a closed system, and that the total entropy of the Universe continually increases and never decreases. However, some portions of the Universe may experience a decrease of entropy, as long as there is at least as much increase elsewhere.

Irreversibility Again

We started with a 90-10 distribution of the energy but found the most probable final distribution to be 60-40. This change is effectively irreversible. The 60-40 distribution is so enormously more probable than the 90-10 arrangement that an observer will essentially *never* see a 90-10 distribution spontaneously recur. Closed macroscopic systems (such as two blocks inside an insulating box) always evolve toward the most probable arrangement, and stay there (or so close that you can hardly tell the difference).

12.6 SPECIFIC HEAT CAPACITY OF A SOLID

In Problem 12.P.71 you calculated the relationship between the temperature of a block and its energy. How can we compare these calculations with experiment, when we don't have a good way to measure the total energy in a solid block? It is, however, possible to measure *changes* in energy and temperature. Suppose that we add a known amount of energy ΔE to a system, and we measure the resulting rise in temperature ΔT. The ratio of the internal energy change to the temperature rise of an object is called the heat capacity of that object. If we express this on a per-gram or per-mole or per-atom basis, it is called "specific heat capacity." We will work with specific heat capacity on a per-atom basis:

$$C = \frac{\Delta E_{atom}}{\Delta T}$$

where

$$\Delta E_{atom} = \frac{\Delta E_{system}}{N_{atoms}}$$

Different materials have different specific heat capacities. For example, at room temperature the specific heat capacity of water (on a per-gram basis) is 4.2 J/K/gram, while the specific heat capacity of iron at room temperature is 0.84 J/K/gram. Specific heat capacity is an important property of a material for two reasons. It has practical consequences in science and engineering because it, along with the thermal conductivity, determines the thermal interactions that the material has with other objects. Also, specific heat capacity is an experimentally measurable quantity that we can compare with theoretical calculations, to test the validity of our models.

How would we measure the heat capacity of a block? One way is to enclose the block inside an insulating box and use an electric heater inside the box to raise the temperature of the block (Figure 12.31). The electric power (in watts) times the amount of time the current runs (in seconds) gives us the input energy ΔE (in joules), which is equal to the internal energy change ΔE_{int}, and we can use a thermometer to measure the temperature rise ΔT of the block. The specific heat capacity (on a per-atom basis) is $\Delta E_{atom}/\Delta T = (\Delta E_{block}/\Delta T)/N_{atoms}$. The heater itself should have a small mass compared to the sample of material whose heat capacity we're measuring, so that the heater isn't a significant part of the material that is being warmed up.

The following exercise illustrates another way to measure specific heat capacity, on a per-gram basis:

12.X.10 A 100 gram block of metal at a temperature of 20° C is placed into an insulated container with 400 grams of water at a temperature of 0° C. The temperature of the metal and water ends up at 2° C. What is the specific heat capacity of this metal, per gram? Start from the Energy Principle. The specific heat capacity of water is 4.2 joules/K/gram.

Experimental Results for Specific Heat Capacity

Measurements of the specific heat capacity of aluminum (circles) and lead (squares) are displayed in Figure 12.32 (the numerical data are given in a table

Figure 12.31 Schematic diagram of an apparatus for measuring the heat capacity of a solid.

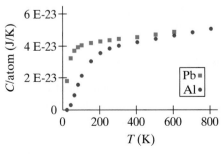

Figure 12.32 Measured heat capacities of lead and aluminum, shown on a per-atom basis.

accompanying Problem 12.P.72). They are plotted as heat capacity (J/K) per atom to facilitate comparison. Note that at low temperatures the specific heat capacity of both substances changes dramatically with temperature. It is not initially obvious why this should be the case, and in fact this temperature dependence was not predicted by classical theory developed in the 1800s. Note also that at room temperature and above, the specific heat capacity (on a per-atom basis) of both materials is about the same.

Will our simple statistical model of a solid be able to predict the temperature dependence of the specific heat capacity of a solid? In Problem 12.P.71 you determined the relationship between energy and temperature for a block (in fact, you did the calculation for two different blocks). This calculation can easily be extended to predict the heat capacity per atom, which is $\Delta E_{atom}/\Delta T$, where E_{atom} is the average energy per atom.

Effective Spring Stiffness

Figure 12.33 In our simple model of a solid, each atom experiences forces from two interatomic "springs." Each of these springs is half as long as an interatomic bond.

In Chapter 4 we determined the interatomic spring stiffness $k_{s,i}$ of aluminum and lead by relating it to the macroscopic stress–strain relationship (Young's modulus). In Problem 12.P.72 you will use the Einstein independent-oscillator model of a solid to predict the specific heat capacity of aluminum and lead. We need to use a value for $k_{s,i}$ in this model, in order to convert the number of quanta to a value for energy in joules. Presumably the spring stiffness $k_{s,i}$ that we use in $\hbar\sqrt{k_{s,i}/m_a}$ should be related to the spring stiffness $k_{s,i}$ determined from Young's modulus. However, we shouldn't expect our predictions to agree exactly with experimental data, because our model ignores the fact that the atoms are not actually isolated from each other.

Moreover, the effective spring stiffness for oscillations in the x, y, or z directions can be expected to be larger than the value of $k_{s,i}$ estimated in Chapter 4 for two reasons.

First, each atom is attached to two "springs" (Figure 12.33). When an atomic core is displaced a distance s_x from its equilibrium position, the spring to the left and the spring to the right each exert a force of $k_{s,i}s_x$, so the combined force is $2k_{s,i}s_x$. This implies that the "effective spring stiffness" for oscillations would be $2k_{s,i}$.

Second, in this simple model we divide the solid into "cubes" surrounding each independent atomic oscillator (Figure 12.33), so each of the two springs to the left and right of an atomic core is effectively half the length of a full interatomic spring, and each half-spring would have a spring stiffness of $2k_{s,i}$.

These considerations suggest that the effective spring stiffness for thermal oscillations in the context of the Einstein model might be a factor of 4 larger than the spring stiffness we determined in Chapter 4 from stretching a wire.

We emphasize that the Einstein model is clearly deficient, since an atom is not actually enclosed in rigid walls. However, having chosen this model we follow through in a consistent way, including using a spring stiffness k_s appropriate to the model. There are more sophisticated models for determining heat capacity that fit experimental data even better than the Einstein model does, but the Einstein model captures the main features.

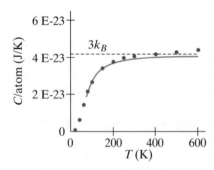

Figure 12.34 Computed specific heat capacity of copper, compared with experimental values.

Specific Heat Capacity as a Function of Temperature

Figure 12.34 displays the result of a calculation of the specific heat capacity of copper on a per-atom basis. The colored dots represent actual experimental data, and the solid line represents the computed values. An outline of the calculation that produced this curve is as follows:

- Use the mass of a copper atom and the effective interatomic spring stiffness for copper, calculated in Chapter 4 and modified by the factor discussed above, to find the value of one quantum of vibrational energy for this system.

■ At this point it is important to do Problem 12.P.72, in which you compute the specific heat capacity of aluminum and lead, and compare with actual experimental data.

■ For a small particle of solid copper containing a given number of atoms, calculate the entropy of the system for increasing values of energy added to the system, in amounts of one quantum (as in Problem 12.P.70).

■ Calculate the change in entropy due to the addition of each quantum of energy, and use the definition of temperature ($1/T \approx \Delta S/\Delta E$) to find the temperature of the particle for each value of energy (as in Problem 12.P.71).

■ Calculate the change in temperature of the particle due to the addition of each quantum of energy, to find the specific heat capacity ($C \approx \Delta E/\Delta T$), and plot values of C (per atom) versus T (in kelvins) as in Problem 12.P.72.

Our simple model of a solid as a collection of independent harmonic oscillators does a surprisingly good job of fitting experimental data over a wide range of temperatures. If we were able to include more oscillators in our calculations (by using double-precision arithmetic to handle larger numbers, or by using mathematical techniques such as Stirling's approximation for factorials), we could extend our predictions to lower temperatures.

The deviation of the experimental data from the prediction at very high temperatures suggests that at these temperatures a simple harmonic oscillator is not a good model of the atoms in this solid. At high quantum levels, the harmonic oscillator potential energy curve is a poor approximation to the actual potential energy curve describing interatomic interactions (see Chapter 7).

Note an interesting aspect of the graph: at room temperatures, the predicted specific heat capacity on a per-atom basis approaches a constant value of $3k_B$ (three times the Boltzmann constant, 1.38×10^{-23} J/K). This value agrees quite well with the measured specific heat capacity of a variety of substances at ordinary temperatures.

Energy Quantization and Specific Heat Capacity

The key difference between our model of solids (the Einstein model) and earlier classical models that did not predict a temperature dependence of specific heat capacity lies in the quantization of energy in atomic oscillators. Statistical mechanics was originally developed in the 1800s, before the beginning of quantum theory. The classical theory did predict that at high temperatures the energy of each atom in a solid would be $3k_BT$ (k_BT per nonquantized one-dimensional oscillator), so that the heat capacity per atom would be $3k_B$.

If you examine your calculations, you'll find that the high-temperature limit (heat capacity per atom = $3k_B$) corresponds to temperatures high enough that k_BT is significantly larger than one quantum of energy. In a situation where the average energy per one-dimensional oscillator is about k_BT and is large compared to one quantum of energy, the quantization of the energy doesn't make much difference in the analysis. That is, if energy quanta are small compared to the energies of interest, mathematical analysis can be carried out adequately by considering energy to be continuous rather than discrete. In such cases pre-quantum and quantum calculations will give the same results, as is the case here at high temperatures.

On the other hand, at low temperatures the average energy per one-dimensional oscillator is comparable to or smaller than one quantum of energy, and the continuous, nonquantum calculations are not valid. The classical theory provided no explanation for the discovery that as materials were cooled down to very low temperatures, the specific heat capacity decreased with decreasing temperature.

In 1907 Einstein carried through the analysis we have just done and predicted the curve we have just plotted. The good agreement with both low-temperature *and* high-temperature measurements of specific heat capacity was strong additional evidence for the hypothesis that the energy of oscillators is indeed quantized. (At extremely low temperatures the model of a solid must be refined, taking into account the electrons in the metal, among other things, to achieve full agreement between theory and experiment.)

The fact that the specific heat capacity for all materials decreases at low temperatures has practical consequences. For example, it makes it difficult to cool a sample to a very low temperature. Cooling a sample depends essentially on putting the sample in contact with a "sink," a large object that is already at a lower temperature, so that there is transfer of energy Q due to the temperature difference out of the sample into the sink, lowering the temperature of the sample and not raising the temperature of the sink very much. At very low temperatures, however, the sink has a low specific heat capacity, so this is difficult to achieve.

> **12.X.11** In an insulated container a 100-watt electric heating element of small mass warms up a 300 gram sample of copper for 6 seconds. The initial temperature of the copper was 20° C (room temperature). Predict the final temperature of the copper, using the $3k_B$ specific heat capacity per atom.
>
> **12.X.12** Since $\Delta T = \Delta E/C$, what will happen at low temperatures to the temperature of the sink when some energy ΔE is transferred to it from the sample? Why is this unfortunate?

Which of Our Results Are General?

We have analyzed simple models of solid matter. Nevertheless, the basic conclusions are quite general. For example, if our two model blocks were made of different materials, so that the energy quanta were of different size in the two blocks, this would complicate the procedures for evaluating the number of ways Ω to arrange the energy, but the basic conclusion would remain, that the entropy will increase to a maximum.

EXAMPLE

A Lead Nanoparticle

In Chapter 4, from Young's modulus for lead we found that the effective interatomic "spring" stiffness was about 5 N/m.

(a) For a nanoparticle consisting of 3 lead atoms, what is the approximate temperature when there are 5 quanta of energy in the nanoparticle?

(b) What is the approximate specific heat capacity (per atom) at this temperature? Compare with the approximate high-temperature specific heat capacity (per atom) for lead.

Solution

We'll use the Einstein model of a solid, though with only three atoms our conclusions will be very approximate.

(a) We model the three atoms as 9 independent quantized oscillators. One quantum of energy in one of these oscillators is this many joules:

$$\hbar\omega_0 = \hbar\sqrt{k_s/m} = (1.05 \times 10^{-34}\,\text{J}\cdot\text{s})\sqrt{\frac{4(5\,\text{N/m})}{(207 \times 1.7 \times 10^{-27}\,\text{kg})}}$$

$$= 7.92 \times 10^{-22}\,\text{J}$$

We need to calculate the number of ways to arrange q quanta in the neighborhood of $q = 5$, and the associated entropy (Figure 12.35).

Since $1/T = \partial S/\partial E$, $T \approx \Delta E/\Delta S$. Take differences from $q = 4$ to $q = 6$, in order to approximate the slope at $q = 5$; energy increase is 2 quanta:

q	Ω	$\ln\Omega$
4	$\dfrac{12!}{4!8!} = 495$	6.20
5	$\dfrac{13!}{5!8!} = 1287$	7.16
6	$\dfrac{14!}{6!8!} = 3003$	8.01

Figure 12.35 Calculation of entropy ($S = k_B \ln \Omega$).

$$T \approx \frac{N\hbar\omega_0}{k_B\Delta(\ln\Omega)} = \frac{2(7.92 \times 10^{-22}\,\text{J})}{(1.38 \times 10^{-23}\,\text{J/K})(8.01 - 6.20)} \approx 63.7\,\text{K}$$

(b) Find T_1 at the midpoint of the 4 to 5 interval, T_2 at the midpoint of the 5 to 6 interval; energy increase from T_1 to T_2 is one quantum:

$$T_1 \approx \frac{(7.92 \times 10^{-22}\,\text{J})}{(1.38 \times 10^{-23}\,\text{J/K})(7.16 - 6.20)} \approx 60.1\,\text{K}$$

$$T_2 \approx \frac{(7.92 \times 10^{-22}\,\text{J})}{(1.38 \times 10^{-23}\,\text{J/K})(8.01 - 7.16)} \approx 67.7\,\text{K}$$

As expected, these two temperatures bracket the temperature of 63.7 K at $q = 5$. We want the specific heat capacity on a per-atom basis, and there are 3 atoms:

$$C_{\text{per atom}} = \frac{1}{3}\frac{\Delta E}{\Delta T} \approx \frac{1}{3}\frac{(7.92 \times 10^{-22}\,\text{J})}{(67.7 - 60.1)\,\text{K}} \approx 3.44 \times 10^{-23}\,\text{J/K}$$

The approximate high-temperature specific heat capacity for a solid is $3k_B$ per atom, which is $3(1.38 \times 10^{-23}\,\text{J/K}) = 4.14 \times 10^{-23}\,\text{J/K}$. This suggests that at a temperature of about 63.7 K, lead is not quite at the high-temperature limit.

The actual experimental value for lead at 63.7 K (interpolation of the data accompanying Problem 12.P.72) is 22.6 J/K/mole, which on a per-atom basis is

$$(22.6\,\text{J/K/mole})\left(\frac{1\,\text{mole}}{6 \times 10^{23}\,\text{atoms}}\right) = 3.8 \times 10^{-23}\,\text{J/K}$$

Our three-atom calculation is not a very accurate model of a macroscopic block of lead, in part because with only three atoms it is not a very good model to say that each atom is connected by springlike bonds to six neighboring atoms. Despite the failings of our simple model, our result is rather close to the actual value.

12.7 THE BOLTZMANN DISTRIBUTION

How does the density of Earth's atmosphere change as altitude increases? How does the distribution of the speeds of molecules in a gas change as the temperature of the gas increases? So far we have mainly been concerned with the thermal equilibrium of two blocks, and how thermal equilibrium arises as that particular distribution of energy between the two blocks that has (by far) the largest number of ways to arrange the quanta. What can we say about the probability of observing a particular amount of energy associated with one particular atom or oscillator? Addressing this question will lead us to the "Boltzmann distribution," which provides insight into the behavior of a very wide variety of physical, chemical, and biological phenomena.

A Constant-Temperature Reservoir

Consider a large system in contact with a small system, as shown in Figure 12.36. The two systems are isolated from their surroundings and share a fixed amount of energy E_{tot}. The large system's energy and temperature cannot change very much, so we call the large system a nearly constant-temperature "reservoir," which keeps the small system always at nearly the same temperature.

The entropy of the total system (reservoir plus small system) will increase rapidly if energy is transferred from the small system to the reservoir, because in the large system that energy can be distributed among a very much larger number of microstates than were available in the small system. Since we expect

A large "reservoir"

Ω_R

Ω_s
A small system

Figure 12.36 A small system in contact with a very large system (a "reservoir").

the total entropy to increase, we expect that most of the time, most of the total energy E_{tot} will be found in the reservoir, and little energy in the small system. We will show how to make this idea quantitative.

Let $\Omega_{res}(E_{res})$ be the number of microstates in the reservoir when the amount of energy in the reservoir is E_{res}. Similarly, let $\Omega(E)$ be the number of microstates in the small system when it has an amount of energy E, where E is much less than E_{res}, because the reservoir is so big. The total energy of the combined system is $E_{tot} = E_{res} + E$, which is a fixed number because the two systems are isolated from their surroundings. The total number of microstates for the combined system corresponding to E_{tot} is $\Omega_{tot}(E_{tot})$, which is also a fixed number.

Probability of a Particular Division of Energy

The number of ways of arranging E_{res} in the reservoir and E in the small system is $\Omega_{res}(E_{res})\Omega(E)$, while the total number of ways of arranging E_{tot} in the combined system is $\Omega_{tot}(E_{tot})$. Therefore the probability $P(E)$ of finding the energy split between the reservoir and the small system so that there is energy E in the small system is this:

$$P(E) = \frac{\Omega_{res}(E_{res})\Omega(E)}{\Omega_{tot}(E_{tot})}$$

The most probable value of the energy E to be found in the small system is zero, because the more energy E that is taken away from the big system and put into the small system, the fewer the microstates in the big system (without a comparable increase in the small system), which would mean a decrease in the total entropy.

We're interested in how fast $P(E)$ decreases as we move more and more of the total energy into the small system. We take logarithms and multiply by k_B to express everything in terms of entropy, then see how the expression varies with E:

$$k_B \ln P(E) = k_B \ln(\Omega_{res}(E_{res})) + k_B \ln(\Omega(E)) - k_B \ln(\Omega_{tot}(E_{tot}))$$

Consider the term $k_B \ln(\Omega_{res}(E_{res}))$, which is the entropy $S_{res}(E_{res})$ of the reservoir when the energy in the reservoir is $E_{res} = E_{tot} - E$ due to some energy E having gone into the small system. In Figure 12.37 we show the familiar calculation of the entropy of the reservoir as a function of the energy in the reservoir. On an expanded scale we show a portion of a plot of the entropy vs. E_{res} near the maximum possible value of E_{res}, which is the total energy E_{tot} of the combined system.

We need to evaluate the entropy of the reservoir for small values of $E = (E_{tot} - E_{res})$. We can obtain this by the following argument. The nearly straight line on the graph has a slope dS_{res}/dE_{res} and goes through $S_{res}(E_{tot})$, where the energy in the reservoir is E_{tot}. Therefore in this region $S_{res}(E_{res})$ can be represented by

$$S_{res}(E_{res}) = S_{res}(E_{tot}) - \frac{dS_{res}}{dE_{res}} E$$

The entropy of the reservoir decreases as more energy E is shifted into the small system.

However, $dS_{res}/dE_{res} = 1/T$, where T is the temperature of the reservoir, and the fact that the slope is nearly constant in this region reflects the fact that removing a small amount of energy from this large object hardly changes its temperature. The large object is a nearly constant-temperature reservoir, which keeps the small object at that temperature.

Making the substitution $dS_{res}/dE_{res} = 1/T$, we have

$$S_{res}(E_{res}) = k_B \ln(\Omega_{res}(E_{res})) = S_{res}(E_{tot}) - \frac{E}{T}$$

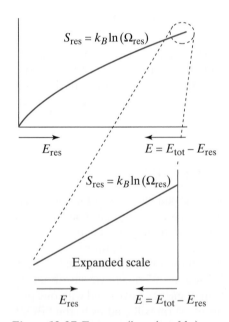

$S_{res} = k_B \ln(\Omega_{res})$

$E_{res} \qquad E = E_{tot} - E_{res}$

$S_{res} = k_B \ln(\Omega_{res})$

Expanded scale

$E_{res} \qquad E = E_{tot} - E_{res}$

Figure 12.37 Entropy (in units of k_B) as a function of E_{res}, for the larger object.

Substitute this into the equation for $k_B \ln P(E)$ and then divide by k_B:

$$\ln P(E) = \frac{S_{res}(E_{tot})}{k_B} - \frac{E}{k_B T} + \ln(\Omega(E)) - \ln(\Omega_{tot}(E_{tot}))$$

Note that the terms that don't involve E are all constants because E_{tot} is constant. Take the exponential of this equation—that is, raise e to both sides of the equation:

$$e^{\ln P(E)} = e^{constant} e^{\ln(\Omega(E))} e^{-E/k_B T}$$

Since e raised to the natural logarithm of some quantity is that quantity, we have the following, where A is some constant:

$$P(E) = A\Omega(E)e^{-E/k_B T}$$

We have calculated the probability of finding a small amount of energy E in a small system that is in contact with a large reservoir. This is called the Boltzmann distribution in honor of the Austrian physicist who developed statistical mechanics in the 19th century:

THE BOLTZMANN DISTRIBUTION

The probability of finding energy E in a small system in contact with a large reservoir is proportional to

$$\Omega(E)e^{-E/k_B T}$$

The exponential part, $e^{-E/k_B T}$, is called the "Boltzmann factor." $\Omega(E)$ is the number of microstates corresponding to energy E.

This result is very general and applies to a very wide variety of phenomena. In many situations the number of microstates $\Omega(E)$ changes much more slowly with energy than does the Boltzmann factor, in which case the qualitative behavior of the system can be determined just from the exponential.

QUESTION What is the most likely value of E, the energy to be found in a microscopic system that is in thermal equilibrium with a large system, if the Boltzmann factor is the important factor? Are you likely to find an energy that is much larger than $k_B T$?

Since the exponent is negative, the exponential factor is largest for $E = 0$, so you're most likely to find the system in its ground state. This agrees with our expectation that taking energy out of the reservoir and putting it in the small system will reduce the number of total microstates and represent a decrease in entropy for the combined system, which is unlikely.

Since e^{-1} is 0.37, the probability of the energy being much larger than $k_B T$ is rather small.

If the temperature is so low that $k_B T$ is small compared to the energy E_1 of the first excited state, you are unlikely ever to find the system in one of its excited states. The system is thermally inert, because the probability of taking in *any* amount of energy is very small. In contrast, if the temperature is high enough that $k_B T > E_1$, then you will sometimes find the system in one of its excited states, although it is still true that the system is most likely to be found in its ground state (Figure 12.38).

The Boltzmann distribution has far-reaching consequences. For example, chemical and biochemical reaction rates typically depend strongly on temperature because with higher temperature the reactants are moving faster and may be found in excited states. Physical reaction rates are also affected. At the very high temperatures found in the interior of our Sun, kinetic energies are

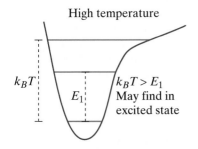

High temperature

$k_B T$

E_1

$k_B T > E_1$
May find in
excited state

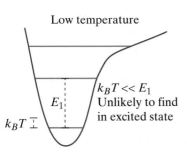

Low temperature

E_1

$k_B T \ll E_1$
Unlikely to find
in excited state

$k_B T$

Figure 12.38 At a low temperature it is unlikely to find a system in an excited state, but at a high temperature the likelihood increases.

high enough to overcome the electric repulsion between nuclei and to allow the nuclei to come in contact, so that they can undergo thermonuclear fusion reactions. A gas becomes a plasma if the temperature is so high that $k_B T$ is comparable to the ionization energy.

> **12.X.13** At room temperature, show that $k_B T \approx \frac{1}{40}$ eV. It is useful to memorize this result, because it tells a lot about what phenomena are likely to occur at room temperature.
>
> **12.X.14** A microscopic oscillator has its first and second excited states 0.05 eV and 0.10 eV above the ground state energy. Calculate the Boltzmann factor for the ground state, first excited state, and second excited state, at room temperature.

Many Observations vs. Many Systems

If a microscopic system (such as a single oscillator) is in contact with a large system, the probability of finding a particular amount of energy in the microscopic system is governed by the Boltzmann distribution.

One can think about this in two ways. The first is to imagine measuring the energy in this particular system repeatedly, over time, and recording the results. The fraction of the results that indicate a particular energy is predicted by the Boltzmann distribution.

A second way to think about this is to imagine assembling a large number of identical systems, each in contact with a large system, and taking one "snapshot" in which the energy of each system is recorded simultaneously. The fraction of systems with a particular energy is predicted by the Boltzmann distribution.

The second approach also gives us a way to predict the distribution of energy in a single large system. For example, consider a system consisting of all the air molecules in a room. Imagine that each molecule is identified by a letter: A, B, C ... One could consider molecule A to be a microscopic system in contact with the large system consisting of all other molecules (B, C, ...). However, one could also consider B to be the microscopic system, in contact with all other molecules (A, C, ...). We can use the Boltzmann distribution to predict the fraction of all molecules in the room that have a particular energy.

These two views of the Boltzmann distribution, one microscopic system observed repeatedly or a large number of microscopic systems observed once, complement each other. Sometimes one view is more helpful, sometimes the other.

12.8   THE BOLTZMANN DISTRIBUTION IN A GAS

The Boltzmann distribution applies to any kind of system—not just a solid. As a major application of the Boltzmann distribution, we will study a gas consisting of molecules that don't interact much with each other. Examples are the so-called "ideal gas" (with no interactions at all), and any real gas at sufficiently low density that the molecules seldom come near each other.

In order to apply the Boltzmann distribution, we need a formula for the energy of a molecule in the gas. We will omit rest energy, and we will also omit nuclear energy and electronic energy from our total, because at ordinary temperatures there is not enough energy available in the surroundings to raise the molecule to an excited nuclear state or an excited electronic state. Also, for simplicity, instead of writing ΔE_{vib} to represent an amount of vibrational energy above the ground vibrational state, we will simply write E_{vib}.

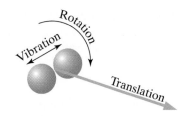

Figure 12.39 Energy of an oxygen molecule.

The energy of a single gas molecule in the gravitational field near the Earth's surface (excluding rest energy, nuclear energy, and electronic energy) is this:

$$K_{\text{trans}} + E_{\text{vib}} + E_{\text{rot}} + Mgy_{\text{CM}}$$

where E_{vib} and E_{rot} are the vibrational and rotational energies relative to the center of mass (if the molecule contains more than one atom, as in Figure 12.39) and y_{CM} is the height of the molecule's center of mass above the Earth's surface. The mass of the molecule is M. For brevity we will speak of "the energy of the gas molecule," but of course the gravitational potential energy really applies to the system of gas molecule plus Earth.

QUESTION   For this expression for the energy to be accurate, why must the gas be an "ideal" gas (or a real gas at low density)?

In a real gas at somewhat high density we may not be able to neglect the potential energy associated with the intermolecular forces, in which case this expression for the energy would not be adequate.

If the temperature T is the same everywhere in the (ideal) gas, the probability that a particular molecule will have a certain amount of energy is proportional to

$$\Omega(E)e^{-(K_{\text{trans}}+E_{\text{vib}}+E_{\text{rot}}+Mgy_{\text{CM}})/k_BT}$$

We mentally divide the gas into two systems: one particular molecule of interest and all the rest of the molecules. These two systems are in thermal equilibrium with each other, because the gas molecules are continually colliding with each other and can share energy. The energy for the one particular molecule is then expected to follow the Boltzmann distribution.

To avoid excess subscripts, in the following discussion we will simply write v for the center-of-mass speed v_{CM} and y for the center-of-mass height y_{CM}.

Separating the Various Factors

It is useful to group the terms of the Boltzmann factor like this:

$$\left[e^{-K_{\text{trans}}/k_BT}\right]\left[e^{-Mgy/k_BT}\right]\left[e^{-E_{\text{vib}}/k_BT}\right]\left[e^{-E_{\text{rot}}/k_BT}\right]$$

The first bracket is associated with the distribution of velocities, the second with the distribution of positions, the third with the distribution of vibrational energy, and the fourth with the distribution of rotational energy. We will discuss each of these individually.

Height Distribution in a Gas

A striking property of Earth's atmosphere is that in high mountains the air density and pressure are significantly lower than at sea level. The air density is so low on top of Mount Everest that most climbers carry oxygen tanks. Can we explain this? In order to get at the main issue, the variation of density with height, we make the rough approximation that the temperature is constant—the same in the mountains as at sea level.

QUESTION   How bad an approximation is it to consider the temperature to be the same at all altitudes?

Even in high mountains the temperature is typically above $-29°C$ $(-20°F)$, which is 244 K, and this is only 17% lower than room temperature of 293 K (20°C or 68°F). So maybe this approximation isn't too bad.

We speak of the probability that the x component of the gas molecule's position lies between some x and $x + dx$, and similarly for y and z. Here dx is considered to be a very short distance, small compared to the size of the container but large compared to the size of a molecule.

Figure 12.40 Number density vs. height in a constant-temperature atmosphere.

Focus just on that part of the Boltzmann distribution that deals with position, where the distribution is proportional to the probability of finding one particular molecule between x and $x + dx$, y and $y + dy$, and z and $z + dz$:

$$e^{-Mgy/k_BT}\,dx\,dy\,dz$$

Evidently there's nothing very interesting about x and z (directions parallel to the ground). However, in the vertical direction there is an exponential fall-off with increasing height for the probability of finding a particular molecule at height y (Figure 12.40).

Looked at another way, our exponential formula tells us how the number density of the atmosphere depends on height, because in telling us about the behavior of one representative molecule, the formula also tells us something about all the molecules.

We shifted gears from thinking about one molecule to thinking about many. Think again about one single air molecule. Suppose that you place it on a table, which is at room temperature. It is not a very unusual event for a thermally agitated atom in the table to hit our air molecule hard enough to send it kilometers high into the air! Actually, our particular molecule will very soon run into another air molecule and not make it very high in one great leap, but on average we do find lots of air molecules very high above sea level rather than finding them all lying on the ground.

Notice again that k_BT is the important factor in understanding the statistical behavior of matter. Here it sets the scale for the variation with height of the atmosphere's number density.

At a height where the gravitational potential energy $U_g = mgy$ is equal to k_BT, the number density has dropped by a factor of $e^{-1} = 1/e = 0.37$. Dry air at sea level is 78% nitrogen (one mole N_2 = 28 grams), 21% oxygen (one mole O_2 = 32 grams), about 1% argon, and 0.03% CO_2. An average mass of 29 grams per mole is good enough for most calculational purposes.

> **12.X.15** Suppose that you put one air molecule on your desk, so it is in thermal equilibrium with the desk at room temperature. Suppose that there is no atmosphere to get in the way of this one molecule bouncing up and down on the desk. Calculate the typical height that the air molecule will be above your desk, so that $Mgy \approx k_BT$.
>
> **12.X.16** Approximately what fraction of the sea-level air density is found at the top of Mount Everest, a height of 8848 meters above sea level?

Distribution of Velocities in a Gas

Next we look at the distribution of molecular speeds. When the gas is confined inside a finite container the momentum (or velocity) and height are quantized, but under almost all conditions the size of the energy quantum associated with momentum (or velocity) and position is so small compared to k_BT that it is appropriate to take a nonquantum approach for these variables.

We speak of the probability that the gas molecule has an x component of velocity within the range between some v_x and $v_x + dv_x$, and similarly for v_y and v_z. Here dv_x is considered to be a very small amount, small compared to the average speed of the molecules.

Since we're explicitly interested in v_{CM}, it will be useful to express translational kinetic energy as $\frac{1}{2}Mv_{CM}^2$ rather than $p^2/(2M)$ in this discussion. Since $v^2 = v_x^2 + v_y^2 + v_z^2$, we can write the formula for the distribution of velocity in a gas as follows (remember that we are simply writing v for the center-of-mass speed v_{CM}):

$$\left[e^{-\frac{1}{2}Mv_x^2/k_BT}\,dv_x\right]\left[e^{-\frac{1}{2}Mv_y^2/k_BT}\,dv_y\right]\left[e^{-\frac{1}{2}Mv_z^2/k_BT}\,dv_z\right]$$

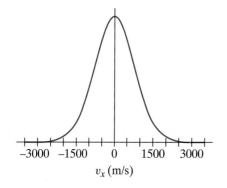

Figure 12.41 Distribution of the x component of velocity for helium at room temperature.

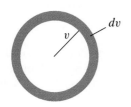

Figure 12.42 The volume of a shell with radius v and thickness dv is $4\pi v^2 dv$.

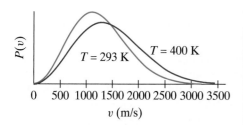

Figure 12.43 Distribution of speeds of helium atoms at two temperatures.

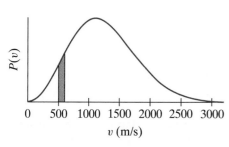

Figure 12.44 Fraction of helium atoms at 293 K with speeds between 500 and 600 m/s.

The distribution for each velocity component is a bell-shaped curve, called a "Gaussian." In Figure 12.41 we show the distribution of the x component of velocity for helium at room temperature.

> **QUESTION** Judging from this graph, what is the average value for v_x? What are the average values of v_y and v_z? Why are these results plausible?

Evidently the average value for the x component of the velocity is zero. This is reassuring, because a molecule is as likely to be headed to the right as it is to be headed to the left. Similarly, the average values for the y and z components of velocity are zero.

It would not be particularly surprising to find a gas molecule with an x component of velocity such that $\frac{1}{2}Mv_x^2 \approx k_B T$. On the other hand, it would be very surprising to find a gas molecule with a value of v_x many times larger. As usual, $k_B T$ sets the scale for thermal phenomena.

Distribution of Speeds in a Gas

By converting from rectangular coordinates to "spherical" coordinates, the velocity-component distribution can be converted into a speed distribution. We won't go into the details, but the main idea is that the volume element in rectangular coordinates in "velocity space" $dv_x dv_y dv_z$ turns into the volume of a spherical shell with surface area $4\pi v^2$, and thickness dv, as shown in Figure 12.42.

Transferring to spherical coordinates, we have the following:

$$e^{-\frac{1}{2}Mv^2/k_B T}dv_x dv_y dv_z = e^{-\frac{1}{2}Mv^2/k_B T}4\pi v^2 dv$$

The only thing missing is a "normalization" factor in front to make the integral over all speeds from 0 to infinity be equal to 1.0 (since our one molecule must have a speed somewhere in that range). Here is the "Maxwell–Boltzmann" distribution for a low-density gas:

MAXWELL–BOLTZMANN SPEED DISTRIBUTION (LOW-DENSITY GAS)

$$P(v) = 4\pi \left(\frac{M}{2\pi k_B T}\right)^{3/2} v^2 e^{-\frac{1}{2}Mv^2/k_B T}$$

The probability that a molecule of a gas has a center-of-mass speed within the range v to $v + dv$ is given by $P(v)dv$.

The distribution function for helium is shown for two different temperatures in Figure 12.43. You can see that the average speed of a helium atom is predicted to be about 1200 m/s at room temperature (20° Celsius, which is 293 Kelvin). The prediction was first made by Maxwell, in the mid-1800s, long before the development of quantum mechanics. In retrospect, this worked because the quantum levels of the translational kinetic energy are so close together as to be almost a continuum, so the classical model is a good approximation.

The average number of helium atoms in a container that have speeds in the range of 415.4 m/s to 415.7 m/s can be calculated by evaluating Maxwell's formula with $v = 415.4$ m/s and multiplying by $dv = 0.3$ m/s, which gives the probability of finding one molecule in this speed range. If there are N atoms in the container, multiplying by N gives the total number of molecules likely to be in that speed range at any given instant. The area of the vertical slice shown in Figure 12.44 represents the fraction of helium atoms that are likely to have

Figure 12.45 Apparatus for measuring the distribution of speeds of gas molecules.

Figure 12.46 Distribution showing fraction of atoms with speeds in a given range.

speeds between 500 m/s and 600 m/s. At higher temperatures the distribution shifts to higher speeds (Figure 12.43).

Measuring the Distribution of Speeds in a Gas

The actual distribution of speeds of molecules in any gas can be measured by an ingenious experiment. Make a tiny hole in a container of gas and let the molecules leak out into a vacuum (that is, a region from which the air is continually pumped out). The moving molecules run through collimating holes and then through a slot in a rapidly rotating drum and strike a row of devices at the other side of the drum that can detect gas molecules (Figure 12.45). Such measurements confirm the Maxwell prediction.

Fast molecules strike soonest, followed by medium-speed molecules, and finally slow molecules, so molecules with different speeds strike at different locations along the rotating drum. The number of molecules striking various locations along the drum is a direct indication of the distribution of speeds of molecules in the gas. (A correction has to be applied to these data to obtain the speed distribution inside the gas, because high-speed molecules emerge through the hole more frequently than low-speed molecules do, even if their numbers are the same in the interior of the gas. Also, the apparatus measures the distribution of v_x rather than v.)

How might we display the results of this experiment? We could present the data as a graph (Figure 12.46) of the fraction $\Delta N/N$ of the N helium atoms that have speeds between 0 and 500 m/s, between 500 and 1000 m/s, between 1000 and 1500 m/s, and so on.

Average Translational Kinetic Energy in a Gas

For a gas confined inside a container, quantum mechanics predicts that the kinetic energy $\frac{1}{2}Mv^2$ is quantized, but if the container is of ordinary macroscopic size, the energy quanta are extremely small compared to the average kinetic energy, even at very low temperatures. Most gases liquefy before the temperature drops so low as to invalidate the pre-quantum analysis for the velocity distribution.

Whenever the energy has a term containing a square of a position or momentum component, such as x^2 or p_x^2, pre-quantum theory predicts an associated average energy of $\frac{1}{2}k_B T$. This result is valid in quantum mechanics for high temperatures, where $k_B T$ is large compared to the quantum energy spacing. This result follows from using an integral to take the average value of a quadratic term when the distribution is a Gaussian involving that quadratic term. That is, using integral tables (or integrating by parts), you can show that the average value of a quadratic term is $\frac{1}{2}k_B T$ (a bar over a quantity means "average value."):

$$\overline{w^2} = \frac{\int_0^\infty w^2 e^{-w^2/k_B T}dw}{\int_0^\infty e^{-w^2/k_B T}dw} = \frac{1}{2}k_B T$$

The average value is obtained by weighting the values of w^2 by the probability of finding that value; the denominator takes care of normalizing the distribution properly.

As an example, when we did quantum-based calculations for the Einstein solid, we found that at high temperatures the specific heat capacity on a per-atom basis was $3k_B$, which implies an average energy of $3k_B T$. We modeled an atomic oscillator as three one-dimensional oscillators, each having an average energy of $\frac{1}{2}k_B T$ corresponding to kinetic energy $p^2/(2M)$ and another $\frac{1}{2}k_B T$

for the potential energy $\frac{1}{2}k_s x^2$. This is a total of six quadratic terms, implying an average energy per atom of $6(\frac{1}{2}k_B T)$, or $3k_B T$. We therefore expect $3k_B T$ the heat capacity per atom in a solid at high temperature to be $3k_B$, which is indeed what is observed.

HIGH-TEMPERATURE AVERAGE ENERGY

If $k_B T \gg$ the energy quantum, the average energy

associated with a quadratic energy term is $\dfrac{1}{2}k_B T$.

The number of quadratic terms in the expression for the energy is often called the "degrees of freedom."

QUESTION   What is the average translational kinetic energy $\frac{1}{2}M\overline{v^2}$ in terms of $k_B T$? What is the associated contribution to the specific heat capacity of the gas?

Since there are three quadratic terms in the translational kinetic energy, $K_{trans} = \frac{1}{2}mv_x^2 + \frac{1}{2}mv_y^2 + \frac{1}{2}mv_z^2$, the average value is three times $\frac{1}{2}k_B T$. We have the following important result:

AVERAGE K_{trans} FOR AN IDEAL GAS

$$\overline{K}_{trans} = \frac{1}{2}M\overline{v^2} = \frac{3}{2}k_B T$$

The contribution to the specific heat capacity for a gas is $\frac{3}{2}k_B$.

Root-Mean-Square Speed

The square root of $\overline{v^2}$ is called the "root-mean-square" or "rms" speed:

$$v_{rms} \equiv \sqrt{\overline{v^2}}$$

With this definition of v_{rms}, we write $\frac{1}{2}m\overline{v^2} = \frac{1}{2}mv_{rms}^2 = \frac{3}{2}k_B T$.

> **12.X.17** Calculate v_{rms} for a helium atom in the room you're in (whose temperature is probably about 293 K).
>
> **12.X.18** Calculate v_{rms} for a nitrogen molecule (N_2; molecular mass 28) in the room you're in (whose temperature is probably about 293 K). Air is about 80% nitrogen.

Average Speed vs. RMS Speed

The way you find the average speed of molecules in a gas is to weight the speed by the number of molecules that have that speed, and divide by the total number of molecules:

$$\overline{v} = \frac{N_1 v_1 + N_2 v_2 + \text{etc.}}{N} = \frac{N_1}{N}v_1 + \frac{N_2}{N}v_2 + \text{etc.}$$

For a continuous distribution of speeds this turns into an integral, where the weighting factors are given by the Maxwell speed distribution, which gives the probability that a molecule has a center-of-mass speed in the range v to $v+dv$. Using integral tables one finds this:

$$\overline{v} = \int_0^\infty 4\pi \left(\frac{m}{2\pi k_B T}\right)^{3/2} e^{-\frac{1}{2}mv^2/k_B T}(v)v^2 dv = \sqrt{\frac{8}{3\pi}}\sqrt{\frac{3k_B T}{m}} = 0.92 v_{rms}$$

This shows that the average speed is smaller than the rms speed ($0.92v_{rms}$). This same calculational scheme can be used to determine any average. For example, if you calculate the average value of v^2 by this method, you find $\overline{v^2} = 3k_BT/m$, as expected.

Earlier we commented that the average speed $\overline{v}$ of helium atoms at room temperature is about 1200 m/s. In Exercise 12.X.17 you found that v_{rms} at room temperature is 1350 m/s. This is an example of the fact that the rms speed is higher than the average speed.

> **12.X.19** The rms speed is somewhat higher than the average speed due to the averaging of squared speeds. Calculate the average of the numbers 1, 2, 3, and 4, then calculate the rms average (the square root of the average of their squares), and show that the rms average is larger than the simple average. Squaring gives extra weight to larger contributions.

■ There's almost no helium in our atmosphere. Where do we obtain helium for party balloons and low-temperature refrigeration and scientific experiments? It is extracted from natural gas when the gas is pumped out of the ground. Heavy radioactive elements such as uranium in the Earth's crust emit alpha particles (helium nuclei), which capture electrons and become helium atoms. These atoms are trapped with natural gas in underground cavities. When the helium-bearing natural gas is pumped to the surface, the helium is extracted (at some cost).

Application: Retaining a Gas in the Atmosphere

Since $v_{rms} = \sqrt{3k_B(T/m)}$, helium atoms typically travel much faster than nitrogen molecules in our atmosphere, due to the small mass of the helium atoms. Some few helium atoms will be going much faster than the average and may attain a high enough speed to escape from the Earth entirely (escape velocity from the Earth is about 1.1×10^4 m/s). This leads to a continuous leakage of high-speed helium atoms and other low-mass species such as hydrogen molecules (Figure 12.47). Other processes may also contribute to the flow of helium away from the Earth.

| QUESTION | Why doesn't the Moon have any atmosphere at all?

The Moon's gravitational field is so weak that escape speed is quite small, and all common gases can escape, not just hydrogen and helium.

> **12.X.20** Calculate the escape speed from the Moon and compare with typical speeds of gas molecules. The mass of the Moon is 7×10^{22} kg, and its radius is 1.75×10^6 m.

Figure 12.47 Low-mass atoms or molecules in our atmosphere may have speeds high enough to escape from the Earth entirely.

Application: Speed of Sound

Sound waves in a gas consist of propagation of variations in density (Figure 12.48), and the fundamental mechanism for this kind of wave propagation involves collisions between neighboring molecules, whose speeds are proportional to v_{rms} (and roughly comparable to the speed of sound). For example, compare the v_{rms} for nitrogen, 510 m/s, which you calculated in Exercise 12.X.18 with the speed of sound in air (which is mostly nitrogen) at 293 K, which is measured to be 344 m/s. (You may know the approximate rule that a 1-second delay between lightning and thunder indicates a distance of about 1000 feet, which is about 300 m.)

Figure 12.48 Periodic variations in gas density make a sound wave, which travels through a gas with speed v (the speed of sound).

12.X.21 Should the speed of sound in air increase or decrease with increasing temperature? What percentage change would result from doubling the absolute temperature? (This effect is readily observed by measuring the speed of sound in a gas as a function of absolute temperature. The excellent agreement between theory and experiment provides additional evidence for our understanding of gases.)

Vibrational Energy in a Diatomic Gas Molecule

We have treated the distribution of velocity and position. Next we discuss the distribution of vibrational energy, with the Boltzmann factor

$$e^{-E_{vib}/k_B T}$$

For a monatomic gas such as helium, there is no vibrational energy term. However, for a diatomic molecule such as N_2 or HCl, the vibrational energy is

$$E_{vib} = \frac{p_1^2}{2m_1} + \frac{p_2^2}{2m_2} + \frac{1}{2}k_s s^2$$

where k_s is the effective spring stiffness corresponding to the interatomic electric force (not to be confused with k_B, the Boltzmann constant), and s is the stretch of the interatomic bond. This formula for the vibrational energy is essentially the same as the formula for a one-dimensional spring–mass oscillator, which we have studied in detail (Figure 12.49).

Since $p_1 = p_2$ for momenta p_1 and p_2 relative to the center of mass, we can write

$$\frac{p_2^2}{2m_2} = \frac{p_1^2}{2m_2} = \left(\frac{m_1}{m_2}\right)\frac{p_1^2}{2m_1}$$

$$\frac{p_1^2}{2m_2} + \left(\frac{m_1}{m_2}\right)\frac{p_1^2}{2m_1} + \frac{1}{2}k_s s^2 = \left(1 + \frac{m_1}{m_2}\right)\frac{p_1^2}{2m_1} + \frac{1}{2}k_s s^2$$

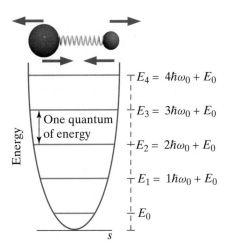

$$
\begin{aligned}
E_4 &= 4\hbar\omega_0 + E_0\\
E_3 &= 3\hbar\omega_0 + E_0\\
E_2 &= 2\hbar\omega_0 + E_0\\
E_1 &= 1\hbar\omega_0 + E_0\\
&\quad E_0
\end{aligned}
$$

One quantum of energy

Figure 12.49 The quantized vibrational energy levels of a diatomic oscillator.

This equation has a form exactly like that for a spring–mass oscillator, with a different mass.

Earlier we modeled a solid as a large number of isolated three-dimensional atomic oscillators (each corresponding to three one-dimensional oscillators, because there are springlike interatomic forces on an atom from neighboring atoms in all directions). This is an overly simplified model of a solid, because in a solid the atoms interact with each other. For example, if an atom moves to the left this affects the atoms to the right and to the left.

In a low-density gas, however, the vibrational oscillators really are nearly independent of each other, because the gas molecules aren't even in contact with each other except when they happen to collide. So the analysis we carried out for the Einstein model of a solid applies even better to the vibrational portion of the energy in a real gas than it does to a real solid.

Among the results that apply immediately are that the specific heat capacity associated with the vibrational energy of one oscillator is k_B at high temperatures. The specific heat capacity decreases at very low temperatures, where $k_B T$ is small compared to the energy spacing of the quantized oscillator energies. Just as it was a surprise when the specific heat capacity of metals was found to decrease at low temperatures, there was a similar surprise in the measurements of the specific heat capacity of diatomic gases at low temperatures, because the contribution of the vibrational motion vanished.

QUESTION There are two quadratic terms in the vibrational energy (kinetic and spring). Therefore, at high temperatures what is the average vibrational energy in terms of $k_B T$? What is the associated contribution to the specific heat capacity of the gas?

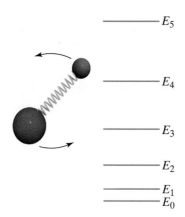

E_5

E_4

E_3

E_2

E_1
E_0

Figure 12.50 Rotational kinetic energy levels of a diatomic molecule.

Figure 12.51 A diatomic molecule can rotate around the x or the y axis. We say there are two rotational degrees of freedom.

At high temperatures the vibrational motion of a diatomic molecule (a one-dimensional oscillator) is expected to have an average energy that is approximately $2(\frac{1}{2}k_BT) = k_BT$, and contributes k_B to the specific heat capacity.

Rotational Energy in a Diatomic Gas Molecule

Finally we consider the rotational-energy portion of the distribution, for which the Boltzmann factor is

$$e^{-E_{\text{rot}}/k_BT}$$

For a monatomic gas there is no rotational term, but for a diatomic gas there is rotational kinetic energy associated with rotation of the "dumbbell" consisting of the two nuclei. The rotational kinetic energy is this (Figure 12.50):

$$E_{\text{rot}} = \frac{1}{2}I\omega_x^2 + \frac{1}{2}I\omega_y^2 = \frac{L_{\text{rot},x}^2}{2I} + \frac{L_{\text{rot},y}^2}{2I}$$

This corresponds to rotational angular momenta $L_{\text{rot},x}$ and $L_{\text{rot},y}$ about the x axis and about the y axis. Note that only the nuclei contribute significantly to the rotational kinetic energy, because the electrons have much less mass.

Rotation about the z axis connecting the two nuclei is irrelevant, for a somewhat subtle reason. The angular momentum is quantized, which leads to energy quantization. For rotations around the z axis (Figure 12.51) the energy is $L_{\text{rot},z}^2/(2I)$, where $L_{\text{rot},z}$ is the z component of the rotational angular momentum. Since I about the z axis for the tiny nuclei is extremely small compared with the moment of inertia about the x and y axes of the diatomic molecule, the rotational energy of the associated first excited state is enormous, and this state is not excited at ordinary temperatures.

Since there are two quadratic energy terms associated with rotation, we conclude that at high temperatures the rotational motion of a diatomic molecule has an average energy that is approximately $2(\frac{1}{2}k_BT) = k_BT$, and contributes k_B to the specific heat capacity.

The spacing between quantized rotational energies for a diatomic molecule is even smaller than the vibrational energy quantum. As a result, the gas must be cooled to a very low temperature before the pre-quantum results become invalid. Many gases liquefy before this low-temperature regime is reached.

Specific Heat Capacity of a Gas

We are now in a position to discuss the specific heat capacity of a gas as a function of temperature. This property is important in calculating the thermal interactions of a gas. Historically, measurements of the specific heat capacity of gases were also important in testing theories of statistical mechanics. We'll concentrate on calculating C_V, the specific heat capacity at constant volume (meaning no mechanical work is done on the gas). The associated experiment would be to add a known amount of energy to a gas in a rigid container and measure the temperature rise of the gas.

Consider the average energy of a diatomic molecule such as N_2 or HCl, consisting of the translational kinetic energy associated with the motion of the center of mass, plus the vibrational energy, plus the rotational energy, plus the gravitational energy of the molecule and the Earth:

$$\left[\frac{1}{2}m\overline{v_x^2} + \frac{1}{2}m\overline{v_y^2} + \frac{1}{2}m\overline{v_z^2}\right] + \left[\left(1 + \frac{m_1}{m_2}\right)\frac{\overline{p_1^2}}{2m_1} + \frac{1}{2}k_s\overline{s^2}\right] + \left[\frac{\overline{L_{\text{rot},x}^2}}{2I} + \frac{\overline{L_{\text{rot},y}^2}}{2I}\right] + Mg\overline{y}_{\text{CM}}$$

QUESTION   At high temperature, what should the specific heat capacity at constant volume be?

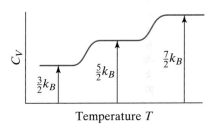

Figure 12.52 Electronic, vibrational, and rotational energy levels for a diatomic molecule.

There are seven quadratic terms in the expression for the energy, so if the temperature is very high, we expect an average energy of $7(\frac{1}{2}k_B T)$ and a specific heat capacity at constant volume $C_V = \frac{7}{2}k_B$.

The discrete energy levels for a diatomic molecule include electronic, vibrational, and rotational energy levels (Figure 12.52). The electronic energy levels correspond to particular configurations of the electron clouds, and the spacing between these levels is typically 1 eV or more. Since at room temperature $k_B T$ is about $\frac{1}{40}$ eV , the electronic levels do not contribute to the specific heat capacity at ordinary temperatures. The vibrational energy spacing is much smaller, and the rotational energy spacing is smaller still. A diatomic gas at high temperature has a bandlike spectrum, since transitions between electronic levels may be accompanied by one or more vibrational or rotational quanta.

What About *Mgy*?

What about the Mgy_{cm} term in the energy? Remember that we are trying to predict the specific heat capacity of a gas at constant volume (so that no mechanical work is done on the system). We could measure C_V by having a known amount of energy transfer Q due to a temperature difference into a closed container of gas and observing the temperature rise of the gas. During this process there is an extremely small increase in y_{CM} of the gas in the container due to the temperature dependence of the height distribution, so the associated Mgy contribution to the specific heat capacity is negligible.

Specific Heat Capacity vs. Temperature

Figure 12.53 is a graph of the specific heat capacity (at constant volume) of a diatomic gas as a function of temperature. In the following exercises, see whether you can explain these values of the specific heat capacity in terms of our analyses of the various contributions to the energy of a diatomic gas molecule.

Remember that $k_B T$ is about $\frac{1}{40}$ eV at room temperature.

Figure 12.53 Specific heat capacity C_V of a diatomic gas vs. temperature.

12.X.22 What is the specific heat capacity on a per-molecule basis (at constant volume) of a monatomic gas such as helium or neon? Why doesn't the specific heat capacity depend on temperature?

12.X.23 Next consider a diatomic molecule.

(a) At high temperatures, what is the specific heat capacity (at constant volume) on a per-molecule basis of a diatomic gas such as oxygen or nitrogen?

(b) Suppose that we lower the temperature of a diatomic gas to a point where $k_B T$ is small compared to the energy of the first excited vibrational state but still large compared to the energy of the first excited rotational state. Now what is the specific heat capacity (at constant volume)?

(c) Lower the temperature even more, so that $k_B T$ is small compared to the energy of the first excited rotational state (some gases liquefy before this low temperature is reached). What is the specific heat capacity (at constant volume) at this low temperature?

(d) As the temperature is decreased, the specific heat capacity (at constant volume) for H_2 eventually decreases to $\frac{3}{2}k_B$, before the gas liquefies. Does this transition to $C_V = \frac{3}{2}k_B$ happen at a higher or lower temperature for D_2 (deuterium, whose nuclei each contain a proton plus a neutron)? Why? (*Hint:* Remember that it is the rotational angular momentum that is quantized.)

SUMMARY

The fundamental assumption of statistical mechanics is that, over time, an isolated system in a given macrostate is equally likely to be found in any of its possible microstates.

Two blocks in thermal contact evolve to that division of energy that has associated with it the largest number of ways $\Omega = \Omega_1\Omega_2$ of arranging the total energy among the atoms (largest number of microstates for a macrostate of given energy).

Second law of thermodynamics: The entropy of a closed system never decreases. Only in a reversible process does the entropy of a closed system stay constant.

Einstein solid: Each atom is modeled as three one-dimensional quantized oscillators.

Number of ways to arrange q quanta of energy among N one-dimensional oscillators

$$\Omega = \frac{(q + N - 1)!}{q!(N - 1)!} \text{(number of microstates)}$$

Entropy: $S \equiv k_B \ln \Omega$, where $k_B = 1.38 \times 10^{-23}$ J/K

Temperature: $\dfrac{1}{T} = \dfrac{\partial S}{\partial E}$

A small flow of energy Q into a system raises entropy by $\Delta S = \dfrac{Q}{T}$.

Specific heat capacity per atom: $C = \dfrac{\Delta E_{atom}}{\Delta T}$, where

$$\Delta E_{atom} = \frac{\Delta E_{system}}{N_{atoms}}$$

The probability of finding energy E in a small system in contact with a large reservoir is proportional to

$$\Omega(E)e^{-E/k_B T}$$

$\Omega(E)$ is the number of microstates corresponding to energy E.

For each energy term involving a quadratic term such as x^2 or p_x^2 ("degree of freedom"), if the average energy is large compared to $k_B T$ the average energy is $\frac{1}{2}k_B T$ and the contribution to the specific heat capacity is $\frac{1}{2}k_B$. This contribution decreases at low temperatures. At high temperatures, specific heat capacity on a per-atom basis in a solid $\approx 3k_B$.

Speed distribution for a gas:

$$P(v) = 4\pi \left(\frac{M}{2\pi k_B T}\right)^{3/2} e^{-\frac{1}{2}mv^2/k_B T} v^2$$

EXERCISES AND PROBLEMS

Sections 12.1–12.2

12.X.24 List explicitly all the ways to arrange 2 quanta among 4 one-dimensional oscillators.

12.X.25 How many different ways are there to get 5 heads in 10 throws of a true coin? How many different ways are there to get no heads in 10 throws of a true coin?

12.X.26 How many different ways are there to arrange 4 quanta among 3 atoms in a solid?

12.X.27 A carbon nanoparticle (very small particle) contains 6000 carbon atoms. According to the Einstein model of a solid, how many oscillators are in this block?

12.X.28 In order to calculate the number of ways of arranging a given amount of energy in a tiny block of copper, the block is modeled as containing 8.7×10^5 independent oscillators. How many atoms are in the copper block?

12.X.29 In Chapter 4 you determined the stiffness of the interatomic "spring" (chemical bond) between atoms in a block of lead to be 5 N/m, based on the value of Young's modulus for lead. Since in our model each atom is connected to two springs, each half the length of the interatomic bond, the effective "interatomic spring stiffness" for an oscillator is 4×5 N/m = 20 N/m. The mass of one mole of lead is 207 grams (0.207 kilograms). What is the energy, in joules, of one quantum of energy for an atomic oscillator in a block of lead?

12.X.30 Consider an object containing 6 one-dimensional os-cillators (this object could represent a model of 2 atoms in an Einstein solid). There are 4 quanta of vibrational energy in the object.

(a) How many microstates are there, all with the same energy?

(b) If you examined a collection of 48,000 objects of this kind, each containing 4 quanta of energy, about how many of these objects would you expect to find in the microstate 000004?

12.P.31 The reasoning developed for counting microstates applies to many other situations involving probability. For example, if you flip a coin 5 times, how many different sequences of 3 heads and 2 tails are possible? Answer: 10 different sequences, such as HTHHT or TTHHH. In contrast, how many different sequences of 5 heads and 0 tails are possible? Obviously only one, HHHHH, and our formula gives 5!/[5!0!] = 1, using the standard definition that 0! is defined to equal 1.

If the coin is equally likely on a single throw to come up heads or tails, any specific sequence like HTHHT or HHHHH is equally likely. However, there is only one way to get HHHHH, while there are 10 ways to get 3 heads and 2 tails, so this is 10 times more probable than getting all heads.

Use the formula 5!/[N!(5 − N)!] to calculate the number of ways to get 0 heads, 1 head, 2 heads, 3 heads, 4 heads, or 5 heads in a sequence of 5 coin tosses. Make a graph of the number of ways vs. the number of heads.

Sections 12.3–12.6

12.X.32 Energy conservation for two blocks in contact with each other is satisfied if all the energy is in one block and none in the other. Would you expect to observe this distribution in practice? Why or why not?

12.X.33 What is the advantage of plotting the (natural) logarithm of the number of ways of arranging the energy among the many atoms (natural logarithm of the number of microstates)?

12.X.34 Which has a higher temperature: a system whose entropy changes rapidly with increasing energy or one whose entropy changes little with increasing energy?

12.X.35 Explain why it is a disadvantage for some purposes that the specific heat capacity of all materials decreases at low temperatures.

12.X.36 Two blocks with different temperatures had entropies of 10 J/K and 35 J/K before they were brought in contact. What can you say about the entropy of the combined system after the two came in contact with each other?

12.X.37 Consider two blocks of copper. Block A contains 600 atoms and initially has a total of 20 quanta of energy. Block B contains 400 atoms and initially has 80 quanta of energy. The two blocks are placed in contact with each other, inside an insulated container (so no thermal energy can be exchanged with the surroundings). After waiting for a long time (for example, an hour), which of the following would you expect to be true?

(A) Approximately 60 quanta of energy are in block A, and approximately 40 quanta of energy are in block B.

(B) Approximately 50 quanta of energy are in block A, and approximately 50 quanta of energy are in block B.

(C) The entropy of block A is equal to the entropy of block B.

(D) The temperature of block A and the temperature of block B are equal.

12.X.38 Object A and object B are two identical microscopic objects. Figure 12.54 below shows the number of ways to arrange energy in one of these objects, as a function of the amount of energy in the object.

E, J	0.8 E–20	1.0 E–20	1.2 E–20	1.4 E–20
# ways	37	60	90	122

Figure 12.54

(a) When there are 1.0×10^{-20} joules of energy in object A, what is the entropy of this object?

(b) When there are 1.4×10^{-20} joules of energy in object B, what is the entropy of this object?

(c) Now the two objects are placed in contact with each other. At this moment, before there is time for any energy flow between the objects, what is the entropy of the combined system of objects A and B?

12.P.39 The interatomic spring stiffness for tungsten is determined from Young's modulus measurements to be 90 N/m. The mass of one mole of tungsten is 0.185 kg. If we model a block of tungsten as a collection of atomic "oscillators" (masses on springs), note that since each oscillator is attached to two "springs," and each "spring" is half the length of the interatomic bond, the effective interatomic spring stiffness for one of these oscillators is 4 times the calculated value given above.

Use these precise values for the constants: $\hbar = 1.0546 \times 10^{-34}$ J·s (Planck's constant divided by 2π), Avogadro's number = 6.0221×10^{23} molecules/mole, $k_B = 1.3807 \times 10^{-23}$ J/K (the Boltzmann constant).

(a) What is one quantum of energy for one of these atomic oscillators?

(b) Figure 12.55 contains the number of ways to arrange a given number of quanta of energy in a particular block of tungsten. Fill in the blanks to complete the table, including calculating the temperature of the block. The energy E is measured from the ground state. Nothing goes in the shaded boxes. Be sure to give the temperature to the nearest 0.1 Kelvin.

q	# ways	E, J	S, J/K	ΔE, J	ΔS, J/K	T, K
20	4.91 E26					
21	4.44 E27					
22	3.85 E28					

Figure 12.55

12.P.40 The interatomic spring stiffness for copper is determined from Young's modulus measurements to be 28 N/m. The mass of one mole of copper is 0.064 kg. If we model a block of copper as a collection of atomic "oscillators" (masses on springs), what is one quantum of energy for one of these atomic oscillators? Note that since each oscillator is attached to two "springs," and each "spring" is half the length of the interatomic bond, the effective interatomic spring stiffness for one of these oscillators is 4 times the calculated value given above.

Use these precise values for the constants: $\hbar = 1.0546 \times 10^{-34}$ J·s (Planck's constant divided by 2π), Avogadro's number = 6.0221×10^{23} molecules/mole, $k_B = 1.3807 \times 10^{-23}$ J/K (the Boltzmann constant).

(a) What is one quantum of energy for one of these atomic oscillators?

(b) Figure 12.56 is a table containing the number of ways to arrange a given number of quanta of energy in a particular block of copper. Fill in the blanks to complete the table, including calculating the temperature of the block and the per-atom heat capacity C. The energy E is measured from the ground state. Nothing goes in the shaded boxes. Be sure to give the temperature to the nearest 0.1 Kelvin.

q	# ways	E, J	S, J/K	ΔE, J	ΔS, J/K	T, K
20	4.91 E26					
21	4.44 E27					
22	3.85 E28					

Figure 12.56

(c) There are 100 atoms in this object. What is the heat capacity on a per-atom basis? (Note that at high temperatures the heat capacity on a per-atom basis approaches the classical limit of $3k_B = 4.2 \times 10^{-23}$ J/K/atom.)

12.P.41 A nanoparticle containing 6 atoms can be modeled approximately as an Einstein solid of 18 independent oscillators. The evenly spaced energy levels of each oscillator are 4×10^{-21} J apart.

(a) When the nanoparticle's energy is in the range $5 \times 4 \times 10^{-21}$ J to $6 \times 4 \times 10^{-21}$ J, what is the approximate temperature? (In order to keep precision for calculating the heat capacity, give the result to the nearest tenth of a Kelvin.)

(b) When the nanoparticle's energy is in the range $8 \times 4 \times 10^{-21}$ J to $9 \times 4 \times 10^{-21}$ J, what is the approximate temperature? (In order to keep precision for calculating the heat capacity, give the result to the nearest tenth of a degree.)

(c) When the nanoparticle's energy is in the range $5 \times 4 \times 10^{-21}$ J to $9 \times 4 \times 10^{-21}$ J, what is the approximate heat capacity per atom?

Note that between parts (a) and (b) the average energy increased from "5.5 quanta" to "8.5 quanta." As a check, compare your result with the high temperature limit of $3k_B$.

12.P.42 For a certain metal the stiffness of the interatomic bond and the mass of one atom are such that the spacing of the quantum oscillator energy levels is 1.5×10^{-23} J. A nanoparticle of this metal consisting of 10 atoms has a total thermal energy of 18×10^{-23} J.

(a) What is the entropy of this nanoparticle?

(b) The temperature of the nanoparticle is 87 K. Next we add 18×10^{-23} J to the nanoparticle. By how much does the entropy increase?

12.P.43 A 50 gram block of copper (one mole has a mass of 63.5 grams) at a temperature of 35° C is put in contact with a 100 gram block of aluminum (molar mass 27 grams) at a temperature of 20° C. The blocks are inside an insulated enclosure, with little contact with the walls. At these temperatures, the high-temperature limit is valid for the specific heat capacity. Calculate the final temperature of the two blocks. Do NOT look up the specific heat capacities of aluminum and copper; you should be able to figure them out on your own.

12.P.44 Young's modulus for copper is measured by stretching a copper wire to be about 1.2×10^{11} N/m². The density of copper is about 9 grams/cm³, and the mass of a mole is 63.5 grams. Starting from a very low temperature, use these data to estimate roughly the temperature T at which we expect the specific heat capacity for copper to approach $3k_B$. Compare your estimate with the data shown on a graph in this chapter.

12.P.45 The goal of this experiment is to understand, in a concrete way, what specific heat capacity is and how it can be measured. You will need a microwave oven, a styrofoam coffee cup, and a clock or watch.

In the range of temperature where water is a liquid (0 C to 100 C), it is approximately true that it takes 4.2 J of energy (1 calorie) to raise the temperature of 1 gram of water through 1 degree Kelvin. To measure this specific heat capacity of water, we need some way to raise a known mass of water from a known initial temperature to a final temperature that can also be measured, while we keep track of the energy supplied to the water. One way to do this, as discussed in the text, is to put water in a well-insulated container within which a heater, whose power output is known, warms up the water. In this experiment, instead of a well-insulated box with a heater, we will use microwave power, which preferentially warms up water by exciting rotational modes of the water molecules, as opposed to burners or heaters that warm up water in a pan by first warming up the pan.

Use a styrofoam coffee cup of known volume in which water can be warmed up. The density of water is 1 gram/cm³.

It is a good idea not to fill the cup completely full, because this makes it more likely to spill.

One method of recording the initial temperature of the water is to get water from the faucet and wait for it to equilibrate with room temperature (which can either be read off a thermostat or estimated based on past experience). After waiting about a half hour for this to happen, place the cup in the microwave oven and turn on the oven at maximum power. The cup needs to be watched as it warms up, so that when the water starts to boil, the elapsed time can be noted accurately.

BE CAREFUL! A styrofoam cup full of hot liquid can buckle if you hold it near the rim. Hold the cup near the bottom. If the cup is full, do not attempt to move the cup while the water is hot. A spill can cause a painful burn.

On the back of the microwave oven (or inside the front door), there is usually a sticker with specifications that says "Output Power = ...Watts" which can be used to calculate the energy supplied. If there is no indication, use a typical value of 600 watts for a standard microwave oven. Using all the quantities measured above and knowing the temperature interval over which you have warmed up the water, you can calculate the specific heat capacity of water.

(a) Show and explain all your data and calculations, and compare with the accepted value for water (4.2 J/K per gram).

(b) Discuss why your result might be expected to differ from the accepted value. For each effect that you consider, state whether this effect would lead to a result that is larger or smaller than the accepted value.

12.P.46 A nanoparticle consisting of four iron atoms (object 1) initially has 1 quantum of energy. It is brought into contact with a nanoparticle consisting of two iron atoms (object 2), which initially has 2 quanta of energy. The mass of one mole of iron is 56 grams.

(a) Using the Einstein model of a solid, calculate and plot $\ln \Omega_1$ vs. q_1 (the number of quanta in object 1), $\ln \Omega_2$ vs. q_1, and $\ln \Omega_{total}$ vs. q_1 (put all three plots on the same graph). Show your work and explain briefly.

(b) Calculate the approximate temperature of the objects at equilibrium. State what assumptions or approximations you made.

12.P.47 Figure 12.57 shows a one-dimensional row of 5 microscopic objects each of mass 4×10^{-26} kg, connected by forces that can be modeled by springs of stiffness 15 N/m. These objects can move only along the x axis.

Figure 12.57

(a) Using the Einstein model, calculate the approximate entropy of this system for total energy of 0, 1, 2, 3, 4, and 5 quanta. Think carefully about what the Einstein model is, and apply those concepts to this one-dimensional situation.

(b) Calculate the approximate temperature of the system when the total energy is 4 quanta.

(c) Calculate the approximate specific heat capacity on a per-object basis when the total energy is 4 quanta.

(d) If the temperature is raised very high, what is the approximate specific heat capacity on a per-object basis? Give a numerical value and compare with your result in part (c).

12.P.48 A box contains a uniform disk of mass M and radius R that is pivoted on a low-friction axle through its center (Figure 12.58). A block of mass m is pressed against the disk by a spring, so that the block acts like a brake, making the disk hard to turn. The box and the spring have negligible mass. A string is wrapped around the disk (out of the way of the brake) and passes through a hole in the box. A force of constant magnitude F acts on the end of the string. The motion takes place in outer space. At time t_i the speed of the box is v_i, and the rotational speed of the disk is ω_i. At time t_f the box has moved a distance x, and the end of the string has moved a longer distance d, as shown.

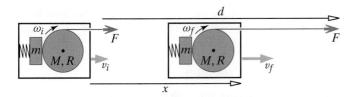

Figure 12.58

(a) At time t_f, what is the speed v_f of the box?

(b) During this process, the brake exerts a tangential friction force of magnitude f. At time t_f, what is the angular speed ω_f of the disk?

(c) At time t_f, assume that you know (from part b) the rotational speed ω_f of the disk. From time t_i to time t_f, what is the increase in thermal energy of the apparatus?

(d) Suppose that the increase in thermal energy in part (c) is 8×10^4 J. The disk and brake are made of iron, and their total mass is 1.2 kg. At time t_i their temperature was 350 K. At time t_f, what is their approximate temperature?

Sections 12.7–12.8

12.X.49 Consider the exponential function e^{-x}. Evaluate this function for $x = 1, 10{,}000$, and 0.01.

12.X.50 At room temperature (293 K), calculate $k_B T$ in joules and eV.

12.X.51 Sketch and label graphs of specific heat capacity vs. temperature for hydrogen gas (H_2) and oxygen gas (O_2), using the same temperature scale. Explain briefly.

12.X.52 Which has more internal energy at room temperature— a mole of helium or a mole of air?

12.X.53 Which of the following are true about the Boltzmann factor $e^{-\Delta E/k_B T}$, where ΔE is the energy above the ground state?

(A) $e^{-\Delta E/k_B T}$ is small when ΔE is large.
(B) $e^{-\Delta E/k_B T}$ is small at low temperature T.
(C) $e^{-\Delta E/k_B T}$ tends to zero as T gets larger and larger.
(D) Even at high temperature, when many energy levels are excited, the ground state ($\Delta E = 0$) is the most populated state.

12.X.54 Explain qualitatively the basis for the Boltzmann distribution. Never mind the details of the math for the moment. Focus on the trade-offs involved with giving energy to a single oscillator vs. giving that energy to a large object.

12.X.55 Many chemical reactions proceed at rates that depend on the temperature. Discuss this from the point of view of the Boltzmann distribution.

12.X.56 Figure 12.59 shows is the distribution of speeds of atoms in a particular gas at a particular temperature. Approximately what is the average speed? Is the rms (root-mean-square) speed bigger or smaller than this? Approximately what fraction of the molecules have speeds greater than 1000 m/s?

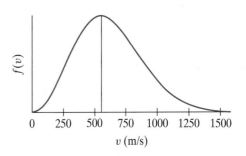

Figure 12.59

12.X.57 The temperature of the surface of a certain star is 7000 K. Most hydrogen atoms at the surface of the star are in the electronic ground state. What is the approximate fraction of the hydrogen atoms that are in the first excited state (and therefore could emit a photon)? The energy of the first excited state above the ground state is $(-13.6/2^2 eV) - (-13.6 eV) = 10.2 eV = 1.632 \times 10^{-18}$ J.

(In this estimate we are ignoring the fact that there may be several excited states with the same energies—for example, the 2s and 2p states in hydrogen—because this makes only a small difference in the answer.)

12.X.58 A gas is made up of diatomic molecules. At temperature T_1, the ratio of the number of molecules in vibrational energy state 2 to the number of molecules in the ground state is measured, and found to be 0.35. The difference in energy between state 2 and the ground state is ΔE. Which of the following conclusions is correct?

(A) $\Delta E \approx k_B T_1$
(B) $\Delta E << k_B T_1$
(C) $\Delta E >> k_B T_1$

At a different temperature T_2, the ratio is found to be 8×10^{-5}. Which of the following is true?

(D) $\Delta E \approx k_B T_2$
(E) $\Delta E << k_B T_2$
(F) $\Delta E >> k_B T_2$

12.X.59 Marbles of mass $M = 10$ grams are lying on the floor. They are of course in thermal equilibrium with their surroundings. What is a typical height above the floor for one of these marbles? That is, for what value of y is $Mgy \approx k_B T$?

12.X.60 Viruses of mass $M = 2 \times 10^{-20}$ kg are lying on the floor at room temperature (about $20°C = 293$ K). They are of course in thermal equilibrium with their surroundings. What is a typical height above the floor for one of these viruses? That is, for what value of y is $Mgy \approx k_B T$?

12.X.61 How does the speed of sound in a gas change when you raise the temperature from $0°C$ to $20°C$? Explain briefly.

12.P.62 Calculate the temperature rise of a gas.

(a) You have a bottle containing a mole of a monatomic gas such as helium or neon. You warm up this monatomic gas with an electrical heater, which inputs $Q = 580$ joules of energy. How much does the temperature of the gas increase?

(b) You have a bottle containing a mole of a diatomic gas such as nitrogen (N_2) or oxygen (O_2). The initial temperature is in the range where many rotational energy levels are excited but essentially no vibrational energy levels are excited. You warm up this diatomic gas with an electrical heater, which inputs $Q = 580$ joules of energy. How much does the temperature of the gas increase?

(c) You have a bottle containing a mole of a diatomic gas such as nitrogen (N_2) or oxygen (O_2). The initial temperature is in the range where not only many rotational energy levels are excited but also many vibrational energy levels are excited. You warm up this diatomic gas with an electrical heater, which inputs $Q = 580$ joules of energy. How much does the temperature of the gas increase?

12.P.63 In studying a voyage to the Moon in Chapter 3 we somewhat arbitrarily started at a height of 50 kilometers above the surface of the Earth.

(a) At this altitude, what is the density of the air as a fraction of the density at sea level?

(b) Approximately how many air molecules are there in one cubic centimeter at this altitude?

(c) At what altitude is air density one-millionth (1×10^6) that at sea level?

12.P.64 At sufficiently high temperatures, the thermal speeds of gas molecules may be high enough that collisions may ionize a molecule (that is, remove an outer electron). An ionized gas in which each molecule has lost an electron is called a "plasma." Determine approximately the temperature at which air becomes a plasma.

12.P.65 100 joules of energy transfer due to a temperature difference are given to air in a 50 liter rigid container and to helium in a 50 liter rigid container, both initially at STP (standard temperature and pressure).

(a) Which gas experiences a greater temperature rise?

(b) What is the temperature rise of the helium gas?

12.P.66 It is possible to estimate some properties of a diatomic molecule from the temperature dependence of the specific heat capacity.

(a) Below about 80 K the specific heat capacity at constant volume for hydrogen gas (H_2) is $\frac{3}{2}k_B$ per molecule, but at higher temperatures the specific heat capacity increases to $\frac{5}{2}k_B$ per molecule due to contributions from rotational energy states. Use these observations to estimate the distance between the hydrogen nuclei in an H_2 molecule.

(b) At about 2000 K the specific heat capacity at constant volume for hydrogen gas (H_2) increases to $\frac{7}{2}k_B$ per molecule due to contributions from vibrational energy states. Use these observations to estimate the stiffness of the "spring" that approximately represents the interatomic force.

12.P.67 In 1988, telescopes viewed Pluto as it crossed in front of a distant star. As the star emerged from behind the planet, light from the star was slightly dimmed as it went through Pluto's atmosphere. The observations indicated that the atmospheric density at a height of 50 km above the surface of Pluto is about one-third the density at the surface. The mass of Pluto is known to be about 1.5×10^{22} kg and its radius is about 1200 km. Spectroscopic data indicate that the atmosphere is mostly nitrogen (N_2). Estimate the temperature of Pluto's atmosphere. State what approximations and/or simplifying assumptions you made.

12.P.68 Buckminsterfullerene, C_{60}, is a large molecule consisting of 60 carbon atoms connected to form a hollow sphere. The diameter of a C_{60} molecule is about 7×10^{-10} m. It has been hypothesized that C_{60} molecules might be found in clouds of interstellar dust, which often contain interesting chemical compounds. The temperature of an interstellar dust cloud may be very low, around 3 K. Suppose you are planning to try to detect the presence of C_{60} in such a cold dust cloud by detecting photons emitted when molecules undergo transitions from one rotational energy state to another.

Approximately, what is the highest-numbered rotational level from which you would expect to observe emissions? Rotational levels are $l = 0, 1, 2, 3, \ldots$.

Computational Problems

It is important to work through the first four computational problems in detail in order to make the ideas concrete.

12.P.69 Calculate the number of ways to arrange energy in an Einstein solid.

(a) Model a system consisting of two atoms (three oscillators each), among which 4 quanta of energy are to be distributed. Write a program to display a histogram showing the total number of possible microstates of the two-atom system vs. the number of quanta assigned to atom 1. Compare your histogram to the one shown in Figure 12.15 (you should get the same distribution).

(b) Model a system consisting of two solid blocks, block 1 containing 300 oscillators and block 2 containing 200 oscillators. Find the possible distributions of 100 quanta among these blocks, and plot number of microstates vs. number of quanta assigned to block 1. Compare your histogram to the one shown in Figure 12.21. Determine the distribution of quanta for which the probability is half as large as the most probable 60-40 distribution.

(c) Do a series of calculations distributing 100 quanta between two blocks whose total number of oscillators is 500, but whose relative number of atoms varies. For example, consider equal numbers of oscillators, and ratios of 2:1, 5:1, and so on. Describe your observations.

12.P.70 Start with your solution to Problem 12.P.69 (b). For the same system of two blocks, with $N_1 = 300$ oscillators and $N_2 = 200$ oscillators, plot $\ln(\Omega_1)$, $\ln(\Omega_2)$, and $\ln(\Omega_1\Omega_2)$, for q_1 running from 0 to 100 quanta. Your graph should look like the one in Figure 12.25. Determine the maximum value of $\ln(\Omega_1\Omega_2)$

and the value of q_1 where this maximum occurs. What is the significance of this value of q_1?

12.P.71 Modify your calculations from Problem 12.P.70 to plot the temperature in kelvins of block 1 as a function of the number of quanta q_1 present in the first block. On the same graph, plot the temperature in kelvins of block 2 as a function of q_1 (of course, $q_2 = q_{tot} - q_1$).

In order to plot the temperature in kelvins, you must determine the values of ΔE and ΔS that correspond to a one-quantum change in energy. Consider the model we are using. The energy of one quantum, in joules, is $\Delta E = \hbar\sqrt{k_s/m}$. The increment in entropy corresponding to this increment in energy is $\Delta S = k\Delta(\ln\Omega)$. Assume that the blocks are made of aluminum. In Chapter 4 you made an estimate of the interatomic spring constant k_s for aluminum. In Section 12.6 we show that the effective k_s for oscillations in the Einstein solid is expected to be about 4 times the value obtained from measuring Young's modulus.

What is the significance of the value of q_1 (and of q_2) where the temperature curves for the two blocks cross (the temperatures are equal)?

12.P.72 Modify your analysis of Problem 12.P.71 to determine the specific heat capacity as a function of temperature for a

$T(K)$	C, Al (J/K/mole)	C, Pb (J/K/mole)
20	0.23	11.01
40	2.09	19.57
60	5.77	22.43
80	9.65	23.69
100	13.04	24.43
150	18.52	25.27
200	21.58	25.87
250	23.25	26.36
300	24.32	26.82
400	25.61	27.45

single block of metal. In order to see all of the important effects, consider a single block of 35 atoms (105 oscillators) with up to 300 quanta of energy. Note that in this analysis you are calculating quantities for a *single* block, not two blocks in contact.

To make specific comparison with experimental data, consider the cases of aluminum (Al) and lead (Pb). For each metal, plot the theoretical specific heat capacity C per atom vs. $T(K)$, with dots showing the actual experimental data given in the following table, which you should convert to the same per-atom basis as your theoretical calculations. Adjust the interatomic spring stiffness k_s until your calculations approximately fit the experimental data.

What value of k_s gives a good fit? Compare with the estimated values of k_s obtained in Chapter 4 based on Young's modulus, but remember that those estimated values are probably 4 times smaller than the effective spring constant for oscillations.

Show from your graph that the high-temperature limit of the specific heat capacity is about $3k_B$ per atom. The rise above this limit at high temperatures may be due to the fact that the assumed uniform spacing of the quantized oscillator energy levels isn't a good approximation for highly excited states. See Chapter 8.

12.P.73 For some examples of your choice, demonstrate by carrying out actual computer calculations that the "square-root" rule holds true for the fractional width of the peak representing the most probable arrangements of the energy. *Warning:* Check to see what is the largest number you can use in your computations; some programs or programming environments won't handle numbers bigger than about 1×10^{307}, for example, and larger numbers are treated as "infinite."

12.P.74 Create a computational model of one atom of the Einstein solid as shown in Figure 12.4. Let the atom move under the influence of the six springlike forces that act on the atom. Use the classical Momentum Principle, although the actual behavior is governed by quantum mechanics, with quantized energies.

(a) Experiment with different initial displacements of the atom away from the equilibrium position.

(b) In many materials the effective stiffness of the interatomic bonds is different in the x, y, and z directions; see how making the stiffnesses different affects the motion.

ANSWERS TO EXERCISES

12.X.1 20
12.X.5 5.4×10^{88} years; 5.4×10^{78} times the age of the Universe!
12.X.7 This looks *very* improbable; the second law of thermodynamics would be violated by this process.
12.X.8 500 K
12.X.9 Yes, there are more ways to arrange energy in liquid water than solid ice. The water has 1.23 J/K more entropy (note that 0°C is 273 K).
12.X.10 1.87 J/K/gram
12.X.11 25.2°C
12.X.12 The temperature of the sink will rise rapidly because the heat capacity is low; you want the temperature of the sink to stay constant.
12.X.14 1; 0.14; 0.02
12.X.15 8.4×10^3 m = 8.4 km! (about 5 miles)

12.X.16 0.35
12.X.17 1350 m/s
12.X.18 510 m/s
12.X.19 2.50; 2.74
12.X.20 2300 m/s; not very large compared with 500 m/s
12.X.21 Increase, because the average speed of the air molecules increases; 40% (factor of $\sqrt{2}$)
12.X.22 $\frac{3}{2}k_B$; no low-lying vibrational or rotational energy levels to be excited
12.X.23 **(a)** $\frac{7}{2}k_B$; **(b)** $\frac{5}{2}k_B$; **(c)** $\frac{3}{2}k_B$; **(d)** transition to $\frac{3}{2}k_B$ occurs at a lower temperature for deuterium because the rotational energy level spacing $L^2/(2I)$ is smaller for deuterium (larger nuclear mass means larger moment of inertia, since the internuclear distance is about the same; $L^2 = l(l+1)\hbar^2$ is the same for the rotational energy levels of hydrogen and deuterium molecules)

Gases and Engines

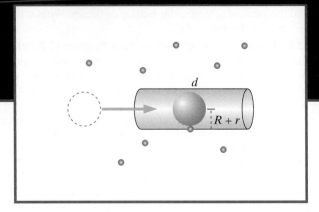

- We can model a gas as tiny balls in rapid random motion.
- Applying statistical ideas to this microscopic model allows us to predict macroscopic relationships among pressure, volume, and temperature, including the ideal gas law.
- The definition of temperature in terms of entropy is consistent with the everyday temperature scale.
- Both work and energy transfer due to a temperature difference can change the temperature and/or volume of a gas in a chamber with a movable piston.
- The attainable efficiency of engines that convert thermal energy into useful work is constrained by the second law of thermodynamics.

13.1 GASES, SOLIDS, AND LIQUIDS

In contrast to a solid, a gas has no fixed structure. The gas molecules are not bound to each other but move around very freely, which is why a gas does not have a well-defined shape of its own; it fills whatever container you put it in (Figure 13.1). Think, for example, of the constantly shifting shape of a cloud, or the deformability of a balloon, in contrast with the rigidity of a block of aluminum. On average, gas molecules are sufficiently far apart that most of the time they hardly interact with each other. This low level of interaction is what makes it feasible to model a gas in some detail, using relatively simple concepts.

In a gas the molecular motion must be sufficiently violent that molecules can't stay stuck together. At high enough temperatures, any molecules that do manage to bind to each other temporarily soon get knocked apart again by high-speed collisions with other molecules. However, at a low enough temperature, molecules move sufficiently slowly that collisions are no longer violent enough to break intermolecular bonds. Rather, more and more molecules stick to each other in a growing mass as the gas turns into a liquid or, at still lower temperatures, a solid.

Figure 13.1 Molecules in a gas.

Liquids Are More Complex

A liquid is intermediate between a solid and a gas. The molecules in a liquid are sufficiently attracted to each other that the liquid doesn't fly apart like a gas (Figure 13.2), yet the attraction is not strong enough to keep each molecule near a fixed equilibrium position as in a solid. The molecules in a liquid can

Figure 13.2 Molecules in a liquid.

slide past each other, giving liquids their special property of fluid flow (unlike solids) with fixed volume (unlike gases).

The analysis of liquids in terms of atomic, microscopic models is quite difficult compared with gases, where the molecules only rarely come in contact with other atoms, or compared with solids, where the atoms never move very far away from their equilibrium positions. For this reason, in this introductory textbook with its emphasis on atomic-level description and analysis we concentrate mostly on understanding gases and solids.

An active field of research, called "molecular dynamics," models liquids, gases, and solids by computational modeling using the Momentum Principle and appropriate forces between the molecules. The difference between this work and the computational models you have made is that sophisticated programming techniques are required to deal with very large numbers of molecules, sometimes as many as a million, in a reasonable amount of computer time.

13.2   GAS LEAKS THROUGH A HOLE

In the previous chapter we used statistical mechanics to determine the average speed of a gas molecule. Here we'll see some interesting phenomena in which the average speed plays a role. We will model the gas molecules as little balls that don't attract each other and interact only in brief elastic collisions.

We will frequently use the symbol $n = N/V$ to stand for the number of gas molecules per unit volume, which we will express in SI units as number per cubic meter:

NUMBER DENSITY: NUMBER PER CUBIC METER

$$\text{Definition: } n \equiv \frac{N}{V}$$

N is the number of gas molecules in the volume V. The units of n are molecules per cubic meter.

Warning: You may be familiar from chemistry with the "ideal gas law" written in the form $PV = nRT$, where n is the number of moles. Here "n" means something else—the number of molecules per cubic meter, N/V.

EXAMPLE   *n* at "Standard Temperature and Pressure"

"Standard Temperature and Pressure" (or STP) are defined for a gas to be $0°C = 273$ K and the average air pressure at sea level. Under STP conditions the ideal gas law can be used to show that one mole of a gas will occupy a volume of 22.4 liters $= 22.4 \times 10^3$ cm^3 $= 22.4 \times 10^{-3}$ m^3. What is the number density n of a gas at STP?

Solution   One mole consists of 6.02×10^{23} molecules, so

$$n = \frac{6.02 \times 10^{23} \text{ molecules}}{22.4 \times 10^{-3} \text{ m}^3}$$
$$= 2.68 \times 10^{25} \text{ molecules/m}^3$$

One-Directional Gas

We will calculate the leakage rate of a gas through a small hole in a container filled with the gas. First we'll consider a simplified one-directional example, in

Figure 13.3 Side view of molecules all traveling to the right with speed v inside a tube. There are n molecules per cubic meter inside the tube.

Cross-sectional area A

Figure 13.4 End view of molecules all traveling the same direction in a tube of cross-sectional area A.

$v \Delta t$

Figure 13.5 A molecule that will leave the tube in a time Δt must be within a distance $v \Delta t$ from the end.

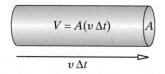

$V = A(v \Delta t)$ A

$v \Delta t$

Figure 13.6 Volume containing the molecules that will leave the tube in time Δt.

order to understand the basic issues before stating the results for a real three-dimensional gas.

The chain of reasoning that we follow is basically geometric. Consider a situation in which many gas molecules are all traveling to the right inside a tube. For the moment, temporarily assume that they all have the same speed v (Figure 13.3). The cross-sectional area of the tube is A (Figure 13.4).

Because eventually we want to be able to calculate how fast a gas will leak through a hole in a container, we will calculate how many molecules leave this tube in a short time interval Δt. For a molecule that is traveling at speed v to be able to reach the right end of the tube in this time interval Δt, it must be within a distance $v \Delta t$ of the end (Figure 13.5).

Since the cross-sectional area of the tube is A, the volume V of the tube that contains just those molecules that will leave the tube during the time interval Δt is simply $A(v \Delta t)$; see Figure 13.6.

There are n molecules per m^3 inside the tube, and $N = nV$ molecules inside the volume V. So the number N of molecules that will leave in the time interval Δt is

$$N = nV = n(Av \Delta t)$$

Dividing by the time interval Δt, we find the following result:

$$\text{number of molecules crossing area } A \text{ per second} = nAv$$

This is for the one-directional case; all molecules have the same speed.

QUESTION   Does this formula make sense? What would you expect if you increased the number of molecules per cubic meter, or the cross-sectional area, or the speed?

The formula does make sense. The more molecules per unit volume ($n = N/V$), the more molecules will reach the right end of the tube per unit time. The bigger the cross-sectional area (A), the more molecules that will pass through that area per unit time. The faster the molecules are moving (v), the more molecules from farther away that can reach the end of the tube in a given time interval. The units are right: (molecules/m^3)(m^2) (m/s) = molecules/s.

Effect of Different Speeds

We need to account for the fact that the gas molecules don't all have the same speed v. Suppose that n_1 molecules per unit volume have speeds of approximately v_1, n_2 molecules per unit volume have speeds of approximately v_2, and so on. The number of molecules crossing an area A per second is

$$n_1 Av_1 + n_2 Av_2 + \cdots$$

The average speed of all the molecules is by definition the following, where we weight each different speed by the number of molecules per cubic meter that have that approximate speed:

$$\bar{v} = \frac{n_1 v_1 + n_2 v_2 + \cdots}{n}$$

Therefore,

$$n_1 Av_1 + n_2 Av_2 + \cdots = nA\bar{v}$$

A horizontal bar over a symbol is a standard notation for "average," and $\bar{v}$ means average speed.

Finally, we have a valid formula for the number of molecules in a one-directional flow leaving the right end of the tube, even in the situation where they have a distribution of different speeds:

NUMBER OF MOLECULES CROSSING AREA *A* PER SECOND (ONE-DIRECTIONAL)

$$\text{1D flow rate} = nA\bar{v}.$$

This is for one-directional flow; molecules have various speeds.

A "one-directional" gas may sound a bit silly, but this formula does apply to real one-directional flows such as the flow of water or gas through a pipe or the flow of electrons through a copper wire in an electric circuit.

A Three-Dimensional Gas

A more realistic model of a gas in a closed container would have approximately equal numbers of molecules heading to the left as well as to the right in the tube, in which case the number of molecules leaving the right end of the tube would be only $\frac{1}{2}nA\bar{v}$. In a real three-dimensional gas, molecules are moving in all directions. Only those molecules that are headed in the $+x$ direction can pass through a hole located to the right, not those moving in the $-x$ direction or in the $\pm y$ or $\pm z$ directions. The molecules are headed randomly in all six directions, so we might expect our formula would have a factor of 1/6.

However, the actual factor is 1/4, which comes from detailed averaging over all directions and is related to our use of the average speed (magnitude of velocity), rather than averages of velocity components v_x or v_y or v_z. We don't want to get bogged down in the rather heavyweight mathematics required to prove this, so we just state that the factor is 1/4 rather than 1/6. (You can find a full derivation in books on statistical mechanics.)

NUMBER OF MOLECULES CROSSING AREA *A* PER SECOND (THREE-DIMENSIONAL)

$$\text{3D flow rate} = \frac{1}{4}nA\bar{v}$$

This is for three-dimensional flow; molecules have various velocities.

EXAMPLE   **Leak Through a Hole in a Balloon**

At standard temperature and pressure, at what rate will helium escape from a balloon through a hole 1 millimeter in diameter? As we saw in the previous chapter, the average speed $\bar{v}$ of helium atoms at ordinary temperatures is about 1200 m/s.

Solution   Earlier we calculated that at STP $n = 2.68 \times 10^{25}$ molecules per cubic meter. Therefore the leak rate is

$$\frac{1}{4}nA\bar{v} = \left(2.68 \times 10^{25}\frac{\text{molecules}}{\text{m}^3}\right)\left(\pi(0.5 \times 10^{-3}\,\text{m})^2\right)(1200\,\text{m/s})$$

$$= 6.3 \times 10^{21}\ \text{molecules/s; about 0.01 moles/s}$$

Cooling of the Gas

There is an interesting and important effect of gas escaping through a hole. If you look back over the derivations of the formulas, you can see that faster molecules escape disproportionately to their numbers. Faster molecules can be farther away from the hole than is true for slower molecules and still escape through the hole in the next short time interval. As a result, the distribution of speeds of molecules inside the container becomes somewhat depleted of high speeds.

13.X.1 What can you say about the temperature of the gas inside the container as the gas escapes? Why?

EXAMPLE   **Leakage from a Balloon**

A party balloon filled with helium is about one foot in diameter (about 30 centimeters). The volume of a sphere is $\frac{4}{3}\pi r^3$, where r is the radius of the sphere. In the previous example we calculated that the initial leak rate through a circular hole 1 millimeter in diameter is 6.3×10^{21} molecules/s.

(a) If the helium were to escape at a constant rate equal to the initial rate, about how long would it take for all the helium to leak out?

(b) As the helium escapes, the balloon shrinks, and the number of helium atoms per cubic meter n will stay roughly constant, in which case our analysis is pretty good. However, remember that the temperature of the gas drops due to a preferential loss of high-speed atoms. Would this effect make the amount of time to empty be more or less than the value you calculated in part (a)?

Solution   **(a)** The leak rate was 6.3×10^{21} atoms per second. Calculate the number of helium atoms in the balloon originally:

$$\frac{4}{3}\pi(0.3 \text{ m})^3 \left(\frac{6 \times 10^{23} \text{ atoms}}{22.4 \times 10^3 \text{cm}^3}\right)\left(\frac{1 \times 10^6 \text{cm}^3}{\text{m}^3}\right) = 3 \times 10^{24} \text{ atoms}$$

Assuming a constant rate:

$$\frac{3 \times 10^{24} \text{ atoms}}{6.3 \times 10^{21} \text{ atoms/s}} = 500 \text{ s}$$

■ When you blow up an ordinary rubber balloon the pressure is higher than one atmosphere, which means higher number density n but also a correspondingly higher leak rate, which is proportional to n. To a first approximation the density doesn't matter in this estimate of the time to empty. Also note that often when you puncture a balloon the balloon rips, creating a large opening; this is not the case we are considering.

(b) With the preferential loss of high-speed atoms, the speed distribution inside the balloon shifts to lower speeds (corresponding to a lower temperature). If the average speed is lower, the leak rate is lower, and it should take longer for the balloon to empty than we calculated in part (a), where we assumed a constant leak rate.

We've implicitly done the analysis in vacuum. If the balloon is in air, air molecules enter the balloon through the same hole.

13.X.2 A spacecraft containing air at STP is struck by a micrometeor that makes a hole 2 mm in diameter. Calculate the initial rate at which air escapes through the hole, in number of air molecules leaving the spacecraft per second. In the previous chapter we found that the average speed of air molecules at STP is about 500 m/s.

13.3   MEAN FREE PATH

Suppose that we place a special molecule somewhere in the helium gas, one whose movements we can trace. For example, it might be a molecule of perfume. On average, how far does this molecule go before it runs into a molecule of the gas? The average distance between collisions is called the "mean free path." It plays an important role in many phenomena, including the creation of electric sparks in air (discussed in Volume II of this textbook).

An approximate calculation of the mean free path depends on a simple geometrical argument. Draw a cylinder along the direction of motion of the special molecule, with length d and radius $R+r$, where R is the radius of the molecule, and r is the radius of a gas molecule, as shown in Figure 13.7 and Figure 13.8.

The geometrical significance of this cylinder is that if the path of the special molecule comes within one molecular radius of a gas molecule, there will be a collision, so any gas molecule whose center is inside the cylinder will be hit. How long should the cylinder be for there to be a collision?

We define d to be the average distance the special molecule will travel before colliding with another molecule, so the cylinder drawn in Figure 13.7 should contain on average about one gas molecule. The cross-sectional area of the cylinder is $A \approx \pi(R+r)^2$ and the volume of the cylinder is Ad. If n stands for the number of gas molecules per cubic meter, we can write a formula involving the mean free path:

Figure 13.7 If a "special molecule" (blue) enters a cylinder of length d and radius $R+r$, it will collide with a gas molecule (green).

Figure 13.8 End view of cylinder.

MEAN FREE PATH

$$d \approx \frac{1}{nA}$$

$A \approx \pi(R+r)^2$; see Figure 13.7.

We are ignoring some subtle effects in this calculation, but this analysis gives us the main picture: the mean free path is shorter for higher density or larger molecular size.

EXAMPLE   **Mean Free Path in Air**

To get an idea of the order of magnitude of a typical mean free path, assume that an N_2 molecule in the air has a radius of approximately 2×10^{-10} m (the radius of one of the N atoms being about 1×10^{-10} m), and calculate approximately the mean free path d of an N_2 molecule moving through air.

Solution

$$n = 2.68 \times 10^{25} \, \frac{\text{molecules}}{\text{m}^3}$$

$$A \approx \pi(2 \times 2 \times 10^{-10} \, \text{m})^2 = 5 \times 10^{-19} \, \text{m}^2$$

$$d \approx \frac{1}{nA} \approx \frac{1}{\left(2.68 \times 10^{25} \, \dfrac{\text{molecules}}{\text{m}^3}\right)(5 \times 10^{-19} \, \text{m}^2)}$$

$$d \approx 7 \times 10^{-8} \, \text{m}$$

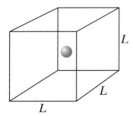

Figure 13.9 The volume of air occupied on average by one molecule.

13.X.3 It is interesting to compare the mean free path of about 7×10^{-8} m to the average spacing L between air molecules, which is the cube root of the volume occupied on average by one molecule (Figure 13.9). Calculate L. You may be surprised to find that the mean free path d is much larger than the average molecular spacing L. The molecules represent rather small targets.

13.4 PRESSURE AND TEMPERATURE

We can use statistical ideas to relate what we know about the motion of molecules in a gas to the pressure that the gas exerts on its container. We'll consider a closed container, with the energy of the gas not changing.

In a party balloon, helium atoms are continually hitting the rubber walls of the balloon. In an earlier example we found that at STP (standard temperature and pressure) about 6×10^{21} helium atoms per second escape through a 1-mm-diameter section of the balloon. That means that in a closed container 6×10^{21} helium atoms per second strike every 1-mm-diameter portion of the container's walls.

The time- and space-averaged effect of this bombardment is an average force exerted on every square millimeter of the container. This average force per unit area is called the "pressure" P and is measured in N/m^2, also called a "pascal." We are going to calculate how big the pressure is in terms of the average speed of the helium atoms, thus building a link between the microscopic behavior of the helium atoms and the macroscopic time- and space-averaged pressure.

On average, a helium atom bounces off the wall with no change of kinetic energy. We emphasize that this is the average behavior. Any individual atom may happen to gain or lose energy in the collision with a vibrating atom in the container wall, but if the container is not being warmed up or cooled down, the velocity distribution in the gas does not change, and on average the helium atoms rebound from the wall with the same kinetic energy they had just before hitting the wall (Figure 13.10).

QUESTION If the kinetic energy doesn't change when an atom bounces off the wall, is there any change in the atom's momentum?

Unlike speed or kinetic energy, momentum is a vector quantity, and there is a large change in the x component of momentum, from $+p_x$ to $-p_x$. (There is no change in the y component.) Therefore the momentum change of the helium atom is $\Delta p_{x,\text{helium}} = -2p_x$.

QUESTION What caused this change in the momentum of the helium atom?

A force is required to change the momentum of an object. In this case the force was applied by the wall of the container (or in more detail, by an atom in the wall of the container). By the principle of reciprocity (Newton's third law), the helium atom must have applied an equal and opposite force to the wall (or more precisely, to an atom in the wall of the container). Therefore the wall must have acquired an amount of momentum $\Delta p_{x,\text{ wall}} = 2p_x$.

If we could calculate the average time Δt between collisions of helium atoms with a small area A of the wall, we could express the pressure (force per unit area) as follows, since $d\vec{p}/dt = \vec{F}$:

$$P = \frac{F}{A} = \frac{1}{A}\frac{\Delta p_x}{\Delta t}$$

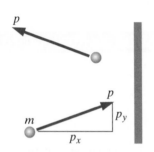

Figure 13.10 On average, an atom bounces off the wall without changing speed.

Assuming One Direction and One Speed

We already know that if there are n_+ helium atoms per unit volume that have an x component of velocity equal to $v_x > 0$ (that is, moving in the $+x$ direction toward the wall), the number of such atoms that hit an area A of the wall in a time Δt is $n_+ A v_x \Delta t$. Therefore the average time between collisions is

$$\Delta t = 1/(n_+ A v_x)$$

and we find

$$P = \frac{F}{A} = \frac{1}{A}\frac{\Delta p_x}{\left(\dfrac{1}{n_+ A v_x}\right)} = (n_+ v_x)(2p_x) = 2n_+\left(\frac{p_x}{m}\right)p_x = 2n_+\frac{p_x^2}{m}$$

An Alternative Argument

Alternatively, we can say that $n_+ A v_x$ helium atoms with this x component of velocity hit the area A every second, and each of them delivers $2p_x$ of momentum to the wall, so that the total momentum transfer to the wall per second is

(number of hits per second) (momentum transfer per hit) $= (n_+ A v_x)(2p_x)$

Divide by A to get the force per unit area, and you again get

$$P = (n_+ v_x)(2p_x) = 2n_+\frac{p_x^2}{m}$$

Taking the Velocity Distribution into Account

Actually, this is just the contribution to the pressure made by those atoms that happened to have this particular value of $v_x > 0$. On average, only half the atoms are headed in the $+x$ direction, so we replace n_+ by $n/2$, where n is the number of atoms per cubic meter (going in either the $+x$ or $-x$ direction).

$$P = 2n_+\frac{p_x^2}{m} = n\frac{p_x^2}{m}$$

We also need to average over the slow and fast atoms:

$$P = n\frac{\overline{p_x^2}}{m}$$

We'll explain what we mean by this average. Suppose that n_1 atoms per unit volume have x components of momentum of approximately p_{x1}, n_2 atoms per unit volume have x components of momentum of approximately p_{x2}, and so on. The pressure is

$$P = n_1\left(\frac{p_{x1}^2}{m}\right) + n_2\left(\frac{p_{x2}^2}{m}\right) + \cdots$$

The average value of p_x^2 for all the atoms is by definition the following, where we weight each different value of p_x^2 by the number of atoms per cubic meter that have that approximate value:

$$\overline{p_x^2} = \frac{n_1 p_{x1}^2 + n_2 p_{x2}^2 + \cdots}{n}$$

so we have $n_1 p_{x1}^2 + n_2 p_{x2}^2 + \cdots = n\overline{p_x^2}$. Therefore we can write this:

$$P = n_1\left(\frac{p_{x1}^2}{m}\right) + n_2\left(\frac{p_{x2}^2}{m}\right) + \cdots = n\frac{\overline{p_x^2}}{m}$$

Taking Direction into Account

We can re-express the pressure in terms of the magnitude of momentum p rather than p_x by the following steps:

First, note that $p = \sqrt{p_x^2 + p_y^2 + p_z^2}$, or $p^2 = p_x^2 + p_y^2 + p_z^2$.

Second, since the atoms are flying around in random directions, there should be no difference in averaging in the x, y, or z direction, so $\overline{p_x^2} = \overline{p_y^2} = \overline{p_z^2}$, which implies that $\overline{p^2} = 3\overline{p_x^2}$.

Taking these factors into account, instead of $P = n\dfrac{\overline{p_x^2}}{m}$ we can write the following important result:

GAS PRESSURE IN TERMS OF ATOMIC QUANTITIES

$$P = \frac{1}{3}n\frac{\overline{p^2}}{m}$$

QUESTION   Look back over the line of reasoning that led us to this result, and reflect on the nature of the argument. Stripped of the details, try to summarize the major steps leading to this result.

We reached this result by combining two effects: the number of molecules hitting an area per second is proportional to v, and the momentum transfer is also proportional to v. Hence the force per unit area is proportional to v^2.

The Ideal Gas Law

It is useful to rewrite our result for the pressure, factoring out the term $\overline{p^2}/(2m)$, which is the average translational kinetic energy K_{trans} of a molecule of mass m:

$$P = \frac{2}{3}n\left(\frac{\overline{p^2}}{2m}\right) = \frac{2}{3}n\overline{K}_{trans}$$

The pressure of a gas is proportional to the number density (number of molecules per cubic meter, N/V), and proportional to the average translational kinetic energy of the gas molecules. This gives us a connection to temperature, because we found in the previous chapter that $\overline{K}_{trans} = \frac{3}{2}k_B T$.

Substituting into the pressure formula, $P = \frac{2}{3}n\overline{K}_{trans} = \frac{2}{3}n(\frac{3}{2}k_B T) = nk_B T$, which relates pressure to temperature and is called the "ideal gas law":

MICROSCOPIC VERSION OF THE IDEAL GAS LAW

$$P = nk_B T$$

13.X.4 We found that at STP, $n = 2.68 \times 10^{25}$ molecules per m^3. Calculate the value of one atmosphere of pressure in units of N/m^2.

The Macroscopic Ideal Gas Law

We can compare our microscopic version of the gas law with experiments that measure the macroscopic properties of gases. For gases with fairly low densities, measurements of the pressure for all gases (helium, oxygen, nitrogen, carbon dioxide, etc.) are well summarized by the macroscopic "ideal gas law," which is probably familiar to you from chemistry, and which describes the observed behavior of any low-density gas:

MACROSCOPIC VERSION OF THE IDEAL GAS LAW

$$PV = (\text{number of moles})RT$$

R is the "gas constant" (8.3 J/K/mole).

To compare with our microscopic prediction, we can convert this macroscopic version of the ideal gas law to a version involving microscopic quantities:

$$P = \frac{(\text{number of moles})\,RT}{V}$$

$$\text{number of moles} = \frac{(\text{number of molecules})}{(6.02 \times 10^{23}\,\text{molecules/mole})} = \frac{N}{6.02 \times 10^{23}}$$

where N is the number of molecules in the gas. Therefore we have

$$P = \left(\frac{N}{6.02 \times 10^{23}}\right)\frac{RT}{V} = \left(\frac{N}{V}\right)\left(\frac{R}{6.02 \times 10^{23}}\right)T$$

This is simply $P = nk_BT$, where $n = N/V$ and $R/6.02 \times 10^{23}$ is the Boltzmann constant k_B:

$$k_B = \frac{R}{6.02 \times 10^{23}} = \frac{8.3\,\text{J/K/mole}}{6.02 \times 10^{23}\,\text{molecules/mole}} = 1.38 \times 10^{-23}\,\text{J/K}$$

Warning about the Meaning of "n"

Sometimes the ideal gas law is written in the form $PV = nRT$, where n is the number of moles rather than the number of molecules per cubic meter, $n = N/V$. On the rare occasions when we need to refer to the number of moles, we'll write it out as "number of moles."

Temperature from Entropy or from the Ideal Gas Law

In the previous chapter we found the relationship $\overline{K}_{\text{trans}} = \frac{3}{2}k_BT$, based on the Boltzmann distribution as applied to a low-density gas. The Boltzmann distribution in turn was based on the statistical mechanics definition of temperature in terms of entropy as $1/T = dS/dE$.

When we inserted $\overline{K}_{\text{trans}} = \frac{3}{2}k_BT$ into the kinetic theory result for pressure, $P = \frac{1}{3}n(\overline{p^2}/m)$, we obtained the molecular version of the ideal gas law, $P = nk_BT$, which we showed was equivalent to the macroscopic version of the ideal gas law, $P = (\text{number of moles})RT/V$.

Low-density gases are described well by the ideal gas law, and for that reason gases are used to make accurate thermometers. You measure the pressure P and volume V of a known number of moles of a low-density gas and determine the temperature from the macroscopic gas law:

$$T = \frac{PV}{(\text{number of moles})\,R} \quad \text{(gas thermometer)}$$

The fact that we could start from $1/T = dS/dE$ and derive the ideal gas law proves that the temperature measured by a gas thermometer is exactly the same as the "thermodynamic temperature" defined in terms of entropy.

Real Gases

Our analysis works well for low-density gases. For high-density gases, there are two major complications. First, the molecules themselves take up a considerable fraction of the space, so the effective volume is less than the geometrical volume, and V in the gas law must be replaced by a smaller value.

Second, at high densities the short-range electric forces between molecules have some effect. In this context these intermolecular forces are called "van der Waals" forces, which are the gradient of the interatomic potential energy discussed in Chapter 7. In a low-density gas almost all of the energy is kinetic energy, but in a high-density gas some of the energy goes into configurational energy associated with the interatomic potential energy. The effect is that in the gas law P must be replaced by a smaller value.

When these two effects are taken into effect, the resulting "van der Waals" equation fits the experimental data quite well for all densities of a gas, although the corrections are different for different gases due to differences in molecular sizes and intermolecular forces.

Energy of a Diatomic Gas

The formula for the average translational kinetic energy $\overline{K}_{\text{trans}} = \frac{3}{2}k_B T$ is valid even for a gas with multiatom molecules such as nitrogen (N_2) or oxygen (O_2). However, as you will recall from Chapter 9, this translational kinetic energy is only a part of the total energy of the molecule. In addition to translational energy, a diatomic molecule can have rotational and vibrational energy relative to the center of mass (Figure 13.11), so the energy contains additional terms. We write the energy of a diatomic molecule in the following way:

$$K_{\text{trans}} + E_{\text{vib}} + E_{\text{rot}} + Mgy_{\text{CM}}$$

The key point, as we discussed in the previous chapter, is that the average energy of a diatomic molecule is greater than the average energy of an atom in a monatomic gas at the same temperature. For a monatomic gas such as helium the average total energy per molecule is just $\overline{K}_{\text{trans}} = \frac{3}{2}k_B T$.

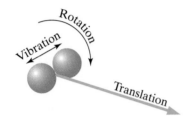

Figure 13.11 Energy of an oxygen molecule.

Figure 13.12 Weighing a box containing a gas.

Application: Weight of a Gas in a Box

The exponential dependence of number density ($n \propto e^{-mgy/k_B T}$) and therefore of pressure on height makes it possible to understand something that otherwise might be rather odd about weighing a box that contains a gas (Figure 13.12). Let's do the weighing in a vacuum so we don't have to worry about buoyancy forces due to surrounding air. The box definitely weighs more on the scales if there is gas in it than if there is a vacuum in it. In fact, if the mass of the gas is M, the additional weight is Mg, because the Momentum Principle for multiparticle systems refers to the net force on a system, and the gravitational contribution to the net force is the sum of the gravitational forces on each individual gas molecule in the box, Mg.

That seems reasonable until you think about the details of what is going on inside the box. At any given instant, the vast majority of the gas molecules are not touching the box! Also, some of the gas molecules are colliding with the top of the box, exerting an upward force on the box. How can these molecules possibly contribute to the weight measured by the scales? What is the mechanism for the Momentum Principle working out correctly for this multiparticle system?

The number density n at the bottom of the box is slightly larger than the number density at the top of the box, if all the gas is at the same temperature, which is a good assumption in most situations. Therefore the pressure, $P = nk_B T$, is larger at the bottom than at the top. Let's calculate the y components of the forces associated with these slightly different pressures, where A is the area of the top (and bottom) of the box, and h is the height of the box (Figure 13.13).

$$F_{\text{net},y} = P_{\text{top}}A - P_{\text{bottom}}A = A\Delta P$$

where ΔP is the pressure difference from bottom to top inside the box. We can calculate ΔP directly, by starting from the fact that for a small height change Δy, we have this:

$$\frac{\Delta P}{\Delta y} \approx \frac{dP}{dy}, \quad \text{so} \quad \Delta P \approx \frac{dP}{dy}\Delta y$$

P slightly smaller

Area *A*

Gas h

P slightly greater

Figure 13.13 The force due to gas pressure on the top of the box differs slightly from the force on the bottom of the box.

If we can evaluate dP/dy, we can determine the small pressure difference ΔP. Since $P = nk_BT$, the height dependence of the pressure is the same as for the number density (for constant temperature), and we can write this:

$$\frac{dP}{dy} = \frac{d}{dy}\left(P_{\text{bottom}}e^{-mgy/k_BT}\right)$$

$$\frac{dP}{dy} = P_{\text{bottom}}e^{-mgy/k_BT}\left(-\frac{mg}{k_BT}\right) \approx -P_{\text{bottom}}\left(\frac{mg}{k_BT}\right)$$

because the factor e^{-mgy/k_BT} is very close to 1. The result is negative because the force the gas exerts on the top of the box is smaller than the force on the bottom. Choosing $\Delta y = h$, and writing $P_{\text{bottom}} = nk_BT$, we have this result for ΔP:

$$\Delta P = \frac{dP}{dy}\Delta y \approx -nk_BT\left[\frac{mg}{k_BT}\right]h = -nmgh$$

Now that we know the pressure difference, we can calculate the y component of the net force:

$$F_{\text{net},y} = A\Delta P = -nmg(Ah)$$

QUESTION Ah is the volume V of the box, and $n = N/V$ is the number of molecules per m^3 in the box, so what does this formula reduce to?

We have the striking result that the air inside the box pushes down on the box with a force equal to the combined weight Mg of all N molecules in the box:

$$F_y = -Nmg = -Mg$$

We have shown that the difference in the time- and space-averaged momentum transfers by molecular collisions to the top and bottom of the box is equal to the weight of the gas in the box, as predicted by the Momentum Principle for a multiparticle system. The pressure difference is very slight, but then the weight of the gas is very small, for that matter. What is surprising is that at any particular instant, relatively few of the molecules are actually in contact with the box, yet the effects of these relatively few molecules is the same as though they were all sitting on the bottom of the box.

You could think of a box full of water in the same way. The water pressure is larger at the bottom than at the top, but with water this difference is quite large, corresponding to the much higher density of water. In fact, a column of water only 10 meters high makes a pressure equal to that produced by the many kilometers of atmosphere.

Consider a column of water 10 meters high with cross-sectional area A. The density of water is 1 gram/cm^3, which is 1×10^3 kg/m^3. The total mass of the column is the density (kg/m^3) times the volume $A(10\,\text{m})$, so the pressure on the bottom of the column is

$$\frac{(1 \times 10^3\,\text{kg}/\text{m}^3)A(10\,\text{m})(9.8\,\text{N}/\text{kg})}{A} \approx 1 \times 10^5\,\text{N}/\text{m}^2$$

This is the same result we found for atmospheric pressure at STP.

QUESTION In the preceding discussion the box was in vacuum, but under normal conditions the box we were weighing would have been surrounded by air. Suppose that the box has thin walls and is initially open to the air. Then we close the lid, trapping air inside it. Consider all forces including buoyancy forces, and determine what the scales will read. Will it be the weight of the box alone or the box plus air?

The scales will measure the weight of the box alone, because the weight of the air is $M_{\text{air}}g$ and the buoyancy force is $M_{\text{air}}g$ upward.

Figure 13.14 The ball has just hit the scale and has rebounded upward.

Figure 13.15 Impacts of the ball on the scale.

Application: Weight of a Bouncing Molecule

There is a related calculation that is amusing. You may have done a related homework problem earlier on collisions, in which case you already went through a similar calculation. Consider a single molecule (otherwise in vacuum) that bounces up and down on a scale without losing significant energy. Figure 13.14 shows a snapshot at the instant that the molecule is just bouncing up after hitting the scale.

What do the scales read? If the scales can respond quickly, we will see brief spikes each time the molecule strikes it (Figure 13.15). Let's determine what the time-averaged force is.

QUESTION How long does it take for the molecule to reach the top of its trajectory, h, starting with a speed v? What then is the time Δt between impacts? How much momentum transfer is there to the scale on each impact?

It takes a time interval v/g to go up (for the speed to decrease from v to 0 with acceleration $-g$) and another time interval v/g to come down, so the time between impacts is $\Delta t = 2v/g$. Each impact transfers an amount of momentum $\Delta P = 2mv$. Therefore the time-averaged force is

$$\frac{\Delta P}{\Delta t} = \frac{2mv}{2v/g} = mg$$

If the scale is sluggish, and can't respond in a time as short as $\Delta t = 2v/g$, the scale will simply register the value mg, just as though the molecule were sitting quietly on the scale.

13.5 ENERGY TRANSFERS

In this section we offer an analysis of energy transfers between a gas and its surroundings. We will address such questions as these: How much energy transfer Q due to a temperature difference is required to raise the temperature of a gas by one degree (the heat capacity of a gas)? How does the temperature of a gas change when you compress it quickly? Do the answers to these questions depend on what kind of gas is involved?

We need a device that lets us control the flow of energy into and out of a gas, in the form of work W or energy transfer Q. We will use a system consisting of a cylindrical container containing gas that is enclosed by a piston that can move in and out of the cylinder with little friction but which fits tightly enough to keep the gas from leaking out (Figure 13.16). This is similar to a cylinder in an automobile engine, into which is sprayed a mixture of gasoline vapor and air. The mixture is ignited by a spark, and the chemical reactions raise the temperature and pressure very high very quickly. The piston is pushed outward, which turns a shaft that ultimately drives the wheels.

Figure 13.16 A cylinder with a piston that can move vertically.

Force and Pressure

Figure 13.17 shows a cylinder with a vertical-running piston on which we can load varying amounts of sand, in order to be able to control the pressure of the contained gas, and to be able to do controlled amounts of work on the gas. We will also put the cylinder in contact with hot or cold objects and allow energy transfer into or out of the cylinder.

QUESTION Consider the piston plus sand as the system of interest for a moment and think about what forces act on this system.

Figure 13.17 The piston can be loaded with varying amounts of sand.

A free-body diagram for the piston + sand system (Figure 13.18) includes the downward gravitational forces on the piston (Mg) and on the sand (mg), an

Figure 13.18 Forces on the piston.

upward time- and space-averaged force due to the pressure of the enclosed gas on the lower surface of the piston, and a downward time- and space-averaged force due to the pressure of the outside atmosphere on the upper surface of the system.

Since pressure is force per unit area, the upward force is PA, where A is the surface area of the bottom of the piston. Similarly, the downward force contributed by the outside atmosphere is $P_{air}A$. These pressure-related forces are not actually continuous, since they are the result of collisions of individual gas molecules with the piston and sand. However, the rate of collisions is so extremely high over the area of the piston that the force seems essentially constant. For example, in Exercise 13.X.2 you found that on average, at STP (standard temperature and pressure) about 6×10^{21} helium atoms strike a tiny 1-mm-diameter section every second.

> QUESTION In mechanical equilibrium (velocity of piston not changing), solve for the gas pressure inside the cylinder.

The Momentum Principle tells us that in equilibrium the net force on the piston + sand system must be zero, from which we are able to deduce that the pressure of the gas inside the cylinder is

$$P = P_{air} + \frac{Mg + mg}{A}$$

A Sudden Change

By varying the amount of sand (m) we can vary the pressure of the gas under study.

> QUESTION If you suddenly add a lot of sand, what happens?

If you suddenly add a lot of sand to the piston, there is suddenly a sizable nonzero net downward force on the piston + sand system: $F_{net} = AP_{air} + Mg + mg - AP \neq 0$. The piston starts to pick up speed downward. As it does so, it runs into gas molecules and tends to increase their speeds.

> QUESTION What happens to the temperature of the gas in the cylinder?

Since higher average speed means higher temperature, the gas temperature starts to increase at the same time that the volume of the gas is decreasing. This is a double whammy: both increased temperature and decreased volume contribute to increased pressure, since $P = (N/V)k_BT$. Therefore the pressure in the gas quickly rises, and eventually there will be a new equilibrium with a lower piston (supporting more sand) and a higher gas pressure in the cylinder.

However, getting to that new equilibrium is pretty complicated. If there is no friction or other energy dissipation, the piston will oscillate down and up, with the gas pressure going up and down. It is even possible to determine an effective "spring stiffness" for the gas and calculate the frequency of the oscillation. However, in any real system there will be some friction, so we know that the system will eventually settle down to a new equilibrium configuration.

Quasistatic Processes

To avoid these complicated (though interesting) transient effects, we will study what happens when we add sand very carefully and very slowly, one grain at a time, and we assume that the new equilibrium is established almost immediately. This is called a "quasistatic compression" because the system is at all times very nearly in equilibrium (Figure 13.19). Similarly, if we slowly remove one grain at a time, we can carry out a "quasistatic expansion." Note in particular that at no time does the piston have any significant amount of kinetic

Add or remove one grain at a time

Gas

Figure 13.19 By adding or removing one grain of sand at a time we effect a "quasistatic" compression or expansion.

energy, and macroscopic kinetic energy is essentially zero at all times. Of course there is plenty of microscopic kinetic energy in the gas molecules and the outside air molecules and in the thermal motion of atoms in the cylinder walls, piston, and sand.

Suppose we carry out a lengthy, time-consuming quasistatic compression by adding lots of sand, one grain at a time, very slowly. The piston goes down quite a ways, the pressure in the gas is a lot higher, and the volume of the gas is a lot smaller. If we know the pressure and volume, we can calculate the temperature by using the ideal gas law $P = (N/V)k_B T$ (assuming the gas density isn't too high to make this invalid).

Can we predict how low the piston will go for a given amount of added sand? Oddly enough, no—not without knowing something more about this device. There are two extreme cases that are both important in practice and calculable for an ideal gas. If the apparatus is made of metal (a very good thermal conductor) and is in good thermal contact with a large object at temperature T, the process will proceed at nearly constant temperature, with energy transfer from the gas into the surroundings. If the apparatus is made of glass (a very poor thermal conductor), the temperature of the gas inside the cylinder will rise in a predictable way, with negligible energy transfer to the surroundings in the form of energy transfer Q. We will analyze both kinds of processes: constant temperature processes and no-Q processes.

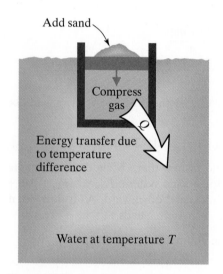

Add sand

Compress gas

Energy transfer due to temperature difference

Water at temperature T

Figure 13.20 A cylinder of gas immersed in a large volume of water.

Constant-Temperature (Isothermal) Compression

Suppose that the cylinder is made of metal (which is a very good thermal conductor) and is sitting in a very big tub of water whose temperature is T, as shown in Figure 13.20.

As we compress the gas, the temperature in the gas starts to increase. However, this will lead to energy flowing out of the gas into the water, because whenever the temperatures differ in two objects that are in thermal contact with each other, we have seen that there is a transfer of energy from the hotter object into the colder object. In fact, for many materials the rate of energy transfer is proportional to the temperature difference—double the temperature difference, double the rate at which energy transfers from the hotter object into the colder one.

The mechanism for energy transfer due to a temperature difference is that atoms in the hotter object are on average moving faster than atoms in the colder object, so in collisions between atoms at the boundary between the two systems it is likely that energy will be gained by the colder object and lost by the hotter object.

Energy transfer out of the gas will lower the temperature of the gas, since the total energy of the gas is proportional to the temperature. Quickly the temperature of the gas will fall back to the temperature of the water. The temperature of the big tub of water on the other hand will hardly change as a result of the energy added to it from the gas, because there is such a large mass of water to warm up.

Therefore the entire quasistatic compression takes place essentially at the temperature of the water, and the final temperature of the gas is nearly the same as the initial temperature of the gas. This is a constant-temperature compression (also called an "isothermal" compression, which just means constant temperature). Similarly, we can slowly remove sand from the piston and carry out a quasistatic constant-temperature expansion.

> QUESTION   Suppose that the original gas pressure and volume were P_1 and V_1, and as the result of a constant-temperature process the final volume is V_2. What is the final pressure P_2?

Since the temperature hasn't changed, from the ideal gas law $P = (N/V)k_B T$ we can deduce that $P_1 V_1 = P_2 V_2$, and therefore $P_2 = P_1(V_1/V_2)$.

Energy in a Constant-Temperature Compression

A more difficult question we can ask (and answer!) is this: How much energy was added to the water in the constant-temperature compression?

> QUESTION   What energy inputs and outputs were made to the gas? What energy change occurred in the gas?

The piston did work on the gas, and there was energy transfer Q out of the gas (and into the water). The Earth's gravitational force did work on the gas (since the center of mass of the gas went down), but this is negligibly small compared to the work done by the piston (the lowering of the heavy piston involves much more gravitational energy than the lowering of the low-mass gas).

> QUESTION   Did the total energy of the gas change (ignoring the small gravitational energy change)?

Since the total energy of an ideal gas is proportional to temperature (including rotational and vibrational energy if the gas is not monatomic), and we made sure that we kept the temperature constant, the total energy of the gas did not change. Therefore we have this energy equation for the open system that is the gas:

$$\Delta(\text{energy of gas}) = (W \text{ by piston}) - (|Q| \text{ that flowed into water}) = 0$$

Symbolically, $\Delta E_{gas} = W + Q = 0$, where Q is negative (transfer out of system). In thermal processes of this kind, the energy equation is called "the first law of thermodynamics."

If we can calculate the work done by the piston, we can equate that result to the amount of energy transfer Q into the water. The piston exerts a (time- and space-averaged) force PA on the gas, where P is the pressure in the gas and A is the cross-sectional area of the piston (Figure 13.21).

> QUESTION   If the piston drops a distance h, is the work it does PAh?

As the piston drops, the pressure in the gas increases, so the force is not constant. We have to integrate the variable force through the distance h in order to determine the work. If we measure x downward from the initial piston position, at each step dx of the way the increment of work done is $PA\,dx$. However, $A\,dx$ is an increment of volume (base area A time altitude dx), and the change in the volume is negative. Putting this all together, we have this:

$$F_{\text{on gas}} = PA$$

Figure 13.21 Force on the gas by the piston.

WORK DONE BY A PISTON ON A GAS

$$W = -\int_{V_1}^{V_2} P\,dV$$

Check the sign: If the volume decreases, the integral will be negative, and the minus sign in front of the integral makes the work done on the gas be positive, which is correct. Conversely, in an expansion of a gas, the integral is positive, and the work is negative, because the gas is doing work on the piston rather than the other way round.

> QUESTION   In a constant-temperature (isothermal) compression, work is done on the gas. Where does this energy flow go? How does the temperature stay constant?

As the piston moves down, it increases the average energy of the gas molecules that run into it. Therefore the input work starts to raise the temperature of the

Figure 13.22 The gas is compressed at a constant temperature T (the cylinder is immersed in a large tub of water at temperature T).

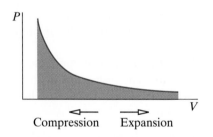

Figure 13.23 If T is kept constant, then the product PV is also constant.

gas (higher speeds), but this higher temperature causes energy transfer Q out of the cylinder, into the (very slightly) lower-temperature water. The net effect is for energy to flow into the gas in the form of mechanical work, and out of the gas into the water in the form of energy transfer Q. There is no change in the total energy of the gas, because the temperature of the gas didn't change. The outward energy transfer brings the gas temperature back to what it was before the falling piston tried to increase the temperature.

Now we can calculate quantitatively the work done by the piston in the constant-temperature compression, which is equal to the energy transfer from the gas to the surrounding water. We add lots of sand, one grain at a time (to maintain quasistatic equilibrium), and we compress the gas from an initial volume V_1 to a final volume V_2 (Figure 13.22).

Replace P by Nk_BT/V, where T is a constant:

$$W = -\int_{V_1}^{V_2} P\,dV = -\int_{V_1}^{V_2} \frac{Nk_BT}{V}\,dV$$

$$W = -Nk_BT\int_{V_1}^{V_2} \frac{dV}{V} = -Nk_BT[\ln V]_{V_1}^{V_2} = -Nk_BT\ln\left(\frac{V_2}{V_1}\right)$$

Since the process was a compression, V_2 is smaller than V_1. Since $-\ln(V_2/V_1) = \ln(V_1/V_2)$, we have

$$W_{\text{by piston}} = |Q_{\text{into water}}| = Nk_BT\ln\left(\frac{V_1}{V_2}\right)$$

Just as the symbol W is normally used for "work," so the symbol Q is normally used for "energy transfer due to a temperature difference." Here Q is negative (energy output).

Graphical Representation

There is a useful graphical representation of this constant-temperature process. We plot the pressure P vs. the volume V, as shown in Figure 13.23, and the process is represented by a curve of constant T (which is also a curve of constant PV, since $PV = Nk_BT$). The area under the process curve represents the work W, which is also the energy transfer Q that flows to the surroundings. In a compression the volume decreases, so we move to the left on the graph. In an expansion the volume increases, and we move to the right on the graph.

$$W = -\int_{V_1}^{V_2} P\,dV$$

The First Law of Thermodynamics

For historical reasons, in thermal processes the Energy Principle is called "the first law of thermodynamics."

THE FIRST LAW OF THERMODYNAMICS
$\Delta E_{\text{sys}} = W + Q$

In the particular case of constant-temperature (isothermal) compression of an ideal gas, we have

$$\Delta E_{\text{int}} = W + Q = 0$$

where E_{int} is the "internal" energy of the gas (the sum of the translational, rotational, vibrational, and other energy terms of all the molecules).

13.X.5 A mole of nitrogen is compressed (by piling lots of sand on the piston) to a volume of 12 liters at room temperature (293 K). The cylinder is placed on an electric heating element whose temperature is maintained at 293.001 K. A quasistatic expansion is carried out at constant temperature by very slowly removing grains of sand from the top of the piston, with the temperature of the gas staying constant at 293 K. When the volume is 18 liters, how much energy transfer Q has gone from the heating element into the gas? How much work W has been done on the piston by the gas? How much has the energy of the gas changed?

(You must assume that there is no energy transfer from the gas to the surrounding air, and no friction in the motion of the piston, all of which is pretty unrealistic in the real world! Nevertheless there are processes that can be approximated by a constant-temperature expansion. This problem is an idealization of a real process.)

Heat Capacity

We were able to calculate the important properties of a constant-temperature compression, where the apparatus was in good thermal contact with its surroundings. Before analyzing the opposite extreme, where the apparatus is thermally insulated from its surroundings and no energy transfer due to a temperature difference is involved, we will review the concept of heat capacity in this context.

Lock the piston in position so that the volume of the gas won't change (Figure 13.24), and so no work is done ($W = 0$). Put the apparatus in a big tub of water whose temperature is higher than the gas temperature, so that energy transfer will go from the water to the gas. Allow an amount of energy transfer Q due to the temperature difference to enter the gas and observe how much temperature rise ΔT has occurred in the gas. We define the "specific heat capacity C_V at constant volume" on a per-molecule basis in the following way, where N as usual is the total number of molecules in the gas:

$$\Delta E_{\text{thermal}} = N C_V \Delta T \text{ (constant volume)}$$

Since $W = 0$, this is also equal to Q. The larger the specific heat capacity, the smaller the temperature rise ΔT for a given energy input Q.

QUESTION What is C_V for a monatomic gas such as helium?

Since the total energy in a monatomic gas is $N\overline{K}_{\text{trans}} = \frac{3}{2}Nk_BT$, the energy transfer Q will increase the total energy of the gas by an amount $\frac{3}{2}Nk_B\Delta T$, so $C_V = \frac{3}{2}k_B$. Experimental measurements of the specific heat capacity of monatomic gases agree with this prediction.

For gases that are not monatomic, such as nitrogen (N_2), C_V can be larger than $\frac{3}{2}k_B$ because the total energy can be larger than $\frac{3}{2}Nk_BT$ due to rotational and vibrational energy relative to the center of mass of the molecule. Before quantum mechanics, theoretical predictions for the contribution of the rotational and vibrational energies to the specific heat capacity of gases did not agree with experimental measurements. This was one of the puzzles that was eventually resolved by quantum mechanics, as we found in the previous chapter.

Figure 13.24 If the piston is locked so the volume of the gas cannot change, we can measure C_V of the gas.

SPECIFIC HEAT CAPACITY AT CONSTANT VOLUME ON A PER-MOLECULE BASIS
$C_V = \frac{3}{2}k_B$ for a monatomic gas (He, etc.)
$C_V \geq \frac{3}{2}k_B$ for other gases (N_2, etc.)

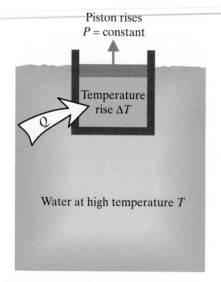

Piston rises
P = constant

Temperature
rise ΔT

Q

Water at high temperature T

Figure 13.25 If the piston is free to move, the pressure inside the cylinder remains the same, and we can measure C_p of the gas.

Heat Capacity at Constant Pressure

If we don't lock the piston but let the gas expand at constant pressure, the incoming energy transfer Q due to the temperature difference not only raises the energy of the gas by an amount $\Delta E_{thermal} = NC_V\Delta T$ but also raises the piston, which involves an amount of work W (Figure 13.25). We define the "specific heat capacity C_P at constant pressure" on a per-molecule basis as follows:

$$Q = NC_P\Delta T \text{ (constant pressure)}$$

Evidently C_P is bigger than C_V, because $NC_P\Delta T = NC_V\Delta T + W$. We can calculate the work done on the piston by the gas as the gas expands, which is the negative of the work done by the piston on the gas:

$$W = \int_{V_1}^{V_2} PdV = P\int_{V_1}^{V_2} dV = PV_2 - PV_1 = Nk_BT_2 - Nk_BT_1 = Nk_B\Delta T$$

Therefore for an ideal gas there is a simple relationship between C_V and C_P:

$$NC_P\Delta T = NC_V\Delta T + Nk_B\Delta T$$
$$C_P = C_V + k_B$$

13.X.6 What is the specific heat capacity at constant pressure on a per-molecule basis C_P for helium?

Molar Specific Heat Capacity

Specific heat capacity is often given on a per-mole basis rather than a per-molecule basis. The molar specific heat capacity at constant volume for an ideal gas is $C_V = \frac{3}{2}R$, where R is the molar gas constant (8.3 J/K), which is $6\times10^{23}k_B$. The molar specific heat capacity at constant pressure is $C_P = C_V + R$ for an ideal gas.

No-Q (Adiabatic) Compression

Initial volume
V_1

More sand

Final volume
V_2

Figure 13.26 If the cylinder and piston are made of insulating material, no energy transfer Q can occur between the gas and its surroundings. This is called an adiabatic compression.

With the apparatus made of metal and sitting in a big tub of water, the temperature during the compression didn't change. Now we analyze the opposite extreme. We make the cylinder and piston out of glass or ceramic (which are poor conductors for energy transfer due to a temperature difference) to minimize energy transfer out of the gas as the gas gets hotter during the compression. In fact, we make the approximation that there is no energy transfer at all (Figure 13.26). Such a no-Q process is also called "adiabatic" (which means "no energy transfer due to a temperature difference").

How realistic is such a no-Q approximation? For many real situations this can be a rather good approximation if the compression or expansion is fast, so that there isn't enough time for significant energy transfer due to a temperature difference. Such a flow of energy from one object to another is a rather slow process. For example, a cup of coffee sitting on a table may stay quite hot for ten minutes or more.

However, didn't we say that we were going to carry out compressions and expansions very slowly, "quasistatically"? Yes, so there is a contradiction. However, it is often the case that motion may be slow enough to be a good approximation to a quasistatic process and nevertheless may also be fast compared to the time required for significant energy transfer due to a temperature difference. Again, think of how long it takes a cup of coffee to cool off.

We again consider the gas in the cylinder as the system of interest, and the work done by the piston is equal to the increase in energy of the gas:

$$W = - \int_{V_1}^{V_2} PdV = NC_V \Delta T$$

QUESTION   Why did we use C_V in this equation? The volume isn't constant in this compression!

An ideal gas is unique among materials in that its total energy is entirely determined by the temperature, independent of volume or pressure. So the change of energy of the gas itself is just $NC_V \Delta T$, even when the volume is not constant. As a practical matter, real gases behave this way as long as the density is not too high. At high densities the energy of the gas includes a significant term associated with the interatomic forces.

We can work through the integration, starting from a "differential" form of the equation shown above:

$$-PdV = NC_V dT$$

Use the gas law $P = (N/V)k_B T$ to substitute for the pressure:

$$-Nk_B T \frac{dV}{V} = NC_V dT$$

Divide through by $k_B T$ and rearrange:

$$\left(\frac{C_V}{k_B} \right) \frac{dT}{T} + \frac{dV}{V} = 0$$

$$\left(\frac{C_V}{k_B} \right) \int \frac{dT}{T} + \int \frac{dV}{V} = 0$$

Now integrate:

$$\left(\frac{C_V}{k_B} \right) \ln T + \ln V = \text{constant}$$

Rewrite, using the properties of logarithms:

$$\ln T^{(C_V/k_B)} + \ln V = \ln \left(T^{(C_V/k_B)} V \right) = \text{constant}$$

Therefore we have

$$T^{(C_V/k_B)} V = \text{constant}$$

13.X.7   Use the ideal gas law to eliminate T from this expression, and show that $PV^\gamma = \text{constant}$ in a no-Q process, where the parameter γ is defined as the ratio of the constant-pressure specific heat capacity to the constant-volume specific heat capacity, $\gamma \equiv C_P/C_V$. (Since C_P is greater than C_V, γ is greater than 1.) In particular, if the initial pressure and volume are P_1 and V_1, and the final pressure and volume are P_2 and V_2, then $P_2 V_2^\gamma = P_1 V_1^\gamma$. Alternatively, we also have $T_2^{(C_V/k_B)} V_2 = T_1^{(C_V/k_B)} V_1$.

Graphical Representation

Again, there is a useful graphical representation of this no-Q process. In Figure 13.27 we plot the pressure P vs. the volume V, and the process is represented by a curve along which $PV^\gamma = \text{constant}$. The area under the process curve represents the work W. For comparison we also show the curve for a constant-temperature process (T and PV constant). The no-Q curve is much steeper than the constant-temperature curve.

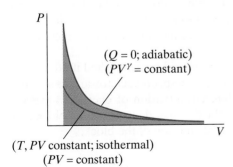

(Q = 0; adiabatic)
($PV^\gamma = \text{constant}$)

(T, PV constant; isothermal)
(PV = constant)

Figure 13.27 *PV curves for a no-Q (adiabatic) process and a constant temperature (isothermal) process.*

Work vs. Q

We have studied the response of a gas to energy inputs and outputs, both mechanical W (work) and thermal Q (energy transfer due to a temperature difference). Work represents organized, orderly, macroscopic energy input. Energy transfer due to a temperature difference represents disorganized, disorderly, microscopic energy input. There is randomness at the atomic level in the collisions of atoms, which is the basic mechanism for energy transfer due to a temperature difference. In the previous chapter we saw deep consequences of the distinction between work W and energy transfer Q.

Figure 13.28 An engine converts net energy transfer Q into net work W.

13.6 FUNDAMENTAL LIMITATIONS ON EFFICIENCY

An important technology involves the conversion of energy transfer Q due to a temperature difference into useful work W (Figure 13.28). For example, in a steam-powered electricity generating plant, coal is burned to warm up water that drives a steam engine to turn a generator, which converts the work done by the steam engine into electric energy. Energy conservation of course puts limits on how much useful work you can get from the burning of the coal—but there is a further limitation due to the second law of thermodynamics. It turns out that in a practical generating plant only about one-third of the energy of the coal can be turned into useful work! The fundamental problem is that energy transfer due to a temperature difference is a disorderly energy transfer, and the second law of thermodynamics takes that into account.

We will find that the most efficient processes are "reversible" processes—that is, processes that produce no change in the entropy of the Universe (a reversed movie of such a process looks possible). In the next sections we discuss two processes, mechanical friction and finite-temperature-difference energy transfer, which are major contributors to entropy production and whose effects must be minimized in order to obtain the most efficient performance in a mechanical system. Where we are headed is to establish limits on how efficiently thermal energy can be turned into mechanical energy in an engine, as a consequence of the second law of thermodynamics. This is a revealing example of the power of the second law of thermodynamics to set limits on possible phenomena.

Friction and Entropy Production

The second law of thermodynamics says that the entropy of the Universe never decreases. Portions of the Universe may experience a decrease of entropy, but only if the entropy of other portions increases at least as much (and typically more). An example of a process that increases the total entropy of the Universe is sliding friction. A block sliding across a table slows down as kinetic energy associated with the overall motion of the block is dissipated into random (thermal) energy inside the block and table, with an increase in entropy.

This friction process is irreversible. We would be astonished if after coming to rest the block suddenly started picking up speed back toward its starting position, although this would not violate conservation of energy. However, the probability is vanishingly small that the dissipated energy in the block and table could concentrate back into an orderly motion of the block.

Energy Transfer and Entropy Production

Mechanical friction contributes to increasing the entropy of the Universe. We will show that energy transfer between two objects that have different temperatures also makes the total entropy of the Universe increase. This is then a process to avoid, if possible, in an efficient engine.

Connect a metal bar between a large block at a high temperature T_H and another large block at a low temperature T_L (Figure 13.29). We write the rate of energy transfer from the high-temperature block (the "source") to the lower-temperature block (the "sink") as $\dot{Q}$. The dot over the letter Q means "rate" or "amount per second." Since Q has units of joules, $\dot{Q}$ has units of joules/s, or watts. This energy transfer rate is

- proportional to the "thermal conductivity" σ of the bar (metals have higher thermal conductivity than glass or plastic),
- proportional to the temperature difference (twice the temperature difference, twice the rate of energy transfer),
- proportional to the cross-sectional area A of the bar (twice the cross-sectional area is like having two bars), and
- inversely proportional to the length L of the bar (twice the length of the bar, half the energy transfer rate):

RATE OF ENERGY TRANSFER DUE TO A TEMPERATURE DIFFERENCE

$$\dot{Q} = \sigma A \frac{(T_H - T_L)}{L}$$

Units are J/S, or watts. $\dot{Q}$ is a positive quantity and is the absolute value of the rate of energy flow into or out of a system due to a temperature difference.

The quantity $(T_H - T_L)/L$ is called the "temperature gradient." The larger the gradient (the more rapidly the temperature changes with distance along the bar), the larger the number of joules per second of energy transfer.

We want to show you that energy transfer between different temperatures increases the total entropy of the Universe. Suppose that the source and sink are so large that for a while the energy transfer doesn't change their temperatures much. Since $1/T = (dS)/(dE)$, it follows that a small energy input (dE) into a system, of amount Q, leads to an entropy change of that system:

$$\Delta S = \frac{Q}{T}$$

Suppose that the energy transfer *rate* is $\dot{Q}$. In a short time interval Δt, the high-temperature source at temperature T_H loses an amount of energy $\dot{Q}\Delta t$.

QUESTION In this short time interval Δt, does the entropy of the source increase or decrease? By how much?

Since $\Delta S_H = (\Delta E)/T_H$, and the high-temperature source loses energy, we have $\Delta S_H = -(\dot{Q}\Delta t)/T_H$.

In this same time interval, the low-temperature sink at temperature T_L gains the same amount of energy, $\dot{Q}\Delta t$.

QUESTION In this short time interval Δt, does the entropy of the sink increase or decrease? By how much?

Since the low-temperature sink gains energy, we have $\Delta S_L = +(\dot{Q}\Delta t)/T_L$.

There is a crucial and perhaps not entirely obvious point: Does the entropy of the bar change during this time interval? No. The blocks are so large that their temperatures don't change much in a short time interval, so the temperatures at the ends of the bar, and all along the bar, are not changing. Every second, energy enters the bar, but the same amount of energy leaves the bar,

Figure 13.29 Energy flows at a rate $\dot{Q}$ from high temperature to low.

Cross-sectional area A

T_H

L

T_L

$\dot{Q}$

$\dot{Q}$

so the energy of the bar isn't changing. Nothing about the bar is changing. It is merely a conduit for the energy transfer, but its state is not changing.

Therefore the change in the total entropy of the Universe in a time interval Δt is

$$\Delta S_{\text{source}} + \Delta S_{\text{sink}} + \Delta S_{\text{bar}} = -\frac{\dot{Q}\Delta t}{T_H} + \frac{\dot{Q}\Delta t}{T_L} + 0 = \dot{Q}\Delta t \left(\frac{1}{T_L} - \frac{1}{T_H} \right)$$

> QUESTION Is this quantity positive or negative? Is this consistent with the second law of thermodynamics?

Since $T_L < T_H$, $1/T_L > 1/T_H$, and the entropy change of the Universe in this process is positive. This is consistent with the second law of thermodynamics, which states that the entropy of the Universe should never decrease.

Reversible and Irreversible Processes

An increase in the total entropy of the Universe is associated with an irreversible process, because to return to the earlier state would require that the total entropy of the Universe actually decrease, which won't happen with macroscopic systems. We would be astonished if the energy transfer suddenly reversed and ran from the colder block thermally back "uphill" to the hotter block!

There is only one way to carry out energy transfer due to a temperature difference reversibly or nearly so (that is, with little or no change in the total entropy of the Universe)—do it between systems whose temperatures are very nearly equal to each other ($T_H \approx T_L$). There's a practical problem, however.

> QUESTION If the two temperatures are nearly equal (to avoid increasing the entropy of the Universe), at what rate is energy transferred from the (slightly) hotter source to the (slightly) cooler sink?

Energy transfer will flow extremely slowly, because $\dot{Q}$ is proportional to the temperature difference (or more specifically to the temperature gradient). If there is hardly any temperature difference, the energy transfer rate will be very small. So we can carry out such energy transfers nearly reversibly, hardly changing the entropy of the Universe, but only if we do it so slowly as to be of little practical use in driving some kind of mechanical engine that converts thermal energy into work.

> **13.X.8** The thermal conductivity of copper (a good thermal conductor) is 400 watts/K/m (for comparison, the thermal conductivity of iron is about 70 watts/K/m, and that of glass is only about 1 watt/K/m). One end of a copper bar 1 cm on a side and 30 cm long is immersed in a large pot of boiling water (100°C), and the other end is embedded in a large block of ice (0°C). It takes about 335 joules to melt one gram of ice. How long does it take to melt one gram of ice? Does the entropy of the Universe increase, decrease, or stay the same?

13.7 | A MAXIMALLY EFFICIENT PROCESS

Despite the serious practical problem that reversible energy transfer due to a nearly zero difference in temperature proceeds infinitesimally slowly, we will analyze reversible "engines" in which (nearly) reversible energy input with tiny temperature differences is used (very slowly!) to do something mechanically useful, such as lift a weight or turn an electric generator. We will also assume that we can nearly eliminate sliding friction. The idea that we will pursue is to see how efficiently we can convert disorderly thermal energy into orderly mechanical energy (a lifted weight). We expect that reversible processes, which don't increase the total entropy of the Universe, should be the most efficient processes, though we recognize that these most efficient processes must proceed excruciatingly slowly.

Since we will use nearly reversible processes to do work, we could run the engine backward. Such a backward-running engine turns out to be a refrigerator—an engine in which work input to the engine can lead to extracting thermal energy from something to make it colder or to keep it cold.

The conception of a theoretically most efficient possible engine, and the recognition that this ideal engine would have to be reversible and not increase what we now call the total entropy of the Universe, was due to a young French engineer, Sadi Carnot, in 1824. His achievement is all the more remarkable because it came before the principle of energy conservation was established!

After using the second law of thermodynamics to determine the maximum possible efficiency of infinitely slow engines, in the last part of the chapter we will analyze engines that run at useful speeds. We will find as expected that they are even less efficient than the infinitely slow engines.

A Cyclic Process of a Reversible Engine

Some of the first practical engines drove pumps to pump water out of deep mines in England, repeatedly lifting large amounts of water large distances. The energy came from burning coal, which boiled water to make steam in a cylinder that pushed up on a piston and operated the pump. People started wondering how efficient such an engine could be. What limits the amount of useful work you can get from burning a ton of coal?

We describe a scheme for running an engine in a reversible way, which ought to be as efficient as possible. For concreteness, our engine is a device consisting of a cylinder containing an ideal gas, with a piston and some sand on the piston to adjust the pressure on the gas (Figure 13.30). This is a simple and familiar device that permits energy exchanges in the form of work and energy transfer due to a temperature difference. As we'll see, the actual construction details of the engine don't matter for the theoretical question regarding the maximum possible efficiency, though they matter very much in the actual construction of a useful engine.

A Constant-Temperature (Isothermal) Expansion

In addition to the engine itself, we need a large high-temperature source, large enough that extracting some energy from it will cause only a negligible decrease in its temperature. Start by arranging that the engine (the gas cylinder with its piston) has a temperature very slightly lower than the temperature T_H of the high-temperature source. We connect the engine to the high-temperature source and perform a reversible constant-temperature (isothermal) expansion of the engine, doing some work on the surroundings in the process (Figure 13.30). We lift the piston by slowly shifting some sand sideways from the piston onto nearby shelves. This requires almost no work on our part but results in the lifting of some weight. We make sure that the temperature of the gas doesn't change during this process. There is energy transfer into the engine because it is very slightly cooler, but the temperature difference is so slight that there is negligible change in the entropy of the Universe.

Entropy and Energy in Lifting the Piston

Remember that the entropy change of a system when there is energy transfer Q is $\Delta S = Q/T$. In lifting the piston the entropy of the high-temperature source has changed by an amount $\Delta S_H = -Q_H/T_H$, where Q_H is a positive

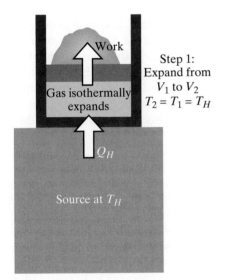

Figure 13.30 Constant-temperature expansion; gas temperature just slightly lower than T_H.

Step 1: Expand from V_1 to V_2 $T_2 = T_1 = T_H$

quantity. The entropy of the gas has changed by an amount $\Delta S_{gas} = +Q_H/T_H$. The entropy of the Universe hasn't changed at all. (The increased entropy of the gas is associated with the fact that there are more ways to arrange the molecules in a large volume than in a small volume.)

Because the temperature hasn't changed, there is no change in the energy of the ideal gas. Recall that the energy of an ideal gas, or a low-density real gas, is a function solely of the temperature, not the volume. (A dense real gas is more complicated, because the electric potential energy for pairs of molecules depends on distance, and therefore the energy of the gas depends on volume as well as temperature.)

We have succeeded in converting all of the energy transfer from the high-temperature source into useful work (lifting the piston), because none of the input energy went into changing the energy of the gas. This is 100% efficiency in converting thermal energy into useful work on the surroundings. What's the problem?

The Need for a Cycle

The problem is this: We need to be able to do this again, and again, and again. For example, we want to keep pumping water out of the deep mine, over and over. However, to repeat the lifting process, we have to return to the original state, with the gas compressed. We could simply reverse the process, doing work on the gas (constant-temperature compression), with energy transfer from the gas into the high-temperature source. The net effect, though, would be that there was zero net energy transfer to the gas, and zero net work done on the surroundings.

We need to run our engine in a nontrivial repetitive "cycle," where we can lift the piston repeatedly. Each time we lift the piston and do useful work, we need to get the piston back down again without having to do the same amount of work to push it down. If we can get the gas back to its original state with net work done on the surroundings, we will have a useful device.

One possibility for bringing the piston down would be to let the gas cool down. However, to lower the temperature we would have to place the gas cylinder in contact with a large object at some low temperature T_L, called the "sink" because, as we'll see, we will dump some energy into it. We can't place the gas cylinder in contact with the sink immediately, because the gas is at a high temperature T_H, and placing the hot gas in contact with the cold sink would mean that there would be a large temperature difference, and there would be a large increase in the entropy of the Universe associated with the energy transfer from the hot gas to the cold sink.

A No-Q (Adiabatic) Expansion

To avoid this large production of entropy, we need to bring the temperature of the gas down almost to T_L *before* making contact with the sink. To do this, we disconnect the gas cylinder from the high-temperature source and perform a reversible no-Q (adiabatic) expansion, which does some more work on the surroundings and lowers the temperature of the engine (Figure 13.31). This is accomplished by slowly removing some weight from the piston, allowing the gas to expand and the temperature to fall. We stop the expansion when the temperature of the gas is just slightly higher than the low temperature T_L of the sink.

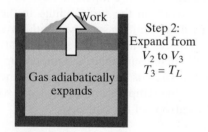

Step 2:
Expand from
V_2 to V_3
$T_3 = T_L$

Figure 13.31 No-Q expansion; the gas temperature drops to T_L.

A Constant-Temperature (Isothermal) Compression

We can now safely place the gas cylinder in contact with the low-temperature sink. This is okay as far as entropy production is concerned, because we made

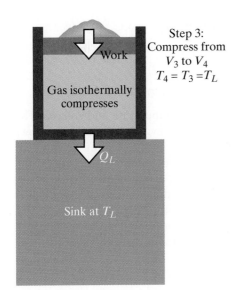

Step 3:
Compress from
V_3 to V_4
$T_4 = T_3 = T_L$

Figure 13.32 Constant-temperature compression; the gas temperature is just slightly higher than T_L.

sure that the temperature of the gas is only very slightly higher than the sink temperature T_L. The piston is now quite high, and we need to bring it down. We do this by slowly adding some weight to the piston, performing a constant-temperature (isothermal) compression (Figure 13.32). There is energy transfer into the sink. Because the temperature of the ideal gas is nearly T_L at all times, there is no energy change in the gas. Therefore the work that we do is equal to the energy transfer Q_L into the sink.

There is an increase in the entropy of the sink $\Delta S_L = +Q_L/T_L$, and there is an entropy decrease in the entropy of the gas $\Delta S_{gas} = -Q_L/T_L$. The entropy of the Universe doesn't change. The decreased entropy of the gas is associated with the fact that there are fewer ways to arrange the molecules in a small volume than in a large volume.

A No-Q (Adiabatic) Compression

We stop the compression at a particular state of the gas chosen with care so that we can do the following: We disconnect the engine from the low-temperature sink and perform a no-Q compression that raises the temperature of the gas to a temperature just barely less than T_H, the temperature of the high-temperature source. We do some work to perform this compression (Figure 13.33).

This compression brings the system back to its original state (density, pressure, temperature), so we can repeat the cycle of four processes all over again. The possible usefulness of this engine is that we can run it repeatedly, over and over. What remains to be analyzed is how much work is done on the surroundings in one cycle, and how much energy input we need to supply. Of course if we find that the net work is zero, the engine won't be useful. However, explicit calculations for an ideal gas show that at least in that case there is net work done on the surroundings. We will find that this is true for any reversible engine run in such a cycle.

This reversible cycle of two constant-temperature (isothermal) processes and two no-Q (adiabatic) processes running between a high-temperature source and a low-temperature sink is called a "Carnot cycle."

Step 4:
Compress from
V_4 to V_1
$T_1 = T_H$

Figure 13.33 No-Q compression; gas temperature rises to T_H. The engine is now back to its original state, ready to begin another cycle.

Summary of the Cycle

Figures 13.34 and 13.35 summarize the four processes of the engine cycle.

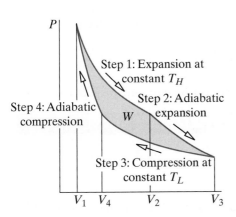

Figure 13.34 P vs. V for the engine cycle. The shaded area represents the net work W done by the engine on the surroundings.

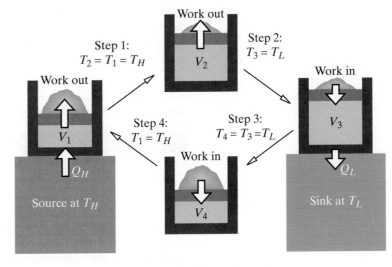

Figure 13.35 The engine cycle.

Step 1: Isothermal expansion in contact with T_H

Step 2: Adiabatic expansion (no contact); temperature falls to T_L

Step 3: Isothermal compression in contact with T_L

Step 4: Adiabatic compression (no contact); temperature rises to T_H

The four processes return the gas to its original temperature and volume. The question is, did we get any net work out of this cycle? Figure 13.34 shows the pressure vs. volume during the cycle. Earlier in this chapter we saw that the work done on the gas is $-\int_{V_1}^{V_2} P dV$, so the work done by the gas on the surroundings is $+\int_{V_1}^{V_2} P dV$, which is the area under one of these curves. Because V increases during steps 1 and 2, the work done on the surroundings is positive, but this work is negative in steps 3 and 4 because V decreases. The net work done on the surroundings is therefore represented by the shaded area shown in Figure 13.34. There is net work, related to the fact that the PV curves for the adiabatic processes (shown in blue and given by PV^γ = constant) are steeper than the PV curves for the isothermal processes (shown in red and given by PV = constant).

Entropy in a Cycle of a Reversible Engine

Let Q_H be the (absolute value of the) energy extracted from the source at temperature T_H during the constant-temperature expansion, and let Q_L be the (absolute value of the) energy dumped into the sink at temperature T_L during the constant-temperature compression. Remember that the entropy change of a system when there is energy transfer Q is $\Delta S = Q/T$.

$$\text{The entropy change of the high-temperature source in one cycle} = -\frac{Q_H}{T_H}$$

$$\text{The entropy change of the low-temperature sink in one cycle} = +\frac{Q_L}{T_L}$$

QUESTION   Is there any entropy change *in the engine* in one cycle?

The entropy change of the engine in one cycle is zero, because the gas is brought back to its original state.

QUESTION   Therefore, in one cycle, how much entropy change is there in the surroundings as a result of the work done on and by the surroundings?

Only the energy transfer due to a temperature difference affects the entropy during one cycle (work doesn't contribute), and we calculated these through the formula $\Delta S = Q/T$. We were careful to run the engine in a reversible way, avoiding making any increase in the total entropy of the Universe, so in one cycle we have

$$-\frac{Q_H}{T_H} + \frac{Q_L}{T_L} + 0 = 0$$

Therefore we can write this important relationship for one cycle of the engine:

ONE CYCLE OF A REVERSIBLE ENGINE
$$\frac{Q_H}{T_H} = \frac{Q_L}{T_L}$$

This is a surprisingly simple result for such a complicated process. What does this simple result depend on? It depends solely on the second law of thermodynamics, that statistically the total entropy of the Universe will essentially never decrease.

We illustrated the processes in the cycle with a cylinder filled with an ideal gas, but the result above doesn't actually depend at all on the details of how the engine is constructed. In particular, it doesn't matter whether the engine contains helium gas, or a solid block of rubber, or a liter of liquid alcohol. There would be an easily visible difference between using a solid or liquid rather than a gas in the engine, because the distance through which the piston would move would be much smaller than with a gas. However, the simple result shown above would remain the same.

Energy in a Cycle of a Reversible Engine

Let's review the energy changes in the engine in one cycle. The source inputs an amount of energy Q_H. The sink removes an amount of energy Q_L. The working substance returns to its original state, so it undergoes no change of energy. Is there anything else? Yes, it may be that there has been net work W done on the surroundings. How can we tell? We can use the fact that the internal energy of the engine does not change in one complete cycle:

$$\Delta E_{\text{engine}} = Q_H - Q_L + W = 0$$

QUESTION   Use the result that $Q_H / T_H = Q_L / T_L$ (because the total entropy of the Universe doesn't change), to determine whether W is positive or negative.

Because $Q_H = (T_H / T_L) Q_L > Q_L$, we have $W = Q_L - Q_H < 0$. The fact that W is negative means that our engine does net work on the surroundings in one cycle. Figure 13.36 outlines the basic scheme. In one cycle the net effect is that the engine takes in energy Q_H from a hot source, does some work on something (for example, turns an electric generator), and exhausts the remaining energy Q_L into a cold sink. The magnitude of the work done is $|W| = Q_H - Q_L$.

The crucial issue is that the engine will not run in repeatable cycles without exhausting some energy to the low-temperature sink. You *cannot* convert all of the energy Q_H into useful work; a portion of Q_H goes into the low-temperature sink (Q_L). We need the low-temperature block to allow us to begin the constant-temperature compression phase of the cycle in a way that avoids any energy transfer with significant temperature difference, which would increase the total entropy of the Universe.

We have to pay for the high-temperature energy Q_H that we use, and the loss of some of the input energy in the form of Q_L is an unfortunate fact of life. We pay for coal or fuel oil or electricity to warm something up to the high temperature T_H, from which we can draw energy to do work for us.

For example, in an old railroad steam engine a fire maintained water at a high temperature to constitute the high-temperature source. In an automobile engine we burn gasoline to create a high temperature and push the pistons. In an electricity generating station we burn coal or fuel oil, or use nuclear fission reactions, to create a high-temperature source from which to drive the generators. We'd like to get our full money's worth (Q_H), but we don't. We only get $|W| = Q_H - Q_L$, having "lost" some of the energy to a low-temperature sink, which typically is our ordinary surroundings at around 20°C (293 K).

Efficiency

This leads to the question, what fraction of the energy Q_H that we pay for turns into useful work W? This fraction is called the "efficiency" of the engine:

$$\text{efficiency} \equiv \frac{|W|}{Q_H}$$

Figure 13.36 Conservation of energy lets us determine the net work output.

QUESTION   Given this definition of efficiency, prove the following result:

THE EFFICIENCY OF A REVERSIBLE ENGINE

$$\text{Efficiency} = 1 - \frac{T_L}{T_H}$$

We have $|W|/Q_H = (Q_H - Q_L)/Q_H = 1 - Q_L/Q_H = 1 - T_L/T_H$. This may seem surprising. The efficiency of a reversible engine depends solely on the ratio of the absolute temperatures of the source and sink. We emphasize that it doesn't matter how the engine is designed or what kind of materials it is made of. This is the highest efficiency we can ever achieve for converting thermal energy into useful work.

> **13.X.9**   What is the efficiency of a reversible engine if the source is a large container of boiling water and the sink is a large block of melting ice?
>
> **13.X.10**   An actual electricity generating plant is powered by a nuclear reactor, with a high temperature of 300°C and low temperature of 25°C (near room temperature). What is the efficiency, assuming that we can treat the processes as being reversible?

No Other Engine Can Be More Efficient

We can show that no engine running between temperatures T_H and T_L can be more efficient than a reversible engine.

The proof is a "proof by contradiction." Suppose that an inventor claims to have invented some cleverly designed new kind of (cyclic) engine, with a higher efficiency than $(1 - T_L/T_H)$ when run between these same two temperatures. In that case, for a given Q_H the new engine would exhaust less Q_L than is exhausted from a reversible engine, and the entropy change of the Universe with this new engine would be negative instead of zero:

$$-\frac{Q_H}{T_H} + \frac{Q_L}{T_L} < 0 \quad \text{(Impossible!)}$$

This would be a violation of the second law of thermodynamics, so it is impossible. We can be quite sure that the inventor's claims are not valid. Note carefully that the inventor's claims don't violate energy conservation. Nothing about energy conservation forbids converting 100% of the input energy Q_H into work W in a cycle. The impossibility lies, rather, in the massive improbability of seeing the total entropy of the Universe decrease.

We took great pains to make a reversible engine, for which the total entropy change of the Universe was zero. No other engine can be more efficient. In fact, any real engine will have some friction and some energy transfer across differing temperatures, in which case the entropy change of the Universe will be positive. So the second law of thermodynamics leads to the following, where "= 0" applies only to ideal, reversible engines, and "> 0" applies to real engines:

ENTROPY CHANGE OF THE UNIVERSE FOR REAL ENGINES

$$\Delta S_{\text{Universe}} = -\frac{Q_H}{Q_H} + \frac{Q_L}{T_L} \geq 0$$

We should not underestimate the need for the ingenuity of inventors, however. The reversible engine made with an ideal gas cylinder is pretty useless for practical purposes. The challenge to ingenuity is to design engines that are practical, and this involves many kinds of engineering design decisions and trade-offs. However, no matter how ingenious the design, the second law of thermodynamics puts a rigid limitation on how efficient *any* engine can be.

QUESTION If $\Delta S_{\text{Universe}} > 0$ in one cycle of an engine, show that the efficiency is less than the ideal efficiency $(1 - T_L/T_H)$ obtained with a reversible engine.

$$\Delta S_{\text{Universe}} > 0$$

$$\frac{Q_L}{T_L} > \frac{Q_H}{T_H}$$

$$\frac{Q_L}{Q_H} > \frac{T_L}{T_H}$$

$$W = Q_H - Q_L = Q_H\left(1 - \frac{Q_L}{Q_H}\right) < Q_H\left(1 - \frac{T_L}{T_H}\right)$$

$$\text{efficiency} = \frac{W}{Q_H} < \left(1 - \frac{T_L}{T_H}\right)$$

Running the Engine Backward: A Refrigerator or Heat Pump

We could run our engine backward, since we took care to make all aspects of the cycle reversible. Starting from our original starting point, we would disconnect from the hot block, perform a no-Q expansion to lower the temperature to that of the cold block, connect to the cold block, do a constant-temperature expansion, then disconnect from the cold block. Next we do a no-Q compression to raise the temperature to that of the hot block, connect to the hot block, and do a constant-temperature compression back to the original state.

The net effect in one cycle is that the low-temperature block gives some energy Q_L to the engine, the high-temperature block absorbs some energy Q_H from the engine, and we do some (positive) work on the system (instead of the engine doing work on the surroundings). Here is the Energy Principle when we run this reversed engine:

$$\Delta E_{\text{engine}} = Q_L - Q_H + W = 0$$

QUESTION Use the result that $Q_H/T_H = Q_L/T_L$ (which is still valid for our reversed engine, as you can convince yourself) to determine whether W is positive or negative.

Since $Q_H = (T_H/T_L)Q_L > Q_L$, we have $W = Q_H - Q_L > 0$. This seems an odd kind of "engine": it doesn't do any work for us—we have to do work on it. Is that of any use? Yes! At the cost of doing some work, we extract energy from a low-temperature block and exhaust it into a high-temperature block. This is a refrigerator (Figure 13.37).

Consider how we keep food cold in an ordinary home refrigerator. A reversed engine maintains the inside of the refrigerator at a low temperature T_L, while the room is at a higher temperature T_H. Although the door and walls are heavily insulated, some energy does leak through from the room, and this energy, Q_L, must be removed to keep the food at a constant low temperature T_L. We exhaust an amount of energy Q_H into the room at temperature T_H. It takes an amount of work W to achieve this effect of moving energy out of the cold region and into the warm region.

In the most favorable case (no increase in total entropy) we have the following two conditions stemming from entropy and energy considerations (the

Figure 13.37 Run the engine backward and you have a refrigerator.

high-temperature exhaust energy must equal the sum of the low-temperature energy plus the work we put in to drive the cycle):

$$\frac{Q_H}{T_H} = \frac{Q_L}{T_L} \quad \text{and} \quad Q_H = Q_L + W$$

Here our view of "efficiency" shifts a bit. What we care about is how much work W we have to do to remove an amount of leakage energy Q_L from the inside of the refrigerator.

QUESTION   Show that the energy removal we get per amount of work done is as follows:

$$\frac{Q_L}{W} = \frac{1}{(T_H/T_L) - 1} > 1$$

Isn't this result somewhat surprising? Are we getting something for nothing when Q_L is bigger than the work W that we do? No, there's no violation of energy conservation. It's just that Q_L plus W equals the output energy Q_H, and we only have to add a modest amount of work to drive the energy "uphill."

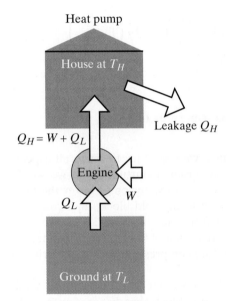

Figure 13.38 Run the engine backward and you have a heat pump.

Heat Pump

A related device is the "heat pump" used in some areas to warm homes by driving low-temperature energy "uphill" into a higher-temperature house. Figure 13.38 shows a house whose interior is maintained at a temperature T_H despite continual leakage Q_H into the outside air. Energy Q_L at low temperature T_L is pumped out of the ground into the house, by the addition of some work W that we do in an engine called a heat pump. This is somewhat similar to the refrigerator situation, except that now what we care about is how much high-temperature energy Q_H we get per amount of work W done by the heat pump.

QUESTION   Show that this ratio is given by the following formula for the heat pump:

$$\frac{Q_H}{W} = \frac{1}{1 - (T_L/T_H)} > 1$$

Again, this feels like we're getting something for nothing, since Q_H is greater than the work W that we do. After all, if you warm the house directly with gas or oil or electric energy, you have to pay for every joule of leakage, not some small fraction of the leakage. So why aren't heat pumps much more commonly used than they are? Basically because we've been calculating the best deal we can possibly obtain (the case of the reversible engine—that is, one whose operation leads to no increase in the total entropy of the Universe). Real engines running either forward or reversed don't attain the theoretical maximum performance, as we will see in the next sections.

13.X.11 Suppose that there is a leakage rate of 50 watts through the insulation into a refrigerator, which we maintain at 3°C. What is the minimum electric power required to continually remove this leakage energy, to maintain the low temperature? In that case, what is the rate at which energy is exhausted into the room? Room temperature is about 20°C.

13.X.12 Suppose that the temperature underground from where we draw low-temperature energy for a heat pump is about 5°C, and we want to keep the house at a temperature of 20°C. How many joules of work must our heat pump supply for every joule of leakage of energy there is out of the house?

*WHY DON'T WE ATTAIN THE THEORETICAL EFFICIENCY?

In Exercise 13.X.10 you calculated the reversible-engine efficiency of an actual nuclear-powered electricity generating plant to be 48%, but the observed efficiency of this plant is only 30%. Real engines do not attain the efficiency predicted by the second law of thermodynamics for ideal reversible engines. In fact, the real efficiency is often only about half of the theoretical efficiency. The reason for less than optimum performance in real engines is partly due to mechanical friction, but this effect can be minimized by good design and proper lubrication. The main limitation on performance comes from the practical necessity of incorporating energy transfers between parts of the system that are at significantly different temperatures, which leads to sizable increases in the total entropy of the Universe, and to much reduced efficiency.

The problem is speed. As we saw earlier in the chapter, the rate of energy transfer $\dot{Q}$ in joules per second between two objects (such as the hot or cold block and the engine) at temperatures T_H and T_L connected by some conducting material of length L and cross-sectional area A, with thermal conductivity σ is this:

$$\dot{Q} = \sigma A \frac{(T_H - T_L)}{L}$$

QUESTION   What is the rate of energy transfer in a reversible engine when in contact with the hot block or the cold block?

Alas, a reversible engine is totally impractical when it comes to getting anything done in a finite amount of time, because the rate of energy transfer is zero. If an engine is perfectly reversible, a cycle takes an infinite amount of time!

Consider the portion of an engine's cycle where the working substance is in thermal contact with the high-temperature source. In order to carry out the expansion quickly, there must be a high rate of energy transfer from the source into the working substance. That means we need a large contact area A, a short distance L, and a high thermal conductivity σ. Most significant of all, the temperature of the source (T_H) must be considerably higher than the temperature of the engine, leading to irreversibility and increase of the entropy of the Universe.

The Efficiency of a Nonreversible Engine

It is actually possible to calculate the effect of such nonreversible energy transfer, as was pointed out in an article titled "Efficiency of a Carnot engine at maximum power output," by F. L. Curzon and B. Ahlborn, *American Journal of Physics,* volume 43, pages 22–24 (January 1975).

Suppose that when the engine is in contact with the high-temperature block at temperature T_H, the engine is at a lower temperature T_1, so that there is a finite energy transfer rate into the engine of $\dot{Q}_H = b(T_H - T_1)$ joules per second, where b is a constant that lumps together the factors of thermal conductivity, cross-sectional area, and length ($\sigma A / L$). Similarly, assume that when the engine is in contact with the low-temperature block at temperature T_L, the engine is at a higher temperature T_2, so that there is a finite energy transfer rate into the working substance of $\dot{Q}_L = b(T_2 - T_L)$ joules per second. We're making the factor b be the same for both the high-temperature and the low-temperature contacts. This simplifies the algebra, and it turns out that using different b factors doesn't affect the final result anyway.

$\dot{Q}_H = b(T_H - T_1)$

T_1

Intermediate temperatures

Engine

T_2 $|\dot{W}| = \dot{Q}_H - \dot{Q}_L$

$\dot{Q}_L = b(T_1 - T_L)$

Sink at T_L

Figure 13.39 An engine that runs at a nonzero rate.

Very slow; no input or output power

Straight connection; no output at all!

Source at T_H Source at T

$\dot{Q}_H$

Q_H

Engine

Q_L $|W| = Q_H - Q_L$

$\dot{Q}_L = \dot{Q}_H$

Sink at T_L Sink at T_L

$\dot{Q}_H = 0$ $|\dot{W}| = 0$
$\dot{Q}_L = 0$
$|\dot{W}| = 0$

Figure 13.40 Two extreme designs where there is zero power output.

Now the energy flow diagram looks like Figure 13.39, with an ideal reversible engine running between the temperatures T_1 and T_2, but with irreversible finite-rate energy transfer between this reversible engine and the source and sink at temperatures T_H and T_L.

Energy enters the engine from the high-temperature source at a rate $\dot{Q}_H$, work is done at a rate $\dot{W}$, and energy is dumped into the low-temperature sink at a rate $\dot{Q}_L$. All of these quantities are measured in watts (joules per second). From energy conservation we have

$$\dot{Q}_H = \dot{W} + \dot{Q}_L$$

Curzon and Ahlborn pointed out that although this engine will necessarily be less efficient than a reversible engine (due to the irreversible energy transfers), it might be useful to ask the question, "What is the *maximum* output power $\dot{W}$ per input power $\dot{Q}_H$?" This question comes down to the question of what value of b will maximize the output power. This is essentially an engineering design question, yet there is an unusually simple answer to this question about the maximum output power.

There Must Be a Maximum Power Output

By considering two extreme designs (Figure 13.40) it is easy to see that there must be a maximum possible power output (per power input). One extreme set of operating conditions is the reversible engine we've already considered, in which all the energy flow rates are zero, including the output power $\dot{W}$.

Another extreme is to arrange conditions in such a way that $T_1 \approx T_2$, so that we have essentially connected the source and sink by a conducting bar. In that case the input power $\dot{Q}_H$ is equal to $\dot{Q}_L$, and the useful output power $\dot{W}$ is again zero.

Somewhere between these two extremes, for both of which there is zero output power, we expect to find a maximum output power. We now look for the conditions that maximize the output power. The search will be carried out by expressing the output power in terms of T_1, the higher of the two temperatures to which the engine is directly exposed in Figure 13.39. Then we will vary T_1 (by adjusting b in Figure 13.39) until we maximize the power output. The details are given in a derivation at the end of the chapter. We obtain the following surprisingly simple result:

THE EFFICIENCY OF A MAXIMUM-POWER ENGINE

$$\frac{\dot{W}}{\dot{Q}_H} \leq 1 - \sqrt{\frac{T_L}{T_H}}$$

This formula gives us the upper bound on the fraction of the thermal input power that can end up as output mechanical power if we maximize the output power. The formula looks quite similar to the reversible-engine zero-power efficiency, but the square root makes a big difference.

Comparison with the Real World

Curzon and Ahlborn give an example of a plant powered by a nuclear reactor, with high temperature 300° C and low temperature 25° C. Using the formula given above, the calculated efficiency is 28%, and the actual observed efficiency is 30% (see Exercise 13.X.13 at the end of this section). This confirms the estimate that Curzon and Ahlborn made that

- the main source of irreversibility is in finite-temperature-difference energy transfer, and
- power plants will normally be run in such a way as to maximize output power.

There is an economic issue here. Suppose that you are responsible for building and operating an electricity generating plant, and you must supply one megawatt of power (one million joules per second). If the operating expense of buying fuel is extremely high but capital investment in generators is inexpensive, it makes sense to build lots of generators and run them very slowly, nearly reversibly, to make the required megawatt of electric power. If the operating expense of buying fuel is relatively low but capital investment in generators is expensive, it makes sense to build few generators and go for maximum power, even though the fuel is not used very efficiently in producing the megawatt of electric power.

Similar issues of maximum efficiency apply to refrigerators and heat pumps. Energy leaks into a refrigerator at some rate (joules per second) and must be removed by "pushing it thermally uphill" to a higher temperature, the temperature of the room. However, in order to have a nonzero rate of energy flow from the food to the engine, the engine must reach an even lower temperature than the food, with entropy-producing energy transfer from the food into the engine. Also, in order for there to be a nonzero rate of energy flow from the engine to the room air, the engine must reach an even higher temperature than the air, with entropy-producing energy transfer from the engine into the air. Schematically the situation looks like Figure 13.41.

Often a refrigerator or freezer has exposed coils (a "heat exchanger") where the energy transfer occurs between the working substance circulating in the coils and the air. If you touch the coils, you find that they are indeed much hotter than room temperature, in order to drive a sufficiently high rate of energy transfer into the air. Notice not only that the heat exchanger must be hotter than the air, but also that it is a rather large and costly device because it has to have a large surface area to get a large rate of energy transfer.

In the case of heat pumps, where the heat pump picks up low-temperature energy underground, the heat pump must reach an even lower temperature in order to get a nonzero rate of energy flow from the ground into the engine. Also, in order for there to be a nonzero rate of energy flow into the house, the engine must reach an even higher temperature than the air, with entropy-producing energy transfer from the engine into the air. The radiators (heat exchangers) in the house must be considerably hotter than the air. There is considerable expense in all the metal in the ground and in the house that enables adequate energy transfer rate. Schematically the situation looks like Figure 13.42.

So although a heat pump does have a theoretical advantage in warming a house in part from energy in the cold ground, the expense of the heat exchangers and the problems of going to rather low temperatures in the ground are practical limitations, especially in very cold climates. Heat pumps are more useful in climates where the winters are not too severe.

Figure 13.41 A refrigerator with nonzero cooling rate.

Figure 13.42 A heat pump with nonzero energy transfer rate due to a temperature difference.

13.X.13 Curzon and Ahlborn give an example of a plant powered by a nuclear reactor, with high temperature 300°C and low temperature 25°C. In Exercise 13.X.10 you calculated the reversible-engine zero-power efficiency to be 48%. Now calculate the non-reversible-engine maximum-power efficiency for this power plant and compare it with the observed efficiency, which is 30%.

Figure 13.43 A 2D random walk, generated by a computer program, using random numbers and many steps.

If you would find it useful to study one more application of the basic statistical concepts underlying our analysis of a gas, here is an interesting one. If we could watch one special molecule wandering around in the air, colliding frequently with air molecules, its path would look something like Figure 13.43. This kind of motion is called a "random walk." It is somewhat surprising to find that despite the random nature of this motion we can calculate something significant about the motion, using simple statistical reasoning.

One Dimension

For simplicity, first we'll consider just the x component of the motion. Pick the origin of the x axis to be at the original position of the special molecule. The first thing that happens is that the special molecule moves until it collides with an air molecule. We call this first x component of the displacement Δx_1.

This component of the displacement may be to the right ($+x$ direction) or to the left ($-x$ direction). As a result of the collision, the special molecule may change direction, and change speed. The next displacement before another collision we call Δx_2, and so on. After N collisions, the net displacement Δx away from the origin is

$$\Delta x = \Delta x_1 + \Delta x_2 + \Delta x_3 + \Delta x_4 + \cdots + \Delta x_N$$

After N collisions, what is the average (most probable) position of the special molecule? On average, it is just as likely that it has moved to the right as moved to the left, so the average x component of displacement ought to be zero. It is easy to see that this will be the case, by taking the average value:

$$\overline{\Delta x} = \overline{\Delta x_1} + \overline{\Delta x_2} + \overline{\Delta x_3} + \cdots + \overline{\Delta x_N}$$

Each of the individual x displacements are equally likely to be to the left or the right, so the average value of the nth x displacement (for $n = 1, 2, 3, \ldots N$) is zero. Therefore the average value of the net x component of displacement is also zero.

On average, after N collisions the special molecule ends up to the left of the origin as often as it ends up to the right. However, we can ask the question, "On average, how far away from the origin (along the x axis) does the special molecule get, no matter whether it ends up to the left or the right?" A good indicator of this distance is the average value of the square of the net x displacement, $(\Delta x)^2$, because that's always a positive number. We can calculate this average.

Since $\Delta x = \Delta x_1 + \Delta x_2 + \Delta x_3 + \Delta x_4 + \cdots + \Delta x_N$, we have

$$(\Delta x)^2 = (\Delta x_1)^2 + (\Delta x_2)^2 + \cdots + (\Delta x_N)^2 + 2\Delta x_1 \Delta x_2$$
$$+ 2\Delta x_1 \Delta x_3 + 2\Delta x_1 \Delta x_4 + \cdots$$

In this square of the net x displacement, there are two kinds of terms: squares of individual x displacements such as $(\Delta x_2)^2$ and "cross terms" like $2\Delta x_1 \Delta x_3$. Given the random nature of the process, we expect that one-fourth of the time both Δx_1 and Δx_3 are positive (product is positive), one-fourth of the time they're both negative (product is positive), one-fourth of the time Δx_1 is positive with Δx_3 negative (product is negative), and one-fourth of the time Δx_1 is negative with Δx_3 positive (product is negative). Therefore, the average value of each cross term $\overline{2\Delta x_i \Delta x_j}$ $(i \neq j)$ is zero.

As for the other terms, the squares of individual x displacements such as $(\Delta x_2)^2$, the average value of each of these terms is some number d_x^2, related to d^2, the square of the mean free path that we discussed in Section 13.3.

QUESTION Count up how many of these square terms there are for N collisions and calculate the average value of the square of the net displacement:

$$\overline{(\Delta x)^2} = ?$$

There are N of these terms, and the square root of this quantity is the "root-mean-square" or "rms" value of the net displacement:

$$\Delta x_{\text{rms}} = \sqrt{\overline{(\Delta x)^2}} = (\sqrt{N})d_x$$

We can also write this formula in terms of time. If we let v be the average speed of the special molecule between collisions and we let T be the average time between collisions, we have $v = d/T$. Also, the total time t for the N collisions is $t = NT$.

QUESTION Rewrite the formula for the rms displacement in terms of just d_x, v, and t (that is, eliminate N and T): $\Delta x_{\text{rms}} = ?$

We find that $\Delta x_{\text{rms}} = \sqrt{N d_x d_x} = \sqrt{N(vT)d_x} = \sqrt{v d_x}\sqrt{t}$. This is a somewhat curious result. For ordinary motion at constant speed, displacement increases proportional to time: double the time, double the displacement. In contrast, the rms displacement in a random walk grows with the square root of the time: on average, the rms displacement doubles when the time quadruples.

This calculation was done for the x component of the motion, but we can generalize our result to real three-dimensional motion in a gas. Figure 13.44 shows a three-dimensional displacement involving Δx, Δy, and Δz.

You can see in Figure 13.44 a three-dimensional version of the Pythagorean theorem for triangles. We can assume that the motion in each dimension is independent of the motions in the other two dimensions, and we have the following result:

$$\Delta r_{\text{rms}} = \sqrt{\overline{(\Delta x)^2} + \overline{(\Delta y)^2} + \overline{(\Delta z)^2}} = \sqrt{N d_x^2 + N d_y^2 + N d_z^2} = (\sqrt{N})d$$

(Here, d_x is the component in the x direction of the three-dimensional mean free path d.)

Writing the rms displacement in terms of the average speed v, the time t, and the mean free path d, we have

$$\Delta r_{\text{rms}} = \sqrt{v d}\sqrt{t}$$

This result is somewhat unusual, because it predicts (correctly, it turns out) that for this random process the displacement from the starting location is proportional to the square root of the time rather than to the time.

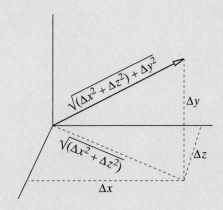

Figure 13.44 Pythagorean theorem in three dimensions.

13.X.14 As you have calculated, the average speed of an air molecule at room temperature is about $500\,\text{m/s}$, and we saw in Section 13.3 that the mean free path is about 7×10^{-8} m. What is Δr_{rms} for an air molecule after one second?

13.10   *DERIVATION: MAXIMUM-POWER EFFICIENCY

In this appendix we find the conditions under which the power output is maximized for an engine that connects to high- and low-temperature reservoirs through finite-temperature-difference connections.

Expressing the Power in Terms of T_1

We will express the power output in terms of T_1, the higher temperature to which the reversible engine is subjected. We assume that the engine, the part that runs between the temperatures T_1 and T_2, can be considered to be a reversible engine because there are no different-temperature energy transfers within it, and we're assuming that we can make mechanical friction negligibly small.

> **QUESTION**   In terms of the temperatures T_1 and T_2, what does the second law of thermodynamics say about the relationship between $\dot{Q}_H$ and $\dot{Q}_L$?

Since we have a reversible engine, we know that $\dot{Q}_H/T_1 = \dot{Q}_L/T_2$. Using this relationship, we can write a formula for the energy transfer into the low-temperature sink in two different ways:

$$\dot{Q}_L = b(T_2 - T_L), \quad \text{and also} \quad \dot{Q}_L = \frac{T_2}{T_1}\dot{Q}_H = \frac{T_2}{T_1}b(T_H - T_1)$$

Therefore we have

$$b(T_2 - T_L) = \frac{T_2}{T_1}b(T_H - T_1)$$

With a bit of manipulation we can solve this equation for T_2 in terms of T_1:

$$T_2 = \frac{T_L T_1}{2T_1 - T_H}$$

For the inner, reversible engine, the input power is of course $\dot{Q}_H = b(T_H - T_1)$, and using the result for the efficiency of a reversible engine, we have

$$\dot{W} = (\dot{Q}_L)(\text{efficiency}) = b(T_H - T_1)\left(1 - \frac{T_2}{T_1}\right)$$

Using our solution for T_2 in terms of T_1, this becomes

$$\dot{W} = b(T_H - T_1)\left(1 - \frac{T_L}{2T_1 - T_H}\right)$$

Varying the Thermal Conduction to Maximize the Power

We consider the source and sink temperatures T_H and T_L to be fixed, so the only temperature that is unspecified in this equation is T_1, which we will vary until we maximize the output power $\dot{W}$. Physically, this corresponds to varying the parameter b that determines the energy transfer rate. As usual, we maximize $\dot{W}$ by differentiating it with respect to T_1, then set this derivative equal to zero (corresponding to a maximum, where the slope is zero):

$$\frac{d\dot{W}}{dT_1} = b\left[-1 + \frac{T_L}{2T_1 - T_H} + (T_H - T_1)\frac{2T_L}{(2T_1 - T_H)^2}\right] = 0$$

Solving for T_1 and simplifying, we get a quadratic equation:

$$4T_1^2 - (4T_H)T_1 + (T_H^2 - T_L T_H) = 0$$

There are two solutions to the quadratic equation:

$$T_1 = \frac{T_H \pm \sqrt{T_L T_H}}{2}$$

Only the "+" sign makes physical sense in this situation, because if T_L were nearly as large as T_H, the solution with the "−" sign would make T_1 nearly 0 instead of lying between T_L and T_H.

We conclude that if $T_1 = (T_H + \sqrt{T_L T_H})/2$, the output power $\dot{W}$ will be maximized. Since $\dot{Q}_H = b(T_H - T_1)$, our value for T_1 determines the value of b ($= \sigma A/L$) that will make T_1 come out right to maximize the power output for a given $\dot{Q}_H$.

Finally, we can plug our value for T_1 back into the expression for $\dot{W}$ to find the efficiency at maximum power:

$$\frac{\dot{W}}{\dot{Q}_H} = \frac{\dot{W}}{b(T_H - T_1)} = 1 - \frac{T_L}{2T_1 - T_H} = 1 - \frac{T_L}{\sqrt{T_L T_H}}$$

Here is the final result, where we write "$\leq$" because we have calculated an upper bound on the efficiency (having taken into account issues of energy transfer rate due to a temperature difference but not mechanical friction):

THE EFFICIENCY OF A MAXIMUM-POWER ENGINE

$$\frac{\dot{W}}{\dot{Q}_H} \leq 1 - \sqrt{\frac{T_L}{T_H}}$$

This gives us the fraction of the thermal input power that ends up as output mechanical power if we maximize the output power. The formula looks quite similar to the reversible-engine zero-power efficiency, but the square root makes a big difference.

For example, in the case where the absolute temperatures in kelvins of T_L and T_H are not very different, it's easy to see that the maximum-power engine has an efficiency only one-half that of the reversible engine:

$$1 - \frac{T_L}{T_H} = \left(1 + \sqrt{\frac{T_L}{T_H}}\right)\left(1 - \sqrt{\frac{T_L}{T_H}}\right)$$

$$\approx 2\left(1 - \sqrt{\frac{T_L}{T_H}}\right)$$

$$\left(1 - \sqrt{\frac{T_L}{T_H}}\right) \approx \frac{1}{2}\left(1 - \frac{T_L}{T_H}\right)$$

In the case of the nuclear power station running between 25°C and 300°C, we saw that a reversible engine would have had an efficiency of 48%, but the efficiency of a maximum-power engine would be only 28% (the actual power plant efficiency was 30%).

SUMMARY

Mean free path d: $n[\pi(R+r)^2(d)] \approx 1$

Root-mean-square speed: $v_{\text{rms}} \equiv \sqrt{\overline{v^2}}$

Work done on a gas: $W = -\int_{V_1}^{V_2} P\,dV$

First law of thermodynamics: $\Delta E_{\text{sys}} = W + Q$

Molecular specific heat capacity at constant volume C_V:
$Q = NC_V\Delta T$

Molecular specific heat capacity at constant pressure C_P:
$Q = NC_P\Delta T$

Number of gas molecules hitting an area A per second $= \frac{1}{4}nA\overline{v}$ (3-D; various speeds)

$$P = \frac{2}{3}n\left(\overline{\frac{p^2}{2m}}\right) = nk_BT$$

where $n = N/V$ (number of molecules per cubic meter)
For a multiatom gas molecule,

$$\Delta\overline{E}_{\text{tot}} = \Delta\left(\frac{3}{2}k_BT + \overline{E}_{\text{rot}} + \overline{E}_{\text{vib}}\right)$$

$C_V = \frac{3}{2}k_B$ for a monatomic gas (He, etc.)
$C_V \geq \frac{3}{2}k_B$ for other gases (N_2, etc.)
$C_P = C_V + k_B$ (molecular specific heat capacity)
$C_P = C_V + R$ (molar specific heat capacity)
In a constant-temperature (isothermal) compression,

$$W_{\text{by piston}} = Q_{\text{into surroundings}} = Nk_BT\ln\left(\frac{V_1}{V_2}\right)$$

In a no-Q (adiabatic) compression,
$PV^\gamma = \text{constant}$, where $\gamma = C_P/C_V$, and also

$$T^{(C_V/k_B)}V = \text{constant}$$

$k_B = 1.38 \times 10^{-23}$ J/K

$R = (6.02 \times 10^{23})k_B = 8.3$ J/K

At "standard temperature and pressure" (STP, which means a temperature of $0°$ Celsius or 273 Kelvin, and a pressure of 1 atmosphere), one mole of a gas occupies 22.4 liters (a liter is 1000 cubic centimeters).

Sea-level pressure (1 atm) is 1×10^5 N/m^2.

Rate of energy transfer due to a temperature difference:

$$\dot{Q} = \sigma A\frac{(T_H - T_L)}{L}\text{ (J/s)}$$

For a cyclic engine running between high temperature T_H and low temperature T_L we have:

Entropy change of the universe for real engines

$$\Delta S_{\text{Universe}} = -\frac{Q_H}{T_H} + \frac{Q_L}{T_L} \geq 0$$

The efficiency of a real engine

$$\text{Efficiency} = \frac{W}{Q_H} \leq 1 - \frac{T_L}{T_H}$$

The "=" sign applies only if the engine is reversible (extremely slow processes, no friction). A heat engine run in reverse is a refrigerator (or a heat pump).

If we maximize power output, we have the efficiency of a maximum-power engine

$$\frac{\dot{W}}{\dot{Q}_H} \leq 1 - \sqrt{\frac{T_L}{T_H}}$$

Random walk root-mean-square displacement:

$$\Delta r_{\text{rms}} = (\sqrt{N})d = \sqrt{v}d\sqrt{t}$$

EXERCISES AND PROBLEMS

Sections 13.1–13.3

13.X.15 Gas leaks at a rate L (in molecules per second) through a small circular hole. If the density of the gas is doubled, the Kelvin temperature is doubled, and the radius of the hole is doubled, what is the new leak rate?

13.X.16 Suppose that we make a circular hole 4 millimeters in diameter in a balloon. Calculate the initial rate at which chlorine escapes through the hole (at 0 C), in number of chlorine atoms leaving the balloon per second.

13.X.17 How does the mean free path of an atom in a gas change if the temperature is increased, with the volume kept constant?

13.X.18 What is the approximate time between collisions for one particular air molecule?

13.P.19 In an example problem in Section 13.2 we considered leakage from a flexible party balloon. If the leakage is from a rigid container (a metal storage tank, for example), the number of atoms per cubic meter, n, will decrease with time t. If the total volume of the tank is V, there are $N = nV$ atoms in the tank at any instant, so we can write the following "differential" equation (that is, an equation that involves derivatives):

$$\frac{d}{dt}(nV) = -\frac{1}{4}nA\overline{v}$$

This says, "the rate of change of the number of atoms in the tank is equal to the (negative) of the rate at which atoms are leaving the tank."

(a) Show that the differential equation is satisfied if the number density

$$n = n_{\text{initial}}e^{-(A\overline{v}/4V)t},$$

where t is the time elapsed since the hole was made. Just plug this function of n, and its derivative, into the equation and show that the two sides of the equation are equal for all values of the time t. Also show that the initial particle density n is equal to $n_{initial}$.

(b) Despite the fancy math, this solution is really only approximate, because the average speed isn't a constant but is decreasing. Suppose, however, that we use a heater to keep the container and the gas at a nearly constant temperature, so that the average speed does remain nearly constant. Suppose that the (rigid) container is again a sphere 30 cm in diameter, with a circular hole 1 millimeter in diameter. About how long would it take for most of the helium to leak out? Explain your choice of what you mean by "most."

13.P.20 Natural uranium ore consists mostly of the isotope U-238 (92 protons and 146 neutrons), but 0.7% of the ore consists of the isotope U-235 (92 protons and 143 neutrons). Because only U-235 fissions in a reactor, industrial processes are used to enrich the uranium by enhancing the U-235 content.

One of the enrichment methods is "gaseous diffusion." The gas UF_6, uranium hexafluoride, is manufactured from supplies of natural uranium and fluorine (each of the six fluorine atoms has 9 protons and 10 neutrons). A container is filled with UF_6 gas. There are tiny holes in the container, and gas molecules leak through these holes into an adjoining container, where pumps sweep out the leaked gas.

(a) Explain why the gas that initially leaks into the second container has a slightly higher fraction of U-235 than is found in natural uranium.

(b) Estimate roughly the practical change in the concentration of U-235 that can be achieved in this single-stage separation process. Explain what approximations or simplifying assumptions you have made to obtain your estimate.

(c) A typical nuclear reactor requires uranium that has been enriched to the point where about 3% of the uranium is U-235. Estimate roughly the number of stages of gaseous diffusion required (that is, the number of times the gas must be allowed to leak from one container into another). Note that the effects are multiplicative.

This is why a practical gaseous diffusion plant has a large number of stages, each operating at high pressure, which makes this an expensive process. The first large gaseous diffusion plant was constructed during World War II at Oak Ridge, Tennessee, and used inexpensive Tennessee Valley Authority electricity.

13.P.21 You are on a spacecraft measuring 8 m by 3 m by 3 m when it is struck by a piece of space junk, leaving a circular hole of radius 4 mm, unfortunately in a place that can be reached only by making a time-consuming spacewalk. About how much time do you have to patch the leak? Explain what approximations you make in assessing the seriousness of the situation.

Section 13.4

13.P.22 A rigid, thermally insulated container with a volume of 22.4 liters is filled with one mole of helium gas (4 grams per mole) at a temperature of 0 Celsius (273 K). The container is sitting in a room, surrounded by air at STP.

(a) Calculate the pressure inside the container in N/m^2.

(b) Calculate the root-mean-square average speed of the helium atoms.

(c) Now open a tiny square hole in the container, with area 1×10^{-8} m^2 (the hole is 0.1 mm on a side). After 5 seconds, how many helium atoms have left the container?

(d) Air molecules from the room enter the container through the hole during these 5 seconds. Which is greater, the number of air molecules that enter the container or the number of helium atoms that leave the container? Explain briefly.

(e) Does the pressure inside the container increase slightly, stay the same, or decrease slightly during these 5 seconds? Explain briefly. If you have to make any simplifying assumptions, state them clearly.

13.P.23 A roughly spherical meteor made mainly of iron (density about 8 grams per cubic centimeter) is hurtling downward through the air at low altitude. At an instant when its speed is 1×10^4 m/s, calculate the approximate rate of change of the meteor's speed. Do the analysis for two different meteors—one with a radius of 10 meters and one with a radius of 100 meters.

Start from fundamental principles. Do not try to use some existing formula that applies to a very different situation. Follow the kind of *reasoning* used in this chapter, applied to the new situation, rather than trying to use the *results* of this chapter.

A major difference from our earlier analyses is that the meteor is traveling much faster than the average thermal speed of the air molecules, so it is a good approximation to consider the air molecules below the meteor to be essentially at rest, and to assume that no air molecules manage to catch up with the meteor and hit it from behind. The meteor drills a temporary hole in the atmosphere, a vacuum, that gets filled explosively by air rushing in after the meteor has passed.

There is good evidence that a very large meteor, perhaps 10 kilometers in diameter, hit the Earth near the Yucatan Peninsula in southern Mexico 65 million years ago and caused so much damage that the dinosaurs became extinct. See the excellent account in *T. rex and the Crater of Doom*, by Walter Alvarez (Princeton University Press, 1997). Alvarez is the geologist who made the first discoveries leading to our current understanding of this cataclysmic event.

Section 13.5

13.X.24 If you expand the volume of a gas containing N molecules to twice its original volume, while maintaining a constant temperature, how much energy transfer due to a temperature difference is there from the surroundings?

13.X.25 If you compress a volume of helium containing N atoms to half the original volume in a well-insulated cylinder, what is the ratio of the final pressure to the original pressure?

13.P.26 Two moles of nitrogen is compressed (by piling lots of sand on the piston) to a volume of 11 liters at room temperature (293 K). The cylinder is placed on an electric heating element whose temperature is maintained at 293.001 K. A quasistatic expansion is carried out at constant temperature by very slowly removing grains of sand from the top of the piston, with the temperature of the gas staying constant at 293 K. (You must assume that there is no energy transfer due to a temperature difference from the gas to the surrounding air, and no friction in the motion of the piston, all of which is pretty unrealistic in the real world! Nevertheless there are processes that can be approximated by a constant-temperature expansion. This problem is an idealization of a real process.)

(a) When the volume is 20 liters, how much energy transfer Q has gone from the heating element into the gas?

(b) How much work W has been done on the piston by the gas?

(c) How much has the energy of the gas changed?

13.P.27 A horizontal cylinder 10 cm in diameter contains helium gas at room temperature and atmospheric pressure. A piston keeps the gas inside a region of the cylinder 20 cm long.

(a) If you *quickly* pull the piston outward a distance of 12 cm, what is the approximate temperature of the helium immediately afterward? What approximations did you make?

(b) How much work did you do, including sign? (Note that you need to consider the outside air as well as the inside helium in calculating the amount of work *you* do.)

(c) Immediately after the pull, what force must you exert on the piston to hold it in position (with the helium enclosed in a volume that is 20 + 12 = 32 cm long)?

(d) You wait while the helium slowly returns to room temperature, maintaining the piston at its current location. After this wait, what force must you exert to hold the piston in position?

(e) Next, you very *slowly* allow the piston to move back into the cylinder, stopping when the region of helium gas is 25 cm long. What force must you exert to hold the piston at this position? What approximations did you make?

13.P.28 Atmospheric pressure at sea level is about 1×10^5 N/m^2, which is about 15 pounds per square inch (psi). A bicycle tire typically is pumped up to 50 psi above atmospheric pressure (psi "gauge"), for an actual pressure of about 65 psi. In rapidly pumping up a bicycle tire, starting from atmospheric pressure, about how high does the temperature of the air rise? Explain what approximations and simplifying assumptions you make.

Sections 13.6–13.8

13.X.29 An aluminum bar 30 cm long and 3 cm by 4 cm on its sides is connected between two large metal blocks at temperatures of 135°C and 20°C, and energy is transferred from the hotter block to the cooler block at a rate $\dot{Q}_1$. If instead the two blocks were connected by an aluminum bar that is 20 cm long and 3 cm by 2 cm on its sides, what would be the energy transfer rate?

13.X.30 A (nearly) reversible engine is used to melt ice as well as do some useful work. If the engine does 1000 joules of work and dumps 400 joules into the ice, what is the temperature of the high-temperature source?

13.X.31 For engines that have high and low temperatures not very different from each other, show that the efficiency of a maximum-power engine is about half of the efficiency of a reversible engine.

13.P.32 In one cycle of a reversible engine running between a high-temperature source and a low-temperature sink, a consequence of the second law of thermodynamics is that $Q_H/T_H = Q_L/T_L$. This result is independent of what kind of material the engine contains. It is instructive to check this general result for a specific model where we can calculate everything explicitly. Consider a reversible cycle of an ideal gas of N molecules, starting at high temperature T_H and volume V_1.

(a) Perform a constant-temperature (isothermal) expansion to volume V_2, and calculate the associated energy transfer Q_H into the gas.

(b) Next perform a no-Q (adiabatic) expansion to volume V_3 and temperature T_L.

(c) Next perform a constant-temperature (isothermal) compression to volume V_4, and calculate the associated energy transfer Q_L out of the gas.

(d) Finally perform a no-Q (adiabatic) compression back to the starting state, temperature T_H and volume V_1.

A N S W E R S   T O   E X E R C I S E S

13.X.1 T decreases because average v decreases.

13.X.2 1×10^{22} molecules per second

13.X.3 $L \approx 3 \times 10^{-9}$ m

13.X.4 1×10^5 N/m^2

13.X.5 986 J; 986 J; 0

13.X.6 $\frac{3}{2}k_B + k_B = \frac{5}{2}k_B$

13.X.7 $\left(\dfrac{PV}{Nk_B}\right)^{c_v/k_B} V = $ constant

$P^{C_V/k_B} V^{C_V/k_B+1} = $ new constant

Take (C_V/k_B) root :

$PV^{1+k_B/C_V} = PV^{(C_V+k_B)/C_V} = PV^{C_P/C_V}$

13.X.8 25 seconds; the entropy of the Universe increases

13.X.9 0.27

13.X.10 0.48

13.X.11 3 watts; 53 watts

13.X.12 0.05 J

13.X.13 28%, which is close to but less than the observed 30%. This implies that the power plant is not run at maximum power.

13.X.14 6 millimeters

Index

A

aberrations, 1021
absorption spectra, 330
  and temperature, 334
AC, 817
  phase shift, 818
  power, 818
  rms voltage, 818
AC circuits, 973
acceleration, 27, 124, 155
  and Momentum Principle, 125
  parallel, 28
  perpendicular, 28
  toward center of kissing
    circle, 195
Achimedes principle, 164
adiabatic compression, 531
Ahlborn
  maximum-power engine, 545
air
  ionization of, 872
air resistance
  dependence on area, 311
  drag coefficient, 312
  falling coffee filter, 310
  force, 312
  mechanism, 313
aluminum
  speed of sound, 163
ammeter, 811
  low resistance, 812
Ampere's law, 915, 918
  applied to long thick wire, 919
  applied to solenoid, 920
  applied to toroid, 921
  differential form, 927
  formal proof, 917
  which currents are inside the
    path, 918
Ampere-Maxwell law, 988, 989
amusement park ride, 207
analytical solutions
  existence, 160

angles and unit vectors, 18
angular distribution, 397
angular frequency, 158
  of a wave, 1043
angular half-width
  of diffraction maximum,
    1069
angular momentum, 415
  atomic, 444
  choosing a location, 419
  comets, 429
  conservation, 430
  in Bohr model of atom,
    735
  moment of inertia, 424
  multiparticle, 422
  nonzero torque, 440
  nuclear, 444
  of atomic electron, 731
  orbital, 415
  quantization, 444, 449
  rotational, 422
  spin, 422
  systems with zero torque, 437
  translational, 415
Angular Momentum Principle,
    The, 427
  multiparticle, 431
  preview, 186
angular resolution, 1067
angular velocity
  right-hand rule, 423
anharmonic oscillator, 160
annihilation, 262
antiderivatives, 273
antiproton, 555
applications
  curving motion, 197
area
  cross-Sectional, 144
atmosphere
  constant-density model, 166
  pressure, 165

atom
   matter composed of, 556
   neutral, 556
   plum pudding model, 396
   size, 556
   size of, 2, 141, 142
   size of nucleus, 556
   structure, 591
atomic angular momentum, 444
atomic structure
   magnet, 730
atoms
   Feynman quote, 138
   force on atomic magnetic moment
      (Stern-Gerlach experiment), 863
   STM images, 2, 139
attractive forces
   potential energy, 252
average velocity, 58

B

ball and spring model
   energy, 297
   potential energy, 294
ball-and-spring model, 139
   compression, 149
battery
   conventional symbol, 806
   emf, 769
   internal resistance, 771, 809
      measurement, 824
   mechanical, 759
      charge separation, 769
      Coulomb and non-Coulomb
         forces, 769
   non-ideal, 809
   power, 808
   role in a circuit, 769
beta decay, 266
Big Bang theory, 36
binding energy, 263
   nuclear, 265
Biot-Savart law, 714
   for a current element, 721
blue sky
   why is it blue, 1011
body
   charging and discharging, 607
Bohr model
   angular momentum, 735
   energy levels, 446
   hydrogen atom, 444
   photon emission, 447
   radii, 446

Boltzmann distribution, 494
   diatomic gas, 497
bosons, 449
bound states, 257
bucket
   vertical swing, 199
buoyancy, 163
buoyant force, 164

C

calculus
   and iteration, 71
   velocity, 26
capacitance, 799
capacitor
   capacitance, 799
   charging, 823
      effect of resistance, 797
      explanation, 796
      role of fringe field, 796
   conventional symbol, 795
   discharging, 795, 823
      explanation, 796
   electric field, 644, 646
   electric field of
      fringe field, 646
   energy in charging and discharging,
      798
   in circuit (quantitative), 814
   in circuits, 794
      effect of insulating layer, 826
      effect of larger plates, 797
      effect of thinner gap, 797
   mechanical construction, 795
   usefulness in circuits, 824
   work and energy, 833
capacitors
   parallel, 798
Carnot cycle, 538
Cavendish, 128
center of mass, 120, 348
   and total momentum, 348
   point particle system, 362
   reference frame, 405
   reference frame
      center of mass frame, 403
   skater, 349
   velocity, 121, 348
center of momentum reference
      frame, 405
chain reaction
   laser, 341
change of identity, 236
change of momentum, 32

change of rest energy, 236
chaos, 116
charge
  accelerated
    radiative electric field, 997
  amount of charge on a tape, 614
  amount on invisible tape, 589
  distributed
    electric field of, 629
  fraction of surface sites charged, 590
  interaction with neutral matter, 591
  net, 610
  on conductors, 604
Charge conservation, 586
charged objects
  interactions, 587
charged particles, 555
  antiproton, 555
  electron, 555
  muon, 555
  pion, 555
  positron, 555
  proton, 555
charging
  a solid object, 587
  by induction, 608
  can't transfer nuclei or protons, 588
charging by contact
  body, 607
  metals, 617
charging by induction, 616
chemical bond
  length, 141
chemical bond is springlike, 139
child on merry-go-round, 197
circuit
  RC
    time constant, 816
    time dependent current, 815
Circuits
  effect of thickness of nichrome wire, 782
  mechanical battery, 759
  surface charge, 760
  two bulbs in parallel, 782
  two bulbs in series, 781
circuits
  current driven by three kinds of
      forces, 962
  effect of length of copper wires, 771
  energy conservation, 772
  feedback, 765
  initial transient, 763
  LC
    time dependent current, 971
  macroscopic analysis, 801

  macroscopic vs. microscopic analysis, 804
  multi-loop circuit, 820
  oscillations in LC circuits, 970
  power, 808
  surface charge
    narrow resistor, 766
    parallel circuits, 774
    wide resistor, 768, 913
  thin filament bulb and thick filament bulb in series,
      775
  with capacitor, 823
  with inductance, 969
  with magnetic induction
    non-uniform ring, 975, 977
    non-uniform ring , 975
    shrinking ring, 975
    two bulbs near solenoid, 974
circular motion in magnetic field, 838
circular pendulum, 205
closed systems, 305
coefficient of friction, 150
coffee filter, 310
coherence, 1046
coherence length, 1055
collisions, 126, 383
  elastic, 385, 387, 393
  equal masses, 387, 393
  head-on, equal masses, 386
  impact parameter, 394
  inelastic, 385, 402, 403
  large internal forces, 383
  maximally inelastic, 386
  reference frames, 403
  relativistic, 399
  scattering, 392
  sticking, 403
  unequal masses, 389
  without contact, 384
comets, 429, 430
components of a vector, 10
compression
  adiabatic, 531
  constant-temperature, 527
  in curving motion, 202
  isothermal, 527
  no-Q, 531
compression force, 148
Compton scattering, 1058
computer iteration, 70
Conductivity, 802
conductivity, 801
  of carbon, 830
  of copper, 802
  ohmic, 804
  with two kinds of carriers, 802

conductor, 587
    energy bands in, 930
conductors
    definition, 587, 608, 609
    polarization, 598
conservation
    momentum, 118
conservation of angular momentum, 430
Conservation of charge, 586
conservation of current
    maintained by feedback, 765
    only in steady state, 799
Conservation of Energy
    potential difference
        round trip, 685
conservation of energy, 221, 245
    in circuits, 772
constants, 555
constraints
    on motion, 206
contact forces, 178
contradiction
    proof by, 599
conventional current, 717, 720
converging lens, 1018
conversion
    units, 142
cosmic microwave background, 36
cosmic rays, 872
    role in ionization of air, 872
coulomb, 555
Coulomb constant, 561
Coulomb force
    in mechanical battery, 769
Coulomb's law, 109, 555
    compared with Gauss's law, 910
    invalid at high speeds, 575
counting states, 476
    permutation formula, 480
cross product, 419, 714
    and unit vectors, 420, 716
    right-hand rule, 416
cross section, 398
cross-Sectional area, 144
crossed electric and magnetic fields, 844
CRT, 838
crystal structure, 2
curl (of a vector field), 927
current
    conventional, 717, 720
    distributed
        magnetic field of, 721
    driven by three kinds of forces, 962
    electron, 712, 718
    electrons move in metals, 819

measuring around a circuit, 754
    mobility, v = uE, 602
    not used up in a light bulb, 754
    series circuit, 754
current density, 801
    and conductivity, 801
current node rule, 755, 773
    in a capacitor circuit, 799
curving motion, 189
    additional examples, 204
    amusement park ride, 207
    applications, 197
    compression force, 202
    kissing circle, 190
    perpendicular component of momentum
        change, 193
    turning car, 191
Curzon
    maximum-power engine, 545
cycle
    of an engine, 536
cyclotron, 840

D

Davisson and Germer, 1061
DC, 817
Delta, 17
delta
    symbol for change, 17
density, 141, 142
    micro, 142
    number, 514
derivative form of the Momentum Principle, 154
detecting interactions, 3
    change of direction, 3
    change of speed, 4
    change of velocity, 4
    uniform motion (no interaction), 5
determinism, 115
deuteron, 400
diagrams
    energy, 253
diatomic gas
    average rotational energy, 505
    Boltzmann distribution, 497
diatomic molecule, 338
    average vibrational energy, 504
    kinetic energy, 351
    quantized energy, 338
dielectric constant, 692
diffraction, 1051
    electron, 1060
    neutron, 1062
    single-slit, 1069

diffraction gratings, 1064
dimensional analysis, 162
diode, 939
dipole, 565
  drawing conventions, 593
  electric field of, 565
  field far from dipole, 567
  in uniform field, 572
  interaction with point charge, 570
  permanent, 593
dipole moment, 572
direction cosines, 18
discharging a capacitor, 795
discharging an insulator, 608
discharging by contact
  body, 607
disk
  electric field, 641
distribution of velocities in a gas, 499
divergence
  standard notation, 926
divergence (of a vector field), 922
diverging lens, 1018
diverging lenses, 1028
Doppler shift, 36
dot product
  scalar, 232
  using angles, 231
  work, 228
drag coefficient, 312
drift speed
  of electrons in circuit, 601
  of free electrons in a spark, 876
Drude model, 601, 755

E
Earth
  satellite, 198
Earth's magnetic field, 728
efficiency
  maximum, 541
  of engines, 533
  of reversible engine, 541
Einstein, 204
  General Relativity, 113
  model of a solid, 475
  special relativity, 35
Einstein and Jack and Jill, 856
elastic collisions, 385, 387
electric dipole
  permanent dipole
    water molecule, 609
electric dipole moment, 572

Electric field
  non-Coulomb, due to changing magnetic
    field, 949
electric field, 556
  and speed of light, 575
  changing, 988
  choice of system, 574
  curly non-Coulomb electric field, 949
  dipole, 565
  direction, 560
  direction in a circuit, 757
  due to surface charge gradient, 761
  energy density associated with, 693
  far from dipole, 567
  gradient of potential, 674
  in a circuit
    location of source charges, 758
  in a conductor at equilibrium, 603
  in a metal at equilibrium, 603
  in a moving reference frame, 876
  in a p-n junction, 936
  inside a hole in a conductor, 912
  inverse square, 560
  magnitude, 561
  near point charges, 561
  non-Coulomb, 950
  nonzero in metal not in equilibrium, 604
  of a capacitor, 646
    calculating, 644
  of a charged disk, 641
    calculating, 641
  of a charged sphere, 646
    calculating, 652
  of a charged thin ring, 640
    calculating, 639
  of a hollow cylinder, 637
  of a point charge, 560
  of a solid sphere of charge, 650
    calculating, 649
  of a sphere
    calculating using potential, 696
  of a thin rod
    calculating, 630
  of accelerated point charge
    quantitative formula, 997
  of distributed charge, 629
  of large plate, 909
  of uniform cube, 911
  of uniform spherical shell, 910
  patterns
    Gauss's law, 899
  procedure for distributed charges, 637
  radiative, 997
  reality of, 574
  related to magnetic field, 878

Electric field (*Contd.*)
  required to ionize molecule, 871
  retardation, 575
  summary of ways to produce, 999
  superposition, 564
  two ways to produce, 950
  uniformly charged sphere, 563
  vectors, 558
  when concept less useful, 609
electric field , 558
electric flux, 901, 902
electric force, 109, 260
electric force and magnetic force, 990
electric generators
  work and energy, 866
electric interactions
  act through matter, 596
  between charged objects, 587
electric motor, 864
electric potential, 672
  arbitrary constant, 690
  compared with emf, 769
  in insulator, 691
  ring, 688
  shifting the zero, 689
  sign, 676
  voltmeters measure potential difference,
    812
electric potential energy, 260
  Morse approximation, 293
  neutral atoms, 262, 293
electromagnetic radiation, 995
  and Gauss's law for magnetism, 992
  direction of propagation, 992, 995–997, 1005
  effect on a charge, 1003
  effect on an atom, 1008
  effect on matter, 1008
  energy and wavelength, 1057
  energy carried by, 1004, 1006
  energy loss, 998, 1008
  intensity, 1044
  magnetic force exerted by, 1005
  momentum in, 1005
  pattern of electric field, 997
  polarization, 1008
  quantitative formula for electric field, 997
  radiation pressure, 1005
  re-radiation or scattering, 1007
  relation of E and B, 995
  resonance, 1010
  sinusoidal
    color vision, 1011
    transmitters, 1003
electromagnetic spectrum
  energy, 326

electromagnetic wave
  coherence, 1046
electromotive force, 769
electron, 555
  mobile charge in metals, 819
  mobile in metal, 600
  spin, 448
electron current, 712, 718
electron diffraction, 1060, 1061
electron excitation, 331
  Franck-Hertz experiment, 332
electron mobility, 602
electron positron annihilation, 262
electron volt, 236
electronic energy levels, 327
electrostatic screening, 913
elliptical orbit, 108
emf, 770
  and magnetic force, 852
  and potential difference across a battery,
    769
  compared with potential difference, 769
  definition, 769
  due to changing magnetic field, 951
  due to changing magnetic flux, 954
  motional, 961
    due to magnetic forces, 851
    quantitative, 853
  motional emf, 851
emission spectra, 329
  and temperature, 334
energy
  potential energy (review), 665
  alpha + carbon makes O*, 268
  and wavelength of light, 342
  average at high temperature, 501
  ball and spring model, 297
  change of rest energy, 236
  choice of system, 306
  conservation, 221
  contrast with momentum, 227, 250
  diagrams, 253
    how to draw, 255
  electric potential, 260
  electronic bands in solids, 929
  fields, 271
  graphs and types of orbits, 255
  gravitational, 246, 247
  gravitational multiparticle, 354
  in a circuit, 768
  internal, 296
  ionization, 261
  kinetic, 222
  kinetic multiparticle, 351
  multiparticle, 347

multiparticle rest, 262
nonrelativistic kinetic, 223
nuclear, 237
other transfers (mass, electromagnetic, etc.), 302
oxygen molecule, 263
particle energy (review), 666
particle kinetic energy (review), 666
quantized electronic, 327
quantized hadronic, 339
quantized nuclear, 339
quantized rotational, 338
quantized vibrational, 336
quantum oscillator, 336
reflection, 266
reflection on forms of, 304
relativistic particle, 225, 274
rest, 222
rest energy, 222
rotational, 353
single particle, 221
spring-mass, 290
thermal, 296
translational kinetic, 351, 361
vibrational, 353
work (review), 667
energy conservation
in a circuit, 780
loop rule, 769, 780
energy density
in electric and magnetic fields, 969
in electric field, 693
in magnetic field, 968
energy dissipation, 310, 315
energy levels
hydrogen atom, 328
solid metal, 929
Energy Principle, The, 220
at high speed, 235
change of rest energy, 236
choice of system, 306
derivation, 238, 274
identifying initial and final states, 267
systems of charged objects, 668
update form, 233
energy transfer
due to a temperature difference, 301
mechanical, 226
thermal, 474
work, 226
energy transfers
gas, 525
mass, electromagnetic, etc., 302
engine, 536
and entropy, 539

energy, 540
maximum power output, 545
maximum-power, 545
engineering statics, 181
entropy, 483
additive, 485
and temperature, 486
and thermal energy transfer, 488
and work, 488
change for real engines, 541
definition, 484
in cycle of reversible engine, 539
of Universe, 486
produced by friction, 533
production by thermal energy transfer, 533
second law of thermodynamics, 485
equilibrium, 753
and feedback, 766
mechanical (statics), 181
potential difference, 678, 689
vs. momentarily at rest, 188
escape speed, 256
for gas molecules, 503
estimating
interaction times, 78
eV (electron volt), 236
external forces, 50
eyeglasses, 1030

F
$F=ma$, 155
falling coffee filter, 310
falling object, 74
Faraday's law, 952, 954
and motional emf, 960
and moving magnets, 957
and non-Coulomb electric field, 954
for multi-turn coil, 956
Lenz's rule, 954
with iron core, 960
feedback in circuits, 765
fermions, 448
Feynman
atoms, 138
fiber optics, 1017
field
electric, 556
in mathematical physics, 559
first law of motion, 6
first law of thermodynamics, 485, 529
fission, 265
flux
electric, 901, 902
magnetic, 953

focal point, 1023
force
    approximate gravitational force,
        101
    buoyant, 164
    compression, 148
        curving motion, 202
    constant, 72
    contact, 178
    electric, 109, 260
    gravitational, 96
    identifying, 180
    magnetic, 837
    measuring, 50
    negative gradient of potential energy,
        264
    net, 50
    normal, 148
    of point charge on neutral atom, 594
    parallel and perpendicular, 77
    parallel component, 192
    perpendicular component, 192
    reciprocity, 103
    string, 181
    tension, 140
    units, 50
force laws
    electric, 260
forces
    external, 50
    internal, 50
    spring, 63
forward reasoning, 204
Fourier analysis, 1076
Franck-Hertz experiment
    electron excitation, 332
free-body diagrams, 54
frequency, 158
friction, 149, 207, 368
    coefficient of friction, 150
    dry sliding, 370
    effective distance, 372
    entropy production, 533
    lubricated, 372
    paradox, 315
    point particle, 369
    sliding, 150
    spring-mass, 314
    static, 150
    work, 370
fundamental assumption of statistical mechanics,
        477
fundamental interactions, 95
fundamental principles of mechanics, 436
fundamental principles:Momentum, 48

G

galaxy, 3
gamma, 29, 221
gas
    adiabatic compression, 531
    Boltzmann distribution, 497
    constant-temperature compression,
        527
    distribution of velocities, 499
    energy transfers, 525
    height distribution, 498
    ideal gas law, 521
    isothermal compression, 527
    leak, 514
    limits of ideal gas law, 522
    mean free path, 517
    measuring speed distribution, 501
    molecules per second per area, 515
    nature of, 513
    no-Q compression, 531
    pressure in atomic quantities, 521
    quasistatic compression, 526
    quasistatic processes, 526
    random walk, 547
    specific heat capacity at constant
        volume, 531
    weight of gas in box, 523
    work by piston, 528
gas law, 521
    limitations, 522
gases, 163
Gauss's law, 903, 907
    applied to circuits, 913
    applied to large plate, 909
    applied to metals, 911
    applied to spherical shell, 910
    applied to transistor, 943
    compared with Coulomb's law, 910
    differential form, 922
    for magnetism, 915
    formal proof, 903, 907
    patterns of electric field, 899
    proof, 904
    relativistically correct, 924
General Relativity, 113, 204
generator, 864
    work and energy, 866
gradient, 248
    of potential, 674
    of potential energy, 272
gravitational constant, 98, 128
gravitational energy
    multiparticle, 354
gravitational field
    energy in field, 271

gravitational force
    approximate, 66, 101
    of sphere, 128
gravitational forced, 96
gravitational potential energy, 246, 247
    near Earth's surface, 259
    three bodies, 258
ground state, 328, 337
grounding, 608
gyroscope, 449

H

hadronic energy levels, 339
Hall effect, 846
harmonic oscillator, 156, 160, 167
    and anharmonic oscillator, 160
Harrison
    sliding friction, 315
heat
    energy transfer due to a temperature difference,
        302
    energy transfer due to a temperature difference
        (Q), 301
heat capacity, 299
    and energy quantization, 492
    calculating, 299
    experimental results, 490
    gas, 505
    solid, 490
    temperature dependence diatomic gas, 506
    water, 299
heat pump, 542, 543
height distribution in a gas, 498
Heisenberg uncertainty principle, 117, 337
hockey pucks, 350
hole
    as mobile charge, 848, 931
hollow cylinder
    electric field, 637
humidity
    effect on charged object, 609
hydrogen
    Bohr model, 444
hydrogen atom
    electronic energy levels, 328

I

ideal gas law, 521
    limitations, 522
idealization, 82
images, 1021
    real, 1021
    virtual, 1022

impact parameter, 394
impulse, 51
    contrast with work, 227, 250
index of refraction, 1014, 1054
Inductance
    of solenoid, 967
inductance, 966
    in circuits, 969
    oscillations in LC circuits, 970
inelastic collisions, 385, 402
    maximally inelastic, 386
inertial frame, 84
inertial frame of reference, 37
infinitesimals
    and integrals, 651
initial conditions
    mass-spring system, 161
    sensitivity, 115
    spring-mass system, 168
instantaneous velocity, 25
insulator, 587
    energy bands in, 930
    location of excess charge, 597
insulators
    definition, 587, 608, 609
    dielectric constant, 692
    electric potential, 691
integral
    and infinitesimals, 651
    infinite sum of infinitesimals, 633
    work, 240
integrals, 273
intelligent plug and chug, 238
intensity
    and amplitude of a wave, 1045
    of an electromagnetic wave, 1044
interaction
    charged objects and neutral matter, 595
    detecting, 3
        additional interaction, 8
        change of direction, 3
        change of identity, 7
        change of shape, 7
        change of speed, 4
        change of temperature, 8
        change of velocity, 4
        indirect evidence, 8
        uniform motion, 5
interaction energy
    multiparticle, 243
interaction of point charge and neutral atom, 594
interatomic bond
    length, 141
    speed of sound, 152
    stiffness, 143, 145

interatomic potential energy, 294
    quantization, 338
interatomic spring stiffness
    Young's modulus, 147
interference, 1045
    constructive, 1046
    destructive, 1047
    maxima and minima, 1049, 1050
    thin film, 1053
    two-slit
        particle model, 1060
interferometry, 1050
internal energy, 296
internal forces, 50
internal resistance, 771
invariant, 225
inverted population, 341
invisible tape, 611
    L tape, 613
    U tape, 611
ionization energy, 261
irreversibility, 486, 489, 535
isothermal compression, 527
isotopes, 2
iterative prediction
    and calculus, 71
    block on vertical spring, 66
    by computer, 70
    of gravitational motion, 104
    of motion, 62

J

Jack and Jill and Einstein, 856
Joule, 299
jumping
    and point particle system, 362
Jupiter, 3

K

kelvin degree, 299
Kepler
    planetary orbits, 431
kinetic energy
    average in gas, 502
    low-speed approximation, 273
    multiparticle, 351, 373
    nonrelativistic, 223
    particle, 222
    relative to cm, 351
    rotational, 355, 425
    translational, 351, 361
Kirchhoff loop rule, 780
Kirchhoff node rule, 755

kissing circle
    curving motion, 190

L

L tape
    attraction by U tape, 613
    preparing, 613
    repulsion by L tape, 613
Laplace's equation, 926
laser, 340
    chain reaction, 341
    coherent light, 1046
    inverted population, 341
lattice, 556
LC circuits, 970
lead
    speed of sound, 163
leak
    gas, 514
lens formula, 1020
lenses, 1017
    aberrations, 1021
    diverging, 1028
    eyeglasses, 1030
    focal point, 1023
    images, 1021
    magnifying glass, 1028
    sign conventions, 1029
    thin, 1019
    thin lens formula, 1020
lenses:converging and diverging,
        1018
Lenz's rule, 954
levitation, 966
Light
    speed of, 574
light
    energy and wavelength, 1057
    in a medium, 1012
    particle model, 326
    relation of E and B, 995
    speed of, 35
    speed same for all observers, 35
light bulb
    cross section, 739
linear systems, 169
liquid, 2
    boiling point, 2
    difficulty of analysis, 513
loop
    magnetic field of, 724, 725
loop rule, 772, 780
Lorentz force, 844, 990
lubrication, 372

M

macrostate, 477
magnet
    atomic structure, 730
    atomic theory, 733
    domains, 733
magnetic dipole moment, 726
    atomic electron, 730
    potential energy, 859
    related to atomic angular momentum, 731
    torque on, 858
magnetic domains, 733
magnetic field
    bar magnet, 727
    Biot-Savart law, 714
    circular motion in, 838
    depends on reference frame, 716
    detecting with a compass, 712
    experiment with coil, 741
    experiment with long straight wire, 741
    in a moving reference frame, 876
    long straight wire, 721, 723
    of a current element, 721
    of a loop, 724, 725
    of a moving charge, 714
    of a solenoid, 738
    of Earth, 728
    of long thick wire, 919
    of solenoid, 920
    of toroid, 921
    procedure for calculating, 721
    related to electric field, 878
    relativity, 716
    summary of ways to produce, 999
    under a current-carrying wire, 713
    uniformity under compass needle, 780
magnetic flux, 953
magnetic force
    compare with electric force, 838
    in a moving a reference frame, 854
    on a current element, 842
    on a current-carrying wire, 841
    on a moving charge, 837
    on current element, 841
    on magnetic moment
        measurement, 881
        quantitative, 861
    on parallel wires, 842
    on wire (experiment), 881
    on wire involves electric forces, 849
    polarization by, 851
    quantitative torque on magnetic moment, 859
    twist of current loop, 858
    twist of magnetic moment, 858
    with electric force, 844

magnetic force and electric force, 990
magnetic induction, 949
magnetic moment
    potential energy of, 859
    twist in magnetic field, 858
magnetic monopole, 728
    fields it would make, 1000
magnetic monopoles, 914
magnetic resonance imaging, 452
magnetic torque, 858
magnifying glass, 1028
mass
    hot object, 264
mass-spring system
    initial conditions, 161
massless particles, 225
maximally inelastic collisions, 386
maximum power output, 545
maximum-power engine, 545
Maxwell's equations, 987, 990
    differential form, 1033
    Faraday's law, 963
    incomplete version, 921
Maxwell-Boltzmann distribution, 500
mean free path, 517
    and sparks, 868
mechanical battery, 759
mechanical energy transfer, 226
Meissner effect (in superconductors),
        965
mesons
    spin, 448
metal
    electric field zero in equilibrium, 603
    polarization of, 602
metals
    equilibrium
        potential difference, 678, 689
    excess charge on surface, 597, 911
    free electrons, 601
    polarization, 601
    polarization process, 599
    proof that electrons are mobile, 819
microstate, 477
microstates
    logarithm, 484
mirrors, 1031
mobile electrons in metal, 600
mobility, 602
model
    ball-and-spring, 140
    solid, 138
modeling
    friction, 368
moment of inertia, 355, 424

momentarily at rest, 187
    vs. equilibrium, 188
momentum, 6
    and center of mass, 348
    approximate formula, 30
    average rate of change of, 33
    center of mass velocity, 121
    change
        parallel and perpendicular, 35
    change of, 32
    change of direction, 33
    change of magnitude, 32
    contrast with energy, 227, 250
    definition, 29
    direction, 31
    fast proton, 29
    gamma, 29
    rate of change, 34, 194
        both magnitude and direction, 201
    rate of change in curving motion, 194
    relativistic, 29, 225
    updating position, 31
    vector change, 32
momentum change
    with changing force, 60
momentum conservation, 118
    relativistic, 122
Momentum Principle, 48
    and acceleration, 125
    components, 53
    derivative form, 154
    limited form (F=ma), 155
    multiparticle, 123
    update form, 53
Momentum Principle, The
    identifying forces, 178
    multiparticle, 347
    parallel and perpendicular components, 208
monopole, 728
    fields it would make, 1000
monopoles, 914
Morse potential energy, 293
motion
    change of direction, 194
    change of speed, 194
    speed and direction changing, 194
motional emf, 851, 961
    and Faraday's law, 960
    small magnitude, 853
motor, 864
MRI, 452
multiparticle
    angular momentum, 422
    interaction energy, 243
    kinetic energy, 351

Momentum Principle, 123
    potential energy, 243
    rest energy, 262
    total energy, 263
multiparticle angular momentum, 431
multiparticle system
    potential energy, 241
multiparticle systems, 347
    energy, 347
muon, 555
    lifetime, 873
musical tones, 1075

N

net charge
    definition, 610
net force, 50
neutral matter
    interaction with charge, 591
neutralizing a tape, 612
neutrino
    discovery of, 401
    proposal and discovery, 245
neutron
    decay, 236, 401
neutron decay, 117, 119
neutron diffraction, 1062
Newton's first law of motion, 6
    and friction, 6
Newton's second law, Momentum
        Principle 95
Newton's second law of motion, 155
NMR, 452
no-Q compression, 531
non-Coulomb electric field, 949,
        950, 962
non-Coulomb force
    in mechanical battery, 769
non-Coulomb forces, 962
non-ohmic resistors
    light bulbs, 804
    semiconductors, 805
nonharmonic oscillations, 1075
nonlinear systems, 1076
nonreversible engine
    efficiency, 544
normal force, 148
nuclear angular momentum, 444
nuclear binding energy, 264, 265
nuclear energy, 237
nuclear energy levels, 339
nuclear fission, 265
nuclear interaction, 111
nuclear waste, 266

nucleus
    discovery of, 395
    size of, 2, 556
number density
    definition, 514
numerical integration, 633
nutation, 449

O

Oersted effect, 712, 741
ohmic resistors, 804
ohmmeter, 811, 813
open systems, 305
orbital angular momentum, 422
orbits
    types and energy graphs, 255
oscillations
    nonharmonic, 1075
oscillations in LC circuits, 970
oscillator
    harmonic, 156

P

p-n junction, 935
parallel capacitors, 798
parallel circuit, 771
parallel component
    of force, 77
parallel component of force, 192
parallel resistance, 807
particle
    massless, 225
    model of light, 326
    point, 361
    spin, 448
pascal
    unit of pressure, 519
path independence, 295
    electric potential, 683
period, 158
periodic motion, 69
periodic table, 2
perpendicular component
    of force, 77
perpendicular component of force, 192
phase shift, 168, 1044
photoelectric effect, 1055
    and particle model of light, 1056
    and wave model of light, 1055
photon, 326
pion, 555
Planck's constant, 336, 342
plane waves, 1012

plasma, 867
plug and chug, 238
plum pudding model, 396
point charge, 555
    electric field of, 560
point particle, 555
    energy, 374
    friction, 369
point particle system, 361
    how to analyze, 366
    jumping, 362
    spring, 364
Poisson's equation, 943
polarizability, 593
    and dipole moment, 593
    low-density approximation, 597
polarization, 591
    conductors, 598
    diagrams
        insulators, 597
        metals, 603, 607
    metals, 601
        process, 599
    not same as charged, 603
    of electromagnetic radiation, 1008
    of metal, 602
    or atoms, 592
    salt water in body, 600
    with nonzero charge, 603
position
    predicting new, 23
position update, 24
    relativistic formula, 39, 40
    with changing momentum, 57
position vector, 10
positron, 555
potential, 672
    at one location, 687, 690
    distributed charges, 695
        uniform disk, 695
        uniform sphere, 696
    electric, 672
    potential difference, 672
        across multiple regions, 676
        and dielectric constant, 692
        definition, 681
        dot product notation, 673
        in a metal in equilibrium, 678
        in a varying field, 679
        in an insulator, 692
        in nonuniform field, 676
        round trip, 685
        sign of potential difference,
            675, 698
    shifting the zero, 689

potential difference
   across a battery, 780
   across p-n junction, 937
   independent of path, 683
   metal not in equilibrium, 686
   round trip
      Conservation of Energy, 685
   round trip is zero, 685
potential energy, 241
   attraction, 252
   change of shape, 247
   electric, 260
   electric and neutral atoms, 293
   gravitational, 246, 247
   gravitational near Earth's surface,
      259
   macroscopic spring, 294
   Morse for neutral atoms, 293
   negative gradient is force, 264
   nuclear, 264
   of magnetic dipole moment, 859
   of magnetic moment, 859
   path independence, 295
   periodic, 929
   properties, 251
   quantized interatomic, 338
   repulsion, 253
   review, 665
   round trip, 295
   springs, 289
   three bodies, 258
   zero at large separation, 252
power, 304
   in battery, 808
   in resistor, 808
   instantaneous, 304
power in circuits, 808
Poynting vector, 1005
precession, 449
   astronomy, 453
   of the equinoxes, 453
pressure, 165
   1 atmosphere, 165
   and temperature, 519
   in atomic quantities, 521
   unit is pascal, 519
principle of relativity, 35, 879
principles, 436
prisms, 1017
problem solving
   forward reasoning, 204
projectile motion, 76
   and air resistance, 312
   graphs, 77
   spin, 312

proof by contradiction
   field inside conductor zero in equilibrium, 599
propagation of a disturbance, 153
proton, 555
   size of, 2

Q
   energy transfer due to a temperature difference, 301
   sign, 302
quantized vibrational energy, 335
quantum oscillator, 336
quarks
   spin, 448
quasistatic compression, 526
quasistatic processes, 526

R
radiation, 995
radiation pressure, 1005
radioactive waste, 266
radioactivity, 266
random walk, 547
rate of change
   momentum, 186
      curving motion, 194
      direction, 194
      magnitude, 194
   of momentum, 34
   of velocity (acceleration), 155
rate of change of direction
   proof, 195
RC circuit, 794
   quantitative, 814
RC circuits, 823
re-radiation, 1007
real images, 1021
real system
   how to analyze, 366
reciprocity, 103
reference frame
   affects observed magnetic field, 716
   inertial, 37
   magnetic force in, 854
reference frames, 391, 403
reflection, 1015, 1016
   from a metal, 1053
   of visible light, 1053
refraction, 1014
   marching band analogy, 1014
refrigerator, 542
relative position vector, 18
relativistic energy, 225
relativistic momentum, 225

relativistic motion, 56
relativistic velocity
    and momentum, 39
relativity
    and electric field, 575
    and time intervals, 856
    field transformations, 876
    magnetic force and reference
        frames, 854
    mass, 262
    principle of, 35, 879
    retardation, 574
        and electric field, 996
        circuit transient, 764
    special theory of, 35, 37
repulsive forces
    potential energy, 253
resistance, 802, 803
    and conductivity, 803
    definition, 802
    ohmmeters measure resistance, 813
    parallel, 807
    series, 806
    series and parallel resistors, 806
resistivity (also see conductivity), 803
resistor
    conventional symbol, 806
resistors
    complex circuit, 820
    non-ohmic, 804
    ohmic, 804
    parallel, 807
    power, 808
    series, 806
resolution, 1070
    telescope, 1070
resonance, 316
    and electromagnetic radiation, 1010
    with electromagnetic radiation, 1010
rest energy, 222
    multiparticle, 262
retardation
    circuit transient, 764
reversibility, 474, 535
reversible engine, 536
    efficiency, 541
    energy, 540
    entropy, 539
    lower efficiency if nonreversible, 544
    run backwards, 542
right-hand rule, 416
    for a wire, 723
    for curly non-Coulomb electric field,
        950
    for current loops, 726

ring
    electric field, 639, 640
    electric potential of, 688
RL circuit, 969
rms speed, 502
rms voltage, 818
rod
    electric field, 630
root mean square speed, 502
rotating vector
    derivative as cross product, 455
rotational angular momentum, 422
    moment of inertia, 424
rotational energy, 351, 353
    quantized, 338
rotational energy levels, 338
rotational kinematics, 442
rotational kinetic energy, 425
rotational kinetic energy , 355
rotational motion, 442
Rutherford
    experiment, 396

S

satellite, 198
scalar, 10
    magnitude, 13
scalar product
    work, 228
scaling a vector, 23
scanning tunneling microscope, 2
scattering, 392, 395
    and details of interaction, 398
    angular distribution, 397
    Compton, 1058
    cross section, 398
    of electromagnetic radiation, 1007
    of light, 1062
    of x-ray and electron, 1058
screening, electrostatic, 913
second-order differential equation, 169
self-force, 559
self-inductance, 966
semiconductor
    diode, 939
    doped, 933
    energy bands in, 930
    n-type, 933
    p-n junction, 935
    p-type, 934
semiconductors, 805
sensitivity to initial conditions, 115
series circuit
    current, 754

series resistance, 806
SI units, 20, 85
single-slit diffraction, 1069
size
    of atoms, 2
    of nucleus, 2
sky
    why is it blue, 1011
sliding friction, 150
    Harrison simulation, 315
Snell's law, 1015, 1016
Solar System, 3
solenoid
    inductance, 967
    magnetic field of, 738
solid
    electronic energy bands, 929
    speed of sound, 152
sound wave
    in solid, 152
spark, 867
    chain reaction, 873
        electric field required, 873
    charge transfer through ionized
        air, 869
    duration, 867
    ionization of air, 872
    light given off in, 870
special relativity
    See Relativity, 574
specific heat capacity
    calculating, 299
    copper, 299
    ethanol, 299
    gas at constant volume, 531
    molar, 531
    water, 299
spectrum
    absorption, 330
    emission, 329
    temperature effects, 334
speed
    average, 20
    instantaneous, 21
speed of light, 35
    from Maxwell's equations, 994
    same for all observers, 35
speed of sound
    aluminum, 163
    analytical expression for a solid, 154
    analytical expression in a solid, 161
    in gas, 503
    lead, 163
    modeling in a solid, 153
    solid, 152

sphere
    electric field, 646
        integrating, 652
    solid
        electric field, 649
    solid volume charge
        electric field of, 650
spheres
    act gravitationally like points,
        128
spin
    particle, 448
spin angular momentum, 422
    moment of inertia, 424
spring
    and Young's modulus, 292
    macroscopic, 288
    massless idealization, 160
    point particle system, 364
    stiffness in Einstein model of
        solid, 491
spring constant, 64, spring
        stiffness 95
spring force, 63
    as a vector, 64
spring stiffness, 64
spring–mass system
    energy, 290
spring-mass system, 156
    analytical solution, 156
    applying The Momentum
        Principle, 157
    harmonic oscillation, 167
    initial conditions, 168
    phase shift, 168
    vertical, 167
    vertical oscillations, 159
    with friction, 314
springs
    end-to-end, 143
    helical, 290
    ideal, 289
    potential energy, 289
    side-by-side, 144
standing waves, 1071
    and quantum mechanics,
        1076
    quantized, 1073
states
    bound, 257
    unbound, 257
static friction, 150
statics, 181
    torque, 186
statics course, 181

statistical mechanics, 474
    Boltzmann distribution, 494
    entropy, 483
    fundamental assumption, 477
    thermal equilibrium, 481
    width of distribution, 483
steady state, 753
    maintained by feedback, 765
    mechanical, 303
steel
    Young's modulus, 147
step size, 69
Stern-Gerlach experiment, 863
stick-slip, 151
stiffness of interatomic bond,
        143, 145
    speed of sound, 152
stimulated emission, 340
STM (scanning tunneling
        microscope), 2
strain, 146
stress, 146
string force, 181
strong interaction, 111
suction, 166
superconductors, 963
    effect of magnetic field
        on, 964
        Meissner effect, 965
    levitation, 966
superposition
    electric field, 564
superposition principle
    and electric field, 913
    electric field, 596
    field, 600
    no blocking of field, 596
Surface charge, 760
surface charge
    amount, 762
    demonstration in high voltage circuit, 778
    feedback mechanism, 765
    gradient makes electric field, 761
    initial transient, 763
    with narrow resistor, 766
    with parallel resistors, 774
    with wide resistor, 768
surroundings, 48
system, 48
    energy accounting, 306
    identifying forces, 180
    momentarily at rest, 187
    multiparticle, 347
    open and closed, 305
    point particle, 361

T

tape
    attraction between U and L tapes, 613
    is an insulator, 616
    neutralizing, 612
    preparing an L tape, 613
    repulsion between L tapes, 613
Tarzan and the vine, 138, 188, 200
telescope resolution, 1070
temperature, 298
    and entropy, 486
    and pressure, 519
    definition in terms of entropy, 488
    effects on spectrum, 334
    from entropy or from gas law, 522
temperature gradient, 534
tension, 140
    microscopic, 141
tension force
    string, 181
terminal speed, 310
thermal conductivity, 534
thermal energy, 296
thermal energy transfer, 474
    and entropy, 488
    rate, 534
thermal equilibrium, 481
thermal expansion, 299
thermometers, 298, 299
thin lens formula, 1020
thin lenses, 1019
Thomson
    discovery of electron, 396
three-body problem, 114
tides, 454
Tolman and Stewart experiment, 819
torque, 186, 428
    calculating, 428
    magnetic, 858
    systems with nonzero torque, 440
    systems with zero torque, 437
total internal reflection, 1016
transformation
    of fields in a moving reference frame, 876
transformers, 968
transistor, 942
translation
    and vibration, 352
traveling fields
    and Gauss's law for magnetism, 992
    moving pulse, 991
turning car, 191
turning point, 187
twist, 186
two-slit interference, 1045, 1050

U

U tape
   attraction by L tape, 613
unbound states, 257
uniform motion, 5, 184
unit conversion, 142
unit vector, 13, 420, 716
   and cross product, 420, 716
unit vectors
   direction cosines, 18
units
   SI, 20
update
   energy, 233
update position
   using momentum, 31
uranium fission, 265

V

Van de Graaff generator, 759
vector
   3D coordinates, 9
   addition, 15
      graphical addition, 16
   basic properties, 9
   components, 10
   cross product, 419, 714
   differs from scalar, 10
   direction, 9, 13
      unit vector, 13
   drawing a vector, 10
   equality of vectors, 15
   illegal operations, 12
   magnitude, 9, 11
   mathematical operations, 12
   multiplied by a scalar, 12
   position, 10
   relative position, 18
   scaling, 23
   sign, 11
   subtraction, 16
      graphical subtraction, 16
   unit vector
      calculating, 14
   velocity, 21
   zero vector, 17
vector cross product, 419, 714
velocity, 4, 20
   and relativistic momentum, 39
   average
      calculating, 22
      definition, 22
      position update formula, 24
   center of mass, 121

   change of, 28
   connection to calculus, 26
   instantaneous, 25
   predicting new position, 23
   vector, 21
velocity distribution in a gas, 499
velocity selector (cross electric and magnetic
      fields), 844
vertical spring-mass system, 159, 167
vibration, 352
   and translation, 352
vibrational energy, 351, 353
   quantized, 335
virtual image, 1022
viscous friction, 314
voltage
   rms, 818
voltmeter, 811
   high resistance, 813

W

water
   heat capacity, 299
water molecule, a permanent dipole, 609
wave
   angular frequency, 1043
   traveling, 1044
wave model vs. particle model of
      light, 1055
wave phenomena, 1043
wavefronts, 1012
wavelength of light
   and energy, 342
waves
   standing, 1071
weight, 203, 204
weightlessness, 203
why the sky is blue, 1011
wire
   magnetic field of, 721, 723
woman, barbell, and Earth, 306
work, 226
   and entropy, 488
   as an integral, 240
   by piston on gas, 528
   calculating, 232
   contrast with impulse, 227, 250
   done by a spring, 241
   dot product, 228
   evaluating, 227
   force at an angle to displacement, 231
   friction, 370
   negative, 222, 227, 228
   nonconstant force, 240

parallel and perpendicular forces, 231
review, 667
sign of, 228, 230
zero work, 230

X

x-ray diffraction, 1051
    condition for constructive interference, 1052

Y

Young's modulus, 146
    interatomic spring stiffness, 147
    range of validity, 147
    steel, 147
    wire acts like a spring, 292

Greek Alphabet and Its Uses in this Textbook

Alpha	A		α	alpha particle (helium-4 nucleus)	Nu	N	ν	neutrino; frequency (= f)
Beta	B		β		Xi	Ξ	ξ	
Gamma	Γ		γ	relativistic factor, photon	Omicron	O	o	
Delta	Δ	change of; small quantity of	δ		Pi	Π	π	circumference/ diameter of a circle
Epsilon	E		ϵ	a small quantity	Rho	P	ρ	density, resistivity
Zeta	Z		ζ		Sigma	Σ sum	σ	conductivity
Eta	H		η		Tau	T	τ	torque
Theta	Θ		θ	angle	Upsilon	Υ	υ	
Iota	I		ι		Phi	Φ flux	ϕ	phase angle
Kappa	K		κ		Chi	X	χ	
Lambda	Λ		λ	wavelength	Psi	Ψ	ψ	
Mu	M		μ	micro (1×10^{-6}), muon, coefficient of friction	Omega	Ω # of microstates of a system, ohm, precession angular frequency	ω	angular frequency, angular speed

Units of Measurement

kg = kilogram   m = meter   s = second   C = coulomb

N = newton (kg·m/s^2)   J = joule (N·m)   W = watt (J/s)

A = ampere (C/s)   V = volt (J/C)   T = tesla (N/C)/(m/s)   eV = electron volt = 1.6×10^{-19} J

Important Prefixes

pico (p) = 1×10^{-12} nano (n) = 1×10^{-9} micro (μ) = 1×10^{-6} milli (m) = 1×10^{-3}

centi (c) = 1×10^{-2} kilo (k) = 1×10^{3} mega (M) = 1×10^{6} giga (G) = 1×10^{9}

Unit Conversions

1 kg = 2.2 pounds   1 inch = 2.54 cm   1 foot = 30.5 cm   1 mile = 1.6 km

1 liter = 1.1 quart $T_{Kelvin} = T_{Celsius} + 273.15$ $T_{Fahrenheit} = (9/5)\,T_{Celsius} + 32$

1 cal = 4.2 J   1 kcal = 4.2×10^{3} J

1 kcal is 1 food calorie, also called 1 "large calorie"